Infectious Disease

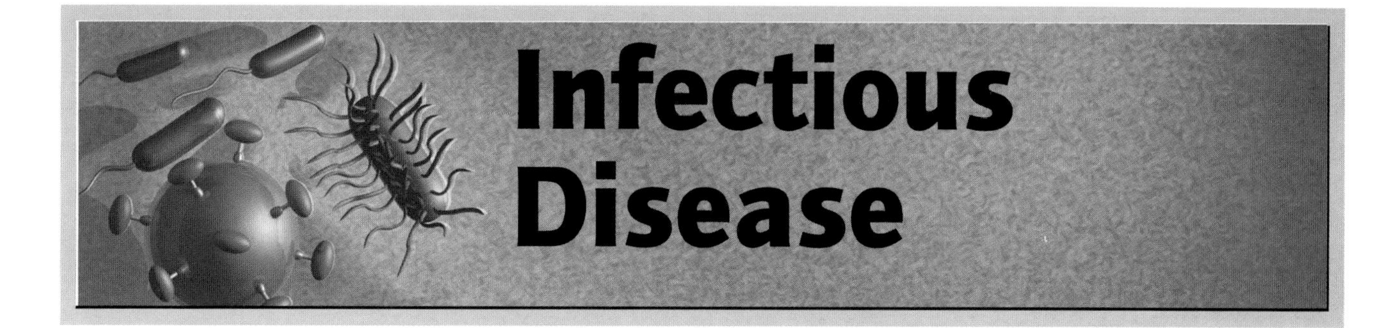

Barbara A. Bannister

MSc, FRCP
Consultant in Infectious and Tropical Diseases
Royal Free Department of Tropical and Infectious Diseases
Coppetts Wood Hospital
London

Norman T. Begg

Formerly
Consultant Epidemiologist
DTM&H FFPHM
Public Health Laboratory Service
Communicable Disease Surveillance Centre
London

Stephen H. Gillespie

MD, FRCP (Edin), MRCPath
Senior Lecturer in Medical Microbiology
Royal Free Hospital School of Medicine
London

SECOND EDITION

Blackwell
Science

Editorial Offices:
Osney Mead, Oxford OX2 0EL
25 John Street, London WC1N 2BL
23 Ainslie Place, Edinburgh EH3 6AJ
350 Main Street, Malden
MA 02148-5018, USA
54 University Street, Carlton
Victoria 3053, Australia
10, rue Casimir Delavigne
75006 Paris, France

Other Editorial Offices:
Blackwell Wissenschafts-Verlag GmbH
Kurfürstendamm 57
10707 Berlin, Germany

Blackwell Science KK
MG Kodenmacho Building
7–10 Kodenmacho Nihombashi
Chuo-ku, Tokyo 104, Japan

First published 1996
Second edition 2000

Set by Excel Typesetters Co., Hong Kong

Printed and bound by G. Canale & C. SpA, Turin

A catalogue record for this title is available from the British Library

ISBN 0-632-05319-4

Library of Congress
Cataloging-in-publication Data

Bannister, Barbara A.
Infectious disease/
Barbara A. Bannister, Norman T. Begg, Stephen H. Gillespie—2nd ed.
p. cm.
Includes bibliographical references and indexes.
ISBN 0-632-05319-4
1. Communicable diseases.
I. Begg, Norman T. II. Gillespie, S. H. III. Title.
[DNLM: 1. Communicable Diseases. WC 100 B219i 2000]
RC111.B364 2000
616.9—dc21 00-023121

DISTRIBUTORS

Marston Book Services Ltd
PO Box 269
Abingdon, Oxon OX14 4YN
(*Orders*: Tel: 01235 465500
Fax: 01235 465555)

USA
Blackwell Science, Inc.
Commerce Place
350 Main Street
Malden, MA 02148-5018
(*Orders*: Tel: 800 759 6102
781 388 8250
Fax: 781 388 8255)

Canada
Login Brothers Book Company
324 Saulteaux Crescent
Winnipeg, Manitoba R3J 3T2
(*Orders*: Tel: 204 837 2987)

Australia
Blackwell Science Pty Ltd
54 University Street
Carlton, Victoria 3053
(*Orders*: Tel: 3 9347 0300
Fax: 3 9347 5001)

For further information on Blackwell Science, visit our website:
www.blackwell-science.com

Contents

Preface, vi

Infection, Pathogens and Antimicrobial Agents

1 The Nature and Pathogenesis of Infection, 1

2 Structure and Classification of Pathogens, 23

3 Laboratory Techniques in the Diagnosis of Infection, 35

4 Antimicrobial Chemotherapy, 51

Systematic Infectious Diseases

5 Skin, Mucosal and Soft-tissue Infections, 81

6 Upper Respiratory Tract Infections, 111

7 Lower Respiratory Tract Infections, 132

8 Gastrointestinal Infections and Food Poisoning, 157

9 Infections of the Liver, 191

10 Infections of the Urinary Tract, 215

11 Childhood Infections, 225

12 Infections of the Cardiovascular System, 243

13 Infections of the Central Nervous System, 254

14 Bone and Joint Infections, 286

Genital, Sexually Transmitted and Birth-Related Infections

15 Genital and Sexually Transmitted Diseases, 296

16 HIV Infection and Retroviral Diseases, 310

17 Congenital and Perinatal Infections, 320

Disorders Affecting More Than One System

18 Tuberculosis and Other Mycobacterial Diseases, 337

19 Bacteraemia and Sepsis, 361

20 Pyrexia of Unknown Origin, 379

21 Postinfectious Disorders, 393

Special Hosts, Environments and the Community

22 Infections in Immunocompromised Patients, 404

23 Hospital Infections, 416

24 Imported and Travel-associated Diseases, 433

25 Some Systemic Zoonoses, 461

26 Control of Infection in the Community, 473

Index, 487

Index of Organisms, 503

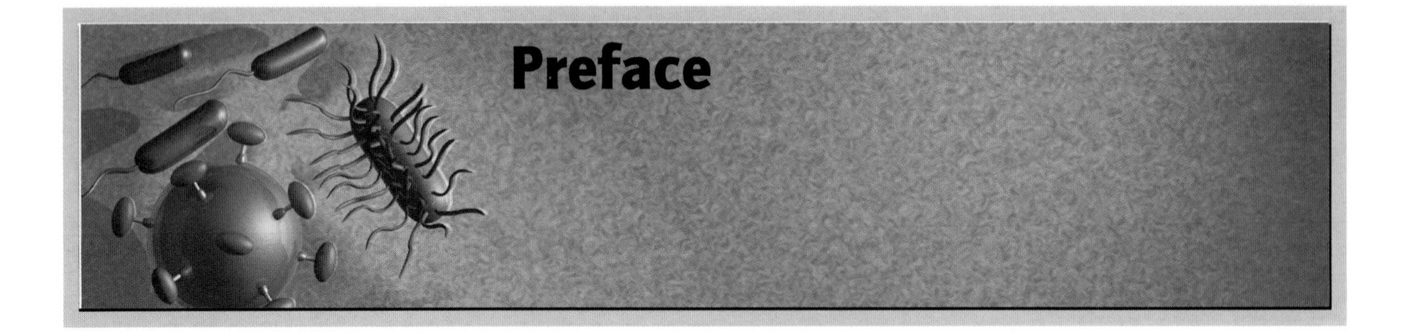

Preface

Infectious Disease encompasses the clinical practice of infectious diseases in hospital and the community. It has been designed as a problem-solving text which will be equally useful to those preparing for examinations or working in the clinical setting.

The introductory chapters set out the important definitions required for the understanding of infectious diseases and describe the nature and presentation of infection, its pathogenesis and scientific investigation. The diagnosis, management and control of individual infectious diseases are described in the systematic chapters. Each chapter introduces the range of diseases affecting the relevant system and a list of pathogens responsible for each disease in the approximate order of importance. For each individual disease the microbiology and epidemiology of each pathogen, the diagnosis and management of the disease and the strategies for their prevention and control, are described.

This textbook is designed to be used either as a basic learning text or as a practical textbook in the clinical setting. It should therefore be most useful for senior medical students and for doctors preparing for the MRCP (UK) and the Primary FRCS. It would also be a useful revision text for those studying for MRCPath and MFCM and a companion textbook for physicians, surgeons and public health specialists early in their career. It contains much information which will be useful to infection control nurses, community nurses and environmental health officers.

Since the publication of the first edition, the understanding of the pathogenesis of infectious diseases has continued to advance. Several new pathogens have been described, and an infectious aetiology for conditions previously considered non-infectious has been proposed or confirmed. Therapeutics have made a major advance. The last few years have seen the introduction of highly active anti-retroviral therapy (HAART) and new classes of drugs for the treatment of influenza and of enteroviral infections. DNA amplification and other molecular diagnostic methods have developed from research techniques to be applied in routine laboratories. Molecular pathogenesis research has borne fruit with the introduction of several vaccines.

Infectious diseases is an exciting and ever-evolving discipline. We hope that you find that this new edition will provide you with the information and insight required for addressing infection-related problems in the varied settings of modern medicine.

We would welcome and value any reader feedback. Let us know how you think this approach could be improved by emailing us at authors@blacksci.co.uk.

1 The Nature and Pathogenesis of Infection

Introduction, 1
The nature of infection, 1
Pathogen, 1
Infection, 2
Communicable disease, 3
Pathogenicity, 3
Virulence, 3
Infectiousness, 3

Epidemiology of infections, 3
Epidemiology, 3
Interaction between host, agent and environment, 4
Sources and reservoirs of infection, 4
Routes of transmission of infection, 5
Definition and types of outbreaks and epidemics, 6

Mechanisms of resistance to infection, 7
Non-specific immunity, 7
Classical and alternative complement systems, 8
Phagocytosis, 8
Specific immune responses, 8

Pathogenesis of infection, 14
Characteristics of successful pathogens, 14
Effects of microbial toxins, 16
Microbial synergy, 17

Manifestations of infection, 17
Fever, 17
Inflammation, 18
Rashes, 20
Harmful effects of immune responses, 21

Dynamics of colonization and infection, 21

Introduction

The nature of infection

The terms used in discussion of infectious diseases are constantly changing to keep pace with changes in our knowledge and understanding. We will start by defining some of the common terms used in this book.

Pathogen

Pathogen
An organism that can invade the body and cause disease.

A pathogen is defined as any organism capable of invading the body and causing disease. Such an organism is said to be pathogenic. Koch isolated and identified organisms such as *Mycobacterium tuberculosis* and *Bacillus anthracis*, where the isolation of the organism only occurred in the presence of disease. It is easy to define these bacteria as pathogens. They also fulfil Koch's further definition of a pathogen, that introduction of a pure culture of the organism into a healthy host can cause the disease. This definition works well for many bacteria, but does not fully describe the complex interactions between microbes and humans which more recent understanding has revealed. For instance, *Escherichia coli* is found in huge numbers in the healthy human bowel, and could therefore be defined as non-pathogenic. *E. coli* is also an important cause of diarrhoeal diseases, and potent enterotoxins and other pathogenicity determinants have been described in some strains. *E. coli* can therefore behave as a pathogen or as a colonizer, depending on various circumstances. A broader definition of a 'biological agent' used in European Union legislation is: 'any microorganism, cell culture or toxin capable of entering the human body and causing harm'.

Changes in medical practice mean that increasing numbers of patients are immunocompromised as the result of either disease or treatments. In such patients organisms which are usually non-pathogenic, such as saprophytic fungi, may act as pathogens. Intensive therapy medicine, with insertion of intravascular cannulae, allows *Staphylococcus epidermidis*, a normal part of the skin flora, to enter the cannula and cause blood-borne infection: behaving as a pathogen.

Parasites as pathogens

The true definition of a parasite is an organism that lives on or in another organism, deriving benefit from it but providing nothing in return. Not all 'parasites' cause disease; organisms such as *Entamoeba dispar*, a protozoan, live in the human gut without causing disease and are thus colonizers. The closely related species *E. histolytica* is capable of invading the bowel wall, causing colitis and abscesses in the liver, brain and other tissues. It is thus a pathogen, and this term will be applied to it, and to other parasitic pathogens, in this book. Multicellular parasites such as schistosomes may also be pathogens. In the past, diseases caused by metazoan parasites, such as schistosomiasis, were sometimes called *infestations*. Nowadays all parasitic diseases are called infections.

Infection

> **Infection**
> A disease caused by a pathogen.

An infection is a disease caused by a pathogen. The human body is colonized on the skin and mucosal surfaces with numerous microorganisms that form the normal flora of the body. These organisms, far from causing disease, often provide benefit to the host, by competing with potential pathogens for attachment sites and nutrients, and by producing antimicrobial substances toxic to pathogens. Thus, the mere presence of microorganisms multiplying in the human body does not constitute an infection (Fig. 1.1). It is

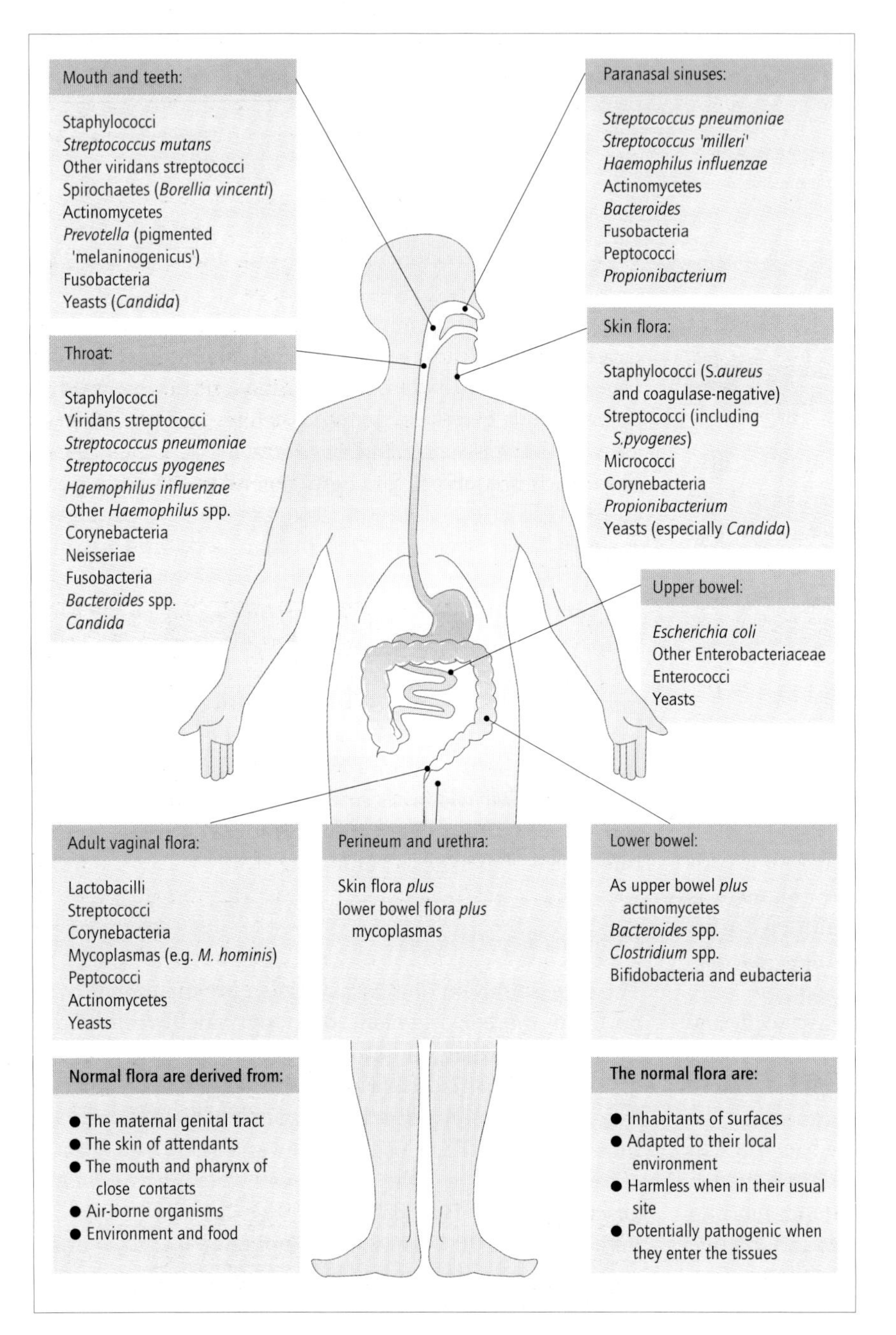

Fig. 1.1 Normal human flora.

the presence of the replicating organism, **associated with tissue damage**, that defines the condition as an infection. *Clostridium tetani* may multiply in a puncture wound, elaborating the neurotoxin tetanospasmin. Because the organism is multiplying in the host's tissues, the resulting disease can be called an infection. In contrast, adult botulism, caused by *C. botulinum*, develops when food is ingested in which this organism has grown and elaborated a neurotoxin; the organism itself does not replicate in the human host. Botulism is therefore an intoxication (poisoning), rather than an infection. *C. difficile* may be isolated from the faeces of 5–20% of normal subjects. However, it is only when conditions within the large bowel are altered by antibiotic therapy that this organism produces its toxin, causing pseudomembranous colitis. In this case, colonization, the presence of microorganisms in the human host in the absence of disease, has developed into an infection.

Communicable disease

Communicable disease
An infection that is capable of spreading from person to person.

A communicable disease is an infection that is capable of spreading from person to person. Not all infections are communicable diseases. A patient with infective endocarditis caused by *Streptococcus sanguis* is suffering from an infection. This is not, however, a communicable disease as it is unable to spread from this patient to another. Communicable diseases may be transmitted by many routes: direct person-to-person transfer; respiratory transmission; parenteral inoculation; by way of fomites (inanimate objects); sexual or mucosal contact; and by insect vectors.

Pathogenicity

Pathogenicity
The ability to cause disease.

Pathogenicity is the ability to cause disease. *Neisseria gonorrhoeae* is the causative organism of gonorrhoea. It is a small Gram-negative diplococcus, some strains of which bear surface projections called pili, while some do not. Those organisms with pili can attach to the urethral epithelium and cause disease. Those that lack this feature cannot, and are non-pathogenic. In this example, pili confer pathogenicity. Mechanisms of pathogenicity are numerous, and will later be discussed more fully (see pp. 14–17).

Virulence

Virulence
A pathogen's power to cause severe disease.

Virulence is defined as a pathogen's power to cause severe disease. When a pathogen causes infection, the resulting disease may be asymptomatic or mild, but can sometimes be severe. This variation may be due to host factors or to virulence factors possessed by the organism. Influenza virus is constantly able to modify its antigenic structure, on which its virulence depends. The difference in the attack rate and the severity of disease in succeeding epidemics is related to the antigenic structure of the causative virus.

Pathogenicity and virulence are not necessarily related. This is illustrated by *Streptococcus pneumoniae*, which depends on its polysaccharide capsule for its pathogenicity. However, it is the biochemical nature of the polysaccharide that determines the virulence of the organism. Type 3 and type 30 pneumococci both produce much capsular material, and both are therefore pathogenic. Infection with type 3 is often associated with severe disease, whereas infection with type 30 is rarely severe. For a further discussion of pneumococcal pathogenicity, see pp. 142–3.

Infectiousness

Infectiousness
The ease with which a pathogen can spread in a population.

Infectiousness is the ease with which a pathogen can spread in a population. Some organisms always spread more readily than others. For example, measles is highly infectious and mumps very much less so. A measure of infectiousness is the intrinsic reproduction rate (IRR), which is the average number of secondary cases arising from a single index case in a totally susceptible population. The IRR for measles is 10–18, while for mumps it is 4–7.

Epidemiology of infections

Epidemiology

Epidemiology
The study of the distribution and determinants of diseases in populations.

This is the study of the distribution and determinants of diseases in populations. The distribution of diseases may be described in terms of time (day, month or year of onset of symptoms), person (age, sex, occupation) or place (region or country). Determinants of diseases are those factors that are associated with an increased or decreased risk of disease. Their effects are usually identified by analytical studies such as case-control or cohort studies. For example, the epidemiology of meningococcal meningitis is characterized by its distribution (commonest in winter, peak incidence in young children, worldwide occurrence but especially in sub-Saharan Africa) and its determinants (close contact with a case, passive smoking).

Interaction between host, agent and environment

The behaviour of a pathogen in a population depends upon the interaction between the pathogen, host and environment. Changes in any one of these three factors will affect the likelihood of transmission occurring, and of disease resulting.

Host factors

Host factors affect both the chance of exposure to a pathogen and the individual's response to the infection. Important host factors include travel, sexual behaviour, hygiene, occupation, crowding, previous immunity, nutrition and underlying disease.

Agent factors

Agent factors include infectiousness, pathogenicity, virulence and ability to survive both in human and animal hosts and under different environmental conditions. Other important factors, such as the ability to resist vaccine-induced immune responses, or drugs, also play a large part in the effect of some diseases.

Environmental factors

Environmental factors such as temperature, dust and humidity, and the use of antibiotics and pesticides affect the survival of pathogens outside the host.

The spread of malaria is a good example of the interaction between host, pathogenic agent and environment. *Plasmodium* sp. is a protozoan parasite transmitted by the bite of an infected female anopheline mosquito. Subsequent asexual development of the organism takes place within the human hepatocytes and erythrocytes. In some forms of malaria (e.g. *P. vivax*) organisms may remain dormant in hepatocytes to mature months later and produce relapses. This does not occur in infections due to *P. falciparum*.

Many host factors affect the transmission of malaria: individuals who live in endemic areas develop partial immunity from repeated exposure and rarely suffer severe disease. This immunity is lost after 1 or 2 years away from endemic exposure. Newcomers to endemic areas will usually suffer severe disease if infected. Certain genetic factors also affect the outcome of infection. For example, individuals with sickle-cell trait have a relatively low parasitaemia when infected with *P. falciparum*, because the organism cannot derive effective nutrition from haemoglobin S.

The agent of *P. falciparum* malaria has developed resistance to an increasing range of prophylactic drugs, making it harder for travellers to protect themselves from infection. The differing antigenic structures of the several stages of the parasite life cycle have, so far, prevented the development of an effective vaccine.

Environmental factors are particularly important in the spread of malaria. Transmission occurs predominantly (although not exclusively) in tropical zones, especially during the rainy season. The anopheline mosquito breeds in stagnant fresh-water environments, and malaria is particularly common in these areas. Drainage of ponds and tanks is an effective means of reducing malaria transmission. Residual insecticides have been used to control adult mosquito vectors; however, this measure has had limited success due to the emergence of insecticide-resistant mosquitoes.

Sources and reservoirs of infection

Pathogens are either endogenous, arising from the host's own flora, or exogenous, arising from an external source. The reservoir of infection is the human or animal population, or environment in which the pathogen exists, and from which it can be transmitted. Infection can be transmitted from carriers of an organism as well as from those suffering active infection.

Person-to-person transmission is the most common method of spread. Horizontal spread is between individuals in the same population, as in the case of whooping cough. Vertical spread is also possible, from mother to fetus during gestation or birth, as in the case of congenital rubella or hepatitis B. Many pathogens can cross the placenta, but only a few cause fetal damage. The consequences of vertical transmission are usually, but not always, most serious when infection occurs during early pregnancy (see Chapter 17).

Animal diseases which spread to humans are called zoonoses. The normal infectious cycle between animals is accidentally entered by humans, most frequently where there is close contact between humans and animals, for instance in occupations such as farming or veterinary work or in recreations such as breeding fancy animals or birds.

Zoonosis
An animal disease that can spread to humans.

Many pathogens are environmental organisms, for example *Listeria monocytogenes*, *Legionella pneumophila* and *Clostridium tetani*. Spread from environment to humans can occur by ingestion (*Listeria monocytogenes*), inhalation (*Legionella pneumophila*) or inoculation (*C. tetani*).

Routes of transmission of infection

Fomites

Occasionally, inanimate environmental objects act as intermediaries, transporting pathogens from source to host. Fomites such as towels or bedding may transmit *Staphylococcus aureus* between hospital patients. Make-up applicators, towels and ophthalmic equipment have all been shown to carry bacterial or viral pathogens from eye to eye when shared without adequate cleaning between uses.

Vectors

Vector
A living creature that can transmit infection from one host to another.

Vectors are living creatures that transmit infection from one host to another. Many arthropod species are able to transmit pathogens (Table 1.1).

Arthropod	Diseases
Mosquito	Malaria, dengue fever, filariasis, yellow fever
Sandfly	Leishmaniasis, sandfly fever
Fly	Trypanosomiasis, onchocerciasis
Flea	Plague, rickettsial infection
Tick	Relapsing fever, rickettsial infection
Mite	Rickettsial infection
Louse	Relapsing fever, typhus

Table 1.1 Arthropod vectors of medical importance.

Direct contact

Where pathogens are present on the skin or mucosal surfaces, transmission may occur by direct contact. Skin infections such as impetigo spread by this means. More fragile organisms cannot survive in a dry, cool environment, but can spread via sexual contact. In children, among whom direct contact is greater than in adults, pathogens in respiratory secretions may also be transmitted by direct contact. A few environmental pathogens can penetrate the skin and mucosae directly. An example of this type of spread is leptospirosis in which organisms excreted in the urine of infected animals penetrate the mucosae or broken skin of a human. Exposure usually occurs via contact with contaminated fresh water, e.g. during swimming, diving and water sports.

Inhalation

Droplets containing pathogens from the respiratory tract are expelled during sneezing, coughing and talking. Droplet nuclei (1–10 μm in diameter) are formed by partial evaporation of these droplets, and they remain suspended in air for long periods of time. Inhalation of droplet nuclei is the principal route of transmission for many human respiratory pathogens, e.g. influenza and measles. Transmission of pathogens by inhalation can also occur from animal to humans, as with *Chlamydia psittaci*, which is present in the droppings and secretions of infected birds. Inhalation of the organism usually occurs when infected birds are kept in a confined space.

Environmental pathogens can also be transmitted by inhalation. The most important example is *Legionella pneumophila*, the causative agent of legionnaires' disease, which is present in aerosols generated from air-conditioning cooling towers, cold-water taps, showers and other water systems. Depending upon wind speed, these aerosols can travel up to 500 m and infect large numbers of individuals.

Ingestion

Enteric pathogens are usually transmitted via contaminated food, milk or water. Many foods are produced from animals, thus ingestion commonly results in animal-to-person transmission. The two most important pathogens causing bacterial food poisoning (*Salmonella* and *Campylobacter*) are both zoonotic pathogens readily transmitted to humans. Food-borne transmission is most likely if food is eaten raw or undercooked, as the pathogens are killed by heat. Many milk-borne infections are also zoonoses acquired by ingestion of unpasteurized milk products from infected cows, sheep or goats.

Spread by ingestion can occur when pathogens discharged in faeces, vomit, urine or respiratory secretions contaminate the hands of an infected individual or fomites such as handkerchiefs, clothes and cooking and eating utensils. Subsequent spread to food or water is favoured by conditions of poor sanitation. A common form of transmission by ingestion occurs by direct contact between faecally contaminated hands and oral mucosa (faecal–oral transmission). Hepatitis A can spread by all of these means. *Salmonella typhi* may also be transmitted by these routes; a few organisms deposited in food will multiply to achieve an infective dose.

Transmission from environment to humans by ingestion is less common. However, the soil- and sewage-borne bacterium *Listeria monocytogenes* is an example, as it can contaminate food and cause invasive disease following ingestion.

Inoculation

Transmission can occur when a pathogen is inoculated directly into the body via a defect in the skin. Contaminated transfusions, blood products or material from non-sterile needles and syringes can transmit viruses such as hepatitis B and human immunodeficiency virus (HIV). Malaria may also be transmitted by contaminated blood transfusions.

Animal-to-person transmission occurs when an infected animal bites, scratches or licks an individual. Rabies is usually spread by this route. Alternatively, the skin may be broken by sharp animal bristles or rough bone meal containing pathogens such as *Bacillus anthracis*.

Environment-to-person spread by inoculation also occurs: *Clostridium tetani* is usually introduced through a puncture wound contaminated with soil, dust or animal faeces.

Definition and types of outbreaks and epidemics

The terms outbreak and epidemic are usually synonymous. The word outbreak is used to describe a localized epidemic. An outbreak is defined as an occurrence of a disease clearly in excess of normal expectancy. Outbreaks occurring in animal populations are called epizootics.

Outbreak
An occurrence of a disease clearly in excess of normal expectancy.

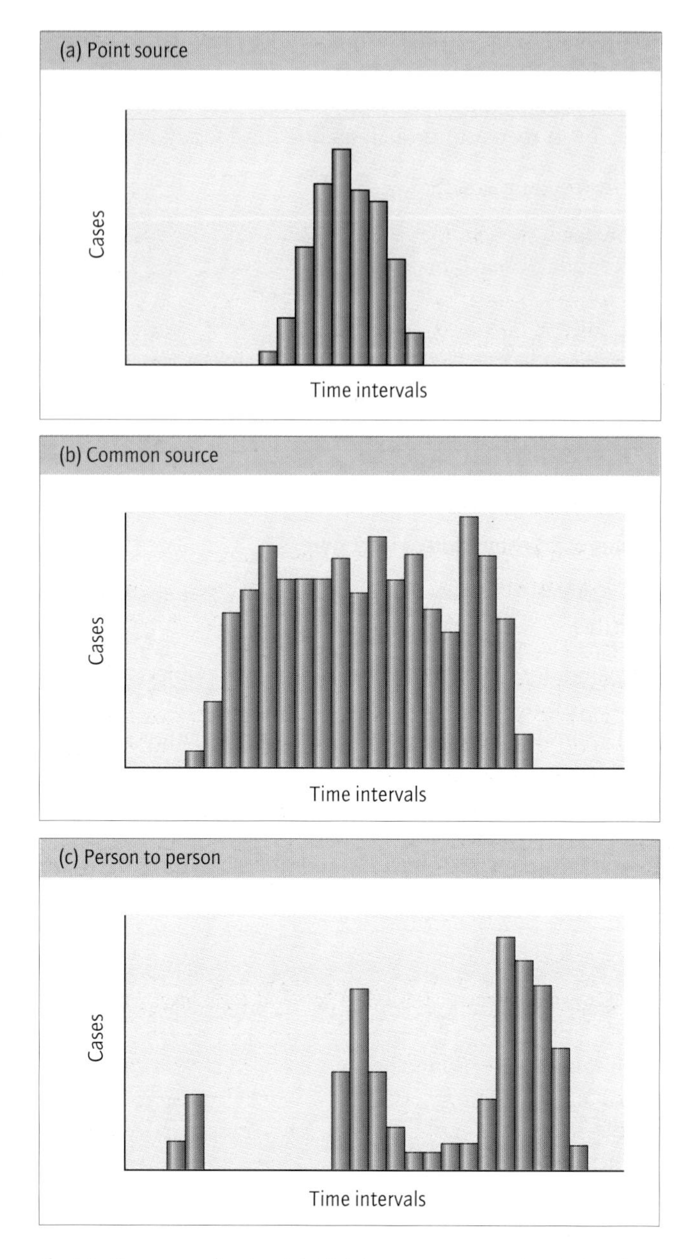

Fig. 1.2 Three types of outbreak. (a) Point-source cases occur in a cluster after a single exposure, for example to a contaminated meal. (b) Common-source cases occur over a period of time after continuing exposure, for example to a commercial distributed food. (c) Person-to-person cases occur in clusters, separated by an incubation period.

There are three main types of outbreaks—point source, common source and person to person (Fig. 1.2).

Point-source outbreak

A point-source outbreak occurs when a group of individuals is exposed to a single source of infection at a defined point in time. An example of this is a group of wedding guests who consume a contaminated food item at the reception. All those affected develop symptoms within a few days of each other.

Common-source outbreak

A common-source outbreak occurs when a group of individuals is exposed to a single source over a period of time. An example is an outbreak of hepatitis B associated with a tattoo parlour using contaminated equipment that is inadequately sterilized between customers. Common-source outbreaks may extend over long periods of time.

Person-to-person outbreak

In a person-to-person (propagating) outbreak there is no common source: the outbreak is maintained by chains of transmission between infected individuals, for example a *Shigella sonnei* outbreak in a school.

Two other terms which sometimes cause confusion are endemic and pandemic. Endemic refers to a disease that occurs commonly all the year round, for example malaria in West Africa. A pandemic is an epidemic that affects all or most countries in the world at the same time. Only four communicable diseases have caused pandemics: influenza, plague, cholera and acquired immunodeficiency syndrome (AIDS).

Mechanisms of resistance to infection

The defence of the human host to infection can be classified into two parts: the non-specific or innate immune system, and the specific immune system. Each is vital for survival against the continuous pressure of microorganisms.

Non-specific immunity

Many components of this function are normal mechanical and physiological properties of the host. They include the skin, the mechanical flushing activity of urine and intestinal contents, ciliary removal of mucus and debris, the enzymatic action of lysozyme in tears, the phagocytes and the normal flora.

The skin

Natural defences of the skin

1 Keratinous surface.
2 Antibacterial effects of sebum.
3 Effect of normal flora.

The skin forms a mechanical barrier to invasion. Also, sebum secreted by the skin inhibits the multiplication of microorganisms. The skin's resident bacterial flora competes with potential invaders, and may produce metabolic products inhibitory to other species. This combination of effects is called colonization resistance and is also important in the pharynx and the bowel. Scratches, ulcers and other defects in the skin surface can bypass its protective mechanisms and permit the entry of skin pathogens such as staphylococci or herpes simplex viruses, as well as environmental organisms such as *Leptospira* spp.

Parasites such as hookworm larvae and schistosome cercariae are capable of penetrating intact skin. Cercariae, which hatch from infected water snails, swim towards potential hosts. The head part of the cercaria secretes proteolytic enzymes which break down the skin (and, incidentally, cause a local dermatitis or 'swimmer's itch').

The physical barrier of the skin can also be breached by biting arthropods. Many types of pathogen are transmitted by this route. Intravenous access devices enable coagulase-negative staphylococi and corynebacteria from the skin to enter the bloodstream and cause septicaemia and endocarditis. HIV and hepatitis B virus infection are transmitted by this route, through the use of contaminated needles.

Mucosal defences against infection

Natural defences of mucosae

1 Mechanical washing by tears or urine.
2 Lysozyme or antibody in surface fluid.
3 Surface phagocytes.
4 Ciliary action moving mucus and debris.

Many bacteria and viruses can invade through intact mucosal surfaces, which are not keratinized, and are often only one cell thick. Nevertheless, mucosal surfaces also have natural defences. They are usually moistened by tissue fluids, which contain lysozymes capable of destroying microbial peptidoglycan. If previous immunization has occurred, secretory immunoglobulin A (IgA) may also be present at the mucosal surface. Phagocytic neutrophils are often expelled at mucosal surfaces, and can ingest foreign material, including pathogens. The mucosae of the gut and the urinary tract are 'washed' by the constant throughput of liquid contents. The respiratory tract can move material upwards towards the pharynx by the action of mucosal cilia. These defence mechanisms are so effective that the urinary tract and respiratory tract are bacteriologically sterile, except for areas near the exterior, such as the mouth and the lower urethra. Sites that possess a colonizing flora, such as the bowel, also benefit from the additional colonization resistance that this confers.

Obstruction of a bronchus, ureter or bile duct will overcome many of these defences, permitting pathogens to accumulate at sites of stagnation, thus predisposing to infection. Similarly, insertion of a urinary catheter or an endotracheal tube will alter clearance mechanisms, encourage stagnation of mucus secretions around the tube and provide an inanimate substrate, encouraging the entry and establishment of a bacterial flora.

Defences against infection via the gut

Natural defences of the gut
1 Gastric acid.
2 Chemical environment produced by normal flora.
3 Bacteriocins.

The gut provides an important route for the acquisition of microbes. The first barrier to infection is gastric acid which inhibits the survival of many intestinal pathogens. Patients with achlorhydria are more susceptible to infections which are transmitted by the faecal–oral route. The normal flora also provides defence against invaders by competing for nutrients and attachment sites. Facultative and obligate anaerobes produce potent inhibitors of bacterial growth called bacteriocins. These protein antibiotics act to inhibit the growth of competing organisms. In addition many obligate anaerobes secrete free fatty acids and alter the local redox potential, making the environment less supportive to other microorganisms. This delicate competitive balance can be upset by disease or by antimicrobial therapy. The most dramatic example of this is pseudomembranous colitis, when patients treated with antibiotics have an overgrowth of *Clostridium difficile* in the intestine. This organism produces a toxin which causes severe ulcerative disease of the large bowel.

Classical and alternative complement systems

The complement system is a complex of plasma enzymes which, by sequential activation, are capable of lysing bacterial cell walls and infected cells (Fig. 1.3). Activation of the classical pathway depends on a specific immune response: C1q binds to the Fc component of bound immunoglobulin at cell surfaces. This pathway can also be initiated by binding to C-reactive protein, an acute-phase plasma protein which is elevated during acute inflammation.

The alternative complement pathway depends on the slower, spontaneous breakdown of C3 at bacterial cell surfaces. In both pathways the generation of active C3b initiates the formation of the 'attack complex' of C6–C9, which perforates and disrupts cell membranes. The alternative pathway is initially slower acting than the classical, but it can act in the absence of a specific antibody and it provides early defence against such severe infections as meningococcal septicaemia. Complement can act to enhance resistance to bacterial and parasitic infection by the action of breakdown products such as C3a and C5a which promote capillary permeability and are chemotactic to neutrophils and macrophages. C3b deposited on the surface of bacteria will opsonize them for phagocytosis.

Patients with congenital deficiencies of the early complement components are more susceptible to pneumococcal infections, in which activation of the alternative complement pathway is important in resistance to infection. Deficiencies in components of the alternative pathway, such as properdin, render the individual highly susceptible to invasive meningococcal infection.

Sialic acid inhibits the natural breakdown of C3. Successful pathogens, such as meningococci, have sialic acid on their surface, which probably reduces the effectiveness of the alternative complement pathway.

Phagocytosis

Phagocytosis of invading organisms is an important nonspecific defence mechanism (Fig. 1.4). Neutrophils and macrophages are attracted to the site of inflammation by mediators such as complement components. The efficiency of phagocytosis is enhanced when organisms are 'opsonized' by attached complement or specific antibody, which provide receptors for the attachment of phagocytes. Organisms are taken up into phagosomes which fuse with the lysosomes containing free radicals and lytic enzymes, resulting in killing.

Patients with deficiencies in phagocyte function suffer repeated pyogenic infections, and develop chronic suppurative granulomata (see p. 404).

Specific immune responses

The specific immune response is a series of adaptive changes whereby the host develops defensive responses to individual microorganisms. This is based on recognition of, and response to, unique antigens which the pathogen possesses. Antigens are made up of individual amino acid or sugar residues linked together to form short sequences. The sequences are displayed in an array on the pathogen's surface because of the way they are integrated into the tertiary structure of the surface chemicals. These short sequences are called epitopes. Immunogenic epi-

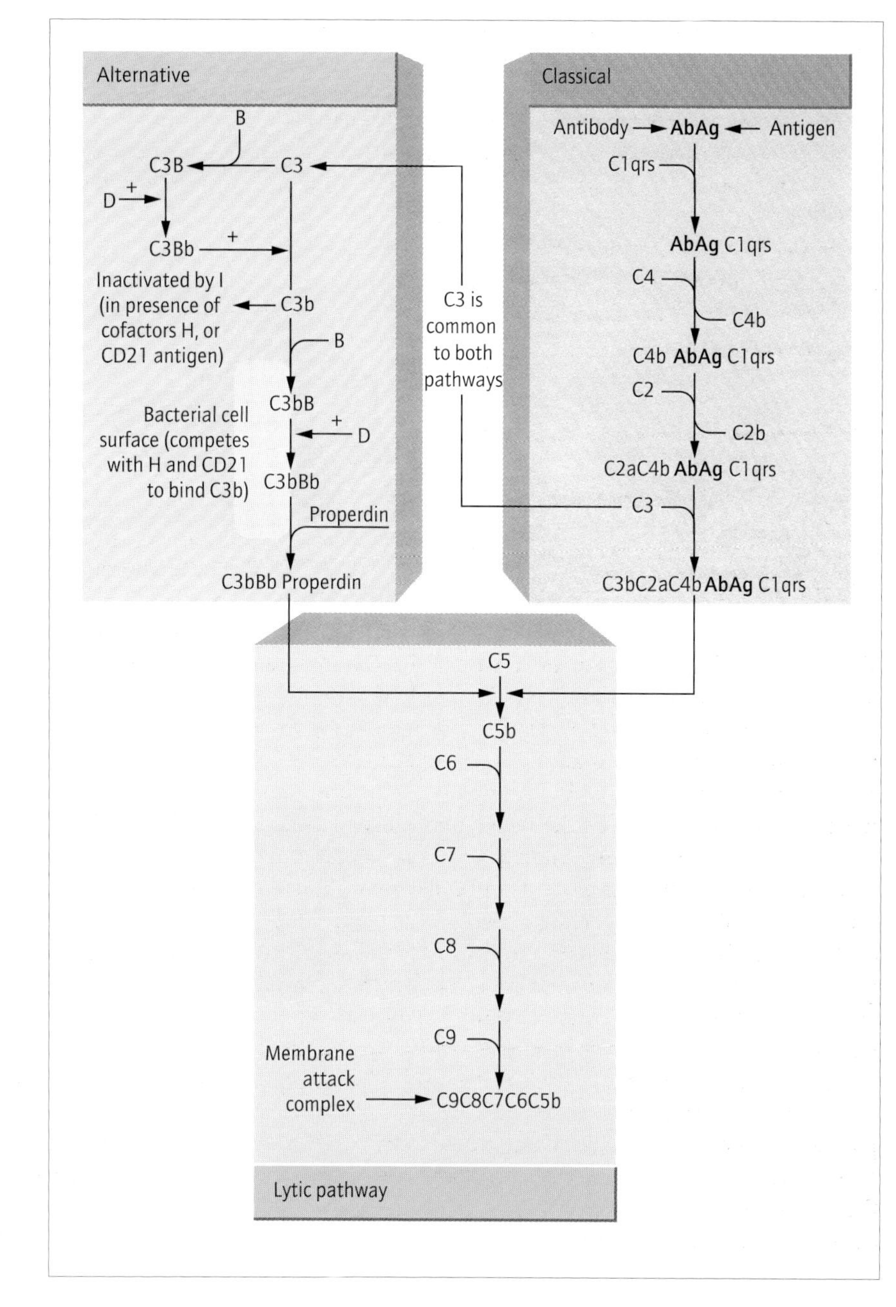

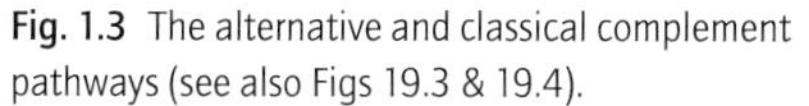
Fig. 1.3 The alternative and classical complement pathways (see also Figs 19.3 & 19.4).

topes may also be parts of toxin molecules, or the abnormal surface proteins displayed on virus-infected host cells. Different epitopes stimulate T and B cell immunity. Organisms each contain many antigens, within which are many different epitopes.

Antigen presentation by macrophages

As well as acting simply as phagocytes, macrophages have an essential role in the development of immunity to individual pathogens. Like other phagocytes, the macrophage takes up foreign material into phagosomes and fuses these with lysosomes to form phagolysosomes.

Substances secreted during the resulting burst of metabolic activity include cytokines (chemicals which modulate the activity of other reticuloendothelial cells). Interleukin-1 (IL-1) and IL-6 modulate the activity of lymphocytes and of other macrophages. They also act on the hypothalamus, causing an increase in body temperature.

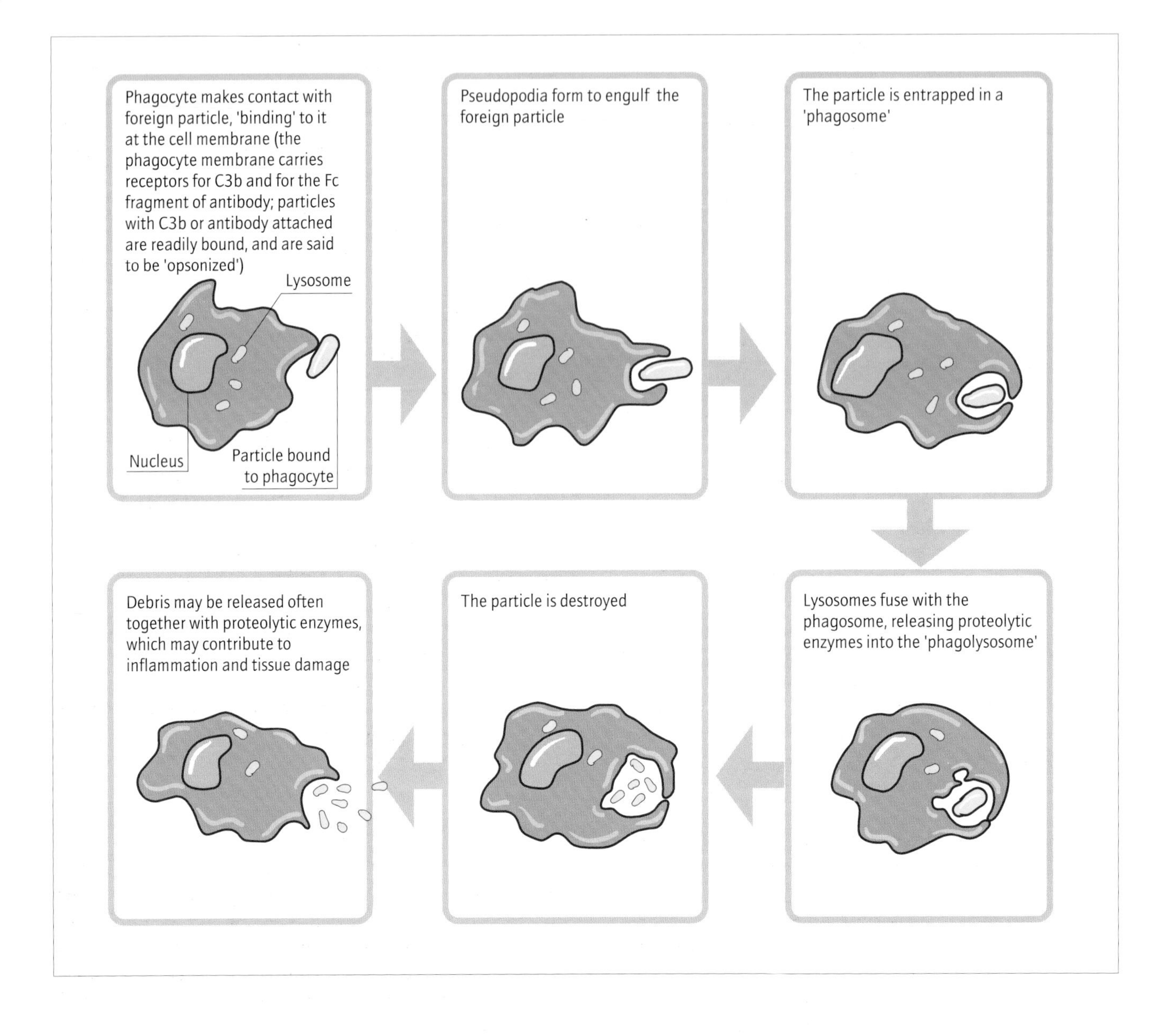

Fig. 1.4 Phagocytosis.

Tumour necrosis factor (TNF) is also released, and acts as an important mediator in the action of endotoxin.

Some of the ingested antigen is broken into fragments by enzymes (processed), and the various fragments are displayed at the cell surface, bound in the molecular groove, or in close association with the cell's own class II human leucocyte antigens (HLAs) of the major histocompatibility complex. The double structure of microbial antigen and class II antigen is recognized by helper T lymphocytes with the same class II antigens, which become activated and capable of displaying chemicals with helper function. Helper T cells also display the CD4 lymphocyte antigen.

Activation of the immune response also sets in motion a mechanism which causes the gradual degeneration and death of activated T cells. This process is called apoptosis. It is important in limiting the duration and extent of an immune reaction, avoiding the progressive tissue damage that might result from an uncontrolled response.

B lymphocytes express immunoglobulin molecules at their surface, which act as antigen-binding receptors, and can recognize the combination of antigen and helper function on helper T cells, which in turn activates them to proliferate. Many enlarge and mature into plasma cells, which secrete immunoglobulin. Others replicate as clones of cells which will recognize the antigen in future, and respond by proliferation and immunoglobulin secretion. This is the basis of immunological memory.

T cells have antigen-binding receptors which are similar in structure and function to immunoglobulins, but which are not secreted into the body fluids. Helper cells are important to both cell-mediated and humoral immune responses (see below). Patients who lack helper-cell function suffer severe cell-mediated immunodeficiency, and also mount a poor humoral immune response to new antigens.

Cell-mediated immune responses

Cell-mediated immunity is a system of lymphocyte-mediated attack, which is not based on major production of antibody. It depends on a system of antigen recognition and cytokine production among macrophages and T cells (Fig. 1.5). T cells with different functions display different cluster differentiation (CD) antigens.

When a macrophage ingests and presents antigen, it can interact with a CD4 cell displaying the same HLA class II identity and a receptor specific for the antigen. This stimulates the macrophage to produce IL-1 which has several actions, including a hypothalamic action leading to fever, and a lymphocyte-stimulating action increasing the density of IL-2 receptors. The CD4 cell is also stimulated, and produces IL-2. When a stimulated T cell encounters free antigen, it, too, will produce IL-2. The system not only causes IL-2 production, but also increases the sensitivity of T cells to its effects. The effects of IL-2 are to cause proliferation of T cells which recognize the antigen, and to mediate interactions between CD4 (helper) cells and other lymphocytes (including the stimulation of B cells: see below).

Among T cell populations are cytotoxic cells (with CD8 antigens). These recognize antigen on cells bearing HLA class I tissue-type antigens, and kill the cells, lysing them by an unknown mechanism. There is also a population of CD8 suppressor T cells, which have an inhibitory or modulating role in lymphocyte interactions.

Patients who lack effective cell-mediated immunity are unable to combat viral infections, particularly those caused by enveloped viruses, such as the herpesvirus family, and infections with intracellular bacteria such as mycobacteria. They may also suffer from yeast infections, such as candidiasis and cryptococcosis (see Chapter 22).

Humoral immune responses

Humoral immunity is the system by which clones of antibody-secreting plasma cells are produced when B lymphocytes respond to 'their' specific antigen (Fig. 1.6). Unstimulated B cells possess large amounts of surface-bound antibody existing in a wide range of apparently random affinities. Some B cells will recognize and bind to the various antigenic epitopes of an invading pathogen. In the presence of helper cells activated by the same antigen, these B cells proliferate, and are transformed into clones of memory cells and antigen-secreting plasma cells. In this way, humoral immunity develops with many clones of B cells which all have differing affinities for a range of epitopes possessed by a pathogen. This is a polyclonal humoral response.

Early in the humoral immune response IgM is produced, but later, in the so-called mature immune response, large amounts of IgG are released. The early IgM response is called the primary response; the prompt and large IgG response to continued or subsequent exposure is called the secondary response. It represents the activation and proliferation of memory cells to form IgG-secreting plasma cells. IgM may also be produced on second or subsequent exposure to a pathogen, but a large memory response of IgM does not occur. Inactivated vaccines are often administered in carefully timed multiple doses to obtain an optimal and long-lasting secondary response.

Antibodies can enhance cell-mediated cytotoxicity in the antibody-dependent cell-mediated cytotoxicity (ADCC) response. In this case, killer lymphocytes (K cells that carry the CD16 antigen) recognize the antibody of bound antigen–antibody complexes on cell surfaces, and kill the antigen-bearing cells.

T cell-independent antigens

Some antigens can induce a humoral immune response without the involvement of T-helper cells. They are usually large polymeric molecules, composed of multiply repeated subunits. It is thought that they bind to many adjacent antigen receptors on the B cell surface, accidentally influencing intervening receptors concerned with recognizing helper function.

The primary immunoglobulin response to T cell-independent antigens is usually a rather small and short-lived IgM response. The secondary response is similarly small, and is usually a further IgM response, without significant IgG production. The absence of a mature IgG response is associated with rather poor and short-lasting immunological memory. This is important as many polysaccharide antigens, such as pneumococcal capsular antigens, act as T cell-independent antigens, especially in young children.

Superantigens

These antigens do not require processing before they are presented to lymphocytes. Lymphocyte activation is

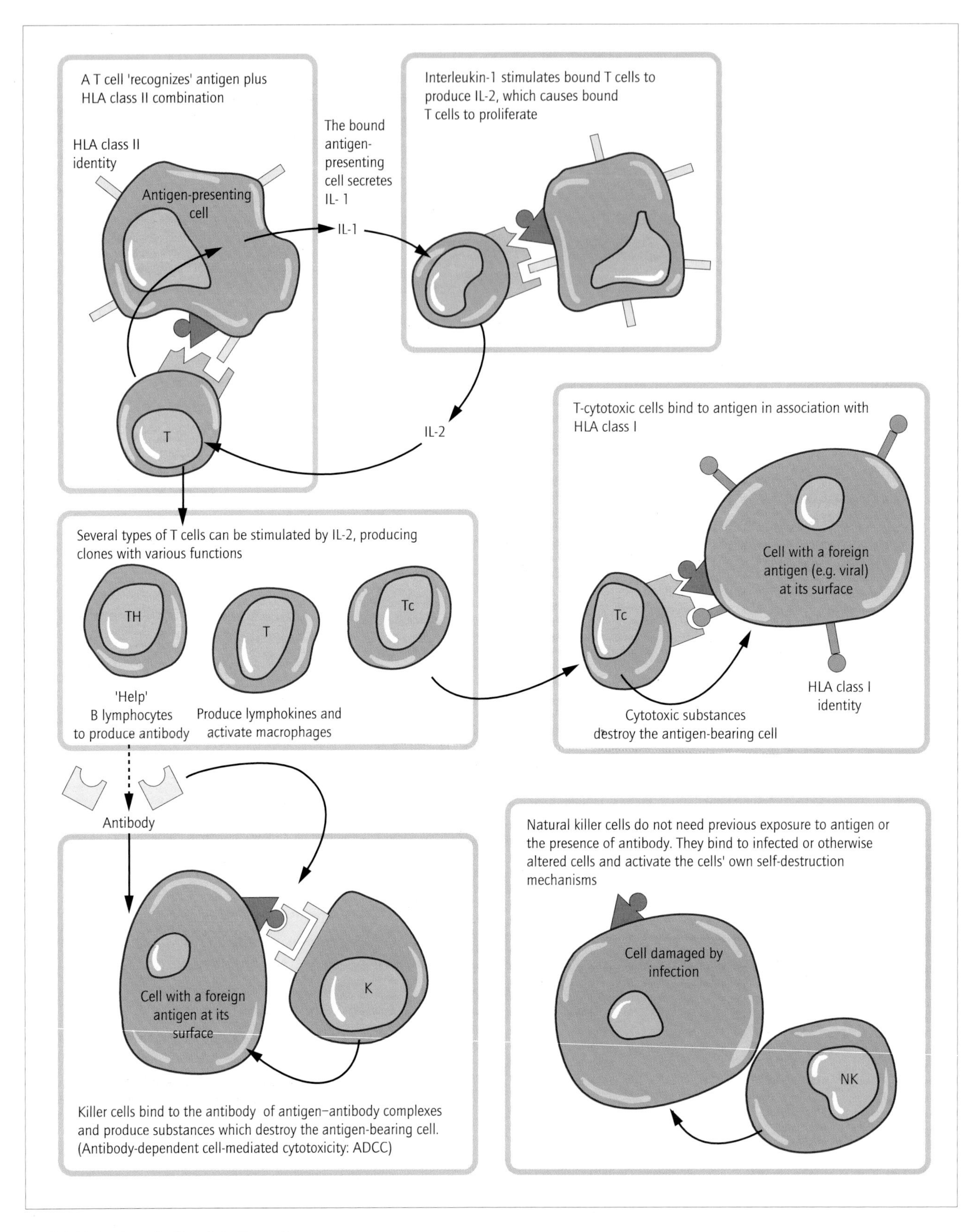

Fig. 1.5 Mechanisms of cell-mediated immunity.

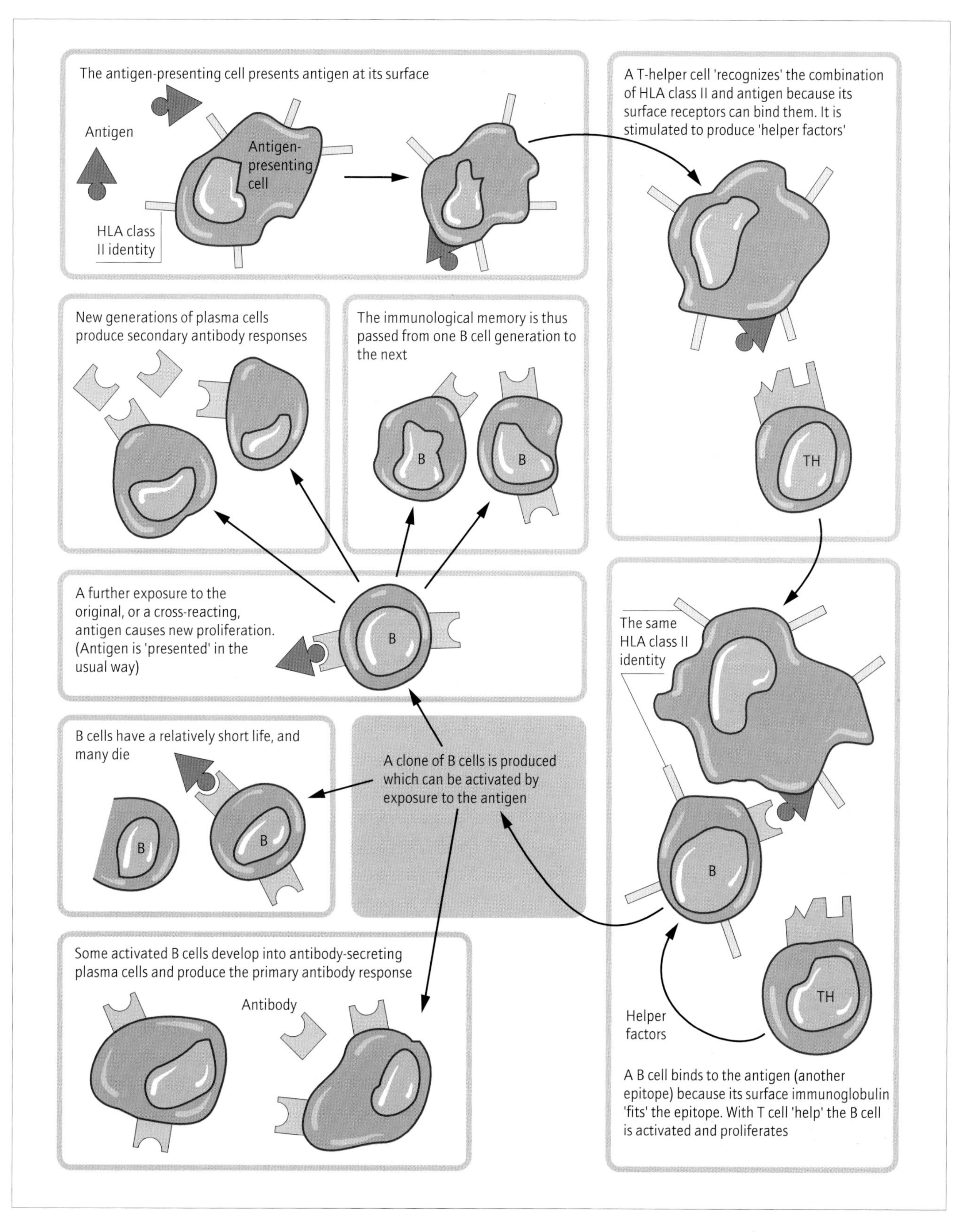

Fig. 1.6 Humoral immune responses.

mediated by an alternative mechanism, using V-beta receptors which are possessed by up to 30% of all lymphocytes, and does not require the association of HLA class II antigens. Superantigen binding does not activate the process of apoptosis. Superantigens therefore cause a very intense immune reaction which is not limited or terminated in the usual way. Staphylococcal enterotoxins, toxic shock syndrome toxins (TSSTs) and some viral antigens act as superantigens.

Pathogenesis of infection

While the spread of a disease is influenced by interaction between the host, the agent and the environment, the effects of disease in the host depend on a number of factors particular to the agent or pathogen. These factors are important determinants of pathogenicity. They are termed the pathogenicity factors of the organism.

The pathogen exploits its host to best advantage if it achieves optimum levels of survival and multiplication. The death of the host is not an advantage unless this contributes to the transmission of microbial genes. Pathogenicity factors are genetically maintained, by natural selection, because they facilitate survival or transmission via the host.

Characteristics of successful pathogens

Features of a successful pathogen

1 Survival and transmission in the environment.
2 Attachment to the surface of the host.
3 Overcoming the body defences against infection.
4 Ability to damage the host, e.g. by toxin production.

Survival in the environment

Many microorganisms are killed by drying, ultraviolet light and variation from their optimum temperature for growth. To overcome these difficulties, organisms have developed many strategies. Organisms which are predominantly environmental have developed survival mechanisms such as the bacterial endospore. This is a structure which contains a single copy of the bacterial DNA in a keratinous protective 'shell', which has little retained water and a very low metabolic rate. Adverse environmental conditions are the stimulus for sporulation, for example the rise in pH in the duodenum for *Clostridium perfringens*. Bacterial spores are capable of survival for many years, 'germinating' to form vegetative cells when conditions are more favourable. The Scottish island of Gruinard was contaminated with anthrax spores early in the Second World War, and was only declared free of infection 50 years later.

Other organisms have found protected ecological niches in the environment: *Legionella pneumophila* inhabits fresh water and can survive in the protected environment within the cytoplasm of free-living amoebae.

Some organisms have such a close relationship with an animal host that they can exist reversibly, as either pathogen or commensal. To increase the chances of survival, some organisms will infect a wide range of species, for example rabies virus, which is able to infect all mammals. This diversity provides a large and adaptable reservoir of infection, which will ensure survival of the pathogen if one host group is eliminated.

Transmission

The problem of transmission between hosts is related to survival. Spore-forming organisms can survive and be transmitted by many routes. Organisms with moderate survival potential can be transmitted by spreading in the air on droplet nuclei (see p. 5).

Bacteria such as *Neisseria gonorrhoeae* are extremely delicate and are unable to survive outside the host. Sexual transmission overcomes this difficulty by depositing the pathogen directly on to the genital mucosa of the new host, and in addition ties the organism's life cycle into an essential part of the host's life cycle, ensuring the survival of the pathogen.

Attachment of organisms to body surfaces

For organisms to gain access to the body via the mucosal surfaces they must first attach themselves. They have to overcome the natural defence mechanisms present in each area.

Organisms may gain attachment by specialized organelles of attachment, or more simply with attachment molecules. Uropathic *Escherichia coli*, which has to overcome the flushing action of urine, uses fimbriae to attach to the urinary epithelium. These fimbriae are pathogenicity determinants. Influenza virus adheres to the host's respiratory mucosal cells via its haemagglutinin molecule.

Microbial defence against immunological attack

From the moment the pathogen enters a new host, it must defend itself against immunological attack. Against organisms which invade via the mucosal surfaces, host secretory IgA is an important defence mechanism. Many respiratory tract pathogens, including *Streptococcus pneu-*

moniae, Haemophilus influenzae and *Neisseria meningitidis*, elaborate a protease which selectively destroys IgA.

Bacterial capsules are an important defence against phagocytosis by neutrophils or macrophages. Capsulate organisms resist phagocytosis unless opsonized by the attachment of specific antibody. Resistance to phagocytosis probably depends on the negative charge on the capsular polysaccharide molecules.

For organisms such as the pneumococcus which activate the alternative complement pathway at their cell wall, the capsule acts as a physical barrier, preventing attached C3b on the cell wall being recognized by phagocytes. Some organisms are able to exploit phagocytes to enhance their life cycle. Once they have entered the phagocyte they are protected from antibodies, and can survive for very long periods. *Mycobacterium tuberculosis* is an intracellular pathogen which can survive inside macrophages and from this site reactivate if the host's immune defences become compromised.

Intraphagocytic survival depends on safe entry into the phagocyte and subsequent avoidance of enzymic degradation by lysozymes. Phagocytic ingestion of a particle is usually accompanied by a 'respiratory burst' which produces intensely toxic oxygen radicals. *Leishmania* spp. overcome this by utilizing alternative mannose/fucosyl and C3b receptors which, unlike the Fc receptors, do not trigger a respiratory burst. Within the phagocyte, there are three main mechanisms by which pathogens can survive. The organism may prevent phagolysosomal fusion, avoiding contact with lysozyme, and continuing to multiply within the phagosome. *Toxoplasma gondii* and *Chlamydia* act in this way. *Leishmania* survives inside the phagolysosome by metabolic adaptation to the hostile environment, and by excreting a factor which scavenges the normally lethal oxygen radicals. Mycobacteria escape from the phagolysosome into the cytoplasm where they are partly protected from digestion by their high lipid content. This effect has been demonstrated by coating staphylococci with the phenolic glycolipid of *M. leprae*. Although successfully ingested, these coated staphylococci are not killed, while uncoated control staphylococci are destroyed.

Viruses may contain genes which code for cytokine-like molecules or cell receptors, thus interfering with the natural functions of cells. Epstein–Barr virus produces proteins which switch off programmed cell death in infected lymphocytes, helping to maintain the population of productively infected cells.

Antigenic variation

For organisms which are obliged to live extracellularly, antibody attack poses a major problem. Some organisms are able to evade the humoral immune system by varying the antigenic make-up of their surface. The major surface antigen of *Trypanosoma brucei* var. *rhodesiense* is the variable surface glycoprotein (VSG). As a result of a complex series of molecular events, the trypanosomes are able to express a different VSG every few days. Thus, as a humoral immune response is produced and parasite numbers are falling, a new clone of trypanosomes emerges with a different VSG and is able to multiply unhindered by the immune system. This process is continued through a preprogrammed set of variations, which are reflected by the episodic nature of symptoms in the early phases of the infection. *Borrelia recurrentis* and *B. duttoni* also undergo antigenic variation, producing a characteristic, relapsing fever.

Influenza virus survives as a pathogen by antigenic variation because its genome can undergo antigenic 'drift' and 'shift'. Drift is the process of gradual changes in the genes coding for viral surface haemagglutinin, enabling the virus partly to escape the effects of population immunization by previous epidemics. Shift is a major change in the antigenic structure of the virus, producing a novel strain to which nobody has any immunity. Antigenic shift may initiate a worldwide epidemic (pandemic). A variation of this approach is adopted by adult schistosomes which absorb host proteins to their surfaces, thus evading detection by the immune system.

Immune suppression

Pathogens employ various strategies to evade or impair the host's immune response.

Many parasitic infections cause overstimulation of the humoral immune system. High concentrations of ineffective antibodies are produced at the expense of normal antibody responses. African trypanosomiasis and leishmaniasis are examples of this; not only is the parasitic infection uncontrolled, but many sufferers die of intercurrent bacterial infections such as acute pneumonia.

The cellular immune response can also be depressed. This occurs in severe tuberculosis and in lepromatous leprosy, where the infection induces a specific cell-mediated immune defect limiting T cell responses to the mycobacteria. Acute viral infections, such as measles and infectious mononucleosis, cause temporary suppression of cell-mediated immune responses.

Immune suppression can be broad spectrum, when a whole arm of the system is impaired by the action of a pathogen. HIV causes a selective depletion of CD4 cells, resulting in susceptibility to tuberculosis, *Pneumocystis* infections and toxoplasmosis. Additionally, reduced T helper-cell function causes an increased susceptibility to many other pathogens, including pyogenic bacteria.

Ability to damage the host

Toxin production

Toxins are responsible for many of the damaging effects of infection. They are also excellent vaccine targets as chemically modified toxin vaccines (toxoids) stimulate strong immune responses. These vaccines have helped in the virtual elimination of diseases such as tetanus and diphtheria. Toxins are often essential for the life cycle of the pathogen; their potential for pathogenicity may be coincidental. Diphtheria toxin, for example, mediates the pharyngeal, cardiac and neurological damage in diphtheria. The gene coding for diphtheria toxin exists in a betaphage, and only organisms carrying this lysogenic phage are toxigenic. In this symbiosis, the phage requires *Corynebacterium diphtheriae* as a host and *C. diphtheriae* is given a biological advantage in colonization of the human host by possession of the toxin gene.

Bacterial toxins are conventionally classified as exotoxins and endotoxins. Exotoxins are toxic substances excreted by organisms. The word endotoxin is usually used to describe the lipopolysaccharide antigen of Gram-negative bacterial cell walls. However, it is now known that many bacterial structural antigens can have toxic effects.

Endotoxin

The lipopolysaccharide of Gram-negative bacteria is an important pathogenicity factor. Lipopolysaccharide is made up of three main parts: (i) the core region, lipid A, which is responsible for the main toxic effects; (ii) an oligosaccharide region which contains heptoses (Hep) and hexoses linked to lipid A via the unusual sugar ketodeoxyoctanoic acid (KDO); and (iii) attached to this a long polysaccharide chain which is the somatic antigen ('O' antigen) of the individual organism. This polysaccharide partly protects Gram-negative bacteria such as salmonellae against the bactericidal activity of serum (Fig. 1.7).

Lipid A acts by stimulating cells of the macrophage series to produce cytokines, such as IL-1 and TNF. These are important factors in activating the complement and clotting cascades, in causing endothelial damage, and in mediating fever and other metabolic and physiological changes.

Toxic activity also resides in the cell wall of some Gram-positive bacteria. The C-polysaccharide and F (Forssmann) antigen of *Streptococcus pneumoniae* are released into the host's tissues, and activate the alternative complement pathway. The products of the complement cascade are responsible for increased capillary permeability and leucocyte migration to the site of infection.

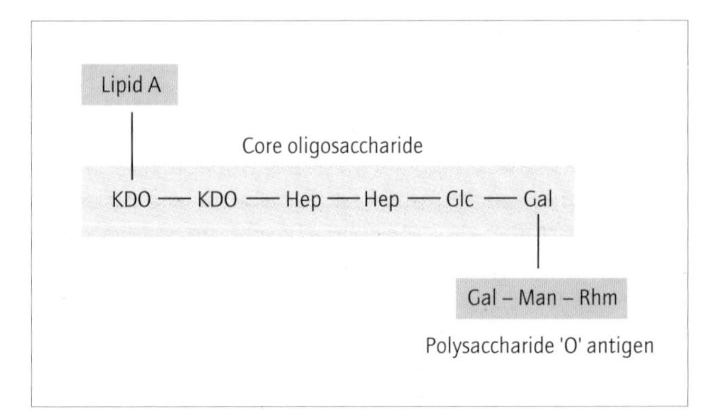

Fig. 1.7 An example of the structure of endotoxin.

Type	Examples
Extracellular cytotoxins (directly poison cells)	Streptococcal hyaluronidase *Pseudomonas aeruginosa* exotoxin A
Transmembrane cytotoxins (enter cells via implanted receptor/transporting molecule)	*Escherichia coli* verotoxin Shiga toxin Diphtheria toxin
Membrane-damaging toxins (cause haemolysis or cytolysis)	Streptolysin O *Clostridium perfringens* alpha toxin *Staphylococcus aureus* P-V leukocidin
Deregulating toxins (cause overactivity of secretory mechanisms)	*E. coli* heat-labile toxin Cholera toxin
Competitive inhibitors (competitive blockers of natural transmitters)	Botulinum toxin Tetanus toxin

Table 1.2 Actions of bacterial exotoxins.

Exotoxins

Bacterial exotoxins are diverse in function and clinical effect. They may be classified either according to the symptoms produced—enterotoxin, neurotoxin or cytotoxin, or according to their mode of action, when this is known (Table 1.2).

Effects of microbial toxins

Some toxins cause important features of a disease, and it is convenient to consider a few examples here.

Streptococci and staphylococci can produce pyrogenic exotoxins (SPEs), toxins which damage the skin and sometimes the internal organs. The erythrogenic toxin of *Streptococcus pyogenes* can cause the rash of scarlet fever. *Staphylococcus aureus* may produce TSSTs, causing toxic shock syndrome with a scarlet fever-like rash. TSSTs also behave like the enterotoxins of *S. aureus* and cause diarrhoea. Some organisms produce haemolytic toxins; *Clostridium perfringens* can cause severe haemolysis by this mechanism. Diphtheria toxin directly damages myocardial cells and causes demyelination of nerve axons, while botulinum toxin inhibits neuromuscular conduction of nerve impulses.

Microbial synergy

Microbes can act together to establish infection, facilitate tissue invasion, reduce the host's immune response and enhance the virulence of pathogens. *Streptococcus pneumoniae* cannot bind to intact respiratory epithelium, but can bind to basal membrane. Influenza virus causes damage to, and shedding of, respiratory epithelium, exposing the underlying basement membrane to attack.

Many infections are polymicrobial, with obligate and facultative pathogens multiplying together and creating the conditions for each other to survive. In synergistic gangrene the metabolic products of facultative organisms reduce the redox potential sufficiently to enable obligate anaerobes to multiply and cause extensive tissue necrosis.

Infection with HIV is an example where one infectious agent reduces the immune response of the host, allowing other organisms to invade. Many parasitic infections (leishmaniasis, trypanosomiasis and malaria) can trigger polyclonal activation of B cells with overproduction of antibody, impairing the host's response to intercurrent bacterial infections.

Chronic schistosomiasis is associated with recurrent *Salmonella* infection. Salmonellae can bind to schistosome eggs, which provide a niche for salmonellae to cause persisting colonization and recurrent infection.

Manifestations of infection

Fever

This is the most common accompaniment of infection, occurring in all but the most trivial or unusual cases.

The body temperature of a healthy person is set and maintained by the hypothalamus. It follows a circadian cycle in which the temperature is lowest in the early morning and highest at about 10 PM, varying by 0.5°C or more. Also, in women who ovulate, a monthly variation in temperature can be detected, with an abrupt step at the time of ovulation.

Infection, in common with a number of other events, can cause a resetting of the hypothalamus to a higher body temperature. This change is initiated by the release of cytokines, particularly IL-1, TNF and alpha-interferon, by activated mononuclear phagocytes. The cytokines act on specialized endothelial cells in the hypothalamic blood vessels, causing the release of prostaglandins which act on the hypothalamic cells (Fig. 1.8).

A raised body temperature is probably useful in combating infection. Many pathogens replicate best at temperatures at or below 37°C. These include respiratory viruses, pneumococci and other bacteria, and many agents of tropical skin infections. Such pathogens are adversely affected by higher temperatures. Even if the pathogen is unaffected by temperature change or, like some campylobacters, replicates well at higher temperatures, fever can still help by accelerating immune reactions such as phagocytosis, antibody and cytokine production, and the complement cascades.

Adverse effects of fever

Delirium

Delirium is a state of confusion, often with agitation. It is most common in children and the elderly, occurring when the temperature is at its highest, especially at night, when it may cause distressing dreams. Although sometimes caused directly by the disease process (toxaemia or cerebral infection), it can often be improved or cured simply by reducing the temperature.

Febrile convulsions

Febrile convulsions affect children between the ages of 6 months and 6 years. They are rarely a sign of true epilepsy, and usually cease spontaneously as the child reaches the age of 4–6. The convulsions most often happen as the temperature is rising.

Treating fever and its complications

Treatment of fever

1 Tepid sponging.
2 Paracetamol.
3 Aspirin.

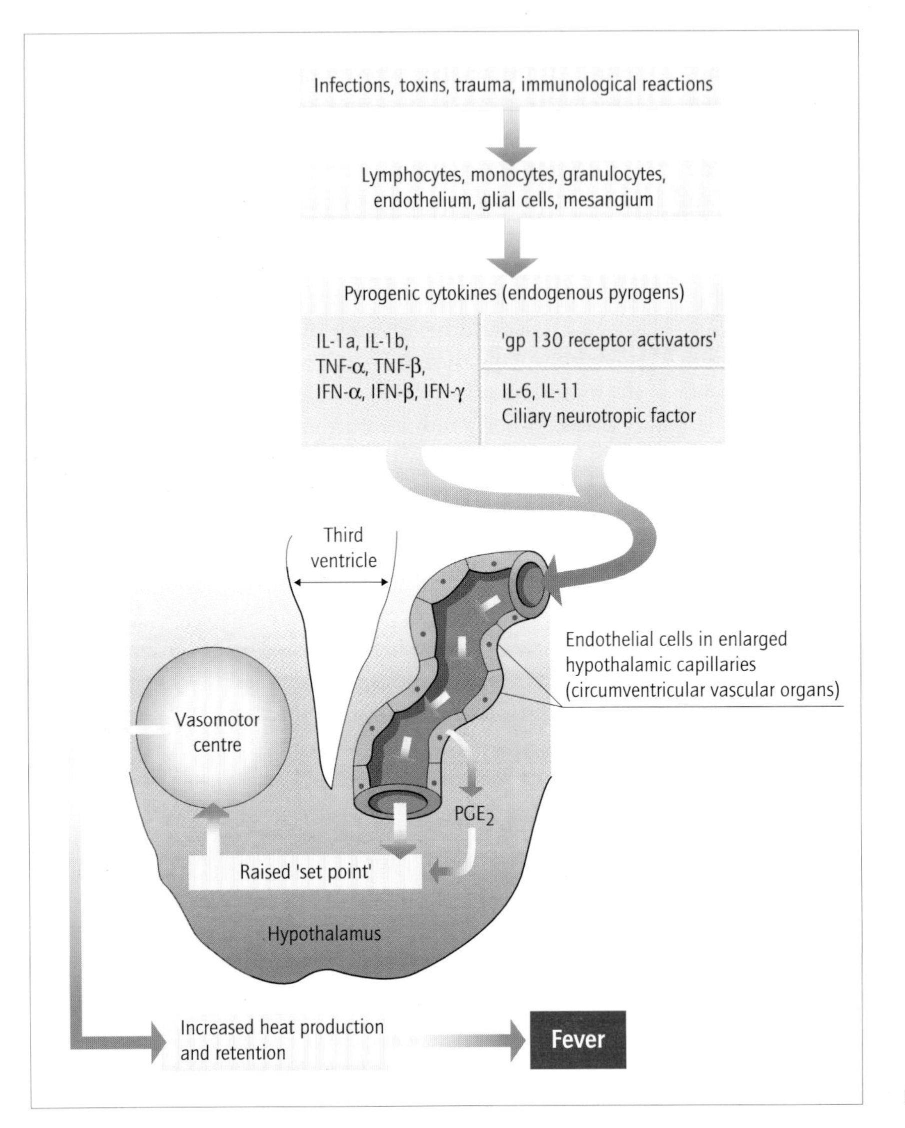

Fig. 1.8 The mechanism of fever.

This is done by sponging the patient with tepid (not cold) water, or by giving antipyretic drugs. Aspirin is an effective antipyretic, but should not be given to children under 12, because of the association of aspirin treatment with Reye's syndrome. Paracetamol is equally effective and is the drug of choice for children.

Convulsions can usually be terminated by lowering the body temperature. If this is unsuccessful anticonvulsant treatment is given. Often a once-only dose of lorazepam or diazepam is enough to avert further attacks during a brief illness. On rare occasions a short course of regular anticonvulsant dosage is required; in this case valproate or carbamazepine is the drug of choice, as in true childhood epilepsy.

Pre-existing disorders

Pre-existing disorders may be adversely affected by fever. Epilepsy may become poorly controlled, and is better managed, if possible, by treating fever than by altering established drug routines. The neurological deficits of multiple sclerosis are reversibly exacerbated by fever. Transient cerebral ischaemic events may recur during fever.

Inflammation

Inflammation is a complex combination of events, whose pathogenesis is still poorly understood. Several important components can be recognized:

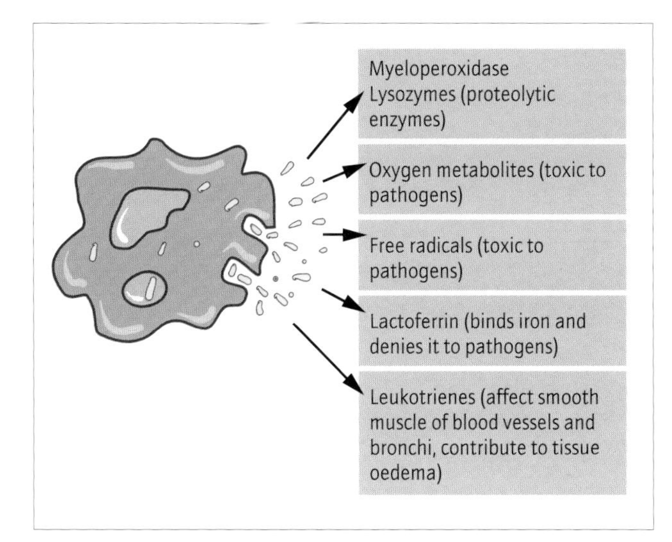

Fig. 1.9 Active substances released by neutrophils.

1 Vasodilatation at the affected site.
2 Exudation of tissue fluid from dilated capillaries.
3 Accumulation of neutrophils and macrophages at the site.
4 Release of active chemicals from neutrophils (Fig. 1.9).

These events combine to cause local heat and redness, sometimes with the advantage of adversely affecting the responsible pathogen. The tissue exudate contains complement components, which comprise one of the immediate defences against bacteria. Some complement activation products are chemotaxins and attract phagocytes to the site.

The phagocytes ingest bacteria and debris, becoming activated and releasing chemicals which both attack the pathogens and contribute to inflammation. These include enzymes which promote rapid synthesis of prostaglandins, a variety of chemicals that are vasoactive and also affect platelet activation. Prostaglandins are important initiators of inflammation. The effects of non-steroidal anti-inflammatory drugs are due to their strong inhibition of prostaglandin synthesis.

Useful anti-inflammatory drugs
1 Ibuprofen (adult and child preparations).
2 Aspirin (over 12s only).
3 Intramuscular or rectal diclofenac.

Detecting inflammation

The signs and symptoms of inflammation are pain, heat, redness and swelling. They are helpful in indicating the site of a localized infection, for instance an abscess or an infected joint, and should always be sought during clinical examination when assessing a patient with suspected infection.

Neutrophils which collect at sites of inflammation may be shed, e.g. in the urine, or discharged from the tissues as pus. Microscopical examination of the appropriate specimen will therefore reveal evidence of infection, even when the patient cannot indicate the affected site.

The swelling of inflammation may be deep in the body and undetectable by surface examination. An X-ray may show the tell-tale soft-tissue shadow, isotope scans may demonstrate the site of hyperaemia, and other imaging procedures such as computed tomography or nuclear magnetic resonance scans are excellent for demonstrating oedema.

Acute phase proteins

Several plasma proteins show a large rise in concentration in the presence of inflammation. Notable among these are caeruloplasmin, ferritin, haptoglobin, alpha$_1$-antitrypsin, alpha$_1$-glycoprotein (orosomucoid) and C-reactive protein (CRP). Levels of transferrin, fibronectin and albumin tend to fall.

The function of these changes, which can be induced by prostaglandins, interferon-alpha or IL-1, is unknown. Caeruloplasmin and haptoglobin will bind to oxygen radicals, perhaps inhibiting their damaging potential in blood. Alpha$_1$-acid glycoprotein can inhibit platelet aggregation, possibly protecting against platelet activation and thrombus formation.

C-reactive protein

C-reactive protein is produced in the liver, and synthesis is greatly increased in acute inflammation. It is a disc-shaped pentameric molecule which readily binds a number of substances, including the C fraction of pneumococcal lysates (from which it gets its name). In its bound form it strongly activates the classical complement pathway, possibly acting as a non-specific defence against infection. It is elevated in many acute bacterial and viral infections and other inflammatory conditions. Because of its rapid response to inflammation it is useful for monitoring responses to treatment in conditions such as endocarditis.

Plasma viscosity and erythrocyte sedimentation rate

The protein changes in inflammation alter the viscosity of the plasma. This can be measured directly, but is more often inferred from changes in the erythrocyte sedimentation rate (ESR: the rate at which red blood cells settle in anticoagulated blood on standing). The normal ESR is not more than about 20 mm/h. This rises to 30–50 mm/h in acute infections, but may reach 70–100 mm/h in some atypical pneumonias and chronic conditions such as abscess formation or immunological disease. It has non-specific diagnostic value, like CRP levels, but responds less rapidly than CRP to changes in the degree of inflammation.

Rashes

Rashes are a particular form of inflammation or tissue damage, affecting the skin. The causative pathogen may be present in the lesions. They can be generalized or localized. The rash of an infectious disease often evolves in a predictable way, starting at a particular site, spreading in a particular direction and containing typical types of skin lesions. Some of the skin lesions that occur in rashes are illustrated in Fig. 1.10.

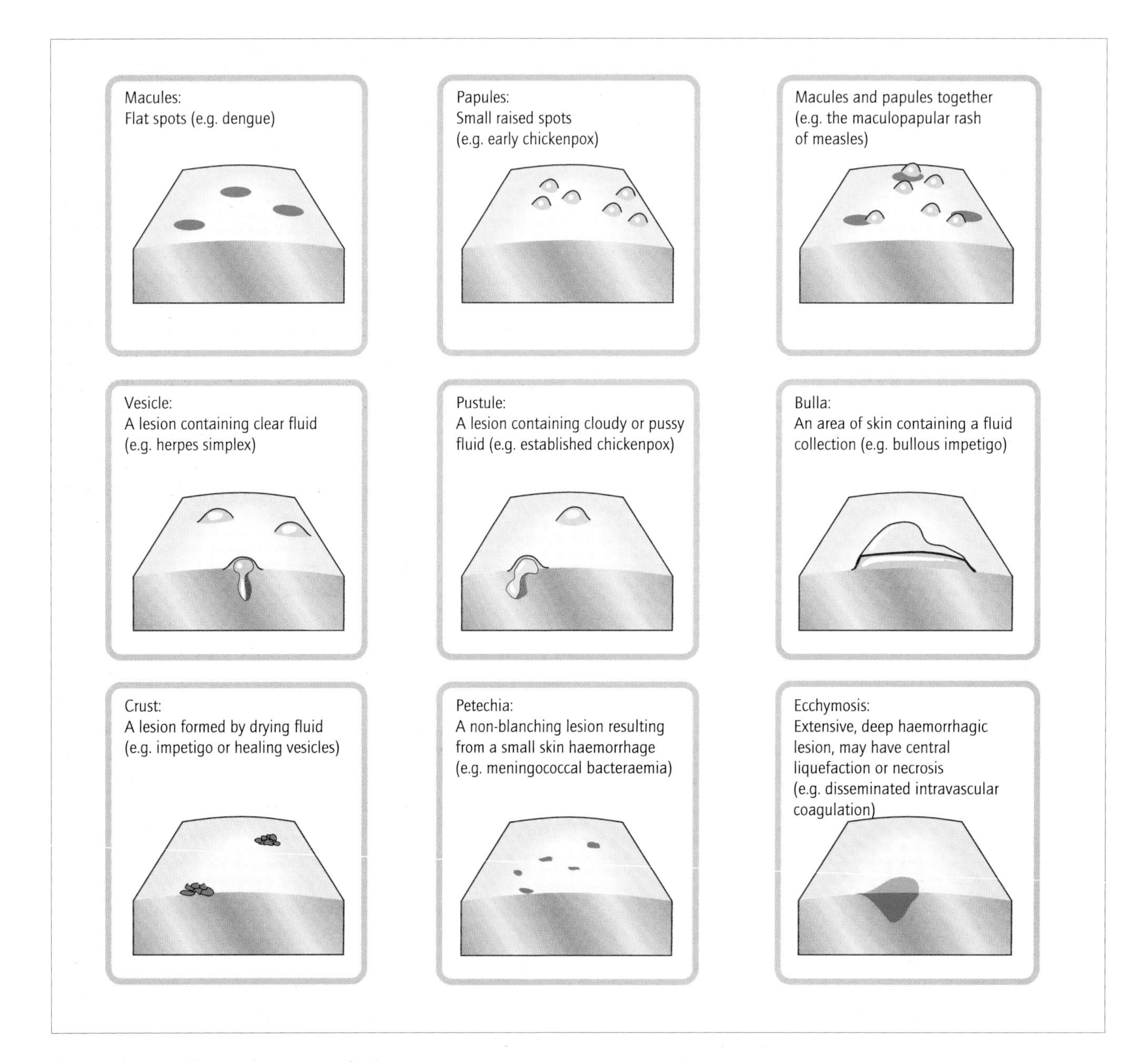

Fig. 1.10 The nomenclature and appearance of rashes.

The rashes of infectious diseases, unlike those of hypersensitivity reactions, are rarely painful or irritating. The lesions of chickenpox may itch quite severely, but this is not so in every patient. However, in rashes caused by severe tissue damage, e.g. the meningococcal rash caused by intravascular coagulation, the more necrotic lesions can be painful.

Harmful effects of immune responses

Immune reactions can be clinically detectable as part of the acute disease, or as a late effect of the disease. This is described as the immunopathology of the disease.

The rashes of some viral infections such as scarlet fever are the manifestation of an immune vasculitis of the skin. The lung damage of respiratory syncytial virus infection is immunopathological, and can be made worse in experimental conditions by immunization against the virus.

Antibodies which accidentally damage human tissues may be manufactured in the course of an infection. Examples include immune thrombocytopenia after rubella and other viral infections, and red-cell agglutination in *Mycoplasma pneumoniae* infections.

Interferon is a lymphokine with many effects, including the inhibition of viruses and reduction of the metabolic activity of virus-infected cells. It also causes the symptoms of fatigue, malaise and myalgia that are typically seen in acute viral infections. High concentrations may contribute to the neutropenia of some viral diseases by a toxic effect on the bone marrow.

The primary function of cell-mediated immunity is to destroy infected cells. Occasionally a very vigorous response can cause severe tissue damage, such as hepatic necrosis in viral hepatitis. It is thought that a similar but slower-onset mechanism is responsible for postviral encephalitides.

Antibody–antigen complexes often form during immune reactions to infection. Most are harmlessly destroyed or cleared, but some may lodge in tissues such as glomerular capillaries or synovial membranes. If they combine with complement there is a risk that the complement will be activated, causing local inflammation and tissue damage. This takes time to develop, but the late effects can produce autoimmune-like postinfectious disorders. Rheumatic fever after *Streptococcus pyogenes* infections is a classic example of this, but postinfectious arthritis, nephritis and neuritis are nowadays more common (see Chapter 21).

These relatively rare complications of infection probably depend also on genetic factors in the patient. A good example of this is the predisposition of HLA B27-positive individuals to develop Reiter's syndrome. Other recognized factors include secretor status—non-secretors appear to be at greater risk of acquiring disease from organisms which they carry and also of developing some postinfectious disorders.

An immunocompromised patient may lack the immunopathological features of a disease. Thus, a child with leukaemia may have severe respiratory features of chickenpox with little or no rash, or a patient with AIDS may fail to develop granulomata in organs infected by mycobacteria (see Chapter 18).

Dynamics of colonization and infection

When a microbe encounters a potential host, a sequence of events takes place. On making contact with the host's mucosa or skin, an organism may be able to adhere to and colonize this surface. If successfully established, colonization often continues without ill effect for a variable length of time. During this period the host may develop immunity to the organism. This is a common means of development of immunity to a number of pathogens such as *Haemophilus influenzae* and *Neisseria meningitidis*. This process is also important since organisms which have little capacity to cause disease may share some antigenic markers with human pathogens. Antibodies developed to these agents of low pathogenicity may provide immunity to powerful pathogens. This effect of cross-immunity is exploited when bacillus Calmette–Guérin (BCG) vaccine is given to confer protection against tuberculosis or leprosy.

Alteration of the host–microbial interaction may permit a change from colonization to invasion of local tissues or the whole body. For many infectious agents the majority of interactions are restricted to colonization.

For other agents, such as poliomyelitis viruses, invasion of the host may take place as part of the life cycle of the organism. In an unimmunized population, newly exposed to the virus, many individuals will be infected. The great majority suffer only a mild bowel or throat infection and become immune to further attacks. In a few cases this infection is complicated by self-limiting viral meningitis, and a minority of meningitis cases develop anterior horn cell infection and paralysis. Paralysis is most likely to affect older children and adults. In populations where the virus is common almost all adults are immune, so most infections occur in young children, who rarely develop paralysis. The disease therefore exists in equilibrium with the population, where the combination of host factors and microbial pathogenicity does not favour the occurrence of symptomatic or severe disease. This situation is often

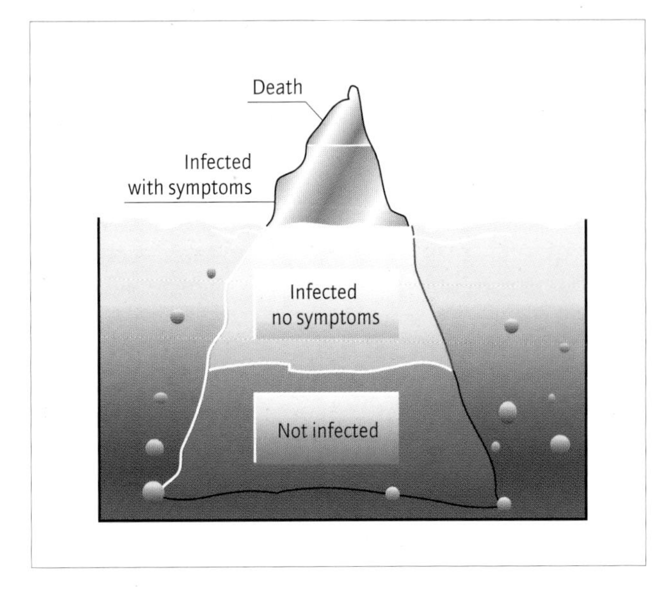

Fig. 1.11 The epidemiological iceberg.

described as 'the iceberg of infection' where the majority of host–microbial interactions are colonization–clearance episodes and only a small proportion result in morbidity or mortality (Fig. 1.11).

The manifestation of an infectious disease is a complex balance between the direct effect of the pathogen or its toxins, and the response of the affected patient. The patient's response depends on several factors, including immune competence, previous experience of the same or similar pathogens and his or her own genetic structure.

Case 1.1: Tuberculosis and the 'iceberg of infection'

History: Two 6-year-old classmates developed tuberculous meningitis within 3 weeks of one another. Both were English children, born and brought up in a prosperous town in East Anglia, where tuberculosis is extremely uncommon.

Question: How should this unusual event be followed up?

Epidemiological investigation: Both cases were notified to the Proper Officer of the Local Authority, as required by statute. The consultant in communicable disease control visited the school to gather more information about the teacher and classmates. The children belonged to a class of 25 6- and 7-year-olds. None had received BCG vaccination, as it would not normally be offered before the age of 11–13. On enquiry, it was found that their teacher had a 10-week history of persistent cough, and the teacher was found on subsequent investigation to have infectious pulmonary tuberculosis.

The 23 classmates all had tuberculin skin testing, and 17 of them tested tuberculin positive (usual prevalence in this community <2%). Four of these had X-ray evidence of pulmonary or pleural tuberculosis.

Question: What action should be taken for the different groups of children?

Management and progress: The six children with tuberculosis were all treated and recovered. The tuberculin-positive children received prophylactic chemotherapy with daily isoniazid for 6 months, as recommended for tuberculin-converters under the age of 16 years (who are at high risk of developing disease); non-infected children were offered immunization with BCG.

This clearly demonstrates the iceberg of infection—8% of exposed children had severe infection; 16% had milder clinical disease; 52% had immunological evidence of infection without demonstrable clinical disease; 28% escaped infection and remained susceptible.

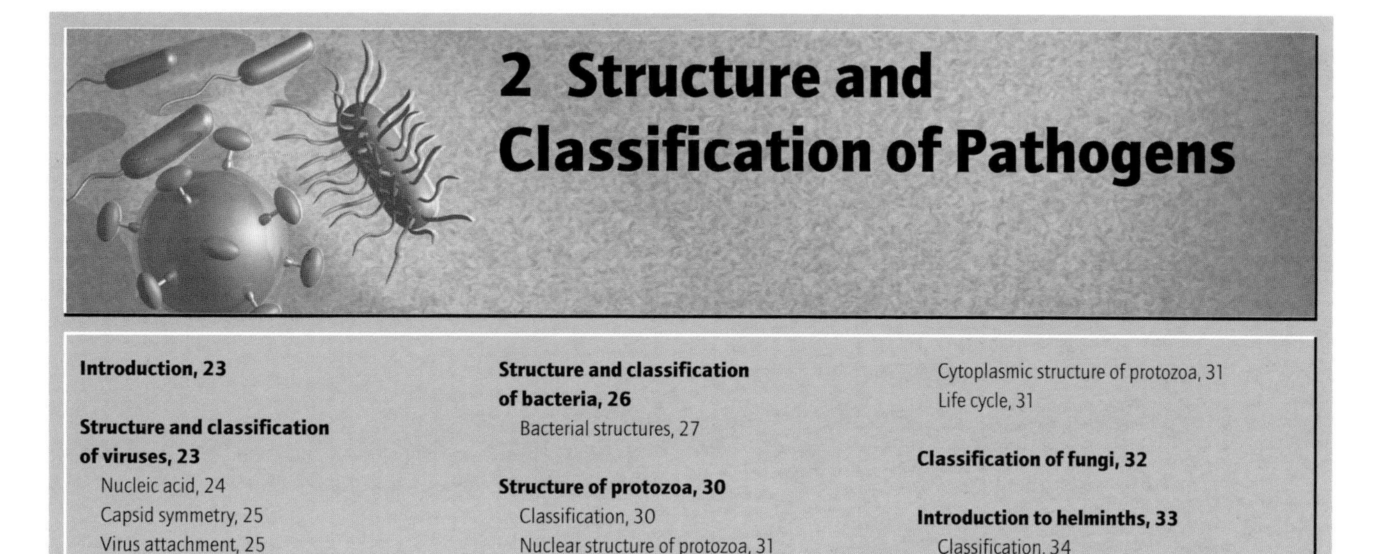

2 Structure and Classification of Pathogens

Introduction, 23

Structure and classification of viruses, 23
Nucleic acid, 24
Capsid symmetry, 25
Virus attachment, 25

Structure and classification of bacteria, 26
Bacterial structures, 27

Structure of protozoa, 30
Classification, 30
Nuclear structure of protozoa, 31
Cytoplasmic structure of protozoa, 31
Life cycle, 31

Classification of fungi, 32

Introduction to helminths, 33
Classification, 34

Introduction

Taxonomy is the process of classifying living organisms into groups (taxa) of related individuals or species. While taxonomists weigh individual characteristics equally, clinical microbiologists do not, as they consider ability to cause disease more important than, say, the ability to ferment a particular sugar. The microbiologist aims to classify microorganisms into clinically relevant groups and, by classifying them, to predict their behaviour.

Pathogens can be classified into five main groups: viruses, bacteria, fungi, protozoa and metazoa (usually helminths). Viruses are smaller than bacteria and consist of a piece of either DNA or RNA supported by nucleoprotein, some enzymes for replication and a 'casing' of structural protein. Some viruses possess an envelope which is derived from host cells (Fig. 2.1). Viruses cannot replicate independently but grow inside host cells, taking control of cellular biochemical processes and subverting them for virus production.

Classification of microorganisms
Viruses, bacteria, protozoa, fungi and metazoa.

Bacteria are single-cell organisms with a single circular DNA chromosome, which is not enclosed in a nucleus (such organisms are called prokaryotes). They have a plasma membrane, and a cell wall of characteristic composition. They replicate by binary fission. Specialized surface structures serve special functions: pili and fimbriae are for attachment and flagella for motility (Fig. 2.2), but they have no internal organelles. Some bacteria can form extremely durable spores. The detailed structure of bacteria is discussed below.

Mycoplasmas are small bacteria, lacking peptidoglycan-containing cell walls. They are the smallest organisms able to live and replicate independently.

Rickettsiae and chlamydiae are also small, and structurally resemble Gram-negative bacteria. They lack some enzymes needed for independent existence, and must replicate intracellularly, borrowing host enzyme systems.

Protozoa are single cells whose name is derived from the Greek words for first animals. They are diploid, possessing paired chromosomes, and as they also possess a nucleus they are eukaryotes. They have several specialized organelles (see below). Protozoan life cycles may be simple, involving only one species, or require intermediate hosts or vectors.

Fungi are also eukaryotes with cell walls containing chitin, cellulose or both. They reproduce by sexual or asexual processes, forming germinative spores.

The final group of organisms that must be considered are the metazoa or multicellular organisms. Virtually all metazoa pathogenic to humans are helminths.

Structure and classification of viruses

The classification of pathogenic viruses developed later than the classification of other organisms. Original classifications of viruses simply described the diseases caused (e.g. rubella), the site of viral shedding (e.g. enteroviruses) or the means of transmission (arthropod-borne; arboviruses). Modern classification is based on genetically determined structures (Table 2.1). Three determinants are considered in the classification of viruses: (i) the type of nucleic acid and its means of transcription; (ii) the structure and symmetry of the structural proteins (capsids); and (iii) the presence or absence of an envelope.

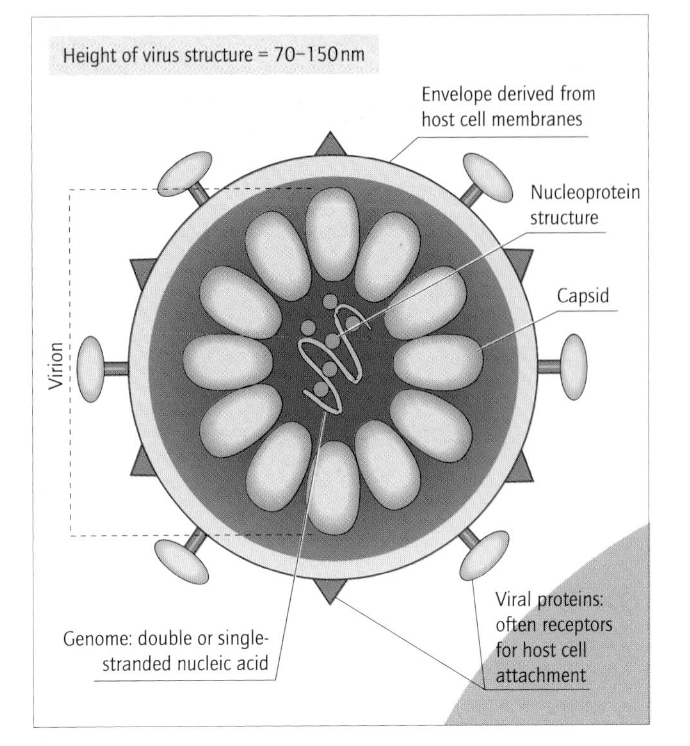

Fig. 2.1 General structure of a virus.

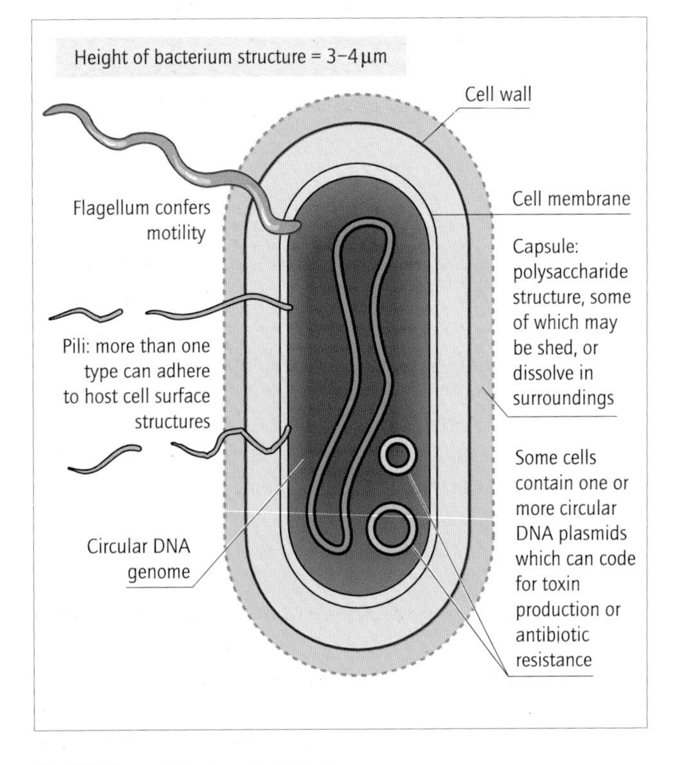

Fig. 2.2 General structure of a bacterium.

Genomic structure in viruses

DNA viruses
Double-stranded DNA.
Single-stranded DNA.

RNA viruses
Positive single-stranded RNA (sense).
Negative single-stranded RNA (antisense).
Double-stranded RNA.
Positive single-stranded RNA which cannot act as messenger (retroviruses).

Nucleic acid

Both the type of nucleic acid and its method of transcription are considered important, as there is considerable variation in the mechanisms that viruses employ in their reproductive process.

DNA viruses

The DNA of viruses can be either double- or single-stranded (ds or ss). Double-stranded DNA viruses include a number of important families that cause human disease. The poxvirus family, possessing the largest viral genomes, includes the agents of smallpox, molluscum contagiosum and also vaccinia virus. Vaccinia virus was the essential factor in the smallpox eradication programme, and may in the future be used as a vaccine vector. The herpesvirus family includes herpes simplex, varicella zoster, cytomegalovirus and Epstein–Barr virus. Adenoviruses also possess double-stranded DNA. Hepatitis B virus is double stranded with a single-stranded portion. The papilloma and polyomaviruses are small viruses containing double-stranded DNA. These viruses are associated with benign tumours (warts) and malignant tumours (cervical, genital and laryngeal cancer).

There are two single-stranded DNA viruses: parvovirus, responsible for 'fifth disease' or 'slapped cheek syndrome' and TT virus, which is a blood-borne virus transmissible by transfusion.

Viral DNA is usually replicated in the nucleus of host cells, using viral DNA polymerase. Newly formed viral DNA is not inserted into host chromosomal DNA, but is assembled with the nucleoprotein and capsids to form new virus particles.

RNA viruses

Single-stranded RNA viruses adopt one of three repro-

Name of family	Type of nucleic acid	Genome size (kB)	Envelope	Examples: genera of medical importance
Poxviridae	ds-DNA	130–280	No	Molluscum contagiosum
Herpesviridae	ds-DNA	120–220	Yes	Herpes simplex, Epstein–Barr virus, etc.
Adenovirus	ds-DNA	36–38	No	Adenovirus
Papovaviridae	Circular ds-DNA	8	No	Human wart virus
Parvoviridae	ss-DNA	5	No	Parvovirus
Hepadnaviridae	ds-DNA with ss portions	3	Yes	Hepatitis B virus
Picornaviridae	ss(+)RNA	7.2–8.4	No	Poliovirus
Togaviridae	ss(+)RNA	12	Yes	Rubella virus
Flaviviridae	ss(+)RNA	10	Yes	Yellow fever virus
Caliciviridae	ss(+)RNA	8	No	Norwalk agent
Rhabdoviridae	ss(−)RNA	13–16	Yes	Rabies virus
Paramyxoviridae	ss(−)RNA	16–20	Yes	Measles virus
Orthomyxoviridae	ss(−)RNA	14	Yes	Influenza virus
Reoviridae	ds segmented RNA	16–22	No	Rotavirus
Arenaviridae	ss(−)RNA	10–14	Yes	Lassa
Retroviridae	ss(+)RNA	3–9	Yes	HIV-1

Table 2.1 Classification of viruses.

duction strategies, depending on whether the RNA is positive or negative (sense or antisense.) If it is positive, or sense, it can serve directly as messenger RNA (mRNA) and be translated into protein, which includes structural proteins and an RNA-dependent RNA polymerase which is used to replicate the viral RNA.

If the RNA is negative or antisense, a different strategy must be adopted. These viruses encode an RNA polymerase (or transcriptase) which transcribes the viral genome into positive RNA, which then acts either as a template for further viral genomic (negative-strand) RNA or as mRNA for translation into proteins (Fig. 2.3).

The third RNA-based strategy is that of the retroviruses, which have single-stranded positive (sense) RNA that cannot act as mRNA. The RNA is transcribed into DNA by an RNA-dependent DNA polymerase (reverse transcriptase, or RT). The DNA enters the host nucleus and is inserted into host DNA. In the host DNA molecule, it is under the control of host transcriptase enzymes which make mRNA and viral genomic RNA.

The reoviruses, which include rotaviruses, possess a segmented, double-stranded RNA genome. Their complex reproductive strategy includes the use of a double-stranded RNA–single-stranded RNA polymerase which produces positive (sense) RNA from the double-stranded portion of viral RNA using the antisense strand as a template. The positive (sense) RNA is extruded from the virus, serving both as mRNA and also as a template to make further antisense RNA, which is then annealed with the complementary strand to form double-stranded RNA.

Capsid symmetry

Viral nucleic acid is contained within a protein coat made up of repeating units known as capsids arranged in either icosahedral or helical structures. In RNA viruses with helical symmetry the capsids are bound around the helical nucleic acid (Fig. 2.4a). Icosahedral symmetry occurs in both DNA and RNA viruses, the capsids forming an approximately spherical polyhedral structure (Fig. 2.4b). Using simple, repeating structures minimizes the amount of nucleic acid devoted to viral coat production and simplifies the process of viral assembly.

Envelopes

In some viruses the nucleic acid and capsid (the nucleocapsid) are surrounded by a lipid envelope derived from the host cell or nuclear membrane (Fig. 2.5). The host membrane is altered by virus-encoded proteins or glycoproteins which may fuse several host cells together, thus facilitating the passage of viruses from cell to cell. Viruses possessing an envelope are sensitive to ether and other substances that dissolve the lipid membrane.

Virus attachment

Virus attachment to host cell membrane is a critical step in the process of infection, and in determining tissue tropism. Viruses have evolved specific antigens that target receptors on the host cells. These are sometimes known as virus attachment proteins or VAPs. The VAP of influenza virus is a haemagglutinin, whereas that of human

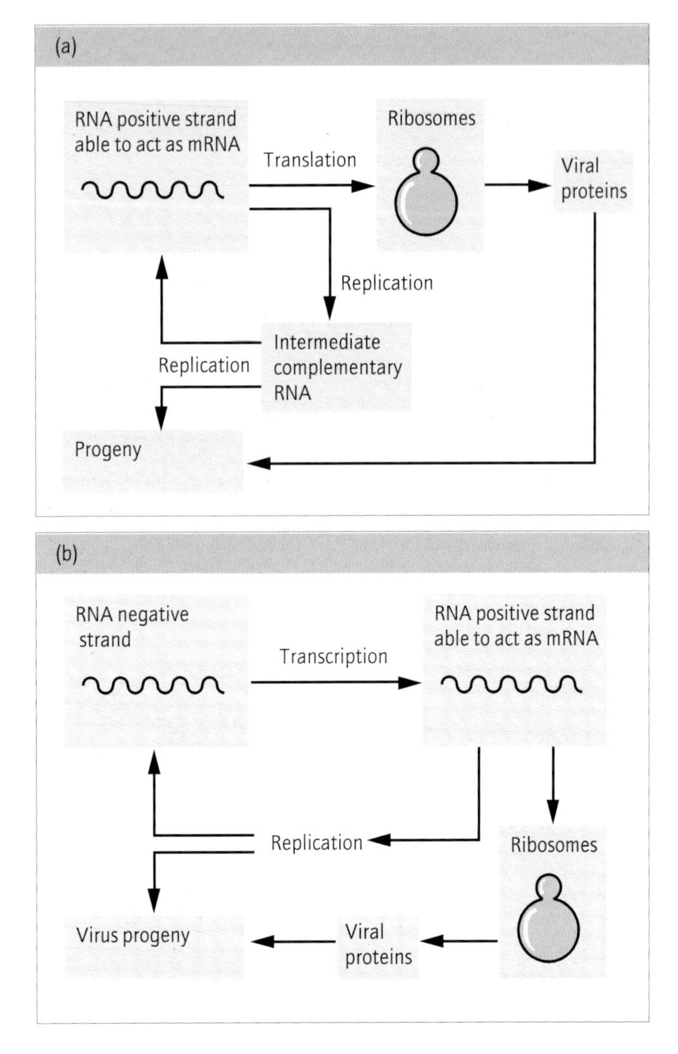

Fig. 2.3 (a) Replication of RNA virus with positive polarity (able to be read as messenger (m)RNA). (b) Replication of RNA virus with negative polarity.

immunodeficiency virus (HIV) is a glycoprotein which binds to the CD4 antigen of T cells.

Structure and classification of bacteria

For descriptive purposes bacteria are often grouped by four main characteristics: the Gram reaction, shape, atmospheric requirements for respiration and the presence of spores.

Classification of bacteria
Gram reaction, shape, atmospheric requirement, presence of spores.

The Gram reaction or Gram stain uses the ability of stains such as crystal violet to bind to the cell wall teichoic acids found in Gram-positive cells, and thereby resist decoloration by alcohol or acetone. It is a simple way of identifying bacteria with different cell wall structures (Fig. 2.6). This is important, as Gram-positive and Gram-negative organisms have different pathogenic potential and different antibiotic susceptibilities.

Bacterial cell walls determine the organisms' shape which is consistent for individual genera. Cocci are spherical and include important human pathogens such as streptococci and staphylococci. Bacilli are rod-shaped and may be short (coccobacilli) or long. They may also vary their shape throughout their length, as do *Fusobacterium* spp. Spiral bacteria are a diverse group, including *Treponema pallidum*, the causative organism of syphilis, which has a very short wavelength, and *Borrelia* spp., which have a longer wavelength, and include the organisms of Lyme disease and relapsing fever. Other spiral

(a) 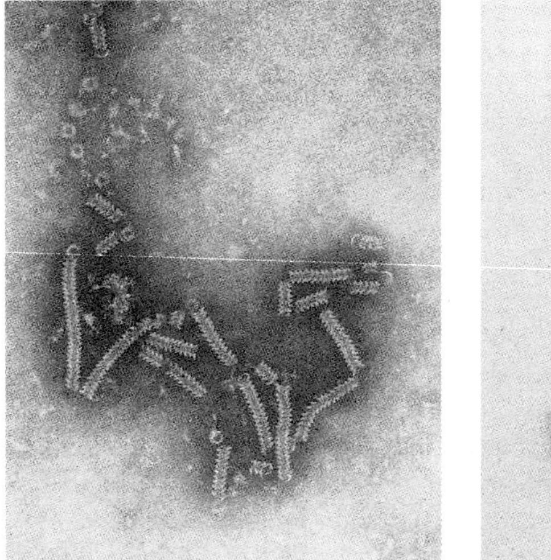

(b) 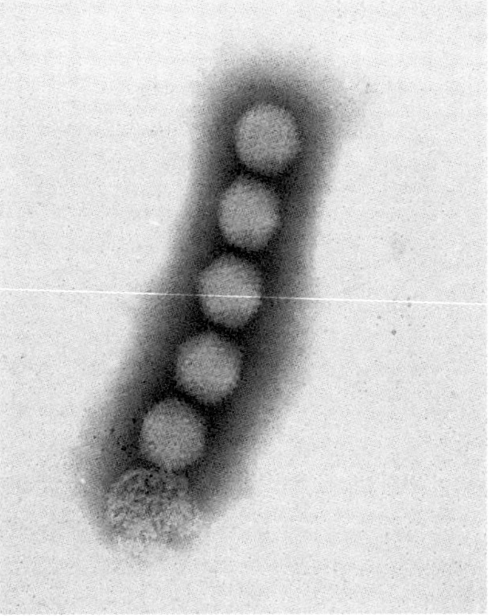

Fig. 2.4 (a) Electron micrograph of a virus with helical symmetry. Parainfluenza type 3 virus × 100 000. (b) Electron micrograph of a virus with icosahedral symmetry. Adenovirus × 100 000.

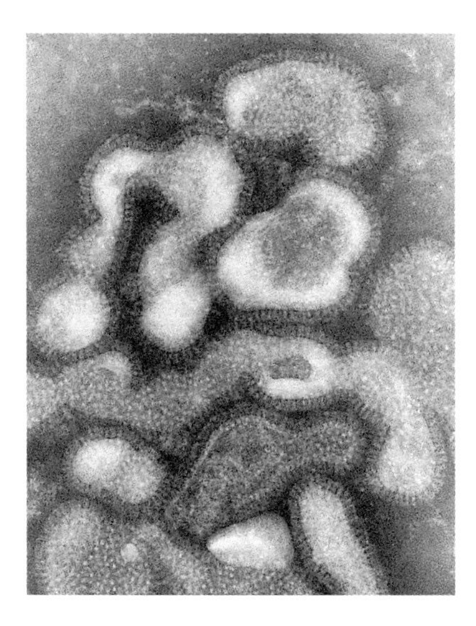

Fig. 2.5 Electron micrograph of an enveloped virus. Influenza virus × 100 000.

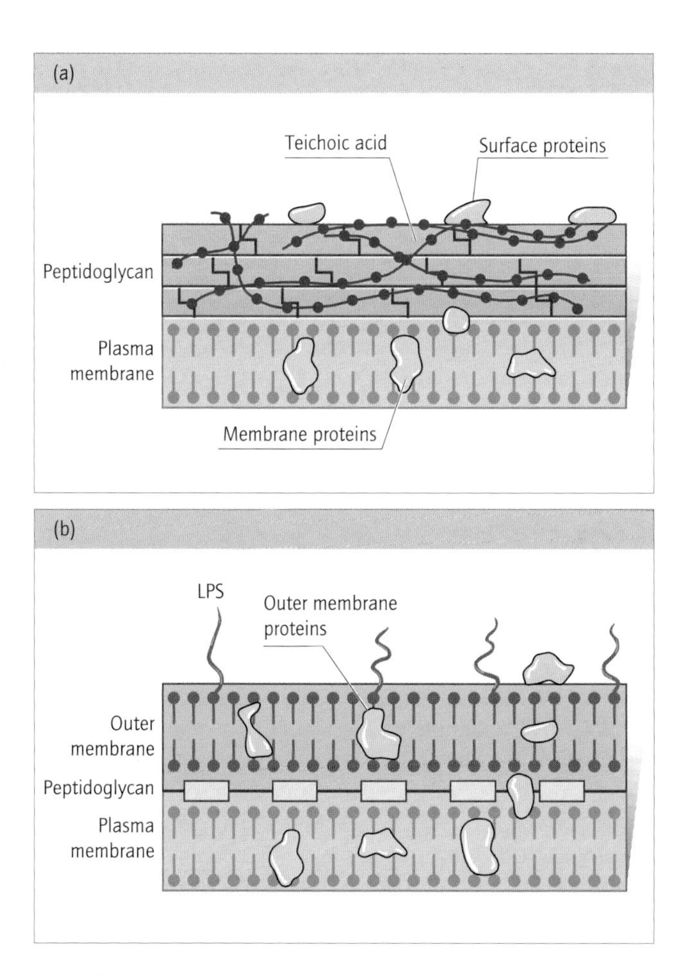

Fig. 2.6 Structures of (a) Gram-positive and (b) Gram-negative bacterial cell walls. LPS, lipopolysaccharide.

organisms, such as the intestinal pathogens *Campylobacter* and *Helicobacter*, have only a few turns.

Bacteria may be classified by their atmospheric requirements into five groups. **Obligate aerobes** are obliged to use oxygen as a terminal electron acceptor, e.g. *Bordetella pertussis*. **Microaerophilic** organisms use oxygen as the terminal electron acceptor but only grow under conditions of reduced oxygen tension, e.g. *Campylobacter* spp. **Capnophiles** are organisms, e.g. *Brucella*, that have enhanced growth in an atmosphere with an increased carbon dioxide concentration. Many human pathogens are **facultative anaerobes** capable of growing aerobically or anaerobically, e.g. staphylococci, streptococci and enteric organisms. **Obligate anaerobes** only grow in the absence of oxygen. They are divided into aerotolerant and strict anaerobes, depending on their sensitivity to oxygen. This difference has practical importance as strict anaerobes do not survive long in specimens and are much more difficult to isolate in the laboratory.

Atmospheric requirements of bacteria
Obligate aerobes, microaerophiles, facultative anaerobes, capnophiles, strict and aerotolerant obligate anaerobes.

Two genera of bacteria found in humans, *Clostridium* and *Bacillus*, possess a bacterial endospore. The shape, size and position of this spore may be helpful in the determination of species identification, especially in the genus *Clostridium* (Fig. 2.7).

Bacterial structures

The cell wall

The bacterial cell wall is essential for survival, as it must withstand the immense osmotic pressure difference between the interior and exterior of the cell. Its strength and rigidity depends on peptidoglycan, a polymer of muramic acid and *N*-acetylglucosamine, cross-linked by peptide bridges.

The Gram-positive cell wall has two layers, consisting of the plasma membrane and a thick peptidoglycan layer. In contrast, the Gram-negative wall has three layers: the plasma membrane, a thinner peptidoglycan layer and an outer membrane (see Fig. 2.6). Cell walls are important targets for beta-lactam and glycopeptide antibiotics (see Chapter 4).

Many bacterial antigens present in the cell wall have an important role in pathogenesis (see Chapter 1). Lipopolysaccharide (endotoxin) is an integral part of the Gram-negative outer membrane. Teichoic acid from *Strep-*

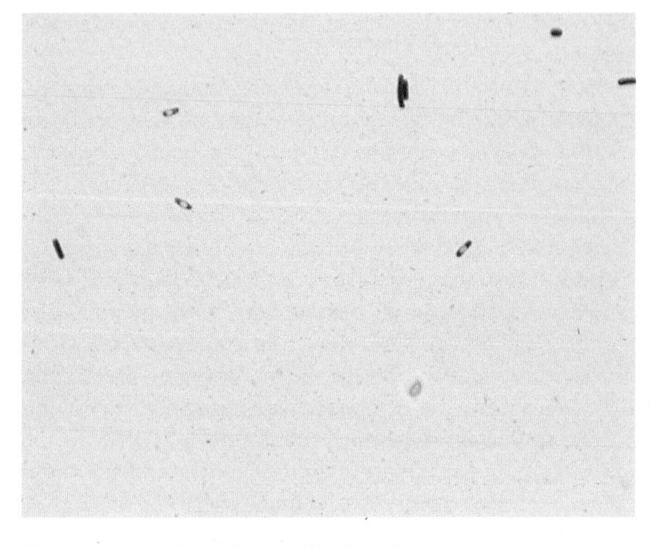

Fig. 2.7 Structure of *Clostridium* sp. showing endospore.

Species	Associated diseases
Streptococcus pneumoniae	Meningitis, pneumonia, septicaemia
Neisseria meningitidis	Meningitis, septicaemia
Haemophilus influenzae	Meningitis, septicaemia
Streptococcus agalactiae	Neonatal septicaemia and meningitis
Klebsiella pneumoniae	Pneumonia, septicaemia, wound infection
Pseudomonas aeruginosa	Pneumonia in cystic fibrosis patients
Escherichia coli	Meningitis, septicaemia
Bacillus anthracis	Anthrax

Table 2.2 Bacterial pathogens whose capsules are important in pathogenesis, and used to identify pathogenic strains.

tococcus pneumoniae causes complement activation and is responsible for attraction of neutrophils to the site of infection. The cell wall of mycobacteria contains approximately 40% lipid, including lipoarabinomannan, mycolic acid and phenolic glycolipids. It will not stain satisfactorily with conventional Gram stain. Vigorous staining and decolorizing methods are required to demonstrate it (see Chapter 18). Several of the surface lipids of mycobacteria are implicated in inhibiting macrophage function (e.g. lipo-arabinomannan) or phagocytosis (phenolic glycolipid from *Mycobacterium leprae*).

The plasma membrane

The bacterial plasma membrane, as in other cells, consists of a trilaminar membrane with two outer hydrophilic layers and an inner lipid core. In the absence of internal membrane-bound structures the plasma membrane is an important site of bacterial metabolism.

Bacterial capsules

These are polysaccharide structures surrounding some of the most important pathogens, including *Streptococcus pneumoniae, Haemophilus influenzae, Neisseria meningitidis, Streptococcus agalactiae, Salmonella typhi, Bacillus anthracis* and *Klebsiella pneumoniae*. *Pseudomonas aeruginosa* sometimes elaborates such abundant alginate capsular material that pus appears gelatinous at infected sites or on culture plates (Table 2.2). Capsules protect pathogens from phagocytosis and humoral immune attack, and inhibit activation of the alternative complement pathway (see Chapter 1).

Capsulation is not, in itself, a pathogenicity determinant as many organisms not usually recognized as human pathogens possess a polysaccharide capsule. The bacterial capsule may have evolved as a mechanism for enabling organisms to survive during conditions of desiccation.

Immune response to polysaccharide capsular antigens

The human host does not respond efficiently to polysaccharide antigens. Such antigens are called T-cell-independent, as T cells are bypassed in the development of antibody to them, and immunological memory is not therefore well established. T cells are, however, involved in the modulation of humoral responses to these antigens. Antibodies are mainly of the immunoglobulin G_2 (IgG_2) isotype. The multiplicity of capsular types expressed by some organisms makes repeated infections possible, and complicates the design of vaccines.

The spleen has an important role in the control of infection with capsulate organisms, as intrasplenic phagocytes can ingest particles which have not been opsonized; thus patients with splenectomy are especially susceptible to pneumococcal infection (see Chapter 22).

Protein 'capsules'

The M antigen of *Streptococcus pyogenes* is a fibrillar protein forming an extracellular capsule-like structure, which protects against phagocytosis by polymorphonuclear leucocytes. Strains of *S. pyogenes* cannot be ingested unless opsonized with specific anti-M antibody. There are more than 80 antigenically distinct M types, and immunity to one does not protect against strains carrying different M antigens. This accounts for the recurrent attacks of streptococcal tonsillitis that some individuals may experience throughout their life.

The M antigen may cross-react with a number of human tissue antigens, with important pathological consequences. Cross-reaction with myocardial antigens is found in many different M types and is thought to be responsible for the pancarditis of rheumatic fever. Infection with a second type of *S. pyogenes* may cause a recrudescence of rheumatic fever. Poststreptococcal glomerulonephritis is confined to a few M types, including 4, 12 and 49, resulting from cross-reaction with the glomerular basement membrane. Infection with another strain of *S. pyogenes* will not result in a recrudescence unless it too is a 'nephritogenic' strain (Table 2.3).

Glycolipid capsules

Many species export extracellular glycolipid material which appears to be important to the organism as a pathogenicity determinant. *Mycobacterium leprae* can survive inside macrophages. It has been shown that the phenolic glycolipid capsule inhibits the activity of the macrophage myeloperoxide–halide microbicidal system, possibly by scavenging free radicals.

Bacterial extracellular material
1 Polysaccharide capsules.
2 Extracellular slime.
3 M protein.
4 Phenolic glycolipid.

Bacterial adhesins
1 Capsular polysaccharide.
2 Extracellular slime.
3 Fimbriae.
4 Lectins.

Extracellular slime

Many organisms of low virulence such as *Staphylococcus epidermidis* colonize intravascular prosthetic devices. This causes fever or septicaemia in patients with intravenous cannulae or long-term intravascular devices such as Hickman catheters. Slime-producing organisms adhere better to the cannulae and more readily colonize and cause infection.

Pseudomonas aeruginosa is an important pathogen of children with cystic fibrosis. It can express a mucoid phenotype in which copious amounts of exopolysaccharide alginate is produced. Alginate is a non-repeating copolymer of beta-D-mannuronate and its C5 epimer, alpha-D-glucuronate, similar to an alginate produced by brown seaweed. Alginate production assists the organism because it enables the formation of microcolonies protected from humoral and cellular immune responses. Paradoxically, alginate-producing strains are often exquisitely sensitive to antimicrobial agents.

Fimbriae or pili

Fimbriae (pili) have an important role in adhesion and assist bacteria in allowing access to a new host or locating the organism close to a nutrient supply. Pili are filamentous proteins capable of binding to host antigens by acting as lectins (proteins that bind to carbohydrate residues). One bacterial strain may express several different pili and may up- and down-regulate different types as required. For some organisms such as *Neisseria gonorrhoeae*, pili are necessary for pathogenesis. Possession of pili may assist the adhesion of organisms to epithelia. Uropathic *Escherichia coli* express not only type 1 fimbriae, in common with many Enterobacteriaceae, but also P fimbriae, which bind to a digalactoside found in the urinary tract. The role of fimbriae is discussed in more detail on p. 215.

Flagella

The flagellum is the main organ of bacterial motility. It is made up of a globular protein, flagellin, arranged in a multistrand helix. The flagellum is bound to the plasma membrane via the motor unit, where energy provided by an ion gradient across the membrane is converted into rotary movement. Motility *per se* is not often a pathogenicity determinant but *Vibrio cholerae* use their motility to

	No. of isolates			
M type	Scarlet fever	Acute glomerular nephritis	Rheumatic fever	Puerperal infections
1	21	24	5	10
2	10	3	0	2
3	85	6	3	11
4	75	10	2	9
9	0	2	1	20
12	9	20	0	10
R28	8	3	0	48
49	0	8	0	2
75	3	0	0	15

Table 2.3 Number of *Streptococcus pyogenes* isolates from different sites from 1980–1990 by M type. Data courtesy of the Streptococcal Reference Laboratory, Public Health Laboratory Service.

burrow through instestinal mucus to the epithelium and strains lacking flagella are non-pathogenic.

Flagella are useful in the laboratory as an identification feature. Flagellar antigens are frequently used to classify *Salmonella* and *Shigella* serologically.

Pathogenicity islands

These are collections of genes that code for pathogenicity and virulence determinants. These genes may include toxins, adhesins, invasins or other virulence factors. Pathogenicity islands are found in both Gram-positive and Gram-negative pathogens and are large genomic regions often containing up to 200 kb of DNA. The G+C content (proportion of guanine plus cytosine in the DNA bases) often differs from the rest of the genome, indicating that the organism has acquired this section of DNA from another organism. In many instances the genetic segments are flanked by specific DNA sequences such as direct repeats or insertion sequence (IS) elements. They may also contain bacteriophage attachment sites and cryptic genes coding for bacteriophage integrases or origins for plasmid replication. These features indicate that pathogenicity islands are transmissible among bacterial populations by horizontal gene transfer, and probably contribute to rapid evolution in bacterial populations.

Structure of protozoa

Protozoa are unicellular microorganisms. They are found in almost every type of environment, over a wide range of pH (3–9.5), temperature, salinity and redox potential.

Classification

The classification of protozoa can be very complicated but may be simplified by subdivision into four main groups: spore forming, flagellate, amoeboid and ciliate.

Classification of protozoa
1 Sporozoa (e.g. *Microsporidium*, *Plasmodium*, *Toxoplasma*).
2 Flagellates (e.g. *Leishmania*, trypanosomes, *Trichomonas*).
3 Amoeboid (*Entamoeba histolytica*, *Acanthamoeba* spp.).
4 Ciliates (e.g. *Balantidium coli*).

Spore-forming protozoa

These can be divided into two groups.

1 Microspora have recently been recognized as human pathogens, particularly in acquired immunodeficiency syndrome (AIDS) patients (see Chapter 15). *Encephalitozoon cuniculi*, *Enterocytozoon beneusii*, *Encephalitozoon intestinalis*, *Pleistophora*, *Nosema connori* and *Vittaforma corneum* are the most important species reported.

2 Sporozoa. This group includes many important human pathogens. The sporozoa possess a complex of organelles known as the apical complex which is important in cell invasion. In the Eimeridia (which include *Toxoplasma gondii*), sexual reproduction occurs in the intestine of the definitive host (species of cat for *Toxoplasma*). In contrast, the sexual stage of haemosporidians, such as *Plasmodium* spp., occurs in mosquitoes, or in ticks for piroplasms such as *Babesia*.

Eimeridia:	*Isospora belli*, *T. gondii*, *Cryptosporidium parvum* *Sarcocystis* sp.
Haemosporidians:	*Plasmodium* spp.
Piroplasms:	*Babesia* spp.

Flagellate protozoa

These comprise the intestinal/genital flagellate protozoa and the blood flagellates. Intestinal flagellates include *Giardia intestinalis* and *Trichomonas* spp.

The blood flagellates are the Trypanosomatidae and are characterized by the possession of a kinetoplast (a dense body at the base of the flagellum). They are transmitted to humans by the bite of arthropod vectors: sandflies in the case of *Leishmania*, *Glossina* (tsetse) flies for African trypanosomiasis, and triatomid bugs for South American trypanosomiasis. *Leishmania* spp. exist in a flagellate form in the arthropod vector (Fig. 2.8) but after injection into the human host they live intracellularly as non-flagellate amastigotes.

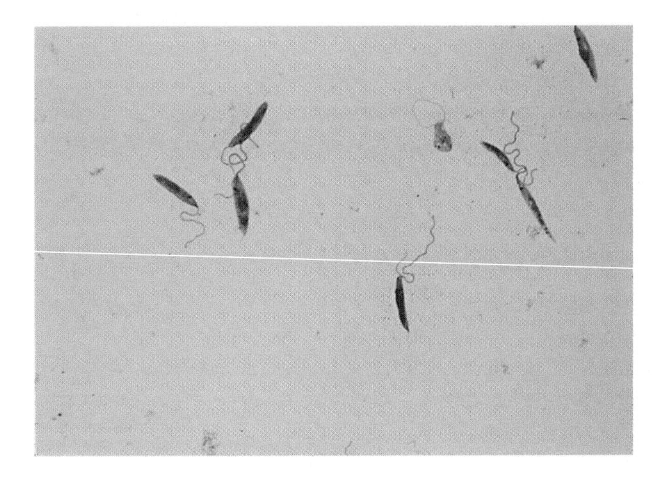

Fig. 2.8 Flagellate appearance of *Leishmania* sp.

Amoeboid protozoa

The most important genus of human amoebae is *Entamoeba*. The main recognized pathogenic species, *E. histolytica*, could be subdivided by electrophoresis of certain enzymes into pathogenic and non-pathogenic subgroups (or zymodemes). It is now known that these groups can be separated by monoclonal antibodies, lectin binding and serum sensitivity and these differences now define two different species: *E. histolytica*, which includes the pathogenic organisms, and *E. dispar*, which contains the non-pathogenic strains.

Naegleria fowleri causes a rare but usually fatal form of meningitis. It has an amoeboid form in the tissues, and a cyst and a flagellate form in the environment. It could therefore be classified with either group. *Acanthamoeba* can cause meningitis, and also a severe form of conjunctivitis associated with contaminated contact lenses.

Blastocystis hominis is an organism of uncertain pathogenic potential but it has been associated with diarrhoeal disease. Its taxonomic position is also uncertain but it is often classified with the amoebae. *Dientamoeba fragilis*, traditionally classified as an amoeba has now been reclassified as a trichomonad.

Ciliate protozoa

Balantidium coli is the only ciliate protozoan regularly found in human specimens. It is sometimes identified in faeces, but is rarely implicated in disease.

Nuclear structure of protozoa

Protozoa are eukaryotes, whose DNA exists as chromosomes within a nucleus. The nucleus is surrounded by a tough nuclear membrane which has multiple channels connecting the nucleoplasm with the endoplasm. Some organisms have a single nucleus, and others may be binucleate (e.g. *Giardia*) or have multinucleate cysts (e.g. *Entamoeba histolytica*). The nucleus may contain single or multiple linear chromosomes. Reproduction is by both sexual and asexual mechanisms.

Cytoplasmic structure of protozoa

The endoplasm is the inner portion of the cytoplasm immediately surrounding the nucleus. It contains chromatidoid bodies, endoplasmic reticulum, Golgi bodies, mitochondria, food vacuoles and microsomes. The ectoplasm is the metabolically active portion of the cell, and is involved in locomotion, respiration, osmoregulation and phagocytosis. The cell is limited by the plasma membrane which controls the intake and output of food, secretions and metabolic waste products. It may vary in shape and can often form pseudopodia for locomotion.

Protozoa have a number of structures external to the plasma membrane. These include an external glycocalyx (as in *Leishmania donovani*), or a tough cyst wall to enable the organism to survive in the external environment (as in *Giardia intestinalis*). Encystation is triggered by alteration in food or oxygen supply, excess of catabolic components, pH changes or desiccation. Some protozoa have specialized organelles: *Leishmania* spp. possess flagella which arise from a specialized kinetoplast containing its own DNA, and metabolic processes. Others, such as *B. coli*, express cilia, which beat in formation to propel the organism.

Life cycle

Protozoa have reproductive cycles which may involve a number of different hosts and environments. Vectors may be included, acting to transmit the infection between hosts and reservoirs of infection. An example of this is the life cycle of *Plasmodium* sp., the cause of malaria (Fig. 2.9). Vectors can also act as semipermanent reservoirs of infection by sustaining vertical transmission, for example certain viral encephalitides in generations of ixodid ticks.

Intermediate hosts

Examples of protozoa that may be transmitted by vectors

1 Mosquitoes (*Plasmodium*).
2 *Glossina* spp. flies (*Trypanosoma brucei*).
3 Sandflies (*Leishmania*).
4 Ticks (*Babesia*).
5 Bugs—triatomid (*Trypanosoma cruzi*).

Protozoa may also be transmitted by water, air, sexual intercourse, food and the faecal–oral route routes (Table 2.4).

Sexual stages of protozoal life cycles

Unlike bacteria which have only limited sexual function, based on transfer of genome fragments by various means, protozoa have well-developed sexual reproduction. In amoebae, simple exchange of DNA is thought to take place, and in ciliates conjugation occurs and microgametocytes are exchanged. Sporozoa have asexual and sexual generations in their life cycle; in coccidia, including *Toxoplasma*, the essential stages of schizogony and gametocyte and zygote formation take place in one host. In sporozoa, including malarial parasites, schizogony and gametocyte production occur in the vertebrate host and gametogeny is completed in the invertebrate host.

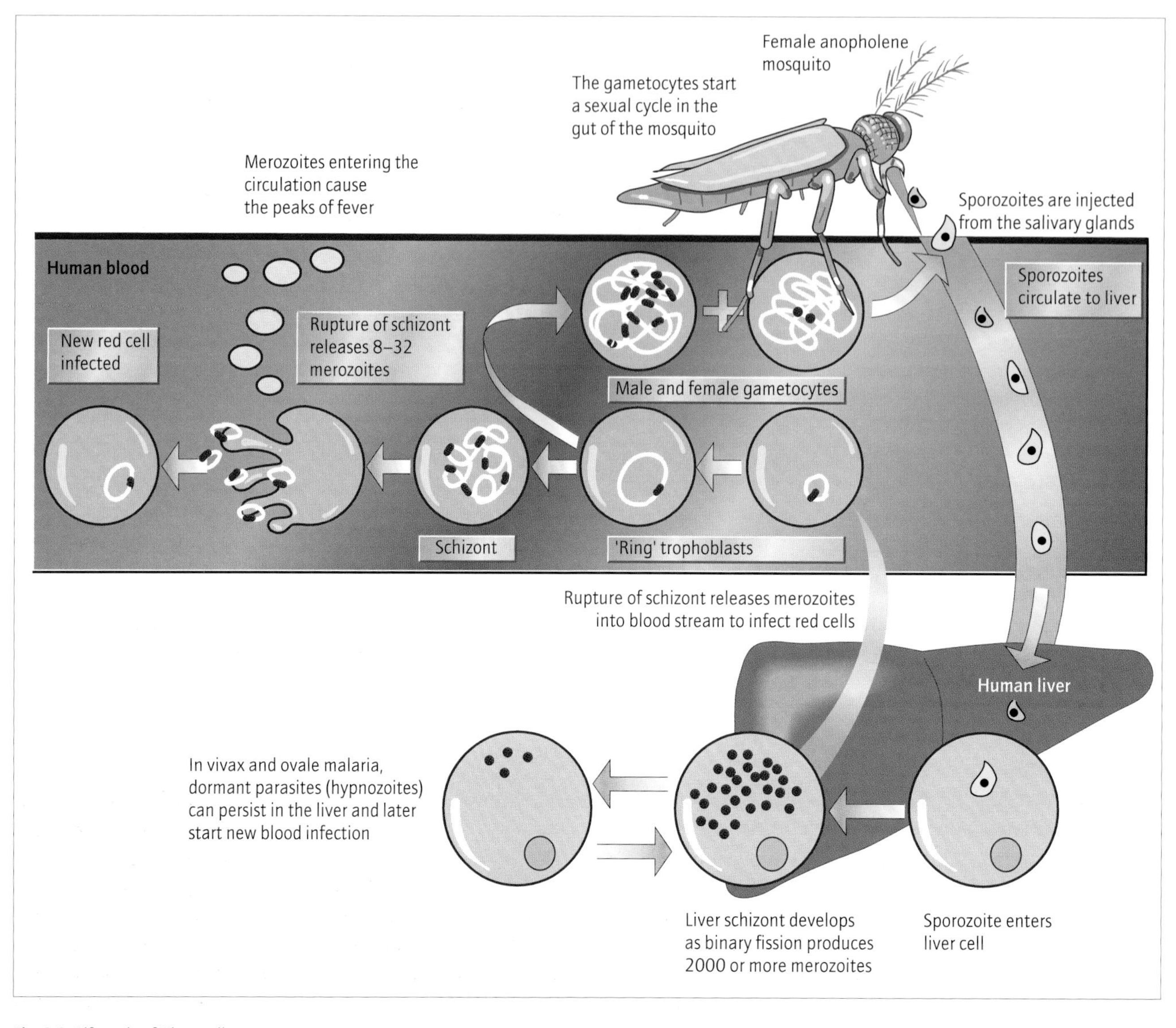

Fig. 2.9 Life cycle of *Plasmodium* sp.

Classification of fungi

Clinical classification of fungi of medical importance
Cutaneous, e.g. dermatophytes.
Systemic, e.g. *Coccidioides*, *Histoplasma*.
Fungal infections of immunocompromised hosts, e.g. *Aspergillus*.

Route	Example
Faecal–oral	*Entamoeba histolytica*
Water	*Cryptosporidium parvum*
Food	*Toxoplasma gondii*
Air	*Pneumocystis carinii*
Sexual	*Trichomonas vaginalis*
Vector	*Plasmodium vivax*

Table 2.4 Transmission of protozoa.

The formal classification of fungi is based upon means of reproduction and morphology of sexual and asexual stages (Fig. 2.10). Unfortunately, the schemes derived have little clinical relevance so a simplified clinical classification is described. This divides pathogenic fungal species into three groups: (i) the superficial and subcutaneous fungi; (ii) the systemic fungi; and (iii) the fungi associated with immunocompromised patients.

Superficial fungal infections are common. They include infection with dermatophytes, such as *Microsporum* and *Trichophyton*. The yeast-like fungus *Malassezia furfur* is

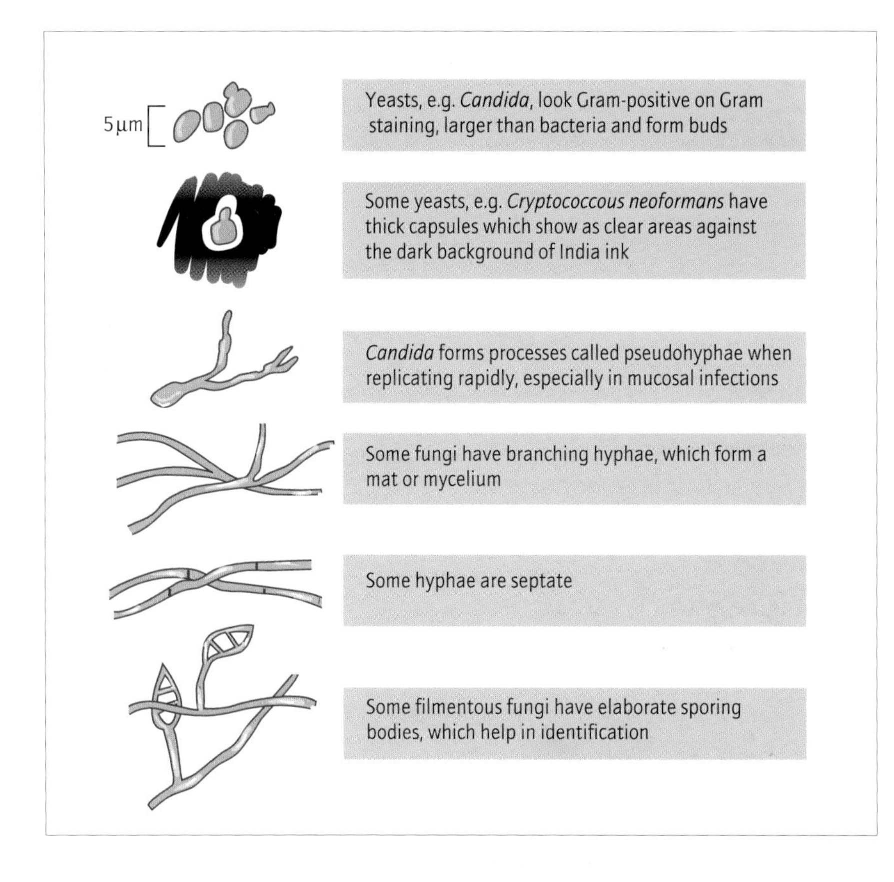

Fig. 2.10 The morphologies of pathogenic fungi.

the causative organism of pityriasis versicolor. Rarer cutaneous fungal pathogens include *Sporothrix schenckii*, the cause of sporotrichosis, and the agents of piedra.

The systemic fungi include *Histoplasma capsulatum*, *Coccidioides immitis* and *Paracoccidioides braziliensis*. These are dimorphic fungi, which have both yeast-like and filamentous forms. They are environmental organisms which usually enter the human body via the respiratory tract. Infection is geographically localized and often clinically mild. Severe disease can occur, however, particularly in immunocompromised subjects.

The main fungi that infect immunocompromised patients are the yeasts *Candida albicans* and related species such as *C. krusei* and *Torulopsis glabrata*. *Aspergillus* species are important filamentous fungi, usually causing pulmonary or disseminated infection. The commonest species implicated in human infection are *A. niger*, *A. fumigatus* and *A. flavus*. The yeast *Cryptococcus neoformans* was a rare cause of chronic lymphocytic meningitis in patients with deficient cell-mediated immunity, and is now an important problem in HIV seropositive patients.

Introduction to helminths

The term helminth is derived from a Greek word meaning worm. It was initially applied to roundworms, but now encompasses all metazoan internal parasites (Tables 2.5 & 2.6). Helminths are complex multicellular organisms with developed organs. Adult worms may possess an alimentary canal whose morphology can be helpful in identification (e.g. hookworms). Female helminths have uteri and produce large numbers of eggs daily. Many pathogenic helminths possess specialized organs of attachment, especially the hookworms and the tapeworms.

In some helminth diseases, humans are the only host and the pathogen produces eggs which are excreted, and survive in the environment. The life cycle is completed when eggs, or free-living larvae from excreted eggs, are ingested by a new human host and a new infection develops. Such infections include those caused by threadworms and roundworms.

Some helminths have complex life cycles, involving an intermediate host (e.g. the pig or cow in tapeworm diseases) or a vector (e.g. the mosquito in filariasis). The eggs of schistosomes are deposited by humans into water, hatch, invade snails, are released as infectious cercariae and reinfect humans by penetrating intact skin. Humans can be infected as an essential part of the life cycle in the case of filariasis or ascariasis, but hydatid disease is a zoonosis, in which humans replace the sheep in a dog–sheep–dog life cycle.

Family	Species	Disease
Trichuroidea	*Trichinella spiralis*	Trichinosis
	Trichuris trichuria	Trichuriasis
	Capillaria hepatica	
	C. phillipinensis	
Rhabditoidea	*Strongyloides stercoralis*	Strongyloidiasis
	S. fuelleborni	
Ancylostomatoidae	*Ancylostoma duodenale*	Hookworm disease
	A. ceylonicum	Ancylostomiasis
	Necator americanus	
Metastronguloidea	*Angiostrongylus cantonensis*	Eosinophilic meningitis
Oxyurida	*Enterobius vermicularis*	Threadworms
Ascarididae	*Ascaris lumbricoides*	Ascariasis
	Toxocara canis	Visceral larva migrans and ocular disease
	T. cati	
Dracunculoidea	*Dracunculus medinensis*	Guinea worm
Filaroidea	*Wuchereria bancrofti*	Lymphatic filariasis
	Brugia malayi	
	B. timori	
	Onchocerca volvulus	River blindness
	Loa loa	Loiasis

Table 2.5 Nematodes that cause human disease.

Family	Species	Disease
Schistosomatoidea	*Schistosoma mansoni*	Bilharzia
	S. haematobium	
	S. japonicum	
Echinostomatoidae	*Fasciola hepatica*	Liver flukes
	Taenia solium	Tapeworms and cysticercosis
Taeniidae	*T. saginata*	
	Echinococcus granulosus	Hydatid disease
	E. multilocularis	

Table 2.6 Most important platyhelminths causing human infection.

Classification

Helminths are classified into the nematodes or roundworms, the platyhelminths or flatworms, the cestodes or tapeworms, and the flukes. The nematodes are divided by the main site of infection: the intestinal nematodes such as *Ascaris* and hookworms; the blood nematodes such as *Filaria*; the tissue nematodes, e.g. *Trichinella*; and skin nematodes such as *Onchocerca volvulus*. The flatworms are further divided into the cestodes (tapeworms) and the trematodes (flukes). The flukes are subdivided by the place where the adult fluke is found, i.e. blood, liver and lung.

PATHOGENIC HELMINTHS

Nematodes (roundworms)

Ascaris lumbricoides, Ancylostoma duodenale, Strongyloides stercoralis (intestinal)
Wuchereria bancrofti, Brugia malayi (blood filariae)
Onchocerca volvulus, Trichinella spiralis (tissue forms)

Platyhelminths (flatworms)

Trematodes (flukes)
Fasciola hepatica, Opisthorchis sinensis (liver flukes)
Schistosoma mansoni, S. japonicum (blood flukes)
Paragonymus westermanii (lung fluke)

Cestodes (tapeworms)
Taenia solium, Diphyllobothrium latum (intestinal form)
Echinococcus granulosus (tissue cyst form)

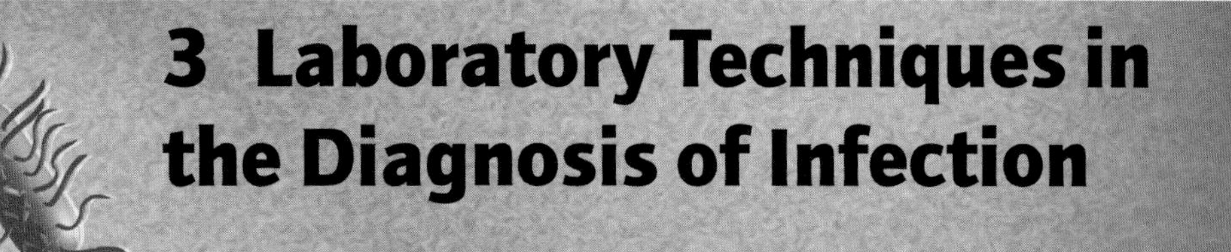

3 Laboratory Techniques in the Diagnosis of Infection

Introduction, 35

Collection of specimens, 35
Laboratory safety and universal precautions, 35

Direct microscopic examination, 35
Microscopy of unstained preparations, 36
Microscopical examination of preparations stained with simple stains, 36
Immunofluorescence, 38

Cultural methods, 38
Media, 39
Limitations of culture, 40
Screening, 40
Automation, 40
Antimicrobial susceptibility testing, 41
Typing microorganisms, 41
Methods of typing microorganisms, 41
Culture of protozoa and helminths, 44
Tissue culture, 44
Virus neutralization, 44

Serology in the detection of infection, 45
Precipitation and agglutination tests, 46
Complement fixation tests, 46
Indirect fluorescent antibody tests, 47
Radioimmunoassay, 47
Enzyme-linked immunosorbent assay, 47
Western blotting (immunoblotting), 49

Molecular diagnostics, 49
Molecular amplification methods, 49
Applications of molecular amplification methods, 49

Summary, 50

Introduction

The microbiology laboratory plays a crucial role in the diagnosis of all infectious diseases. Because there are many different tests available for the diagnosis of individual patients, there are several ways of making a microbiological diagnosis.

1 Microscopical methods.
2 Cultural methods.
3 Serological methods.
4 Molecular methods.

Collection of specimens

Body fluids, secretions and biopsy material can all be examined microbiologically. Samples from the environment, e.g. water, food or soil may also be examined. Some samples must be collected at a particular time; for example malaria parasites are best sought at the peak of fever and a short time afterwards, whereas blood for bacterial culture should be taken as the fever begins to rise. Special precautions must often be taken to ensure survival of the pathogen and exclude contaminants, e.g. cleaning of the perineum before a midstream specimen of urine is collected. Anaerobic species may die if exposed to atmospheric oxygen and, ideally, samples of pus should be obtained, rather than swab specimens. Many pathogens die quickly outside the body, and must be transported to the laboratory without delay. *Neisseria gonorrhoeae* are susceptible to drying, so specimens likely to contain this organism should be inoculated onto microbiological medium near to the patient.

Laboratory safety and universal precautions

Specimens may contain hazardous pathogens and must be handled with care. Concerns about the risk of transmitting blood-borne viruses have led to the introduction of a system of universal precautions which defines the personal protective measures to be taken in collecting and examining specimens irrespective of their source. The idea is to handle specimens on the assumption that they may contain a transmissible pathogen rather than relying on clinical suspicion or written clinical details, which may be faulty or absent. For example, any sputum specimen, not just those where the clinician suspects it, must be assumed to carry the risk of tuberculosis.

Direct microscopic examination

Antony van Leeuwenhoek first saw microscopic 'animalcules', and Alexander Ogston, a surgeon, described the characteristic microscopical morphology of staphylococci in pus and discovered their role in pyogenic sepsis. The light microscope has since been indispensable in the study of microorganisms. The equipment required is cheap, reagent costs are low and early results can be obtained. The diagnosis of malaria or vaginal trichomoniasis, for instance, can be made while the patient waits at the clinic. The organism sought need neither multiply nor even be alive. Microscopy is especially useful for detecting organisms that are difficult or dangerous to grow.

Types of microscopy for the diagnosis of infections
1 Unstained preparations.
2 Simple stains—Gram, Giemsa.
3 Special stains—Ziehl–Nielsen, Gomori–Grocott, India ink.
4 Immunofluorescence—direct and indirect.
5 Electron microscopy.

Microscopy of unstained preparations

Direct microscopy of unstained preparations
Faecal protozoa and helminths, vaginal discharge, urine for bacteria and pus cells.

Direct examination of unstained 'wet' preparations is suitable for rapid diagnosis in the laboratory and the outpatient setting. Many pathogens have a characteristic appearance, e.g. hookworm eggs, or may appear in diagnostic circumstances, for instance bacteria, together with white cells, in the urine from a case of acute urinary tract infection (Fig. 3.1).

Viral specimens can also be examined directly by electron microscopy, using a negative-staining technique. The viruses are concentrated by vigorous centrifugation and then suspended in a heavy metal solution. As it dries, the heavy metal salt fills in the spaces between the viruses, providing an electron-dense background which outlines the virus. This technique is useful for detecting viruses with distinctive morphology, such as poxviruses in scrapings from the lesions of orf and molluscum contagiosum, herpesviruses in vesicle fluid, or one of the many gastrointestinal viruses such as rotavirus or calicivirus in diarrhoea stools (Fig. 3.2).

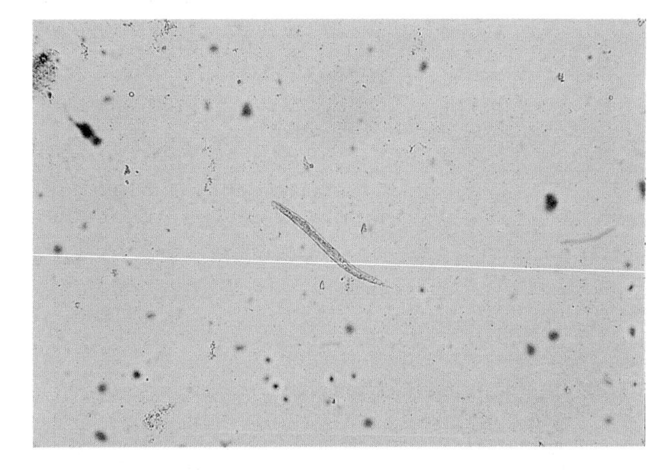

Fig. 3.1 Unstained wet preparation of faeces showing larva of *Strongyloides stercoralis*. ×94 approximately.

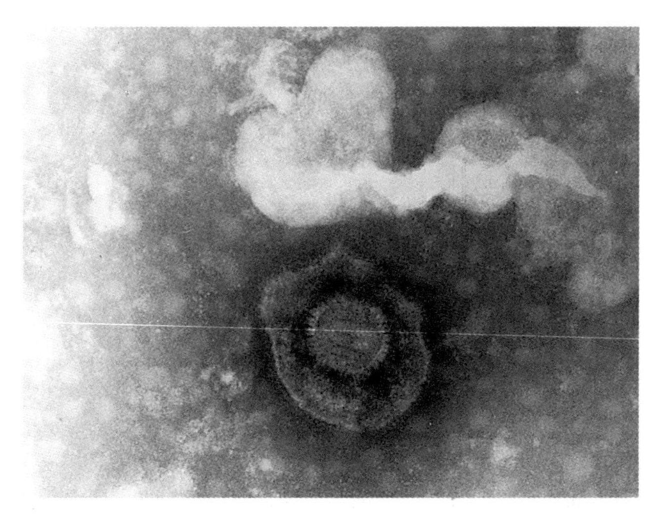

Fig. 3.2 Negatively stained electron micrograph of herpesvirus. Courtesy of Dr David Brown, Central Public Health Laboratory.

Microscopical examination of preparations stained with simple stains

Gram stain

Dried fixed preparations of specimens can be examined using simple stains, such as Gram stain, which dye the bacteria. This technique can demonstrate the shape of the bacteria and their ability, or not, to retain the blue Gram dye (Fig. 3.3). It provides a rapid answer to the clinical question: 'are there any organisms present?' It is therefore most useful when sterile fluids such as cerebrospinal fluid (CSF) or pleural fluid are examined. However, the sensitivity of Gram stain is relatively low; more than 100 000 organisms per millilitre must be present for a diagnosis to be made.

Gram stain
Sterile fluids (cerebrospinal fluid, ascites, pleural fluid), sputum (to exclude poor-quality specimens), pus from any site, urethral discharge.

Gram staining alone can rarely answer the question: 'what organism is present?' because the morphology of bacteria is rarely diagnostic. Important exceptions to this include the finding of Gram-positive or -negative diplococci in CSF from a patient with meningitis, or the charac-

Fig. 3.3 Gram stain of a smear from the mouth, showing a mixture of Gram-positive and Gram-negative cocci and bacilli.

Fig. 3.4 India ink-stained preparation of cerebrospinal fluid, showing *Cryptococcus neoformans* with a thick, clear capsule.

teristic appearances of *Borrelia* and *Fusobacterium* in Vincent's angina. Similarly, a Gram-stained preparation of urethral pus showing Gram-negative intracellular diplococci is sufficiently characteristic to allow a presumptive diagnosis of gonorrhoea.

Other simple stains

Other simple stains include acridine orange (which is more sensitive in demonstrating organisms than Gram stain, but more prone to confusing artefactual effects), lactophenol blue to demonstrate the morphology of fungi, or India ink, which is used to detect the presence of *Cryptococcus neoformans* in the CSF by negative staining (Fig. 3.4).

Special stains

Stains such as Ziehl–Nielsen (ZN) are used to demonstrate specific features of organisms which simple stains will not demonstrate. Specimens are stained with carbolfuchsin, destained with an acid–alcohol solution and then counterstained with methylene blue. The lipid-rich mycobacterial cell wall retains the pink dye and organisms are seen as pink bacilli against the blue background (Fig. 3.5). The number of acid-fast species is limited and this technique is therefore useful in the diagnosis of mycobacterial infection, including tuberculosis and leprosy, and parasitic infections such as cryptosporidiosis.

A variation of this technique uses the naturally fluorescent substance auramine to stain the organisms. The specimen is processed in a similar way to the ZN method, and acid-fast organisms fluoresce bright yellow under an ultraviolet light. Auramine microscopy is used for screening large numbers of specimens but, because it lacks the specificity of the ZN stain, all positive specimens must be overstained by the ZN method, and re-examined to confirm the findings.

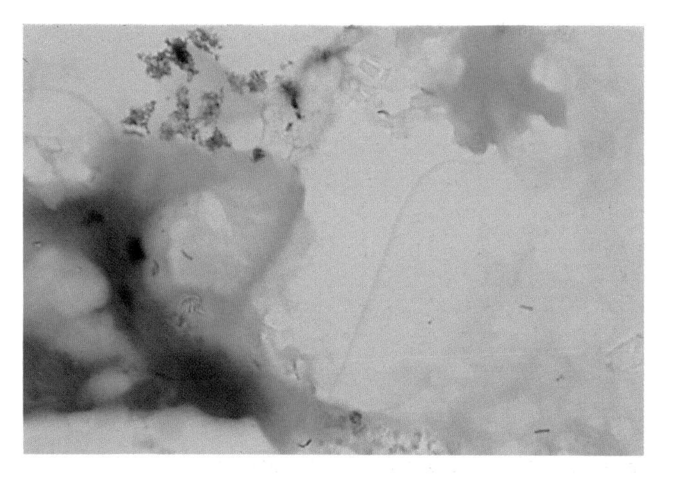

Fig. 3.5 Ziehl–Nielsen-stained smear, showing acid-fast bacilli.

Romanowsky stains

Romanowsky stains colour cytoplasm and chromatin, and are normally used to demonstrate blood cells. Stains such as Giemsa are used in the diagnosis of blood parasites. Taking malaria or filariasis as examples, Giemsa-stained smears not only demonstrate the presence of the organisms, but permit speciation by demonstration of morphological details (Fig. 3.6).

Special stains used in microbiology

1 Ziehl–Nielsen	*Mycobacterium* spp.
2 Gomori–Grocott	Fungi, *Pneumocystis carinii*.
3 Giemsa	Malaria, *Filaria*.

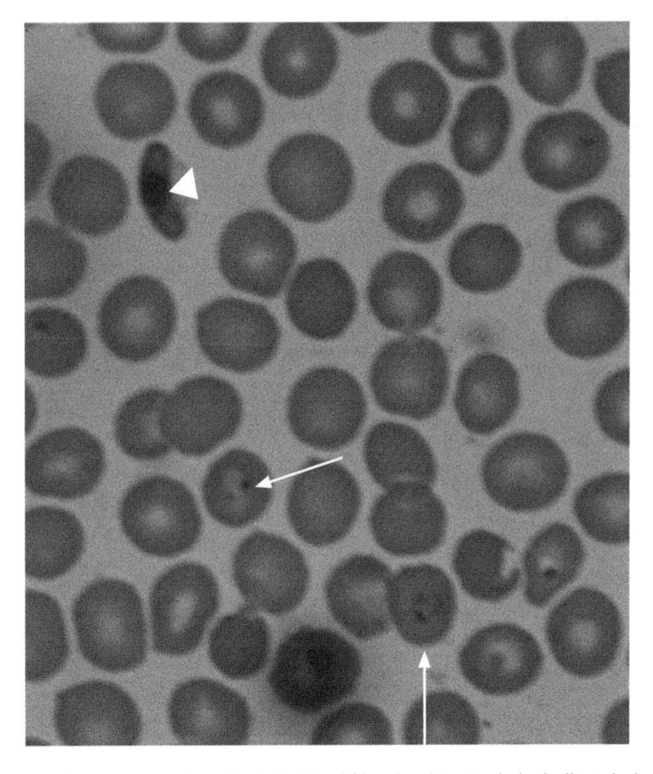

Fig. 3.6 Romanowsky-stained thin blood film, showing morphologically typical trophozoites of *Plasmodium falciparum* (arrows) and a gametocyte (arrowhead).

Immunofluorescence

Direct immunofluorescence techniques detect organisms by their binding with fluorescence-labelled antibodies. Specimens are dried on a multiwell slide, together with control positive and negative specimens. A specific fluoroscein-labelled antibody is then added. The slides are washed and examined microscopically under ultraviolet illumination. Where an antibody specific to the pathogen has bound there is an apple-green fluorescence (Fig. 3.7). This technique is both sensitive and specific and provides a rapid, presumptive diagnosis. It can be applied to a wide range of specimens and is used in the diagnosis of chlamydial urethritis, influenza, parainfluenza virus, respiratory syncytial virus, measles and lyssavirus infections.

Some organisms detectable by direct immunofluorescence	
1 Viruses	Parainfluenza viruses, respiratory syncytial virus.
2 Bacteria	*Legionella*, *Treponema pallidum*.
3 Protozoa/fungi	*Giardia intestinalis*, *Pneumocystis carinii*.

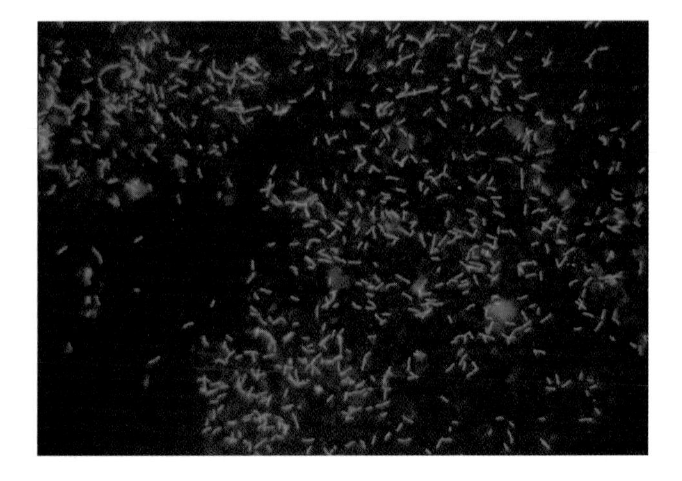

Fig. 3.7 Immunofluorescence-stained preparation of legionellae. Courtesy of Dr Tim Harrison, Central Public Health Laboratory.

Immunofluorescence techniques are also used for the detection of specific antibody. This will be described in more detail below.

Cultural methods

Culture can aid diagnosis in bacterial, parasitic and viral diseases. In bacteriology, culture allows isolation of bacteria on solid media, and recognition by their morphological and biochemical characteristics. It is the main means of obtaining pure cultures for antimicrobial susceptibility testing.

Modern bacterial culture is made possible by the use of agar, a gelatinous substance derived from seaweed, which melts at 90°C but solidifies at 50°C. It is highly stable, rarely affected by organisms in cultures and can be mixed with nutrients such as blood, serum and protein digests to make solid media. Koch introduced agar to microbiology to replace gelatin, which melts at a lower temperature, near to that used for incubation.

Bacteriological culture on solid agar is usually performed in Petri dishes—plastic dishes, 90 mm in diameter, with a vented lid. When prolonged culture is necessary, as in the diagnosis of mycobacterial or fungal infection, it is usually performed in sealed containers to prevent desiccation and the entry of contaminating organisms. Various specimens are usually incubated in a range of atmospheric conditions to optimize the growth of organisms with differing atmospheric requirements (see Chapter 2).

Media

Three types of bacteriological media are used: enrichment, selective and indicator.

Bacteriological media
1 Enrichment.
2 Selective.
3 Indicator.

Enrichment media

Enrichment media are required when fastidious organisms such as *Streptococcus* spp., *Haemophilus influenzae* or *Bacteroides fragilis* are being sought. Their main purpose is to ensure that small numbers of fragile pathogens will multiply sufficiently to be detected.

Simple base media may be enriched by the addition of blood, yeast extracts, tissue infusions, meat, etc. Enrichment media can be solid, for example blood agar, or liquid, as with Robertson's cooked meat broth. Liquid media are especially valuable for investigating body fluids which are normally sterile. Such specimens often contain very small numbers of organisms which will multiply in the highly nutritious medium, and can then be subcultured on to solid media for identification and susceptibility testing.

Enrichment medium aims to amplify the organism by growth, e.g. for *Neisseria* spp., *Streptococcus* spp. or *Vibrio* spp.

Selective media

Selective media are used when a pathogen must be separated from a mixture of organisms. Many body sites, such as the upper respiratory tract or the gut, have a normal resident flora, and pathogens must be isolated from this bacterial competition. Selective media contain compounds which may be chemicals (e.g. selenite F), dyes (crystal violet) or mixtures of antibiotics (such as lincomycin, amphotericin B, colistin and trimethoprim in New York City medium), which selectively inhibit the normal flora, enabling the pathogen to grow through (Table 3.1). These media can be complex mixtures requiring careful preparation.

Although the selective agents have their maximum effect on the unwanted organisms, some inhibition of the target organism inevitably occurs. Therefore, an enrichment medium should also be inoculated so that small numbers of pathogens can be detected.

Selective cultures, e.g. sputum, stool or throat-swab specimens, aim to identify, for instance, *Haemophilus* spp., *Salmonella* spp. or *Streptococcus* spp. from among normal flora. Antibiotics, dyes, antiseptics, chemicals and bile salts are examples of substances used in selective bacteriological media.

Culture of pathogens from sites with normal flora requires selection, and from sterile sites requires enrichment.

Indicator media

Indicator media are used to identify colonies of pathogens among the mixture of organisms able to grow on the selective medium. Commonly used indicator media are usually selective as well. An example of a selective indicator medium is MacConkey's agar, which uses bile salts to select for bile-tolerant enteric organisms. It also contains lactose and the indicator neutral red. Colonies of lactose-fermenting organisms produce lactic acid, and are coloured red by the neutral red indicator. Pathogens are usually non-lactose fermenters, which produce colourless colonies in this medium (Fig. 3.8).

Indicator medium (often combined with selective function)
MacConkey
Selects with bile salts.
Indicates lactose fermentation with pH indicator.

Medium	Selective agent	Specimen	Organism/s sought
Crystal violet–blood agar	Crystal violet	Throat swab	*Streptococcus pyogenes*
Desoxycholate citrate	Desoxycholate	Faeces	*Salmonella, Shigella*
New York City medium	Lincomycin, colistin, amphotericin, trimethoprim	Urethral/cervical smear	*Neisseria gonorrhoeae*
Selenite broth	Selenite F	Faeces	*Salmonella*
Sabouraud's agar	High dextrose content	Many	Fungi

Table 3.1 Examples of selective media for bacteriological culture from sites containing a normal flora.

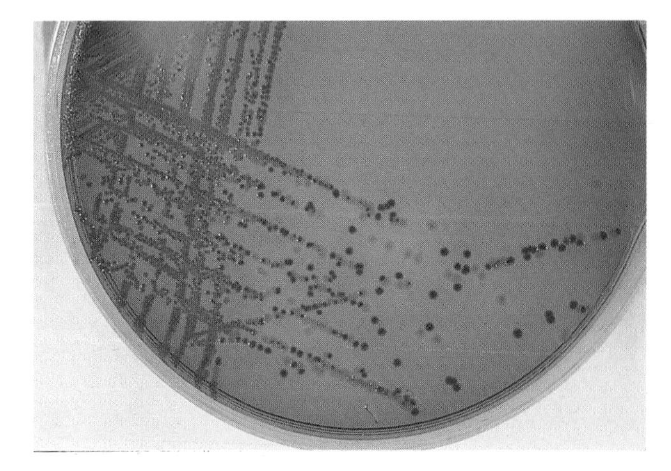

Fig. 3.8 Mixed growth of Enterobacteriaceae on MacConkey's agar plate, showing pink, lactose-fermenting colonies of *Escherichia coli* among the colourless non-lactose fermenters.

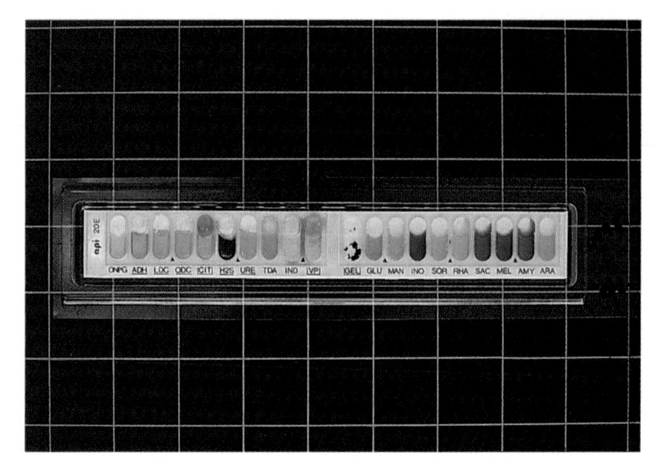

Fig. 3.9 A typical bank of biochemical tests for the identification of species of Enterobacteriaceae.

There are many liquid indicator media, containing different sugars or other substrates such as urea and citrate. They also contain indicator dyes which change colour when bacterial metabolic products alter the pH of the medium. Thus an organism which ferments a particular sugar lowers the pH of the medium, or one which metabolizes urea produces ammonia and raises the pH. These indicator media are usually used in banks of tests which allow identification of an organism by its biochemical profile. Preprepared sets of miniculture vials are commercially available (Fig. 3.9). Automated systems linked to computerized identification programmes are also coming into use.

Limitations of culture

The main limitation of bacterial culture is the incubation time it takes. Often, infected patients are ill and require immediate treatment. The decision to initiate therapy must be made based on the clinical features of the case, along with rapid microscopical examination and haematological or biochemical results. Culture is valuable for confirmation or rebuttal of a diagnosis. It is particularly useful for revealing unusual or unexpected organisms, or showing that organisms have unusual antibiotic susceptibility patterns. The initial therapy in such cases may need modification when the result of culture is available.

Culture has an important epidemiological purpose, not related to the treatment of individual patients. Most cases of acute diarrhoea could be managed properly without microbial culture. The isolation of the same *Salmonella* sp. from several individuals, however, might prompt a search for a common source such as an infected food-handler at a restaurant. Similarly, the isolation of a toxigenic strain of *Corynebacterium diphtheriae* from a throat swab should prompt clinicians to initiate surveillance and control measures among the patient's contacts.

Screening

Microbiological culture can be used to screen patients and healthcare workers for colonization with pathogens such as *Streptococcus pyogenes* or epidemic methicillin-resistant *Staphylococcus aureus* (EMRSA). It can allow a prompt response to the presence of dangerous or difficult-to-treat organisms in a hospital setting.

Automation

Bacterial culture is time-consuming and labour intensive. Conventional methods of blood culture, for example, require manual subculture of each bottle after 12, 24 and 48 h, and later subcultures as necessary. Fortunately, techniques for automatic detection of bacterial growth are now available. They detect the microbial production of carbon dioxide, or changes in the electrical impedance of the medium, to indicate the need for subculture. Continuous monitoring is possible permitting early subculture, shortening significantly the time taken to detect bacterial growth. Adaptations of this approach are available for mycobacterial culture. Computer analysis of results from multiple biochemical tests systems allows rapid identification of bacterial isolates.

> Detection of carbon dioxide produced by bacteria or changes in the electrical impedance of liquid media allow automatic detection of bacterial growth.

Antimicrobial susceptibility testing

Testing antimicrobial susceptibility is an important function of the microbiology laboratory. Apart from confirming the expected behaviour of pathogens, it is valuable when the infecting organisms have unpredictable antimicrobial susceptibilities. This is especially true of Enterobacteriaceae, among which multidrug resistance may develop and spread in the hospital environment. In some geographical areas *Salmonella* may be resistant to most commonly used therapeutic agents. Some strains of *Streptococcus pneumoniae*, previously sensitive to many antimicrobials, have recently developed resistance to penicillin and other commonly used drugs, making empirical therapy difficult.

Almost all currently available methods of antimicrobial susceptibility testing depend on isolation of the organism by culture, followed by reculture of the pure growth, in the presence of antimicrobial agents. This is described in more detail on pp. 56–60.

Typing microorganisms

Typing is the use of further identification methods to distinguish between strains of organisms within the same species.

For most clinical purposes it is sufficient to identify the genus or species of the infecting organism. Further characterization is desirable when for example, a group of *Salmonella typhimurium* infections could be caused by different strains of this common pathogen. The only way to prove that the cases represent an outbreak is to show that the strains are identical by typing.

The principal characteristic of a typing system is that it divides a species of microorganism into sufficient different groups to be epidemiologically discriminating. For example, a system separating a species into only six groups would not be useful. Nor would a system that placed all the common strains in one group. Typing methods should be reproducible so that results from one laboratory can be compared with those of another. The technique should also be simple to perform, as delay in receiving results hinders early control of an outbreak. Finally, it should be inexpensive.

Methods of typing microorganisms

Methods of typing are numerous, ranging from simple biochemical to complex genetic characterization. It is common for two or more methods to be used, to provide adequate distinction of strains; for instance, *Legionella pneumophila* is serotyped into nine groups, and then further biotyped for more accurate identification. *Listeria monocytogenes* is serotyped and phage typed.

Simple laboratory typing

Simple phenotypic markers are often sufficient to demonstrate the identity of isolates and indicate that an outbreak has occurred. It can also be useful to type successive isolates of the same organism from an individual patient; for instance, in a case of *Staphylococcus epidermidis* bacteraemia the organisms isolated from blood or intravascular devices on different occasions could represent a variety of contaminants from the patient's skin. They could be considered identical, and representative of a true infection, if a range of biochemical characteristics and the antibiogram were identical (Fig. 3.10). This approach could not be adopted for *S. aureus*, however, as many strains exist which share the same resistance patterns and biochemical phenotypes. For this species phage typing is most appropriate.

Biotyping

Pathogens are often assigned to a species by means of biochemical testing. Biotyping uses the results of additional biochemical tests to assign members of the same species into different groups or biotypes. The tests employed may be sugar fermentation tests, or tests for the action of enzymes such as urease. Strains of *Legionella pneumophila* and *Corynebacterium diphtheriae* are often identified by biotyping (Fig. 3.11).

Auxotyping

This biochemical method tests the ability of an organism to grow on minimal medium and use single chemicals such as arginine as a source of, for example, nitrogen. The profiles detected by the tests, called auxanograms, are most suitable for typing fastidious organisms with complex nutritional requirements, for example *Neisseria gonorrhoeae* or *Haemophilus influenzae*.

Serological typing

Many organisms can be typed by testing a range of serological reactions. Many laboratories type enteric pathogens such as *Shigella flexneri* or *Salmonella* using a series of antisera raised in mice or guinea-pigs. Serological typing is performed by suspending a pure growth of the organism in a drop of antiserum on a glass slide and

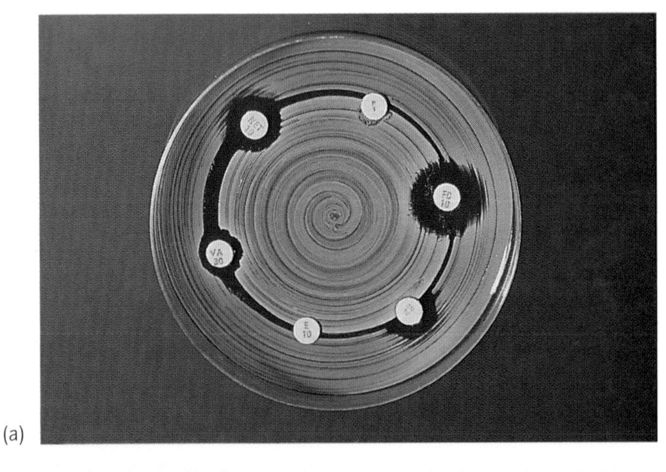
(a)

(b)

Fig. 3.10 Antibiograms of successive isolates of *Staphylococcus epidermidis* from the blood of a patient in intensive care; although the density of growth is different in the two tests (the centre of the plate) the pattern of inhibition by antibiotic discs is identical, indicating that they may be the true cause of a persisting fever.

mixing them with a rocking movement of the slide. The development of agglutination indicates that the organism carries the antigen specific to the antiserum used.

For organisms where it is difficult to generate sets of typing sera, or the test is rarely required, reference laboratories have been set up to provide a central expert service.

Phage typing

Many bacteriophages lyse the bacteria they infect as part of their life cycle (this is known as a lytic cycle). Different strains of an organism have different sensitivities to the phages which commonly infect the species. This property is useful in typing staphylococci, *Listeria monocytogenes*

Fig. 3.11 Biotyping of *Corynebacterium diphtheriae*: the sugar tests are typical of the gravis strain.

and some species of *Salmonella*. A lawn inoculum of the test organism is made, and a battery of suitable phages is inoculated at fixed points. The plates are then incubated overnight and the pattern of lysis indicates the phage type. A *Staphylococcus aureus* may be identified as phage type 88/24 because it was lysed only by the 88 and 24 phages (Fig. 3.12). Some strains are not lysed by any phage, and are called phage untypable.

Bacteriocin typing

Many microorganisms produce protein antibiotics directed against other bacteria competing for the same ecological niche. These bacteriocins often have different inhibitory effects on different strains within a species, and this characteristic is utilized in bacteriocin typing. A test strain is inoculated as a streak on a suitable agar plate and grown overnight. The growth is then scraped off with a glass slide, and the plate exposed to chloroform to kill any remaining organisms. A battery of indicator strains is then inoculated as a set of streaks at right angles to the test strain, and incubated overnight. The next day some indicator strains will have gaps in their growth where they have been inhibited by the test strain's bacteriocin. The pattern of inhibition is used to assign strains to a bacteriocin type. This method has been used in typing *Shigella sonnei* and *Pseudomonas aeruginosa*.

Molecular methods

Many molecular typing methods are now entering into clinical and research practice, much improving our ability to follow outbreaks and to study the epidemiology of infection.

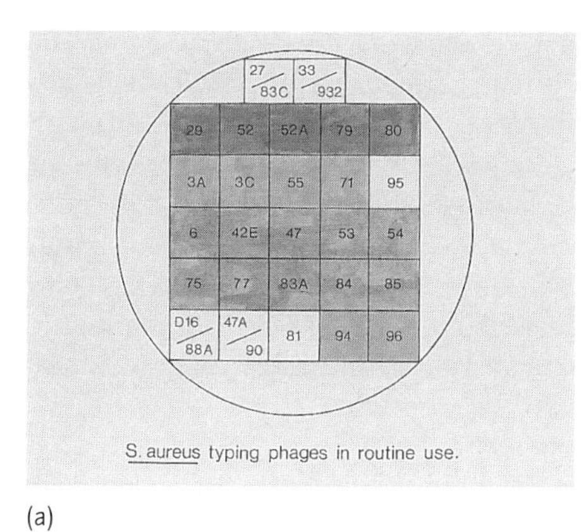

(a)

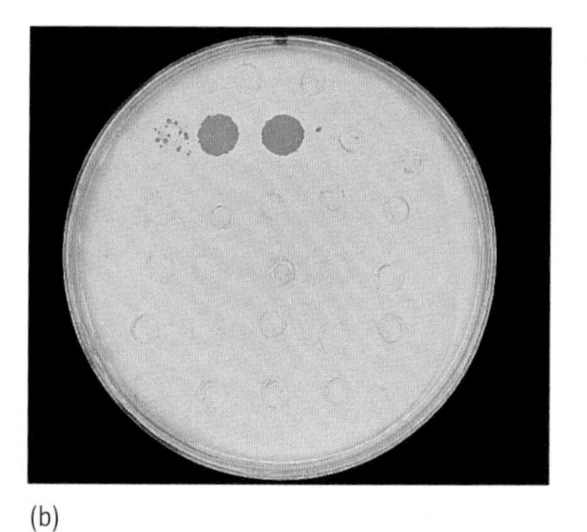

(b)

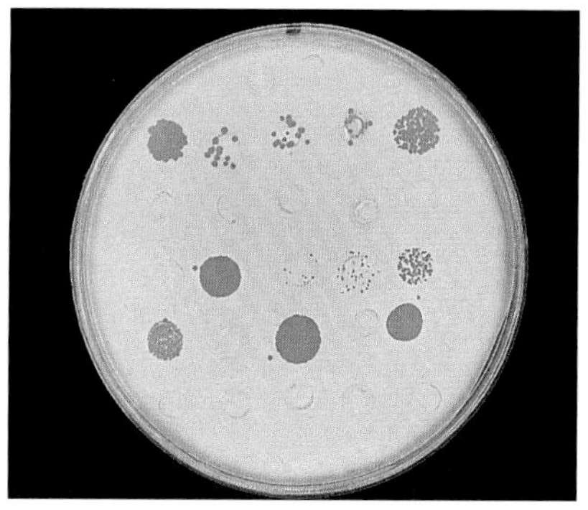

(c)

Fig. 3.12 Phage typing of *Staphylococcus aureus*. (a) Template of currently used phages; (b) lysis by phages 52/52a; (c) complex type. With permission of Central Public Health Laboratory.

Protein typing

This is a relatively crude method of typing in which the proteins of an organism are simultaneously extracted and suspended in a detergent solution (usually sodium dodecyl sulphate, SDS). The pattern of proteins is demonstrated by polyacrylamide gel electrophoresis (PAGE). SDS-PAGE typing techniques have largely been superseded by nucleic acid profiling.

Multilocus enzyme electrophoresis (MLEE)

In this technique cell contents are subjected to electrophoresis and the position of several metabolic enzymes is detected. Small variations in the amino acid sequence of these enzymes results in differences in their final position on the gel, and these differences are used to define the electrophoretic type of the organism. MLEE is used to type bacteria such as *Streptococcus pneumoniae*. When used for protozoa, e.g. *Entamoeba* and *Leishmania*, the technique is then known as zymodeme analysis.

Nucleic acid typing

Many nucleic acid typing methods use the activity of endonuclease enzymes to split the genome into a characteristic range of different-sized fragments. Genomic or plasmid DNA, or ribosomal RNA, is harvested from the test strains and digested with restriction endonucleases. The resulting fragments are then separated by electrophoresis or pulsed-field gel electrophoresis (PFGE), to produce a pattern of bands. The variation in band patterns is called restriction fragment length polymorphism (RFLP; Fig. 3.13). This method can be varied by using a frequently cutting enzyme, and Southern blotting the DNA fragments with a repetitive sequence. This approach is used in ribotyping or insertion sequence typing and can be applied to most bacterial species.

The problem of gel-to-gel variation makes long-term comparisons both within and between laboratories difficult but has been addressed by agreed protocols and computerized gel scanning and analysis. An example is the international method for *Mycobacterium tuberculosis* (see Chapter 18).

Polymerase chain reaction-based methods

Polymerase chain reaction (PCR, see p. 49) can be used to generate DNA for RFLP typing if a variable gene or genes is selected for study. For example, the outer membrane proteins of *Neisseria* are variable and their genes may be analysed in this way. Bacterial genomes contain many repetitive nucleotide sequences and PCR protocols can be designed to amplify parts of them. Electrophoresis of the PCR products produces a ladder-like pattern allowing strains to be differentiated. Examples of this approach are ERIC-PCR (enterobacterial repetitive intergenic consensus sequence) used for a number of different species including *S. aureus*, Enterobacteriaceae and *M. tuberculosis*.

Methods of typing bacteria
1 Biotyping and auxotyping.
2 Serotyping.
3 Phage typing.
4 Bacteriocin typing.
5 Multilocus electrophoresis.
6 Restriction fragment length polymorphism typing.
7 PCR-based methods.

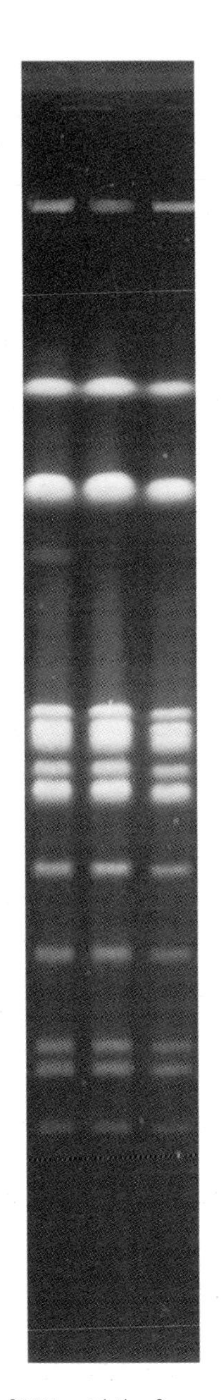

Fig. 3.13 Electrophoresis of DNA restriction fragments of three cytomegalovirus isolates, showing that the infected patients had acquired the same strain.

Culture of protozoa and helminths

Culture of protozoa and helminths is often difficult. In many examples, such as *Cryptosporidium parvum*, only one stage in the life cycle may be culturable. In others, such as *Strongyloides stercoralis*, the conditions of the natural environment can be reproduced in the laboratory. Using activated charcoal as a culture medium, larvae in faecal specimens will mature and multiply.

The asexual life cycle of *Plasmodium falciparum* can be reproduced by culture in banked red blood cells. Sensitivity to antimalarial agents can thus be tested. Although this technique is relatively simple it is inappropriate for routine diagnosis, because direct microscopy is sufficiently sensitive to detect clinically significant parasitaemias. Similarly, amoebae can be cultured in solid media if certain bacteria are included, but simple microscopy is an accurate and rapid means of diagnosis, and culture remains a research procedure.

The promastigote stage of *Leishmania* can be cultured in artificial media. To achieve culture of the amastigote, or tissue stage, material must be inoculated into isolated peripheral blood macrophages or macrophage cell lines.

Tissue culture

Viruses are obligatory intracellular pathogens, thus viral culture must be performed in living cells or tissues. Cells are usually prepared as monolayers on to which the specimen is inoculated. Some cells used for culture are 'primary cultures', extracted directly from an organ, as with fetal lung cells. Primary cell lines can be propagated for up to 50 generations, and are essential for isolation of some viruses. Alternatively, continuous cell lines can be subcultured indefinitely. Cell cultures are maintained at 37°C in an atmosphere of increased humidity and carbon dioxide. Cultured viruses infect the cells and take over their metabolic processes as they do in the human host. The cell monolayer is inspected at intervals for the presence of viral damage to the cells—the cytopathic effect (CPE). Some viruses may be presumptively identified by their distinctive CPE (Fig. 3.14); for example, measles virus produces multinucleated giant cells. Some viruses produce no visible CPE but growth of the virus is detected by, for example, adherence of red cells to viral haemagglutinin molecules expressed on the cells. Viral antigens can be directly demonstrated by fixing the cell monolayer and using immunofluorescence staining techniques (Fig. 3.15).

Evidence of virus growth in cell cultures

1 A cytopathic effect.
2 Haemagglutination.
3 Antigen detection.
4 Electron microscopy.

Virus neutralization

Identification of virus isolates can be achieved by lysing the cells of the monolayer and visualizing the virus

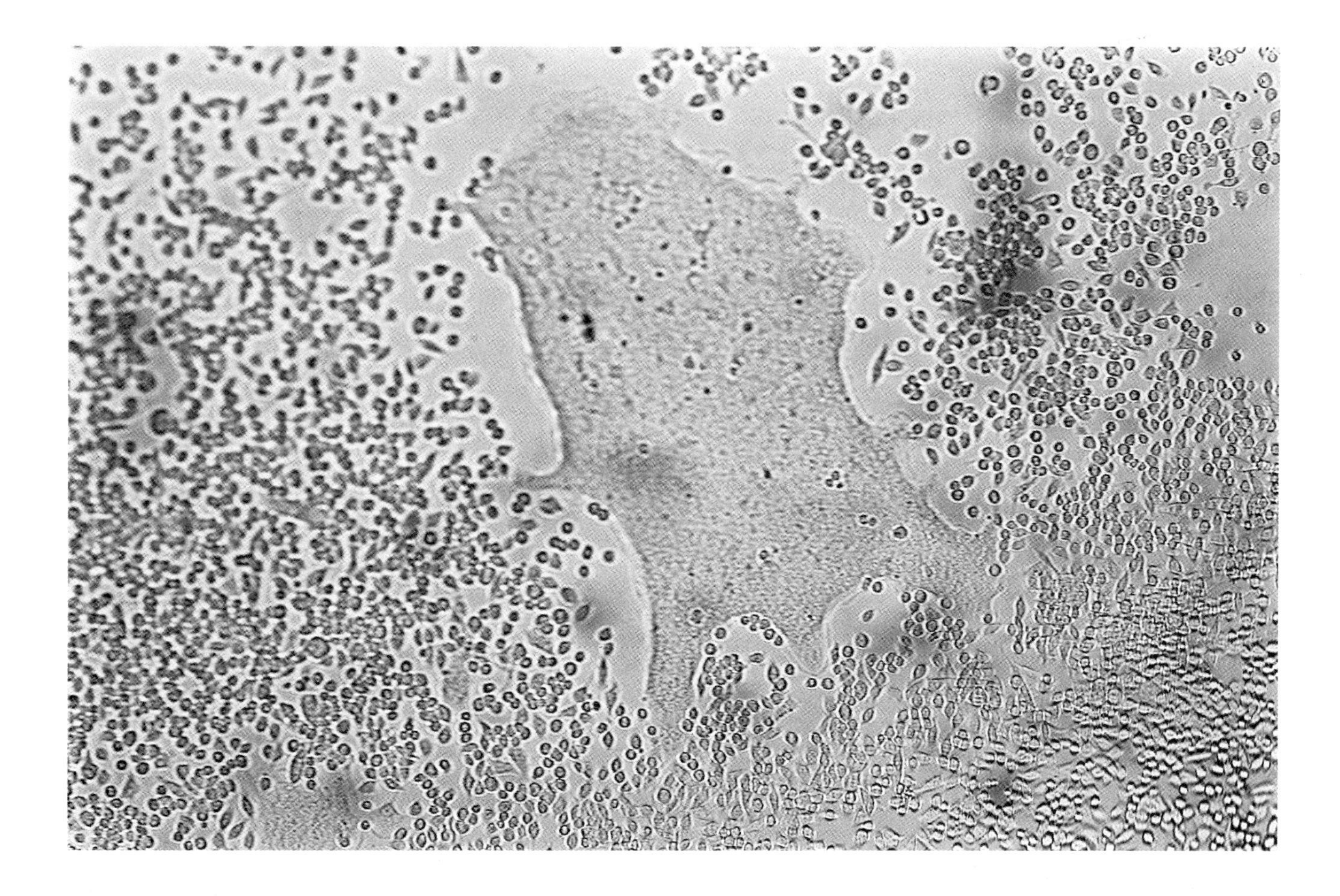

Fig. 3.14 Cytopathic effect of measles virus on lymphoblastoid cells; progressive coalescence of infected cells into a syncytium has formed a giant cell. Such giant cells may contain up to 50 nuclei.

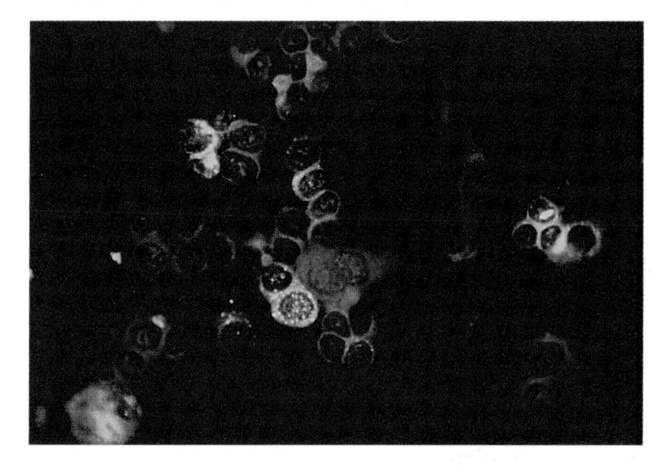

Fig. 3.15 Fluorescence micrograph of Lassa virus-infected vero cells, stained with immunofluorescent-labelled anti-Lassa serum. Courtesy of Dr David Brown, Central Public Health Laboratory.

under the electron microscope. Alternatively, inhibition of subculture by specific antibodies can be used to identify the species. This is known as virus neutralization.

Serology in the detection of infection

Serological techniques depend on the interaction between antigen and specific antibody. They are of particular value when the pathogen is difficult or impossible to culture, or is dangerous to handle in hospital laboratories.

The process can be divided into two parts: (i) the antigen–antibody interaction; and (ii) the demonstration of this interaction by a testing process. The antigen–antibody reaction depends on the specific binding between epitopes on the pathogen and the antigen-binding sites on the immunoglobulin molecules. The sensitivity of a serological test depends partly on the specificity and strength of the antigen–antibody reaction, but mostly on the ability of the test system to detect the reaction.

In older tests antibody–antigen binding was detected

by observing a natural consequence of this interaction: precipitation, agglutination or the ability of the antigen–antibody complex to bind and activate (fix) complement. Some laboratory tests based on these reactions are still in daily use. Newer tests use 'labelled' immunoglobulin molecules to facilitate detection. The main methods employed are labelling with fluoroscein, radioiodine or enzymes. Examples of each of these techniques are described in detail below.

Precipitation and agglutination tests

Precipitation

The simplest serological technique is precipitation, in which an insoluble complex is formed between antigen and antibody. The test is performed by mixing soluble antigen with the serum being tested, and observing any precipitate formation. The antigen and the serum to be tested are usually placed in separate wells, cut in an agar gel a few millimetres apart. Antibody and antigen diffuse from the wells and, at an optimal concentration ratio, will form a precipitin line. This takes up to 48 h, but can be speeded up by applying an electrical field to the gel. This technique, called countercurrent immunoelectrophoresis (CIE) is suitable for rapid diagnosis. It has been used in the diagnosis of acute pyogenic meningitis and fungal infections. The simple equipment and small quantities of reagents employed make the test very inexpensive.

Slide agglutination

Agglutination tests are used to identify the species or serotype of an infecting organism, by observing the aggregation of a suspension of bacteria in the presence of specific antibody. They can be performed on glass slides and are in daily use in speciating faecal pathogens such as *Salmonella* or *Shigella*.

Tube agglutination

In tube agglutination, particulate antigens form a lattice with specific antibody in the serum being tested, and fall to the bottom of the test vessel as a fine mat (a positive result). In a negative test, the particulate antigen falls quickly to the bottom to form a condensed button. This technique has been widely used in *Brucella* standard agglutination tests. A rapid microagglutination test (RMAT) is still the standard test for antibodies to *Legionella pneumophila* (Fig. 3.16).

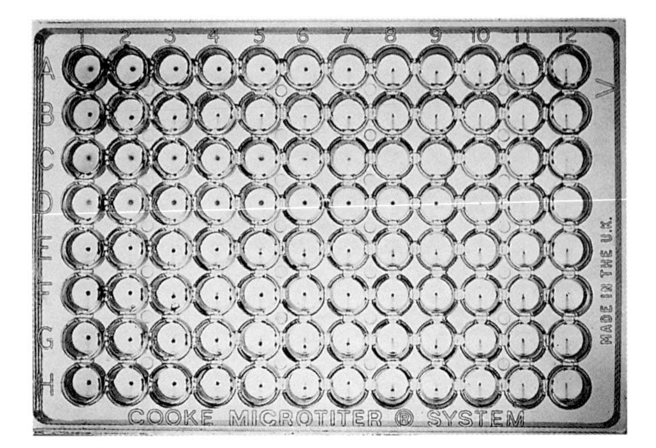

Fig. 3.16 Rapid microagglutination test for the detection of antibodies to *Legionella pneumophila*. Each row is composed of doubling dilutions of an individual patient's serum, mixed with dead organisms. The microtitre plate has conical wells, so that centrifugation 'jams' the agglutinated organism at the bottom of the well. Unagglutinated organisms form a streak when the plate is tipped on its edge: a dot is therefore positive and a streak is negative.

Coagglutination

Specific antibodies can be attached to uniform latex particles, or killed protein A-possessing staphylococci. These particles will be agglutinated when they attach in large numbers to antigen molecules. In this way, otherwise soluble immune complexes may be detected in an agglutination reaction. Such latex and coagglutination techniques are used to detect the presence of the polysaccharide antigens of *Streptococcus pneumoniae*, type b *Haemophilus influenzae*, *Neisseria meningitidis* and *Cryptococcus neoformans* in CSF, or *S. pyogenes* in throat swabs.

Complement fixation tests

In a complement fixation test defined antigen is added to a patient's serum (which has been heated to destroy naturally occurring complement) and supplemented with a measured amount of guinea-pig complement. When specific antibody is present in the serum, it interacts with the antigen which activates (or fixes) complement. Sheep red cells sensitized with rabbit antisheep red cell antibody are added and if complement has already been fixed none remains to lyse the sensitized sheep cells and no lysis will occur (a positive result). In contrast, if no antigen–antibody reaction has taken place in the first reaction, the complement is still available to lyse the red cells (a negative result). The test is technically difficult and requires careful standardization. It is therefore being superseded by those described below.

Indirect fluorescent antibody tests

Specific antigen is fixed on to a multiwell microscope slide and patient's serum added. The slides are incubated, then washed. Fluorescein-labelled antihuman immunoglobulin is then added, followed by further incubation. After a final wash the slides are examined under ultraviolet illumination. Where antibodies from the patient's serum have bound to the antigen, the antihuman globulin will bind, and is indicated by apple-green fluorescence. Individual positive sera may be titrated. Indirect immunofluorescence is both sensitive and specific, but rather time-consuming. It is used in the diagnosis of a number of infections, especially where the throughput of specimens is small, for instance in the diagnosis of syphilis (FTA, fluorescent treponemal antibody test) or parasitic disease such as leishmaniasis or amoebiasis.

Radioimmunoassay

Specific antigen is radiolabelled, then mixed and incubated with the patient's serum. The specimen is then centrifuged to separate the bound from the unbound radioactivity. The heavier antibody–antigen complexes collect at the bottom of the tube, where the radioactivity is then counted. The quantity of antibody is determined by reference to a standard curve. Radioimmunoassays are highly sensitive and specific. However, they require expensive radiolabelled reagents which have a short shelf-life and require additional safety measures and disposal arrangements. The processing of each specimen is time-consuming because of the multiple wash stages required to achieve good results. The gamma counters required are expensive and usually beyond the means of individual microbiology laboratories, but use is usually low, and allows the sharing of equipment by neighbouring laboratories.

Enzyme-linked immunosorbent assay

This technique is similar to radioimmunoassay, but differs in that either the antigen or antibody in the reaction is allowed to bind to a solid phase, such as the walls of microtitre wells. This eliminates the need for centrifugation. There are many variations of enzyme-linked immunosorbent assay (ELISA) but four will be described here and illustrated in the figures.

Antibody-detecting ELISA tests

In an antibody-detecting ELISA, specific antigen is coated on to the wells of a microtitre ELISA plate. The patient's serum is added and any specific antibody binds to the antigen. The plates are then washed and enzyme-labelled antihuman immunoglobulin is added. Plates are washed again and substrate for the enzyme is added. The enzyme–substrate reaction generates a colour which indicates the specific antibody interaction (Fig. 3.17). The optical density of the wells is measured by an ELISA reader. A positive result can be determined by reference to control values: for example a positive result may be defined as a well with an optical density three standard deviations above the mean of a series of negative controls. Alternatively, control positive and negative sera can be examined in parallel and a positive result is reported if the optical density is significantly higher than that of the control negative.

As in other serological techniques, sera can be titrated but this is time-consuming and defeats the main advantages of ELISA, which are simplicity of performance and automation. It is difficult to use an antibody capture assay to answer any question other than: 'is there antibody present?' If a diagnosis is to be made on a single specimen, specific immunoglobulin M (IgM) must be detected. This can be achieved by purifying an IgM fraction from the serum and retesting this in an antibody detection test. A simpler alternative is the IgM antibody-capture ELISA described below.

IgM antibody-capture ELISA

By placing antihuman IgM antibody on the solid phase, it is possible to capture all IgM antibody. After washing, labelled antigen is placed in the well so that if the serum contains antigen-specific IgM the labelled antigen will bind. A positive result is detected by adding substrate, when a colour will be produced by the enzyme-labelled antigen.

Antigen-capture ELISA

Antibody to a specific antigen is bound on the solid phase. If antigen is present in the specimen, it will bind to the antibody. After washing, the bound antigen is then detected by enzyme-labelled specific antibody (Fig. 3.18). The amount of antigen present can be quantified by reference to an antigen standard curve.

Competitive ELISA

Competitive ELISA is a technique which is particularly useful for the measurement of antigen concentrations and is also used to detect antibodies or antibiotic concentrations. Microtitre plates are coated with antibody and the

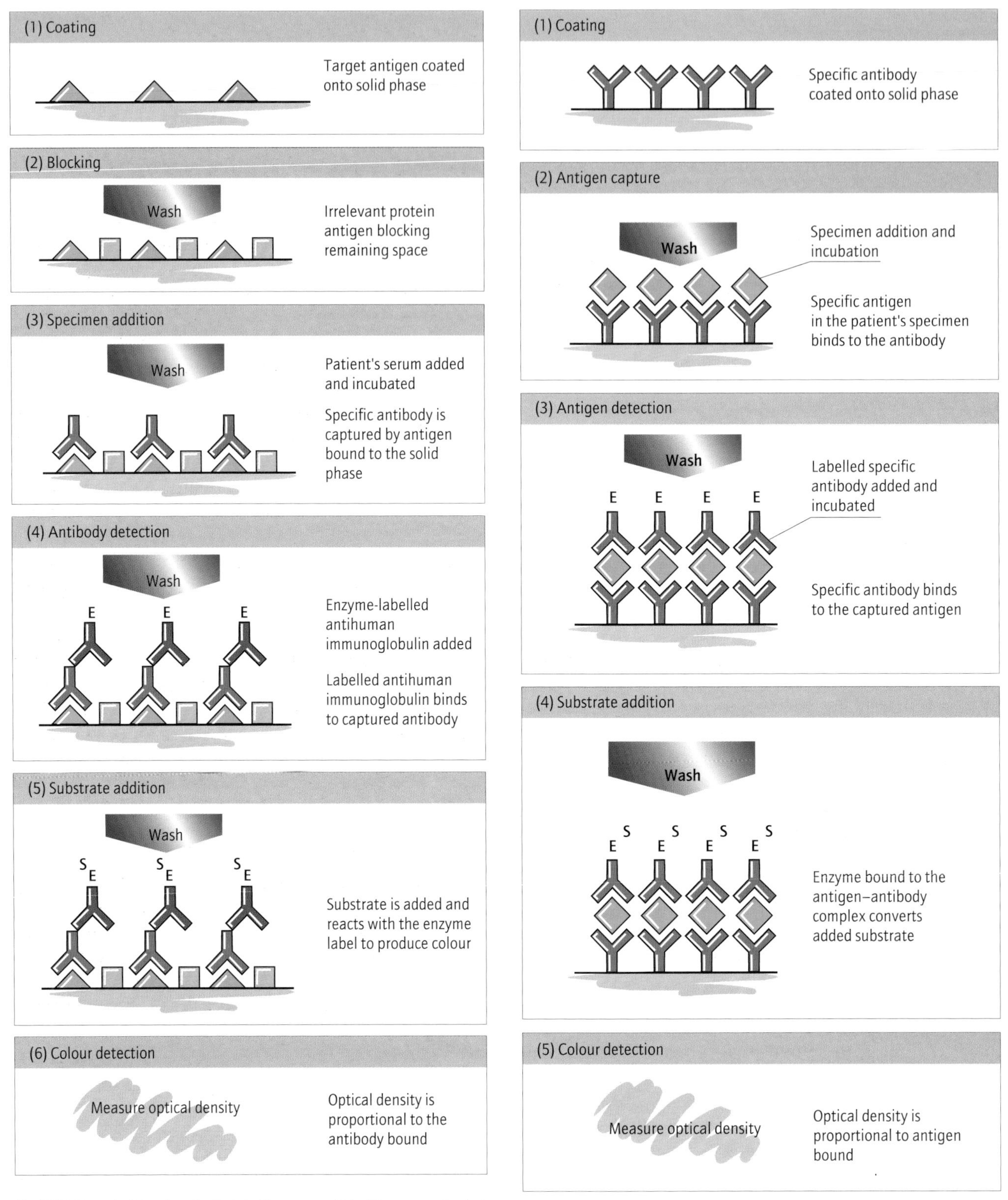

Fig. 3.17 The principle of antibody-detection enzyme-linked immunosorbent assay.

Fig. 3.18 Antigen-capture enzyme-linked immunosorbent assay, e.g. used for demonstrating the presence of hepatitis B surface antigen in the test serum.

patient's serum is added together with a known quantity of labelled antigen. Labelled, and any unlabelled, antigen compete for binding sites on the solid phase. The quantity of labelled antigen bound is then determined by the addition of substrate and measurement of the optical density as before. In other words, if antigen is present in the specimen, little labelled antigen will bind and there will be no colour change. Results are computed in comparison with a control antigen curve.

ELISAs utilize relatively inexpensive reagents and do not require expensive detection systems. The ELISA techniques readily lend themselves to automation and the reagents have a long shelf-life. As a result, ELISAs have been widely applied in the diagnosis of bacterial, parasitic and viral infections.

Western blotting (immunoblotting)

Microbial proteins can be separated by SDS-PAGE and transferred electrophoretically to a nitrocellulose membrane. Strips of the membrane are exposed to the patient's diluted serum, so that any antibodies specific to microbial proteins are bound, and can be detected using enzyme-labelled antihuman antibodies. The pattern of antibody recognition can be used to confirm a diagnosis, and to demonstrate the stage of disease by the repertoire of antibody specificities that the patient has developed. This is useful in the diagnosis of Lyme disease, in which simpler serological techniques are unreliable. It is also used in the study of human immunodeficiency virus (HIV) seroconversion illnesses.

Molecular diagnostics

The development of molecular techniques has opened up a new range of diagnostic methods. These include the detection of specific sequences of DNA from pathogens by Southern blotting. In this technique the patient's specimen is dried on a nitrocellulose filter, and is then heated in the presence of a specific DNA sequence which has been labelled (this may be a radioactive or a biotin-avidin label). The heating separates the double strands of the DNA. On cooling, some of the labelled DNA will anneal with any corresponding sequence in the specimen. After thorough washing of the nitrocellulose filter to remove leftover single-stranded test DNA, demonstration of the label's fluorescence or colour reaction will confirm the presence of the pathogen's DNA. Southern blotting is not readily applicable to routine diagnosis, but is useful when large numbers of specimens have been collected in epidemiological surveys.

Molecular amplification methods

Polymerase chain reaction

In polymerase chain reaction (PCR) a reaction mixture consists of the specimen together with a pair of primers (short sequences of nucleotides specific for a specific nucleic acid sequence of the pathogen sought). Nucleotides and Taq polymerase (an enzyme which catalyses the construction of DNA but is stable at high temperatures) are added. The reaction mixture is heated to separate all DNA strands. The primers bind to specific sequences in the specimen, and are then annealed by cooling the mixture. The Taq polymerase adds nucleotides to make double-stranded DNA fragments. At the end of one cycle there are therefore two copies of the nucleic acid sequences bound by the primers. The cycle of temperature manipulations is repeated, allowing exponential multiplication of these sequences. Positive specimens are detected by performing chromatography of the PCR products on an agarose gel. The presence of the 'target' DNA sequence can be confirmed by its characteristic size. Added sensitivity and specificity can be obtained by Southern blotting the amplified DNA with a specific probe.

PCR can be modified to demonstrate viral RNA by using reverse transcriptase to make DNA, which can then be amplified in the usual way. It has been applied to the diagnosis of many viral diseases, including HIV, cytomegalovirus and some, such as hepatitis C, which still cannot be cultivated. It is also finding a place in the diagnosis of *Mycobacterium tuberculosis* infection.

Alternative amplification strategies

PCR is a patented process. Commercial kits are widely available and have brought it out of the research laboratory into routine use. Alternative strategies now exist for amplification of small quantities of nucleic acid to detectable levels. These techniques include the ligase chain reaction (using DNA ligase), and strand displacement PCR.

Applications of molecular amplification methods

Susceptibility testing

The molecular basis of drug resistance is now becoming understood. For some organisms the presence of a particular gene is associated with the presence of resistance to antimicrobial agents, as in the case of the *Pfmdr* gene in multidrug-resistant *Plasmodium falciparum*. HIV readily mutates to resistance, and the nucleotide sequence of the affected genes (protease and reverse transcriptase genes) can confirm the resistance and suggest which other drugs may also be affected (see Chapter 4).

Response to treatment

It is often important to judge the early success of treatment. Amplification techniques can be useful for this. Several approaches are used to follow the response to anti-tuberculosis therapy; the number of organisms can be estimated by performing limiting dilution PCR (as the number of organisms falls, negative results are obtained after fewer dilutions). A quicker alternative is to detect the presence of mRNA, which is short lived and only present in viable cells. The presence of specific mRNA implies the presence of viable organisms allowing the success or failure of therapy to be judged.

In managing HIV and hepatitis C infection PCR is used to quantify the number of viral genome copies in the serum allowing treatment success or failure to be followed by serial measurements of viral load. The latest techniques will detect as few as 50 copies/ml of HIV genome.

Summary

An important skill in clinical infectious diseases and microbiology is the choosing of appropriate diagnostic investigations. While the latest techniques offer new advantages, elegant and timely results are still obtainable using less dramatic methods, as will be seen in the following chapters. Timely and accurate diagnosis depends on obtaining the correct specimen, at the optimum time of collection. It is important that the specimen is transported to the laboratory quickly and in conditions that maintain the viability of the organisms present or the integrity of the antigen sought. In the laboratory the correct diagnostic method must be used and the results effectively communicated to the doctor managing the case. This process depends on close co-operation between clinician, microbiologist and laboratory scientists.

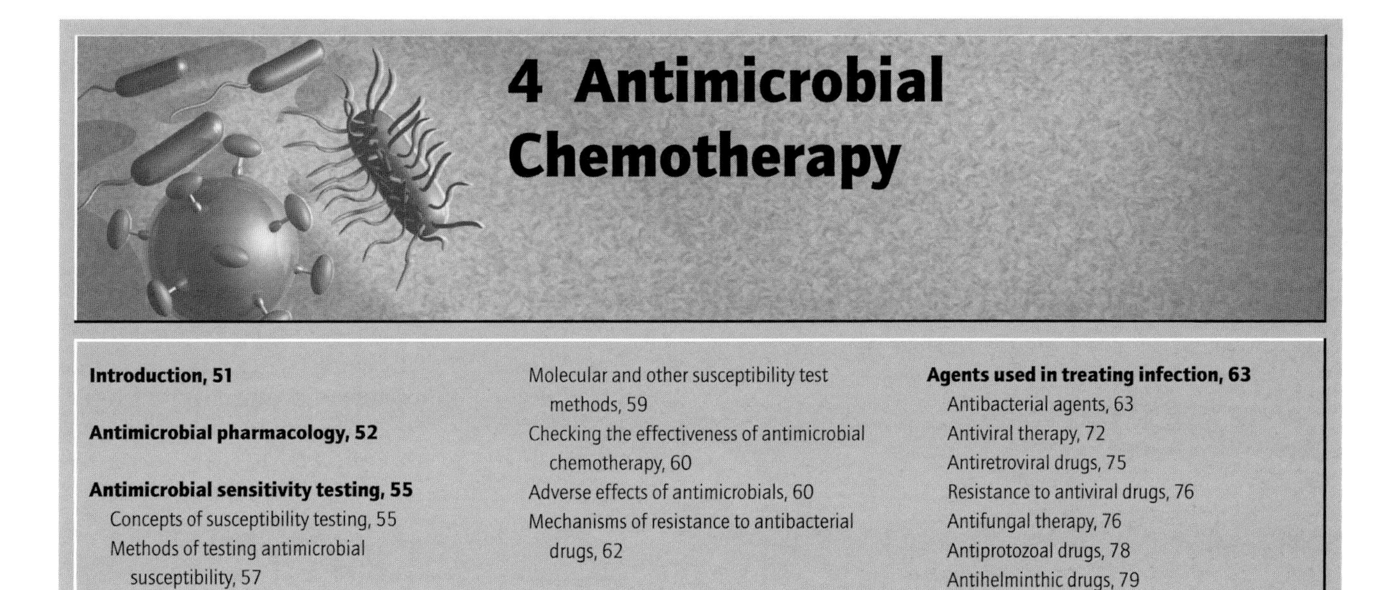

4 Antimicrobial Chemotherapy

Introduction, 51

Antimicrobial pharmacology, 52

Antimicrobial sensitivity testing, 55
- Concepts of susceptibility testing, 55
- Methods of testing antimicrobial susceptibility, 57
- Molecular and other susceptibility test methods, 59
- Checking the effectiveness of antimicrobial chemotherapy, 60
- Adverse effects of antimicrobials, 60
- Mechanisms of resistance to antibacterial drugs, 62

Agents used in treating infection, 63
- Antibacterial agents, 63
- Antiviral therapy, 72
- Antiretroviral drugs, 75
- Resistance to antiviral drugs, 76
- Antifungal therapy, 76
- Antiprotozoal drugs, 78
- Antihelminthic drugs, 79

Introduction

Most communities have a long tradition of using herbal and other medicines for the treatment of fevers. The underlying religious or scientific theory has varied, but even very ancient treatments may still be valid. The use of cinchona bark originated in the theory that where hazards (marshy land, the bringer of malaria) existed, the natural remedy would be found also. This led to the development of quinine, which is still obtained from its natural source. More recently, the artemesinin family of antimalarial drugs have been derived from species of wormwood plants.

Ehrlich thought that the selective staining of microorganisms by dyes might be used to target and kill microbes. From this he developed salvarsan, the first specific antimicrobial. All modern antimicrobial agents have their effect because of their selective toxicity against microbial metabolism or function with minimal effect on the host.

> Antimicrobial agents have selective toxicity: they severely damage microorganisms but have much less effect on human metabolism.

Various strategies have been used in the design of antimicrobials. Some interrupt microbial metabolic pathways which are not possessed by humans. Such antimetabolites are often false substrates or competitive inhibitors of microbial enzymes. The first example of this was the development of sulphonamides by Domagk in 1935. These compounds mimic para-aminobenzoic acid (PABA) and competitively inhibit the conversion of PABA into dihydropteroic acid, an essential precursor of folate. Bacteria rely on this metabolic pathway, as they are unable to take preformed folate from the host, and folate is an essential substrate in the synthesis of DNA. Further down the same pathway, trimethoprim and pyrimethamine inhibit the activity of dihydrofolate reductase, which converts dihydrofolate into tetrahydrofolate (Fig. 4.1). This depletes tetrahydrofolate concentrations. The two effects can be combined in the treatment of bacterial infections (trimethoprim–sulphamethoxazole) or protozoan infections such as toxoplasmosis (pyrimethamine and sulphadiazine).

The bacterial cell wall protects the organism from lysis caused by the osmotic gradient between the interior and exterior of the cell. The major structural component of the cell wall is peptidoglycan, composed of long polysaccharide chains with alternating *N*-acetylglucosamine and muramic acid molecules. The peptidoglycan chains are cross-linked between short peptide side-chains by an amide linkage (Fig. 4.2). Beta-lactam antibiotics (penicillins, cephalosporins, monobactams and penems) work by inhibiting transpeptidation and preventing cross-linking (Fig. 4.3). Each of these antibiotics possesses a beta-lactam ring which mimics the shape of the amide bond. Inhibition of cross-linking weakens the cell wall and the bacteria are killed by lysis.

Vancomycin and teicoplanin, glycopeptide antibiotics, affect bacterial cell wall synthesis by a different mechanism. They bind to the D-alanine of the peptide chain and the lipid II precursors and inhibit the cross-linking process. This activity is limited only to Gram-positive cell walls.

In order to contain the DNA strand in the bacterial cell, the DNA must be supercoiled. It must be uncoiled and recoiled during the processes of replication and transcription. Quinolone antimicrobials, which include nalidixic acid, ofloxacin and ciprofloxacin, work by interfering with the enzyme topoisomerase, or DNA gyrase, which supercoils bacterial DNA (Fig. 4.4).

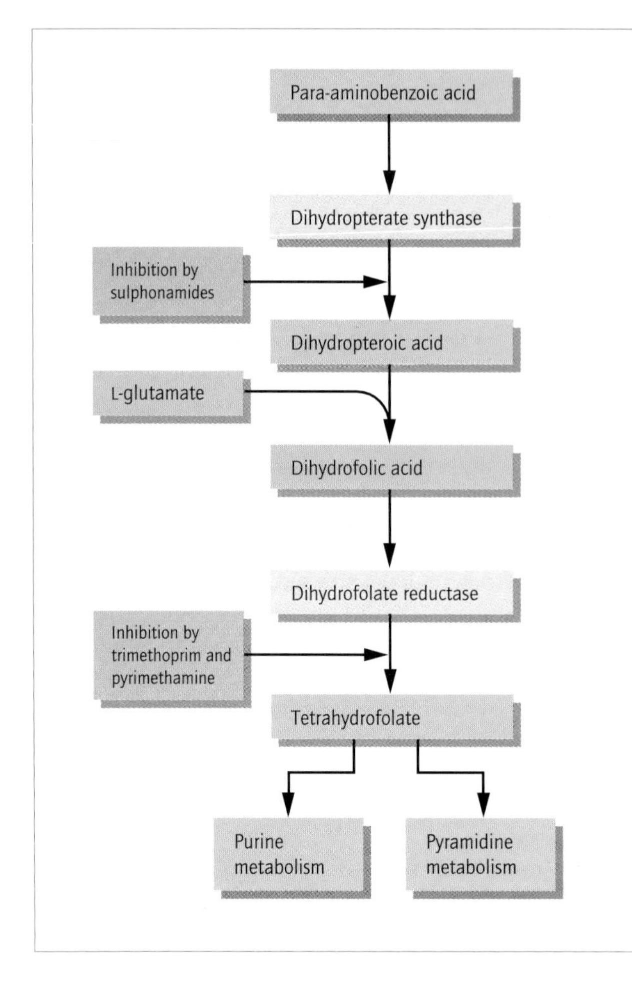

Fig. 4.1 The actions of sulphonamides and trimethoprim.

Rifampicin and rifabutin act by inhibiting DNA-dependent RNA polymerase, preventing transcription of the genetic code to mRNA.

Bacterial protein synthesis (Fig. 4.5) can be inhibited by a number of different mechanisms. Tetracycline inhibits the binding of transfer RNA (tRNA) to the 30S ribosome, whereas macrolides inhibit RNA-dependent protein synthesis at the 50S ribosome. Chloramphenicol prevents binding of tRNA to the 50S subunit of the ribosome. Aminoglycosides bind to both subunits of the ribosome, interfering with protein synthesis and causing the genetic code to be misread.

Streptogramins

Members of this class of antibiotic (e.g. quinupristin/dalfopristin) are being introduced to deal with emerging Gram-positive infections. The combination is produced by *Streptomyces pristinaespirilis*. The two drugs bind to the 50S ribosome at different sites and cooperate to inhibit protein synthesis. They are active against streptococci, *Mycoplasma*, *Chlamydia* and *Legionella*, and also against *Staphylococcus aureus* (including methicillin-resistant *S. aureus*, MRSA) and resistant enterococci. Various resistance mechanisms have already been described including target modification, efflux and modifying enzymes. Resistance encoded by the *erm* gene provides cross-resistance to quinupristin but does not affect the bacteriostatic activity of the streptogramins.

Oxazolidinones

These act by binding to the 50S ribosome and inhibit the initiation phase of protein translation. They are a novel class of antibiotic with activity against Gram-positive organisms and some Gram-negative organisms. They show no cross-resistance with other classes of antibiotic.

Everninomycin

This is an old antibiotic that has been reconsidered for clinical use to meet the challenge of resistant Gram-positive infections. It is a polysaccharide for which the mode of action is not known. It is active only against Gram-positive pathogens.

Antimicrobial action

1 Antimetabolites interrupt microbial chemical pathways.
2 Cell wall agents prevent the construction of bacterial cell walls.
3 Protein synthesis inhibitors interrupt the transcription and/or translation of microbial genes.
4 DNA gyrase inhibitors damage the tertiary structure of bacterial DNA.

Antimicrobial pharmacology

Effective treatment of infection depends on knowledge of the likely infecting organisms, their antimicrobial susceptibility, the site of infection, the spectrum of action of antimicrobial agents and their absorption and distribution within the body.

Knowledge of the pharmacology of antimicrobial agents is important in ensuring that an adequate antibiotic concentration is delivered to the site of infection (bioavailable). Infection in special sites such as the meninges cannot be treated by antibiotics that do not cross the

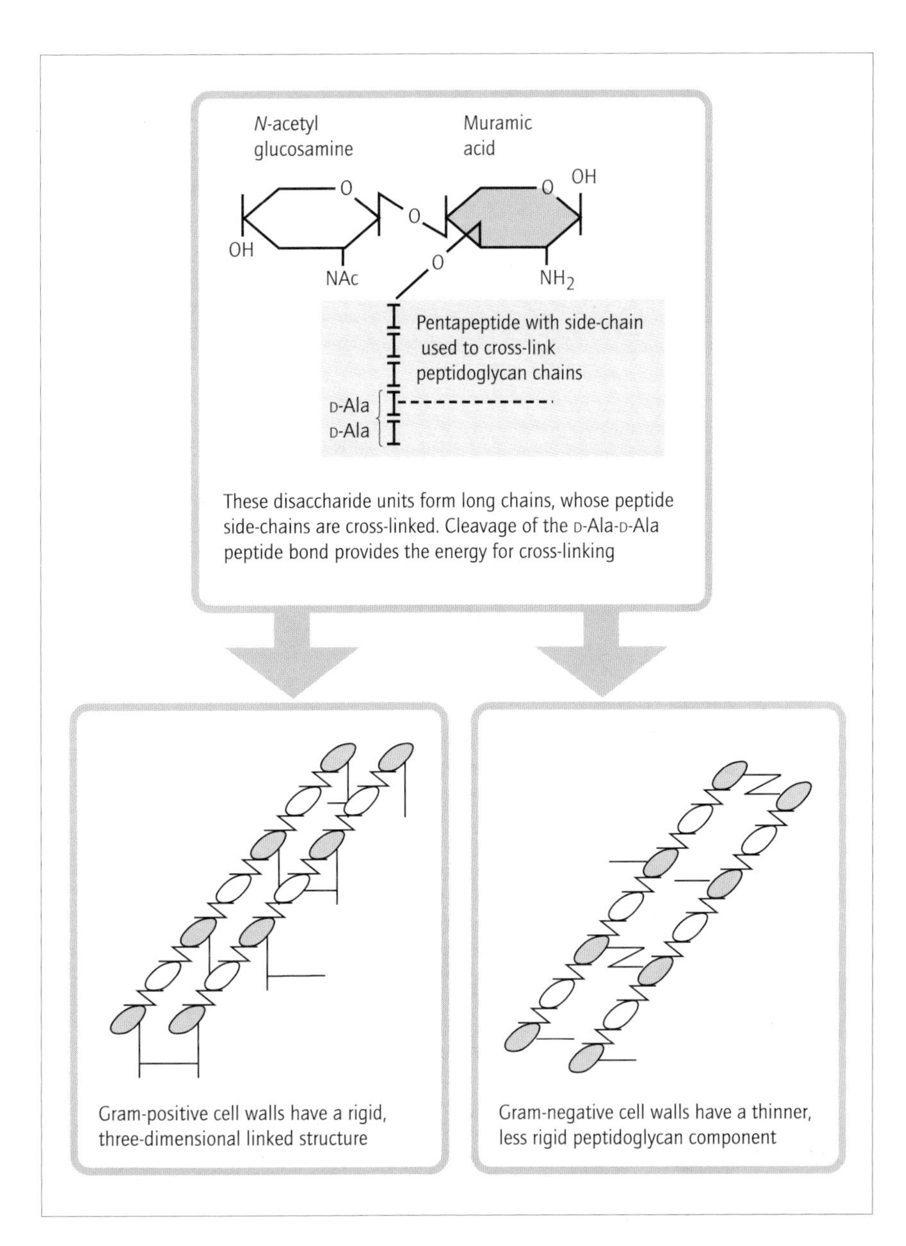

Fig. 4.2 The peptidoglycan cell wall structures in Gram-positive and Gram-negative organisms.

blood–brain barrier. Certain antibiotics also penetrate very poorly into abscesses.

The normal rules of pharmacology apply to the absorption and distribution of antibiotics within the body. Non-polar (lipid-soluble) agents such as chloramphenicol are well absorbed and cross the blood–brain barrier easily, whereas the more polar agents, penicillins and cephalosporins, are less well absorbed and are confined to extracellular compartments. Protein binding also affects the duration of action and the bioavailability of antimicrobial agents.

Conditions at the site of infection may greatly modify the effect of an antimicrobial agent. A concentration gradient of antimicrobial may be set up in an abscess, with lower concentrations at its centre than in the serum. Also, the presence of cellular or bacterial debris and the products of metabolism lower the pH and redox potential, interfering with the action of antimicrobials such as aminoglycosides (gentamicin-like agents) and macrolides (erythromycin-like agents).

The route of excretion also influences the effect of antimicrobial therapy. Agents excreted by the kidney

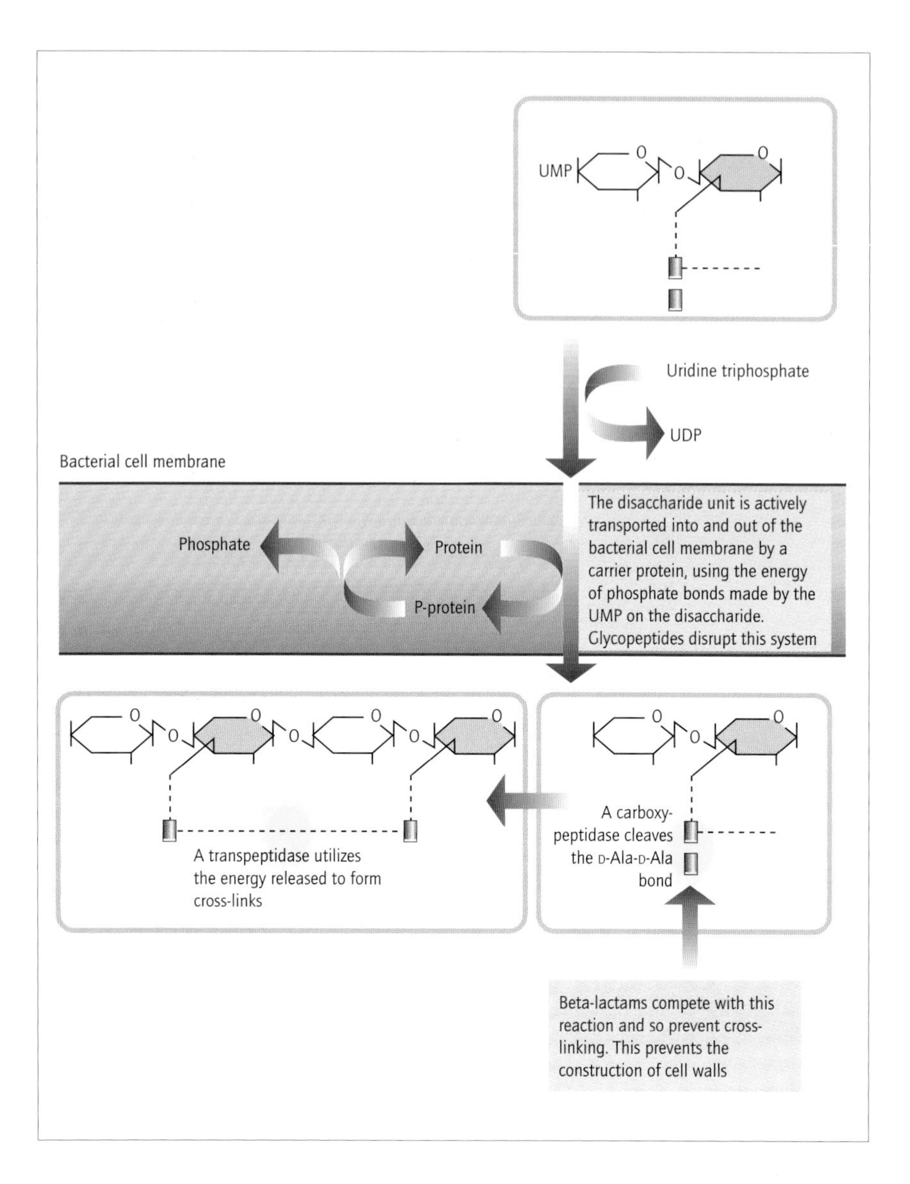

Fig. 4.3 The actions of beta-lactams and glycopeptides. UDP, uridine diphosphate; UMP, uridine monophosphate.

are likely to be effective in pyelonephritis and cystitis as high concentrations will be found in renal tissue, the bladder and urine. However, they will accumulate in cases of renal failure, and the dose must be reduced to avoid toxicity. Antibiotics excreted in the bile are likely to be effective in acute cholangitis, as high concentrations are available at the site of infection.

Many antimicrobials are metabolized by the liver, and hepatic disease may interfere with their excretion, causing an increased risk of side-effects and often necessitating reduction in dosage.

Factors affecting the effect of antimicrobials at the sites of infections

1 The concentration of the antimicrobial (influenced by its ease of access and rate of excretion).
2 Local pH and redox potential.
3 Ability of the pathogen to destroy the antibiotic.
4 Destruction of antimicrobial by host lysozymes and proteases.
5 Renal and/or hepatic damage, which can impair antibiotic excretion.

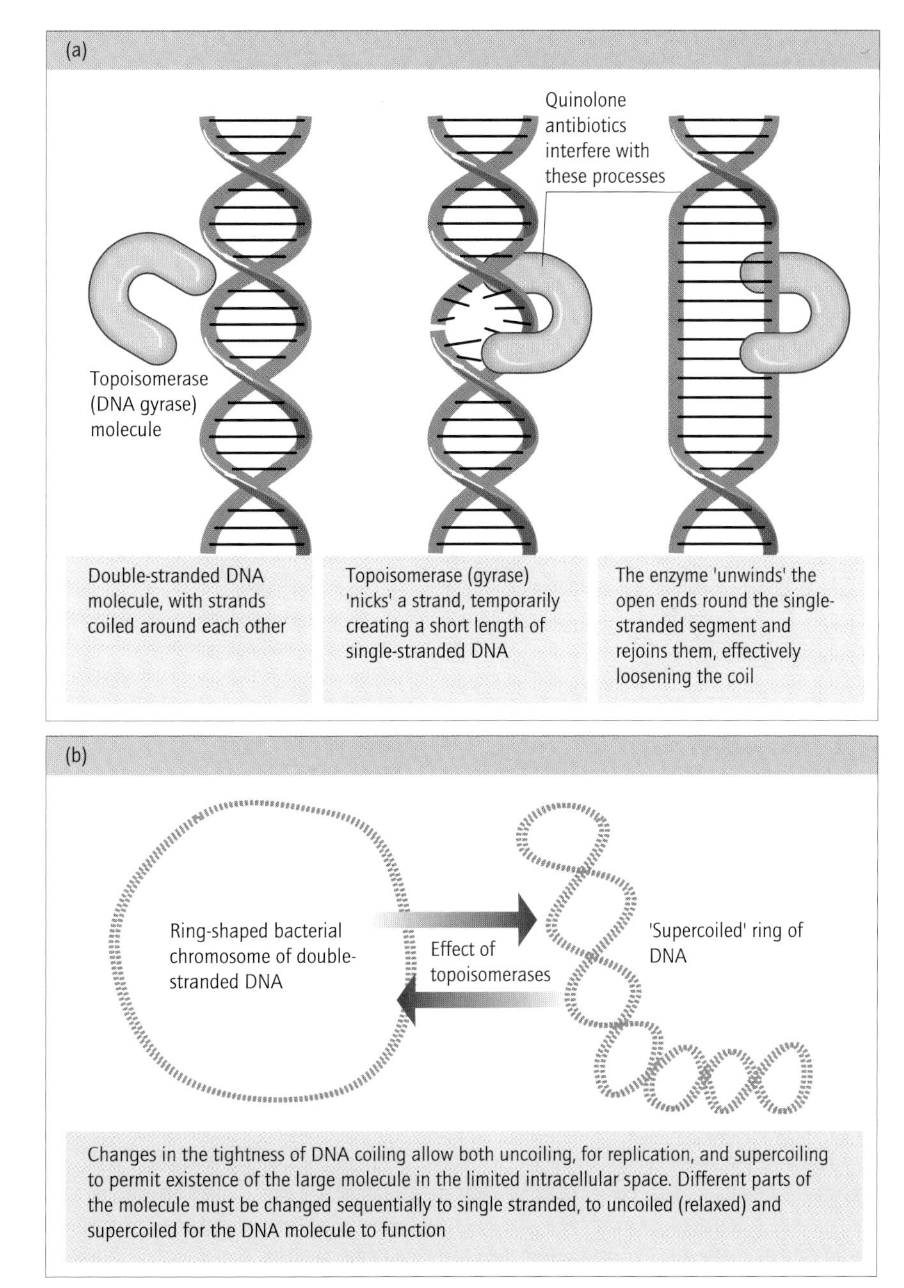

Fig. 4.4 The actions of quinolone antimicrobials. (a) Topoisomerase tightens (supercoils) or loosens (relaxes) the coiling of the DNA molecule by opening and reclosing small gaps in a DNA strand. DNA must be relaxed for its strands to part during replication or the production of RNA. (b) A bacterial DNA molecule is approximately 1 m long. It must be coiled many times to enable it to fit inside a bacterial cell. It must then be uncoiled, part by part, to allow its code to be read. Quinolone antibiotics prevent these changes in configuration, disrupting the function of the DNA.

Antimicrobial sensitivity testing

Concepts of susceptibility testing

Definitions

An organism is defined as sensitive or resistant to an antimicrobial, depending on whether treatment with the usual dose is likely to be successful. A moderately resistant organism is likely to respond to an increased dose.

An organism is considered sensitive if treatment with an antibiotic at standard dosage is likely to be successful.

Treatment for a moderately sensitive (or moderately resistant organism is likely to be successful if an increased dose of antibiotic is used.

An organism which is resistant is unlikely to be successfully treated with a given antibiotic, irrespective of dosage.

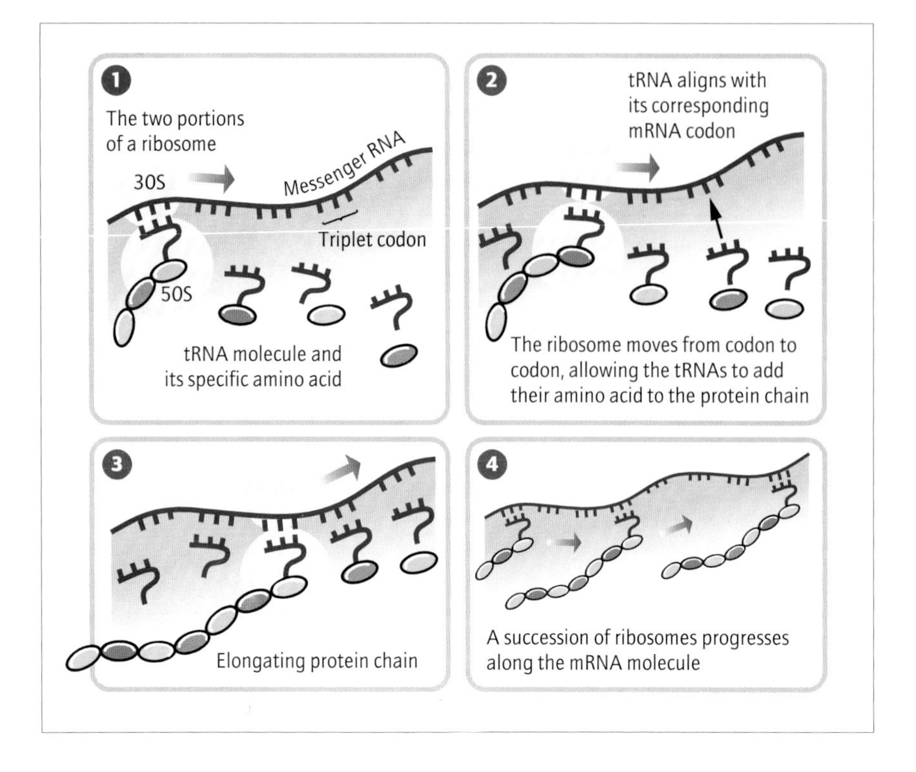

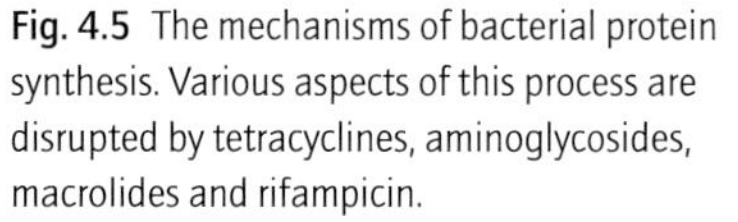
Fig. 4.5 The mechanisms of bacterial protein synthesis. Various aspects of this process are disrupted by tetracyclines, aminoglycosides, macrolides and rifampicin.

These definitions depend on the assumption that the results of laboratory tests reflect what is happening in the patient. Laboratory tests of susceptibility measure the activity of pure drug against bacteria growing exponentially in artificial culture. This is nothing like the situation at the site of an active infection. Antimicrobials, once administered, are absorbed and distributed throughout the body according to their lipid solubility and protein binding. They may also be metabolized by the liver and other tissues, and excreted or secreted by the kidney. Their breakdown products may or may not have antimicrobial activity. The site of infection may be devascularized, or an abscess with a fibrous cavity may have developed making penetration difficult. Organisms often are not 'free' in the tissues but exist inside macrophages or other cells. The site of infection may be at a low pH, at which some antimicrobials are relatively inactive. Organisms at the site of infection are not multiplying exponentially, as they do in the test tube where nutrients are not limited. In some infections bacteria may be in a stationary phase because of limited nutrients or, in some cases, such as tuberculosis, they may be dormant. In active infection the immune system is a vital contributor to recovery and may enhance the apparent activity of antimicrobials.

Laboratory tests measure directly or indirectly, and under controlled conditions, the inhibitory or killing effect of the antimicrobial on the pathogen isolated from the patient. The results of this, together with the known pharmacokinetics of the antibiotic, may predict the likely outcome of treatment with that agent.

Minimal inhibitory concentration and minimal bactericidal concentration

Two concepts are important in this context. The minimal inhibitory concentration (MIC) is defined as the lowest concentration at which growth of the organism is completely inhibited. The minimal bactericidal concentration (MBC) is the lowest concentration at which the organism is killed (actually defined as a 99.9% kill). In planning chemotherapy, the aim is to achieve a concentration of antimicrobial exceeding the MIC at the site of the infection. In some cases, where the multiplying organisms are inaccessible to the additive effect of phagocytosis and antibody, it is highly desirable that the MBC should be exceeded. This is especially true in endocarditis, in which bacteria are buried in thrombus, and in infections near to artificial material such as implants and prostheses.

> The minimum inhibitory concentration (MIC) is the lowest concentration of antibiotic that completely inhibits the growth of an organism.
>
> The minimum bactericidal concentration (MBC) is the lowest concentration of antibiotic that kills an organism.

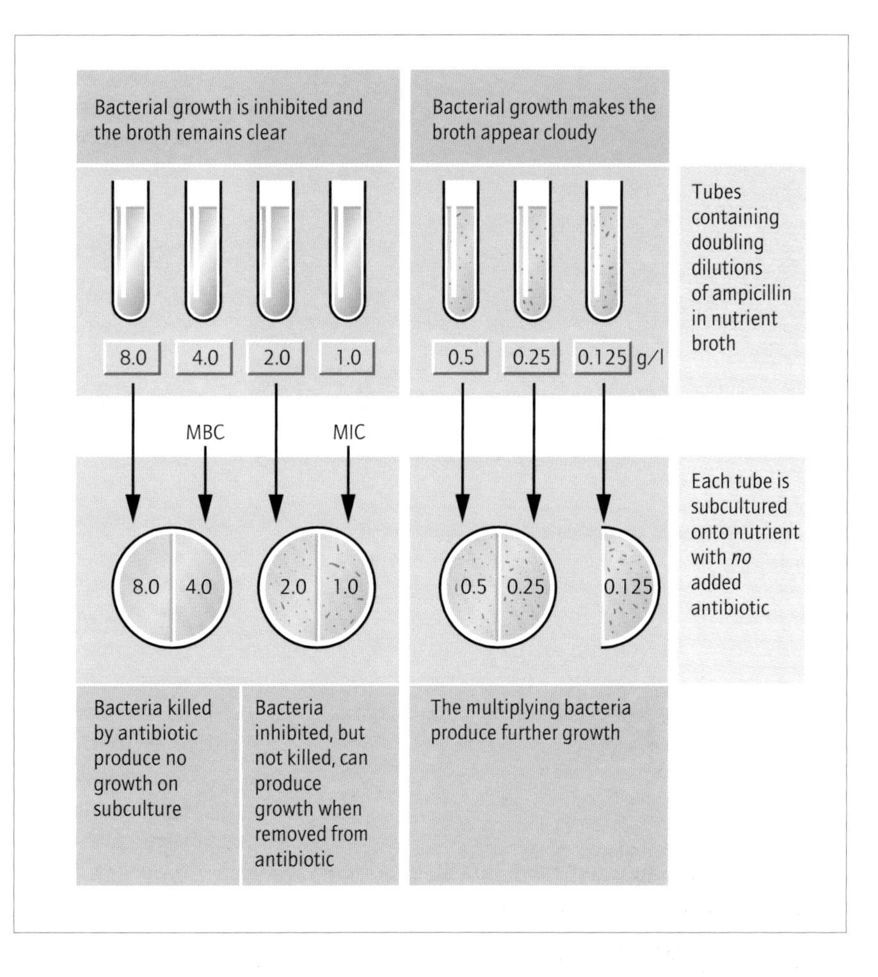

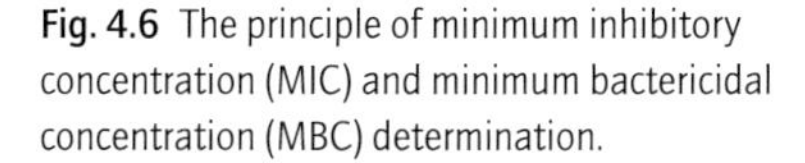
Fig. 4.6 The principle of minimum inhibitory concentration (MIC) and minimum bactericidal concentration (MBC) determination.

The MIC can be determined by cultivating organisms in broth cultures or on agar plates which incorporate a range of concentrations of the desired antimicrobial. By subculturing the broth cultures which show no bacterial growth, the death of the organisms can be confirmed, and the MBC can be determined (Fig. 4.6).

Methods of testing antimicrobial susceptibility

MIC determination is too cumbersome to use routinely in a clinical laboratory. Quicker, indirect methods are therefore used, which allow a large number of isolates to be tested against many antimicrobials. The most popular methods are based on filter-paper discs impregnated with antimicrobials. These are placed on agar plates which have been seeded with bacteria. The antimicrobials diffuse from the disc into the medium, inhibiting bacterial growth. At a certain distance from the edge of the disc, the concentration falls below that which will inhibit the growth of the organism, and a zone of demarcation can be seen. The diameter of this zone is related to the MIC of the organism.

The size of the zone may be altered by a number of factors unrelated to the susceptibility of the organisms tested. These include the composition of the medium, particularly the concentration of divalent cations; the pH of the medium; the molecular size of the antimicrobial; and incubation conditions. These variables must be controlled if reproducible results are to be obtained.

Each of the disc susceptibility methods uses a slightly different way of relating zone size with MIC, and controlling for variation in the conditions of individual tests.

Stokes method

The Stokes method is popular in UK laboratories for its simplicity and flexibility. In this method the test organism is directly compared with a control organism of known MIC, grown on the same plate. This controls for variation in the composition and characteristics of different media and the conditions of an individual test. It also relates the result (the zone size) to the zone size for a susceptible organism (Fig. 4.7). A sensitive *Staphylococcus* is chosen as the control organism for most isolates from sterile sites

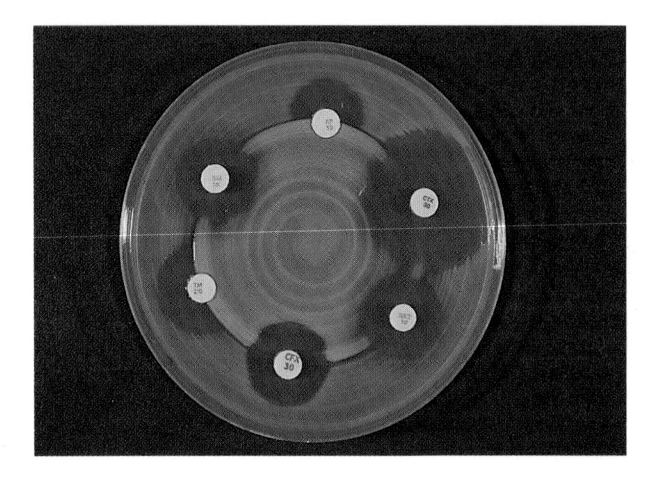

Fig. 4.7 Use of the Stokes method for determining antimicrobial sensitivity by disc diffusion. Outer band, susceptible control organism; central band, test organism.

and an *Escherichia coli* for urinary isolates. When organisms have special growth characteristics, for example *Haemophilus influenzae* or *Pseudomonas aeruginosa*, controls of the same species are used.

NCCLS method

The National Committee for Clinical Laboratory Standards (NCCLS) method is the standard technique used in the USA. The medium, inoculum (which must be confluent) and incubation conditions are strictly defined in this method. The zone sizes are directly related to MIC using a regression line, a graph which plots the known MIC of a series of organisms against zone size. Control strains are regularly tested and must be within defined limits of variation if the results are to be accepted as valid. The diameters of the zones are measured and compared with the previously defined ranges for susceptible organisms.

BSAC standardized disc sensitivity testing method

The British Society of Antimicrobial Chemotherapy (BSAC) introduced a method to standardize disc susceptibility testing. This uses a standard medium, Iso-Sensitest Agar, that can be supplemented with 5% whole horse blood with or without 20 mg/l beta-nicotinamide adenine dinucleotide (NAD). The only exception to this is the use of Mueller–Hinton agar for detection of methicillin resistance in staphylococci. A semiconfluent inoculum is used, defined discs are applied and the plates incubated within 15 min. Plates are incubated for 18–20 h (except for enterococci, which require 24 h, and coagulase negative staphylococci, which may require incubation for 48 h when testing for methicillin resistance). Zone sizes are measured and compared with preprepared tables of zone size versus MIC.

Agar incorporation method

In this method concentrations of antibiotic are incorporated into a solid medium in Petri dishes. The technique has the advantage that many tests can be performed on the same plate, using a multipoint inoculator. It is difficult, however, to obtain an estimate of the MBC in this test.

Breakpoint method

The agar incorporation MIC method can be simplified for routine use by using only two carefully chosen dilutions or 'breakpoints'. Organisms inhibited by the lower concentration of antibiotic would be reported as sensitive to conventional doses of the antimicrobial. If inhibition occurred only at the higher concentration, successful therapy would require higher doses and this result is reported as moderately sensitive. Organisms not inhibited by either concentration would be reported as resistant, and therapy would be predicted to fail.

High- and low-concentration antibiotic-impregnated discs are sometimes used to test organisms in this way. *Neisseria gonorrhoeae*, for example, is often sensitive to benzylpenicillin, and shows a large zone of inhibition around a low-concentration penicillin disc. Some gonococci have altered penicillin-binding characteristics, and are inhibited only by high-concentration penicillin discs. Penicillinase-producing *Neisseria gonorrhoeae* (PPNG) are not inhibited even by high-concentration penicillin discs. Placing the two types of disc on the original selective medium plate therefore provides an early indication of antibiotic sensitivity (Fig. 4.8), but this screening test must be confirmed by formal testing of the purified organism.

E-test

The E-test uses a semiconfluent inoculum and a strip impregnated with varying concentration of antibiotic throughout its length. The point a which growth meets the calibration on the strip is equivalent to the MIC. A wide range of organisms can be tested against most antimicrobials (even including difficult-to-test mycobacteria).

(a)

(b)

Fig. 4.8 (a) The organism on the upper half of the plate is a penicillin-sensitive gonococcus with zones to the penicillin 0.25 μg and 1.0 μg disc. Below is a strain resistant to penicillin. (b) The organism on the upper half of the plate is a non-beta-lactamase-producing penicillin-resistant gonococcus as there is only a small zone to co-amoxiclav (right-hand disc). The strain in the lower half is also penicillin-resistant: no zone to the ampicillin disc on the left but a wide zone to co-amoxiclav, which negates this bacteria's beta-lactamase.

Methods of routine sensitivity testing

1 Stokes method.
2 NCCLS method.
3 BSAC method.
4 Agar incorporation method.
5 Breakpoint method.
6 The E-test

Molecular and other susceptibility test methods

Beta-lactamase testing

Neisseria gonorrhoeae commonly becomes resistant to penicillin by elaborating a beta-lactamase enzyme. It is possible, therefore, to test for penicillin resistance by testing for the action of the beta-lactamase enzyme. This is done by using a chromogenic cephalosporin that changes colour when the beta-lactam bond is hydrolysed by beta-lactamase. This test takes only a few seconds to perform.

Molecular methods of susceptibility testing

The molecular mechanism of antibiotic resistance is now known for many organisms. It is possible for some organisms to detect the presence of gene mutations associated with resistance phenotypes. This is done by amplifying the target gene and determining the DNA (or RNA) sequence. Defining the susceptibility of organisms in this way is useful in investigating the antiviral susceptibility of human immunodeficiency virus (HIV) as the gene sequence can also indicate which drugs might be successful. A similar approach can be used to determine susceptibility of *Mycobacterium tuberculosis*, especially for rifampicin. *M. tuberculosis* becomes resistant to rifampicin through mutations in the RNA polymerase gene, which can be amplified by polymerase chain reaction (PCR) and resistance-associated mutations detected. A number of different techniques have been applied including direct nucleotide sequencing.

An alternative is to perform single-strand conformation polymorphism electrophoresis, where the two strands of bacterial DNA are separated and run on a gel. Point mutations alter the conformation of the DNA, resulting in a different migration speed compared to the non-mutated gene. Point mutations can also be detected by Southern blotting; commercial kits using this approach are available. Several PCR-based methods (e.g. heteroduplex PCR) have also been described. *Streptococcus pneumoniae* develops resistance to penicillin by acquisition of small portions of DNA in the penicillin-binding protein genes. PCR amplification and gel electrophoresis of these genes allows susceptible and resistant isolates to be distinguished.

Plasmodium falciparum becomes resistant to several antimalarial agents by acquisition of the *Pfmdr* gene. A PCR that amplifies this gene enables multidrug-resistant organisms to be identified much more simply than by *in vitro* cultivation. A similar approach can be used to detect, for example, the *mecA* gene of *S. aureus* coding methicillin resistance or the *VanA* gene of enterococci encoding vancomycin resistance.

Checking the effectiveness of antimicrobial chemotherapy

Therapeutic monitoring

Antibiotic concentrations are nowadays rarely monitored except for the following reasons.

1 To ensure that adequate concentrations of antibiotic are found in the serum (or in the cerebrospinal fluid (CSF) in difficult cases of meningitis).

2 To ensure that antibiotic concentrations have not reached toxic levels. This is important for antibiotics with a narrow therapeutic index such as aminoglycosides or vancomycin, or when the patient is in renal failure.

3 To monitor patient compliance. This is important in managing infections that require long-term therapy as an outpatient, e.g. tuberculosis.

4 To confirm adequate absorption of oral antibiotics after a shift from parenteral administration, for example in the case of infective endocarditis.

5 To study the pharmacokinetics of new antibiotics.

A number of different methods are used in the laboratory. The simplest methods are the automated commercial immunoassays available for measuring the concentrations of aminoglycosides ad vancomycin. For many other antibiotics high-performance liquid chromatography (HPLC) must be used. Specimens must be taken at appropriate times: specimens for trough concentration should be taken just before the next dose is given, and peak concentrations should be taken 1 h after the parenteral dose.

Adverse effects of antimicrobials

The basic principle of antimicrobial chemotherapy is selective toxicity, to damage the pathogen's metabolism with a minimal effect on the host. Most antimicrobial agents have very favourable therapeutic indices, meaning that the ratio of toxic level to therapeutic level is high. Adverse effects fall into two major groups: those which are dose-dependent, as in the bone marrow toxicity of sulphonamides or renal toxicity of aminoglycosides; and idiosyncratic reactions, such as penicillin-induced anaphylaxis or chloramphenicol-induced aplastic anaemia.

Adverse antimicrobial effects
1 Dose-dependent toxicity.
2 Idiosyncratic reactions.

Renal toxicity

Renal damage can occur for a number of reasons during antimicrobial chemotherapy. Not all are direct effects of the antimicrobial drug; for instance, renal failure can follow an anaphylactic reaction, because the accompanying hypotension causes acute tubular necrosis. Obstructive nephropathy can occur if high concentrations of drugs such as long-acting sulphonamides or first-generation cephalosporins form microcrystals in the tubules or larger crystals in the collecting ducts.

Direct renal cell damage

Direct renal cell damage can be caused by nephrotoxic antimicrobials. Aminoglycosides are toxic to cells of the proximal renal tubules, causing accumulation of membrane structures within the cell. This occurs at serum levels close to those necessary for effective antimicrobial action; the therapeutic index is narrow. There is evidence of mild, self-limiting tubular effects in all patients, and small reductions in the glomerular filtration rate occur in approximately 80% of patients.

Aminoglycoside toxicity is more likely in elderly patients and those with pre-existing renal compromise, including renal failure caused by sepsis. Toxicity is also increased by coadministration of diuretics, other renal toxic drugs or recent aminoglycoside therapy. A rapid decline in renal function can occur in patients treated with aminoglycosides for infective endocarditis, because of the additive effects of aminoglycoside toxicity, uncontrolled sepsis and the nephritis accompanying the infective endocarditis syndrome. In the ear aminoglycosides may damage the hair cells of the organ of Corti, or the type I hair cells at the summit of the ampullar cristae, leading to impaired hearing. To prevent these consequences the serum aminoglycoside concentration must be closely monitored, and the dose adjusted if toxic levels are approached.

Renal toxicity is common in patients treated with amphotericin. The mechanism is not known but may be associated with distal tubular lesions and a decrease in glomerular filtration rate. This problem has been largely overcome by using the less toxic liposomal preparations.

Tetracycline can cause renal damage by several different mechanisms. Its catabolic action produces an increase in nitrogen degradation products, including urea. Some tetracycline degradation products may be directly toxic to the kidneys.

Sulphonamides are a diverse group of agents, some of which can crystallize in the renal tubules, causing obstructive uropathy. Occasionally they cause direct renal tubular cell damage. Co-trimoxazole, which contains sulphamethoxazole, may accumulate in renal failure, producing a fall in the glomerular filtration rate. Patients with a small urinary volume may develop crystalluria, even with this modern sulphonamide preparation.

Liver toxicity

Damage to the liver may take the form of acute or chronic hepatitis, cholestasis, fatty degeneration or a granulomatous hepatitis.

Hepatocellular damage

This is the commonest form of liver damage. It is often associated with the antituberculous agents isoniazid and rifampicin. Among patients given isoniazid prophylaxis, acute hepatitis occurs in approximately 1%. The risk is age-related, being much greater in patients aged over 35. Toxicity is probably caused by a toxic intermediate metabolite, acetylhydrazine. Hepatitis is more likely in patients with previous liver disease, or a history of alcohol abuse.

Rifampicin may also induce hepatitis. A transient elevation of transaminases is common when dosing is started, but levels return to normal despite continued drug use, and jaundice does not develop. Severe hepatitis with jaundice is occasionally seen. Rifampicin may also produce transient hyperbilirubinaemia due to competition for hepatic excretion. The combination of rifampicin and isoniazid in the treatment of tuberculosis causes hepatitis in about 5% of patients. Hepatitis can complicate the use of erythromycin estolate, parenteral tetracycline, pyrazinamide and ethionamide.

Rare cases of granulomatous hepatitis have been reported following high-dose ampicillin or flucloxacillin treatment, or prolonged quinine therapy.

Cholestatic jaundice

Cholestatic jaundice is a dose-related effect of fusidic acid, induced by doses above 2 g/day. Tetracycline may induce fatty change in the liver if high doses are given intravenously or if patients are predisposed by pregnancy or renal failure.

Effects on the haemopoietic system

Bone marrow toxicity can be caused by many different agents, either as a result of the normal action of the drug, as an idiosyncratic reaction, or by immune mechanisms. It may affect granulocytes alone or the whole haematopoietic system.

Aplastic anaemia

Aplastic anaemia can follow chloramphenicol therapy from 10 days after therapy until up to 6 months. This idiosyncratic reaction must be distinguished from a mild, reversible dose-dependent bone marrow depression which occurs in many patients on higher doses. True aplasia has an incidence of 1 in 40 000–1 in 100 000, and is often irreversible.

Metabolic bone marrow depression

Metabolic bone marrow depression can complicate treatment with antimicrobials affecting nucleic acid synthesis, either directly, as with ganciclovir and zidovudine, or by reducing folate availability, as with sulphonamides.

Granulocytopenia

Granulocytopenia is usually idiosyncratic. It is commonest after sulphonamide or sulphone therapy, but rarely follows high-dose benzylpenicillin therapy (>12 MU/day). Penicillin-associated agranulocytosis is always associated with the same reaction to cephalosporins, and will recur following even small oral doses of these drugs.

Immunologically mediated cytopenias

Immunologically mediated cytopenias are rare. Antibiotics, like other drugs, may act as haptens and induce antibodies to red blood cells or platelets, resulting in a Coombs-positive haemolytic anaemia or immune thrombocytopenia.

Cutaneous adverse reactions

These vary from fixed drug eruptions, urticarial and maculopapular eruptions to erythema multiforme or even life-threatening Stevens–Johnson syndrome. By far the commonest causes of severe skin reactions are sulphonamides, and the risk increases dramatically in the elderly. Mild, itching rashes are common with sulphonamides, penicillins (Fig. 4.9) and cephalosporins. Although not strictly an allergic response, 95% of patients with infectious mononucleosis develop a rash if treated with ampicillin, but this does not recur on re-exposure to the drug after convalescence.

Anaphylaxis occurs when an antibiotic generates a type I hypersensitivity reaction, usually as a result of

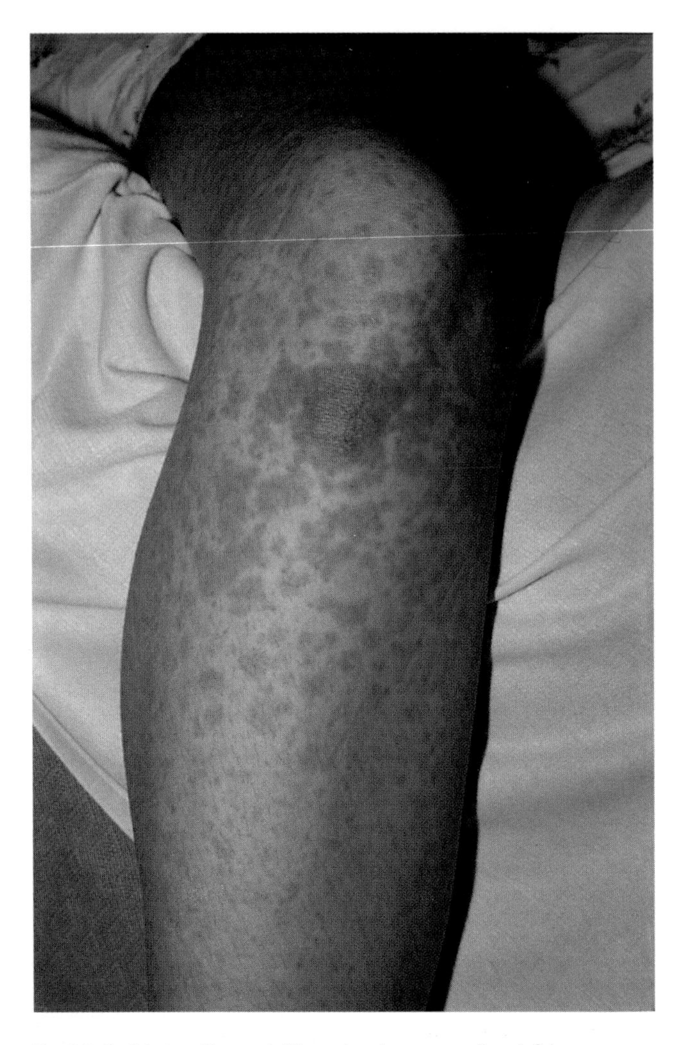

Fig. 4.9 Rash induced by ampicillin; such rashes are usually painful or irritating.

hapten–drug sensitization. This complication is uncommon, perhaps affecting 1 in 100 000 treatments. Penicillins are most often responsible. Cephalosporins may cross-react, but in less than 10% of penicillin-sensitive patients. New monobactam agents have even lower cross-sensitivity rates.

Mechanisms of resistance to antibacterial drugs

Each antimicrobial agent is effective against a limited range of organisms, usually because the metabolic process with which they interfere does not occur in all species. Vancomycin prevents peptidoglycan cross-linking only in Gram-positive organisms. All Gram-negative organisms are naturally resistant. Metronidazole is active at such low redox potentials that only anaerobic bacteria are susceptible. Amphotericin is active against fungi because it inhibits ergosterol synthesis, necessary for cell wall production in fungi, but not in bacteria.

Reasons why organisms may be naturally resistant to antimicrobials

1 They are naturally impermeable to the antimicrobial agent.
2 They lack the target binding site.
3 They lack the target metabolic pathway.
4 They naturally produce antibiotic-destroying enzymes.
5 They produce more of the target to overcome resistance.

Antibiotic-inactivating enzymes

Some organisms naturally produce enzymes that inactivate certain antimicrobials. When penicillin was first introduced, 95% of *Staphylococcus aureus* were sensitive, but 5% produced a beta-lactamase which destroyed the beta-lactam bond needed for activity. Over a year or two an increasing proportion of strains acquired beta-lactamase. Now almost all *S. aureus* isolated in hospital are penicillin-resistant. *S. aureus* has acquired resistance due to the selective pressure applied by antibiotics, favouring survival of beta-lactamase-producing strains. In the presence of similar selection pressure, other organisms expressing a beta-lactamase may also be favoured. The genes coding for these enzymes may even be transferred to other organisms, passing on the selective advantage. By this means, antibiotic-destroying enzymes can become widespread. For example a beta-lactamase emerged in enteric organisms and has spread widely to include *H. influenzae* and *N. gonorrhoeae*.

Other antibiotics are inhibited by different bacterial enzymes. Bacteria become resistant to aminoglycosides by developing inactivating enzymes which adenylate, hydroxylate or acetylate the aminoglycoside molecule at various sites. Resistance to chloramphenicol depends on possession of an acetyltransferase enzyme.

Altered antibiotic-binding sites

The action of antibiotics may be inhibited by altered binding of the antibiotic to its target. Organisms resistant to macrolides and aminoglycosides have small alterations in binding sites on the ribosomal target. Resistance to streptomycin, for example, may develop in a one-step process in which the ribosomal binding site is methylated, preventing the binding of the antibiotic. Penicillin resistance in *Streptococcus pneumoniae* depends on an alteration in the affinity of penicillin-binding protein (PBP). This has developed in a multiple stepwise fashion under the selective pressure of antibiotics available in the human environment. Recombination has occurred between PBP genes in *S. pneumoniae* and commensal organisms at low frequency (10^{-13}). This eventually results in penicillin-binding proteins that do not bind penicillin, but can still support cell wall synthesis. Resistance to sulphonamides and trimethoprim occurs due to alterations in the target enzymes; dihydropterate synthetase and dihydrofolate reductase, respectively. Resistance to rifampicin arises by alterations in the peptide sequence of RNA polymerase. Many viruses can develop resistance to antiviral agents by point mutations in the genes encoding the antiviral target protein; for example, HIV becomes resistant to lamivudine by a single point mutation.

Factors affecting antibiotic penetration

Alteration of the permeability of bacterial outer membranes may result in high-level resistance, often affecting all drugs of a class. This form of resistance is common in *Pseudomonas* spp. Some bacteria become resistant to tetracyclines by actively transporting the antibiotic out of the cell (the efflux mechanism). This mechanism results in resistance to all compounds in this class.

The major resistance mechanism for tetracycline among Gram-negative organisms is an alteration of an inner membrane protein which inhibits accumulation of tetracycline within the bacterial cell.

Glycopeptide-resistant *S. aureus* has a much thicker cell wall and subtle alteration to its peptide structure. This results in an increase in the amount of glycopeptide absorbed to the surface, preventing the antibiotic reaching the site of action.

Methods by which bacteria can acquire resistance to antimicrobial agents

1 An alteration in permeability to the antibiotic.
2 An alteration in the target binding state.
3 Use of an alternative metabolic pathway.
4 Switching on of a gene for an antibiotic-destroying enzyme.
5 Acquisition of a new gene for an antibiotic-destroying enzyme.

Bacterial genes coding for antibiotic resistance can be transmitted not only between organisms of the same species but even between different genera. Transmission of resistance to previously sensitive organisms means that resistance to commonly used antibiotics becomes widespread. Bacterial resistance genes, like all other DNA fragments, are transmitted by four main mechanisms: transformation, transduction, conjugation and transposons (Fig. 4.10). Transformation is the process whereby bacteria take up segments of naked DNA and incorporate them into their own genetic make-up. Transduction takes place when fragments of DNA are taken up by a bacteriophage and transmitted when the phage infects a new bacterial cell. Conjugation involves transfer of chromosomal and/or plasmid DNA between bacterial cells. Transposons are movable genetic elements capable of transferring from one bacterial strain to another, between plasmids and between plasmids and the bacterial chromosome.

In the presence of antimicrobial selection pressure, organisms possessing resistance determinants will be favoured. Some strains possess multiple resistance determinants and present significant therapeutic problems in the selection of effective chemotherapy. Control of antimicrobial prescribing may be effective in limiting the development of these strains, but must be very strict to eradicate multidrug-resistant strains, as a positive selective pressure exists if any antimicrobial to which the organism is resistant is found in the hospital environment.

Mechanisms for gene transfer between bacteria

1 Transformation (transfer of naked DNA).
2 Transduction (where the gene is acquired with a bacteriophage infection).
3 Conjugation (DNA transfer between conjoined bacteria).
4 Transposons (transposition).

Agents used in treating infection

The main types of anti-infection agents will be reviewed here briefly. Antiviral, antifungal and antiparasitic agents will also be discussed with their main therapeutic targets in the systematic chapters.

Antibacterial agents

Beta-lactam antibiotics

Beta-lactam antibiotics include penicillins and cephalosporins which have similar structures.

Penicillins can be conveniently divided into classes on the basis of their antibacterial activity:

1 Natural penicillins (penicillin G).
2 Penicillinase-resistant penicillins (cloxacillin).
3 The aminopenicillins (ampicillin-like agents).

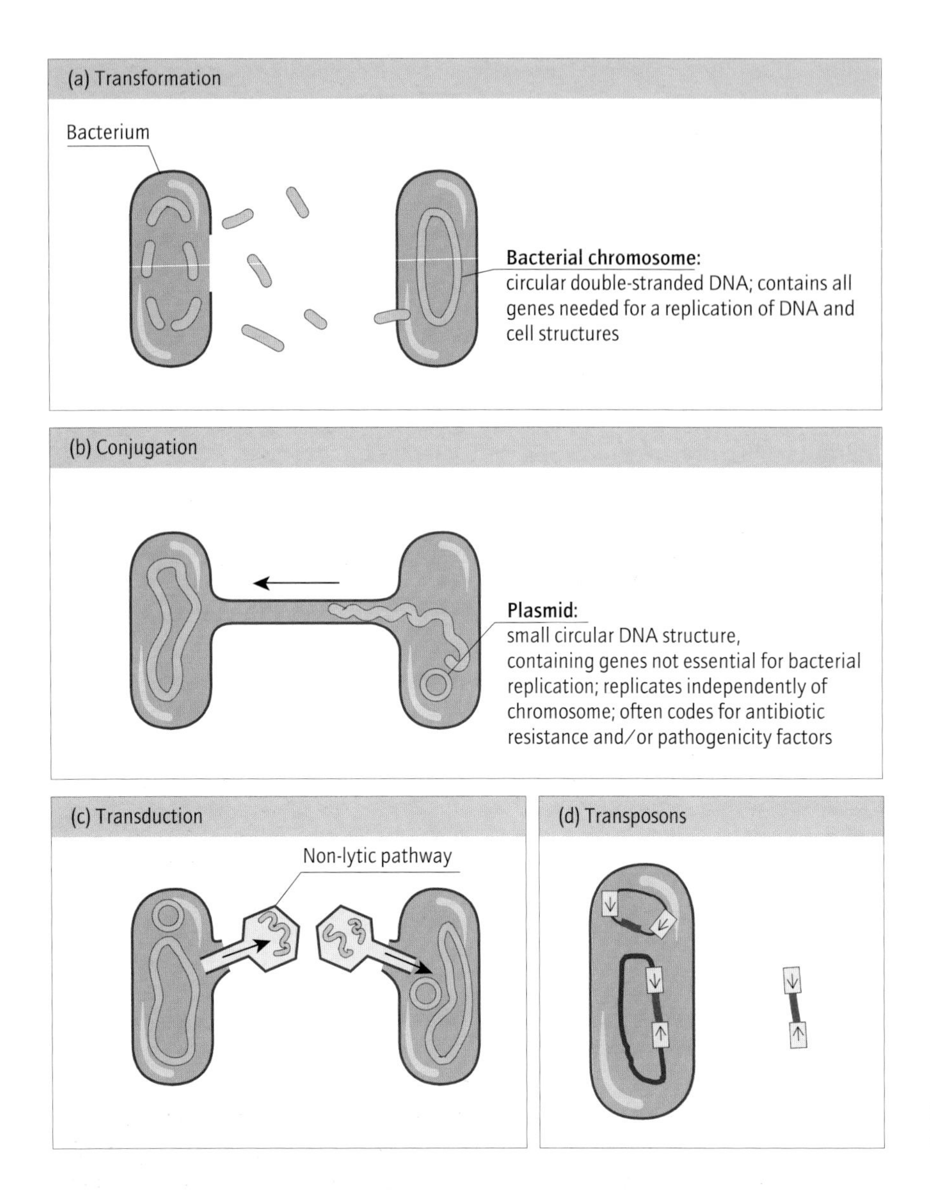

Fig. 4.10 The mechanisms of DNA transfer between bacteria.

4 Expanded-spectrum penicillins.
5 Carbapenems.
6 Monobactams.
7 Beta-lactamase inhibitors.

Spectrum of activity

Penicillin G is derived from cultures of *Penicillium chrysogenum*. It is active against Gram-positive bacteria, including streptococci, staphylococci (excluding most *Staphylococcus aureus*), clostridia and corynebacteria, and some Gram-negative cocci, including *Neisseria meningitidis*. *Bacteroides fragilis* is resistant, but *Prevotella* sp., *Porphyromonas* sp. and anaerobic cocci are susceptible.

Isoxazolyl penicillins such as flucloxacillin, oxacillin and cloxacillin were developed to deal with the problem of penicillin-resistant *S. aureus*. Their bulky side-chain sterically hinders the binding of penicillinase to these drugs (Fig. 4.11). These agents retain activity against streptococci and *Neisseria*. They are well absorbed orally.

The aminopenicillins include ampicillin, its esters and amoxycillin. They are well absorbed orally, and have an expanded spectrum. *Haemophilus influenzae* is susceptible, as are many species of the Enterobacteriaceae, including *Escherichia coli*, *Salmonella* sp. and *Shigella* sp. Following the introduction of ampicillin, hospitals experienced cross-infection problems with Gram-negative pathogens which were naturally resistant to this agent. These included *Klebsiella* spp. and *Pseudomonas aeruginosa*. Newer penicillins were developed to address this

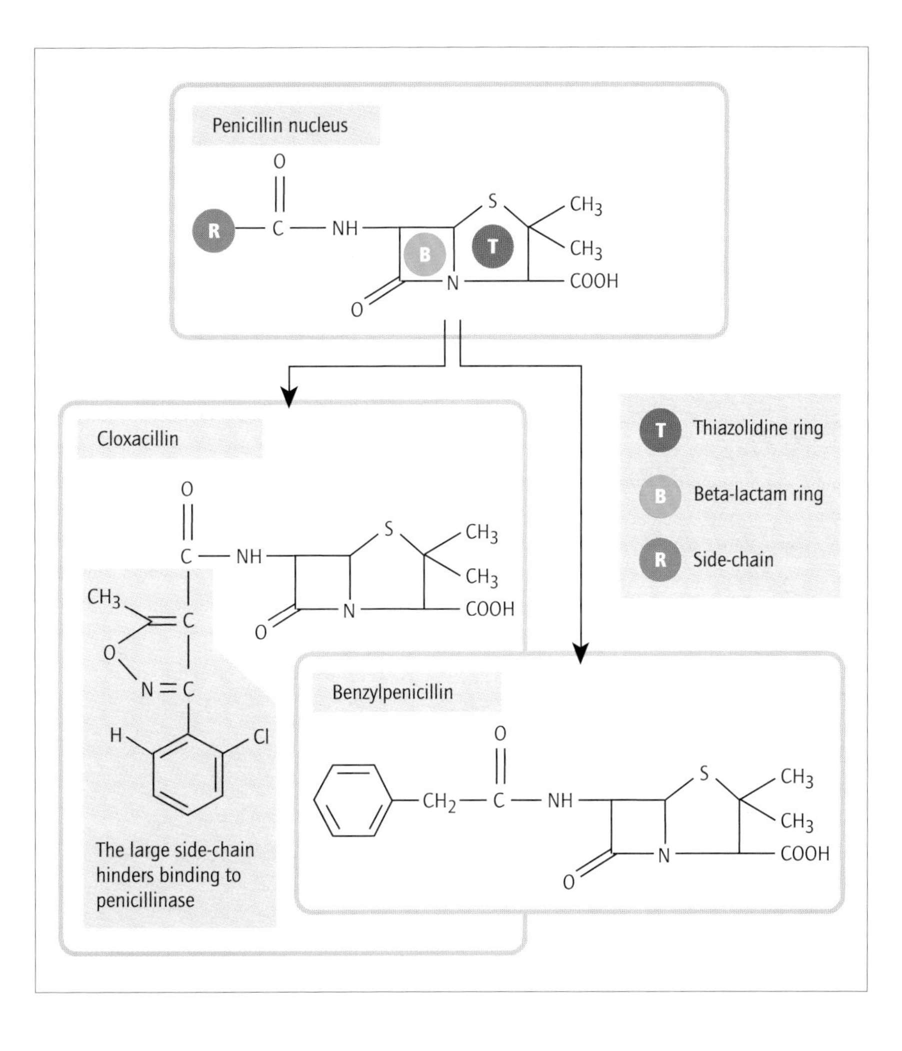

Fig. 4.11 The structure of penicillin G and cloxacillin.

problem and these include the acylureidopenicillins; azlocillin and piperacillin, and the carboxypenicillins; carbenicillin and ticarcillin (Fig. 4.12). All of these agents must be administered parenterally. They are active against some ampicillin-resistant Enterobacteriaceae and *P. aeruginosa* but are susceptible to plasmid-mediated beta-lactamases, which limits their effectiveness.

The carbapenems, imipenem (always given with cilastatin) and meropenem, have a broad spectrum of activity against Gram-positive and -negative bacteria, including beta-lactamase-producing Gram-negative organisms and *Bacteroides fragilis*. The monobactams, of which aztreonam is the best example, are synthetic compounds which have little activity against Gram-positive organisms but are very active against Gram-negative species, including beta-lactamase-producing strains and *Pseudomonas* sp.

As one of the main reasons for resistance to penicillins is the elaboration of beta-lactamases, coadministration of a beta-lactamase inhibitor is a logical step. These are beta-lactamase-stable beta-lactam compounds with minimal antimicrobial activity. It is essential that these compounds have similar pharmacokinetic properties to their companion active agent. Several are available for clinical use, including clavulanic acid and sulbactam. These have been combined with different active agents—clavulanic acid with amoxycillin or ticarcillin, sulbactam with ampicillin or piperacillin.

Pharmacology

Penicillins vary markedly in their oral absorption. Penicillin G is not stable to gastric acid and must be given intravenously. Long-acting derivatives of benzylpenicillin such as procaine and benzathine penicillin are available in some situations for intramuscular administration. Penicillin V (phenoxymethylpenicillin), in contrast, is stable to gastric acid and can be given by the oral route. The aminopenicillins and isoxazolylpenicillins are

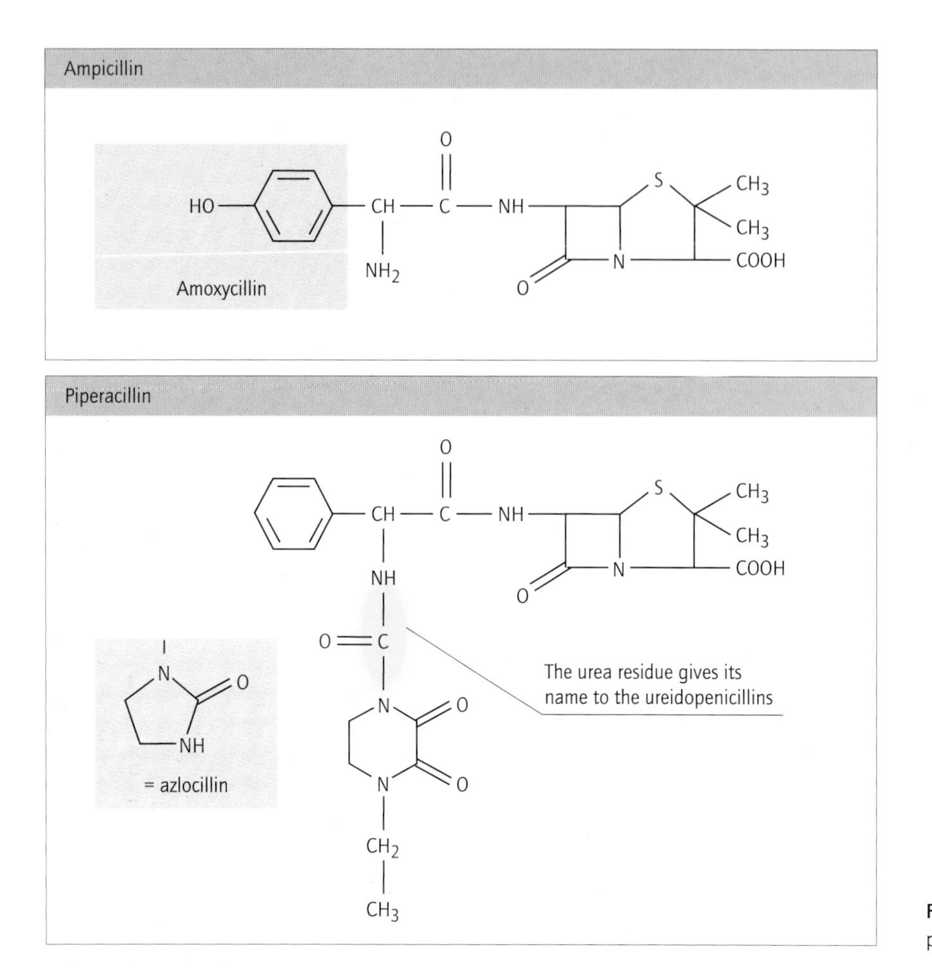

Fig. 4.12 The structure of ampicillin and piperacillin.

also absorbed orally. The remainder must be administered by the intravenous route.

All the orally active penicillins yield peak levels in the serum at 1–2 h after ingestion and this peak is delayed by taking the dose with food. Penicillins are excreted and secreted by the kidney. The half-life of penicillins can be increased by coadministration of probenecid which competes with penicillin for the renal tubular secretory mechanism. The consequence of rapid excretion is a very short half-life, ranging from 30 to 72 min. Protein binding is variable, from 17% for aminopenicillins to 97% for dicloxacillin. Penicillins are distributed in extracellular fluid to most of the body tissues, including lung, liver, kidney, muscle, bone and placenta. They do not cross the blood–brain barrier unless the meninges are inflamed. High concentrations of penicillin are found in the urine. Penicillins are actively secreted into the bile. Imipenem is broken down by renal dihydropeptidase I, so cilastatin, an inhibitor, must be coadministered for this compound to be effective. Meropenem is not broken down, and can be given alone.

Unwanted actions

Penicillins are noted for the infrequency of their adverse effects, and this has contributed to their wide application. The most serious side-effect is acute anaphylaxis which is uncommon but is most likely to complicate treatment with benzylpenicillin. Less severe allergic manifestations include angioneurotic oedema, pruritus and urticaria. In some patients there is a delayed 'serum sickness' reaction with urticaria and arthralgia. Other side-effects include Stevens–Johnson syndrome, dermatitis, morbilliform eruptions and allergic vasculitis. Neutropenia occasionally occurs with prolonged high dosage of benzylpenicillin. There is a high salt load associated with the sodium salts and especially the disodium salts of the expanded-spectrum penicillins. High doses in patients with renal failure may result in hypernatraemia, hypokalaemia and convulsions.

Ampicillin can precipitate antibiotic-associated diarrhoea and, occasionally, pseudomembranous colitis due to its activity against obligate anaerobic organisms

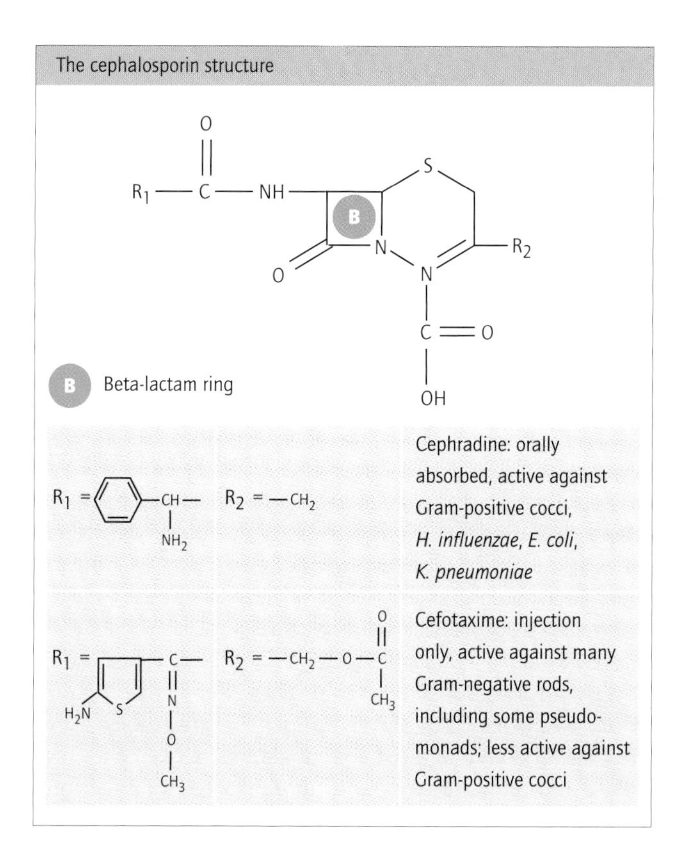

Fig. 4.13 The structure of cephalosporins.

and the consequent effect on the balance of the intestinal flora.

Cephalosporins

Cephalosporins are naturally occurring compounds which are closely related to penicillins (Fig. 4.13). They are usually classified according to the route of administration and spectrum of activity.

The first group is the orally active cephalosporins, which all have a strong Gram-positive spectrum, initially introduced in response to the spread of penicillinase-producing *S. aureus*. They are, however, less effective than isoxazolyl penicillins against this organism. Some agents such as cefaclor also have useful activity against *H. influenzae*, making it particularly valuable in the treatment of community-acquired pneumonia.

The second group contains older, injectable agents, such as cefazolin, cefamandole and cefuroxime. These all retain a good spectrum of activity against Gram-positive organisms, but show strong activity against *E. coli* and some species of *Proteus*. They have established a role in surgical prophylaxis (e.g. cefazolin plus metronidazole before large bowel surgery). Esters of cefuroxime, such as cefuroxime axetil, are absorbed orally.

Among the newer injectable cephalosporin agents (third-generation cephalosporins), cefotaxime is active against most Gram-negative organisms and retains good activity against streptococci, but is less active against *S. aureus* species. These agents are widely used in the management of septicaemia, serious community-acquired pneumonia and neonatal meningitis. Some have only weak activity against *P. aeruginosa*. However, cephalosporins such as ceftazidime and cefepime are active against this species. Ceftriaxone has an extended half-life, making single-dose treatment of penicillinase-producing *N. gonorrhoeae* and once-daily therapy of Gram-negative infections possible. More recently, orally active cephalosporins such as cefixime have become available, with a spectrum of activity similar to the newer injectable cephalosporins.

Cephalosporins are eliminated by the kidney with a half-life of 1–2 h (note the exception of ceftriaxone). They are distributed widely in the extracellular fluid, and penetrate well into tissues. The earlier agents did not cross the blood–brain barrier but the newer injectable cephalosporins such as cefotaxime and ceftriaxone do. Hepatic metabolism is important for some compounds, including cefotaxime.

Cephalosporins have a low incidence of anaphylaxis, but cross-reactivity occurs in approximately 10% of penicillin-allergic patients. Rashes may also occur, as with the penicillins. The earlier agents cephaloridine and cephalothin are toxic to the kidney but more recent agents have not had this side-effect.

Aminoglycosides

The first aminoglycoside, streptomycin, was isolated from *Streptomyces griseus* in 1943. Since then, a number of different agents have been developed, including kanamycin, gentamicin, tobramicin, netilmicin and amikacin (Fig. 4.14). Aminoglycosides are defined by the presence of amino sugars linked by glycosidic bonds to aminocyclitol. Aminoglycosides inhibit protein synthesis by preventing ribosomes from translating messenger RNA codes. They are active against aerobic and facultative anaerobic Gram-negative bacilli and *S. aureus*. In addition they are active against mycobacteria and *Brucella*.

Aminoglycosides are not absorbed from the gastrointestinal tract and must be administered parenterally. They are polar, and distributed in the extracellular fluid. They are not metabolized by the liver but are excreted unchanged in the urine. Although they have a low thera-

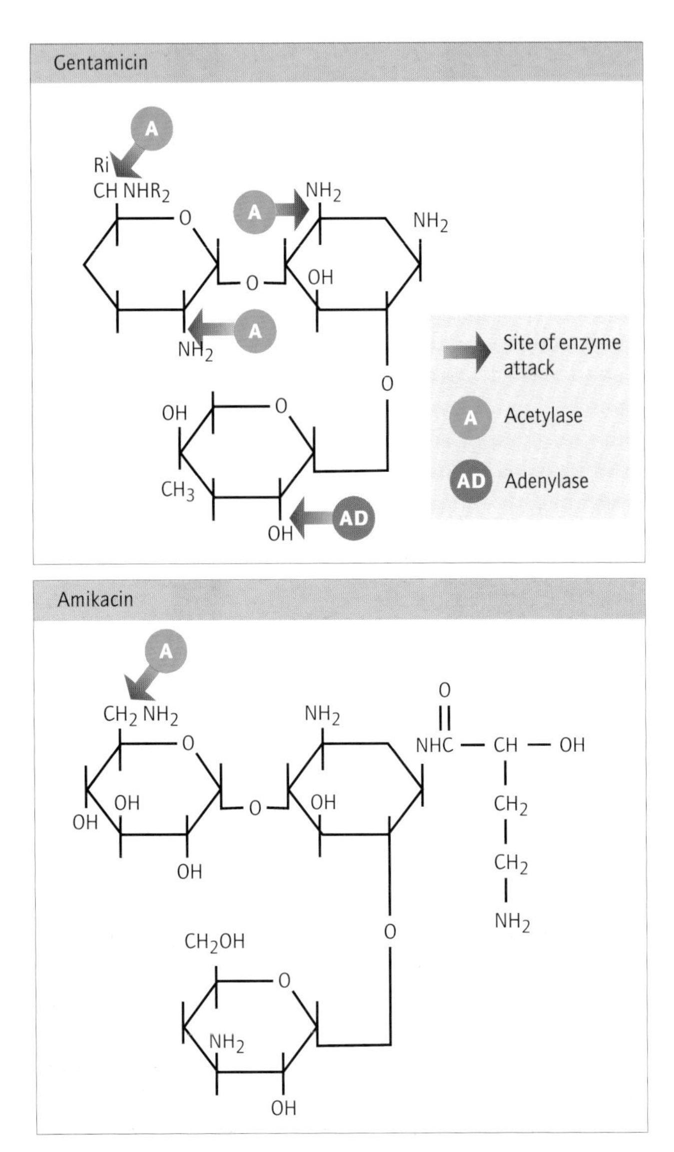

Fig. 4.14 Structure of gentamicin, indicating sites of antimicrobial enzyme attack. Amikacin, by contrast, lacks all but one of these vulnerable sites.

peutic index, they are valuable for their effectiveness in treating Gram-negative septicaemia.

The three main adverse events associated with aminoglycoside use are ototoxicity, renal toxicity and neuromuscular paralysis.

Bacteria become resistant to aminoglycosides by three main mechanisms. Methylation of the ribosome prevents the aminoglycoside from binding to its substrate. This mode of resistance is limited to streptomycin. The bacterial cell membrane may become impermeable to aminoglycoside antibiotics. This affects all of the drugs in the aminoglycoside class. Bacteria can also become resistant by developing aminoglycoside-inactivating enzymes. There are three main enzyme groups: *N*-acetyltransferases, *O*-phosphotransferases and *O*-nucleotidyltransferases. Although a wide range of enzyme-mediated resistances can occur, bacterial strains with aminoglycoside resistance are uncommon in hospitals with well organized infection control and antibiotic policies.

Tetracyclines

Tetracyclines were first isolated from *Streptomyces* species and are based around four fused rings (a hydronaphthacene nucleus). They act by binding to the 30S ribosomal subunit, locking band A of transfer RNA to the septal site on the messenger RNA–ribosomal complex. Many tetracycline compounds have been described, all with similar spectra of activity but varying in their pharmacokinetics. Tetracyclines are active against *S. aureus*, streptococci, including *Streptococcus pneumoniae*, *N. gonorrhoeae*, *N. meningitidis*, *H. influenzae* and some species of Enterobacteriaceae. They are active against *Bacillus* species and many species of anaerobes. They are also effective against mycoplasmas, rickettsiae, coxiellae and chlamydiae, and in spirochaete infections, including those caused by *Treponema pallidum*. Some species of mycobacteria are also susceptible. The activity of tetracyclines is not limited to bacteria but also includes protozoa, such as *Plasmodium* spp. and *Entamoeba histolytica*.

> Tetracyclines act by arresting the sequence of transfer RNA molecules.

Tetracyclines are well absorbed orally but vary widely in their half-life. Some agents, such as doxycycline, have a long half-life, and adequate therapeutic levels may be obtained by once-daily dosage. Tetracyclines are distributed to many tissues including the lung, liver, kidney, brain and respiratory tract. They are concentrated in the bile but do not cross the blood–brain barrier into the CSF. They should be used with caution in renal or hepatic sufficiency.

Gastrointestinal intolerance is the commonest adverse event. Skin reactions include photosensitivity and exfoliation. Tetracyclines are concentrated in developing bones and teeth, and are therefore contraindicated in children and pregnant women. Fatty degeneration of the liver, with fatal liver failure, has been reported in pregnant patients. Tetracycline treatment causes increased protein breakdown and an apparent deterioration in the biochemical status of patients with renal impairment. Irreversible renal failure has developed in some patients treated with tetracyclines.

Because of their wide range of side-effects, tetracyclines are first-line treatment for relatively few infections. They are the treatment of choice for severe rickettsial infections, and for Q fever.

Chloramphenicol

Chloramphenicol was first isolated from *Streptomyces venezuelae* but is now produced synthetically. It inhibits protein synthesis by reversibly binding to the 50S ribosomal subunit, preventing the attachment of aminoacyl transfer RNA, and subsequent peptide bond formation. It is active against most Gram-positive and Gram-negative aerobic bacteria, and many obligate anaerobes, including *Bacteroides*. It is also active against spirochaetes, rickettsiae, chlamydiae and mycoplasmas. It is bacteriostatic against most microorganisms but is bactericidal against *H. influenzae* and *Neisseria* spp.

> Choramphenicol inhibits protein synthesis by blocking the transfer of amino acids from their respective transfer RNAs to the protein chain.

Chloramphenicol may be given parenterally but is well absorbed orally. It is widely distributed in the tissues and crosses the blood–brain barrier. It is metabolized in the liver and excreted by the kidney. A special oily preparation is available and can be given by the intramuscular route. This has proven valuable in developing countries for single-dose treatment of bacterial sepsis.

The principal adverse effect of chloramphenicol is in preterm infants in the neonatal period, when they lack sufficient hepatic enzymes to conjugate chloramphenicol efficiently: toxic levels build up, eventually causing the 'grey baby' syndrome, characterized by circulatory collapse. The dose must be limited to prevent this. In adults the principal toxic effect is on the bone marrow. There is a reversible dose-dependent bone marrow depression due to the inhibition of mitochondrial protein synthesis. A rare form of idiosyncratic aplastic anaemia occurs in approximately 1 in 40 000 patients. This can occur even after the completion of therapy and is not dose-related. It is usually irreversible.

Despite the wide spectrum of activity of chloramphenicol, fear of toxicity has limited its use to a number of specific indications, including alternative therapy of bacterial meningitis, typhoid fever and invasive salmonellosis, brain abscess and rickettsial diseases.

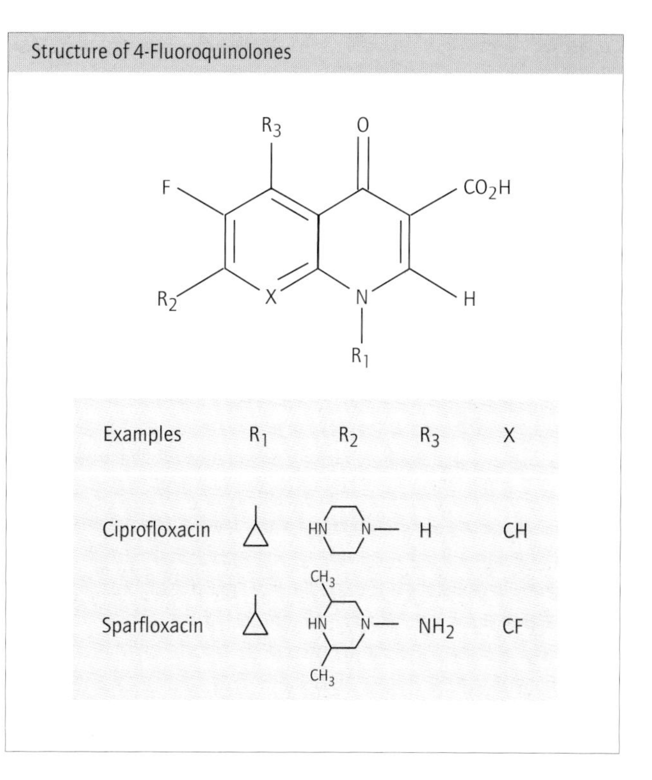

Fig. 4.15 The structure of quinolone antibiotics.

Quinolones

All members of the quinolone group of antimicrobial agents (Fig. 4.4) have a similar action. When DNA is transcribed, the supercoiled molecule must be unwound for transcription to occur. Quinolones interfere with the action of DNA topoisomerase A, the bacterial enzyme responsible for this process. The first agent in clinical use was nalidixic acid, active mainly against Enterobacteriaceae and used for the treatment of Gram-negative urinary tract infections, as high concentrations could be reached in the urine, *Haemophilus* and *Neisseria*. *Pseudomonas* spp. and *Serratia* spp. are resistant to nalidixic acid, as are staphylococci and streptococci.

> Quinolones interfere with DNA function by preventing the uncoiling and recoiling of the molecule necessary for polymerization and transcription.

Chemical modification of the quinolone nucleus by the introduction of a fluorine at the sixth position produced the fluoroquinolones, with antibacterial activity much greater than nalidixic acid. Fluoroquinolones such as ciprofloxacin and ofloxacin are highly active against the Enterobacteriaceae, *Pseudomonas* spp., *Haemophilus* spp.,

Neisseria spp., *Chlamydia* spp. and also mycobacteria. Some newer quinolones have excellent activity against anaerobic bacteria and streptococcal species.

The fluoroquinolones are well absorbed orally, reaching peak concentrations in the serum 90 min after oral administration. Protein binding is low and they are distributed widely in the tissues, including 'difficult' areas such as the prostate. They do not cross the blood–brain barrier. Quinolones penetrate cells, including macrophages and polymorphs. They vary in their degree of metabolism and renal excretion. Ciprofloxacin is partly metabolized and excreted in the urine. The dosage should be reduced in patients with renal insufficiency.

Quinolone antibiotics are generally well tolerated but nalidixic acid often causes nausea, vomiting, diarrhoea, abdominal pain and skin reactions, including photosensitivity. Fluoroquinolones are associated with mild gastrointestinal side-effects, dizziness, tiredness, restlessness and depression. They increase the tendency to seizures in epilepsy, and should be avoided or used with caution in known epileptics. They also have important interactions with some drugs, including aminophylline, increasing their toxic effects. Photosensitivity has proved a problem with some agents, leading to their withdrawal from use.

Fluoroquinolones are valuable in a wide range of systemic bacterial diseases, including the management of hospital Gram-negative septicaemia, especially when *P. aeruginosa* is a likely pathogen, as in neutropenic patients. They reach high concentrations in bronchial secretions, and are useful in treating hospital-acquired Gram-negative pneumonia, and chest infections in children with cystic fibrosis (who are subject to recurrent *Pseudomonas* infection). They are widely used in the management of typhoid, invasive *Salmonella* infections and sexually transmitted diseases, and may be valuable in the management of mycobacterial infections. Their usefulness is somewhat limited by the gradual emergence of resistance, particularly in *Salmonella typhi* and *Neisseria gonorrhoeae*.

Glycopeptides

Vancomycin is a glycopeptide antibiotic first isolated from *Streptomyces orientalis*. It inhibits peptidoglycan synthesis by interfering with chain-lengthening. Glycopeptides are large polar molecules and cannot penetrate the outer membrane of Gram-negative bacteria. Resistance among Gram-positive organisms was rare but is now increasingly reported in enterococci and even staphylococci. Vancomycin and teicoplanin are not absorbed orally and must be administered parenterally, or occasionally locally into the peritoneal cavity. Vancomycin is distributed in the extracellular fluid. It does not cross the blood–brain barrier unless there is meningeal inflammation. More than 80% is excreted unchanged by the kidney.

Glycopeptides inhibit Gram-positive cell wall construction by blocking cross-linking of peptide bridges.

Vancomycin causes thrombophlebitis, so administration via a central venous line is preferred. An idiosyncratic vasodilatation–hypotension ('red man') syndrome, probably related to histamine release, develops in up to 15% of patients given bolus doses. To avoid this, the drug is always given by slow infusion. Teicoplanin may be given by a peripheral line, or by intramuscular injection.

Renal toxicity is important and is more likely in patients receiving concomitant aminoglycosides. Deafness may develop, particularly in elderly patients or those with renal impairment. These toxic effects can be minimized by careful attention to dosage schedules and regular monitoring of serum levels.

Vancomycin is indicated for the treatment of severe Gram-positive infections unresponsive to beta-lactam antibiotics, or for patients who are sensitive to beta-lactams. It has an important role in the therapy of invasive methicillin-resistant staphylococcal infection and infections with ampicillin-resistant enterococci. Oral therapy is indicated for the treatment of pseudomembranous colitis. Gram-positive infections in chronic ambulatory peritoneal dialysis may be treated by the addition of the drug in the dialysis fluid. The dosage of vancomycin should be modified in patients with renal failure, to avoid accumulation and toxicity.

Resistance was once uncommon but has now emerged in enterococci and staphylococci.

Teicoplanin

Teicoplanin is a new glycopeptide antibiotic isolated from *Actinoplanus teichomyceticus*. It is similar to vancomycin in its spectrum of activity. However, some strains of *Staphylococcus haemolyticus* are naturally resistant. The use of teicoplanin is not associated with the severe adverse reactions reported with vancomycin and serum monitoring is unnecessary. The clinical indications for the use of teicoplanin are similar to those of vancomycin.

Metronidazole

Metronidazole is a nitroimidazole drug active against anaerobic bacteria, and the parasites *Giardia intestinalis*, *Trichomonas vaginalis* and *Entamoeba histolytica*. The drug acts as an electron acceptor at low redox potentials, allowing the formation of toxic metabolites which damage bacterial DNA.

Metronidazole acts at low redox potential by generating highly toxic metabolites.

Metronidazole is indicated for the treatment of serious infections in which anaerobic bacteria may play a part: abdominal sepsis, brain abscess, lung abscess, anaerobic lung infections and dental infections. It may also be effective in oral or parenteral therapy of pseudomembranous colitis. It is the treatment of choice for giardiasis, intestinal amoebiasis and amoebic liver abscess and in the treatment of vaginal trichomoniasis. It is also valuable for surgical prophylaxis for abdominal and gynaecological surgery (see Chapter 23).

Metronidazole is well absorbed orally and rectally, and can also be administered parenterally. It is not strongly protein bound. It is widely distributed throughout all tissues and body compartments. It crosses the blood–brain barrier and achieves high concentrations in brain abscesses. Approximately 30% of the dose is metabolized in the liver and 70% is excreted unchanged in the urine. Initial therapy of serious sepsis and abscesses is via the intravenous route but oral or rectal administration can be commenced as soon as practicable.

Metronidazole is usually well tolerated, but many patients complain of minor side-effects, including fever, dizziness, dry mouth and a metallic taste. More serious side-effects include encephalopathy and peripheral neuropathy, more likely with extended therapy. A disulfiram (Antabuse) reaction can develop in patients who drink alcohol while taking metronidazole.

Metronidazole has mutagenic activity in prokaryotic systems (the Ames test) but there is no evidence of carcinogenicity in humans, despite extensive use over a number of years.

Macrolides

The macrolides are a group of related antimicrobials. They have a common macrocyclic lactam ring with 14, 15 or 16 members. The members include erythromycin, clarithromycin, azithromycin and spiramycin. All the agents have a similar antimicrobial spectrum, including Gram-positive organisms, *Neisseria*, *Haemophilus* and *Bordetella*, and some Gram-negative anaerobes. They are also active against *Mycoplasma* sp., *Rickettsia* sp. and *Toxoplasma gondii*. Their mechanism of activity is uncertain but occurs early in the course of RNA-dependent protein synthesis. The macrolides bind to the 50S ribosome. Bacteria become resistant to macrolides by mechanisms which decrease cell permeability and by the alteration of the ribosomal binding site. Resistance may be inducible or non-inducible (the latter confers resistance to both macrolides and the lincosamide clindamycin).

Macrolides inhibit protein synthesis by an unknown action against ribosomal function.

Erythromycin

Erythromycin is derived from *Streptomyces erthythreus*. Erythromycin base is poorly soluble and inactivated by gastric acid. To improve absorption it may be protected by enteric coating but this may lead to delayed or incomplete absorption. Esters of erythromycin with stearate and ethyl succinate are better absorbed. A water-soluble lactobionate preparation is available for intravenous use. Erythromycin is absorbed orally and is distributed throughout the total body water. It is concentrated in alveolar macrophages and polymorphs. It does not cross the blood–brain barrier but does cross the placenta. It is concentrated in the liver and excreted in the bile.

Erythromycin is well tolerated but some patients complain of nausea and gastric irritation. Thrombophlebitis may follow intravenous use. Cholestatic jaundice may develop, but was most common with older oral preparations, such as the estolate salt. Allergic reactions are rare. Erythromycin is often indicated as a alternative to penicillin in allergic subjects. It is a first-choice treatment of primary atypical pneumonia, including legionellosis. It may shorten the clinical course of *Campylobacter* enteritis. It is the treatment of choice in neonatal chlamydial infections. It also used in treating diphtheria and whooping cough, and may be valuable in some mycobacterial infections.

A number of new macrolide antibiotics have been developed, including clarithromycin and azithromycin. These have a similar antimicrobial spectrum to erythromycin, but improved pharmacokinetic properties. Azithromycin may be given in a single daily dose for the treatment of respiratory tract infections and clarithromycin is often used for the treatment of *Mycobacterium avium-intracellulare* infection. Azithromycin reaches very high intracellular concentrations and may be particularly useful in treating persisting chlamydial infections, such as trachoma.

Fusidic acid

Fusidin or fusidic acid is a fermentation product of *Fusidium coccineum*. It is active against Gram-positive bacteria, including most *S. aureus* strains, and against Gram-negative cocci. Streptococci and pneumococci are relatively resistant and Gram-negative bacilli are highly resistant. It has some activity against a variety of protozoa, including *Giardia intestinalis*. Resistance arises naturally and readily develops during therapy, unless a second agent is coadministered to prevent this. It is well absorbed orally and widely distributed. It does not reach the CSF but does penetrate into cerebral abscesses and bone. When given orally, sodium fusidate is well tolerated, although mild gastrointestinal upset and rashes have been reported. With intravenous use, cholestasis and jaundice can develop but usually resolve following withdrawal of therapy.

Fusidic acid is indicated for the treatment of severe staphylococcal infection, especially infections of bones and joints. It is also used for the treatment of recurrent furunculosis and for topical therapy of some superficial staphylococcal infections.

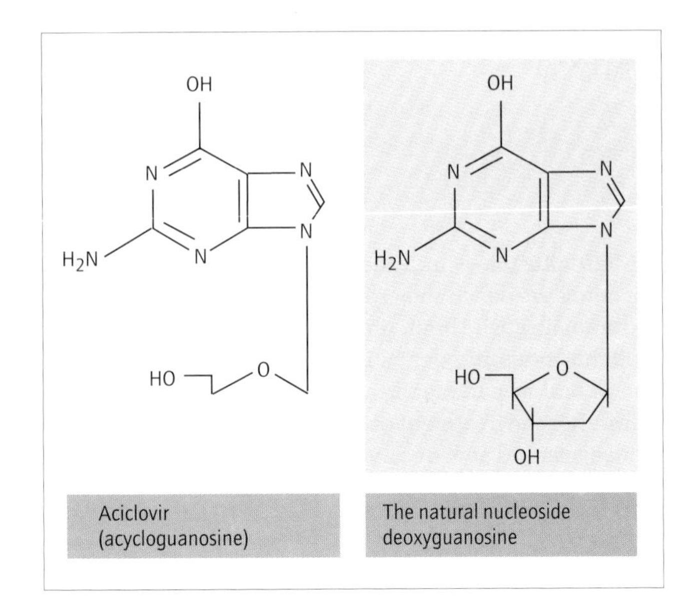

Fig. 4.16 The structure of aciclovir.

Antiviral therapy

The scope of antiviral therapy is rapidly broadening, in parallel with increasing understanding of the mechanisms of viral replication.

Amantadine

Amantadine is a symmetrical 10-carbon tricyclic amine. It interferes with the replication of influenza type A virus, probably by an effect on cell penetration and viral uncoating. It is used for the prophylaxis and treatment of influenza A during outbreaks. Prophylaxis is usually given to patients with underlying chronic conditions who are at increased risk from severe disease. Resistant influenza A strains do occur and are particularly likely to arise if the drug is given as postexposure prophylaxis.

Amantadine is well absorbed orally and concentrations in respiratory secretions are equivalent to those in plasma. Side-effects include insomnia, poor concentration, nervousness, dizziness and headaches. In overdose convulsions, psychoses and cardiac dysrhythmias can develop. Amantadine is contraindicated in pregnancy, as teratogenicity has been observed in animals.

Nucleoside analogues

These drugs are all chemical analogues of natural purine or pyrimidine nucleosides. They require phosphorylation to an active form in host cells, and then interact with viral DNA or RNA polymerase, either becoming incorporated into the nucleotide chain and distorting it, or terminating further elongation.

Aciclovir

Aciclovir is a synthetic acyclic purine nucleoside (Fig. 4.16) active against herpes simplex virus (HSV) types 1 and 2 and varicella zoster virus (VZV). It is activated by monophosphorylation by virally encoded thymidine kinases (TKs) followed by further phosphorylation by cellular thymidine kinases to form aciclovir triphosphate (Fig. 4.17). The need for viral thymidine kinase to perform the first phosphorylation means that the drug only becomes active in virally infected cells. The activated drug inhibits HSV DNA polymerase and is a DNA chain terminator. Resistant virus mutants arise naturally and are readily induced in the laboratory but they tend to be of low pathogenicity and rarely cause clinical problems.

Aciclovir is absorbed orally, with 15–20% bioavailability, and can be administered parenterally. It is widely distributed and adequate concentrations are found in the CSF. It is well tolerated, although deterioration in renal function occurs if the drug is given in a rapid bolus or in overdosage. It is excreted by the kidneys, so the dose must be reduced in patients with renal impairment.

Aciclovir is used parenterally for the treatment of severe HSV and VZV infections in normal and immunocompromised subjects. Topical preparations are available

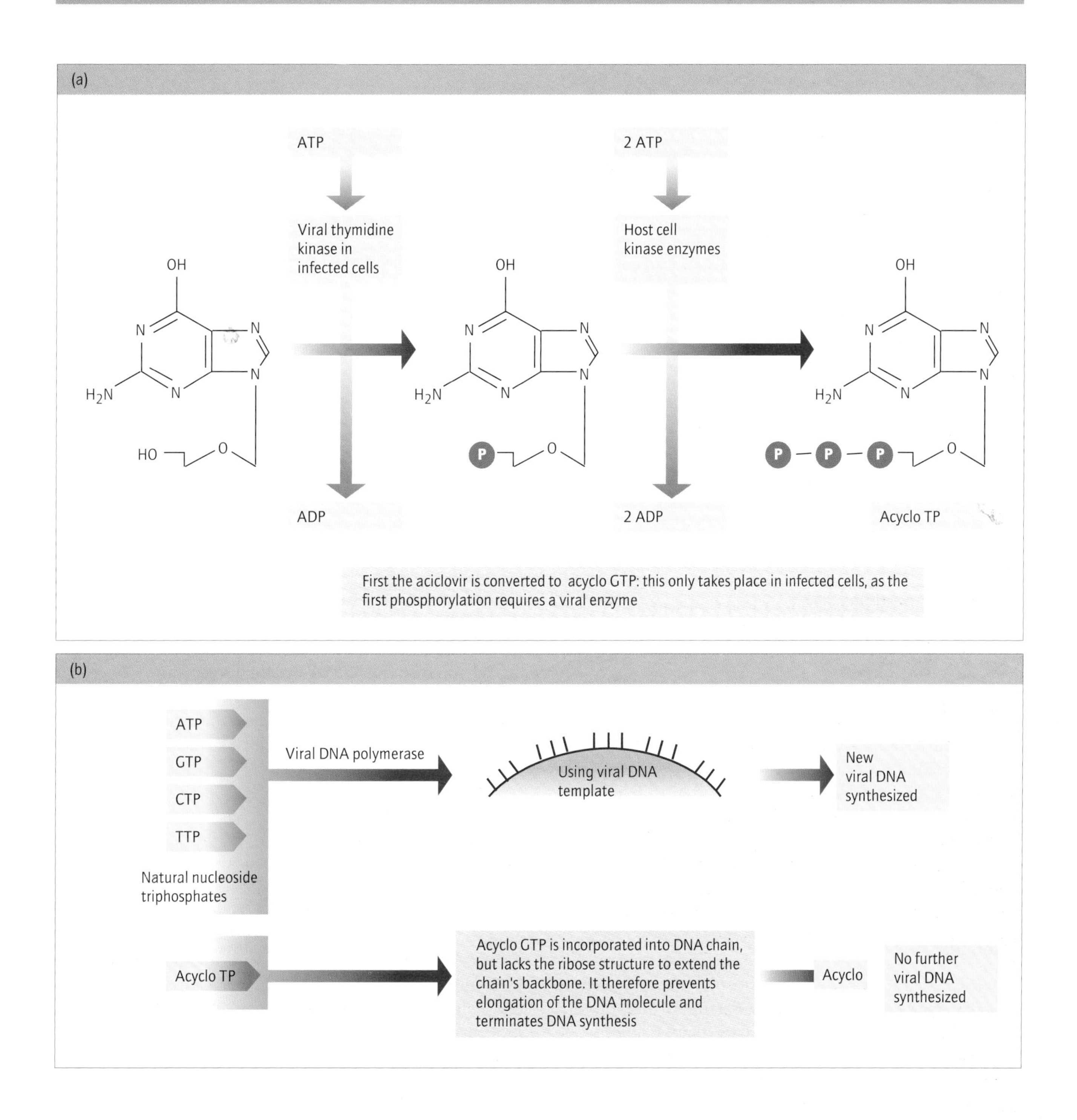

Fig. 4.17 The mechanism of action of aciclovir. (a) Stages in the phosphorylation of aciclovir to a 'false' nucleoside triphosphate. (b) Incorporation of aciclovir into the DNA molecule prevents further chain extension.

for the treatment of herpes simplex infections of the skin and mucous membranes but these have only a weak, local action.

Valaciclovir, the valyl ester of aciclovir, is well absorbed orally, then rapidly and completely converted to aciclovir by hepatic enzymes, increasing bioavailability approximately threefold compared with aciclovin. Another ester compound is famciclovir, a prodrug of peniciclovir. It has excellent bioavailability after oral dosing. Like aciclovir it depends on virally encoded TK for activation.

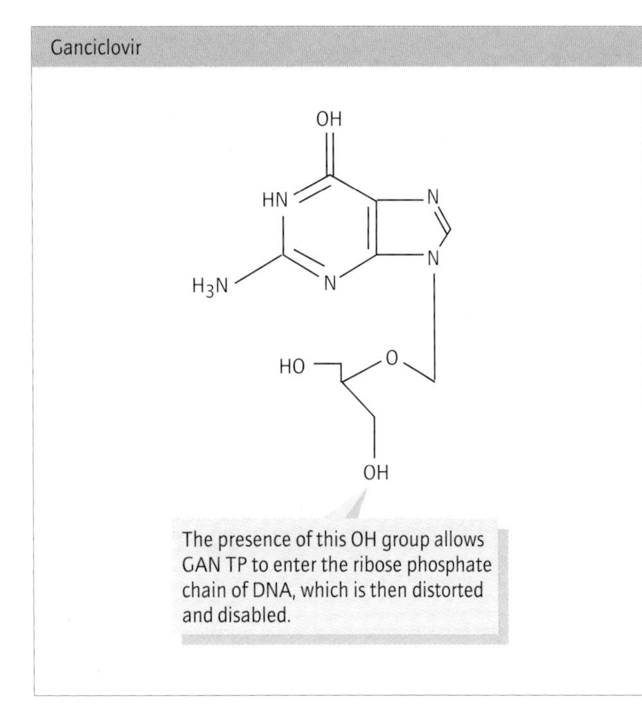

Fig. 4.18 The structure of ganciclovir.

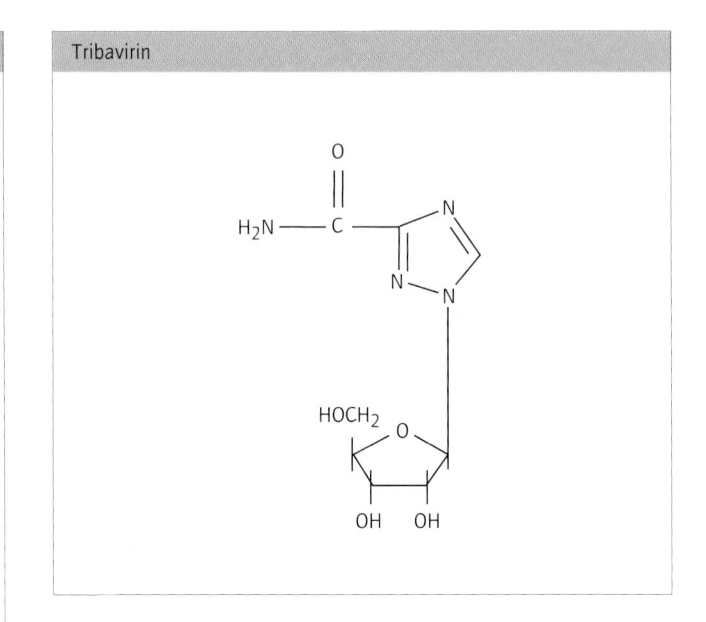

Fig. 4.19 The structure of tribavirin.

Ganciclovir

Ganciclovir (Fig. 4.18) is more active than aciclovir against cytomegalovirus (CMV) (which does not contain the same viral thymidine kinase as herpes simplex and varicella zoster). It is not well absorbed orally and must usually be given intravenously. It is used in the treatment of life- or sight-threatening CMV infections in immunocompromised patients, but is too toxic to the bone marrow to be used for less severe indications.

Cidofovir

Cidofovir is an acyclic nucleoside phosphonate that is phosphorylated only by cellular enzymes. Consequently it is active against strains of herpesvirus deficient in TK or with an altered enzyme. It also has some action against poxviruses. Mild or moderate cutaneous adverse reactions are commonly reported and renal tubular toxicity is an important adverse effect. Adefovir is a similar drug.

Tribavirin

Tribavirin is a synthetic nucleoside derivative (Fig. 4.19). It is active against DNA viruses such as the herpesviruses and some human RNA viruses such as the orthomyxoviruses, paromyxoviruses, arenaviruses including Lassa virus, Bunyaviridae including Rift Valley virus, and hantaviruses. It is absorbed orally and can be administered by aerosol for the treatment of respiratory syncytial virus (RSV) infections. The drug is well tolerated, although a dose-related fall in haemoglobin is common, due to haemolysis. It is used for the treatment of RSV infections in infants and is effective in the treatment of Lassa fever, Crimean-Congo haemorrhagic fever and hantavirus infections if given early in the course of disease.

Other types of antiviral drug

Foscarnet

This is a non-nucleoside pyrophosphate analogue. It inhibits the DNA polymerases of herpesviruses, including CMV. It must be given by intravenous infusion. It is indicated in CMV retinitis in acquired immunodeficiency syndrome (AIDS) patients, and mucocutaneous herpes simplex in immunocompromised patients unresponsive to aciclovir. It is relatively toxic, causing impaired renal function, including acute renal failure, hypokalaemia, granulocytopenia and thrombocytopenia. It should be used with care in renal impairment and electrolyte disturbance.

Zanamivir

Zanamivir is a sialic acid analogue that inhibits neuraminidase. It has been shown to inhibit influenza viruses A and B in cell culture, in infected animals and humans. It has proved to be effective in clinical trials.

Protease inhibitors for other viruses

Protease inhibitors are being investigated as therapeutic agents in a number of viral infections including hepatitis C, CMV, enterovirus and rhinovirus. Some of these agents will enter clinical trial and use over the next few years.

Drugs in development

Drugs are being developed which inhibit the binding of enteroviruses to their host-cell receptor. They are given orally and widely distributed in the body, including the CSF. They are intended for the treatment of viral meningitis.

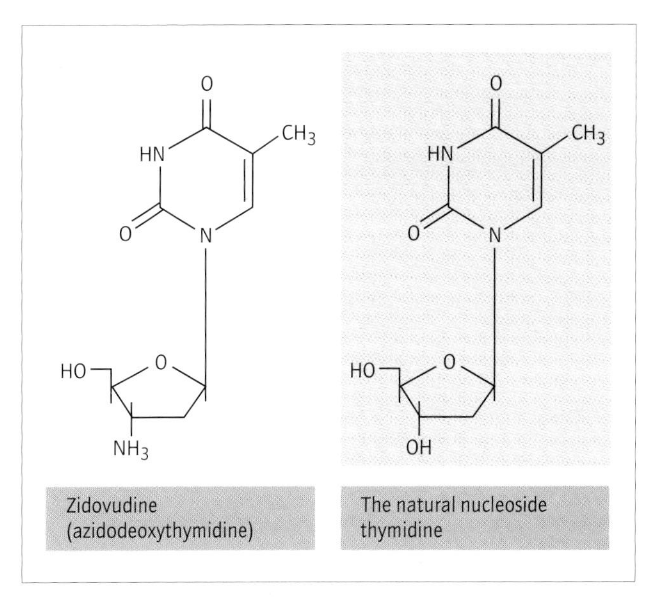

Fig. 4.20 The structure of zidovudine.

Antiretroviral drugs

The introduction of highly active antiretroviral therapy (HAART) has transformed the benefits of HIV care by producing dramatic reductions in the HIV viral load and improvement in the CD4 count. This has led to a fall in the incidence of opportunistic infections, with much improved survival and quality of life. Antiretroviral drugs used to achieve this are divided into three classes and these are described below.

Nucleoside reverse transcriptase inhibitors (NRTIs) or nucleoside analogues

These are nucleoside analogues that inhibit reverse transcriptase, the enzyme responsible for the conversion of viral RNA into a DNA copy. These include the longest established antiretroviral drugs zidovudine (AZT), lamivudine (3TC), stavudine (d4T), didanosine (ddI) and zalcitabine (ddC), and also abacavir. They are the mainstay of antiretroviral therapy and are used in combination in initial therapy (see below). Lamivudine is highly active but selects for resistance to other NRTIs, and should only be used in completely suppressive regimens.

Zidovudine

Zidovudine (azidodeoxythymidine or AZT) is a dideoxynucleoside analogue originally developed as an anticancer agent (Fig. 4.20). It is phosphorylated to AZT monophosphate by viral thymidine kinase and is subsequently converted to diphosphate and triphosphate by cellular thymidylate kinase. It inhibits retroviral reverse transcriptase to which it binds preferentially.

It is well absorbed orally and crosses the placenta and the blood–brain barrier. It is metabolized in the liver. The main side-effects are anaemia and neutropenia which usually occur after about 6 weeks of therapy, particularly in patients with pre-existing cytopenias. The drug should be discontinued if the haemoglobin falls below 7.5 g/l or the neutrophil count below 0.7×10^9/l. These effects are reversible. Other adverse events reported include nausea, headache, rash, abdominal pain, fever, myalgia and, rarely, metabolic acidosis.

Protease inhibitors

Protease inhibitors are the antiretroviral compounds which produce the greatest fall in viral load when used as single agents. They are central to HAART. They include indinavir, ritonavir, saquinavir and nelfinavir. The activity of these drugs appears to be similar.

Protease inhibitors can cause significant toxicity. This includes unusual complications such as renal stones, crystalluria, a 'buffalo hump' and gynaecomastia. Lipodystrophy is a recently recognized complication of protease inhibitor therapy and is characterized by lowered body fat and raised cholesterol and triglygeride concentrations. Fat loss occurs in all regions other than the abdomen and it may be associated with insulin resistance Some of the adverse events associated with antiretroviral drugs are listed in Table 4.1.

Agent and class	Main adverse events
Nucleoside reverse transcriptase inhibitors	
Abacavir	Severe hypersensitivity reactions
Didanosine (ddI)	Pancreatitis, peripheral neuropathy, hyperuricaemia,diarrhoea
Lamivudine (3TC)	Nausea, vomiting, diarrhoea, pancreatitis
Stavudine (d4T)	Peripheral neuropathy, pancreatitis, nausea, vomiting
Zalcitabine (ddC)	Peripheral neuropathy, oral ulcers, nausea, vomiting, dysphagia, diarrhoea, abdominal pain
Zidovudine (AZT)	Anaemia, leucopenia, nausea, vomiting, neuropathy, pancytopenia
Protease inhibitors	
Indinavir	Renal stones, triglyceridaemia, raised blood glucose
Nelfinavir	Diarrhoea, raised blood glucose and triglycerides
Ritonavir	Gastrointestinal disturbance, oral paraesthesia, drug interactions
Saquinavir	Diarrhoea, mucosal ulceration, hyperlipidaemia
Non-nucleoside reverse transcriptase inhibitors	
Nevirapine	Rash common, also Stevens–Johnson syndrome, toxic epidermal necrolysis.
Delavirdine	Rash, fever

Table 4.1 Antiretroviral agents currently available.

Agent	Class
Efavirenz (DMP266)	Non-nucleoside reverse transcriptase inhibitor
Amprenavir (141W94)	Protease inhibitor
ABT-378	Protease inhibitor
PNU-140690	Non-peptide protease inhibitor
T-20	Membrane fusion inhibitor

Table 4.2 Some new antiretroviral agents in various stages of development.

Non-nucleoside reverse transcriptase inhibitors (NNRTIs)

These drugs inhibit reverse transcriptase by an alternative mechanism to NRTIs. They have been shown to be effective agents in combination regimens. Because of their susceptibility to viral resistance after a single mutation event, they should only be used in regimens designed to be maximally suppressive. For new drugs see Table 4.2.

Resistance to antiviral drugs

RNA viruses have a highly variable genome because they have no proof-reading mechanism to preserve genetic stability during genome replication. Many genetic variants exist in a population of virus (quasi-species). Mutations may be neutral, having no effect on fitness, or deleterious where the variants replicate less well than the 'wild' type. Selective mutations are those where the mutation yields a replicative advantage in the face of immunological or therapeutic pressure. In HIV infection approximately 10 billion virus particles are produced daily and contain one mutation in 9200 nucleotides. Thus every possible single drug-resistant mutant is likely to be generated every day. Spontaneous double mutations are extremely unlikely. Resistant virus clades can be detected in patients not previously exposed to treatment. When selective pressure is applied by therapy pre-existing minor quasi-species rapidly emerge as the dominant type. Resistance to some drugs can arise by a single mutation event (e.g. lamivudine and certain NNRTIs). When these drugs are given in partially suppressive regimens resistance will emerge within weeks. For other drugs, for example zidovudine and certain protease inhibitors, three or more mutations must occur before a resistant phenotype is expressed.

Rapidly emerging resistance is increasingly recognized in the antiviral therapy of chronic hepatitis B, a DNA virus infection. Treatment with multiple drugs, or with agents that require multiple mutations to resistance should be preferred, and to prevent the accumulation of resistant mutants, viral replication should be kept as low as possible. Previous treatment may have selected mutants which are no longer detectable but will reappear rapidly if treatment is restarted with that drug.

Antifungal therapy

Many antifungal drugs exist, but most are highly toxic, and their use is limited to topical applications. The polyenes are the oldest drugs, but among them, amphotericin B is still the drug of first choice for many systemic mycoses. Modern formulations are designed to maximize its efficacy and minimize its significant toxicity.

Polyenes

The polyenes consist of a macrolide ring with a variable number of hydroxyl groups along the hydrophilic side and numerous repeating double bonds on the hydrophobic. They bind to sterols and eukaryotic cell membranes, causing leakage of cellular contents leading to cell death. Some polyenes are toxic only to fungal membranes but others are equally toxic to fungal and mammalian cells.

Nystatin

This drug is used as topical treatment, mostly for mucosal *Candida* infections, and for intestinal candidiasis. It is not absorbed by the oral route and is too toxic for parenteral use. It has largely been replaced by the more convenient and highly active imidazole drugs in the treatment of vaginal candidosis.

Amphotericin B

This important polyene drug is a heptaene, with seven double bonds. It is active against most of the fungi pathogenic for humans and is used as parenteral treatment for systemic mycoses. It also has activity against *Leishmania* spp. and can be used in the treatment of kala-azar and cutaneous leishmaniasis. Acquired resistance is rare, but has been described in some *Candida* species, notably *C. tropicalis*, *C. parapsilosis*, *C. lusitaniae* and *C. krusei*. Amphotericin B is administered parentally and is highly protein-bound, thus it does not penetrate well into the CSF. Treatment causes fever, rigors, headache, vomiting and thrombophlebitis, which may be ameliorated by the concurrent administration of corticosteroids. Hyperkalaemia and anaemia are also common. The incidence of serious side-effects has been reduced by incorporating amphotericin B into liposomes or other lipid carriers, producing liposomal AMB (ambisome), or amphotericin B lipid complex. This allows much higher doses to be administered with fewer toxic reactions.

Azoles

The azoles are a large group of synthetic agents which have an imidazole or triazole ring with an N-carbon substitution. These agents act by blocking the 14 alpha-D-methylation step in the biosynthesis of ergosterol. This leads to ergosterol depletion and interference with fungal membrane function. As a group they have a wide spectrum of activity against *Candida* and dermatophytes, and some are active against *Aspergillus* sp.

Imidazoles

Topical imidazoles. These are powerful topical agents used for treating oral, vaginal and skin candidosis, and dermatophyte infections. They include clotrimazole, econazole, fenticonazole, isoconazole, miconazole, sulconazole and tioconazole. Miconazole can be given by mouth for treatment of intestinal candidosis. Adverse effects are rare; irritation caused by the pharmaceutical vehicles or preservatives in creams is relatively common.

Ketoconazole. This drug can be used parenterally, as in the treatment of histoplasmosis, but is more toxic than the triazoles (see below). In oral use, side-effects include severe hepatitis (1 case per 15 000 patients). Topical formulations are used to treat seborrhoeic dermatitis and other fungal infections of the skin.

Triazoles

Fluconazole. This drug is active against yeast-like fungi, including *Candida*, *Histoplasma* and *Cryptococcus* spp. It is effective in dermatophytosis but not against aspergillosis. It is well absorbed orally and is well tolerated. Elevation of transaminases may occur, and warfarin activity may be increased. It is useful in the oral treatment of vaginal candidosis, and treatment of mucosal and systemic candidosis in immunocompromised patients.

Itraconazole. This agent, closely related to fluconazole, additionally has some activity against *Aspergillus* spp. It is orally absorbed and achieves high concentrations in the stratum corneum and hair. It is used in the treatment of superficial mycosis including dermatophytosis and pityriasis versicolor, and in oral and vaginal candidosis. It may also be useful in the treatment of aspergillosis, histoplasmosis and cryptococcosis as an alternative to amphotericin.

Allylamines

Terbinafine is a recently introduced allylamine drug used for the oral treatment of ringworm and dermatophyte nail infections. It acts by inhibition of squalene epoxidase resulting in accumulation of aberrant and toxic sterols in the fungal cell wall. It is indicated for the oral treatment of conditions such as toenail infections which have failed to respond to local therapy or are unlikely to respond.

Stevens–Johnson syndrome, toxic epidermal necrolysis, loss of the sense of taste and hepatic toxicity are reported adverse effects. Treatment should be continued for up to 6 weeks for skin infections and for 3 months or longer for nail infections.

Other antifungal drugs

Flucytosine (5-fluorocytosine)

This is a synthetic fluorinated pyrimidine with activity limited to *Candida* spp., *Cryptococcus neoformans* and some fungi causing chromomycosis. It is incorporated into fungal mRNA in the place of uracil, causing disruption to protein synthesis. It also blocks thymidylate synthetase causing interruption of DNA synthesis. It is given either orally or parenterally and is excreted by the kidney. Dosage must be reduced when the creatinine clearance falls. Resistance is quite common and may arise during treatment. The main side-effects are bone marrow aplasia, enteritis and hepatotoxicity. Flucytosine is used in combination with amphotericin for the treatment of systemic candidosis and cryptococcal infections, when it allows the dosage of amphotericin to be reduced, thereby reducing side-effects.

Griseofulvin

This is an antibiotic derived from penicillium species, such as *Penicillium griseofulvum*. Its mechanism of action is unknown but it interferes with nucleic acid synthesis. Its activity is restricted to the dermatophytes. The drug is absorbed orally and it appears in keratinized tissues at concentrations sufficient to prevent further invasion by the fungus. It is useful for treating nail infections, but must be given for extended periods until all infected keratinized structures are shed and replaced. Serious adverse events are uncommon but headache and rashes occasionally occur. Griseofulvin reduces the anticoagulant effect of warfarin.

Antiprotozoal drugs

Antimalarials

Quinine

Quinine is the principal alkaloid of chinchona bark first used in European medicine in the mid 17th century. It is a 4-aminoquinoline drug, a potent schizonticide thought to act by inhibiting haem polymerase. It can be given orally or parenterally, and crosses the blood–brain barrier with concentrations reaching 7% of serum levels in the CSF. It is indicated for the treatment of *P. falciparum* malaria, especially in severe or complicated infection. Quinine has a narrow therapeutic index, and tinnitus, headache and nausea are common with normal doses. The drug also aggravates hypoglycaemia in severe malaria. Patients with cardiac diseases should be closely monitored during therapy, as quinine depresses cardiac function. It causes blindness and encephalopathy if taken in overdosage.

Chloroquine

Chloroquine, a 4-aminoquinoline drug, is schizonticidal against all four human species of *Plasmodium* and against amoebae (though resistance to chloroquine is now widespread in *P. falciparum*). It is thought to act by inhibiting the enzyme that polymerizes and detoxifies ferriprotoporphyrin IX. It is approximately 90% bioavailable after oral administration, and has a long half-life. It is highly bound to blood constituents such as thrombocytes and granulocytes and plasma proteins, and also to melanin-containing cells in the retina. Retinal damage can follow prolonged dosing. Chloroquine is the treatment of choice for non-falciparum malaria, but is not now used for the treatment of *P. falciparum* infections, though it retains a role in prophylaxis. Minor adverse effects are common, including nausea and vomiting, diarrhoea, dizziness, and pruritus in dark-skinned individuals. It exacerbates psoriasis and is contraindicated in this condition.

Mefloquine

This drug was developed after over 300 000 compounds were synthesized and tested by the American Army following their experience of drug-resistant malaria in Vietnam. It is a quinolone carbinol possessing a quinine-like ring structure. Mefloquine is administered orally and absorbed from the gut, with peak concentrations 2–12 h later. It is 98% protein-bound. The average half-life is 21 days. Convulsions, dizziness, altered balance, vomiting, diarrhoea, and psychological reactions have been reported and are relatively common in patients on therapeutic doses.

Mefloquine is effective in treatment and prophylaxis of all species of human malaria, and is effective for prophylaxis of chloroquine-resistant *P. falciparum*, but mefloquine resistance already exists in some areas of the Far East. Adverse effects, however, limit its tolerability. Weekly dosage with 250 mg gives a steady-state serum concentration of approximately 1 mg/l (2 mg/l inside red cells). The drug is mainly excreted in the bile and faeces.

Mefloquine should not be prescribed for patients with renal compromise or a past history of psychiatric disease or epilepsy. It is teratogenic to rats and mice in early pregnancy and relatively contraindicated for pregnant and lactating women.

Primaquine

This 8-aminoquinoline drug is the only effective agent for eradicating long-term liver infection with the hypnozoite forms of *P. vivax* and *P. ovale*. It is given after treatment of the feverish acute infection with other antimalarials. It causes haemolysis in individuals with glucose-6-phosphate dehydrogenase (G6PD) deficiency, and must not be given to patients whose G6PD status is unknown. This problem prevents its routine use for malarial prophylaxis.

Artemisinins

Artemisinin is derived from the plant *Artemisis annua* or qinghaosu and is a sesqueterpinoid. This plant has a long history of antimalarial use in traditional Chinese medicine. The mode of action is not known, but is believed to depend on activation by cleavage of the endoperoxidase bridge, generating free radicals which form covalent bonds with parasite proteins. All artemisinin derivatives are converted *in vivo* into the active antimalarial compound dihydroartemisinin.

In China and Vietnam, artemesinin has replaced quinine in treatment of *P. falciparum* malaria. It has a rapid action, shortening fever clearance time by 8 h. It is therefore useful for treating serious and complicated malaria, including cerebral malaria. It causes mild cerebral irritation and does not terminate coma as rapidly as quinine, but this is not clinically important.

This drug can be administered intravenously, rectally, intramuscularly and orally. Artusenate is a water-soluble form for intravenous and oral use. Artemether is a methyl derivative for intramuscular injection or oral use. In severe cases of malaria a dose of 2 mg/kg followed by 1 mg/kg in 12 h later and each 12 h until the patient is able to take oral medication has been shown to be effective.

Other antimalarial drugs

Proguanil is used in combination with chloroquine for malarial prophylaxis, and may be used alone for prophylaxis of non-falciparum malaria. Halofantrine is related to quinine, but carries a significant risk of cardiac arrythmias, and its use is restricted to treating *P. falciparum* infections when other drugs cannot be used. Some drug combinations are used against malaria: Fansidar® is a combination of pyrimethamine plus sulphadoxine; Malarone® combines atovaquone plus proguanil.

Other antiparasitic drugs

Pyrimethamine

This is an antifolate drug which is usually used as part of a drug combination (as in Fansidar®). It is used in combination with sulphadimidine for treating toxoplasmosis.

Atovaquone

Hydroxynaphthoquinones are inhibitors of mitochondrial electron transport, competing with the biological electron carrier ubiquinone. They inhibit the ubiquinone-linked dihydro-oroate dehydrogenase which is central to pyrimidine biosynthesis. Atovaquone binds to the ubiquinoyl–cytochrome c reductase region (complex III). It is active against chloroquine-sensitive and -resistant *P. falciparum*, *Toxoplasma gondii* and *Pneumocystis carinii*.

Atovaquone is a second-line drug for the treatment of parasitic infections in HIV patients. It is combined with proguanil in Malarone®, which is licensed for the oral treatment of uncomplicated *P. falciparum* infections.

Antihelminthic drugs

Benzimidazoles

Three members of this group are commonly used in the treatment of nematode infections, namely mebendazole, thiabendazole and albendazole. They interfere with beta-tubulin function, resulting in a failure to form microtubules and leading to paralysis of the worm.

Mebendazole

This drug acts against *Ascaris*, hookworm, *Trichuris* and *Enterobius*, and is the drug of choice for these infections. Although a single dose of 100 mg is sufficient for the eradication of threadworm infections, *Ascaris*, hookworm and *Trichuris* infections require a dose of 100 mg twice daily for 3 days. The drug is usually well tolerated but side-effects such as abdominal pain, diarrhoea and headache are more likely with a heavy worm load. It is not sufficiently well absorbed for use in systemic parasite infections. Serum concentration is lowered by phenytoin or carbamazapine, but increased by cimetidine. An oral suspension is available for the treatment of children.

Thiabendazole

This drug is absorbed after oral dosage and is used worldwide for the treatment of strongyloidiasis and visceral larva migrans, and sometimes for intestinal helminth infections. It can be made into a topical paste for the treatment of cutaneous larva migrans. It can cause significant gastrointestinal irritation and skin hypersensitivity rashes.

Albendazole

This drug achieves better tissue concentrations and is less toxic than thiabendazole. It crosses the blood–brain barrier to achieve a CSF concentration 20% that of plasma, and also achieves useful concentrations inside parasitic cysts. It is metabolized quickly and completely, with an approximate plasma half-life of 9 h.

It is active against *Ascaris*, hookworm, *Trichuris* and *Enterobius*, and also against *Strongyloides*, *Capillaria*, *Trichinella*, and the tissue stages of hydatid disease and cysticercosis (*Taenia solium*). It is well tolerated, with only transient side-effects such as epigastric pain and diarrhoea following single-dose treatment.

Other antihelminthic drugs

Piperazine

This drug causes paralysis of *Ascaris* and *Enterobius*. Very little bioavailability data exist, though it is thought to be well absorbed and eliminated rapidly. It is highly effective, with success rates of over 95% but side-effects, including nausea, diarrhoea and urticaria, are described. Severe neuropsychiatric reactions occur rarely.

Levamisole

Levamisole is well absorbed and widely distributed through the body tissues. It is metabolized in the liver and mainly excreted by the kidney giving a plasma half-life of 4–5 h. It is active against *Ascaris* but less effective against hookworm, and is therefore used only for *Ascaris* mono-infections. Nausea and vomiting, abdominal pain and headache have been reported. Influenza-like reactions, blood disorders, photosensitivity and renal failure, only develop at higher doses used for immunomodulation in some non-infectious diseases. (It is not currently available in the UK.)

Ivermectin

Ivermectin is an antiparasitic agent used for single-dose treatment of systemic filariasis. It is the treatment of choice for suppression and cure of onchocerciasis.

Niclosamide

This chlorinated salicylanilide acts against *Taenia saginata* and *T. solium* by interfering with the energy metabolism of the worm. The pharmacokinetics are unknown but treatment is associated with a high cure rate (90–100%). Side-effects are rare, but nausea, vomiting and diarrhoea are encountered. An antiemetic should be given before treatment and drugs which cause vomiting avoided, as retrograde ingestion of eggs can cause cysticercosis. In view of this, patients with intestinal tapeworm infection should preferably be referred to a specialist for treatment with praziquantel.

Praziquantel

Originally developed for the treatment of schistosomiasis, praziquantel has a wider spectrum of activity against blood, lung and liver flukes and tapeworms. It acts by altering the calcium permeability of membranes, resulting in tetanic contraction and disruption in the parasite surface which allows the host's immune system to mount an effective immune response. This mechanism is most important in killing adult schistosomes, which survive in the host because they conceal their surface antigens by absorbing host antigens onto their surface.

The drug is well absorbed and distributed to all the body tissues. It crosses the blood–brain barrier and levels in the CSF may reach a concentration 25% that of serum. It is metabolized in the liver and excreted rapidly in the urine, having a half-life of 1–1.5 h. The concentration of praziquantel is reduced by phenytoin, carbamazepine and dexamethasone. There is no information on human teratogenicity; clinically insignificant amounts are excreted into breast milk. Side-effects, such as dizziness, headache, lassitude and limb pain, are few and mainly dose related. A single dose of 40 mg/kg is adequate for treating beef and pork tapeworms and simple schistosomiasis. Higher or multiple doses are required for liver flukes and *Schistosoma japonicum*.

5 Skin, Mucosal and Soft-tissue Infections

Introduction, 81
Structural considerations, 81
Skin changes in systemic disease, 81
Natural defences of the skin, 82

Viral infections of the skin and mucosae, 82
Papillomavirus infections (warts), 83
Herpes simplex infections, 83
Herpes zoster, 86
Hand, foot and mouth disease, 88
Cutaneous poxvirus infections, 89

Bacterial infections of the skin and mucosae, 91
Impetigo and furunculosis, 91
Erysipelas, 96
Cellulitis, 98
Necrotizing infections of skin and soft tissue, 99
Otitis externa, 100
Erythema chronicum migrans (cutaneous Lyme disease), 100
Erythrasma, 101
Erysipeloid, 101
Cat-scratch disease, 101
Actinomycosis, 102
Cutaneous mycobacterial infections, 102
Propionibacterium acnes and acne, 102

Fungal infections of the skin and its appendages, 103
Introduction, 103
Candidiasis, 103
Dermatophytoses (tinea infections), 104
Pityriasis versicolor, 106
Sporotrichosis, 106
Rarities, 107
Pityriasis rosea, 107

Parasites of the skin, 108
Scabies, 108

Introduction

Structural considerations

The skin consists of the superficial epidermis, and the deeper dermis (Fig. 5.1). The epidermis is composed of layers of squamous cells, arising from a basal cell layer, with a thick 'prickle-cell' layer overlaid by the increasingly keratinized and flattened granular, lucid and horny layers. The keratinous surface is further protected by a film of sebaceous secretion which resists penetration by fluids and microorganisms.

Beneath the basal cell layer lies the dermis. This is a sensitive and vascular structure supported by loose connective tissue, which is well hydrated and supple, allowing movement of the skin surface over deeper structures.

The epidermis is penetrated by hair follicles and the ducts of sweat glands arising in the dermis. These can provide a niche for colonizing organisms and, if occluded by keratinous debris or inspissated secretions, they can become sites of loculated or spreading infection. The hair and nails are modified epidermal structures, consisting mainly of keratin.

The epidermis, hair and nails can be locally invaded by organisms which digest keratin or sebum. These are usually fungi. They cause damage and flaking, sometimes with a minimal inflammatory reaction. Bacterial infection of the epidermis is more invasive and inflammatory. Some infections spread along the epidermal surface, producing a macerated, weeping lesion, which advances to adjacent areas. Such open lesions are infectious, often spreading by contact to other sites or individuals. Intraepidermal infections are less common. They advance within the epidermal cell layers, sometimes separating them to form fluid-filled bullae. Erysipelas is an example of this.

Infections of the dermis spread widely in the loose connective tissue, and the advancing inflammation has an indistinct edge. This is the typical appearance of cellulitis. The dermis can be greatly expanded by the oedema and cellular infiltrate which accompany infection. Scarring or fibrosis afterwards can tether the skin to underlying structures such as fascia or muscle. In severe cases the underlying tissue can become infected (for instance, when a deep skin ulcer erodes underlying bone; see Chapter 14).

Skin changes in systemic disease

The skin often displays abnormalities in systemic diseases. Rashes, jaundice, and vasculitic, embolic and haemorrhagic lesions are valuable physical signs, and may be so typical as to be diagnostic. They may be due to infectious lesions as in the chickenpox rash, to a vasculitic reaction as in erythema nodosum, or to damage by toxins as in scarlet fever. Vasculitis or intravascular coagulation may produce petechial or haemorrhagic lesions, seen in rickettsiosis or meningococcal disease. Chemical infiltration with bilirubin or methaemalbumin may cause typical discoloration.

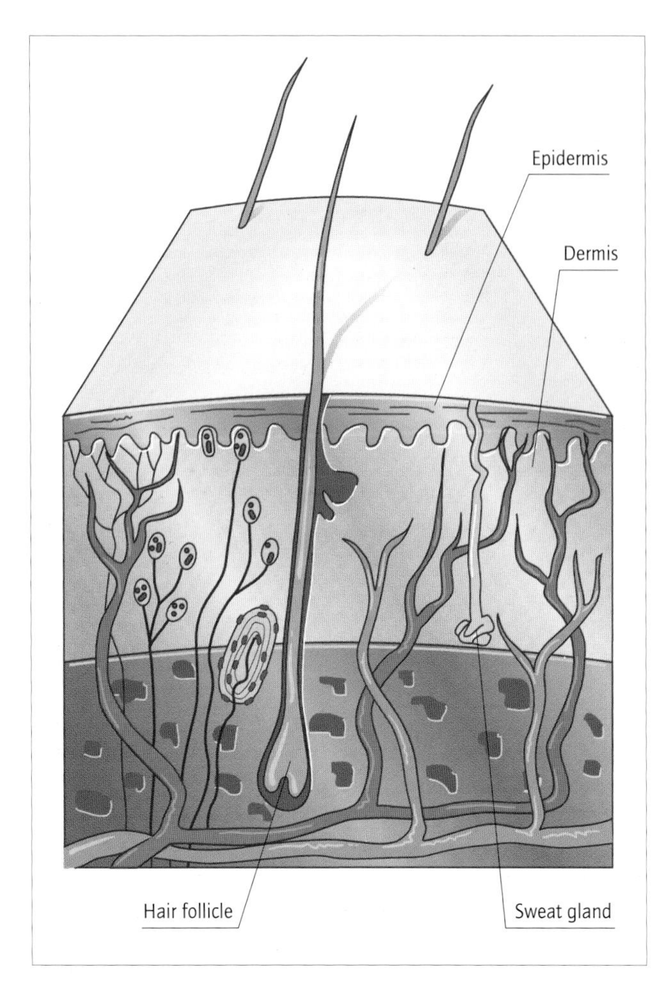

Fig. 5.1 Structure of the skin.

Natural defences of the skin

The surface structure and acidic sebaceous secretions are hostile to many pathogens. There is a dense population of normal flora inhabiting the surface, ducts and follicles. These organisms are adapted to the skin environment, which they may further modify by adjusting the pH, redox potential or local concentration of bacteriostatic substances. The rich blood and lymphatic supply of the dermis ensure that both specific and non-specific immune responses can quickly be recruited when pathogens enter the skin.

Important normal skin flora
1 Coagulase-negative staphylococci.
2 *Staphylococcus aureus.*
3 *Streptococcus pyogenes.*
4 Many species of *Corynebacterium.*
5 *Propionibacterium* spp.
6 *Candida* spp.

Skin defences are compromised if the surface is penetrated by injury or thinned and excoriated by inflammatory processes. Sustained wetness of the skin causes maceration of the keratinous epidermis, which can then be invaded by pathogens. Patients who suffer from eczema are highly susceptible to skin infections caused by *Staphylococcus aureus* or herpes simplex.

Natural defences of the skin
1 Keratinous surface.
2 Antibacterial effects of sebum.
3 Effect of normal flora.

Mucosal structure and defences

Mucosae do not have a keratinized surface; they are always moist and some are only one cell thick. They are much more susceptible than skin to invasion by surface pathogens. For defence they rely on the washing action of secretions, which may contain lysozymes or specific secretory immunoglobulin A (sIgA). Mucosae adjacent to skin, as in the nose and mouth, are colonized and protected by their own normal flora, which merges into that of the skin at the mucocutaneous margin.

Natural defences of mucosae
1 Mechanical washing by tears or urine.
2 Lysozyme or antibody in surface fluid.
3 Surface phagocytes.
4 Ciliary action moving mucus and debris.

Viral infections of the skin and mucosae

ORGANISM LIST

Papillomaviruses
Herpes simplex virus type 1
Herpes simplex virus type 2
Varicella zoster virus
Coxsackie A viruses
Molluscum contagiosum
Cowpox
Orf/paravaccinia
Vaccinia

Papillomavirus infections (warts)

Introduction

There are many types of human papillomaviruses (HPVs), of which the commonest are those associated with common warts, plantar warts (verrucae) and genital warts. Unlike some genital papillomaviruses, particularly HPV-16 and HPV-18, but including other types (see Chapter 15), HPV infections of the skin are rarely associated with malignancy. Exceptions include rare squamous cell transformation of condylomata acuminata, and the uncommon Bowenoid papulosis and epidermodysplasia verruciformis.

Common (plane) warts

These are usually seen in children, as immunity to the viruses is gained by adulthood. They spread by contact, the viruses entering tiny fissures, and infecting the cells of the prickle-cell layer. The infected cells hypertrophy and multiply, producing a keratinized, nodular papilloma. Curetting the lesion confirms its identity by revealing the small bleeding points of dermal capillaries supplying the lesion (in involuting lesions, thrombosed capillaries appear as black dots). While solitary on the palms, plane warts may be multiple in the looser skin on the wrists or dorsa of the hands.

Plantar warts may become large, a centimetre or more in diameter. They are pressed into the foot by the body weight and often develop an eroded central area. Local pressure can cause severe pain on weight-bearing. They can be distinguished from corns by the bleeding points of dermal capillaries revealed on paring the lesion; corns are composed entirely of keratin, with no capillaries.

Treatment of warts is rarely required in children, as immunity gradually develops and the lesions involute. Warts in adults can be curetted, or preferably ablated by freezing with a cryoprobe or with a small swab dipped in liquid nitrogen. Plantar warts may be reduced in size by paring. Salicylate and lactic acid ointment, painted on to the lesion, allowed to dry and covered with an adhesive dressing, softens the keratin and inhibits the viruses.

Plantar warts are infectious, especially during swimming, when the uncovered skin is softened by immersion. If the lesions are temporarily covered with a waterproof plaster or latex sock, however, swimming need not be forbidden.

Herpes simplex infections

Introduction and epidemiology

Herpes simplex viruses are colonists and invaders of mucosae. They can also infect skin if entry is available via small punctures, fissures or macerated areas. Typically, infections of the mouth and upper body have been caused by herpes simplex type 1, while genital infections have been due to the genetically distinct type 2 virus. In recent years the epidemiology of the two viruses has become less distinct, and either virus may be found in either site.

Like other herpesviruses, herpes simplex has the property of latency; after causing a primary infection it persists in the immune host as a dormant form. Subsequent reactivations may cause either asymptomatic virus excretion or a clinical lesion, usually less severe than the primary infection.

Herpes simplex infection is spread by direct contact with mucosa or skin from which virus is being shed. In infancy primary infection is often buccal (Fig. 5.2), acquired by oral contact or shared eating utensils. Nailfold or finger pulp infections may follow contamination of the finger with the saliva, and often occur in infants who suck their thumb or nurses who provide mouth care for debilitated patients (Fig. 5.3). Inoculation into the eye may cause conjunctival (Fig. 5.4) or corneal infection.

Reactivation lesions may be precipitated by acute feverish illness and are therefore called cold sores (Fig. 5.5). Severe herpetic lesions often accompany pneumococcal or meningococcal infections. Overexposure to sunlight is also a common precipitating event.

Sites of skin and mucosal herpes simplex infection

1. Primary gingivostomatitis.
2. Lip margin (cold sore).
3. Nailbed (felon or whitlow).
4. Finger pulp.
5. Facial skin.
6. Eczema-affected skin (eczema herpeticum).
7. Cornea (dendritic ulcer).
8. Conjunctiva.
9. Genitalia (see Chapter 15).
10. Central nervous system infections (see Chapter 13).

Effective cell-mediated immunity is essential for the control of herpes simplex infection. Patients treated with cytotoxic drugs, who are undergoing bone marrow grafting, or those with human immunodeficiency virus (HIV) infection are at risk of severe infection.

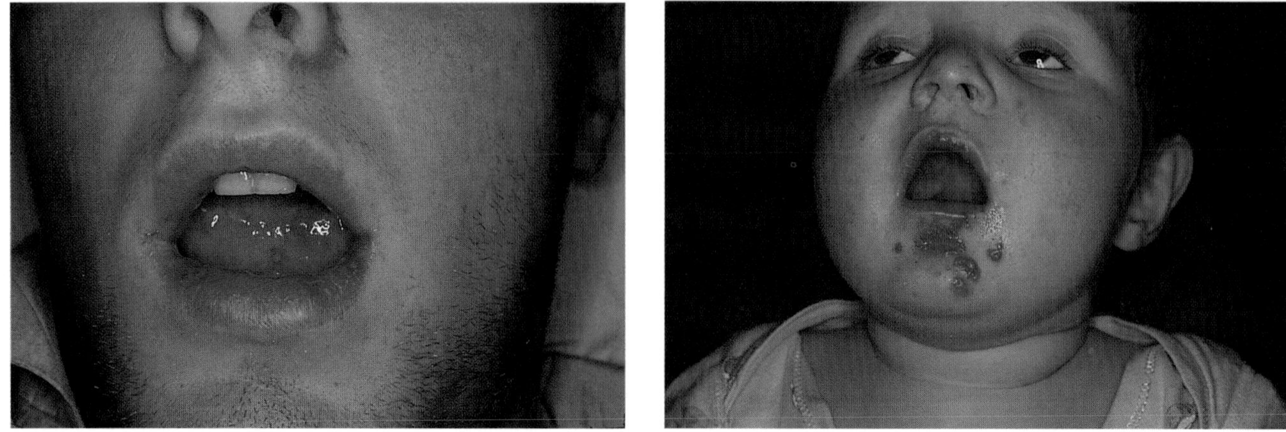

Fig. 5.2 Primary herpetic gingivostomatitis showing: (a) lesions on the tongue (unusual in an adult); (b) in an infant, herpetic lesions on skin macerated by infected saliva.

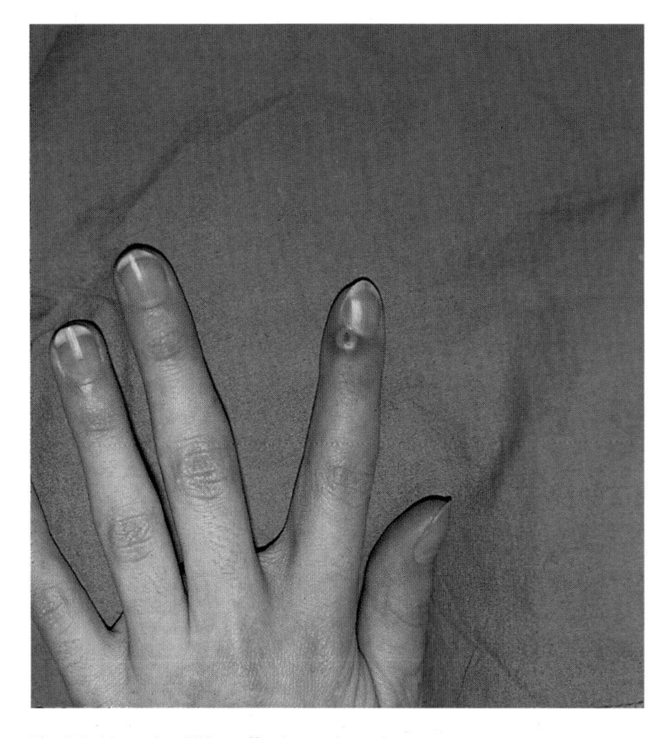

Fig. 5.3 Herpetic whitlow affecting an intensive care nurse.

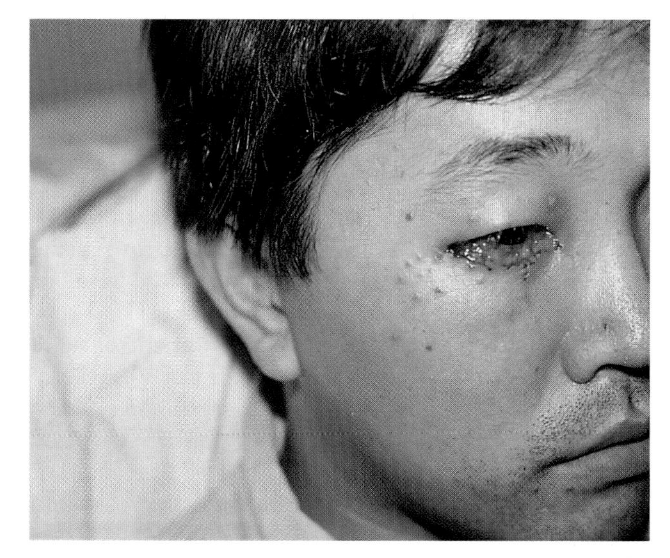

Fig. 5.4 Herpes simplex conjunctivitis.

Pathology

Herpes simplex is a cytolytic virus that causes degeneration and 'ballooning' of epidermal prickle cells, leading to vesicle formation. Keratinized cells form the vesicle roof, but in the mouth and at other mucous membranes the roof is unstable and sloughed away. The vesicle contains cellular debris with fused keratinocytes containing typical inclusion bodies (Tzanck cells) (Fig. 5.6).

Virus spreads from cell to cell and also up the sensory neurones via the axon, establishing latent infection in the sensory neurones. Herpes simplex virus can be isolated from the trigeminal ganglia of approximately 50% of patients tested. If the virus spreads to the brain, encephalitis may develop.

Latency is a dynamic equilibrium with a low-grade viral infection which does not cause cell lysis. Intermittently, the virus reactivates and travels down the sensory neurone to cause a secondary infection. The mechanisms of latency and reactivation are not clearly understood, but specific antisense DNA sequences from the herpes simplex virus are detectable in latently infected cells.

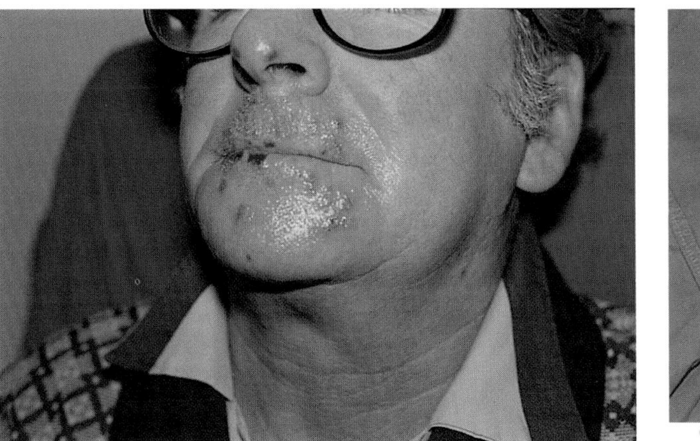

(a)

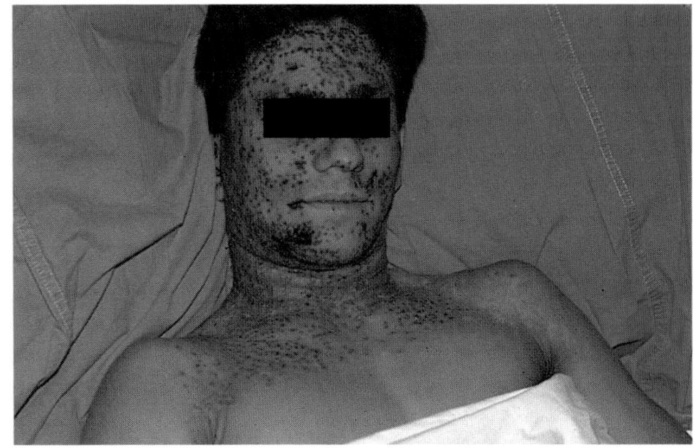

(b)

Fig. 5.5 (a) Herpes simplex reactivation lesion (cold sore); (b) eczema herpeticum originating from a cold sore.

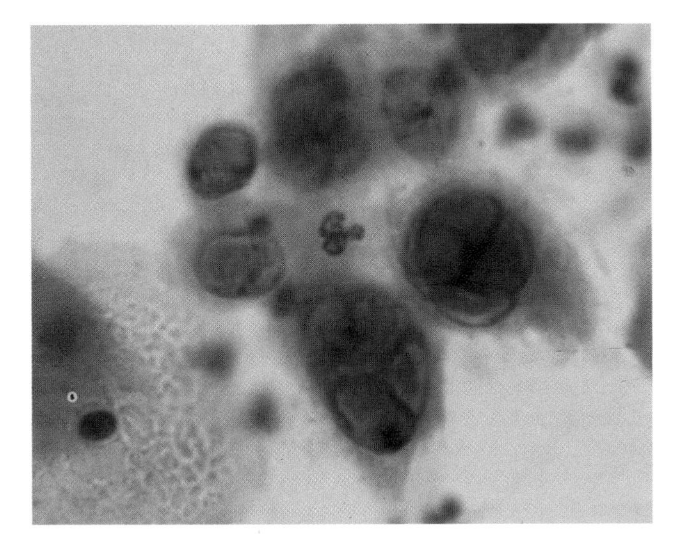

Fig. 5.6 Microscopical appearance of cells from a herpetic vesicle.

Clinical features

Primary herpetic gingivostomatitis

Primary herpetic gingivostomatitis is common in infants, though some people present in their teens or later. After a day or two of fever and tender enlargement of cervical lymph nodes painful, oval, whitish vesicles appear in the mouth. They may cover the tongue and are common on the soft palate, the gums and inside the lips. Soreness often prevents suckling or taking solid food. Dribbling of infectious saliva often leads to secondary lesions on the fingers, chin or chest.

Few new lesions appear after 2 or 3 days of eruption, and most have deroofed and healed within 5–7 days. There is risk of spread to affected skin or, rarely, generalized skin infection in individuals with eczema, even if the disease is currently inactive. This condition is called eczema herpeticum.

Diagnosis

The diagnosis is usually clinically obvious. Oral blisters are seen in hand, foot and mouth disease, but these are accompanied by a typical cutaneous rash. Oval plaques of oral candidiasis sometimes occur in infants, but without associated fever and lymphadenopathy.

Laboratory diagnosis

Light microscopy of scrapings from the base of lesions may reveal cells with typical intranuclear inclusions surrounded by a clear halo (Tzanck cells) and, in skin specimens, fused cells (polykaryocytes). Electron microscopy demonstrates virus particles in scrapings and is the investigation of choice in early cases. These features are common to other herpesvirus infections. Vesicle fluid may give false-negative results, but polymerase chain reaction (PCR) can demonstrate specific herpes simplex DNA. Herpes simplex antigen can be detected by immunofluorescence staining of cells from herpetic lesions.

Herpes simplex virus is readily cultivated. Fresh vesicle fluid, saliva or scrapings should be inoculated as soon as possible. A cytopathic effect can often be seen after 24 h but may be delayed for up to 7 days. It is characterized by grape-like clusters of refractile cells, usually more pronounced with herpes simplex virus type 2.

Techniques for diagnosing herpes simplex infections of skin and mucosae
1 Light microscopy (cytology).
2 Electron microscopy.
3 Antigen detection.
4 PCR for herpes simplex DNA.
5 Cell culture.
6 Serology.

Serology

Herpes virus-specific IgG and IgM are detectable by immunofluorescence, radioimmunoassay (RIA) or enzyme-linked immunosorbent assay (ELISA). Specific IgM antibodies in serum indicate active infection but do not reliably differentiate between primary infection and reactivation. Seroconversion in primary infection can be detected serologically. A fourfold rise in antibody titres demonstrated by these methods often indicates reactivation–infection.

Management

Maintenance of hydration is important. Mild analgesia with paracetamol may help. In severe cases an analgesic gel can be applied sparingly to the gums and anterior tongue. Older children and adults usually cope by taking frequent cool or tepid drinks.

Treatment of herpes simplex

For skin and mucosal infections
1 Aciclovir orally 200 mg five times daily for 5 days.
2 Valaciclovir orally 500 mg twice daily for 5 days.
Doses may be doubled in severe cases or immunosuppressed patients.

For severe infections or extensive eczema herpeticum
1 Aciclovir i.v. 5 mg/kg 8-hourly.
Dose may be doubled in immunosuppressed patients.

Prevention and control

Transmission may be reduced by avoiding direct contact with herpetic lesions, although asymptomatic infections and reactivations, with transient viral shedding, are probably common. To avoid herpetic whitlow those attending dental and tracheostomy patients should always wear gloves. Patients with herpetic lesions should avoid contact with immunosuppressed patients, newborns and patients with severe eczema or burns.

Herpes zoster

Introduction

Herpes zoster is the condition caused by reactivation of latent varicella zoster virus (VZV) whose genomes persist in the sensory root ganglia of the brainstem and spinal cord. Reactivation may be spontaneous or may follow a physical or emotional insult, such as a fever, injury or bereavement.

Infection is common and debilitating in middle or old age, but children and teenagers can develop zoster if they had chickenpox as babies, and neonatal zoster occasionally occurs after intrauterine varicella infection.

The disease begins with inflammation in the affected dorsal root ganglion and the appearance of virus particles in the neurones. There is inflammation in the surrounding meninges and affecting the courses and connections of the neurones in the central nervous system. On reaching the skin the disease causes blistering lesions with the same pathological appearance as those of chickenpox (see Chapter 11).

Clinical features

Illness begins with pain and hyperaesthesia in the affected dermatome, as virus particles migrate down the nerve fibres towards the skin. Virus also spreads centrally, causing intramedullary and meningeal inflammation. A few patients have clinically apparent meningism, and children in particular may present with viral meningitis and cerebrospinal fluid pleiocytosis before they develop a rash (Fig. 5.7).

The rash begins as small, pink patches at the site of penetrating cutaneous nerves. Vesicles quickly develop in these areas, and the lesions spread until the dermatome is filled. The rash is painful, with a disturbing burning, stabbing quality. Unlike chickenpox vesicles, those of zoster may coalesce to form bullae which become flaccid and lose their roofs, forming ulcerated areas (Fig. 5.8). The vesicles heal by drying and scabbing, rarely leaving a scar.

Multidermatomal and bilateral zoster are rare, occurring when more than one ganglion is affected. Severe zoster in a young person may be an early sign of HIV infection or another immunodeficiency. Blood-borne dissemination of virus, with the development of a chickenpox-like rash, also indicates immunodeficiency (Fig. 5.9).

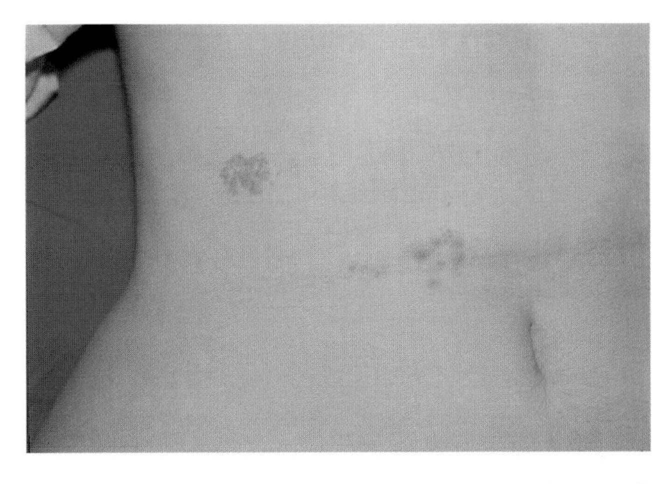

Fig. 5.7 Herpes zoster lesions appearing 2 days after this 8-year-old presented with lymphocytic meningitis; the rash did not extend further.

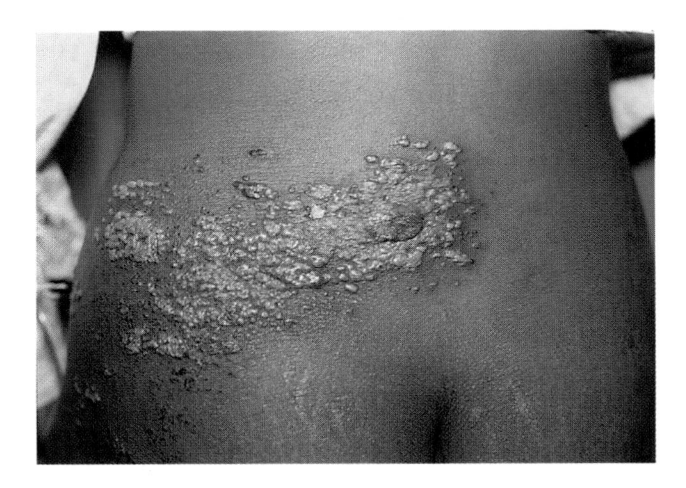

Fig. 5.8 A close-up view of herpes zoster lesions, showing their tendency to coalesce.

Signs of immunodeficiency in herpes zoster
1 Bilateral or multidermatomal dermal rashes.
2 Unusually severe or deep lesions.
3 Recurring herpes zoster.
4 Association with chickenpox-like rash.
5 Association with progressive central nervous system features.

The course of the illness depends on the patient's age. In children the rash may be minimal; in adults the average duration of rash is 5–8 days. In elderly people there can be weeks of illness, with bullous rash and severe central nervous system disruption causing confusion and debility.

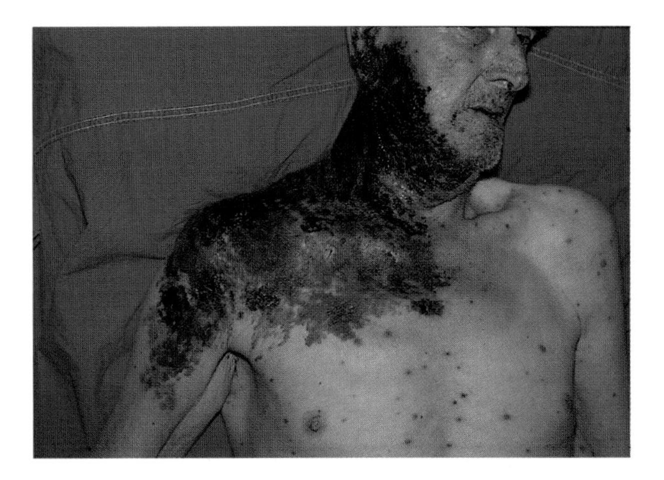

Fig. 5.9 A severe and haemorrhagic herpes zoster rash, with disseminated chickenpox lesions in a patient with chronic lymphocytic leukaemia (the patient recovered fully on treatment).

Diagnosis

The diagnosis is usually clinically obvious. If necessary, herpesvirus particles can be demonstrated by electron microscopy of vesicle fluid or scrapings. The virus grows with a typical cytopathic effect in cell culture. Serological testing usually shows a large rise in IgG antibodies (secondary response; see Chapter 1).

Laboratory diagnosis of herpes zoster
1 Light microscopy (cytology).
2 Electron microscopy.
3 Cell culture.
4 Serology.

Management

Rest and analgesia may be sufficient in young children and mild cases. The illness can be shortened in older people by early treatment with aciclovir, valaciclovir or famciclovir. Oral aciclovir has poor bioavailability and may not be effective; severe cases benefit from intravenous treatment for 4 or 5 days, or longer in the immunosuppressed. Valaciclovir (an ester of aciclovir) or famciclovir, an ester of penciclovir, are well absorbed before being hydrolysed to the parent drug in the circulation. They can be given in lower and less frequent dosage than aciclovir.

Treatment of herpes zoster for uncomplicated infections:
1 Oral aciclovir 800 mg five times daily for 7 days.
2 Oral valaciclovir 1 g three times daily for 7 days.
3 Oral famciclovir 250 mg three times daily for 7 days or 750 mg once daily for 7 days.
For severe infection:
4 Aciclovir infusion 5 mg/kg 8-hourly for 4–7 days.
For immunocompromised patients:
5 Aciclovir infusion 10 mg/kg for 7–14 days.

Complications

Postherpetic neuralgia (zoster-associated pain, ZAP)

Postherpetic neuralgia is causalgic (pain unrelated in quality or severity to its cause), and severely affects the elderly, lasting an average of 60 days in the over-50s. It is often noticed when the rash is healing, and responds poorly to analgesics, including opiates, phenytoin, carbamazepine and vitamin B_{12} injections. Tricyclic antidepressants often give relief, probably by blocking reuptake of 5-hydroxytryptamine in the presynaptic nerve endings of central pain pathways. Gabapentin may be an effective alternative. Either can be combined with standard analgesics. ZAP is approximately halved in duration in those commencing antiviral treament within 48 h of rash onset.

Secondary bacterial infection

Secondary bacterial infection with *Staphylococcus aureus* can affect severe or denuded rashes, especially in the elderly. Oral treatment with cloxacillin, flucloxacillin or trimethoprim is usually effective.

Ascending myeloencephalitis

Ascending myeloencephalitis is an exceptionally rare complication which should be treated with high-dose intravenous aciclovir. Immunosuppressed patients are at higher risk, and respond less well to treatment.

Hand, foot and mouth disease

This is a highly infectious disease of toddlers caused by coxsackie A viruses, most often A16. Family outbreaks are common, occasionally affecting adults as well as children, though adults are usually immune. A brief feverish prodrome is followed by the simultaneous eruption of several blisters on the hands and feet and in the pharynx. The palms and soles are most affected (Fig. 5.10). There is also a papular rash on the buttocks. There is mild discomfort and soreness of the throat, but the illness is short and self-limiting in most cases. A recent epidemic in the Far East was caused by enterovirus type 71, and some cases were complicated by encephalopathy and/or shock.

If microbial diagnosis is needed, the viruses can be iden-

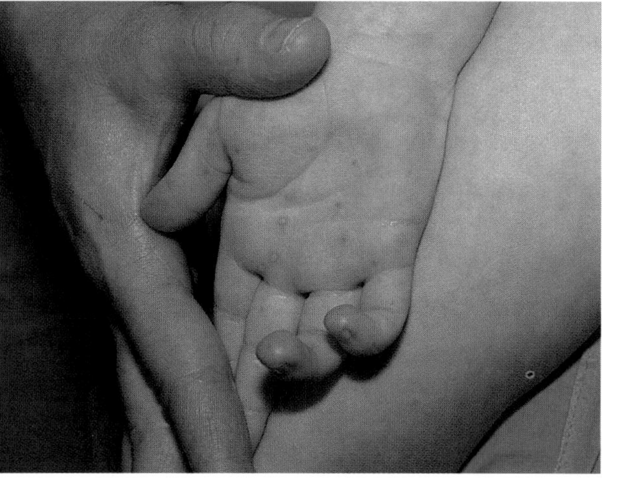
(a)

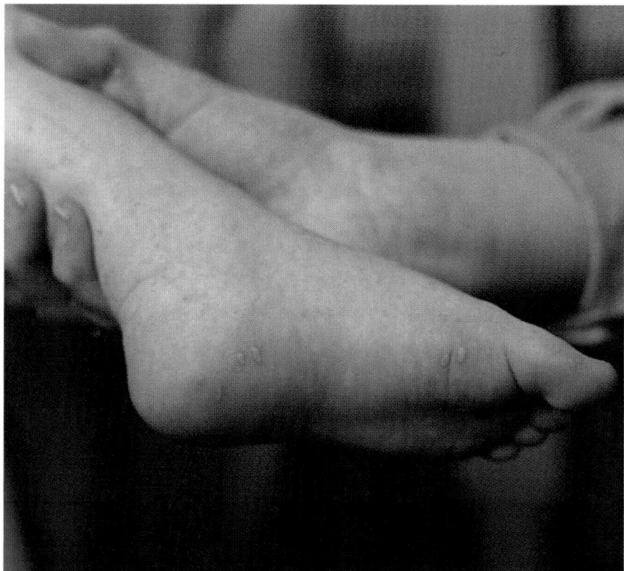
(b)

Fig. 5.10 Hand, foot and mouth disease: the typical blisters.

tified in cell cultures of vesicle fluid, throat swab or faeces. The laboratory diagnosis of enterovirus infections is discussed in Chapter 6.

Cutaneous poxvirus infections

The poxviruses which infect the skin are molluscum contagiosum, cowpox, orf and vaccinia. Monkeypox and tanapox cause systemic infections with cutaneous pocks.

Poxviruses are divided into three groups: (i) orthopoxviruses which include monkeypox, vaccinia and cowpox; (ii) parapoxviruses which include orf; and (iii) unclassified viruses which include molluscum contagiosum and tanapox. The viruses are large with a double-stranded DNA genome coding for more than 100 viral proteins. They have complex symmetry with a brick-like shape 200–250×250–300 nm. Virus uncoating is initiated by host enzymes but completed by viral enzymes. Viral replication takes place in the cytoplasm producing inclusion bodies.

Molluscum contagiosum

This is a wart-like condition caused by a poxvirus infection of the prickle-cell layer of the epidermis. The infected cells proliferate, vacuolate and enlarge, protruding above the surface of the skin as typical, pearly lesions up to 3 mm in diameter. They are always umbilicated, with a small central cavity containing whitish, pulpy material. The vacuolated cells are shed into the lesion, which is highly infectious. Infection is transmitted by skin contact, clothing and towels, and autoinoculation to other skin sites is common.

The lesions are typically in groups, often on the face or arms. HIV-positive patients may have numerous lesions, particularly on the face. Their appearance is diagnostic but, if necessary, poxviruses can be demonstrated by the presence of characteristic molluscum bodies in a scraping stained with, for example, Giemsa or iodine, or by electron microscopy of expressed material (Fig. 5.11).

Spontaneous resolution is uncommon, but adequate treatment is curative. The lesions can be removed by curetting, or 'killed' by inserting the point of an orange-stick dipped in 80% phenol solution into the umbilicated centre. Extensive skin lesions in HIV patients seem to respond well to topical cidofovir ointment.

Cowpox and orf

These are zoonoses caused by poxviruses. Cowpox produces lesions on the udders of cattle, and causes systemic disease with pneumonitis in cats. Human cases have occurred after contact with cattle, domestic cats and big cats. Transmission occurs by direct contact, and causes large, volcano-shaped pocks with vesicular or necrotic centres, usually on the hand (Fig. 5.12). The central crater may appear black, and a history of animal contact may

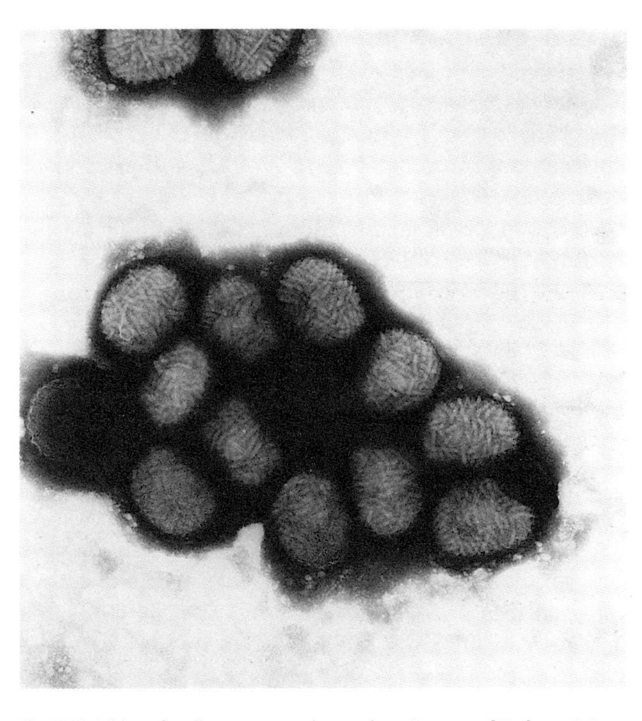

Fig. 5.11 Virion of molluscum contagiosum virus. Courtesy of Professor P. D. Griffiths and Ms G. Clewley, Department of Virology, Royal Free Hospital School of Medicine.

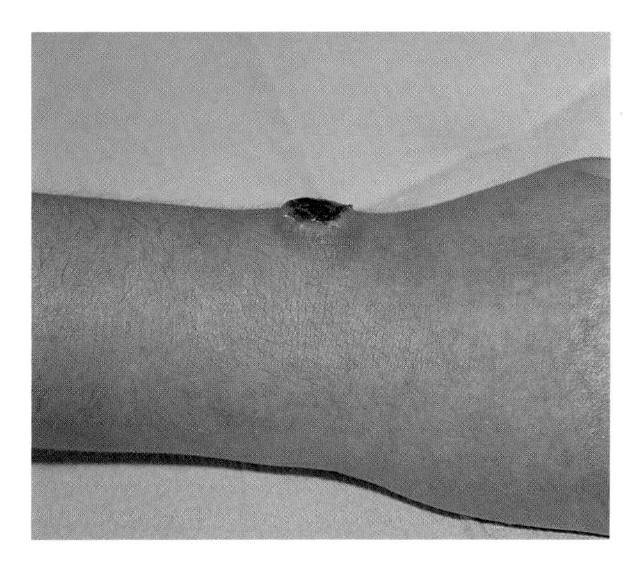

Fig. 5.12 Lesion of cowpox on the wrist of a stable-girl (possibly contracted from the farm cat).

lead to suspicion of anthrax. Cowpox lesions have no halo of vesicles, however, and the surrounding oedema is rarely as severe or extensive as in anthrax.

Orf is a parapoxvirus whose natural host is sheep. Human infections follow the handling of sheep or their carcasses. The hand or wrist is usually affected by one or more indurated papules 1–2 cm in diameter. There may be a central vesicle or crater, or simply a depression.

Viruses can be identified by electron microscopy or culture of biopsy material or curettings from the edge or base of the lesions. Culture of cowpox on the chorioallantoic membrane of hens' eggs produces characteristic lesions (Fig. 5.13).

No specific treatment is required, as the conditions are self-limiting, with a natural history of up to 3 or 4 weeks. Aciclovir and penciclovir are not effective against poxviruses.

Vaccinia

Vaccinia exists in many strains whose origins are uncertain. It may have been derived from subcultures of cowpox or of variola (the agent of smallpox) (Figs 5.14–5.16). It was used for three centuries as an effective vaccine against smallpox (which was declared extinct from nature in 1979). Vaccinia is now an uncommon laboratory organism, used for developing hybrid vaccines, though canarypox and other animal poxviruses are now more favoured. It is rarely given as a vaccine as the indication no longer exists. Researchers should, however, be aware that vaccinia is a pathogenic virus. It can cause severe necrotizing or spreading infection in patients with altered immunity. It has a predilection for eczematous

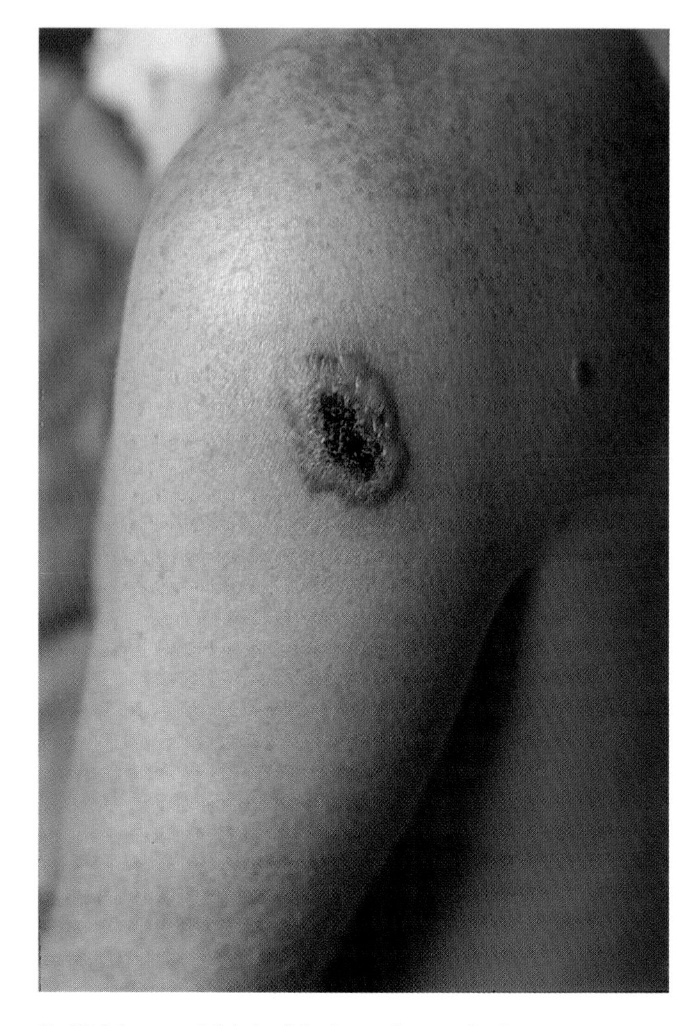

Fig. 5.14 Large vaccinia lesion following smallpox vaccination.

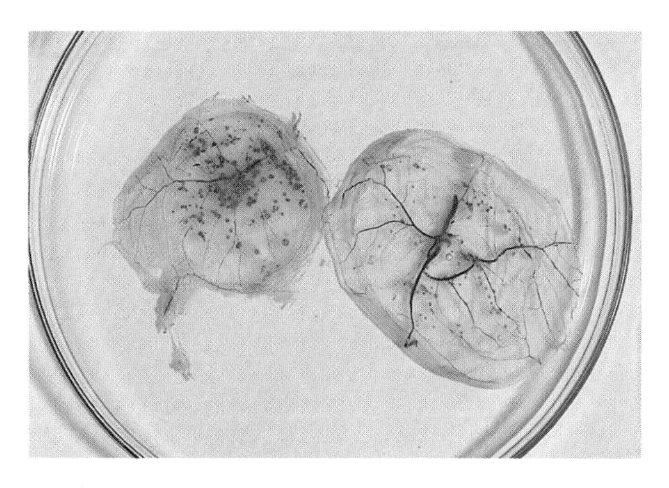

Fig. 5.13 Pocks produced by cowpoxvirus on the chorioallantoic membrane of a hen's egg.

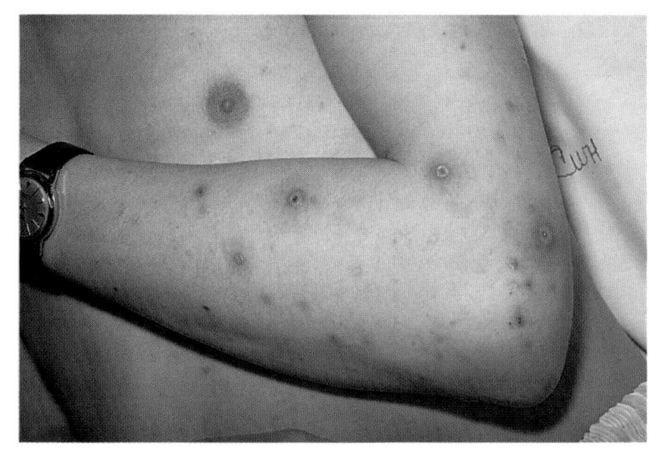

Fig. 5.15 Autoinoculation lesions of vaccinia on the skin.

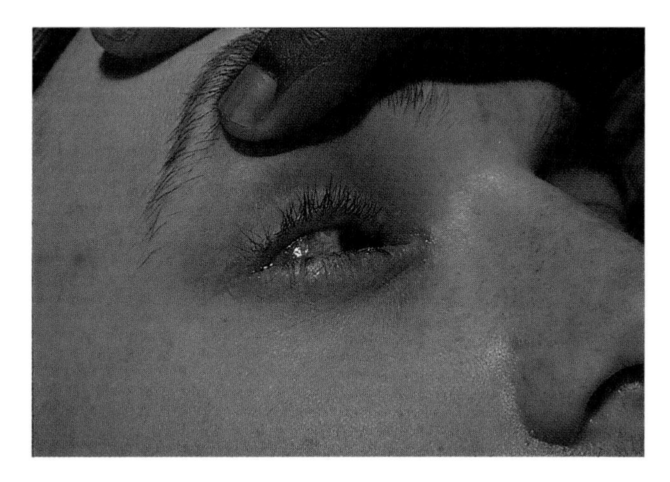

Fig. 5.16 Autoinoculation lesions of vaccinia on the conjunctiva.

skin. It is readily transmitted by the transplacental route. A postinfectious encephalitis can follow primary vaccinia infection. Aciclovir is not effective, but adefovir has significant antivaccinia activity.

Monkeypox

Monkeypox is a zoonotic orthopoxvirus infection which causes smallpox-like disease in humans in central and West Africa. The diagnosis may be suspected clinically because of the typical rash, lymphadenopathy and recent monkey contact. Human-to-human spread is uncommon. Specimens of vesicle fluid are suitable for investigation. The virus is stable and survives well while being transported. In recent cases in Zaire the diagnosis was made by a combination of electron microscopy, virus isolation and serology. Distinctive pocks are produced on chorioallantoic membrane culture in hens' eggs. Antibody to specific monkeypox antigen can be detected in serum absorbed with smallpox and vaccinia antigens. Surveillance of systemic monkeypox is maintained because of the remote possibility of an increase in virulence.

Tana

Tana is a poxvirus infection probably spread by insect bites, and usually acquired in Africa. It is unusual to see cases with more than a single skin lesion. The diagnosis is suggested by the travel history. The virus can be demonstrated by electron microscopy or cultivated in cell culture. Neutralizing antibodies are produced in convalescence.

Bacterial infections of the skin and mucosae

ORGANISM LIST

Staphylococcus aureus
Streptococcus pyogenes
Corynebacterium spp., including *C. minutissimum* and *C. diphtheriae*
Borrelia burgdorferi, B. garinii, B. afzelii
Pasteurella spp.
Bartonella henselae
Erysipelothrix rhusiopathiae
Mycobacteria, including *Mycobacterium tuberculosis* and environmental mycobacteria
Actinomyces spp.

Impetigo and furunculosis

Introduction

Impetigo is a pyogenic infection of the epidermis. Furunculosis is infection of sebaceous glands or sweat glands. Both are usually caused by *Staphylococcus aureus*, and are characterized by an intense local inflammatory response and the production of pus.

S. aureus is distinguishable by phage-typing into many groups, some of which produce powerful toxins. Some toxins act locally, contributing to the pathogenesis of lesions. Others are systemically active, causing severe systemic disease as a consequence of relatively limited local infection. The most important example of this is toxic shock syndrome (see below).

Epidemiology

Pus from skin lesions is highly infectious. Staphylococcal skin infections therefore spread easily, contiguously to adjacent skin sites, by autoinoculation to distant sites or by contact to the skin of other individuals.

Microbiology

Staphylococci are Gram-positive, catalase-producing organisms in the family Micrococcaceae. *Staphylococcus* spp. are facultative, but *Micrococcus* spp. (which rarely cause human disease) are obligate aerobes. The organisms are non-motile and rarely produce capsules.

There are at least 16 *Staphylococcus* species of varying pathogenicity to humans. They can be divided into the pathogenic coagulase-positive *S. aureus*, and the less invasive coagulase-negative staphylococci, on the basis of the coagulase test. *S. aureus* colonies on modern media are not always gold-coloured, and are therefore not morphologically distinguishable from some coagulase-negative staphylococci.

Pathogenesis

The enzyme coagulase catalyses the conversion of fibrinogen to fibrin without the presence of thrombin. It has conventionally been associated with pathogenicity, but not all strains of *S. aureus* are highly pathogenic. Most *S. aureus* produce microcapsules of which there are 11 types although most of the pathogenic strains express type 5 or 8. Some organisms produce a pseudocapsule in host tissues. Many strains of *S. aureus* express a fibronectin receptor at their surface. This may facilitate adhesion to host tissues, where fibronectin is present, and is often produced in increased quantities during the reaction to acute infection. Foreign material such as intravenous cannulae or sutures rapidly become coated with serum components such as fibronectin, to which staphylococci can adhere.

S. aureus also produces extracellular enzymes, such as hyaluronidase, collagenase and lipase, which break down host tissues and may facilitate invasion. Cytolytic toxins, conventionally named haemolysins, are also produced. The way in which these enzymes contribute to the pathology of the disease is not known.

Staphylococcal enterotoxins are important in food-borne disease (see Chapter 8), and toxic shock syndrome toxins 1 and 2 have, among other effects, toxicity for the skin and destroy the desmosomes. Pyrogenic exotoxins are also produced. These are closely homologous with toxins produced by *Streptococcus pyogenes*, and act as superantigens. They induce a vigorous proliferation of activated T cells, resulting in a massive release of cytokines by both macrophages and T cells. Toxin-mediated disease is associated with severe and exfoliative skin rashes.

Staphylococcal peptidoglycan and lipoteichoic acid stimulate the release of tumour necrosis factor-alpha (TNF-α), interleukin (IL)-1, IL-6 and IL-8; this together with cellular activation and activation of the complement and coagulation pathways leads to shock and coagulopathy.

Different stages of staphylococcal infection correspond with the expression of different bacterial antigens. During the initial stages, matrix-binding proteins favour colonization and, later, exoproteases facilitate spread through the tissues. Global regulatory genes that control the expression of groups of staphylococcal genes have been studied, including *agr*, which suppresses the expression of surface protein through a bacterial density-sensing peptide.

Factors affecting pathogenicity of staphylococci

1 Coagulase.
2 Capsules.
3 Fibronectin receptor.
4 Extracellular enzymes.
5 Haemolysins.
6 Pyrogenic exotoxins.
7 Enterotoxins A–E.
8 Toxic shock syndrome toxins 1 and 2.

Clinical features

Impetigo

Impetigo often occurs around the mouth or nose, where skin is easily damaged, or in superficial skin lesions such as scratches, insect bites or broken chickenpox vesicles. It is common in children, but adults can also be affected, particularly by secondary 'impetiginization' of skin lesions. Beginning as a small spot, impetigo extends to form a plaque-like inflamed lesion on which a yellowish exudate forms, and dries into thick scabs. The lesions are irritating and sore; scratching or rubbing contributes to infection of other sites (Fig. 5.17).

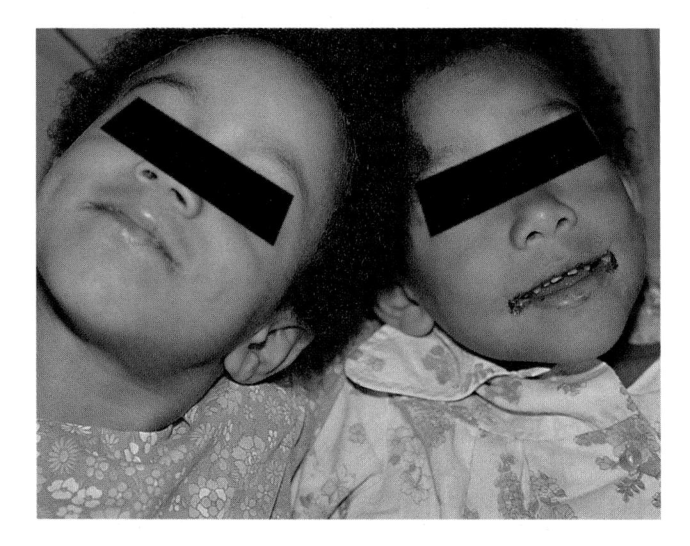

Fig. 5.17 Impetigo in a typical site: both sisters are affected.

Furunculosis

Furunculosis tends to occur in stagnant sebaceous material, and therefore affects children and men more than women. Poor personal hygiene can predispose to furunculosis by allowing blockage of sebaceous ducts. The lesions may be single, when they are often called boils, or multiple. They begin as small papules which increase in size and tenderness to a variable degree. Some will 'point' and discharge yellowish pus before gradually resolving, while others may remain 'blind', gradually healing by becoming less indurated and inflamed.

Concurrent infection of several neighbouring glands causes a composite lesion with several discharging sinuses at the skin surface. These lesions are called carbuncles. They tend to occur in men, often affecting the axilla or the hairline at the back of the neck, and can reach several centimetres in diameter.

Diagnosis

The diagnosis is usually clinically evident. *S. aureus* can readily be recovered from cultures of exudate or pus, allowing confirmation of the antibiotic sensitivities of the organism.

A minority of impetiginous lesions produce *Streptococcus pyogenes* on culture. Such lesions have a tendency to contain bullae, but this is not a reliable distinguishing feature.

Lesions similar to furuncles are occasionally caused by non-staphylococcal infections, such as *Pseudomonas aeruginosa* in swimmers. Microbiological examination is therefore important in managing lesions which do not readily respond to standard treatment.

Microbiological diagnosis

Staphylococci are not nutritionally demanding and grow readily on simple media. Colonies are readily identified on blood agar by their opaque, butyrous consistency. Many colonies elaborate a haemolysin, and are surrounded by a zone of complete haemolysis. Colonies of *S. aureus* are golden-yellow to creamy-white, whereas coagulase-negative colonies are smaller and white. When *S. aureus* must be identified from sites contaminated by other organisms, selective media may be used. Staphylococci can tolerate salt concentrations of 5% or more. Five per cent salt agar will therefore select for these organisms. Mannitol and an indicator are also incorporated, as mannitol is fermented by almost all *S. aureus*, allowing selection of positive colonies for further study. These screening techniques are used to identify carriers of *S. aureus* in outbreaks of infection. An alternative medium is Baird–Parker medium with additional antibiotics, one such as ciprofloxacin is used for selection of Epidemic MRSA that are resistant to ciprofloxacin. The identity of the organisms is confirmed by demonstrating characteristic staphylococcal morphology on Gram staining, catalase production and the presence of coagulase.

Coagulase is either bound to the bacterial cell wall (when it is known as clumping factor) or expressed extracellularly. Isolates are usually screened using a slide agglutination technique to detect clumping factor. This identifies more than 89% of strains of *S. aureus*. Negative isolates are then tested using a tube coagulase technique which detects free coagulase. This identifies more than 99% of *S. aureus*.

More than 95% of *S. aureus* produce deoxyribonuclease (but so do up to 10% of coagulase-negative staphylococci). Once an isolate is identified as a *S. aureus* no further identification is usually necessary. Speciation of the coagulase-negative staphylococci may be indicated when attempting to confirm that multiple isolates from the same patient are the same, and therefore likely to be clinically significant. A number of useful commercial testing kits are available for use in routine laboratories.

Typing

Several typing methods are available to demonstrate the transmission of resistant staphylococci in a hospital environment. The longest established method is phage typing. Molecular methods are now also used, including pulsed field gel electrophoresis and ribotyping.

Antimicrobial susceptibility

S. aureus is susceptible to erythromycin, clindamycin, fucidic acid, ciprofloxacin, aminoglycosides, chloramphenicol, the first-generation cephalosporins and the glycopeptides vancomycin and teicoplanin. Individual strains may acquire resistance to any of these agents. Intermediate glycopeptide resistance has been reported and may be an increasing clinical problem.

S. aureus strains were once almost universally susceptible to penicillin but when penicillin was used widely, penicillinase-producing strains soon predominated. The introduction of methicillin and, later, flucloxacillin, initially solved this problem. However, strains of methicillin-resistant *S. aureus* (MRSA) have become a major problem in many hospitals throughout the world. Methicillin resistance develops by acquisition of the gene *mecA*, which codes for a low-affinity penicillin-binding protein PBP2. This enables the organism to synthesize its cell wall despite the presence of the antibiotic. MRSA is commonly resistant to several other antimicrobials, and some MRSA known as epidemic ('E') strains spread readily in hospitals. Most hospitals have policies to control the spread of MRSA (see Chapter 23).

Methicillin resistance can be difficult to detect in the laboratory and isolates should be inoculated on salt-containing medium and the plates incubated at 30°C. Molecular techniques are being employed for rapid identification of MRSA. PCR-based approaches are used to detect the *mecA* gene responsible for methicillin resistance. As many coagulase negative staphlyococci also possess this gene multiplex PCRs that detect *mecA* and coagulase genes are employed. A simple slide agglutination technique uses a monoclonal antibody to PBP2 and this together with a coagulase test is adequate to confirm a diagnosis of MRSA.

S. aureus with intermediate glycopeptide sensitivity cannot be distinguished from susceptible strains by disc diffusion methods. Broth minimal inhibitory concentration (MIC) and E-test are satisfactory methods.

Laboratory tests used to identify *S. aureus*

1 Slide coagulase test.
2 Tube coagulase test.
3 Tests for deoxyribonuclease.
4 Slide agglutination tests to detect PBP2s.

Management

Impetigo is an extremely superficial infection, which responds readily to modest doses of antibiotics. In British practice oral antistaphylococcal agents flucloxacillin or cloxacillin are usually given, and these are also effective in the minority of impetigo cases caused by *Streptococcus pyogenes*. In penicillin-allergic individuals oral cephalosporins may be given (unless anaphylaxis was the problem, in which case there is approximately a 10% chance of anaphylactic reaction to cephalosporins also). Trimethoprim is also likely to be effective. For methicillin-resistant staphylococci, treatment must be determined by sensitivity testing of the organism.

Topical antimicrobial agents may also be effective, but carry a risk of hypersensitivity, especially with topical penicillins and neomycin. Aminoglycosides may be absorbed from the lesions, with risk of toxicity if used on large areas. Tetracyclines and fusidic acid may be effective, but can encourage the emergence of resistant organisms. The only topical antimicrobials unrelated to commonly used systemic agents (and therefore without risk of devaluing them by encouraging resistance) are mupirocin and silver sulphadiazine. Mupirocin ointment is effective against many bacteria, and promotes healing of superficial infected lesions. It is particularly useful when intolerance or hypersensitivity limits the value of systemic agents. *S. aureus* can develop mupirocin resistance, so prolonged or extensive use is not recommended. Silver sulphadiazine also has a wide spectrum and is used for prophylaxis and treatment of infection in burns.

Treatment of staphylococcal skin infections

First choice
Oral flucloxacillin 250 mg or cloxacillin 500 mg 6-hourly for 5–7 days.

Alternatives
1 Oral cephalosporins 250–500 mg 6–8-hourly (see data sheet) for 5–7 days.
2 Trimethoprim 200 mg 12-hourly for 5–7 days.

Mild furunculosis, with small or moderate pustular lesions, will often heal spontaneously if good skin hygiene is maintained. Large lesions and carbuncles are painful and may be accompanied by spreading inflammation and fever. They should be treated with antistaphylococcal antibiotics, which may need to be given parenterally in severe cases.

Surgical management

Once an abscess has formed, healing is unlikely until the pus is discharged or drained. Continued antibiotic treatment may limit inflammation and abolish systemic features but walled-off pus sometimes remains, with risk of recurrent infection. Large pustules and abscesses should therefore be aspirated or incised. This is usually followed by rapid resolution of the lesion and relief of pain.

Surgical drainage is the most effective treatment for large abscesses and carbuncles.

Complications

Failure to respond to treatment

Skin infections acquired in hot climates may be caused by *Acinetobacter* spp. Vibrios from brackish or marine environments can cause quite severe, sometimes necrotizing, lesions. Both types of organism are resistant to antibiotics commonly used in skin infections. *S. aureus* acquired overseas, even as near as Portugal and Spain, may be resistant to a wide range of agents. Patients whose skin is frequently exposed to disinfectants, for example indoor swimmers, may be colonized with opportunist pathogens such as *Pseudomonas* spp. and develop pseudomonal furunculosis.

Scalded skin syndrome or toxic epidermal necrolysis

This condition, caused by the pyrogenic exotoxins of *S. aureus*, is also called Ritter's or Lyell's syndrome when it affects neonates. Spreading outwards from each infected staphylococcal lesion is an area of epidermal damage. Affected epidermis is loose and can be rubbed off the underlying layers simply by an examining finger (Nikolsky's sign). Early lesions look pale, and often form flaccid, shallow bullae, which may be very extensive. In severe cases the lesions become confluent and the whole surface of the skin separates, leaving a typical scalded appearance (Fig. 5.18).

Prompt treatment with antistaphylococcal agents will abort the lesions. The surface epidermal layers are shed, leaving regenerating skin areas. Severe cases must be nursed in warm, humid conditions to avoid excessive loss of heat and moisture from denuded areas. This is especially important for babies, because of their relatively large surface area.

Toxic epidermal necrolysis is an immunologically mediated skin disorder, sometimes caused by drug reactions. In such cases the lesions resemble those of scalded skin syndrome but there is no evidence of staphylococcal infection. A history of drug ingestion may be obtainable. Commonly implicated drugs include sulphonamides, sulphonylureas, phenytoin, indomethacin and allopurinol. Antistaphylococcal treatment is not indicated in these cases.

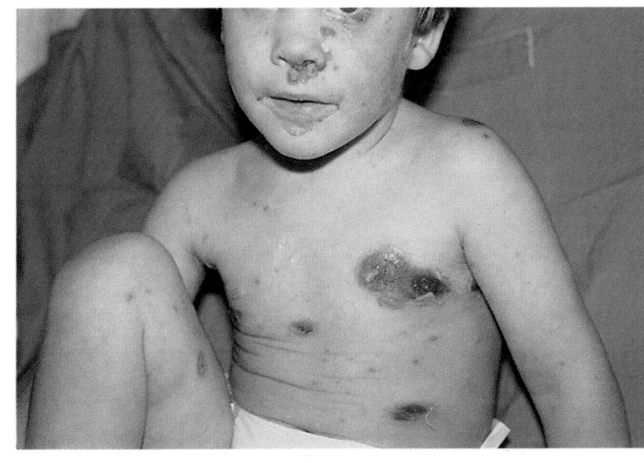

Fig. 5.18 Scalded skin appearance: the child has healing chickenpox, the lesions of which have become infected with *Staphylococcus aureus*.

Toxic shock syndrome

This is a systemic disease caused by the toxic shock syndrome toxins (TSSTs) of *S. aureus*. The patient often has a trivial staphylococcal infection, commonly a skin abscess or, in women, a vaginal infection associated with the use of highly absorbent tampons during menstruation. Occasionally, however, toxic shock syndrome complicates staphylococcal bacteraemia or endocarditis.

The main features of the illness are fever, diarrhoea, myalgia, rash, hypotension and confusion. The rash is similar to that of scarlet fever, but without the rough, punctate effect. It particularly affects the peripheries, where it may contain petechiae, and is exaggerated in the flexures. Conjunctival injection is often prominent (Figs 5.19–5.21). Accompanying laboratory findings are raised creatine kinase and other muscle enzymes, low serum calcium, mild thrombocytopenia and a white cell count which may be normal or high but usually with a predominance of neutrophils.

Definition of toxic shock syndrome

1 Temperature 39°C or greater.
2 Rash: diffuse macular erythema.
3 Desquamation: 1–2 weeks after onset.
4 Systolic blood pressure 90 mmHg or less.

Plus involvement of three or more of the following organ systems:

1 Gut—vomiting or diarrhoea.
2 Muscles—creatine kinase twice upper limit or more.
3 Mucosae—vagina, mouth, conjunctiva inflamed.
4 Renal—creatinine twice upper limit or more.
5 Hepatic—bilirubin or transaminases twice upper limit or more.
6 Platelets—100 000/mm^3 or less.
7 Central nervous system—altered consciousness without focal signs.

The diagnosis can be suspected clinically and treatment should be commenced as soon as possible. Antistaphylococcal antibiotics should be given intravenously. The treatment of choice is flucloxacillin, to which agents such as rifampicin or fusidic acid may be added, as they readily penetrate into inflamed tissues. Fluid balance and haemodynamic support are important in severe cases. Rare cases of toxic shock syndrome are caused by methicillin-resistant staphylococci. It is therefore important to seek the focus of infection and take pus or swab specimens to identify the organism and check its sensitivity spectrum.

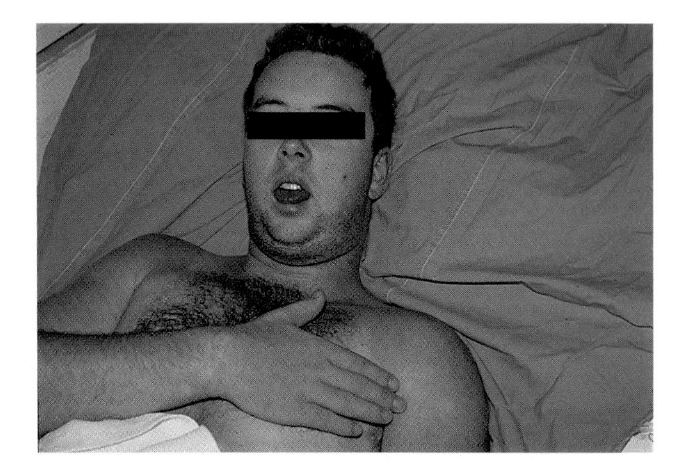

Fig. 5.19 Toxic shock syndrome: erythema and conjunctival injection in a confused and hypotensive patient (the causative infection was a small abscess on the occiput).

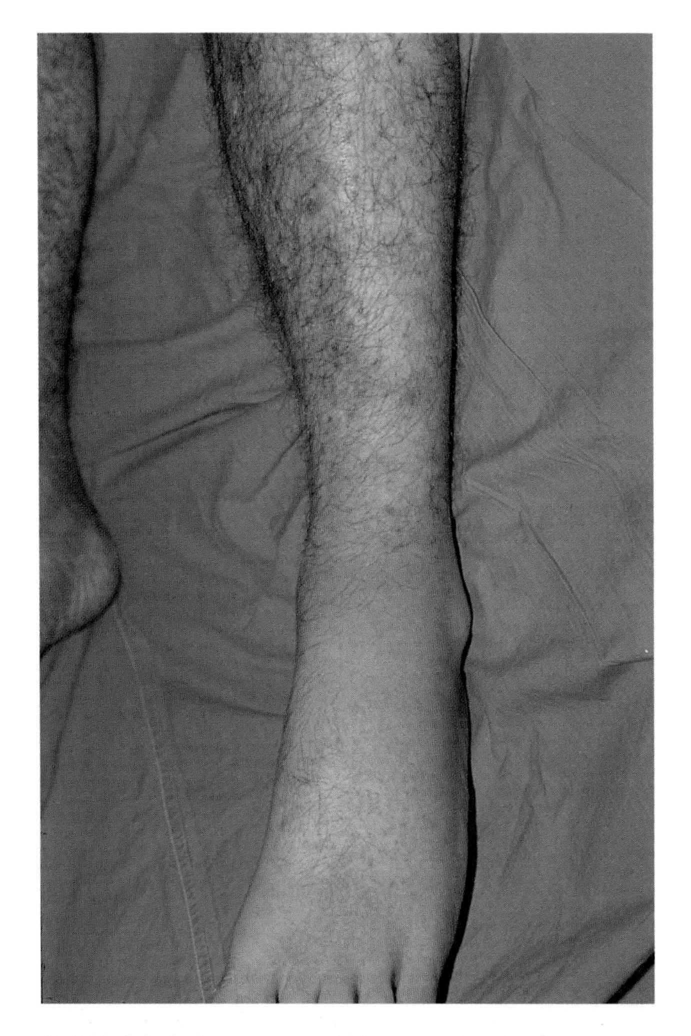

Fig. 5.20 Toxic shock syndrome: petechial component in rash on legs.

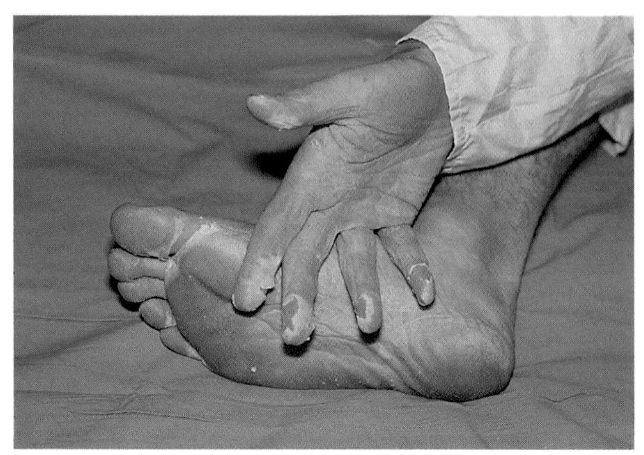

Fig. 5.21 Toxic shock syndrome: typical desquamation of the digits.

It may take some days for diarrhoea and hypotension to resolve. The erythema often heals by desquamation, with characteristic shedding of the whole nailfold and skin overlying the finger pulp.

Streptococcal impetigo can be complicated by surgical scarlet fever (see Chapter 6) and, in young children, by poststreptococcal nephritis (see Chapter 21).

Erysipelas

Introduction

Erysipelas is an intradermal infection caused by *Streptococcus pyogenes*. It is often confused with cellulitis, which is a subcutaneous infection caused by a variety of skin flora and sometimes opportunistic pathogens. The two conditions can often be clinically distinguished from one another, and this is important both in deciding on treatment and in determining which cases may have severe underlying disease, such as bacteraemia.

The origin of the infection is almost always endogenous, from the normal skin flora.

Pathology

The inflammation spreads in the layers of the epidermis, causing an expanding bleb of infection. In severe infection, fluid-filled bullae may form in the epidermal layers. Properties of *S. pyogenes* predisposing to virulence are discussed in Chapter 6.

Clinical features

Erysipelas almost always affects the face or the shin, sites

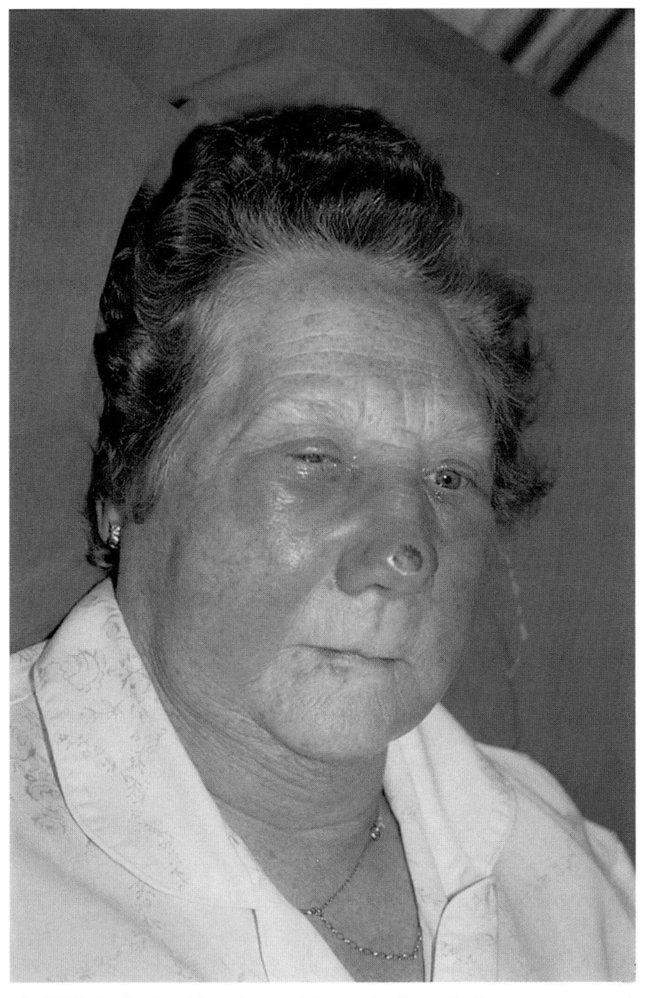

Fig. 5.22 Erysipelas: this rash spread from a tiny fissure in the nose.

where the skin is easily traumatized or fissured; a small lesion between the toes or at the angle of the nose or mouth may afford entry for the streptococci (Fig. 5.22).

Aching, throbbing or tenderness of the skin is followed in a few hours by an indurated, hot, tender, erythematous lesion which is clearly demarcated from the normal skin, both visually and by palpation. The patient can easily feel the boundary between normal and infected skin. There is a variable fever, moderate malaise and a neutrophilia with a total white cell count of $11–13 \times 10^6/l$.

If untreated, the lesion spreads rapidly. Tender enlargement of draining lymph nodes is common, and severe cases may have suppurative lymphadenitis (Fig. 5.23) with surrounding erythema and considerable pain. Breakdown of lymph nodes indicates a danger of secondary bacteraemia (Fig. 5.24). Erysipelas other than on the face or shin is extremely uncommon, and is an important marker of streptococcal bacteraemia.

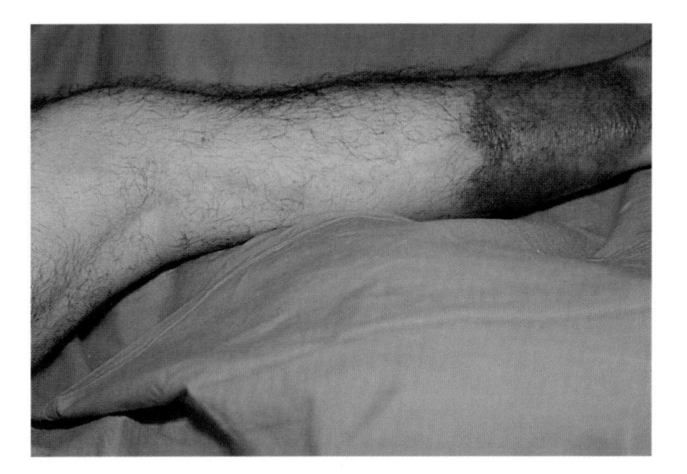

Fig. 5.23 Erysipelas: lymphangitis ascending from the skin lesion (the patient had tender, enlarged inguinal lymph nodes).

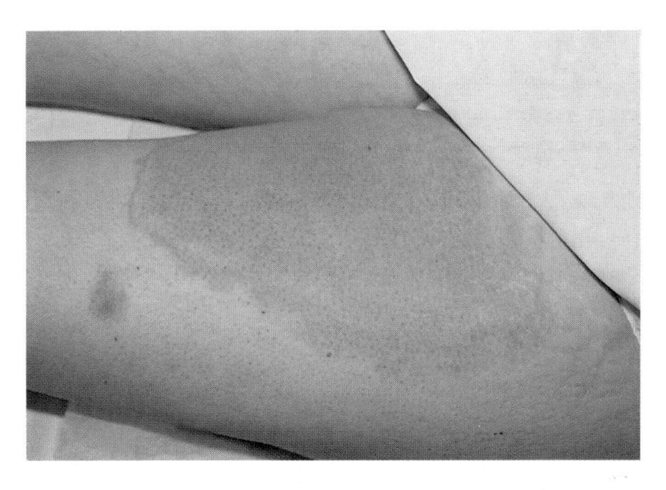

Fig. 5.24 Erysipelas: spreading infection surrounding the draining inguinal lymph node.

Diagnosis

The appearance and site of the lesions are usually diagnostic. As the infection is 'enclosed' in the epidermis, *S. pyogenes* is rarely recovered from swabs, and blood cultures are usually negative. A serological diagnosis can be made by demonstration of a rising antistreptolysin O titre (ASOT), but the rise in titre is modest or delayed in many patients. Anti-DNase or antihyaluronidase titres (reference laboratory tests) may show greater rises.

Management

Early treatment is important to limit the extension of the infection. Many cases are successfully treated in general practice with oral antibiotics such as ampicillin and flucloxacillin or with erythromycin. Failure to respond within 36–48 h should prompt admission for intravenous therapy. The treatment of choice in hospital is benzylpenicillin. Cefuroxime and erythromycin are suitable alternatives; however, some streptococci are resistant to erythromycin, which is also irritant and difficult to give intravenously.

Treatment of erysipelas
1 Oral treatment in mild cases (7–10 days): amoxycillin 500 mg 8-hourly, ampicillin 500 mg 6-hourly, flucloxacillin 500 mg 6-hourly or erythromycin 500 mg–1 g 6-hourly.
2 Intravenous treatment in severe cases (10–14 days): benzylpenicillin 2.4 g 4–6-hourly, cefuroxime 1.5 g 6–8-hourly or erythromycin 1 g 6–8-hourly.

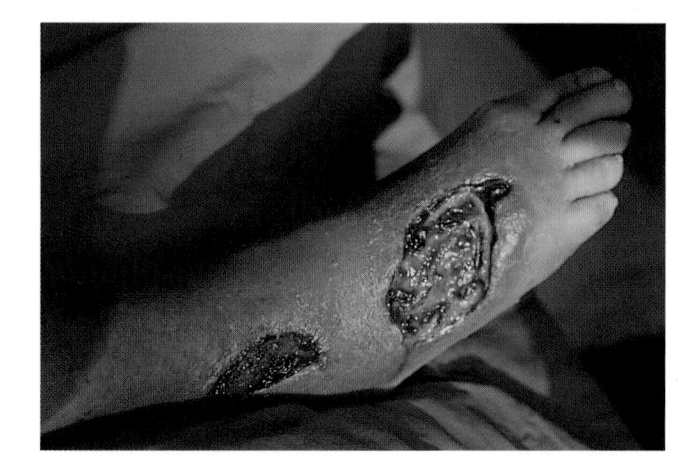

Fig. 5.25 Erysipelas: this patient made a rapid recovery on penicillin treatment, but the affected site sloughed, requiring a full-thickness skin graft.

The skin lesion usually spreads for 12–24 h after treatment is commenced, possibly because of the effect of streptococcal hyaluronidase. Thereafter, the swelling and redness subside and healing is often accompanied by desquamation of the affected skin. Five to 7 days' treatment is usually sufficient, but a few cases prove very difficult to control and may require up to 3 weeks' therapy.

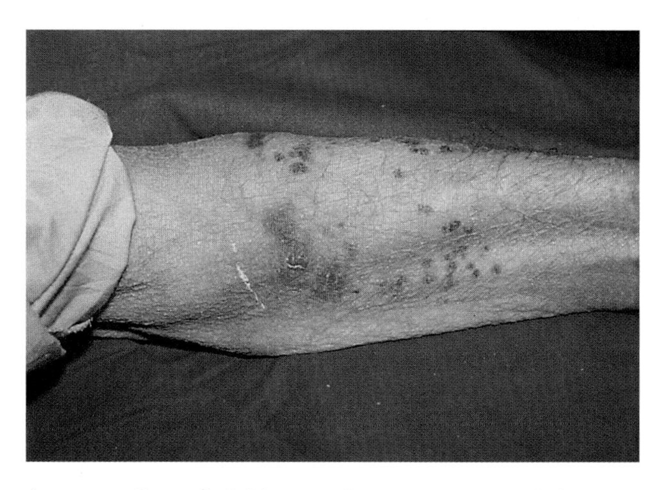

Fig. 5.26 Small area of cellulitis surrounding a venepuncture site (this was caused by a methicillin-resistant *Staphylococcus aureus*).

Complications

Complications are generally rare, the only important one being tissue damage with necrosis. This occasionally occurs even after apparently prompt and effective treatment. Full-thickness sloughing of skin may require referral for grafting (Fig. 5.25). Painful suppuration or sloughing of local lymph nodes indicates risk of bacteraemia, and deserves inpatient treatment with parenteral antibiotics.

Recurrence of erysipelas is common, and may occur weeks or months after the first attack. A few patients suffer repeated attacks, usually in the same site. Prophylactic oral penicillin or erythromycin may prevent further attacks, but is not effective in all cases. If no attack occurs after a year of prophylaxis, it may be possible to discontinue the antibiotic.

Cellulitis

Often confused with erysipelas, this is an infection of the loose subcutaneous tissue, with inflammation of the overlying skin. It often results from a penetrating injury or local lesion which allows ingress of pathogenic bacteria. In hospital practice it is a common complication of indwelling cannulation of veins (Fig. 5.26). In rare cases, apparently spontaneous cellulitis is a blood-borne condition complicating bacteraemia. Unlike erysipelas, the erythema of cellulitis has an indistinct margin; indeed its subdermal or subfascial extent may be much greater than the area of cutaneous erythema. While erysipelas affects the face and lower leg, cellulitis can occur anywhere.

The commonest causes of cellulitis are *Streptococcus pyogenes* and *Staphylococcus aureus*. Other organisms, such as *Pasteurella multocida*, may complicate dog or cat bites. Marine vibrios can enter via scratches or cuts from rocks and coral. Coliforms or enterococci can infect the lower limb in incontinent patients, and pseudomonads may cause hospital-associated cellulitis.

Community-acquired cases often respond well to treatment with oral cloxacillin, flucloxacillin, cephalosporins or trimethoprim. Erythromycin may be effective, but is not a reliable antistaphylococcal drug. Ciprofloxacin is likely to be effective in cases associated with unusual precipitating exposures. In hospital, treatment may need to be guided by culture and sensitivity data. Severe infection, especially in debilitated patients, may require parenteral antibiotics.

Necrotizing infections of skin and soft tissue

Introduction

In general these infections are endogenously acquired but they can also complicate penetrating skin lesions, surgical incisions and wounds and, occasionally, decubitus or diabetic ulcers.

Gas-forming infections of limited extent

These are often polymicrobial infections which arise when devitalized tissue is contaminated by facultative Gram-negative organisms (which often produce gas during carbohydrate metabolism) and various anaerobes. Enterococci or anaerobic cocci may coexist in these infections. Devitalized tissue enables the growth and multiplication of facultative organisms which further lower the redox potential allowing obligate anaerobes to multiply. There is inflammation and moderate gas formation at the affected site, but significant systemic toxaemia is rare. This type of infection is relatively common in diabetic ulcers and decubitus ulcers, and can complicate diabetic or ischaemic gangrene of the toes or feet.

Clinical examination may reveal slight crepitus in the inflamed area, and X-rays may show streaks of gas in tissue planes. In spite of these appearances, treatment with broad-spectrum antibiotics is often successful. Suitable treatment should include cover for Enterobacteriaceae, enterococci and anaerobes. Mild cases may respond to oral metronidazole plus amoxycillin or co-amoxiclav. For parenteral treatment, ampicillin, or a broad-spectrum cephalosporin, such as cefotaxime, plus gentamicin or another aminoglycoside should be combined with metronidazole, which penetrates tissues and abscess walls well, and is highly effective in anaerobic environments. Surgery, other than that demanded by the pre-existing condition (such as ray amputation in the diabetic foot or debridement of necrotic ulcers), is rarely necessary.

Gas gangrene

This is a clostridial infection of subcutaneous tissue, particularly of muscle (clostridial myonecrosis). Most cases follow the inoculation of *Clostridium perfringens* organisms or spores into a wound or incision. Clostridia, which are spore-bearing, anaerobic Gram-positive rods, are common in faeces and soil, so that wounds acquired out of doors, in wars or in field sports are at risk. Operations at or near the perineum or, rarely, gallbladder surgery can also lead to gas gangrene. Rare, apparently spontaneous cases, especially in unusual sites such as the arm or trunk, are often the result of blood-borne spread from a malignancy of the colon or genital system. Initiation of infection is also facilitated by the inoculation of foreign material, such as soil or surgical implants.

Several clostridial species are capable of causing gas gangrene. They all produce copious amounts of gas from the metabolism of either saccharides or proteins. Mainly saccharolytic organisms include *Clostridium perfringens*, *C. septicum* and *C. tertium*. Mainly proteolytic organisms include *C. oedematiens* and *C. histolyticum*.

C. perfringens, in particular, is a highly toxic organism. It produces an alpha toxin, which is a lecithinase, and strongly haemolytic; a necrotizing beta toxin; a similarly necrotizing epsilon toxin; and a theta toxin, which is strongly haemolytic, producing large clear zones around colonies on blood agar plates. Many other toxins are also produced (including a delta toxin, which causes rare cases of necrotizing jejunitis).

Clinical features

There is rapidly advancing swelling and discoloration of a limb or other affected tissue, with obvious crepitus, and the skin often contains fluid- and gas-filled blisters. The infection may produce a characteristic sickly sweet smell. The patient is feverish and hypotensive, and may also be severely anaemic.

Diagnosis

The diagnosis is usually clinically evident. The presence of gas in the tissues is easily demonstrated by clinical and X-ray examination. Microscopy of vesicle fluid or wound swabs shows plentiful Gram-positive rods with surprisingly few neutrophils. The white cell count is elevated, serum methaemalbumin is elevated and haptoglobins are often reduced. Fluid and wound cultures readily produce a growth of clostridia on anaerobic culture. Although *C. perfringens* is common, other clostridial and facultative organisms may be present.

Blood cultures should be performed, as bacteraemia can coexist.

Treatment

Treatment has three important aspects:
1 Antimicrobial chemotherapy: traditionally this is intravenous benzylpenicillin, but metronidazole penetrates tissues better and is an excellent antianaerobe drug—probably both should be given. A broad-spectrum cephalosporin can be substituted for penicillin if Gram-negative rods may be involved.
2 Surgery is important: extensive debridement or amputation may be necessary to halt the spreading infection, and to remove infected, necrotic tissue.
3 Hyperbaric oxygen therapy has been strongly advocated but has never gained a place in routine therapy. It may save critically devitalized tissue and/or make demarcation between salvageable and necrotic tissue more evident to the surgeon.
Anti-gas gangrene serum (AGGS) is available from regional pharmacies. It is directed against *C. perfringens* and its toxins, and may protect tissues from further toxic damage. Its use is unproven.

Prevention

Prevention of clostridial myonecrosis is important. Benzylpenicillin or metronidazole is given prophylactically in at-risk operations such as high amputations of the leg, and after contamination of traumatic or military wounds. Adequate cleaning of wounds, removal of debris and devitalized tissue, and avoiding primary closure of severely contaminated wounds all decrease the risk of anaerobic infection.

Necrotizing fasciitis

This is a rapidly spreading infection predominantly affecting subcutaneous and perimuscular fat. Necrotic liquefaction of fatty tissue is the characteristic pathology. It is usually caused by a mixed bacterial infection, which may include pathogens derived from the skin and bowel. Fournier's gangrene is a full-thickness necrosis of the perineal skin which can leave the testicles denuded. It has a similar aetiology. Treatment is with broad-spectrum antibiotics plus metronidazole. Surgical management with extensive debridement is almost always essential.

Rare cases of necrotizing fasciitis are caused by *Streptococcus pyogenes* infection. The optimum antibiotic treatment in such cases is thought to be clindamycin. Penicillin plus metronidazole is an alternative. Prompt and complete surgical debridement may be life-saving.

Acute pyomyositis

Acute pyomyositis is a rare infection usually seen in the tropics. It is a pyogenic *Staphylococcus aureus* infection of muscle frequently affecting the leg. It is difficult to treat with antistaphylococcal drugs alone. Surgical drainage and debridement are usually required.

Otitis externa

This is a superficial inflammation of the skin of the external auditory meatus. Eczema is a common precipitating condition. It presents with erythema, weeping and often severe local pain, or may be similar to impetigo or furunculosis. The poorly ventilated auditory canal can be colonized not only by common skin-infecting bacteria but also by fungi, including *Aspergillus* spp. Swimmers and divers, whose ears are often wet, are prone to pseudomonal otitis externa caused by water-borne organisms.

Simple cleaning and drying of the ear canal under direct vision may be sufficient to allow resolution. Short courses of corticosteroid drops may eradicate the eczematous reaction. Mild infection is often treated topically with neomycin, framycetin or clioquinol drops. Such preparations should not be given in courses lasting longer than 7 days, as they can predispose to both local sensitization and the establishment of fungi. Topical clotrimazole is useful if fungal infection is present. Staphylococcal infection is best treated with oral cloxacillin or flucloxacillin, as for other skin infections.

Erythema chronicum migrans (cutaneous Lyme disease)

Erythema chronicum migrans (ECM) is the cutaneous manifestation of early Lyme disease, caused by tick-borne borrelias (*Borrelia burgdorferi, B. afzelii, B. garinii* and probably others: see Chapter 25). It consists of a circular or discoid lesion which begins and expands from the site of an infecting tick bite (Fig. 5.27). Approximately 75% of individuals with early localized infection develop ECM. Some lesions are large, disappearing when they have traversed a whole limb, while occasional patients have atypical or multiple lesions.

Typical ECM is pathognomonic of Lyme disease. At this stage, just over half of patients have IgM antibodies to *B. burgdorferi* in the serum. However, false positives are relatively common, and false negatives sometimes occur, particularly after early treatment. Western blotting will confirm true-positive antibody reactions. A firm diagnosis

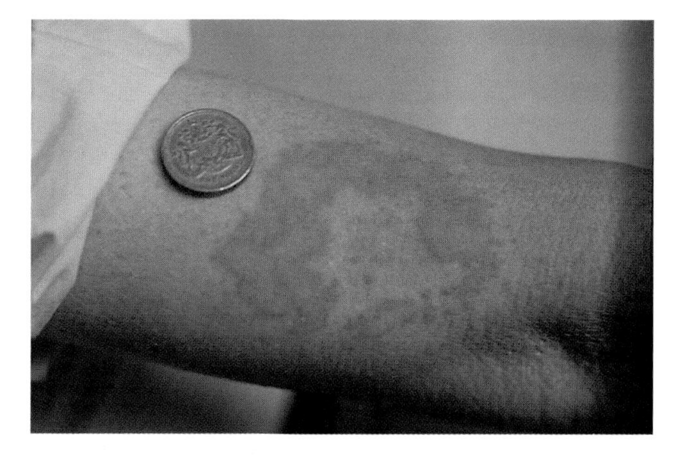

Fig. 5.27 Erythema chronicum migrans expanding from the site of a tick bite on the arm. Courtesy of Dr M. G. Brook.

should therefore rest on a combination of clinical and serological evidence. *B. burgdorferi* can be cultured from blood or biopsies of lesions, but this requires experience and is most reliably performed in reference laboratories. *B. burgdorferi* DNA can be demonstrated in lesions by PCR.

Treatment is important. Although untreated ECM is self-limiting, this leaves the risk of later manifestations of Lyme disease (see Chapter 25). The treatment of choice is oral doxycycline 200 mg, followed by 100 mg daily for 3 weeks. Erythromycin is a second-choice alternative. A small risk of continued infection remains after this treatment. This possibility must be excluded if the patient suffers later, systemic symptoms.

Erythrasma

This is a superficial inflammation of the skin, usually of the flexures, caused by *Corynebacterium minutissimum* infection. The advancing flexural erythema can be mistaken for a fungal infection or for erysipelas, but it is not painful, and antifungal treatment is ineffective. When illuminated by ultraviolet light the lesion shows characteristic, salmon-pink fluorescence.

C. minutissimum may be distinguished from *C. jeikeium* and *C. bovis*, which also do not utilize nitrate, hydrolyse urea or digest gelatin, by its lack of dependence on lipid in the culture medium. It produces small colonies which fluoresce red-orange under Wood's light in serum-containing medium.

Erythromycin treatment will cure most cases; the organism is also sensitive to tetracycline. Toeweb infections may fail to respond to oral antibiotics, but can often be eradicated by topical treatment with compound benzoic acid ointment.

Erysipeloid

Erysipelothrix rhusiopathiae is a zoonotic organism which causes erysipeloid in pigs. Human infections result from inoculation injuries, such as puncture wounds from bone splinters. A dull red erythema advances, often spreading from one finger to another via the web. Underlying joints may become sore. Systemic manifestations rarely occur.

Clinical diagnosis is usually possible from the history of exposure (often occupational) to a source of infection, and from the clinical features. The infection responds rapidly to 5–7 days' treatment with oral penicillins or tetracyclines.

E. rhusiopathiae is a facultative, catalase-negative, non-sporing Gram-positive rod, which produces alpha haemolysis on blood agar. Unlike *Listeria*, it is non-motile and it produces hydrogen sulphide in Kligler's triple sugar–iron medium. The mechanisms of pathogenesis are uncertain but are thought to be related to neuraminidase production.

Cat-scratch disease

This is a lymphocutaneous disease caused by *Bartonella henselae*. The patient usually has a history of cat scratch or cat exposure. After 5–10 days' incubation, a nodular or indurated swelling appears at the site of the scratch, and may discharge a little pus. The local-draining lymph nodes enlarge and become tender, occasionally suppurating and discharging.

Histopathological examination of affected lymph nodes shows a mixture of non-specific inflammatory reactions including granulomata and stellate necrosis. Bacilli may be demonstrated by Warthin–Starry silver staining (Fig. CS. 1, p. 109) and less effectively by Gram staining. The causative organism *Bartonella henselae* has been isolated in culture. Freshly prepared brain–heart infusion, agar containing 5 or 10% rabbit or horse blood, should be used and the plates should be incubated in a humid atmosphere for up to 3–4 weeks. *Bartonella* spp. grow best on semisolid media and do not produce turbidity or convert enough carbon dioxide for ready detection in automated systems. Colonies on blood agar are pleomorphic. PCR-based techniques are now sufficiently sensitive to be very useful in diagnosis.

Immunofluorescence and enzyme immunoassay techniques for the detection of IgM and IgG antibodies to *Bartonella henselae* are available through reference laboratories.

The disease is self-limiting, but may last for 3 weeks or more. Treatment with oral tetracycline or erythromycin may shorten the course.

B. henselae can cause systemic or bacteraemic infection, pelious hepatitis or indolent dermatoses in the immunosuppressed (see Chapter 22).

Actinomycosis

This is an infection of skin and subcutaneous tissue caused by *Actinomyces israelii* or, rarely, by other *Actinomyces* spp. The organism may invade from underlying mucosa, from the mouth, the pleura or the peritoneum. The commonest lesion is an abscess of the cheek.

The lesion begins as a hard, enlarging, nodule which suppurates and discharges greyish pus containing tiny pale or yellow dots. These 'sulphur granules' are spherical colonies of the branching bacteria; they can be crushed between a slide and a cover slip and stained to demonstrate their variable degree of Gram positivity (Fig. 5.28). The organisms can be isolated by anaerobic culture but the laboratory must be informed of the suspected diagnosis as the period of incubation should be extended to at least 2 weeks. Isolation allows accurate speciation and sensitivity testing. The *Actinomyces* sp. is usually accompanied by coexisting bacteria, such as *Actinobacter actinomycetecomitans* or Gram-negative bacteria of intestinal origin, which may require concurrent treatment with additional antibiotics to ensure eradication of the infection.

Penicillin or erythromycin are effective in eradicating the infection, but high doses are needed and the course should last for 6 or 8 weeks. Co-amoxiclav is a useful monotherapy for *Actinomyces* spp. and the combination of organisms often found in lesions.

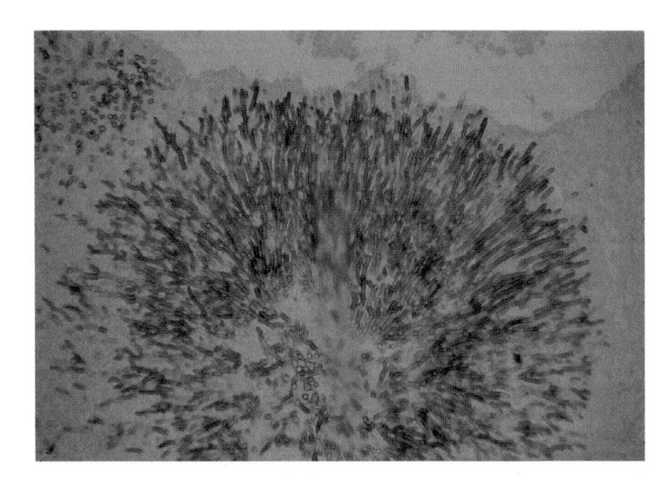

Fig. 5.28 Gram-stained crush preparation of 'sulphur granule', showing a Gram-variable, branching appearance.

Cutaneous mycobacterial infections

Some 'environmental' mycobacteria can cause skin infections which fail to respond to simple antibiotic treatment and which seem sterile on standard bacterial culture.

Mycobacterium marinum is found in pools, rivers and aquaria. Infection, probably of small skin defects, produces indolent, nodular lesions, almost always on the hand. These can spread to affect subcutaneous tissue, fascia and tendons. Extensive lesions may ulcerate and/or discharge pus. Mycobacteria can be seen and cultured in curettings or biopsy material (see Chapter 18). Treatment with rifampicin plus ethambutol or clarithromycin is often successful; debridement is sometimes required.

M. chelonei and *M. fortuitum* occasionally infect inoculation sites, for instance in diabetics taking insulin. The infection usually causes a 'sterile' subcutaneous abscess. Mycobacteria are recovered by culture of pus. Treatment with co-trimoxazole or ciprofloxacin is often curative without the need for drainage.

M. tuberculosis causes lupus vulgaris, a granulomatous lesion which appears slightly nodular, lichenified and sometimes scaling, often with an atrophic centre. It may have a natural history of years, leading to misdiagnosis as chronic dermatitis. Pressing a glass slide on the lesion to blanch it may reveal the granulomata as translucent granular structures heaped together (called the apple-jelly appearance). Mycobacteria may be demonstrable in, and culturable from, biopsy material.

Many patients also have pulmonary or other foci of tuberculosis. Standard antituberculosis treatment will cure the skin lesion as well as any other focus.

Propionibacterium acnes and acne

Acne is a multifactorial skin disorder, in which excessive sebaceous secretions are produced in response to strong androgenic stimulation. The sebaceous glands become engorged and blocked, causing pustules and comedones. Secondary infection may cause severe inflammation and contribute to later scarring. *P. acnes* can be recovered from the lesions. It is not known what contribution the organism makes to the pathology of acne, or whether other skin flora are also involved, but broad-spectrum antibiotics can control the condition to a large extent. Doxycycline is often given once daily in courses of several months. Erythromycin is also effective in many cases.

Fungal infections of the skin and its appendages

ORGANISM LIST

Candida albicans
Dermatophytes
Microsporum spp.
Trichophyton spp.
Epidermophyton spp.
Malassezia furfur
Rarities
Sporothrix schenckii
Blastomyces
Histoplasma capsulatum
Cryptococcus neoformans

Introduction

The fungal infections (mycoses) commonly seen in Europe are superficial mycoses, affecting only the skin and causing superficial inflammation confined to the site of infection. They are usually recognizable by their typical skin lesions, and the organisms are easily identified in swabs or scrapings.

The deep mycoses are rare in temperate countries. Although some are respiratory infections (see Chapter 7), several are infections of subcutaneous tissue, acquired by inoculation through the skin. They tend to produce granulomatous lesions which sometimes invade, either by spread to adjacent tissue or by metastasis to lymph nodes and other body sites. The lesions of deep mycoses must be distinguished from infectious and autoimmune granulomata and from tumours.

Candidiasis

Introduction and epidemiology

Candida albicans is a yeast which is part of the normal flora of the skin, mucosae and bowel. Its balance with other flora and the health of the tissues is important in preventing superficial invasion and infection. Normal skin is rarely affected by candidiasis, but it requires only wetting and slight maceration of the epidermis to allow the establishment of replicating organisms. Antibiotic treatment increases the likelihood of candidiasis. Mild degrees of immunosuppression, including the effects of corticosteroid or cytotoxic therapies, pregnancy, diabetes and other endocrine diseases, can all predispose to candidiasis.

Predisposing conditions to candidiasis
1 Antibiotic treatment.
2 Corticosteroid treatment.
3 Cytotoxic therapy.
4 Diabetes mellitus.
5 Pregnancy.
6 Cell-mediated immune deficiency.

Clinical features

The warm, moist areas of the folds under the breasts, in the natal cleft or under the abdominal 'apron' of the obese are most often affected. Candidiasis is a common infection of the perineum of infants or incontinent adults. Paronychia can affect individuals who constantly wear rubber gloves for washing-up or other tasks, and whose hands are always sweaty or damp.

Skinfold infection (intertrigo) produces reddening and slight thickening of the skin, causing plaque-like lesions which are clearly demarcated from adjacent, normal skin. Moist lesions are dull and may produce a slight, sticky exudate. Drier lesions often appear shiny or flaky, may have circular satellite lesions nearby, and must be distinguished from psoriasis. This type of lesion is common in napkin rash. Both types of lesion are often irritating or itchy.

Nailfold lesions (paronychia) cause bolster-like swelling, with thickened rolls of skin which may bulge over the nail. A 'cheesy' exudate is sometimes seen in the cleft under the swelling. Inflammation of the nailbed causes ridging of the nail, and in rare cases infection of the nail itself produces an opaque greenish or brownish discoloration.

Diagnosis

The site of the lesions and the cheesy exudate, if present, strongly suggest candidiasis. The differential diagnosis includes erythrasma, dermatophyte infections, contact dermatitis and flexural psoriasis. Swabs from lesions or exudate can be Gram stained to demonstrate the diagnostic presence of budding yeasts. Inoculation on to Sabouraud's agar will allow cultural identification of yeasts in cases of doubt. The organism will also grow on blood agar or heated blood agar, but produces tiny colonies which are easily overlooked in mixed cultures of skin organisms. The use of chrome agar allows *C. albicans* to be speciated rapidly as it grows with a characteristic green colour on this agar. Other useful identification tests include germ tube formation, specific for *C. albicans*, and auxanograph testing, which allows speciation of most isolates.

Management

Removing the predisposition, when possible, will assist treatment and reduce recurrences. Most candidal infections respond readily to topical treatment. Water-soluble creams are recommended for the skin, as they do not have the occlusive and macerating effect of ointments. The polyenes nystatin and amphotericin are effective and cheap. The imidazoles, miconazole, clotrimazole and econazole are also effective; the last two are available as sprays, solution or lotion for application to large or hairy areas.

In severe or extensive disease, or persisting predisposition, topical treatment may fail. The orally administered triazoles, itraconazole or fluconazole may be effective in these cases. They are contraindicated in liver disease and should be given with caution during pregnancy and lactation; they have important interactions with terfenadine.

Treatment of severe *Candida* infections of skin or mucosae

1 Itraconazole 100 mg daily for 15 days; or
2 Fluconazole 50 mg daily for 14 days.
(Up to 4 or 6 weeks for foot infection or severe intertrigo.)

Dermatophytoses (tinea infections)

Introduction

Dermatophytes are filamentous fungi that digest keratin. Different species have varying affinities for skin, hair and nails. Although they cannot invade living tissues, their presence in the epidermis can induce an inflammatory reaction in the affected site. Some species are exclusive to humans; others are acquired by contact with infected animals, and these often cause the most severe inflammation.

ORGANISM LIST

Organism	Host	Fluorescence
Scalp infections		
Microsporum canis	Cat, kitten, dog	Positive
M. audouinii	Human	Positive
Trichophyton sulphureum	Human	Negative
T. violaceum	Human	Negative
T. schoenleinii	Human (favus)	Positive
Body infections		
T. mentagrophytes	Animal	Negative
T. verrucosum	Animal	Negative
Groin and foot infections		
T. rubrum	Human	Negative
T. interdigitale	Human	Negative
Epidermophyton floccosum	Human	Negative
Nail infections		
T. rubrum	Human	Negative

Laboratory identification of dermatophytes

Scrapings of infected skin, hair and clippings from nails should be sent dry to the laboratory. These are clarified by gently heating in a solution of potassium hydroxide, and examined under the microscope for the presence of the typical branching hyphal elements.

Dermatophytes grow readily on many microbiological media but a useful selective medium is Sabouraud's dextrose agar. As all are resistant to the action of cyclohexamide, this is incorporated as a selective agent. Chloramphenicol and gentamicin can be added when bacterial contamination is likely. A specialized dermatophyte test medium incorporates all three of these agents together with an indicator which detects the rise in pH which occurs when dermatophytes grow. Cultures are incubated at 30°C for up to 4 weeks.

The identification of dermatophytes is based on colonial morphology, on microscopic appearance of the fungal hyphae and conidia, and on physiological and biochemical testing. Slide preparations of mycelia can be stained with lactophenol cotton blue and examined microscopically for the morphology of the conidia and chlamydospores. These structures are often characteristic for different species and some examples are seen in Fig. 5.29. Other tests which may be employed include the ability of the fungal isolate to penetrate an uninfected hair, to hydrolyse urea and to produce characteristic growth on rice grains.

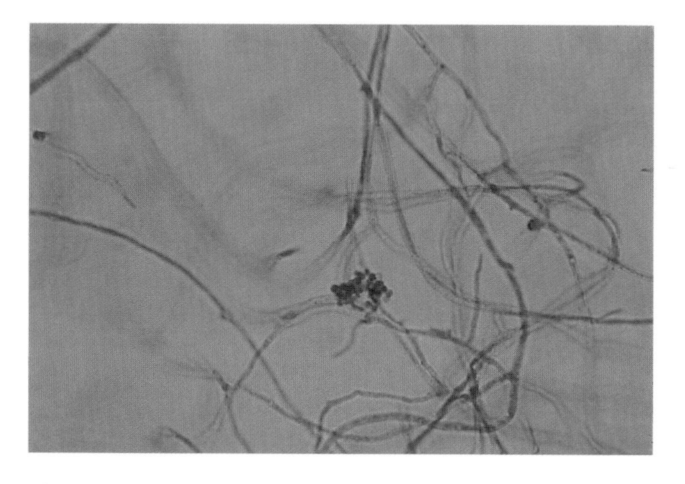

Fig. 5.29 Lactophenol blue-stained preparations of dermatophytes, showing the characteristic morphology of conidia.

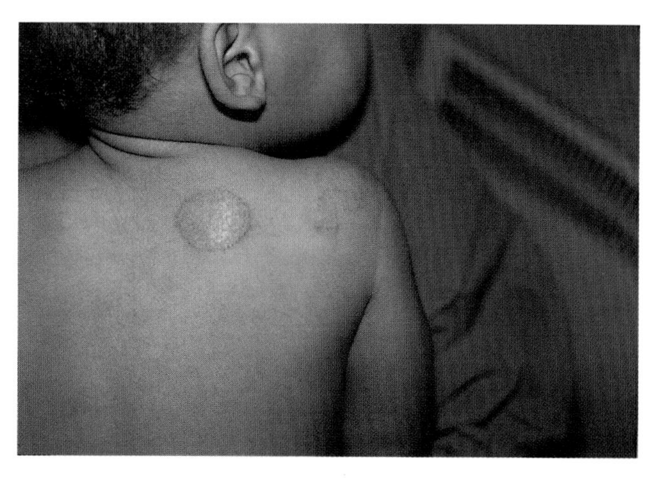

Fig. 5.30 A typical lesion of tinea corporis (the family cat was also affected).

Clinical features

The commonest manifestation of dermatophyte infection is an expanding lesion with a scaly or inflamed advancing edge. This is a typical tinea or ringworm infection. The eruption is usually described by its position on the body, e.g. tinea capitis, tinea corporis or tinea cruris.

Scalp ringworm (tinea capitis)

This is common in children, and is usually caused by a *Microsporum* species. It presents as one or more oval patches of hair loss, which expand steadily and can affect the whole of the scalp. The hairs are damaged, and broken off near to the skin (unlike alopecia, in which there is no scaling, and hairs are absent in the acute stage).

Trichophyton species can infect both adults and children. Swelling of the scalp is often marked and hairs broken off at the skin surface may appear as black dots. Severe inflammation may cause a purulent exudate from the follicles, causing hairs to be completely shed. This terminates the infection, but can leave scarring of the scalp and follicles, with permanent hair loss.

Diagnosis can be made presumptively by clinical features. *Microsporum* infections cause a greenish-blue fluorescence of the affected hairs and skin under ultraviolet light (Wood's light). Scales can be gently scraped off and hair stumps removed by plucking, and both examined by microscopy and culture.

Topical treatment rarely succeeds. The treatment of choice for children is oral griseofulvin which is well absorbed and concentrated in keratin. At least 2 months' treatment is necessary; the need for further treatment may be reduced by applying miconazole or clotrimazole cream. Griseofulvin enhances the effect of alcohol and increases the degradation of warfarin and oral contraceptives. Women should avoid conception for 1 month and men for 6 months after therapy. Oral itraconazole, fluconazole or terbinafine are licensed for treating adults.

Reinfection can be prevented by seeking and treating any reservoir of infection, which may be another child or adult, a family pet or a farm animal.

Ringworm of the body (tinea corporis)

This is usually an obvious round or oval expanding lesion with a scaly, slightly inflamed periphery (Fig. 5.30). It must be distinguished from discoid eczema and isolated lesions of psoriasis. Examination and culture of scrapings will confirm the diagnosis.

Treatment with topical antifungals such as clotrimazole or miconazole cream is adequate for most mild lesions. Very extensive or severely inflamed lesions may be best treated with a 3- or 4-week course of oral griseofulvin. Oral itraconazole is effective in a dose of 100 mg daily for 15 days.

Ringworm of the groin (tinea cruris)

This is often intertriginous, affecting the inguinal folds and adjacent thighs. *T. rubrum* infection, however, can cause extensive lesions spreading down the thighs and posteriorly to the natal cleft and the buttocks. Inadvertent treatment with topical corticosteroids partly inhibits the inflammation causing the lesion margin to disappear, but papulopustular lesions often develop. This modified disease is often called tinea incognita. It can also occur on the face in similar circumstances. Microscopy and culture of scrapings will make the diagnosis. The important differential diagnoses are intertriginous candidiasis or psoriasis, and erythrasma.

Topical antifungal treatment is effective, but often causes skin irritation. If this happens, or the skin is intensely inflamed at presentation, dilute potassium permanganate soaks can be applied twice a day for 3 or 4 days before commencing a weak corticosteroid cream and topical antifungal. The steroid can be discontinued when the inflammation subsides, and the antifungal continued as required. Oral itraconazole offers an alternative approach, especially in extensive lesions. The dose is as for tinea corporis.

Ringworm of the hands, feet and nails

Dermatophyte infections of thick keratin often produce extremely scaly lesions which resist treatment with topical antifungals. The interdigital webs may become fissured and macerated, with scaly infection extending on to the digits or the dorsum of the foot (or hand). *Trichophyton* spp. can cause extensive infection, resembling the shape of a slipper on the dorsum of the foot. Nails become opaque, discoloured and brittle, starting at the tip, gradually affecting the lateral margins and then the whole nail plate, which may flake away. The nailfolds do not swell, as they do in candidiasis. The main differential diagnoses are psoriasis and contact dermatitis.

Treatment must be systemic and prolonged. Terbinafine in 2–4-week courses of 250 mg daily is effective in treating palmar and plantar disease, as well as early nail infections. It is well tolerated, apart from occasional gastrointestinal side-effects. Rare allergic rashes and liver toxicity have been reported. Itraconazole 100 mg daily for 30 days may be effective in treating the hands or feet; for toenail infections the dose is 200 mg daily for 3 months (pulses of 200 mg twice daily for 7 days can be used, repeated after 3 weeks; two pulses for fingers, three for toes). Oral griseofulvin for 6 months is the minimum for nail infections; 12 months or more is usual for toenails. As griseofulvin enhances the ill-effects of alcohol many people prefer not to take it for toenail infections.

Treatment for fungal infection of the nail

1 Terbinafine 250 mg daily for 6 weeks–3 months (not recommended for children).

2 Itraconazole 200 mg daily for 3 months *or* 200 mg twice daily for 7 days, repeated after 21 days; two repeats for fingers, three for toes.

3 Griseofulvin 500 mg (child 10 mg/kg) daily for several weeks or months until cure is complete. Beware of its Antabuse-like effect.

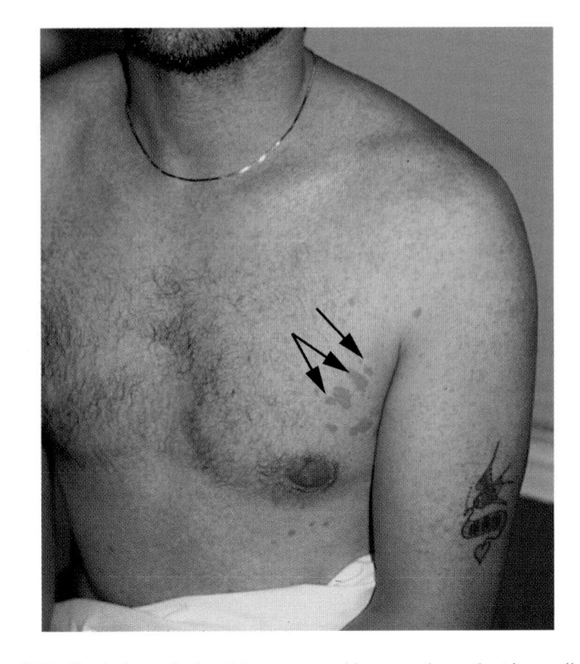

Fig. 5.31 Pityriasis versicolor: this appears red because the patient has a slight fever.

Pityriasis versicolor

This is a very superficial skin infection caused by the filamentous fungus *Malassezia furfur*. Pale brown, fine scaly macules develop on the upper chest or back, forming an irregular pattern which appears slightly brown in a white-skinned person, or slightly pale in a dark skin. There is little or no inflammation, and sensation is normal in the affected areas. When the skin is warm after a bath or during a feverish illness, the plaques may become red or pink (Fig. 5.31).

The clinical appearance is characteristic, but the diagnosis can be confirmed by microscopy. Skin scrapings are placed on a microscope slide, and mixed with a drop or two of 5% potassium hydroxide. This reveals a mycelium in which are scattered groups of rough, round sporing bodies. This typical 'meat balls and spaghetti' appearance is diagnostic (Fig. 5.32).

Almost any topical antifungal cream will clear the lesions, but recurrence is common. Washing all clothes and shampooing the hair with selenium sulphide (Selsun) shampoo may help to remove a reservoir of infection. Oral itraconazole 200 mg daily for a week is also effective.

Sporotrichosis

This is a localized, nodular skin infection caused by

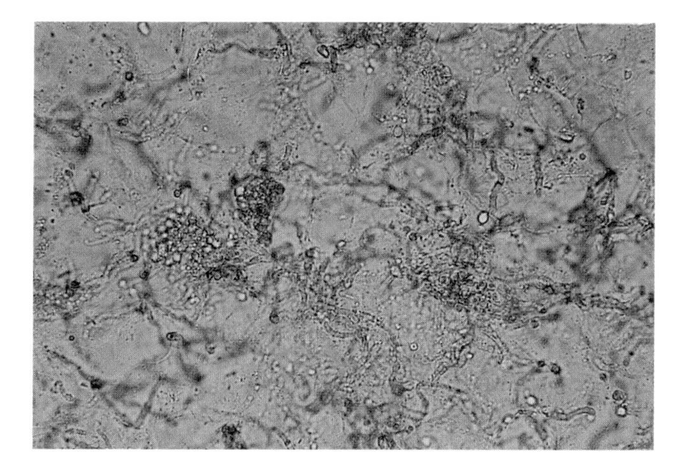

Fig. 5.32 Potassium chloride preparation of scraping from a case of pityriasis versicolor (stained with blue ink (Quink)).

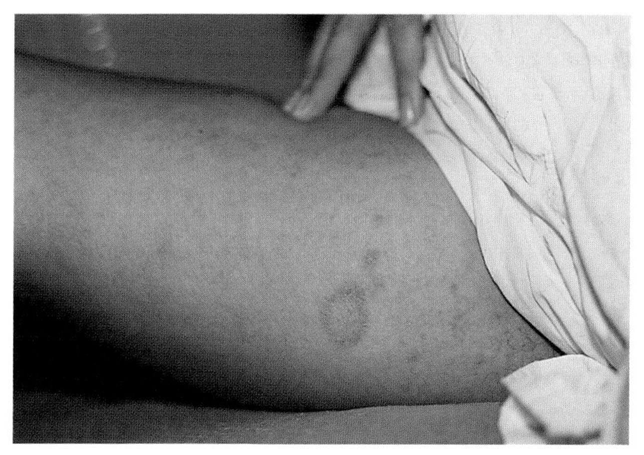

Fig. 5.33 Pityriasis rosea: herald patch on the thigh.

Sporothrix schenckii, a ray fungus which produces characteristic stellate microcolonies *in vitro*. The fungus exists in wood, soil and vegetation. Infection is usually by inoculation, and leads to nodules, abscesses or ulcers which may expand locally. Satellite lesions may appear along lymphatic pathways, and draining lymph nodes may be affected. Haematogenous spread occasionally occurs in debilitated or immunosuppressed individuals.

Differential diagnoses include mycobacterial infections (fishtank granulomata) or rare syphilitic lesions. Histology and culture of biopsy material are diagnostic.

The traditional treatment of sporotrichosis has been oral potassium iodide for courses of several weeks. Amphotericin is also effective but, because of its toxicity, is reserved for severe spreading or systemic disease. Results of treatment with itraconazole or ketoconazole have yet to be critically reviewed.

Rarities

Fungal infections of the skin are more common in the tropics than in temperate countries. Several exotic fungi can produce nodular or granulomatous lesions on the skin, typically of the lower legs. Diagnosis is by histology and culture of biopsy material. A positive complement-fixation test is usual in histoplasmosis, which is often a systemic disease with risk of lymph-node, buccal mucosa and lung involvement. Expert advice should be sought about management, which may be difficult and is always prolonged.

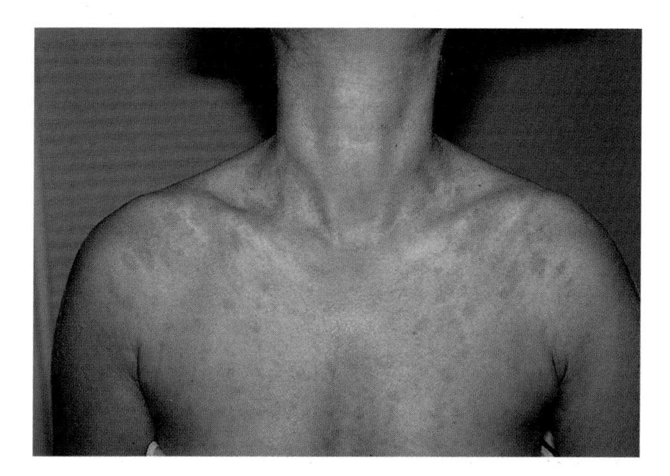

Fig. 5.34 Pityriasis rosea: typical 'vest and pants' rash, with 'Christmas tree' orientation of lesions.

Pityriasis rosea

This is a disease of unknown aetiology, but which has recently been associated with seroconversion to human herpesvirus type 7 (HHV-7). The first sign is the 'herald patch', an inflamed, scaly, rather scabby lesion up to 4 cm in diameter on the trunk or upper leg (Fig. 5.33). After anything from 3 to 14 days there follows a symmetrical eruption of oval plaques and smaller macules and papules, which are sometimes itchy. The plaques affect the trunk and proximal limbs ('vest and pants' distribution), and have their long axes aligned with the skin creases, giving the rash a typical 'Christmas tree' appearance (Fig. 5.34).

Each plaque, or medallion, has an indistinct, slightly raised margin with central arrays of pointed scales, with the points orientated towards the edge of the lesion. The lesions expand up to 5 cm long in about 2 weeks, and then fade over about 2 months.

The differential diagnosis is of tinea or discoid eczema, or occasionally syphilis. There is no specific treatment. The condition is not contagious, and second attacks are extremely rare.

Parasites of the skin

Scabies

Scabies is caused by the mite *Sarcoptes scabiei*, which burrows in the epidermis, the female mites laying eggs along their tracks. The condition is infectious by direct skin contact; mites being attracted by warmth. It is easily transmitted between sexual partners and those who share beds, or to the carers of dependent and infected individuals.

The infection itself is asymptomatic, but hypersensitivity to the mites, their eggs or surface proteins eventually causes irritation, scratching and excoriation of affected skin. The soft skin of the flexures, digital webs, perineum and axillae is most affected; the face is spared, except in small children and severe infections. So-called Norwegian scabies is caused by an aggressive strain of mite which causes severe lesions, even affecting the face in adults. Scabies can cause widespread chronic lesions in immunosuppressed HIV patients.

The diagnosis is suggested by the typically distributed, itchy rash. Burrows may be visible, with a tiny, pearly nodule at the advancing end. The nodule may be teased out and shown to be a mite. Often, however, burrows are destroyed by scratching, but scrapings may still contain round, black dots of mite faeces.

Treatment is usually with topical acaricides. All members of a household should be treated, and the medication should be applied to the whole skin excluding the face. Lindane 1% lotion or cream is inexpensive; malathion 0.5% lotion or liquid may succeed if lindane fails. Permethrin 5% cream is also available, and can be applied to the face if necessary. Monosulphiram solution may be used for treating children. Most cases will respond to one application of acaricide, washed off after 12–24 h. If lindane is used, a second application should be given after 1–3 days. Itching may persist for many days after successful treatment. It is often ameliorated by topical crotamiton cream. Oral albendazole has shown useful additive effects with topical treatment in immunosuppressed patients.

Case 5.1: Complicated cat bite

History: A 31-year-old woman was bitten on the left first metacarpophalangeal joint while removing a stray cat from her garden shed. Rabies is not endemic in Britain, and she had recently had a tetanus toxoid booster immunization, so she cleaned the puncture wounds with proprietary disinfectant solution and covered the site with a waterproof adhesive dressing. Thirty-six hours later she attended the emergency department because of increasing redness, aching and irritation at the site. She was previously well, and took no regular medications.

Physical findings: The finger was swollen from the dorsum of the hand to the first interphalangeal joint, and the overlying skin was shiny and dusky red. Four puncture wounds were present on the dorsal and medial aspect of the metacarpophalangeal joint, and yellowish pus was visible in three of them. There was no associated lymphangitis or lymphadenopathy.

Questions: Are cat bites common?
Do they commonly become infected?
Which bacterial pathogens are likely to be identified?
Are any of these pathogens potentially invasive?
What antimicrobial chemotherapy can be recommended?

Management and progress: Animal bites are extremely common; over 250 000 accident and emergency consultations yearly are caused by them, and many more general practitioner consultations take place. Over 90% are for cat and dog bites. About 80% of cat bites are puncture wounds and two-thirds affect the hand. About half of dog bites are puncture wounds, the remainder are lacerations or mixed injuries, and about half affect the hand. Children are more likely to suffer severe bites, and bites on the face and head. Various studies show that 30–80% of cat bites become infected, compared with 5–20% of dog bites. A mixed aerobic and anaerobic flora can be identified in half of infected bites, in 40% only aerobic or facultative organisms are found, and the remaining infections are purely anaerobic (see Table CS.5).

Table CS.5 Bacteria commonly isolated from infected dog and cat bites.

Aerobic and facultative organisms	
Pasteurella species	*P. septica* and *P. multocida*, most often from cat bites; *P. canis*, mostly from dogs
Streptococci	*S. pyogenes*, *S. milleri* types, and viridans streptococci
Staphylococci	Half of these are *S. aureus*, probably intrinsic to the host, the remainder are coagulase-positive and -negative staphylococci of animal and human origin
Neisseria species	Several species which are pharyngeal commensals of animals
Moraxella species	Pharyngeal commensals of animals
Capnocytophaga species	Pharyngeal commensals of animals
Anaerobic organisms	
	A range of organisms including *Fusobacterium*, *Bacteroides*, *Prevotella*, *Porphyromonas* and *Propionibacterium* species of animal or human origin

Many of these organisms are sensitive to benzylpenicillin or phenoxymethylpenicillin, but *S. aureus Moraxella* and some *Pasteurella* species produce beta-lactamase enzymes. The treatment of choice for infected animal bites is therfore co-amoxiclav. Cefuroxime axetil or clarithromycin are suitable second choices.

Pasteurella multocida is capable of causing invasive infection in debilitated individuals. Bacteraemias and endocarditis have been described in patients with alcoholic liver cirrhosis, those receiving chemotherapy or antirejection medication and, rarely, as an opportunistic infection in AIDS. *Capnocytophaga canimorsus* is also recognized as potentially invasive.

The patient was given a 5-day course of oral co-amoxiclav 250 mg three times daily. She re-presented 2 weeks later, saying that after initial slight improvement, the finger had remained inflamed, and she had developed a tender swelling in the left axilla. Examination revealed improvement in finger swelling, but persistent discharge of pus from the puncture sites and a 3×2.5 cm tender, mobile lymph node in the left axilla.

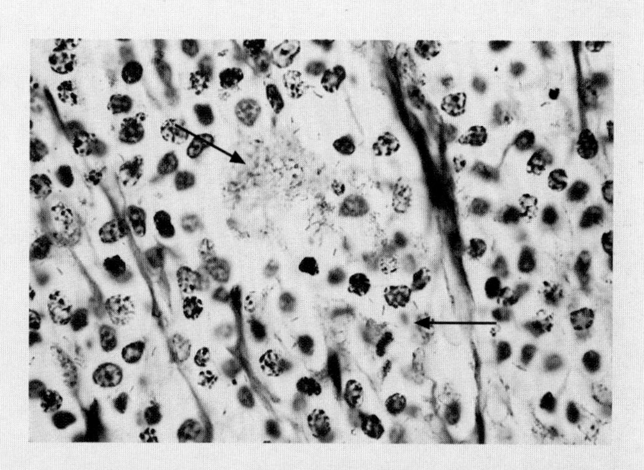

Fig. CS.1 Lymph-node biopsy examined with Warthin–Starry silver staining, showing many active cell nuclei, and groups of intracellular rod-shaped bacteria (arrows).

Questions: What less common organism could cause ulceroglandular infection?
Are debilitated and immunocompromised individuals at high risk from this infection?
What treatment is available?

Further management and progress: Lymph-node biopsy (Fig. CS.1) showed typical inflammatory changes with cords of active mononuclear cells, and intracellular rod-shaped bacteria demonstrated by silver staining (immunological staining techniques are also available). Although rare, cat-scratch disease classically follows a cat scratch or bite. It is caused by *Bartonella henselae*, an obligately intracellular Gram-negative rod, which is a member of the family of Rickettsiaceae. A serum enzyme-linked immunosorbent assay (ELISA) test, available from reference laboratories, is positive in over 85% of infected patients. Ulceroglandular tularaemia, caused by *Francisella tularensis*, can cause a similar presentation. It results from rodent exposure and occasionally from cat bites in wooded or forested terrain, but is rarely seen in Britain. It tends to affect hunters and trappers, particularly in North America.

B. henselae infection can cause indolent skin and

reticuloendothelial disease in the immunosuppressed, and has particularly been described in AIDS, in which an angiomatous dermatitis or pelious hepatitis may occur.

This patient was treated with doxycycline 100 mg daily for 2 weeks, resulting in gradual healing of the skin lesions. The lymphadenitis resolved over the following 3 weeks. Serological tests showed antibodies to *B. henselae* in a titre of 1:256, considered to indicate recent infection.

6 Upper Respiratory Tract Infections

Introduction, 111

Conjunctivitis and keratoconjunctivitis, 111
- Introduction, 111
- Clinical features, 112
- Diagnosis, 112
- Management, 112
- Herpes simplex keratitis (dendritic ulcer), 112
- Bacterial conjunctivitis, 113
- Neonatal ophthalmia, 113
- Chlamydial conjunctivitis and trachoma, 113
- Amoebic keratoconjunctivitis, 114

Infections of the middle ear, 114
- Acute otitis media, 114
- Paranasal sinusitis, 115

Viral infections of the throat and mouth, 115
- The common cold (coryza), 115
- Enteroviral pharyngitis, 116
- Adenoviral sore throats and pharyngoconjunctival fever, 117
- Infectious mononucleosis, 118

Bacterial throat infections, 120
- Streptococcal tonsillitis, 121
- Acute epiglottitis, 125
- Diphtheria, 127
- Ludwig's angina, 130
- Retropharyngeal abscess, 130
- Vincent's angina, 131

Introduction

The upper respiratory tract comprises the conjunctiva, nose, paranasal sinuses, middle ear, nasopharynx, oropharynx and laryngopharynx. It is largely covered with ciliated columnar epithelium. Exceptions are the oropharynx, vocal cords and upper posterior epiglottis which are lined with stratified squamous epithelium. The conjunctiva is also stratified squamous epithelium, continuous with and similar to the epithelium of the cornea. Squamous epithelium is also found in the mastoid antrum of the middle ear.

The adenoids and tonsils are important structures of the upper respiratory tract. They are lymphoid organs whose surfaces are marked by many deep clefts, both macroscopic and microscopic.

The whole upper respiratory tract is colonized by a variety of normal flora.

Normal upper respiratory tract flora

1. *Streptococcus pneumoniae*.
2. Anaerobic and microaerophilic streptococci.
3. *S. 'milleri'* (found in the sinuses).
4. *Haemophilus influenzae*.
5. Other *Haemophilus* species.
6. Diphtheroids.
7. Coagulase-negative staphylococci.
8. *Staphylococcus aureus*.
9. *Moraxella catarrhalis* and *Neisseria* spp.
10. *Prevotella melaninogenicus* and related species.

Temporary colonization of the pharynx, nose or eye by potential pathogens is also common, and may provide an important reservoir of infection, for instance with *Neisseria meningitidis* or *Corynebacterium diphtheriae*. A variety of viruses are intermittently excreted from the pharynx, including rhinoviruses, paramyxoviruses, enteroviruses, adenoviruses and myxoviruses. These are sometimes associated with symptoms, but often replicate asymptomatically.

Finally, the pharynx is a site of intermittent shedding of latent viruses. Herpes simplex virus and Epstein–Barr virus are the most important of these, but others include cytomegalovirus and possibly other human herpesviruses such as HHV-6.

The environment of the upper respiratory tract is varied; different areas are susceptible to infection with different pathogens. While most infections are of surfaces, the middle ear and the paranasal sinuses are hollow structures with narrow outlets (the ostia of the sinuses and the eustachian tubes of the middle ears) whose obstruction leads to loculated infection and abscess formation. The soft tissues of the fauces, surrounding the tonsils, are susceptible to abscess formation if severely inflamed.

Conjunctivitis and keratoconjunctivitis

Introduction

The conjunctiva is often inflamed during infections of the respiratory tract, such as colds, influenza and measles. It is also exposed to many air-borne infections, but is protected to a great extent by the washing action of the tears. Tears contain a number of substances, including lysozymes and immunoglobulins, which inhibit pathogens. Nevertheless, a number of organisms commonly cause primary conjunctivitis. Conjunctival infections can also be transmitted directly from eye to eye by fomites such as ophthalmological instruments, shared cosmetic applicators and, in conditions of poor hygiene, by flies. When the cornea is involved, the condition is called keratitis or keratoconjunctivitis.

Occlusion of the conjunctiva by contact lenses increases the likelihood of infection, and poor lens hygiene can lead to severe pseudomonal or even amoebic infections with the risk of corneal damage.

ORGANISM LIST

Adenoviruses
Enteroviruses (especially type 30)
Herpes simplex virus
Moraxella lacunata
Streptococcus pneumoniae
Haemophilus influenzae
Neisseria gonorrhoeae and *N. meningitidis*
Chlamydia trachomatis
Pseudomonas aeruginosa
Acanthamoeba spp.

Clinical features

The eye feels sore and itchy, and there is a discharge of watery, mucoid or purulent material which may dry, especially during sleep, and glue the eyelids together. In severe cases there is swelling of the eyelids which further inhibits eye-opening and drainage of secretions.

The conjunctiva appears red, often with a thin, clear outline surrounding the iris. Occasionally marked swelling causes it to bulge through the palpebral fissure.

Childhood conjunctivitis

Childhood conjunctivitis (pink eye) is common and often spreads among small children in families and school communities. It is usually caused by a respiratory adenovirus. It begins unilaterally and may spread to the other eye. It is self-limiting and usually mild, with a natural history of a few days.

Shipyard eye

Shipyard eye is a colloquial term applied to keratoconjunctivis spread by ophthalmological equipment. Often caused by adenovirus type 8, it was common in occupational settings where minor eye trauma and frequent examinations took place. Simple hygiene precautions such as hand-washing by staff and adequate sterilization of equipment prevent continuing spread of infection.

Haemorrhagic conjunctivitis

Haemorrhagic conjunctivitis caused many epidemics worldwide in the early 1980s. The agent was enterovirus type 30. The disease was abrupt in onset, moderate to severe, and associated with intense, haemorrhagic inflammation of the conjunctiva.

Diagnosis

The clinical diagnosis is usually evident. Important differential diagnoses for a red, painful eye include herpes simplex keratitis (dendritic ulcer—see below) and acute glaucoma. Both of these conditions can be sight-threatening, and should be considered whenever a red eye persists.

In infants the lacrimal sac may drain poorly, causing swelling and a mucus discharge at the inner canthus. The condition is non-infectious, harmless and self-limiting. Lacrimal sac drainage by digital compression abolishes the 'sticky eye' and can be discontinued after 1 or 2 weeks.

Management

Most viral cases resolve spontaneously in a few days. Failure to respond should prompt investigation with swabs for bacterial culture, a search for chlamydial infection and perhaps viral culture. In difficult cases an early ophthalmological examination is advisable.

Warning

A red eye unresponsive to antibiotic treatment should not be treated with corticosteroids until the possibility of a herpes simplex infection has been ruled out.

Herpes simplex keratitis (dendritic ulcer)

This is a progressive infection of the corneal epithelium which presents as a red, painful eye. It produces a branching ulcer which destroys the corneal epithelium and may damage the underlying tissue, leading to scarring and visual impairment. *It progresses rapidly if treated with topical corticosteroids.*

The diagnosis is suggested by a persistently red eye, unresponsive to topical antibiotic treatment. The branching ulcer can be seen on slit-lamp examination, or by inspection after the instillation of fluorescein drops. Herpes simplex virus can be cultured from corneal scrapings or herpes simplex antigen (by direct fluorescence staining) or DNA (by polymerase chain reaction, PCR) can be demonstrated in corneal scrapings.

The treatment of choice is topical aciclovir ointment. Treatment is continued until healing is complete. Follow-up by an ophthalmologist is important, both to monitor healing and to offer advice if residual scarring remains.

Treatment of herpes simplex keratitis
Topical aciclovir 3% eye ointment five times daily (continue for at least 3 days after complete healing).

Bacterial conjunctivitis

Bacterial conjunctivitis may occur alone, or complicate upper respiratory infections; common causes are *Haemophilus influenzae*, *Streptococcus pneumoniae* and sometimes *Staphylococcus aureus*. A purulent exudate is often apparent, and may form crusts at the inner canthus. The condition is usually mild and often self-limiting.

Unusual bacteria affecting the conjunctiva include *Moraxella lacunata*, which causes indolent or subacute infections (often in outbreaks where spread is by towels, make-up applicators or unwashed hands) and *H. aegyptius*, more common in tropical climates, which causes more acute infection and may also spread by the droplet route. *Pseudomonas aeruginosa* can cause keratoconjunctivitis with blurred vision in contact lens wearers. Infection is derived from unsterile cleaning fluids or from inappropriate use of stored tap water.

Treatment with chloramphenicol drops speeds healing. Chloramphenicol ointment can be used at night. The course should rarely be longer than a week. Repeated or sustained use of chloramphenicol carries a risk of agranulocytosis and should be avoided. Gentamicin drops and ointment are alternatives, which are also useful for treating *Pseudomonas* conjunctivitis. Severe, unresponsive or ulcerating eye infections require specialist management which includes subconjunctival injections of antibiotics.

Treatment of bacterial conjunctivitis
1 Chloramphenicol 0.5% eye drops at least 2-hourly, then four times daily when infection is controlled. Continue for 48 h after healing. Chloramphenicol 1% eye ointment may be used instead of drops at night or alone in a dose of three or four times daily.
Alternatives: gentamicin, neomycin or framycetin drops or ointments. Fusidic acid 1% drops are available for treating staphylococcal conjunctivitis.
2 *Pseudomonas* conjunctivitis should be treated with gentamicin 0.3% eye drops in the same regimen as choramphenicol drops; an alternative is tobramycin 0.3% eye drops.
Alternatives (and for ulcerative conjunctivitis): ciprofloxacin 0.3% eye drops, every 15 min for 6 h, then every 30 min for that day, every hour the second day, then every four hours for 2 more days (ofloxacin drops are also available).

Warning: prolonged or repeated use of chloramphenicol eye drops can predispose to agranulocytosis.

Neonatal ophthalmia

This severe conjunctivitis is usually acquired during birth from the infected maternal genitalia. It is caused by either *Chlamydia trachomatis* or *Neisseria gonorrhoeae* and appears in the first 2 or 3 days of life. Gonococcal ophthalmia responds to topical chloramphenicol. Chlamydial infection is nowadays more common. It is treated with chlortetracycline eye ointment plus systemic erythromycin. Systemic treatment is needed because respiratory chlamydial infection may follow ophthalmia (see Chapter 17). In both cases the parents should be offered follow-up treatment.

Treatment of ophthalmia neonatorum and chlamydial conjunctivitis (see Chapter 17)
1 Chlamydial: tetracycline 1% eye ointment four times daily (systemic erythromycin treatment should be given simultaneously).
2 Gonococcal: chloramphenicol eye drops and/or ointment (systemic penicillin or other appropriate treatment should be given simultaneously).

Chlamydial conjunctivitis and trachoma

Oculogenital strains of *Chlamydia trachomatis* that commonly colonize the genital tract can also affect the eye. Oculogenital chlamydiae can be distinguished from one another and from the agents of trachoma and lymphogranuloma venereum by serotyping. They cause subacute conjunctivitis unresponsive to chloramphenicol treatment which spreads between sexual contacts and from eye to eye. *Chlamydia trachomatis* inclusion bodies can be demonstrated by light microscopy in cells from conjunctival scrapings or from positive cell cultures. Methods of rapid diagnosis include enzyme immunoassay (EIA) to detect chlamydial antigen, and PCR or ligase chain reaction methods to detect chlamydial DNA.

Tetracycline eye ointment is the treatment of choice. Oral treatment, e.g. erythromycin or tetracycline, may be added. Investigation and treatment of the patient and sexual partner for genital infection are also necessary (see Chapter 15).

Trachoma

Trachoma is a disease of crowding and poor hygiene. It is precipitated by persisting or repeated infection with *C. trachomatis*, spread from eye to eye by unwashed hands, and possibly by flies. Untreated infection and repeated superinfections may lead to the formation of a plaque of vascular inflammatory tissue (pannus) which deforms the eyelid. Scarring, leading to entropion and trichiasis, is an important cause of corneal damage, scarring and blindness. Topical chlortetracycline eye ointment and hygienic measures can control this disease.

Amoebic keratoconjunctivitis

This is a rare condition associated with contact lenses. The lens cleaning fluid becomes colonized by free-living amoebae, usually *Acanthamoeba*, but occasionally *Naegleria*, which are then repeatedly inoculated into the eye. The resulting severe keratitis is difficult to treat and often damages vision. The treatment of choice is propamidine isethionate (Brolene) 0.1% eye drops four times daily or 0.15% dibromopropanidine ointment once or twice daily. These may be used in combination with chlorhexidine and neomycin eye drops (specialist supervision required). Control is by the use of only sterile cleaning fluids, which are discarded after use.

Infections of the middle ear

Acute otitis media

ORGANISM LIST

Many respiratory viruses
Streptococcus pneumoniae
Haemophilus influenzae
S. pyogenes
Staphylococcus aureus
Chlamydia pneumoniae
Mycoplasma pneumoniae
Moraxella catarrhalis

Introduction

Infection of the cavity of the middle ear causes pain, reddening, opacification and sometimes rupture of the tympanic membrane with discharge from the ear. Traditionally it has been thought of as a primary or secondary infection caused by respiratory tract organisms ascending the eustachian tube to infect the obstructed or virus-inflamed cavity. However, several studies have suggested that antibiotic treatment is rarely more successful than symptomatic treatment, and furthermore that myringotomy and culture of middle ear contents do not improve the results of treatment.

Many cases of otitis media are probably viral. *Chlamydia pneumoniae* is now also recognized as a cause of mild cases, unresponsive to treatment with ampicillin or co-trimoxazole. However, severe infections are sometimes seen, with early rupture of the eardrum and frankly purulent discharge. These infections may be spontaneous, but are often complications of catarrhal respiratory infections, including the now rare measles. Implicated organisms include *Streptococcus pneumoniae*, *H. influenzae*, *S. pyogenes* and *Staphylococcus aureus*. It seems reasonable to treat these promptly with antibiotics, and to culture any discharge.

Management

1 For mild pain and reddening of drum, with no fluid level or other features, or with general upper respiratory symptoms: offer analgesics, and decongestant if indicated. Review if persistent or worsening.
2 Evidence of *C. pneumoniae* or *M. pneumoniae infection* (see Chapter 7): systemic erythromycin (or tetracycline in an adult) may be beneficial.
3 Severe pain, discharge or important precursor such as measles or influenza: ideally, obtain pus for bacterial culture. Offer antibiotics, which may need to include an antistaphylococcal spectrum. Analgesia is essential.

Complications

Mastoiditis

Mastoiditis is the extension of pyogenic infection into the mastoid antrum. If treated early with antibiotics this may resolve, but loculation in the air cells of the antrum makes the infection difficult to eradicate. There is severe pain behind and within the ear, and often a high fever. There is also a risk of extension into the cranial cavity. Treatment must then be surgical, with opening and debridement of the air cells (mastoidectomy).

Attic infection

Attic infection is damage to the high roof of the middle ear cavity, which can involve the facial nerve. This is more common in chronic or neglected infections. A cholesteatoma (tumour of waxy inflammatory tissue) may form, and can erode the cranial bones, predisposing to intracranial infection. The treatment of cholesteatoma is surgical. Rare cases of chronic middle ear infection can be complicated by the presence of anaerobic pathogens.

Paranasal sinusitis

In this condition the paranasal sinuses fill with exudate, and the draining ostia become blocked. Like otitis media, it is often secondary to a catarrhal infection and may be caused by *S. pneumoniae*, *H. influenzae*, *S.'milleri'*, anaerobes or *S. aureus*.

Clinical features are pain and tenderness over the affected sinus, usually a maxillary or frontal sinus. There is a loss of transillumination, and X-rays show thickening of the soft-tissue wall of the cavity, often with a fluid level.

Treatment includes elevation of the head and offering decongestants to aid drainage. An oral broad-spectrum antibiotic such as co-amoxiclav, cefuroxime axetil or tetracycline will reduce the purulent exudate. In relapsing or chronic cases, the ostia of the frontal sinuses can be surgically enlarged, or false ostia can be made to connect the maxillary sinuses to the buccal cavity, allowing improved drainage. The ethmoid sinus has thin lateral and superior walls, which sometimes rupture. Lateral spread of infection causes orbital cellulitis, while superior spread may lead to meningitis. Inflammation in the walls of the sphenoid sinuses affects the overlying venous sinus. Cavernous sinus thrombosis is a grave complication of severe, untreated sinusitis. It is best diagnosed by computed tomography or magnetic resonance scan.

Viral infections of the throat and mouth

ORGANISM LIST

Rhinoviruses
Coronaviruses
Enteroviruses
Adenoviruses
Epstein–Barr virus
Herpes simplex virus

The common cold (coryza)

Introduction

This disorder is caused by numerous strains of rhinoviruses and sometimes by coronaviruses or other respiratory virus infections. Colds are extremely infectious by the droplet route. The familiar clinical features are mild fever, swelling of the mucosa of the nose, throat and conjunctiva, and finally a copious mucoid exudate. Otitis media in children and sinusitis in adults are common complications.

Pathology

There are more than 100 different serotypes of rhinoviruses, which are members of the Picornaviridae. They are small RNA viruses 28–34 nm in diameter expressing icosohedral symmetry. The genome consists of a single strand of positive-polarity RNA of approximately 7.2 kb. A single polypeptide is produced and cleaved. There are four capsid proteins; VP1–4. The viral capsid consists of 12 pentamers each of which contains a prominent valley in which there is a 'pocket'. In most strains this contains a strongly hydrophobic molecule known as pocket factor. This pocket is a potential site for specific inhibitors of the virus. The virus binds via its VP1 protein to intracellular adhesion molecule 1 (ICAM-1) on the host cells, to gain attachment and entry.

Coronavirus is the only genus in the family Coronaviridae. It is a pleiomorphic, non-enveloped RNA virus ranging in size from 60 to 220 nm. The virus has characteristic club-shaped projections 20 nm in length extending from the surface. There are three main structural proteins: the nucleocapsid protein, the surface projection protein and the transmembrane or matrix protein. The nucleocapsid proteins are species-specific whereas strain-specific antigens are found on the surface projections. The surface projections are a high-molecular-weight glycoprotein (180 kDa) and are readily removed by protease activity.

The main mechanism of pathogenesis is probably a direct cytotoxic effect on respiratory epithelial cells.

Laboratory diagnosis

This is rarely attempted due to the frequency and trivial nature of the infection. Nasal washings may be inoculated into human embryonic lung fibroblasts. A cytopathic effect (CPE) similar to that of enterovirus infection develops within 8 days, although it may not develop until a second passage has been performed. Neutralizing antibody can be detected, but this test is too cumbersome for routine use. An EIA-based method is available but has the disadvantage of being serotype-specific. Viral RNA can be detected by RT-PCR.

Coronaviruses are difficult to isolate, and serological investigations are not readily available. Virus can be isolated from nasal and throat swabs and nasopharyngeal aspirates by inoculation in human embryonic lung fibroblasts. The CPE consists of small granular round cells in the monolayer. Antibodies can be detected by virus neutralization, complement fixation test and haemagglutination inhibition (HAI).

Management

Treatment is symptomatic and includes antipyretic analgesics, mild decongestants such as pseudoephedrine tablets and bed rest in severe cases. Enthusiasm for intranasal interferon therapy has waned because of local irritation and poor evidence of benefit.

Enteroviral pharyngitis

Epidemiology

Pharyngitis due to enterovirus infection is common. Coxsackie type A10 is frequently responsible, although other coxsackie and echoviruses can cause pharyngitis. Infections predominate during summer and autumn; unlike winter viruses whose main target is the respiratory tract (e.g. influenza and adenovirus). Humans are the only known reservoir of infection and transmission occurs by direct contact and through droplet spread. The disease often occurs as epidemics affecting young children in nurseries, playgroups and schools. Crowding and poor hygiene increase the risk of epidemics.

Virology

More than 70 serotypes of enterovirus have been isolated from human sources. All belong to the family Picornaviridae which comprises four genera, of which two, the rhinoviruses and the enteroviruses, cause disease in humans. The enteroviruses can be differentiated into five main groups: (i) polioviruses; (ii) coxsackieviruses group A; (iii) coxsackieviruses group B; (iv) echo (enteric cytopathogenic human orphan) viruses; and (v) five more recently characterized human enteroviruses types 68–72 (type 72 is hepatitis A virus). Each of these has different tissue tropisms (see Chapter 13).

The viruses are 27 nm in diameter, with icosahedral symmetry. The virion contains four proteins—VP1–4. The genome consists of a single strand of positive-sense RNA. Variations in VP1–VP3 are responsible for serological diversity; antiviral antisera neutralize only the homologous virus strain. Enteroviruses bind to host cell-specific receptor sites via VP4 and it is thought that differences in these receptors are responsible for different tissue tropisms among enteroviruses. The VP4 structure has been fully identified for two poliovirus serotypes and in both there is a deep cleft in the protein through which the virus binds to the host cell membrane.

Groups of enteroviruses

1 Polioviruses.
2 Coxsackieviruses group A.
3 Coxsackieviruses group B.
4 Echoviruses.
5 Other recently characterized human enteroviruses.

Pathology

Enteroviruses enter the body through the pharynx and the alimentary tract. The virus multiples locally in the tonsils, Peyer's patches and other bowel-associated lymphoid tissue. A viraemic phase often occurs, and this may be followed by disease in different organs, for example meninges, myocytes, brain or skin. In poliomyelitis the virus multiplies and causes damage in the anterior horn cells, resulting in a flaccid paralysis (see Chapter 13).

Clinical features

The incubation period of about 1 week is followed by sore throat and fever. The severity and duration of symptoms are very variable. In severe cases headache, stiff neck or meningism also occur. Symptoms rarely last more than 5–7 days.

Reddening of the fauces is usual, but does not parallel the intensity of symptoms. Chains of moderately enlarged lymph nodes are often palpable in the anterior and posterior triangles of the neck. Both echovirus and coxsackievirus infections occasionally produce an accompanying rash of sparse macules or small papules, concentrated on the cheeks and trunk.

Coxsackie A infections may cause a faucial rash of blisters with haloes of inflammation (herpangina; Fig. 6.1). Hand, foot and mouth disease of toddlers causes similar lesions in the mouth, accompanied by blisters on the

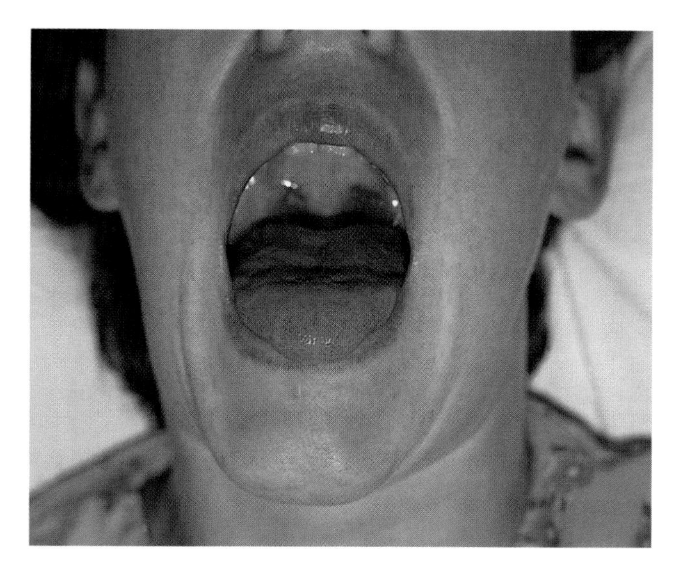

Fig. 6.1 Faucial blisters in a case of herpangina.

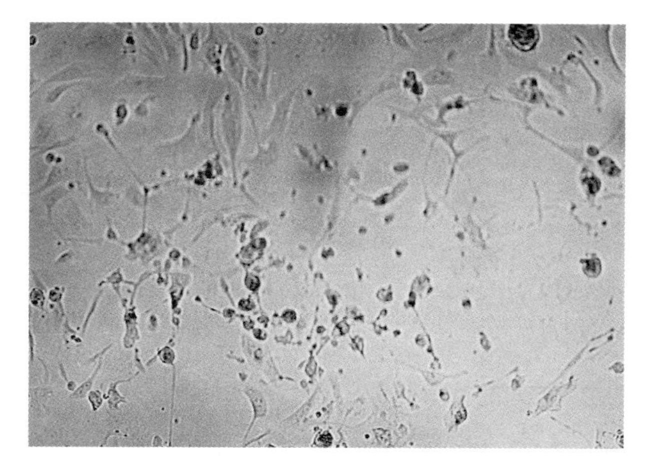

Fig. 6.2 Cytopathic effect of enteroviruses, showing rounding, shrinkage and loss of contiguousness of the cell monolayer. Courtesy of Dr M. Zambon, Central Public Health Laboratory.

palmar aspects of the hands and feet, and a papular rash on the buttocks (see Chapter 5).

Coxsackie B infections can cause pleurodynia (Bornholm disease) in which the trunk muscles are severely tender and painful, mimicking pleurisy. There is usually also a high fever and sore throat. Coxsackie B can also cause maculopapular rashes, mild to moderate orchitis and, rarely, myocarditis.

Management

Analgesics, especially non-steroidal anti-inflammatory agents, are helpful. In patients over the age of 12 years, soluble aspirin gargles may help, and can be swallowed for their systemic effects. Recently developed antiviral drugs, such as pleconaril, which prevent attachment of viruses to host cells, may be useful in enteroviral infections.

Laboratory diagnosis

Enteroviruses are most easily recovered from faeces, but throat swabs and cerebrospinal fluid should also be examined in cases of meningitis. Culture of an enterovirus from a sterile site is diagnostic, whereas isolation from faeces is less certainly so. Enteroviruses grow well in cultures of human embryonic lung fibroblasts. The infected cells become rounded and refractile before separating from the monolayer (Fig. 6.2). Isolates are typed by neutralization using pooled antisera.

Coxsackie group A viruses grow poorly in tissue culture, but growth can be achieved by intracerebral inoculation of mice. Serodiagnosis is available using immunoglobulin M (IgM) antibody capture methods. Enteroviral RNA can be detected by RT-PCR, a useful rapid test done on cerebrospinal fluid samples in meningitis cases.

Complications

The spectrum of enteroviral infections includes lymphocytic meningitis, myositis, pericarditis and acute myocarditis. Patients with significant meningism, precordial pain, dysrhythmias or heart failure require further investigation.

Adenoviral sore throats and pharyngoconjunctival fever

Introduction and pathology

Adenoviruses of types 1–10 are capable of causing severe sore throats, usually with high fever, severe faucial inflammation and painful enlargement of lymph nodes in the upper neck. The illness may last 7–10 days and is often followed by debility in the convalescent period.

The same adenoviruses commonly cause conjunctivitis. When this coexists, the condition is called pharyngoconjunctival fever.

Human adenoviruses form part of the genus *Mastadenovirus* which is part of the family Adenoviridae. Adenoviruses are unenveloped viruses with a double-stranded DNA genome of approximately 35 kDa and icosahedral symmetry. Forty-two serotypes are recognized and are divided into subgenera A–F. The disease caused varies according to the subgenus. Subgenus A is highly oncogenic in animals, B and C are associated with respiratory disease, D is associated with keratoconjunctivitis, E with conjunctivitis and respiratory disease, and F with infantile diarrhoea.

Adenoviruses multiply inside the nuclei of epithelial cells. They are cytopathic for human cells, and this is probably responsible for the tissue damage and symptoms associated with infection. Different target specificity of serotype-specific surface 'fibres' contributes to the different tissue trophisms noted above. A toxin-like activity has been associated with the vertex capsomeres.

Laboratory diagnosis

Adenovirus is most easily isolated from stool but can also be recovered from conjunctival swabs, nasopharyngeal aspirates and cerebrospinal fluid. Human cell lines, e.g. Hep-2, are suitable for virus isolation. A cytopathic effect is seen in 48 h and is characterized by rounded-up cells with refractile intranuclear inclusion bodies. The identification can be confirmed by electron microscopy. Antibodies to a group-specific antigen can be detected by complement fixation test or EIA. PCR diagnosis is available.

Infectious mononucleosis

Introduction

Infectious mononucleosis is caused by primary Epstein–Barr virus (EBV) infection. It is a systemic disease, but in 75% of cases its main feature is sore throat (anginose infectious mononucleosis), and the differential diagnosis and main complications fall into this chapter. However, 15% present as hepatitis and 10% as fever alone (see Chapters 9 & 20).

Clinical presentations of Epstein–Barr virus infection

1 Sore throat 75%.
2 Hepatitis 15%.
3 Fever alone 10%.
4 Rare: viral-type meningitis, mononeuritis or polyneuritis, perisplenic pain.

Although primary disease causes considerable morbidity, the late effects of infection may also be important. Both Burkitt's lymphoma and nasopharyngeal tumours are consequences of EBV infection in early infancy. Cofactors such as early infection with malaria, or ingestion of toxins in, for example preserved vegetables, may also be important. In the immunosuppressed, EBV infection can cause interstitial pneumonitis and B-cell lymphomas.

Epidemiology

EBV is shed in pharyngeal secretions, and transmission occurs via close oral contact, through kissing, shared eating utensils or in some cultures by a mother chewing food for her infant.

EBV infection is common in young children, when it is usually asymptomatic. Clinical disease occurs mainly in teenagers and young adults. The estimated annual incidence, based on general practitioner consultations, is 1 per 1000 population.

Virology and pathogenesis

EBV was first identified in Burkitt's lymphoma tissue. Morphologically identical to other herpesviruses, it is a DNA virus with isocohedral symmetry and enveloped with a membrane derived from the host plasma membrane. The double-stranded DNA is 172 kb in length—large enough to code for 100–200 proteins. These include the Epstein–Barr nuclear antigen (EBNA) complex, the latent membrane protein (LMP), the terminal protein, the membrane antigen complex, the early antigen (EA) complex and the viral capsid antigen (VCA). The function of many of these antigens is unknown. The EBNA complex consists of at least six proteins and is probably important in maintaining the virus in the infected cell. EBNA-1 antigen is expressed on all infected cells but may be lost as infected cells die. The early antigen (EA) complex is expressed only on cells that have entered a lytic phase. Antibodies to EBNA and EA appear early in the infection and are transient. Antibodies to capsid antigens appear in IgM by the time disease is apparent, and may persist for weeks or months. IgG anticapsid antibodies indicate immunity, and persist throughout life.

Virus adheres to the B cells via the major glycoprotein gp350/220 which binds to the host-cell CD21 receptor (the receptor for the C3d component of complement). Adherence is followed by endocytosis and the virus fuses with the vesicle membrane via viral gp85, permitting release into the cytoplasm. Viral DNA persists in host cells as double-stranded DNA episomes organized into nucleosomes similar to chromosomal DNA. During viral latency, 10% of EBV genes are expressed including six EBNAs (1, 2, 3A, 3B, 3C and LP).

The function of these genes in latency and in the immortalization of infected cells is now known. For example LMP-1 is essential for EBV-induced transformation, and is found in many EBV-related lymphomas. A lytic cycle can be stimulated during which genes are expressed that encode structural proteins and those proteins associated with viral release. Latently infected B cells then develop the characteristic cytopathic effect of a lytic herpesvirus infection.

Virus first invades pharyngeal cells, from where B lymphocytes become infected and the infection is spread throughout the body. EBV is capable of activation and immortalization of B cells. This is the probable pathogenic mechanism for many of the 'immunological' effects and complications of infection.

Clinical features

The incubation period is 6–8 weeks. Fever, sore throat and widespread lymphadenopathy develop more or less simultaneously. A white creamy exudate appears on the tonsils, and becomes confluent within 24–36 h. The exudate is rarely discoloured. It may become bulky, but it does not involve the pharyngeal mucosa (Fig. 6.3). The pharyngeal and nasal mucosae are congested and swollen. Gross pharyngeal swelling may make it impossible to swallow saliva and can threaten the airway.

The spleen is palpably enlarged in 25–40% of cases; the liver edge is often palpable. A chest X-ray occasionally shows enlarged mediastinal lymph nodes and, surprisingly, as many as 20–25% of patients have a lung opacity suggesting segmental pneumonitis. The X-ray abnormalities resolve spontaneously as the fever abates.

Infected B cells remain within reticuloendothelial tissues, while a vigorous T-cell response releases activated T cells into the bloodstream. These are the 'atypical mononuclear cells' and may constitute 40% or more of the total lymphocyte count during the acute infection (Fig. 6.4).

The liver function tests show raised transaminases, usually in the range of 60–500 IU/ml. The alkaline phosphatase may rise during convalescence, sometimes approaching 1000 IU/ml, but this rarely causes any clinical consequence, and usually resolves uneventfully.

The activated B cells produce various antibodies, including the heterophile antibodies which agglutinate horse and sheep red blood cells. Other detectable antibodies include haemolysins, platelet antibodies, antinuclear antibodies, rheumatoid factors and anticardiolipin antibodies. These antibodies are occasionally associated with immune cytopenias or with autoimmune-like diseases, but the most common of these effects does not exceed an incidence of 1 in 20 000–30 000 clinical cases.

The duration of fever and exudative pharyngitis varies widely from a few days to as much as 3 weeks. Most patients convalesce steadily after this, but a minority suffer postinfectious fatigue for up to 6 months. Persisting chronic fatigue syndrome (CFS) is a rare complication of EBV infection.

Diagnosis

The clinical picture is often sufficient for diagnosis. Exudate is present in many cases of streptococcal pharyngitis, but this is follicular and rarely confluent. The pseudomembrane of diphtheria is sometimes confluent, but it is discoloured, and tends to spread beyond the tonsillar margin. In both bacterial diseases there is a low-grade neutrophilia, quite unlike the atypical lymphocytosis of EBV infection.

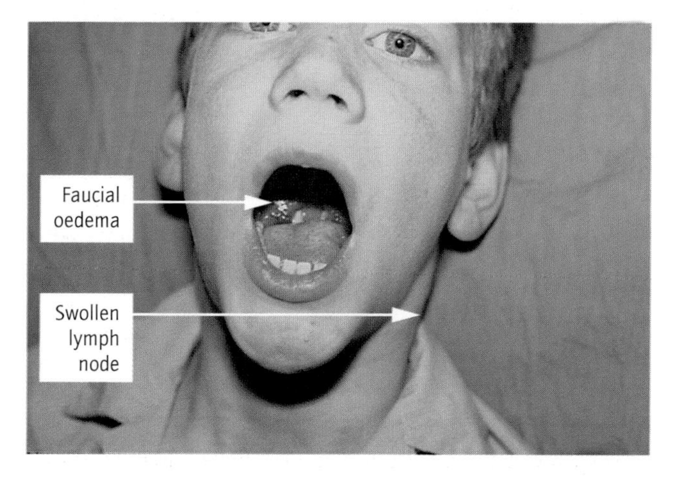

Fig. 6.3 Exudative tonsillitis with marked cervical lymphadenopathy in a case of infectious mononucleosis.

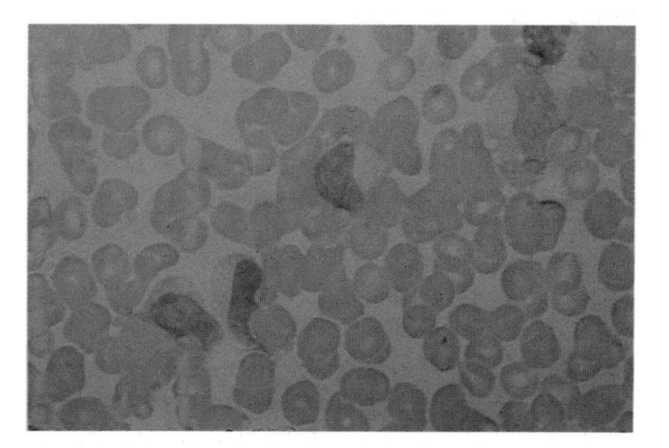

Fig. 6.4 Atypical mononuclear cells in the peripheral blood of a case of infectious mononucleosis.

A commercial slide test, the Monospot, which demonstrates the presence of heterophile antibody is now available. This utilizes horse red cell agglutination, and is positive in about 90% of patients at presentation. Small children, however, do not produce high levels of heterophile antibodies, so the test is positive in about 50% of patients aged less than 5 years. PCR is emerging as a sensitive method of detecting EBV infection.

All patients have IgM anti-EBV capsid antibodies, which are diagnostic of acute infection. In most patients it is not possible to detect rising IgG titres to VCA as concentrations are already high when the patient presents. These antibodies can also be detected by immunofluorescence.

Although EBV can be cultivated, this approach is not used as cultures are likely to be positive in specimens from latently infected asymptomatic individuals.

Management

There is no specific treatment; most cases recover uneventfully. In older children and adults soluble aspirin in standard doses may be gargled and swallowed. In the under-12s paracetamol may not be sufficient; paediatric formulations of ibuprofen may then be helpful. Stronger analgesics may be indicated in some cases with severe pharyngeal pain.

Complications

Threatened respiratory obstruction

Threatened respiratory obstruction is the most common reason for hospital admission. It will often improve with elevation of the head (to encourage drainage of oedema from the pharyngeal tissues), anti-inflammatory analgesics and reassurance. Danger signs are inability to swallow saliva and a rapidly increasing pulse rate; these may be followed by increasing respiratory rate and cyanosis. Respiratory obstruction can often be avoided by reducing oedema with an intravenous bolus of 100 mg hydrocortisone. The patient usually reports improvement within 20–30 min. Fears of adverse effects have been much allayed by successful trials of similar treatment in croup and bronchiolitis. The corticosteroid has no effect on the general progress of the infection, and dosing can be repeated if non-steroidal anti-inflammatory agents do not maintain the improvement. Emergency intubation or tracheostomy are occasionally necessary.

Effects of abnormal antibodies

Effects of abnormal antibodies are rarely clinically important. Thrombocytopenia is the least rare, followed by haemolytic anaemia. A handful of cases of systemic lupus erythematosus-like disease are reported; joint pains occur in convalescence and may persist for some months, often improving when anti-DNA antibodies disappear.

Suppurative complications

Peritonsillar abscess, pharyngeal abscess, ethmoiditis, infection of other intracranial sinuses and periorbital cellulitis occasionally occur, and probably represent secondary infection of inflamed mucosae and obstructed sinuses. They should be investigated and treated as for primary pyogenic infections at these sites.

Rupture of the spleen

Rupture of the spleen is extremely rare. It may present with left upper quadrant and shoulder pain during acute disease or be precipitated by apparently trivial trauma, such as a blow during play, a sudden movement or a cough. An abdominal tap may yield blood-stained fluid. Imaging studies may demonstrate free fluid in the peritoneal cavity, or the disrupted splenic anatomy. Prompt surgical intervention in needed to terminate bleeding. Contact sports or combat sports should therefore be avoided until lymphadenopathy and, by inference, splenomegaly have subsided.

Neurological complications

Neurological complications include lymphocytic meningitis, mononeuritis or brachial plexitis (Fig. 6.5). These are benign and self-limiting. Occasional cases of encephalopathy are reported, of which some are progressive or even fatal.

Bacterial throat infections

ORGANISM LIST

Streptococcus pyogenes
Haemophilus influenzae
Corynebacterium diphtheriae
Other bacteria, including *Neisseria meningitidis*, *H. haemolyticum*, *Chlamydia pneumoniae*, *Staphylococcus aureus* and *N. gonorrhoeae*

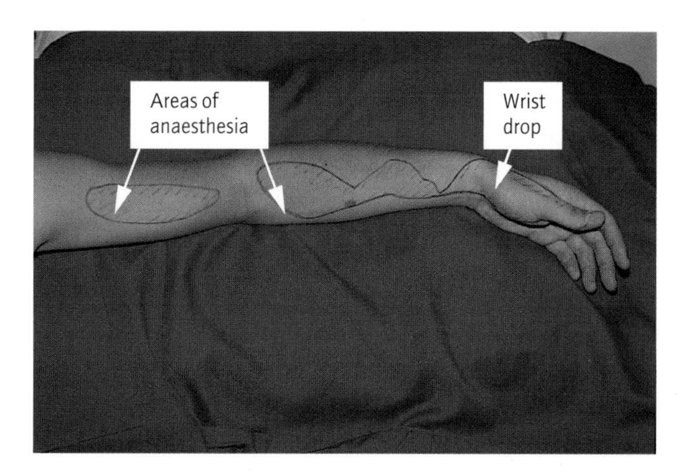

Fig. 6.5 Brachial plexitis complicating infectious mononucleosis: complete recovery occurred within 3 weeks.

Streptococcal tonsillitis

Introduction

This disease is caused by *S. pyogenes* (group A streptococcus, or GAS), and is common worldwide, affecting both adults and children. It causes considerable short- and medium-term morbidity, and can be recurrent, as *S. pyogenes* is a tenacious colonist of the throat, and exists in many serotypes. It is also important because streptococcal throat infections can be followed by scarlet fever, poststreptococcal nephritis or rheumatic fever (see Chapter 21).

Epidemiology

Streptococcal pharyngitis is common in temperate climates, occurring mainly in the winter. The disease is usually spread by direct contact with respiratory excretions, and less commonly by air-borne droplets or indirect contact by hands. Outbreaks due to contaminated food and milk have been described but are uncommon.

The infection occurs mainly in children, up to 20% of whom may be asymptomatic carriers. Disease is commoner in crowded settings such as children's homes and military camps.

The incidence and severity of scarlet fever (and other manifestations of group A streptococcal infection) have declined steadily over the past 50 years. In 1936, when yearly notifications began, there were 104 862 notifications and 440 deaths from scarlet fever in England and Wales—a case fatality ratio of 0.42 per 100. By 1986, this had dropped to 6888 notifications and 3 deaths (0.05 per 100). The incidence (but not the case fatality ratio) increased during 1988 and 1989, but has subsequently declined to around 5000 cases per annum. A similar fall was seen in the USA; however, a reappearance of severe infections complicated by rheumatic fever has been observed since 1986, associated with mucoid strains of *S. pyogenes*.

Pathogenesis

The main pathogenicity determinant is the M-protein antigen of which there are many types (see Chapter 2 & Table 2.3). This protein has a fibrillar structure with a similar function to the polysaccharide capsule of other pyogenic organisms, such as *H. influenzae* and *N. meningitidis*. It inhibits bacterial phagocytosis by neutrophils. *S. pyogenes* can only be ingested when opsonized by anti-M antibody. This antibody is serotype-specific, providing no protection to infection by *S. pyogenes* of other M-protein types. Cross-reactions between these M proteins and host myocardial cells are thought to be responsible for the development of rheumatic fever. Cross-reactions between antigens in the glomerular basement membrane and certain 'nephritogenic' M types of *S. pyogenes* are responsible for poststreptococcal glomerulonephritis.

Pathogenicity factors for *Streptococcus pyogenes*
1 M proteins.
2 Lancefield group antigens.
3 Streptolysin S.
4 Streptolysin O.
5 Hyaluronidase.
6 Collagenase.
7 DNAase.
8 Streptokinase.
9 Pyogenic exotoxins A, (B) and C.

Lancefield group antigens

These are polysaccharide–teichoic acid antigens located in the bacterial cell wall. Lancefield antigens group streptococci into pathologically related groups and species, and constitute the main method of classifying beta-haemolytic streptococci. The Lancefield group antigen has not, however, been directly associated with the pathogenesis of infection.

Lytic enzymes

S. pyogenes elaborates a wide range of lytic toxins, some of which are virulence factors and some of which are used in diagnosis. The organism produces two haemolysins, one of which is oxygen-stable (streptolysin S). An oxygen- and heat-labile streptolysin (streptolysin O) is one of a family of cytolytic toxins found in Gram-positive organisms (see p. 16). The antibody response to anti-streptolysin O (ASO) is used in the serological diagnosis of *S. pyogenes* infections (see p. 97). Hyaluronidase and collagenase may aid tissue invasion by breaking down collagen and hyaluronic acid in connective tissue. The ability to disrupt tissue and to reduce tissue redox potential is thought to be important in the pathogenesis of synergistic gangrene, where mixed infection with Gram-positive cocci, including *S. pyogenes* and/or *Staphylococcus aureus*, and obligate anaerobes can result in destruction of tissue planes.

Streptokinase acts as a plasminogen activator, producing clot lysis, which may enhance the spread of the organism. Other lytic enzymes include four serologically different DNAses (A–D). On its surface, the organism expresses C5a- and immunoglobulin-binding proteins which may interfere with host immune responses. A recently identified potential virulence factor is surface-expressed enolase. This is a metabolic enzyme that acts by binding plasminogen.

Streptococcal pyrogenic exotoxins (SPEs)

There are three distinct SPEs: A, B and C, of which A and C are structurally similar. These toxins are responsible for the rash of scarlet fever (erythrogenic toxins) and stimulate macrophages to produce tumour necrosis factor, with all the resulting metabolic and immunological consequences. They are superantigens, and have close homology with some *Staphylococcus aureus* exotoxins. They are thought to mediate the shock syndrome in severe group A streptococcal infections.

SPE B has a different structure. It is secreted as a zymogen and is converted to a proteinase which is mitogenic and cardiotoxic.

Clinical features

The usual features are fever, pain in the throat, enlargement of the tonsils and tender swelling of the tonsillar lymph nodes at the angles of the jaw. The severity of the symptoms is variable; some patients with severe pain have only the mildest reddening of the throat, while some with moderate pain have alarming tonsillar inflammation.

The classical appearance is of follicular tonsillitis, in which the tonsils are enlarged, very red and dotted with patches of soft white exudate (Fig. 6.6). The throat is extremely painful and a large, tender lymph node swells downwards from beneath the angle of the jaw. It is not unusual for one or other tonsil to be predominantly affected, with proportionately greater swelling of the lymph node on that side.

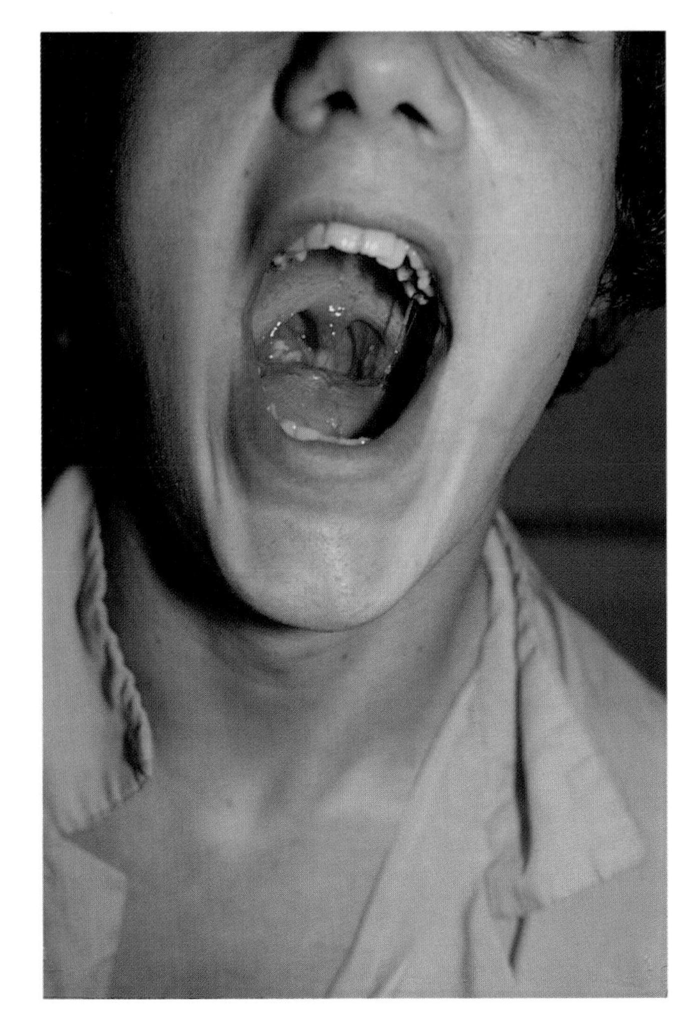

Fig. 6.6 Typical appearance of follicular tonsillitis.

Scarlet fever

Scarlet fever is a hypersensitivity response to erythrogenic toxin produced by the infecting *Streptococcus pyogenes*. There is severe malaise, and many children with the condition vomit at the onset of illness.

An erythema appears on the chest and quickly involves all of the skin. It is more marked in the folds and valleys of the skin (Pastia's sign). The papillae of the skin are swollen, forming tiny conical papules which give the rash a sandpaper texture, and a stippled appearance (often called a punctate erythema). There may be a less affected area around the mouth or 'snout' area, but this is neither a constant nor a diagnostic feature (Fig. 6.7).

Initially the tongue is furred and white (white strawberry tongue). Over about 3 days this clears from the tip backwards, leaving a reddened (red strawberry tongue) appearance (Fig. 6.8).

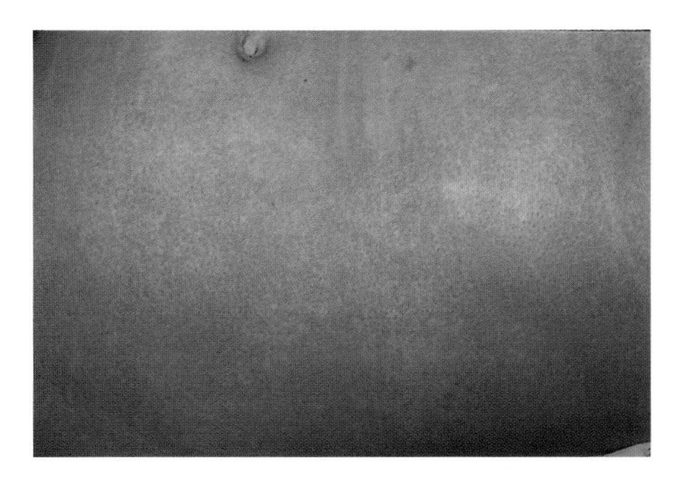

Fig. 6.7 Typical rash of scarlet fever.

Fig. 6.8 Red strawberry tongue of scarlet fever.

Streptococcal toxic shock syndrome

This is a severe version of scarlet fever, with multisystem involvement. Pleural and peritoneal effusions are common, and renal failure is often seen. Shock and thrombocytopenia are other features, but haemorrhage is rare.

Diagnosis

Mild streptococcal tonsillitis is difficult to recognize clinically. However, a neutrophilia in the peripheral blood is a predictor of bacterial rather than viral aetiology (with a sensitivity of 75%), and allows an early decision for antibiotic treatment to be made. Follicular tonsillitis with associated tender enlargement of tonsillar lymph nodes is typical of *S. pyogenes* infection, and is clinically diagnostic.

Recovery of *S. pyogenes* from throat swabs strongly suggests the diagnosis, but streptococcal colonization can coexist with viral and other sore throats. Also, *S. pyogenes* does not transfer well from swab to culture, so that there is a 20–30% false negativity rate in the results of swab cultures.

S. pyogenes is a fastidious organism, growing only on rich nutrient media, usually containing blood. Haemolytic toxins produce the characteristic zone of complete (beta) haemolysis, enabling selection of suspect colonies for identification. The growth of the organism and the haemolysis it produces are enhanced by incubation in an anaerobic atmosphere. When large numbers of specimens are to be screened, a selective medium incorporating antibiotics such as ofloxacin or nalidixic acid, or dyes such as crystal violet, can be used to inhibit Gram-negative and Gram-positive commensal organisms, respectively.

For most routine laboratories, identification of *S. pyogenes* is achieved on the basis of colonial morphology (large clear colonies surrounded by a zone of beta haemolysis) and the presence of the group A Lancefield antigen. The antigen is extracted with an enzyme and its presence detected by agglutination of latex particles coated with antibodies to the different Lancefield group antigens. Lancefield groups, their related species and pathogenicity are listed in Table 6.1. Rapid diagnostic tests exist, based on latex agglutination of throat-swab material. These techniques are sufficient simple for use in a general practitioner surgery. They are highly specific but of variable sensitivity (as low as 55% in some studies).

Laboratory identification of *Streptococcus pyogenes*
1 Chains of Gram-positive cocci growing on blood agar, plus beta-haemolysis, plus Lancefield group A antigen.
2 Rapid antigen detection in throat swabs.
3 Antibody detection: high or rising titres of antistreptolysin O, antihyaluronidase and/or anti DNAse.

Serology

Acute streptococcal infection can be diagnosed by detecting a rise or fall in antibodies to the oxygen-labile streptolysin (ASO), DNAse B or hyaluronidase. ASO titres (ASOT) above 400 I.U. indicate likely recent infection, while a fourfold rise in titre in paired sera is strong evidence of streptococcal disease. Although these techniques are available in many laboratories, they have limited utility as antibody response may be delayed for up to 4 weeks. Their main use is in providing supportive evidence in the diagnosis of complications such as rheumatic

Group	Species	Diseases caused
A	*Streptococcus pyogenes*	Acute pharyngitis, quinsy, otitis media, erysipelas, synergistic gangrene, rheumatic fever, poststreptococcal glomerulonephritis
	S. 'milleri' (minute colony)	Metastatic suppurative infection
B	*S. agalactiae*	Neonatal septicaemia, meningitis and pneumonia
C	*S. dysgalactiae* *S. equi* *S. equisimilis* *S. zooepidemicus*	Rare cause of skin sepsis and endocarditis. Postinfectious glomerulonephritis has been reported
D	*S. bovis*	Endocarditis and bacteraemia associated with colonic neoplasm
F	*S. 'milleri'*	Metastatic suppurative disease, dental sepsis

Table 6.1 Species of beta-haemolytic streptococci and their pathogenicity.

fever or glomerulonephritis. Anti-DNAse B responses usually last longer and may be used as a marker of infection in the late convalescent phase.

Management

The treatment of choice is penicillin. Early and mild cases may respond to oral therapy with ampicillin, but established and severe infections can require inpatient treatment with intravenous benzylpenicillin. Suitable alternative drugs are cephalosporins and erythromycin. However, a small proportion of highly virulent strains are resistant to erythromycin, so failure to respond to this drug should prompt an early change of treatment.

It is not easy to eradicate *S. pyogenes* from the throat. Research shows that 2 weeks of vigorous (often parenteral) penicillin therapy is needed and erythromycin treatment is unreliable. This is important in dormitory- or barracks-associated outbreaks of streptococcal sore throat, especially those associated with rheumatic fever. In a domiciliary setting a course of a least 7–10 days' treatment is probably advisable.

Treatment of streptococcal sore throat

1 Early and mild cases: oral ampicillin 250–500 mg 6-hourly for 10 days, or erythromycin in the same dosage and schedule.

2 Severe cases: benzylpenicillin 1.2–2.4 g 4–6-hourly.

Complications

Peritonsillar abscess

Peritonsillar abscess (quinsy) is the commonest complication of tonsillar sepsis. The tissue of the fauces and soft palate on the affected side becomes boggy and pendulous. The throat is intensely painful and tender, often with severe tenderness and swelling of the draining lymph node. The tonsil may be invisible in the mass of oedematous tissue. A small proportion of cases have bilateral quinsy, which carries a high risk of airway obstruction (Fig. 6.9).

Prompt and vigorous treatment can avert the need for surgical drainage. High doses of penicillin (up to 2.4 g 4-hourly) can be given intravenously; many specialists find the addition of metronidazole aids improvement. (Streptococci grow very well anaerobically, and metronidazole penetrates the oedematous tissue well.) Early abscesses will resolve, others often rupture into the throat, causing a brief rancid taste followed by relief of pain. A few progress and enlarge, requiring surgical drainage.

Streptococcal bacteraemia

Streptococcal bacteraemia is a rare complication with a high mortality rate. Warning signs are high fever, extreme pain in the draining lymph nodes (rarely, even suppuration or sloughing of the painful nodes; Fig. 6.10) or the appearance of erysipelas-like lesions on the skin. Patients with these features should be treated intravenously while blood culture results are awaited (see Chapter 19).

Poststreptococcal disorders

Poststreptococcal disorders include rheumatic fever, nephritis, erythema multiforme and erythema nodosum. These are discussed in Chapter 21.

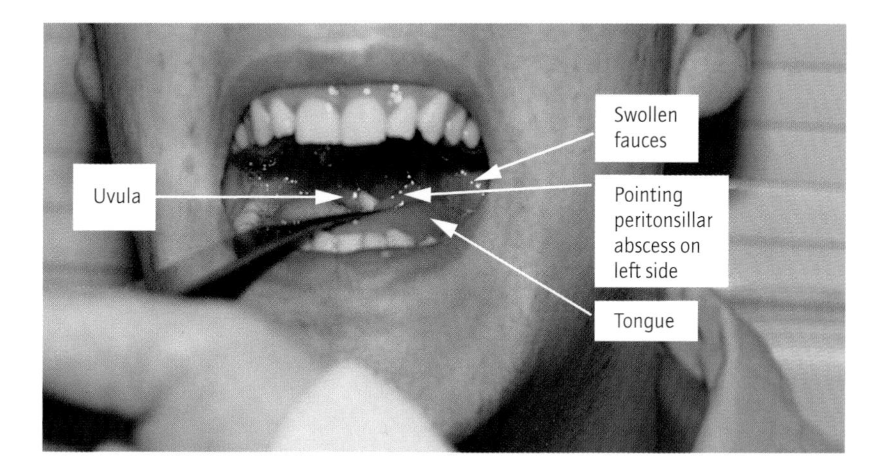

Fig. 6.9 Bilateral quinsies (peritonsillar abscesses).

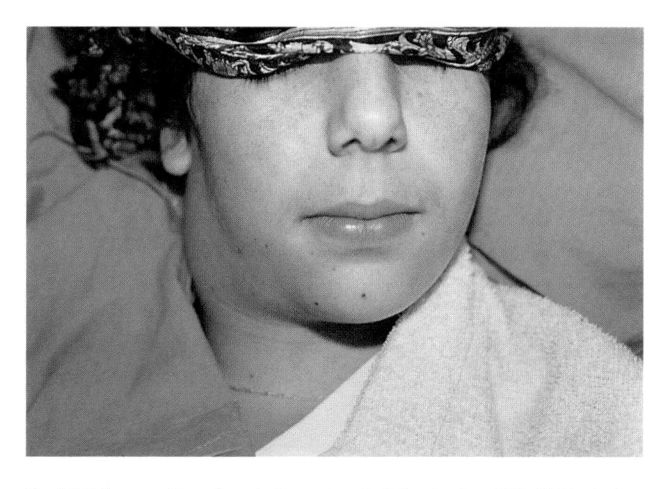

Fig. 6.10 Suppuration of cervical lymph node following tonsillitis (300 ml of pus was drained at surgery).

Acute epiglottitis

Introduction

This is a severe throat infection which causes massive oedema of the epiglottis and threatens the airway. It is a rare but important disease of preschool children; occasional adult cases are reported. It is usually caused by *H. influenzae* although *S. pyogenes* causes a small percentage of cases. The disease must be considered in cases of severe sore throat, as it can cause sudden respiratory obstruction.

Epidemiology

Acute epiglottitis is now rare since the introduction of immunization against *H. influenzae* type b (see Chapter 13). In the prevaccine era, the peak incidence was at about 3 years of age, and most infections occurred during the winter. Epiglottitis due to *S. pyogenes* is uncommon and affects mainly adults.

Clinical features

The illness develops rapidly, with high fever, severe throat pain, swelling and tenderness of the neck and hyoid region, and great difficulty in swallowing. As epiglottic swelling increases the patient drools and then develops stridor.

On examining the throat with the tongue depressed, the red, swollen epiglottis can be seen protruding upwards like a cherry. However, manipulation of the throat can precipitate complete respiratory obstruction and should be avoided if urgent X-ray diagnosis is available. The swollen epiglottis is visible on a lateral X-ray of the soft tissues of the neck, in which it looks like the rounded tip of the thumb, filling the lower oropharynx (Fig. 6.11). The blood count often shows a high neutrophilia, with total white count of $15–25 \times 10^9/l$.

Diagnosis

The diagnosis should be suspected in any patient with severe throat pain and drooling or stridor. Many cases are missed until a late stage because epiglottitis is not considered.

Clinical confirmation is provided by a lateral X-ray of the soft tissues of the neck, or by direct inspection of the throat if there is no alternative. Blood culture should be obtained for microbiological diagnosis. Throat swabs should be deferred until the airway is made safe.

Differential diagnoses

Differential diagnoses include infectious mononucleosis, bilateral peritonsillar abscesses, diphtheria, retropharyn-

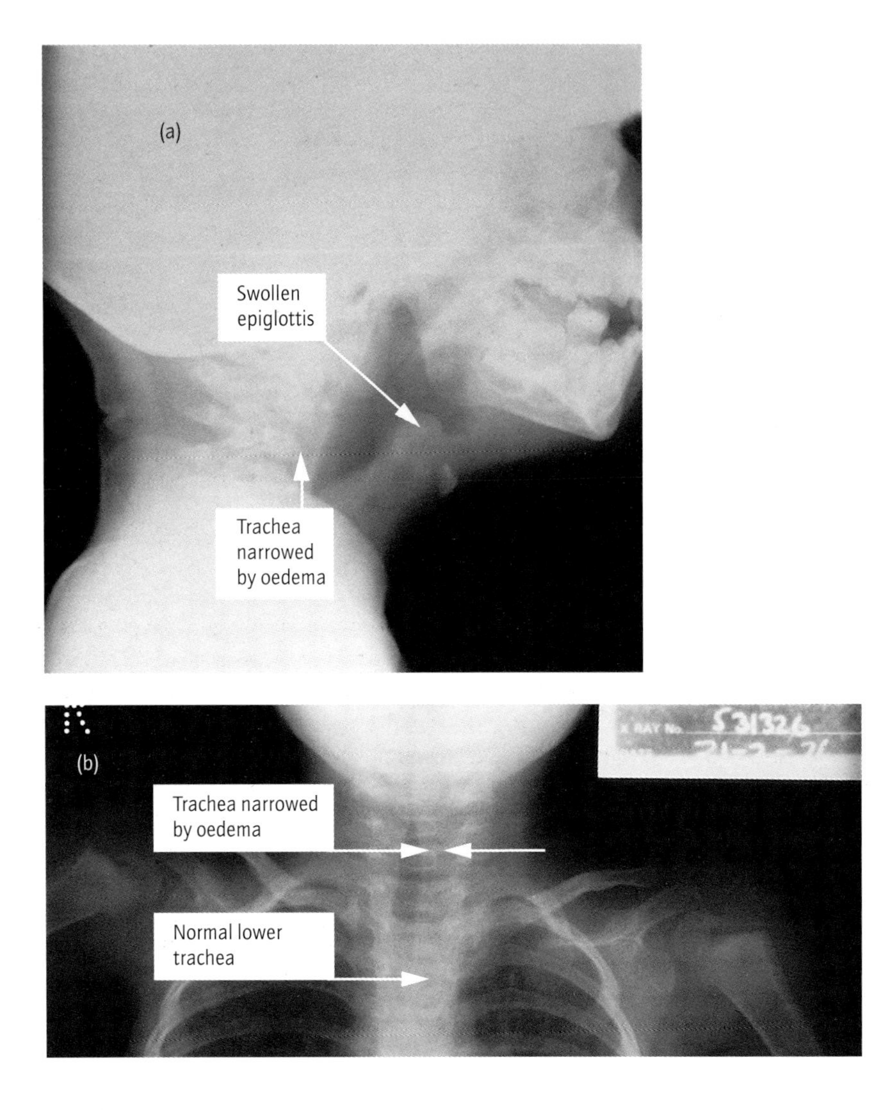

Fig. 6.11 Acute epiglottitis: (a) enlarged epiglottis demonstrated by lateral X-ray of the soft tissues of the neck; (b) oedema tracking downwards narrows the trachea.

geal abscess and Ludwig's angina. A differential white count and heterophile antibody test will exclude infectious mononucleosis (PCR or IgM antibody tests for EBV are more reliable in small children). Cautious examination of the mouth and fauces will exclude the membrane of diphtheria, gross sublingual swelling of Ludwig's angina and the faucial swelling of quinsy. The X-ray will show the site of swelling within, rather than behind, the pharynx.

Treatment of acute epiglottitis

First choice: cefotaxime—child i.v. 150–250 mg/kg daily in two to four divided doses; adult 2–4 g 8-hourly. Or: ceftriaxone—child 50 mg/kg as a single daily dose; adult 2–4 g as a single daily dose. Treatment continued for 10 days.

NB Consider a single bolus dose of corticosteroid to reduce epiglottic oedema.

Management

Acute epiglottitis is a medical emergency. The patient should be allowed to sit up, as this often helps to keep the airway open. High-dose antibiotic treatment should be begun immediately. The treatment of choice is a broad-spectrum cephalosporin such as cefotaxime or ceftriaxone. A course of 10 days is adequate for most cases.

If the airway is critically obstructed, oxygen should be given by mask while urgent tracheostomy is considered. A bolus dose of corticosteroid may reduce oedema and avoid the need for tracheostomy while antibiotic therapy is taking effect. It will not compromise the response to treatment. Many ear, nose and throat departments keep a supply of heliox (80% helium and 20% oxygen), which has an extremely low viscosity and will pass much better than pure oxygen through a tiny airway or a small cannula in the trachea.

Complications

Once the patient reaches medical care the mortality is low and the complications are mainly those of intubation and ventilation: hypostatic lung infections, pneumothorax in young children, and infected tracheostomy sites. Metastatic complications of the bacteraemia are rare, but a few cases also have pneumonia or skeletal infection, which responds to the treatment for epiglottitis.

Prevention and control

Many cases are prevented by childhood immunization against *H. influenzae* type b.

Diphtheria

Introduction

Diphtheria is a rare infection of the respiratory mucosa, and sometimes of broken skin, caused by *Corynebacterium diphtheriae*. Although controlled in most communities by childhood immunization, elderly unvaccinated travellers, or refugees and immigrants from rural areas may suffer from the disease or carry the organism in the nose and throat. Vaccine-induced immunity declines in adult life. Tourists may unwittingly pass through endemic areas, and be unexpectedly infected.

Epidemiology

Humans are the only reservoir of infection. The disease is spread by direct contact with cases or carriers. Patients with cutaneous diphtheria are more infectious than those with pharyngeal and other forms. Following the introduction of routine immunization in 1942 in Britain, the incidence declined rapidly and by the 1960s diphtheria had almost been eliminated. Since then a handful of cases have been reported each year. Most of these are in adults and are acquired abroad, although limited indigenous transmission does occur from time to time. It is noteworthy that the case fatality ratio for diphtheria has increased slightly since 1960 (from 5 to 8 per 100), probably because of failure to recognize the disease now that it is so rare. *C. ulcerans* infection from unpasteurized milk occasionally presents as classical diphtheria.

Immunization also resulted in a decrease in carriers of both toxigenic and non-toxigenic strains of *C. diphtheriae*. This was unexpected, as a toxoid preparation should in theory only protect against the toxic manifestations of disease. The reason for the decline in carriers has not been satisfactorily explained.

In tropical countries where hygiene is poor, diphtheria is still common. Here the usual presentation is the cutaneous form of disease, which is often limited in extent, and confers immunity without a severe, toxic illness. Nasal diphtheria also tends to be mild, but the infected discharge is important in the spread of disease among children.

Pathology

The consequences of infection with *C. diphtheriae* are twofold: the effects of the potent exotoxin; and obstruction of the airway by necrotic debris, which forms a tough pseudomembrane on infected respiratory mucosa (Fig. 6.12).

Diphtheria toxin is the major pathogenicity determinant of *C. diphtheriae*. The genetic sequence encoding the toxin resides on a beta-phage. Bacteria not infected by the phage are non-toxigenic and non-pathogenic. The protein toxin possesses three domains. A receptor portion binds to the target cell and the central portion, which is highly hydrophobic, dissolves in the cell membrane, carrying the toxin portion into the cell. The toxin itself is an adenosine diphosphate ribosylase which ribosylates an amino acid diphthamide present in elongation factor 2. Elongation factor 2 is essential for protein synthesis in the host cell, and its inhibition leads to cell death.

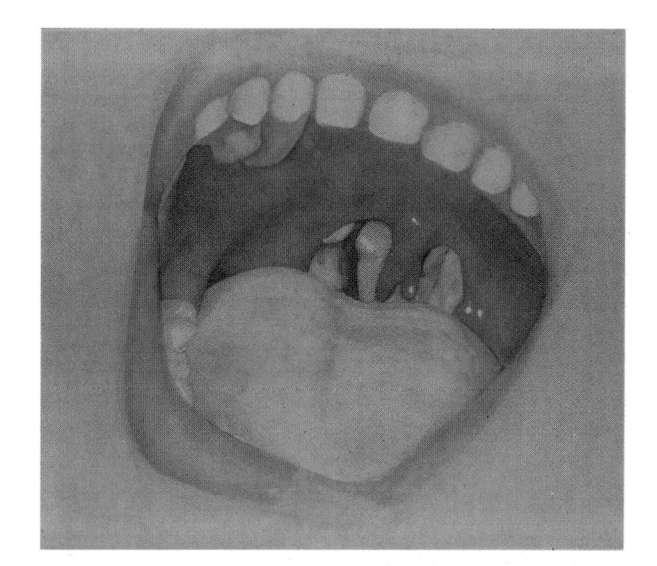

Fig. 6.12 Diphtheria: the off-white, smooth pseudomembrane affects the tonsils and pharynx.

Clinical features

After an incubation period of 3–5 days, the infected site becomes inflamed and gradually covered by a spreading, tough, adherent slough (often called the membrane). The toxic effects of diphtheria are proportional to the extent of the infection, indicated by the area of membrane.

Pharyngeal diphtheria is the commonest respiratory form. This presents with fever, sore throat and marked oedema of the cervical lymph nodes, which may produce a bull-neck appearance (Fig. 6.13). The membrane may affect the tonsils but, unlike the exudate of infectious mononucleosis, also spreads across the pharyngeal mucosa. It is usually greyish and semitransparent, opacified where it contains areas of altered blood, and sometimes blackened and necrotic. It is strongly adherent, and attempts to scrape it away cause pain and bleeding.

Threatened airway obstruction causes stridor. Children often assume a characteristic posture, leaning forward with the neck extended, to hold the airway open.

Diphtheria can also affect the larynx and trachea, and then presents as croup. Only a strong index of suspicion will allow prompt diagnosis in such cases, as the membrane may not be visible in the pharynx.

Other sites which may be affected include the nares, the conjunctiva and small areas of broken skin, such as abrasions or ulcers. These small sites are inflamed, often produce a serosanguineous exudate and may have small adherent patches of membrane. *C. diphtheriae* can colonize sites of other infections and has been found in impetigo, cellulitis and broken chickenpox lesions.

During acute diphtheria there is a modest fever, but disproportionate prostration. There is a neutrophilia in the peripheral blood, but minimal disturbance of renal and liver function.

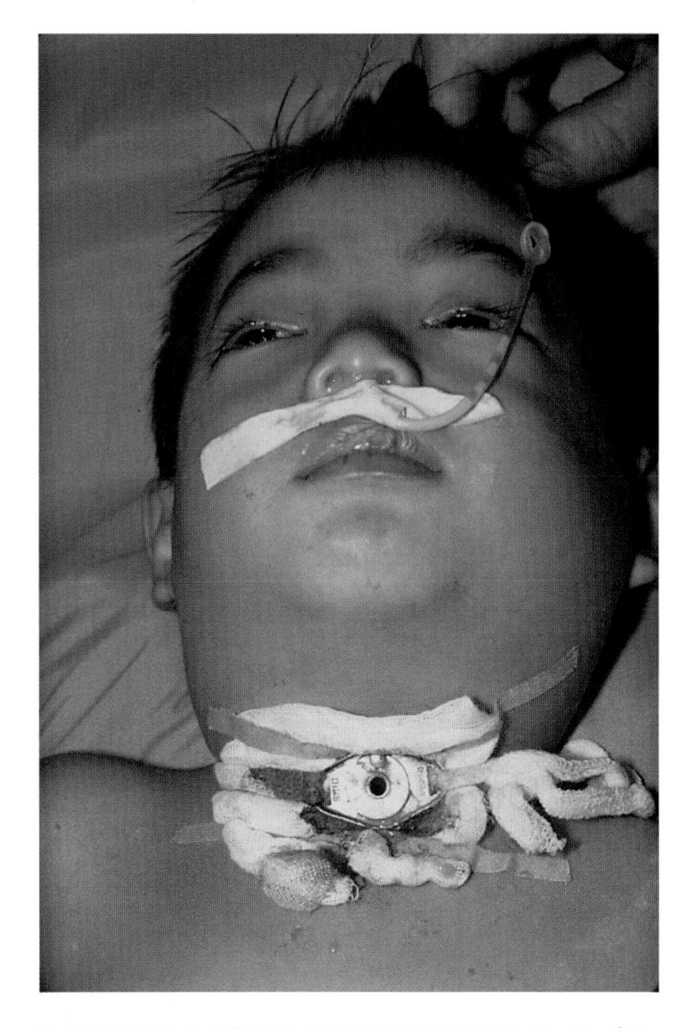

Fig. 6.13 Diphtheria: bull-neck appearance and tracheostomy. Courtesy of the World Health Organization.

Fatalities are usually due to irreversible heart failure. Intrabronchial or tracheal membrane can cause respiratory failure. Very severe cases occasionally die from a Waterhouse–Friderichsen syndrome of adrenal failure and haemorrhagic features.

Diphtheria toxin

The diphtheria toxin causes early cardiac damage and later neurological lesions. In the first week heart failure and conduction defects are likely, and profound heart block is a risk, but the myocardium recovers completely when convalescence is established.

The neurological damage is caused by demyelination, and appears from the second week. It occurs earliest and most severely near to the site of the membrane. Palatal and ocular palsies are common after throat infections. After 3 or 4 weeks some patients develop a generalized weakness or paralysis similar to Guillain–Barré syndrome, but this also is fully reversible.

In rare cases a late nephritis causes impaired renal function.

Diagnosis

Clinical suspicion is aroused by severe throat or pharyngeal swelling and typical membrane with modest fever, severe prostration and neutrophilia.

Laboratory diagnosis

Specimens from the throat, larynx, nose or skin may be examined. Swabs are adequate. It is very important to inform the laboratory that diphtheria is suspected, other-

Organism	Sucrose hydrolysis	Mannitol hydrolysis	Starch hydrolysis	Glycogen hydrolysis	Urease	Nitrate production	Toxins		
							D	Cp	Both
C. diphtheriae var. *gravis*	+	+	+	+		+	+		
C. diphtheriae var. *intermedius*	−	+	−	−		+	+		
C. diphtheriae var. *mitis*	+	+	−	−		+	+		
C. ulcerans	−		+		+	−	+	+	+

D, diphtheria toxin; Cp, toxin produced only by the animal pathogen *C. pseudotuberculosis* and by *C. ulcerans*.

Table 6.2 *Corynebacterium diphtheriae*: tests to identify and distinguish *gravis*, *intermedius* and *mitis* biotypes.

wise special media will not be inoculated and the pathogen may be discarded as a diphtheroid. On tellurite media corynebacteria reduce tellurite to a black substance. The colonies have a black shiny appearance, aiding selection for identification. The Loeffler's slope contains a rich serum medium and the organisms grow rapidly. Sufficient growth is usually available after 6 h to allow staining by Albert's method which demonstrates the volutin granules found in this species. The organism should then be subcultured on to blood agar for biochemical confirmation using sugar tests adapted for corynebacteria (Table 6.2).

Confirmation of toxigenicity is made by polymerase chain amplification which allows rapid detection of the toxin gene in *C. diphtheriae* isolates.

It must be emphasized that the laboratory's role is to confirm the presence of toxigenic *C. diphtheriae* and to alert the public health services. The identification of non-toxigenic *C. diphtheriae* requires no public health control measures.

Management

Antibiotic treatment

Intravenous benzylpenicillin is the treatment of choice. Parenteral erythromycin or a cephalosporin are alternatives. As inflammation subsides in 24–36 h the membrane loosens. Casts of the upper airway or bronchi may be shed, sometimes needing assisted removal by suction or bronchoscopy.

Antitoxin

Antitoxin is given to neutralize circulating toxin and prevent further damage to myocardium and myelin. Dosage ranges from 10 000 I.U. for nasal disease to 120 000 I.U. for aggressive nasolaryngeal diphtheria.

Antitoxin treatment in diphtheria

Type of diphtheria	*Dose (units)*	*Route*
Nasal	10 000–20 000	Intramuscular
Tonsillar	15 000–25 000	Intramuscular or intravenous
Pharyngeal or laryngeal	20 000–40 000	Intramuscular or intravenous
Combined sites or late diagnosis	40 000–60 000	Intravenous
Cutaneous disease	Not widely recommended: wound toilet and antibiotics preferred	

Precede full dose by subcutaneous test dose of antitoxin, e.g. 50–100 U 30 min before main dose. (From WHO (1994) *Manual for the Management and Control of Diphtheria in the European Region*. World Health Organization.)

Human immunoglobulin is not effective, so diphtheria antitoxin is a horse serum preparation. A small intramuscular test dose is therefore given 30–45 min before the main dose, to detect possible hypersensitivity. In severe cases the main dose may be given intravenously.

Elective tracheostomy

Elective tracheostomy avoids possible emergency tracheostomy, which is difficult when the tissues are very oedematous. It also provides airway protection if palatal palsy develops later.

Follow up

Diphtheria cases require prolonged observation to detect cardiographic changes, rhythm disturbances and late neurological complications, which may require airway protection or other support. When patients have recovered, they are likely to be carriers of *C. diphtheriae*. It is usual to obtain two successive negative nose and throat swabs to indicate clearance. Carriage is difficult to eliminate, often requiring one or two cycles of 10–14 days' erythromycin treatment (penicillin will not eradicate carriage).

Prevention and control

Diphtheria vaccine is a formalin-inactivated toxoid preparation. A standard paediatric dose contains at least 30 IU of antigen. In adults and children over 10 years a low-dose (1.5 IU) vaccine is used because of the risk of hypersensitivity reactions. Three doses of vaccine given at monthly intervals, starting at 2 months of age, are recommended for primary immunization in the UK, with a booster dose 3 years later and again before leaving school (see Chapter 26). Nowadays over 95% of children in the UK receive a full course of vaccine. Adults born before 1942, when routine immunization was introduced, have only naturally acquired antibody. Up to 25% of people in this age group have no measurable antibody.

Measuring immunity to diphtheria by antitoxin concentration

1 <0.01 IU/ml: no protection.
2 0.01–0.10 IU/ml: partial protection.
3 >0.10 IU/ml: reliable protection.

Diphtheria is a notifiable disease. Cases and carriers of toxigenic strains should be isolated until three consecutive throat and nose cultures taken 24–48 h apart (the first at least 5 days after completion of antibiotic therapy) are negative. Close contacts should have cultures taken, and also be kept under surveillance for a minimum of 7 days from the last date of exposure to the case. Contacts should also receive a booster dose of vaccine (or a full primary course if unimmunized) and a course of chemoprophylaxis. The regimen for chemoprophylaxis is erythromycin 250 mg four times a day for 7 days. Bacteriological clearance is confirmed by two consecutive negative nose and throat swabs, the first at least 5 days after completion of chemoprophylaxis.

Ludwig's angina

This is a suppurative infection of the hypoglossal tissue planes. It can become an emergency because the oedema and exudate push the tongue upwards and backwards, which may deform the pharynx and threaten the airway.

The origins of the infection are probably the mouth and teeth. Most infections are polymicrobial. Typical implicated organisms include *Streptococcus pyogenes*, *Staphylococcus aureus*, oral streptococci and 'mouth' anaerobes such as *Prevotella melaninogenicus* and *Fusobacterium* spp.

Clinical features develop rapidly. They include pain, fever, difficulty in swallowing and increasing stridor. Examination shows a bull-neck appearance with tenderness of the neck and throat. It is difficult to open the mouth and, when it is open, the tongue is elevated so that its underside is visible above the lower teeth. The fauces may also be swollen.

Treatment should be prompt. Antibiotics should include one active against *S. aureus*. Satisfactory therapy includes penicillin plus cloxacillin or flucloxacillin, or a cephalosporin such as cefuroxime or cefotaxime. Although these agents are effective against mouth anaerobes, metronidazole penetrates oedematous tissues well and may be useful additional treatment. Treatment should usually last for about 10 days.

Oedema may be reduced medically with a bolus dose of corticosteroid, e.g. 200 mg hydrocortisone or 10 mg dexamethasone, intravenously. If this does not provide relief, drainage of pus from the sublingual space may be effective. This is performed by passing perforated drains through the floor of the mouth and out through the skin anterior to the hyoid bone.

Retropharyngeal abscess

This is a suppurative infection in the tissue spaces behind the pharynx. Normally there is only a narrow space between the posterior pharyngeal wall and the anterior ligaments of the spinal column. If oedema and pus expand this space, the posterior pharyngeal wall is pushed forwards, obstructing the airway.

The abnormal position of the pharyngeal wall is difficult to see on inspection, especially if the abscess is low in the throat. Many patients therefore present as emergencies, with neck or throat pain and difficulty in breathing. The diagnosis can be demonstrated by showing a wide soft-tissue space between the vertebrae and the air-filled pharynx on a lateral X-ray of the neck.

Emergency tracheostomy may be life-saving in urgent cases. Medical treatment is identical to that for Ludwig's angina, as the infection is almost always of mouth origin (and only rarely from a spinal infection). Pus can be released by incising the posterior pharyngeal wall. Spinal infection should be excluded by appropriate imaging of the spinal tissues. Rare cases of retropharyngeal abscess result from cervical infection with tuberculosis or (in endemic areas) brucellosis, so pus should be obtained for culture if the spine is involved.

Vincent's angina

This is a synergistic infection of the mouth which particularly affects the gums. It is associated with poor dental health and poor oral hygiene. The patient presents with extreme soreness of the mouth and gums, accompanied by an offensive halitosis.

The microbial cause is the synergistic action of the mouth spirochaete *Borrelia vincenti* and the anaerobes *Fusiformis* spp. If laboratory confirmation of the diagnosis is needed, the organisms can be demonstrated in large numbers by making a Gram stain of the material on a mouth swab.

Treatment with metronidazole will quickly eradicate the anaerobic organisms. Penicillin or ampicillin is also effective. Improved mouth care may be needed to avoid recurrence.

7 Lower Respiratory Tract Infections

Introduction, 132

Laboratory diagnosis, 133
- Introduction, 133
- Sputum examination, 133
- Bronchoalveolar lavage, 133

Viral infections of the lower respiratory tract, 133
- Introduction, 133
- Laboratory diagnosis, 134
- Croup, 134
- Bronchiolitis, 135
- Influenza, 137
- Other viral pneumonias, 140

Pyogenic bacterial respiratory infections, 141
- Chronic obstructive pulmonary disease (COPD, chronic bronchitis), 141
- Bronchiectasis, 142
- Cystic fibrosis, 142
- Pneumococcal (lobar) pneumonia, 142
- *Klebsiella pneumoniae* pneumonia, 144
- *Haemophilus influenzae* pneumonia, 144
- Legionnaires' disease and legionelloses, 147
- Initial assessment and management of community-acquired pneumonias, 149
- Aspiration pneumonia, 149
- Lung abscess, 150
- Anaerobic pneumonia (necrobacillosis), 150

Atypical pneumonias, 150
- Introduction, 150
- Epidemiology, 150
- Microbiology of *Chlamydia* infections, 151
- Clinical features, 151
- Diagnosis of atypical pneumonias, 152
- Management, 152

Pertussis, 153
- Introduction, 153
- Epidemiology and pathology, 153
- Clinical features, 153
- Diagnosis, 153
- Management, 154
- Prevention and control, 154

Fungal infections of the lower respiratory tract, 155
- Introduction, 155
- Aspergillosis, 155
- Histoplasmosis, 155
- Blastomycosis, 155
- Coccidioidomycosis, 155

Introduction

The lower respiratory tract comprises those structures extending downwards from the vocal cords.

The trachea, bronchi and bronchioles are lined with ciliated columnar respiratory epithelium, within which are distributed mucus-producing goblet cells. The airways walls contain smooth-muscle cells, and are elastic, dilating on inspiration and narrowing during expiration. Any pathological narrowing is therefore increased during expiration.

The smallest bronchioles have no muscular coating, and their epithelium is flat and non-ciliated. The alveoli themselves are lined with two types of cells; smooth, flat cells, and slightly thicker cells with a granular cytoplasm. The thicker cells are related to the alveolar macrophages, which are mobile within the alveoli. The alveolar cells are separated from the underlying capillary endothelium by a film of interstitial fluid, in hydrodynamic equilibrium with the alveoli (which usually contain no fluid) and the capillary blood.

The normal lower respiratory tract is bacteriologically sterile. Inhaled particles, including bacteria, are trapped in the mucus which lines the airways and is moved towards the pharynx by the beating of the epithelial cilia. The mucus is then swallowed and the bacteria are destroyed by gastric acid. Particles which reach the alveoli are phagocytosed by alveolar macrophages, which are also expelled by the 'ciliary escalator'. Secretory immunoglobulin A (IgA) in the respiratory epithelium helps to inhibit or destroy invading pathogens.

Common mechanical problems in the respiratory tract include:

1 Paralysis of the cilia by cigarette smoke.
2 Excessive volumes of mucus which cannot be effectively cleared.
3 Mucus too thick to be effectively cleared (in cystic fibrosis).
4 Paroxysmal narrowing of the airways by asthma attacks.
5 Immobilization or damage of alveolar macrophages by particles which they cannot destroy (e.g. silica or asbestos particles).
6 Loss of ability to cough, due to coma or paralysis.
7 Endotracheal intubation, which introduces microorganisms and inhibits effective coughing.

These factors impair the clearance of bacteria and other particles. Larger airways can be cleared by coughing, but this cannot maintain the patency of smaller airways, even when supplemented by vigorous physiotherapy.

The lower respiratory tract is exposed to a variety of inhaled pathogens, which it can often expel before infection becomes established. *Streptococcus pneumoniae* and *Haemophilus influenzae*, from the adjacent pharynx, are common colonists, and are also potential pathogens.

Their rich blood supply makes the lungs vulnerable to infection by blood-borne organisms, of which *Staphylococcus aureus* is the most important.

Non-infectious conditions can mimic lower respiratory infections, by presenting with fever and cough or shortness of breath. These include:

1 Autoimmune disorders such as systemic lupus erythematosus.
2 Granulomatous conditions such as sarcoidosis.
3 Vasculitides, particularly Wegener's granulomatosis or polyarteritis nodosa (both of which can also produce nodular opacities on chest X-ray).
4 Hypersensitivity reactions such as farmer's lung or the more severe Goodpasture's syndrome.
5 Malignancies, such as infiltrating lymphomata.

Laboratory diagnosis

Introduction

Identifying pathogens in respiratory specimens is difficult because of the problems of obtaining uncontaminated specimens. Diagnostic efforts are therefore directed at the common important pathogens. Adequate diagnosis can usually be achieved by a combination of sputum microscopy and culture, and serological tests on blood.

Sputum examination

This is the commonest procedure carried out in the microbiological laboratory, but it is also the most controversial because of the difficulty in obtaining adequate specimens. Physiotherapy is often helpful in obtaining lung secretions, rather than a specimen containing mostly saliva and buccal epithelial cells, and may be additionally aided by prior inhalation of nebulized saline (providing an 'induced sputum' specimen).

The specimen is examined initially by microscopy of a Gram-stained smear. Specimens containing few polymorphs and more than 25 epithelial cells per low-power field are largely salivary, and are likely to yield only upper respiratory tract flora on culture. Alternatively, a specimen containing many neutrophils and ciliated columnar cells is largely of bronchial origin. Some organisms, such as *Klebsiella pneumoniae*, have a characteristic morphology and are seen in high numbers when causing infection.

An important consideration is that the common respiratory pathogens also occur in the normal upper respiratory tract flora. This can be overcome by estimating the numbers of organisms in the specimen, using semiquantitative methods. Organisms present at more than 10^6 CFU/ml are considered more likely to be acting as pathogens. The results of semiquantitative sputum culture are comparable with results from specimens obtained by cricoid puncture or bronchial aspiration.

Bronchoalveolar lavage

Bronchoscopy is valuable in the diagnosis of lower respiratory infections with unusual pathogens or in immunocompromised patients (who rarely produce excess sputum because of their reduced inflammatory responses). The bronchoscope is passed into a lower airway and 100–200 ml of sterile saline is injected and aspirated. Although the material has come directly from the alveoli, semiquantitative methods are advisable, as contamination can still occur from the advancing end of the bronchoscope. Organisms present at more than 10^3 CFU/ml may be pathogens. Bronchoalveolar lavage is of particular value for the diagnosis of infections such as tuberculosis, legionellosis, fungal infections and *Pneumocystis carinii* infection.

Obtaining specimens from the lower respiratory tract

1 Sputum obtained by coughing.
2 'Induced sputum' specimens.
3 Broncheolar lavage.
4 Direct bronchial aspiration.
5 Cricoid puncture.

Viral infections of the lower respiratory tract

Introduction

Viral respiratory infections are among the most common of all infections and many pass unremarked, being part of a systemic disease such as measles, varicella or infectious mononucleosis, or mild cases of influenza or respiratory syncytial virus (RSV) infection. Nevertheless, some viral infections can cause severe pulmonary disease.

Laboratory diagnosis

Nasopharyngeal secretions, which always contain respiratory cells, should be obtained from children under the age of 5 years. A fine-bore catheter is passed through the nostril into the nasopharynx and secretions are aspirated using gentle suction via either a 20- or 50-ml syringe or a low-pressure suction apparatus. In older children and adults, specimens may be obtained by gargling normal saline.

Direct immunofluorescence is a useful rapid laboratory test for some viral antigens, particularly RSV, influenza, parainfluenza or measles. Nasopharyngeal secretions are dried on a Teflon-coated slide, acetone-fixed and incubated with rabbit antibody to the expected virus. If viral antigen is present in the respiratory cells, the rabbit antibody will bind to it, and is demonstrated by staining with fluorescein-labelled antirabbit antibodies.

ORGANISM LIST

Influenza virus
Parainfluenza viruses
Respiratory syncytial virus (RSV)
Adenoviruses
Measles, varicella and Epstein–Barr viruses (EBVs)

Croup

Introduction

Croup is a syndrome of laryngotracheobronchitis, which usually affects children but also occurs occasionally in adults. Its importance is that almost every young child has at least one episode, it is often a child's first severe illness, the physical signs are distressing and there is a risk of respiratory obstruction in the most severe cases, due to gross swelling of the aryepiglottic folds. The common syndrome of croup must be distinguished from rarer causes of partial respiratory obstruction such as epiglottitis or diphtheria.

Epidemiology

Croup is common in infants. The principal causal agents are parainfluenza viruses types 1 and 2 which circulate during the late autumn and winter months. They are highly infectious by the air-borne route. Prodromal measles causes a similar cough; influenza can cause croup in both adults and children.

Organisms that may cause croup

1 Parainfluenza virus.
2 Influenza virus.
3 Measles virus.
4 Respiratory syncytial virus.
5 *Haemophilus influenzae.*
6 *Corynebacterium diphtheriae.*

Pathology

Parainfluenza viruses are classifed in the genus *Paramyxovirus* within the family Paramyxoviridae. Four parainfluenza virus types (types 1–4) are pathogenic to humans. The virus is pleomorphic with a roughly spherical shape, 120–300 nm in diameter, and possesses an envelope. The envelope contains glycoprotein projections of haemagglutinin, and a neuraminidase. The virion contains a helical nucleocapsid, which consists of viral negative-sense RNA attached to the large (L) protein (a polymerase complex), the matrix (M) protein and the P protein (a polymerase complex). Viral RNA is transcribed by virally encoded transcriptase into positive (sense) RNA.

The virus invades epithelial cells throughout the tracheobronchial tree, damaging the columnar ciliated cells and their ciliary function. There is oedema of the mucosa, with an acute neutrophilic infiltrate. In a minority of cases a severe pneumonia develops; in fatal cases there is haemorrhagic bronchiolitis and also loss of ciliated epithelium. Polymorphs and macrophages are found in the alveolar exudate.

Clinical features

The syndrome develops rapidly, with bursts of harsh, barking coughs interspersed with noisy breathing. Many cases are mild, with no other features. In more severe cases there is fever and increasing stridor, sometimes with recession of the intercostal spaces on inspiration. Restlessness and tachycardia increase as respiratory obstruction develops. Auscultation of the chest rarely reveals abnormal sounds, unless the bronchi are sufficiently affected to be loaded with mucus. Cyanosis is a late and grave sign. Most cases resolve within a week, but a croupy cough may return if there are further respiratory infections in following weeks.

Diagnosis

In most cases this is clinical, based on the typical cough. A low or normal white cell count supports a probable viral aetiology. In patients with neutrophilia, other causes of croup or pharyngeal obstruction should be considered. These include epiglottitis and occasionally diphtheria.

A parainfluenza virus may be demonstrable by immunofluorescent staining or reverse transcriptase polymerase chain reaction (RT-PCR) of nasopharyngeal aspirate. Virus may be recovered by cell culture of respiratory secretions although this is too slow for clinical use. Serodiagnosis is rarely undertaken.

Management

The classic management of surrounding the patient with a warm, steamy atmosphere is often effective. It relieves pain and reduces cough, and the heat may inhibit viral replication. Controlled trials have confirmed that short courses of dexamethasone 4–8 mg twice daily or daily (child 1–4 mg daily or twice daily) improve airway function without causing adverse effects. Intravenous doses of 25–50 mg hydrocortisone (100–200 mg in an adult) may be repeated three or four times on the first day.

Bronchiolitis

Introduction

This is similar in its pathology to croup, but the small bronchioles are most affected instead of the upper airways. In this condition respiratory difficulties often develop suddenly in a child with an apparently mild illness. This causes considerable morbidity in infants, but with modern treatment mortality is very low.

Epidemiology

The infection principally affects infants up to the age of 2 years. Epidemics occur during late winter and early spring (Fig. 7.1). These are usually due to RSV, although other viruses may cause epidemics. Epidemics often coincide with increases in the incidence of sudden infant death syndrome, suggesting a possible role of RSV in its aetiology. Preterm infants and those with congenital heart disease are at particular risk of severe infection. Serious

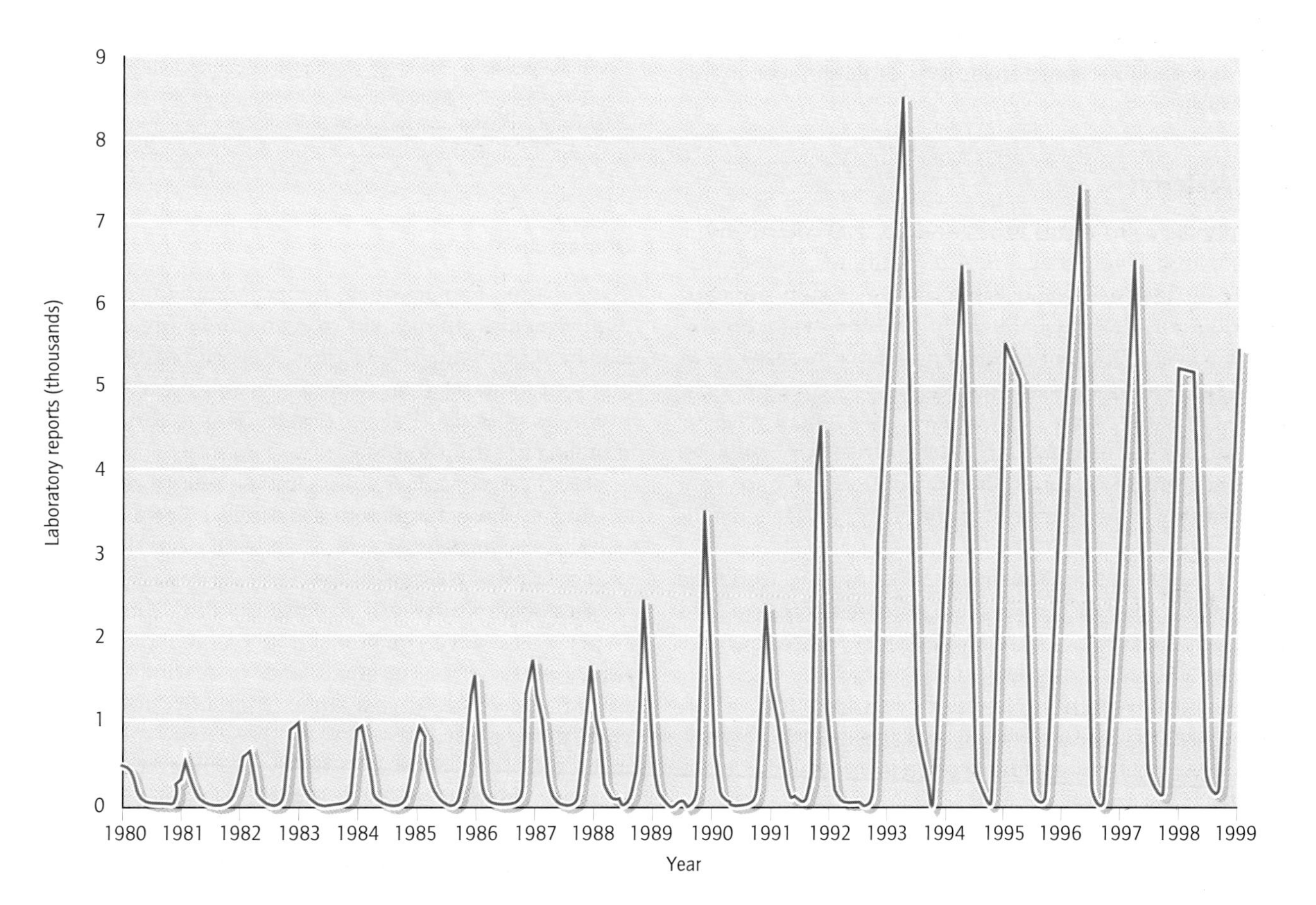

Fig. 7.1 Epidemic curve of respiratory syncytial virus infections, showing regular winter/spring epidemics.

outbreaks have been reported among newborn babies on intensive care units.

Humans are the only source of infection. The incubation period is 5–8 days and the patient remains infectious for about 7 days after the onset of symptoms.

Pathology

RSV is a member of the genus *Pneumovirus* in the family Paramyxoviridae. It is an enveloped virus of 120–300 nm diameter. The helical RNA genome codes for at least 10 polypeptides including F and G envelope-associated glycoproteins. The F or fusion protein is associated with penetration of the virus into cells, and its spread from cell to cell. The larger G protein is responsible for initial attachment of the virus to host cells. There is antigenic variation among strains of RSV but this is not easy to demonstrate, as human sera contain neutralizing antibody which neutralizes all viruses of this genus.

RSV is acquired through the upper respiratory tract and spreads throughout the respiratory epithelium. Infection is characterized by peribronchial inflammation, with oedema of the submucosa and adventitium. There is necrosis of the bronchiolar epithelium and plugging of the airways. When the bronchiolar lumen is incompletely obstructed a ball-valve effect occurs, which results in air-trapping, and the characteristic hyperinflation seen in this condition.

Clinical features

Initially the child seems to have a cold with catarrh and a loose cough. Then after 3 or 4 days the respiratory rate suddenly increases, and intercostal recession appears. Paradoxically, fever usually declines at this stage. Breathing is wheezy, and bottle-feeding is often impossible, as crying or feeding causes cyanosis.

Examination of the chest reveals widespread wheeze, particularly in expiration. Patchy areas of dullness, reduced breath sounds or adventitious sounds are common and may change with coughing, which is repetitive and often constant.

Bronchiolar obstruction causes atelectasis in parts of the lung, while partial obstruction causes air-trapping and hyperinflation elsewhere. The chest X-ray therefore shows varying areas of both opacity and lucency.

Slow improvement begins after 5 or 6 days. It may take a week or two for the child to return to normal, but the capillary oxygen saturation is often subnormal for twice as long.

Diagnosis

Clinical diagnosis is often obvious.

Laboratory diagnosis

Laboratory diagnosis of RSV infection can be made by immunofluorescence, enzyme immunoassay (EIA) or RT-PCR performed on nasopharyngeal secretions. Sensitivity of between 70 and 100% can be expected. RSV can also be cultured in Hep-2 or HeLa cells. The characteristic cytopathic effect is the formation of syncytia, generally seen between 2 and 7 days after inoculation.

A retrospective diagnosis can also be achieved by detecting rising titres of RSV antibodies, by complement fixation, virus neutralization, EIA and indirect immunofluorescence tests.

Laboratory diagnosis of respiratory syncytial virus infection

1 Direct immunofluorescence of respiratory secretions.
2 Reverse transcriptase PCR to demonstrate viral RNA.
3 Cell culture.
4 Serological: complement fixation, enzyme immunoassay or indirect immunofluorescence.

Management

Oxygenation and adequate hydration are the immediate requirements. Humidified oxygen can be given in a cot-sized oxygen tent. The inspired oxygen concentration is adjusted to maintain as near normal arterial oxygen saturation as possible. Pulse oximetry is a useful means of monitoring saturation.

Bronchospasm plays little part in the bronchiolar narrowing, so bronchodilators are rarely helpful. A trial of nebulized salbutamol is worthwhile however, if there is a past or family history of atopy.

Nebulized tribavirin shortens the duration of hypoxia and fever in bronchiolitis, and is licensed for this use. Its effect is small and is of limited benefit in otherwise healthy infants. Morbidity from bronchiolitis is great, however, in those with pre-existing cardiac or lung disorders; tribavirin may be of benefit to these children. A 20 mg/ml solution of tribavirin is nebulized, and administered for 12–18 h daily, for at least 3 days.

Complications

Secondary bacterial infection is uncommon, but can threaten life. High fever, change from mucoid to purulent bronchial secretions, sustained deterioration in oxygen saturation and rising neutrophil count are all warning signs. The chest X-ray may show a segmental opacity or an air bronchogram, suggesting consolidation.

Blood and aspirated secretions should be cultured and intravenous antibiotic treatment started without delay. A cephalosporin such as cefuroxime or cefotaxime will be effective against *Streptococcus pneumoniae* or *Haemophilus influenzae*, which are the likely pathogens, and will also act against occasional *Staphylococcus aureus* infections.

Prevention

No vaccine is currently available; however, a number of promising candidates are undergoing clinical trials, including both subunit and recombinant preparations.

Influenza

Introduction

Influenza is a moderate to severe illness most often caused by influenza A or B viruses. It is highly infectious to susceptibles, and with its short incubation period it can cause overwhelming epidemics. Sudden absence of staff adversely affects commerce, industry and public services. Individuals with the disease can be seriously ill, and severe secondary bacterial infections cause as much morbidity and mortality as the influenza itself.

Pathology

Influenza virus infects ciliated respiratory cells, killing many, and causing shedding of the majority. This leaves denuded airways, susceptible to colonization and invasion by bacterial pathogens. There is a vigorous immune response, with much production of interferon and temporary impairment of cell-mediated immunity (the tuberculin reaction cannot be elicited until convalescence is complete). Although influenza viraemia is difficult to demonstrate, viral genome is present in large quantities within the mononuclear cells of the blood during acute infection, and many body tissues can be invaded.

Virology

Influenza virus is part of the genus *Orthomyxovirus* in the family Orthomyxoviridae. Virus particles are 80–120 nm in diameter. The RNA genome is segmented, with four fragments. The RNA is closely associated with nucleoprotein (NP) in a helical structure. The NP is specific to the three types of influenza virus: A, B and C. The matrix protein or membrane protein, with a mass of 2 kDa, surrounds the nucleocapsid and is the major protein of the virus particle. This is enclosed by the envelope, which contains two virus-encoded glycoproteins: the haemagglutinin, responsible for attachment of the virus to host cell receptors, and the neuraminidase, of molecular weight 200–250 kDa, important for release of virus from infected cells (Fig. 7.2). Both the haemagglutinin (H) and the neuraminidase (N) are antigenically variable, and this is useful in the classification of the viruses.

The virus adsorbs to host cell surface structures which contain sialic acid. After internalization the haemagglutinin molecule is transformed into a fusogenic form facilitating fusion of the virus with the endosome. The M2

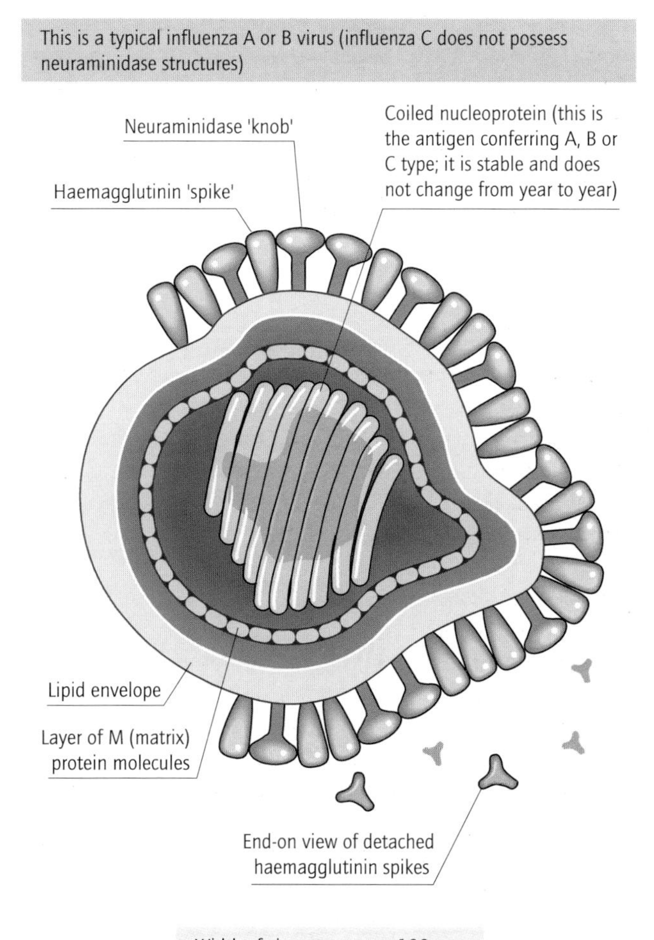

Fig. 7.2 Structure of influenza viruses.

matrix protein forms an ion channel that results in an influx of protons releasing the core from the M1 matrix protein.

Epidemiology

Epidemics of influenza occur in most years during the winter months. These are usually due to type A viruses, although type B and mixed type A and type B epidemics also occur. Type C is occasionally recognized as causing local outbreaks. Attack rates during epidemics seldom exceed 10% in the general community, but may be up to 50% in closed institutions such as boarding schools and nursing homes for the elderly. Extensive pandemics are much less common. This century, pandemics have been recorded in 1918, 1957 and 1968.

These epidemiological findings are mirrored by the changes in the antigenic structure of the virus.

Antigenic drift is responsible for the yearly epidemics as new subtypes of haemagglutinin slowly evolve. Infection with an epidemic strain provides antibody which will cross-react with closely related haemagglutinin types, and confers partial immunity to infection by slowly evolving types from year to year. Antigenic drift arises in the virus because of selection pressure imposed by the partial immunity of the population. Periodically a major change (shift) occurs in the haemagglutinin or neuraminidase antigen. The population has little or no cross-reacting antibody to the 'new' virus, and a pandemic results.

Age-specific attack rates during an epidemic reflect existing immunity from exposure to previously prevalent strains. Attack rates are often highest among school-age children. In contrast, complication rates and mortality are greatest in the elderly and those with underlying chronic conditions. The most severe forms of infection occur during pandemics.

Excess mortality during influenza epidemics

1 During an influenza epidemic, deaths due to respiratory disease may increase by as much as 50%.

2 Deaths due to cerebrovascular and cardiovascular disease also increase.

3 Excess mortality is not usually followed by a deficit during the following year.

4 There were an estimated 26 080 excess deaths attributable to influenza in England and Wales during the 1989/90 epidemic.

5 Between 80% and 90% of excess deaths are in people aged 65 years or over.

Clinical features

After an incubation period of 3 or 4 days there is a sudden onset of severe malaise, arthralgia, myalgia and prostration. Fever may be high, and accompanied by shivering and sweating. Malaise can be so severe that the disease begins with an episode of vomiting. A variable range of associated symptoms includes headache, sore throat, loose stools and shortness of breath. Uncomplicated influenza usually persists for 5–7 days before the fever decreases and gradual convalescence begins.

The respiratory tract is the most important target of the infection. There is always a viral pneumonitis, which may be widespread and relatively mild, or may take the form of interstitial pneumonia or a segmental infection. Severe pneumonitis can cause considerable morbidity and sometimes death. Other respiratory features include tracheitis, croupy cough, laryngitis and sometimes sinusitis and conjunctivitis. Influenza C infections are uncommonly recognized. They are often mild, and conjunctivitis is a prominent feature.

Other body systems can be significantly affected. There is often a subclinical myocarditis with altered electrocardiogram patterns and variable extrasystoles, with postural or exercise-related hypotension.

There is a variable rate of viral encephalitis or meningoencephalitis in influenza epidemics. This coincides with viral excretion from the respiratory tract. Cases often present with headache, meningism, irritability, altered personality and drowsiness. Examination of cerebrospinal fluid (CSF) shows excess lymphocytes and slightly raised protein levels. Recovery is slow and fluctuating.

Laboratory findings

Laboratory findings reflect tissue damage and the response to interferon production. The white cell count is low, often 3 or $4 \times 10^9/l$. The transaminase levels are raised, and are of both liver and tissue origin. Amylase levels are also occasionally raised. Oxygen saturation, measured by pulse oximetry, is often reduced, even when there is no clinical evidence of significant pneumonitis. Some patients with severe myalgia and weakness have elevated creatine kinase levels.

Complications

Secondary bacterial infection

Secondary bacterial infection is a common problem and should be urgently considered in any patient with deteriorating repiratory function. Influenza predisposes not only to *Streptococcus pneumoniae* and *Haemophilus influenzae* secondary infections, but also to severe *Staphylococcus aureus* pneumonia, which often develops quickly, causing purulent sputum and deteriorating lung function, but any increase in the white cell count is slow or late. Nodular or patchy opacities are seen on chest X-ray; small abscess cavities may develop; and staphylococcal bacteraemia may occur.

Blood and sputum cultures should be obtained, and antibiotic treatment promptly commenced, including drugs that are effective against *S. aureus*.

Bacterial sinusitis or otitis media can also complicate influenza in both children and adults. The expected pathogens are the same as for pneumonia.

Postviral complications

Postviral complications occur with variable frequency, depending on the characteristics of both the causative virus and the affected population. Prolonged convalescence with fatigue is common, often lasting from 6 to 12 weeks. Neurological effects include Guillain–Barré syndrome, mononeuropathies and occasionally encephalopathy. Persisting encephalopathy is rare, but the encephalitis lethargica cases in 1918 may have been related to the influenza pandemic of that year.

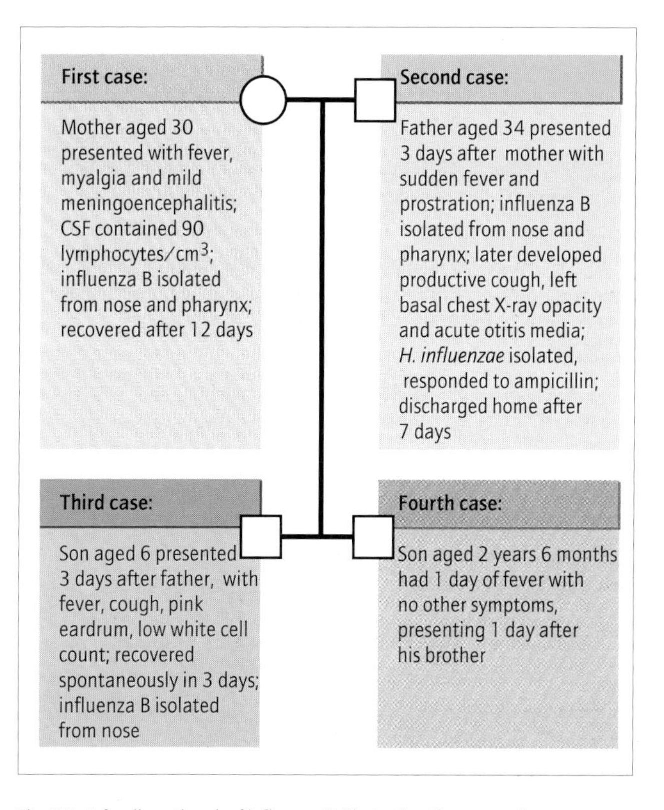

Fig. 7.3 A family outbreak of influenza B, illustrating the range of presentations and their relationship to the patients' ages.

Diagnosis

Mild and sporadic cases may be mistaken for other feverish conditions, although virus may later be identified in cell culture of nose- or throat-swab specimens, nasopharyngeal aspirate, bronchial washings or even CSF. Cases may be identified clinically in epidemics (Fig. 7.3).

Laboratory diagnosis

A rapid diagnosis can be made by demonstrating type-specific antigen in nasopharyngeal cells by direct immunofluorescence. Influenza virus antigen can be detected by EIA and viral RNA by RT-PCR.

Virus isolation should also be carried out so that the antigenic structure of the virus can be defined for future vaccine preparation. The virus may be isolated by inoculating nasopharyngeal secretions on to continuous human diploid fibroblasts incubated at both 33°C and 37°C, as some viruses replicate optimally at a lower temperature. The growth of virus in the cell culture is detected by haemadsorption using guinea-pig and chicken erythrocytes (but guinea-pig erythrocytes do not adhere to cells infected with type C). Virus can also be recovered from nose or throat swabs and occasionally from CSF. After primary isolation, the virus is typed in dedicated reference laboratories.

Rising titres of neutralizing, complement-fixing or haemagglutinin antibodies can be demonstrated in paired sera.

Laboratory diagnosis of influenza

1 Direct immunofluorescence, EIA or RT-PCR of respiratory secretions.

2 Cell culture of respiratory secretions or cerebrospinal fluid.

3 Serologically by complement fixing or haemagglutinating antibodies.

Management

Rest, warmth, adequate hydration and analgesia all give considerable relief. The myocardium is protected from excess work, and oxygen desaturation is limited while the patient rests in bed. Reduction of fever is not usually necessary; indeed virus replication may be reduced when the temperature of the mucosae rises above 35°C.

Antiviral drugs

Amantadine, given within 48 h of onset, has some effect against clinical influenza A (but not other types), shortening the duration of fever by 24–36 h. It is given in a dose of 100 mg twice daily for 5–7 days. Side-effects include drowsiness, confusion and reticulate rash, and can be difficult for elderly patients to tolerate. Rimantidine (not licensed in the UK) is an alternative, to which resistance quickly emerges.

Neuraminidase antagonists, such as zanamavir, are now available and also reduce the severity and duration of illness. They are only effective if given within the first 24–36 h of symptoms.

Tribavirin, given by nebulizer (see above, for RSV) has been used in treating influenza pneumonia, but is not licensed for this indication.

Antibiotics

Antibiotics should be given as soon as culture specimens have been obtained when there is suspicion of secondary bacterial infection. An antistaphylococcal agent should always be included.

Prevention and control

Influenza vaccines are non-replicating virus vaccines with a clinical efficacy of approximately 70%; though less efficacious in the elderly, debilitated or immunosuppressed, they still reduce the severity of disease in these groups. Different strains of antigen are used in different years, to adapt to the changeable antigenic structure of the influenza A virus. Usually there are two type A and one type B virus in the vaccine. Some vaccines are composed of purified viral surface antigens; others, called split vaccines, are made by separating viral core material from surface structures and using the partly purified surface structures as the vaccine antigens.

Vaccination is recommended for all people over 75 years of age, and for children and adults under 75 with chronic respiratory and cardiac disease, chronic renal failure, diabetes and other endocrine disorders and immunosuppression. It should also be given to residents of nursing homes and other institutions for the elderly. In some countries healthcare workers are routinely vaccinated. The vaccine is contraindicated in patients with known anaphylactic hypersensitivity to egg products.

Amantadine hydrochloride may be used prophylactically during an outbreak of influenza type A. It is indicated for patients in high-risk groups for whom vaccination is contraindicated, or in the 2 weeks following vaccination, before protective antibodies have developed.

Amantadine for treatment and prophylaxis of influenza A

1 Treatment: 100 mg twice daily orally for 5–7 days (child 10–15 years 100 mg daily).

2 Prophylaxis: 100 mg daily orally for 7–10 days.

Other viral pneumonias

Many viruses are capable of causing localized or generalized pneumonitis. Segmental pneumonia, with a distinct chest X-ray opacity, can be seen in infections with adenovirus, RSV, parainfluenza virus and EBV. Pneumotropic hantaviruses, excreted by rodents, cause severe and fatal pneumonias in various parts of the world (see Chapter 25), but are not endemic in the UK.

Adenovirus

Adenoviruses, particularly type 5, tend to affect children and young adults. Chest infection may be accompanied by pharyngitis, conjunctivitis, tender lymphadenopathy or, occasionally, a morbilliform rash. Epidemics can affect closed communities in barracks or institutions. Vaccines are used to prevent epidemics in military recruits.

Respiratory syncytial virus

RSV can cause outbreaks of segmental pneumonia, particularly among the elderly. The illness is mild to moderate, with a low mortality.

Pneumonias in other viral infections

Parainfluenza and EBV tend to cause subclinical lung disease, though there may be a cough. The segmental lung disease is often discovered by chance on chest X-ray. EBV can also cause lymphocytic interstitial pneumonitis in immunosuppressed patients.

Diffuse pneumonitis can be caused by parainfluenza virus, measles and chickenpox. The lung disease is rarely clinically significant in immunocompetent individuals. Varicella pneumonia is the exception, being potentially severe or life-threatening, but it is accompanied by the characteristic chickenpox rash (see Chapter 11).

Viruses causing pneumonitis
1 Adenovirus.
2 Respiratory syncytial virus.
3 Parainfluenza virus.
4 Varicella zoster virus.
5 Epstein–Barr virus.
6 Pneumotropic hantaviruses.

Pyogenic bacterial respiratory infections

ORGANISM LIST

Streptococcus pneumoniae
Haemophilus influenzae
Moraxella catarrhalis
Klebsiella pneumoniae
Staphylococcus aureus
Escherichia coli and other Gram-negative rods
Fusobacterium necrophorum
Rare bacterial lung infections: *Streptococcus pyogenes*, enteric fevers (see Chapter 24), anthrax, plague (see Chapter 25), leptospirosis (see Chapter 9)

Chronic obstructive pulmonary disease (COPD, chronic bronchitis)

Introduction

This is a condition characterized by excessive mucus production and poor clearance of the bronchi, with varying degrees of accompanying bronchospasm. The abnormal bronchial tree acquires a persistent colonizing flora, and repeated attacks of acute infection cause progressive damage to the bronchi, leading to further infectious episodes, increasing bronchospasm, and an inexorable decline in lung function.

Epidemiology

COPD is among the commonest causes of hospital admission. The incidence is greatest in adults, although children are also affected. Patients who smoke and those with dust diseases are particularly affected, probably because ciliary clearance of secretions is impaired. Patients with cardiac disease have poor alveolar clearance due to oedema. Most infective episodes occur during the winter. There is evidence that the incidence is related to levels of atmospheric pollution, particularly sulphur dioxide levels.

Pathology

The infections are superficial, affecting the mucosa of the bronchi and the mucus-producing goblet cells. There is oedema and acute inflammatory exudate, both of which acutely exacerbate chronic obstruction and bronchospasm.

Clinical features

The illness develops with gradually increasing cough, respiratory rate, expectoration and sputum purulence. The tendency to bronchospasm increases. Systemic effects are few. There may be a mild or moderate fever and the white cell count may be slightly raised. The chest X-ray rarely shows any definite change unless bronchial obstruction has precipitated a complicating segmental infection.

Diagnosis

This is clinical. The sputum is always colonized with upper respiratory tract flora, and may contain neutrophils. This does not change, except in degree, during acute exacerbations.

Management

General measures are important. Adjustment of bronchodilator treatment, including temporary increases in topical or systemic corticosteroid dosage, will aid ventilation, coughing and drainage of the bronchi. Physiotherapy may be a valuable adjunct. An increase in inspired oxygen concentration may be indicated if hypoxia is severe; inspired oxygen dosage should be sufficient to maintain adequate blood oxygen saturation or PO_2.

Oral antibiotic treatment is often adequate for the superficial mucosal infection. Agents suitable for *S. pneumoniae* and *H. influenzae* include tetracyclines, ampicillin, amoxycillin or clarithromycin. Almost all *Moraxella catarrhalis* isolates produce a beta-lactamase, but are sensitive to co-amoxiclav and most non-beta-lactam antibiotics. Courses of antibiotic should be just long enough to terminate infection, but insufficient to encourage the emergence of new colonizing flora; 5 days is ideal. Patients who have received repeated courses of antibiotics may become colonized with Gram-negative rods such as *Escherichia coli* or *Klebsiella pneumoniae*. These may need treatment with antibiotics such as co-amoxiclav or cefuroxime.

In severe episodes, parenteral antibiotics may be indicated. Ampicillin or cefuroxime is often effective. The use of very broad-spectrum antibiotics, such as third-generation cephalosporins, predisposes to antibiotic-associated diarrhoea.

Bronchiectasis

In bronchiectasis the structure of bronchi is abnormal, either because of congenital conditions (e.g. Kartagener's syndrome) or because of damage from repeated infections. Areas of lung contain saccular or fusiform spaces derived from dilated bronchi, and these discharge copious purulent sputum. Cilary clearance is ineffective in these wide spaces. Indeed, the respiratory epithelium may become squamous after many infectious insults.

In bronchiectasis, long-term complications include amyloidosis, hypertrophic pulmonary osteoarthropathy (HPOA) and late malignant change.

Cystic fibrosis

In cystic fibrosis (CF), abnormal and tenacious mucus accumulates in the bronchial tree. The ciliary clearance process is ineffective, leading to colonization of the bronchi with a resident flora which causes repeated infections.

In bronchiectasis and CF segments of the bronchial tree are chronically obstructed and inflamed, producing large volumes of mucus and mucopus. The effect of chronic infection is debility and, in children, failure to thrive. Progressive respiratory insufficiency adds further debility. The infecting organisms include *S. pneumoniae* and *H. influenzae*, but others are common, including *Staphylococcus aureus*, *E. coli* and other Gram-negative rods.

CF sufferers have poor phagocyte function, which additionally predisposes to chronic infection with *S. aureus*, *Pseudomonas aeruginosa* and *Burkholderia cepacia*. *P. aeruginosa* sometimes produces large quantities of alginate capsule-like material, enclosing microcolonies which are protected from the host's immune defence. Furthermore, it produces a variety of enzymes and toxins, including elastases, proteases, DNAse, lecithinase and exotoxin A. These are damaging to both bronchial structures and defence systems, and contribute to deteriorating lung function. Interestingly, azithromycin can switch off production of these substances, even though it does not kill the organism (which has high minimum inhibitory concentrations for macrolide drugs). This finding has not so far been put to clinical use. Some strains of *Burkholderia cepacia* can colonize the lungs of CF sufferers and cause severe infective episodes. The rate of decline in lung function is directly related to the acquisition of these organisms.

Treatment of both conditions depends on vigorous physiotherapy and postural drainage, usually a lifelong twice-daily burden. This is combined with early and vigorous antibiotic treatment of infections. In CF patients it is usual to treat exacerbations immediately with intravenous drugs which have an antipseudomonal action, e.g. ceftazidime. Many young patients have implanted cannulae with subcutaneous access so that domiciliary treatment can be commenced without delay.

Surgical treatment of localized bronchiectasis consists of removing an affected and devitalized lung segment or lobe. This often permits a marked improvement in general health. Heart–lung transplant may be offered to CF patients. This cures respiratory failure and normalizes lung mucus, but the denervated lungs lack a cough reflex, so the exacting physiotherapy must be continued to avoid aspiration and minimize infection of the transplanted lung.

Pneumococcal (lobar) pneumonia

Introduction

Streptococcus pneumoniae, also called the pneumococcus, attacks previously healthy individuals as well as those with predisposing conditions, and causes severe morbidity. Antibiotic-resistant forms have become common in many parts of the world, and are increasing in the UK.

Epidemiology

S. pneumoniae is the commonest bacterial cause of community-acquired pneumonia. One community-based study estimated that 4% of adults aged 60–79 years developed pneumococcal pneumonia each year. There is a higher incidence in certain occupational groups and geographical areas, for example South African gold miners and in the highlands of Papua New Guinea. Predispositions include sickle-cell disease, anatomical or functional asplenia, chronic cardiac, respiratory, liver and renal disease, diabetes mellitus, alcohol abuse and immunosuppression. The case fatality rate is 5–10% (20% in bacteraemic cases), and is highest in elderly or compromised patients.

Pathogenesis of *S. pneumoniae* infections

S. pneumoniae is a Gram-positive coccus forming part of the *S. oralis* group together with *S. oralis* and *S. mitis*. There are more than 90 different serotypes based on polysaccharide capsular antigens.

The capsule is a major pathogenicity determinant; virulence is related to the capsular polysaccharide type and less to the quantity of capsular material produced. The lethal infective dose for a mouse is reduced from 10^9 to 10 CFU for organisms with a capsule. Infection with type 3, which produces copious amounts of a linear polysaccharide, is associated with high mortality, but type 30, which produces a similar amount of capsule, is rarely associated with clinical infection.

Cell wall C polysaccharide activates the alternative complement pathway giving rise to C3a and C5a which are vasoactive and recruit polymorphs to the infected site.

Toxins produced by *S. pneumoniae* include pneumolysin, a cytotoxin analogous to streptolysin O (in *S. pyogenes*) and to listeriolysin (in *Listeria monocytogenes*), which forms a ring of molecules which dissolve in plasma membranes leaving a hole. This toxin appears to have an important role in developing bacteraemia. Neuraminidase is an enzyme which removes sialic acid from host glycoproteins. Two forms of this enzyme are present in *S. pneumoniae*. Immunization against these toxins partially protects mice from lethal pneumococcal infection. Pneumococcal surface protein A is a surface-exposed protein with several serological variants. Its biological role is yet to be established. *S. pneumoniae*, like *S. pyogenes*, also produces a hyaluronidase and a surface-exposed enolase.

In common with other mucosal pathogens, *S. pneumoniae* possesses an IgA protease. This enzyme assists the organism in achieving colonization in the nasopharynx.

Pathogenicity factors for *Streptococcus pneumoniae*
1 Capsular polysaccharide.
2 Cell wall (C) polysaccharide.
3 Toxins: pneumolysin, neuraminidase.
4 Surface protein A.
5 Immunoglobulin A protease.
6 Hyaluronidase and enolase.

Pathology

Lobar pneumonia is an infection of the alveoli. The affected part of the lung is solidified and red, due to blood-stained acute inflammatory exudate. As the action of phagocytic enzymes liquefies the exudate, some of it is expectorated. This produces blood-stained, 'rusty' sputum, full of neutrophils and Gram-positive diplococci. In patients who recover, the exudate resolves completely, and the intact alveolar epithelium resumes its normal appearance.

Clinical features

The onset is sudden, with high fever and non-specific symptoms including vomiting, loose stools, headache or pleuritic chest pain. Sweating is not usually prominent. The main clue suggesting bacterial illness is neutrophilia, often 15–25 $\times 10^9$/l.

After a variable interval, dry cough suddenly develops, and the classical signs of consolidation develop. The chest X-ray now shows a typical dense opacity filling all or most of a lobe (Fig. 7.4).

A further interval elapses before the solid exudate breaks down, and sputum is produced. If untreated, a proportion of patients suddenly lose their fever and improve (healing by crisis), as effective antibodies appear in the blood. Others develop potentially fatal complications.

Diagnosis

Bacteriological confirmation may not be possible, as few patients produce sufficient sputum to demonstrate neutrophils and Gram-positive diplococci. Several antigen detection and PCR methods are now available for rapid

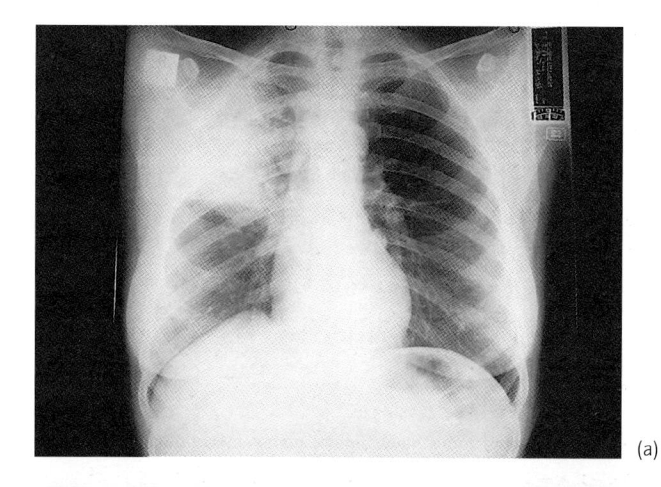

(a)

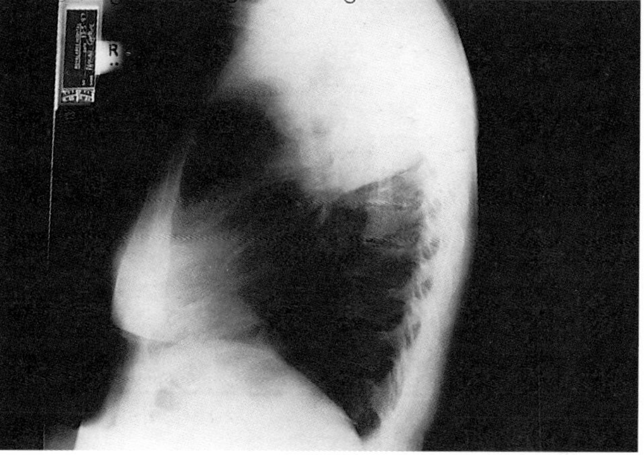

(b)

Fig. 7.4 Lobar pneumonia: chest X-ray showing typical consolidation in the right upper lobe: (a) posteroanterior and (b) right lateral views.

diagnosis. Blood cultures should be taken in all cases as they allow a definitive diagnosis to be made in bacteraemic patients.

Management

In Britain, less than 5% of pneumococci are moderately resistant to benzylpenicillin, but can be treated with adequate doses of this drug (1.8–2.4 g 6-hourly); in other Western countries, about 15% are moderately resistant and approaching 10% are fully resistant. The same remarks apply to amoxycillin and co-amoxiclav. In the USA, cefuroxime resistance varies from 3 to 12%, but cefotaxime and ceftriaxone are effective in 95%. Resistance to macrolides and tetracyclines is seen in 3–10%. Recently, isolates with vancomycin tolerance have been described.

Complications

Sterile pleural effusion

A common problem, this is a sterile exudate, containing few white cells. It tends to occur during treatment, and often responds to aspiration. Rarely, persistent reaccumulation may require operative drainage and pleurodesis.

Extension of infection

Extension of infection to adjacent pleura or pericardium is more likely in late or untreated cases. It is marked by swingeing fever and typical localizing pain.

Empyema should be drained as completely as possible, to avoid loculation. Up to a litre of fluid can be withdrawn each day (withdrawing more may lead to hypotension or reactive pulmonary oedema). An indwelling chest drain can be used if rapid reaccumulation of fluid persists (Fig. 7.5). High-dose intravenous antibiotic treatment should be given. Intrapleural antibiotic offers no advantage, as the antibiotic often becomes trapped at the site of injection, or destroyed by proteases or lysozymes.

Pericardial drainage is hazardous, but must be carried out if there are signs of tamponade. The procedure is best performed by an experienced operator.

Bacteraemia

Bacteraemia is a serious complication, which increases the case fatality rate to 20%. It is more likely in infections with type 3, 6 or 5 pneumococci. Metastatic infection, such as meningitis, peritonitis or brain abscess, may further complicate bacteraemia.

Prevention and control

Pneumococcal vaccine is a non-replicating vaccine containing purified capsular polysaccharide from each of the 23 types of pneumococcus which together account for over 85% of invasive pneumococcal disease in most Western countries. The efficacy is 60–70%, but may be lower in children under 2 years of age and in those with immunosuppression. It is indicated for all those above age 2 with predisposing conditions (see above); in some countries routine vaccination of all elderly people above 60 years of age is recommended. In patients undergoing elective splenectomy, the vaccine should be given 2 weeks before the operation.

Revaccination is not normally advised, because of the risk of adverse reactions, except for patients with asplenia or nephrotic syndrome, whose antibody levels are likely to decline more rapidly. Long-term antibiotic prophylaxis with penicillin or amoxycillin is also indicated for patients with sickle-cell disease or other splenic dysfunction, as they can become infected with uncommon types of pneumococci.

Several conjugate pneumococcal vaccines are currently in clinical trials; a 7-valent preparation is likely to be licensed in the USA soon. These conjugate vaccines will provide superior immunity to their plain polysaccharide counterparts, and are likely to be routinely used in children to prevent septicaemia, meningitis and otitis media.

Klebsiella pneumoniae pneumonia

Klebsiella pneumoniae can cause lobar pneumonia in patients with long-standing asthma, debilitating disease or alcoholism. Unlike pneumococcal infection, there is copious sputum containing many neutrophils and large, capsulated Gram-negative rods, permitting presumptive diagnosis. This organism is resistant to amoxycillin, and is usually treated with a cephalosporin (see also Table 7.1).

Haemophilus influenzae pneumonia

Introduction and epidemiology

H. influenzae non-capsulated (untypable) strains are common pathogens in COPD and other chronic chest conditions. Encapsulated strains, particularly type b (Hib) readily colonize unimmunized infants below the age of 5 years, but rarely affect older individuals, as natural immunity develops early in life. Manifestations of Hib disease include pneumonia, meningitis, epiglottitis, septic arthritis and facial cellulitis.

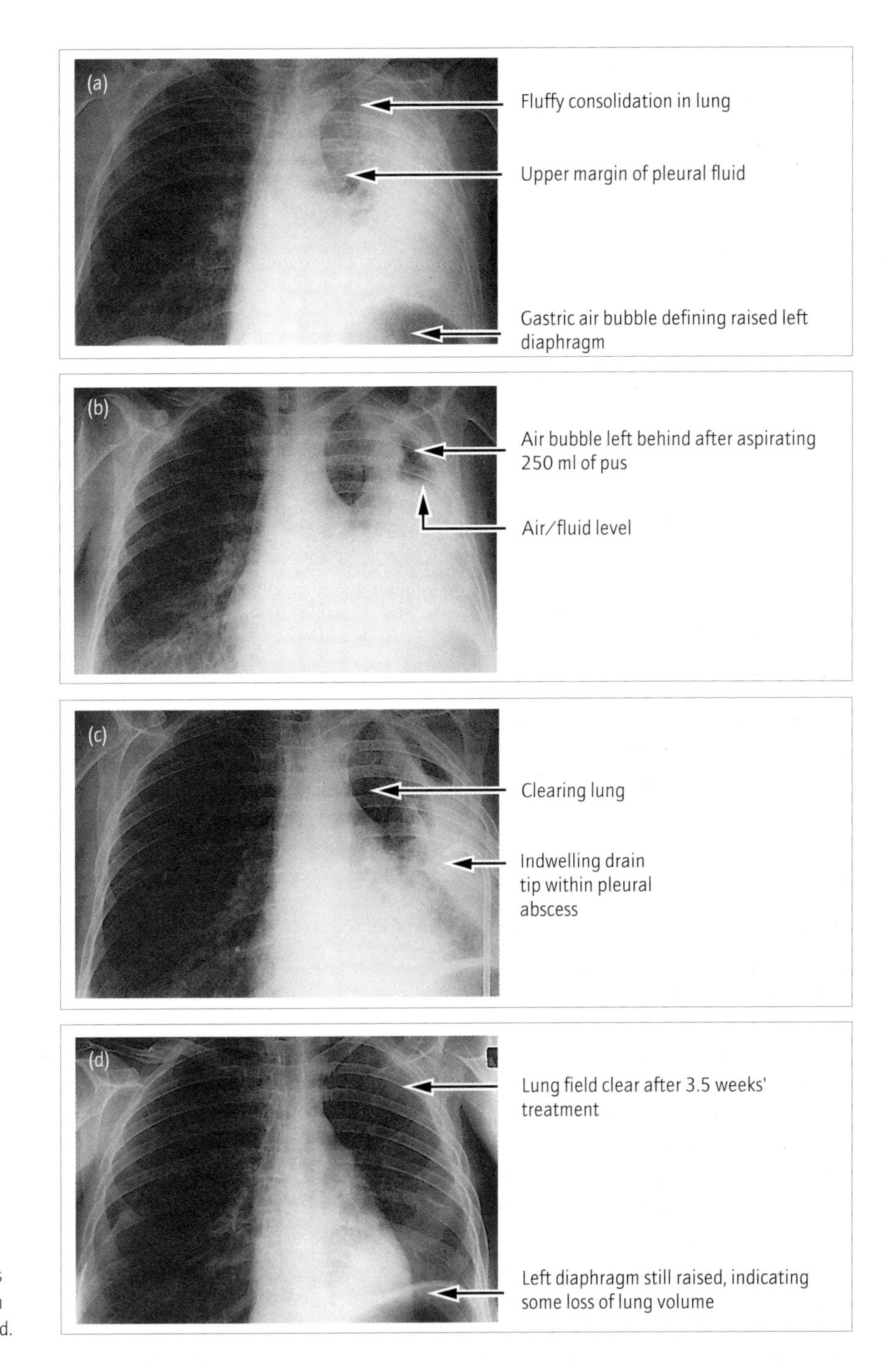

Fig. 7.5 (a–d) Chest X-ray series showing management of pneumococcal empyema complicating lobar pneumonia: the empyema was first aspirated, but later required drainage with an indwelling catheter before resolution was achieved.

Microbiology

Non-capsulate strains of *H. influenzae* are important commensals and secondary pathogens of the upper and lower respiratory tract. There are six capsular serotypes (a–f) differing in their polysaccharide composition. Serotype b is the usual cause of invasive infection; other serotypes are rarely implicated.

Clinical features

Illness develops gradually, over a few days. The child is feverish and listless. Cough may not be prominent, but chest examination usually shows typical features of consolidation. Chest X-ray reveals a segmental, lobar or more widespread opacity, and an air bronchogram is often seen.

Type of disease	Recommended treatment
Uncomplicated	
Adult up to age 60	**1** Oral amoxycillin 500 mg–1.0 g three times daily. **2** Oral erythromycin 250–500 mg four times daily. **3** Intravenous ampicillin 500 mg 6-hourly.
Child	**1** Oral erythromycin: up to 2 years, 125 mg 6-hourly; 2–8 years, 250 mg 6-hourly. **2** Oral clarithromycin: under 1 year, 7.5 mg/kg; 1–2 years, 62.5 mg; 3–6 years, 125 mg; 7–9 years, 187.5 mg; 10–12 years 250 mg – all doses given twice daily. **3** Intravenous ampicillin: under 10 years, 250 mg 6-hourly; over 10 years, 500 mg 6-hourly (may be combined with oral erythromycin to 'cover' *Mycoplasma* etc.).
Coexisting with other conditions (e.g. heart/lung/renal disease or diabetes mellitus)	**1** Oral co-amoxiclav: 625 mg three times daily. **2** Oral clarithromycin: 250–500 mg twice daily. **3** Consider oral levofloxacin. **4** Intravenous co-amoxiclav: 1.2 g 8-hourly. **5** Intravenous cefuroxime: 750 mg 8-hourly.
Severe (see below)	
Adult	**1** Intravenous co-amoxiclav: 1.2 g 8-hourly *or* intravenous cefuroxime: 1.5 mg 8-hourly; *plus* intravenous erythromycin: 0.5–1.0 g 6-hourly. **2** Intravenous clarithromycin: 500 mg 12-hourly. NB: Intravenous rifampicin: 600 mg 12-hourly can be added to **1** or **2**.
Child	**1** Intravenous co-amoxiclav: up to 3 months, 25 mg/kg 8-hourly; 3 months–12 years, 25 mg/kg 8- or 6-hourly; *or* intravenous cefuroxime: 20–30 mg/kg 8-hourly; *plus* intravenous erythromycin: 50 mg/kg daily by continuous infusion or in four divided doses. NB: Intravenous rifampicin: under 1 year, 5 mg/kg 12-hourly; over 1 year 20 mg/kg 12-hourly can be added.
Severe infection coexisting with other conditions or requiring intensive care	
Adult	**1** Intravenous cefotaxime: 1.0 g 8-hourly; *or* intravenous ceftriaxone 2.0 g daily; *or* intravenous imipenem/cilastatin: 500 mg–1.0 g 8-hourly; *plus* intravenous erythromycin: 500 mg–1.0 g 6-hourly. NB: Intravenous rifampicin: 600 mg 12-hourly may be added.
Child	**1** Intravenous cefotaxime: 50–75 mg/kg twice daily; *or* for children above 6 weeks, intravenous ceftriaxone: 50–80 mg/kg daily; *or* intravenous imipenem/cilastatin: for children above 3 months, 15 mg/kg 6-hourly. NB: Intravenous rifampicin: under 1 year, 5 mg/kg 12-hourly; over 1 year 10 mg/kg 12-hourly may be added.

Signs of severe community-acquired pneumonia	Indicators of need for intensive therapy
Clinical signs	Severe respiratory failure and/or need for mechanical ventilation
Respiratory rate > 30 breaths/min.	Blood pH < 7.3
Diastolic blood pressure < 60 mmHg	Systolic blood pressure < 90 mmHg
	Multiorgan failure
Laboratory signs	Severe disseminated intravascular coagulation
Hypoxia (Pao_2 < 8 kPa)	
White cell count < 4×10^9/l or > 20×10^9/l	
Blood urea > 7.0 mmol/l	
Multilobar chest X-ray opacities	

Table 7.1 Initial management of community-acquired pneumonias.

There is usually a neutrophilia of $15–20 \times 10^9/l$ in the peripheral blood.

Diagnosis

Blood cultures often produce a growth of Hib, which may be identifiable by agglutination test in 18–24 h. Small children do not expectorate their sputum, but samples are often obtainable by pharyngeal aspiration (which stimulates coughing), and are also culture-positive.

The differential diagnoses are pneumococcal and *Mycoplasma pneumoniae* pneumonia, both of which are common in toddlers.

Laboratory identification

Haemophilus is a fastidious genus whose growth requires the presence of nicotinamide adenine dinucleotide (NAD) and haematin normally found in blood. These factors are not available in blood agar, but can be released by adding the blood to the medium at 80°C, which gives a chocolate appearance. If a clear medium is required, as for the study of capsulation, a filtered extract from red cells or a peptic digest of meat may be added. Dependence on NAD and haematin can be tested by inoculating the organism on to nutrient agar and placing paper discs containing the factors; growth will only occur around a disc containing both factors. The presence of a capsule can be demonstrated by slide agglutination with serotype-specific antiserum. This should be performed on all isolates where type b disease is suspected.

Antibiotic resistance is common, especially to ampicillin (about 15% of Hib isolates in the UK). The presence of a beta-lactamase, TEM-1, which had its origin in *E. coli*, is usually responsible. It can be detected by rapid tests using a cephalosporin, nitrocephin, which changes colour when its beta-lactam bond is broken.

Management

It is advisable to treat immediately with an intravenous agent while awaiting more diagnostic data. A broad-spectrum cephalosporin such as cefuroxime or cefotaxime is effective against both Hib and pneumococcal infection. Erythromycin is effective against pneumococci and *M. pneumoniae*, but Hib isolates are increasingly resistant, as they also are to amoxycillin. Clarithromycin is effective against 99% of isolates. Quinolone antibiotics are effective against almost all *H. influenzae*, but are not recommended for small children.

Prevention and control

Polysaccharide conjugate vaccines against Hib are included in the routine infant vaccination programmes in many countries (see also Chapter 26). They have greatly reduced the incidence of all invasive forms of Hib disease, including pneumonia. Hib vaccines do not protect against pneumonia caused by non-type b strains, or infections due to non-encapsulated strains.

Legionnaires' disease and legionelloses

Introduction

Legionnaires' disease is important because it is often unrecognized until a late stage, it does not respond well to conventional treatment for community-acquired pneumonia, and it can be life-threatening. It commonly occurs in clusters or outbreaks of cases.

Epidemiology

The infection is transmitted by inhalation of contaminated aerosols generated from hot-water tap and shower outlets, water-cooling towers of office and industrial sites, and domestic or recreational whirlpool spas. Legionellae thrive in water at temperatures between 20–50°C, and the organism can often be isolated from water systems in buildings as well as from ponds, streams and soil. Legionellae can exist within free-living amoebae.

The disease occurs both sporadically and as localized outbreaks. Outbreaks are usually associated with a contaminated water source in a large building. Illness preferentially affects the middle-aged and elderly, particularly males and those who smoke or have underlying chronic respiratory disease. In the UK, there are about 150 cases annually and the case fatality rate is 10%. Nearly 50% of cases are travel-associated, the implicated source often being the water system of a tourist hotel (Fig. 7.6). Person-to-person spread does not occur.

The incubation period is 2–10 days for legionnaires' disease and 1–2 days for Pontiac fever, the milder non-pneumonic form of the infection.

Microbiology and pathogenesis

Legionellae are Gram-negative aerobic non-sporing bacteria. They are catalase- and oxidase-variable and most species of *Legionella pneumophila* produce beta-lactamase. This organism produces an extracellular acid polysaccharide layer which acts as a capsule. At least 12 serotypes are recognized in human infections.

(a)

(b)

Fig. 7.6 (a) Poorly maintained Turkish bath in a hotel which caused an international outbreak of legionnaires' disease: 11 people from four countries contracted the infection. (b) The hotel cooling tower.

Legionella spp. express a lipopolysaccharide which exhibits some cross-reactions with *Salmonella*. The major surface protein is a protease which is thought to play an important role in the pathogenesis of lung damage. Inhalation of organisms contained within environmental amoebae may enhance invasiveness.

Clinical features

Most of the Legionellaceae are capable of producing a mild or moderate flu-like illness, sometimes with a cough. This condition, when recognized as legionellosis, is often called Pontiac fever. A similar illness, accompanied by a rash, was called Fort Bragg fever after an outbreak among office workers there.

Legionnaires' disease is a severe systemic infection with pneumonia, usually caused by a serotype of *L. pneumophila*. After the incubation period of 2–10 days, there is fever, prostration and cough, often misinterpreted as influenza or bronchitis. In spite of treatment with beta-lactam antibiotics, fever continues, loose stools are commonly seen, cough and tachypnoea worsen and the patient becomes increasingly confused.

Physical examination shows tachypnoea and purulent sputum in a morose or confused patient. There is usually a distinct area of consolidation or coarse crepitations in the chest, confirmed by demonstrating one or more large opacities on chest X-ray (Fig. 7.7).

Laboratory findings include neutrophilia, an erythrocyte sedimentation rate of 70 mm/h or more, variably elevated urea and creatinine levels and raised transminases. There may be marked hyponatraemia and/or normocytic anaemia. Severe respiratory and/or renal failure can develop quickly.

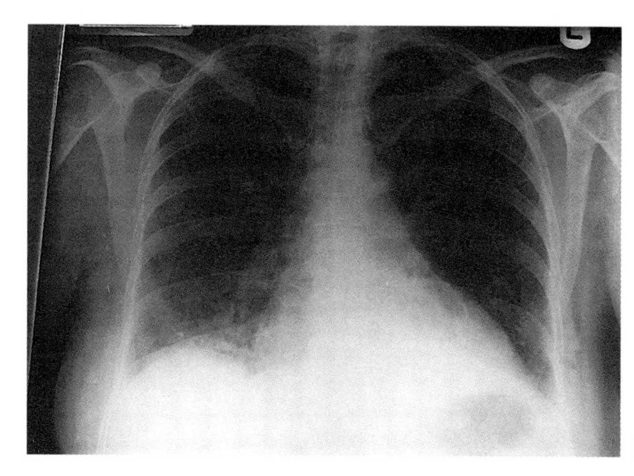

(a)

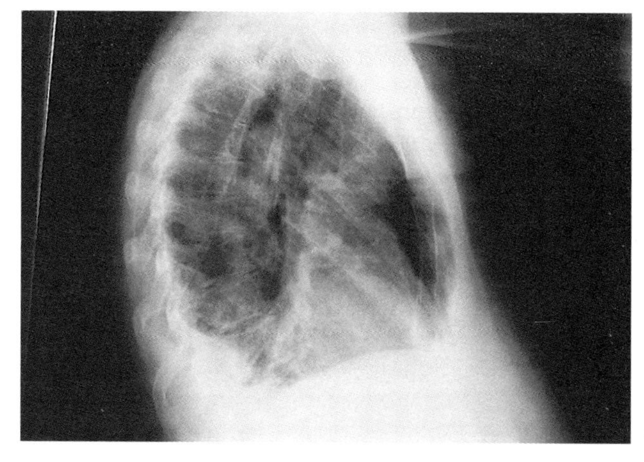

(b)

Fig. 7.7 Legionnaires' disease: extensive lung opacity in a patient with severe symptoms (one of several people infected during a hotel-based group holiday): (a) posteroanterior and (b) right lateral views.

Diagnosis

This may be suspected on epidemiological grounds. The presence of pneumonia with severe confusion and metabolic disturbance in a previously fit patient is highly suggestive.

Laboratory diagnosis

Gram-stained smears of sputum are not useful as legionellae take up the stain very poorly. An enzyme-linked immunosorbent assay (ELISA) method for detection of serogroup 1 antigen in urine is now the diagnostic method of choice as it produces a rapid result. A direct fluorescent antibody method is available for the diagnosis of *L. pneumophila* serotype 1 in sputum. It has relatively poor sensitivity but a high specificity. *Legionella* spp. are fastidious, growing slowly on a cystine-containing charcoal yeast extract agar. This is made selective by the addition of antibiotics which inhibit other respiratory commensals. Legionellae are identified by their cystine dependence and are serotyped by direct immunofluorescence antibody testing. Detection of *Legionella* DNA by PCR has been described.

Legionnaires' disease can be diagnosed by detecting a fourfold rise in antibody titres or a single high value greater than 1 in 128. Methods such as indirect immunofluorescence and commercial rapid microagglutination tests are available.

Management

Appropriate treatment should be given on suspicion of the diagnosis, to avoid the danger of respiratory or renal failure. The treatment of choice is intravenous erythromycin in a dose of at least 3 g daily. The addition of ciprofloxacin or rifampicin may speed the response.

Treatment of legionnaires' disease
Erythromycin i.v. 3–4 g daily in four divided doses *plus* either ciprofloxacin i.v. 200 mg twice daily *or* rifampicin 600 mg twice daily. Oxygen is often required, and intensive respiratory support may be needed. Renal failure in late or severe cases demands intensive management.

Complications

Patients who recover usually do so completely, though they may need many weeks' convalescence. A few develop lasting, patchy cerebral or bulbar deficit, often attributed to severe hypoxia during the illness. Guillain–Barré syndrome is a rare but recognized complication of acute legionnaires' disease.

Prevention and control

Effective maintenance of building water systems and other potential sources of infection is the key to prevention of legionellosis. This should include adequate chlorination of water supplies and stored cold water, and heating (to above 60°C) of hot water. The design of water systems should avoid stagnation of water in peripheral sections, and permit the achievement of a temperature of 55°C at all hot taps within 15 seconds of turning on. Whirlpool spas should be frequently cleaned and continuously disinfected.

Initial assessment and management of community-acquired pneumonias

Management of bacteriologically defined pneumonias is ideally based on knowledge of the likely antimicrobial sensitivity of the causative organism. However, there is a wide clinical overlap between exacerbations of COPD and primary pneumonias of different aetiologies. General guidelines are therefore appropriate for assessing the severity of illness and assigning initial therapy, which can be adjusted later if indicated by microbiological results (see Table 7.1).

Repeated, unexplained community-acquired pneumonias may be a sign of underlying disease: neonates and infants may have undiagnosed cystic fibrosis; patients of any age may have an acquired disorder of immunity, and should be investigated (see Chapter 22).

Aspiration pneumonia

This results from failure to cough and protect the airways. It can occur in unconscious or anaesthetized patients, or in those with neuromuscular disease affecting swallowing or breathing. When the chest is erect, the right main and lower lobe bronchi are nearly vertical, so the right lower lobe tends to be most affected. In a recumbent patient the apical bronchus of the right lower lobe is affected most.

The aspirate contains mixtures of mouth and pharyngeal organisms, mainly penicillin-sensitive streptococci and spirochaetes mixed with anaerobes. These produce a patchy pneumonia, most dense in the vulnerable segment, sometimes leading to abscess formation.

The diagnosis must be suspected in vulnerable patients, and treated with physiotherapy, a penicillin or cephalosporin plus metronidazole. If gastric contents have been aspirated, the inflammatory effect of gastric acid can be inhibited by giving a corticosteroid such as methylprednisolone or prednisolone for the first 24–36 h.

Lung abscess

The commonest cause of a lung abscess is aspiration. Systemic sepsis or primary lung infections are rarer causes. Most lung abscesses are polymicrobial, with facultative and anaerobic organisms occurring equally commonly in most series of cases. Staphylococcal bacteraemia is often accompanied by a nodular pneumonitis which can progress to abscesses. Rare infections such as nocardiosis and aspergillosis (more common in immunosuppressed patients) can cause small abscesses in the lungs.

Hospital-acquired lung abscesses may behave in the same way as community-acquired ones, but in the former, *Klebsiella pneumoniae*, *Pseudomonas aeruginosa* and *Candida* spp. are additional causes, which carry a particularly poor prognosis.

If the abscess can be aspirated, culture and sensitivity testing of the pus are helpful, and healing is promoted. Blood cultures should be obtained before treatment is commenced. Most abscesses can be treated as for aspiration pneumonia, with percutaneous drainage when possible. A minority require surgical drainage.

Anaerobic pneumonia (necrobacillosis)

Introduction and epidemiology

This is a rare condition which tends to affect adolescents and young adults, particularly men. It is a bacteraemic disease, usually arising from infection with *Fusobacterium necrophorum*, which is a colonist of the pharynx and mouth.

Clinical features and diagnosis

Illness begins abruptly with a high fever and very severe sore throat. After 2 or 3 days there is cough and often chest pain. One or more areas of consolidation are detectable on examination and chest X-ray. Standard antibiotic treatment is unsuccessful, the pneumonia extends, renal function deteriorates, hypoxia and hypotension develop. Suppuration of cervical lymph nodes with thrombophlebitis of the underlying jugular vein are recognized features of severe disease. Most cases are fatal unless suspected and treated early.

Blood cultures are often positive, and must always be obtained. Sputum culture may also be positive, but the bacteria are strictly anaerobic, and survive poorly unless maintained in an anaerobic environment. The diagnosis must be considered if the patient is to have a good chance of recovery; the sore throat is a helpful pointer.

Management

Although the pathogen is penicillin-sensitive in laboratory conditions, penicillin has only a slow effect. Metronidazole is more successful, and can be given together with penicillin or broader-spectrum drugs while the diagnosis is confirmed.

Atypical pneumonias

Introduction

The atypical pneumonias were so called because they appeared so unlike typical lobar pneumonia, having a gradual onset, prolonged fever, profuse sweating and slow recovery with no point of crisis, and pathogens were not demonstrable by Gram-staining. They are important because they cause morbidity in adults of working age, and epidemics or outbreaks are not uncommon.

ORGANISM LIST

Mycoplasma pneumoniae
Chlamydia pneumoniae
C. psittaci
Coxiella burnetii

Epidemiology

M. pneumoniae and *C. pneumoniae* are human pathogens whose reservoir of infection is the organism circulating among affected people.

Mycoplasma pneumonia affects all ages but is commonest in school-age children and young adults. Outbreaks occasionally occur, especially in closed institutions. In the UK, epidemics occur at 3- or 4-year intervals. During epidemic years, *M. pneumoniae* competes with *Streptococcus pneumoniae* as the commonest cause of community-acquired pneumonia.

C. pneumoniae has only been distinguished from *C. psittaci* in recent years. The prevalence of seropositivity rises steadily with age until about 60% of individuals have evidence of past infection by the fifth decade. Large-scale epidemics have been recognized in Denmark and Canada, and it appears that both primary infection and reinfection can be symptomatic. Persisting infection with *C. pneumo-*

niae is suspected to be a factor in the development of atherosclerosis.

C. psittaci, the agent of ornithosis, is widespread among birds, which excrete many organisms in their faeces. Infection in humans is acquired by inhaling the organism from desiccated bird droppings and secretions of infected birds. Sporadic cases and limited outbreaks have been associated with keepers and fanciers of parrots and parakeets. Larger-scale outbreaks have been associated with duck or poultry farming, and with poultry packing plants. Person-to-person spread is rare. The incubation period is 4–15 days.

Coxiella burnetii is a pathogen mainly of sheep. Transmission to humans occurs by air-borne spread of organisms in dust contaminated by infected placental tissues and faeces. Rarer sources of infection include parturient cats and unpasteurized milk. The infection mainly affects meat workers, farmers and veterinarians. Infection is severe in pregnant women, particularly affecting the placenta, and often precipiting abortion. No case of infection from a parturient woman has been described.

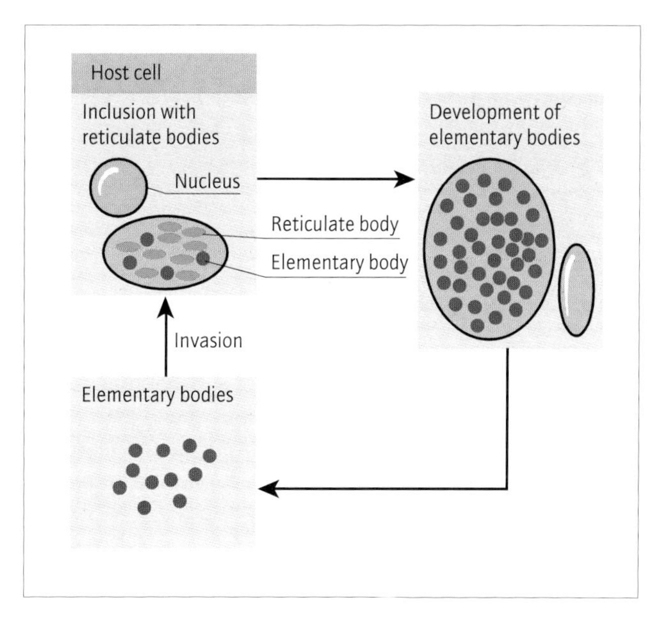

Fig. 7.8 Appearance of intra- and extracellular forms of *Chlamydia* spp.

Microbiology of *Chlamydia* infections

The family Chlamydiaceae consists of one genus including three species: *Chlamydia trachomatis* affects the eye and genital tract and is responsible for trachoma, non-specific urethritis and lymphogranuloma venereum; *C. psittaci* is the causative agent of a zoonotic pneumonia; and *C. pneumoniae* causes upper and lower respiratory tract infection.

The chlamydiae are obligate intracellular bacteria which exist in two forms: the reticulate body, which is the intracellular vegetative form found inside cells of the hosts, and the infective elementary body which is adapted for extracellular survival and is responsible for transmission (Fig. 7.8). The reticulate body divides by a series of fissions to give intermediate forms and finally elementary bodies. The cycle takes place within a cytoplasmic inclusion body, and ends with the release of infective elementary bodies from the host cell.

Lipopolysaccharide is a genus-specific antigen common to all three species. There are also species-specific polypeptide antigens and, in addition, *C. trachomatis* exhibits serotype-specific antigens (see p. 300).

The virulence of *Chlamydia* spp. can be correlated with their *in vitro* growth characteristics; more virulent strains tend to grow more slowly. Cysteine-rich proteins are found in elementary body outer membrane proteins. One of these, a 60 Da protein, is thought to be associated with enhanced invasiveness and virulence. Adhesion molecules have also been recognized in *C. trachomatis* and *C. psittaci*.

Clinical features

The acute feverish part of the illness is similar in all types of atypical pneumonia. After an incubation period of 4 days–2 weeks, there is an insidious onset of irregular, swingeing fever, malaise, fatigue and sweating, especially at night. There is often a dry cough. Physical examination is rarely remarkable, except for a patch of crepitation or dullness in the lung fields. The chest X-ray may show anything from a faint, ill-defined segmental opacity to a large and fairly dense consolidation. Some patients have a small or moderate pleural effusion, often accompanied by pain on the affected side.

Laboratory findings are variable. There may be a modest neutrophilia, but this is rare in *C. pneumoniae* infections. Mild elevation of liver transaminases is common; the erythrocyte sedimentation rate is often raised to 70 mm/h or more. Most patients with *Mycoplasma* pneumonia develop cold agglutinins, but these tend to appear in convalescence, too late for diagnostic use. A few patients develop severe haemolysis in convalescence. Associated clinical features differ in the different infections.

Chlamydia pneumoniae

C. pneumoniae usually causes acute self-limiting respiratory infections, frequently starting with sore throat, often mild, rarely presenting to hospital. Other syndromes include moderately severe atypical pneumonia, otitis media, acute-on-chronic bronchitis, persistent or relapsing pharyngitis and, less commonly, myocarditis. *C. pneumoniae* antigen and genome have been demonstrated, and the organism has been recovered from the foamy cells of atheromatous plaques. There is an epidemiological association between *C. pneumoniae* seropositivity and atheromatous disease.

Mycoplasma pneumoniae

M. pneumoniae usually causes pneumonia, but this may be accompanied by tracheitis, bullous myringitis, mild hepatitis or sometimes by a lymphocytic meningitis. Pericarditis and myocarditis are rare but recognized features. Probably at least 25% of infections are feverish illnesses without pneumonia.

Chlamydia psittaci

C. psittaci causes pneumonia in about 60% of clinically infected individuals. The remainder have fever alone. Infection is debilitating and can be severe and accompanied by respiratory failure. It is a rare cause of culture-negative endocarditis. Occasional cases of myelopathy have been reported, with or without associated pneumonia.

Coxiella burnetii

C. burnetii causes Q fever. This may be an atypical pneumonia, but over half of cases are feverish illnesses with hepatocellular disorder. A small proportion of cases, probably less than 1%, become chronic with swinging fever, granulomatous hepatitis, often clinical jaundice and, in half of cases, endocarditis. In endemic areas this is an important cause of morbidity. Mild to moderate myocarditis may occur in acute or chronic disease.

Diagnosis of atypical pneumonias

The diagnosis of many atypical pneumonias can be inferred from epidemiological circumstances and clinical presentation. RSV and influenza can cause similar illnesses, often in outbreaks.

Although *M. pneumoniae* can be cultivated on serum-based artificial media the organism is slow-growing and the diagnostic yield is low. Serology is the mainstay of diagnosis. In many laboratories particle-agglutination tests are used, and a positive result is taken to indicate the presence of IgM antibodies and therefore recent infection. IgM-detecting ELISA tests are commercially available. Antigen detection by ELISA tests and PCR-based genome detection are available for rapid diagnosis.

Chlamydiae

Chlamydiae are often still detected by finding rising titres of complement-fixing antibodies to whole-cell antigen preparations. There is much cross-reaction between antibodies to different species. Species-specific antibodies are detected by microimmunofluorescence techniques, which detect reactions against elementary body antigens. Culture of chlamydiae is not useful in the diagnosis of pneumonias, although it is useful when a diagnosis of *Chlamydia trachomatis* genital infection is required in cases of sexual abuse or rape. Where a culture diagnosis is required, *C. trachomatis* is readily grown in tissue culture with cyclohexamide-treated McCoy cells.

Q fever

Q fever is diagnosed by finding rising titres of complement-fixing antibodies to phase 2 and (in chronic Q fever) phase 1 antigens. Coxiellae isolated from infected animals express phase 1 antigens, but after passage through embryonated eggs, these are replaced by phase 2 antigens.

In cases of hepatitis, there is a characteristic histological change, with many small granulomata, often with clear centres and always with a halo of eosinophils in their periphery (see p. 205). *Coxiella burnetii* can be demonstrated in silver- or Giemsa-stained smears, or by direct immunofluorescent staining of biopsy specimens.

Management

The antibiotic of choice is tetracycline. Erythromycin is a useful alternative, but may be less effective in Q fever and some cases of psittacosis. Treatment may be given orally in mild and moderate cases, but should be given parenterally in severe or complicated cases (this is easier with erythromycin). The response is not always fast, so courses of 10–14 days are advisable to avoid relapse or recrudescence.

Chloramphenicol may still be the treatment of choice when Q fever occurs in pregnancy. Ciprofloxacin is highly active in laboratory testing and may be a valuable alternative but treatment failures have been reported and there may be a risk of cartilage damage in the fetus. Ciprofloxacin is not reliably effective in chlamydial and mycoplasmal pneumonias; newer quinolones such as levofloxacin may be more effective, but diarrhoea is a common unwanted effect.

Treatment of *Mycoplasma pneumoniae* and *Chlamydia pneumoniae*
1 Erythromycin orally or i.v. 500 mg 6-hourly.
2 Alternative: oxytetracycline orally 250–500 mg 6-hourly.
Both above regimens for 10–14 days.

Treatment of Q fever
1 First choice: tetracycline orally—same dose as in previous text note.
2 In pregnancy: chloramphenicol orally or i.v. 500 mg 6-hourly *or* ciprofloxacin orally 500 mg twice daily (i.v. 200 mg twice daily).
All above regimens for 14 days.

Treatment of psittacosis
1 First choice: tetracycline orally—same dose as for *Mycoplasma pneumoniae* and *Chlamydia pneumoniae*.
2 Alternative: erythromycin orally or i.v—same dose as for *M. pneumoniae* and *C. pneumoniae*.
Both above regimens for 10–14 days.

Care must be taken not to miss endocarditis in Q fever. The presence of phase 1 antibodies is a warning. Echocardiography, physical examination, repeated C-reactive protein estimations and follow-up assessments for 2 or 3 months are advisable in cases of doubt.

Pertussis

Introduction

Pertussis, or whooping cough, is a prolonged, severe and distressing disease of children caused by *Bordetella pertussis*. Although it is now rare, as immunizaton is widespread, large epidemics occurred in the 1980s after concerns about vaccine safety resulted in low rates of acceptance (Fig. 7.9). Morbidity is high in infants and in older children with respiratory or cardiac disease.

Epidemiology and pathology

B. pertussis is readily transmitted by the air-borne route, and is most often seen in infants too young to have been vaccinated or in families who have declined immunization. The reservoir of infection is in adult carriers and subclinical cases, as immunity wanes after childhood. It is a small, Gram-negative rod which is typed by its cell wall proteins, or agglutinogens, designated 1, 2 and 3. Most clinical isolates are of type 1,3 or 2,3. Antibodies to these are protective.

Clinical features

After an incubation of 14–20 days, illness begins with a simple cough and slight fever. Over 4 or 5 days coughs begin to be grouped into paroxysms, which lengthen until up to 20 or 30 coughs occur with no inspirations between them. Large volumes of thick mucus accumulate in the upper airways and are expectorated in ropy strands. Finally the paroxysm ends with a stridulous inspiratory cry, and often with a vomit. Cough, whoop and vomiting may present separately. Adults and a few children may suffer sneezing paroxysms. The paroxysms last for 2 or 3 weeks before gradually improving over 10–14 days.

Cyanosis and, in infants, apnoeic attacks are common. The venous pressure is raised, there is facial lividity and sometimes conjunctival haemorrhage or facial petechiae (Fig. 7.10). In infants there may be intracranial haemorrhage. Secondary or aspiration pneumonia often occurs, but the rare fatalities are usually caused by hypoxic and vascular cerebral damage.

Diagnosis

This is usually clinically obvious, but mild cases may be similar to croup, bronchiolitis or persisting asthma. A lymphocytosis of $15–25 \times 10^9/l$ is strongly suggestive of pertussis.

Microbiology

B. pertussis is a small Gram-negative coccobacillus, related to *B. parapertussis*, which is rarely associated with the whooping cough syndrome, and *B. bronchoseptica*, which is primarily a pathogen of animals.

The pathogenicity of the organism is related to its toxins, including fimbrial haemagglutinin, pertussis toxin and a 67 kDa protein. Pertussis toxin mediates the paroxysmal cough and a unique lymphocytosis, affecting all lymphocyte subtypes. Much research has focused on which of these antigens are immunologically protective. Other toxins include tracheal cytotoxin, which causes ciliostasis and destruction of ciliated epithelium, and adenyl cyclase which destroys polymorphs and stimulates the cough reflex.

The bacterial agglutinogens and fimbrial haemagglutin are the main adhesion antigens.

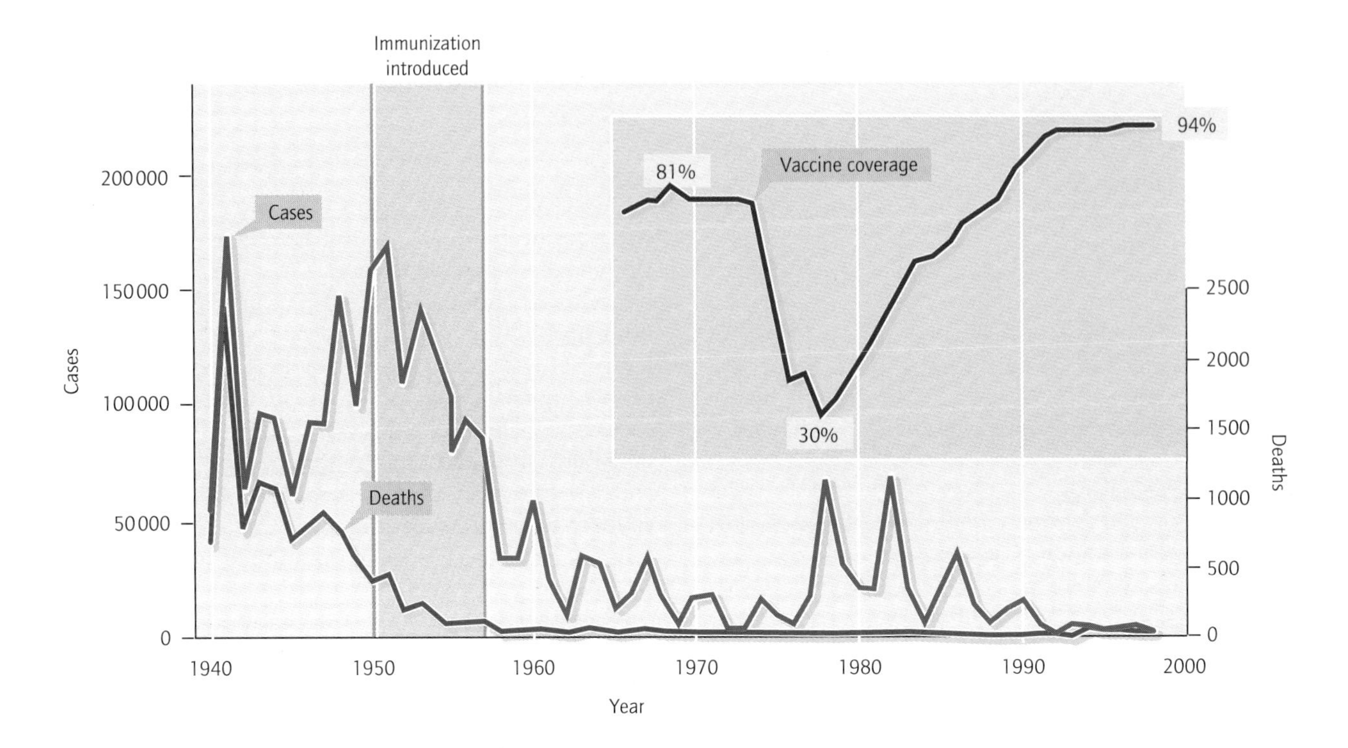

Fig. 7.9 Decline and resurgence of pertussis in England and Wales with changing vaccine coverage.

Laboratory diagnosis

A pernasal swab is the optimal method for obtaining a specimen for culture. The cultivation of the organism is difficult because *B. pertussis* is fastidious and relatively slow-growing, but it may be detected by culture of nasopharyngeal swab specimens on Bordet–Gengou or charcoal yeast extract agar with antibiotic supplements. Genome detection by PCR techniques is a useful rapid test. Serodiagnosis is possible by IgM ELISA.

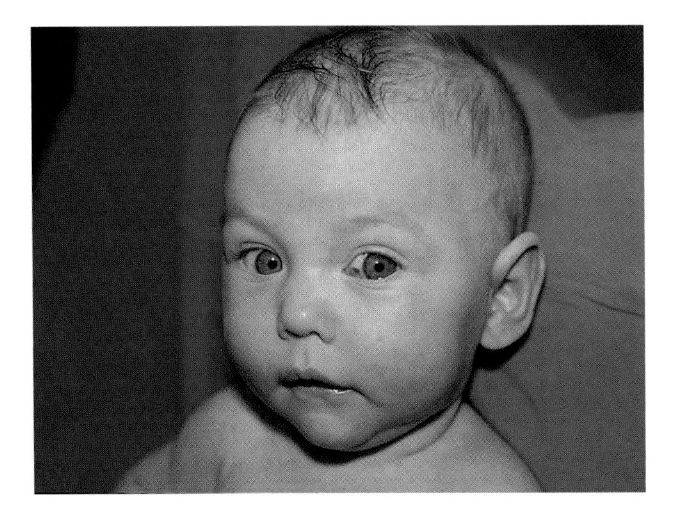

Fig. 7.10 Pertussis in a 6-month-old: subconjunctival haemorrhage produced by severe cough.

Management

Erythromycin is the antibiotic of choice, though it has not produced definite improvement in controlled trials. Supportive treatment is important. The airway should be protected by holding infants face- and head-downwards during paroxysms. Cyanotic or apnoeic attacks can often be helped by suctioning the airways to remove tenacious mucus. A humidified atmosphere may help. Severely affected infants may be nursed in a cot oxygen tent. Assisted ventilation is rarely needed.

Paroxysms are easily precipitated by feeding or by breathing cold air. Most children lose weight during the paroxysmal stage, and frequent feeds are important; food should be offered after coughing or vomiting. Small children may be fearful of paroxysms, and need sympathetic reassurance.

Prevention and control

Erythromycin treatment may reduce the infectiousness of the disease, and is also given for postexposure prophylaxis. Cases remain infectious for up to 3 weeks after paroxysms begin. An effective whole-cell killed vaccine forms part of the universally available triple vaccine in the UK. Subunit vaccines containing combinations of fimbrial haemagglutinin, pertussis toxin, 67 kDa antigen and agglutinogens are now available, and cause fewer febrile adverse reactions.

Fungal infections of the lower respiratory tract

Introduction

Fungal respiratory infections are rare in immunocompetent patients in the UK, mainly because exposure to highly pathogenic fungi is rare. There is a limited list of pathogens, mostly from endemic areas overseas, which cause diseases requiring differentiation from unusual, nodular or cavitating chest infections.

Aspergillosis

Aspergillus spp. have a worldwide distribution. They inhabit dark areas, such as ventilation ducts, cavities in bualdings and spaces behind wall panels. They produce millions of spores which are inhaled and primarily infect the lungs.

Indolent colonization of stagnant bronchioles or old cavities can produce a hypersensitivity reaction with bronchoconstriction. This can precipitate deterioration in patients with pre-existing asthma or chronic bronchitis. Affected patients have precipitin antibodies in their serum, and may respond well to treatment of the *Aspergillus*, although recurrence can be a problem.

In pre-existing cavities fungal hyphae may produce a ball-shaped growth which irritates the cavity wall, causing cough and often haemoptysis (a severe problem in up to 25% of sufferers). A single lesion like this may not produce hypersensitivity, and precipitins may not be present. The X-ray or imaging appearance of this aspergilloma is typically of a round, space-occupying lesion surrounded by a narrow, clear air space. Such lesions are treated surgically by segmental or lobar resection when circumstances permit, but half or more may improve with chemotherapy.

Itraconazole has good activity against the fungus, and is the treatment of choice for disease of moderate severity. Inoperable aspergilloma may respond to this or to intravenous amphotericin B.

Histoplasmosis

Histoplasma capsulatum is yeast, widely distributed in dry, warm areas, and some caves. After exposure, subclinical or mild, self-limiting disease is common. Acute clinical disease appears after 5–18 days, with fever, chills and myalgia. Physical examination is often normal, but mild hepatosplenomegaly, lymphadenopathy and maculopapular rashes or erythema nodosum sometimes occur. Chest X-ray shows midzone opacities and hilar adenopathy.

Antigen detection and serological tests (complement fixation or immunodiffusion) are useful for diagnosis. Bronchoscopic biopsies show interstitial pneumonia with eosinophilic infiltrates.

Itraconazole 200 mg once or twice daily for 10 days is effective treatment. The chest X-ray becomes normal in 2–4 weeks; serology remains positive for many months.

A few cases, or immunosuppressed patients develop disseminated disease, metastatic foci or cavitary pneumonia. These are treated with itraconazole, plus or minus amphotericin (see Chapter 22).

Blastomycosis

This is caused by the yeast *Blastomyces*, whose endemic area is confined to the USA. It causes a disease similar to histoplasmosis. Diagnosis and treatment are as for histoplasmosis.

Coccidioidomycosis

The cause of this disease, *Coccidiodes immitis*, is virtually confined to the San Joachim valley of the western USA. It tends to cause a rapidly progressive bronchopneumonia or pneumonia unresponsive to antibiotics, but mild cases also occur. The fungal hyphae grow rapidly in standard cultures, and produce segmented chains containing millions of spores. The spores are highly infectious, and patients with suspected coccidioidomycosis should always be isolated.

Treatment is with amphotericin (see Chapter 4).

Case 7.1: Chef's exotic cough

History: A 62-year-old chef of Indian origin had not worked for 5 years because of chronic obstructive pulmonary disease. He had a 1-week history of fever, increasing cough and shortness of breath, with increasing production of thick, greenish sputum. He also had a long history of itching of the lower legs. He had entered the UK 12 years previously and never travelled overseas since then. He used regular budesonide and ipatropium inhalers, and took nifedipine for control of moderate hypertension. For the last 6 weeks he had been taking oral prednisolone 5 mg daily for control of worsening asthma.

Physical findings: A distressed, mildly dehydrated patient with shortness of breath at rest. Pulse 110 bpm, blood pressure 130/85 mmHg, heart sounds normal with a third sound in the precordium. Chest expanded, with limited movement, especially on the left; auscultation revealed widespread expiratory wheeze, coarse crepitations at both bases and the left mid-zone, with reduced breath sounds and bronchial breathing in the left axilla. Abdominal and neurological examination were normal.

Investigations: Haemoglobin 15.6 g/dl, white cell count 16.7×10^9/l, with 12.2×10^9/l neutrophils and 2.2×10^9/l eosinophils. Blood urea 8.7 mmol/l, sodium 129 mmol/l, potassium 4.0 mmol/l, ESR 65 mm/h, alkaline phosphatase 255 IU/l, other liver function tests normal. Arterial $P\text{O}_2$ 7.7 kPa, $P\text{CO}_2$ 6.4 kPa. Chest X-ray showed normal lung outlines, widespread ground-glass appearance and a large, ill-defined opacity in the left mid-zone (see Fig. 8.20).

Questions: What diagnosis is fulfilled by the findings?
What treatment is indicated?
What other important pulmonary infection should be excluded?
What unexpected finding should be followed up?

Management and progress: A diagnosis of severe pneumonia was made in this patient with significant comorbidity (see Table 7.1). Treatment was commenced with intravenous cefotaxime, and nebulized salbutamol by inhalation with 30% oxygen. His prednisolone dosage was temporarily increased to 10 mg daily. Sputum and blood cultures were obtained, and tuberculosis was excluded initially by negative acid-fast staining. Moderate eosinophilia is unusual in chronic pulmonary disease; this was further investigated.

Question: What is a common cause of isolated eosinophilia in individuals from third-world countries?

Further management and progress: Strongyloidiasis is a persisting parasite infection in which adult worms in the bowel generate rhabditiform larvae, which may become invasive and reinfect by penetrating the colonic mucosa or perianal skin. The larvae then become filariform, and migrate through the tissues to the lungs, where they cause an inflammatory reaction and are coughed into the pharynx from where they reinvade the gut. Migrating parasites in the blood or tissues stimulate a protective eosinophilia. Serological tests were requested for strongyloidiasis and filariasis. Stool samples were obtained and charcoal culture performed to identify *Strongyloides* larvae. Sputum examination revealed many neutrophils, a few eosinophils and a heavy growth of *Escherichia coli*, sensitive to cefotaxime and co-amoxiclav. Scanty *Strongyloides* larvae were present in the sputum. Oral albendazole 400 mg twice daily was therefore added to the treatment regimen.

The fever and shortness of breath resolved after 3 days' therapy, and oral co-amoxiclav was substituted for intravenous therapy on the fourth day. Oral prednisolone dosage was stepped down over 3 weeks and was then discontinued.

Questions: Why did *Strongyloides* larvae appear in the sputum?
Is the finding of coliform pneumonia of interest in this case?
Is a standard three-day course of albendazole adequate treatment for the strongyloidiasis?

Completion of management: In debilitated patients, *Strongyloides* larvae may migrate, carrying coliform organisms from the patient's bowel in their gut. The resulting pulmonary irritation may be mistaken for asthma and treated with increasing doses of prednisolone, resulting in further immunosuppression and escalating larval migration. Coliform pneumonia or even meningitis may result. In severe immunosuppression, eosinophilia does not occur, and larvae are found in the stool, sputum, liver biopsies and effusions. Fatal invasive disease and/or coliform infections may result. In this mildly immunosuppressed patient a 1-week course of albendazole was given, and a further 3-day course was given on ceasing prednisolone dosage.

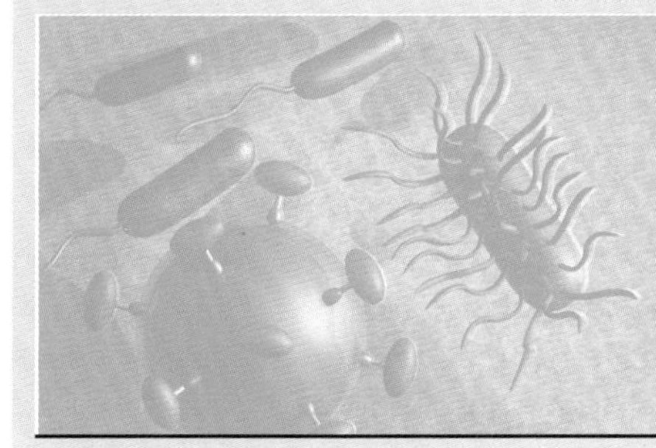

8 Gastrointestinal Infections and Food Poisoning

Introduction, 157
- Structure and environment of the gastrointestinal tract, 157
- Natural defences of the gastrointestinal tract, 158

General principles of managing gastrointestinal infections, 158
- Diagnostic tests in intestinal infections, 158
- Management of gastrointestinal diseases, 159
- Laboratory diagnosis of diarrhoea, 160
- Managing food-borne disease in the community, 162

Viral infections of the intestinal tract, 163
- Introduction, 163
- Rotavirus gastroenteritis, 163
- Other viral infections of the gastrointestinal tract, 164
- Diagnosis of viral gastroenteritis, 165

Bacterial diseases of the gastrointestinal tract, 165
- *Escherichia coli* gastroenteritis, 165
- *Salmonella* infections, 169
- Shigellosis (bacillary dysentery), 173
- *Campylobacter* infections, 175
- Infection with 'food-poisoning' vibrios, 176
- *Yersinia* infections, 177
- *Helicobacter pylori* and acute gastritis, 178
- Diseases caused by bacterial toxins, 178

Non-bacterial toxins and food poisoning, 181
- Bean haemagglutinins, 181
- Scombrotoxin, 181
- Diarrhoetic shellfish poisoning, 181
- Paralytic shellfish poisoning, 181
- Cyanobacterial toxins, 182

Parasitic infections of the gastrointestinal tract, 182
- Laboratory diagnosis, 182
- Giardiasis, 182
- Cryptosporidiosis, 183
- Cyclosporiasis, 184
- Amoebiasis, 185
- Helminth infections of the bowel, 186
- *Ascaris lumbricoides*, 187
- Hookworms, 187
- Strongyloidiasis, 188
- Toxocariasis, 189
- Trichinellosis, 189

Introduction

Infections of the gastrointestinal tract are among the commonest infections in all communities of the world. There is little doubt that they cause the greatest morbidity and mortality. Even in the UK, where water supplies, sanitation and education reach high standards, surveys suggest that intestinal infections are causing increasing illness and loss of working days.

Structure and environment of the gastrointestinal tract

The gastrointestinal tract is a tube, open to the environment at both ends, into which much foreign material is introduced. Microorganisms, which may be pathogens, enter every time fingers, food, drink or utensils are put into the mouth. The local environments vary widely in different parts of the gastrointestinal tract, and this influences the nature of the local flora and of pathogens which may invade.

The stomach is highly acidic and contains variable amounts of air. Few bacteria can survive this environment for long, but many pass quickly through. Boluses of food, especially fats, can protect bacteria during this passage. Chocolate and cheese are thought to enhance survival of *Salmonella* in this way. *Helicobacter pylori* and mycobacteria resist the effect of gastric acid, and can persist in the stomach.

The upper small intestine is alkaline. It has a highly vascular mucosa which is attractive to adherent parasites. It also contains bile, which inhibits the growth of many bacteria, and strongly selects the bile-resistant local flora; almost all flora of the small bowel grow well on bile-containing media, e.g. MacConkey's agar. The highly absorptive mucosa of the jejunum and ileum offers a means of entry for toxins, not only those adapted to adhere to mucosal cells but also systemic toxins which cause non-intestinal diseases such as botulism. Toxins may be elaborated by bacteria in food or be components of the food itself.

The terminal ileum and colon are inhabited by large numbers of anaerobic bacteria, both Gram-positive and -negative, cocci and rods, including sporing organisms. Facultative organisms such as *Escherichia coli* and enterococci can also occupy this environment.

Some organisms, such as the yeast *Candida*, are able to reside throughout the bowel. Others may pass through, being excreted in the faeces for a while, but not causing long-term colonization.

Natural defences of the gastrointestinal tract

Gastric acid

Gastric acid is one of the major first defences of the gastrointestinal tract. While it provides a suitable environment for *Helicobacter* spp., and has no effect on acid-fast organisms, it greatly reduces the numbers of bacteria that go on to enter the intestines. This is practically demonstrated in patients with achlorhydria, who are easily infected by very small numbers of salmonellae or vibrios, and suffer severe illness as a result.

Bile salts

The bile salts of the duodenum inhibit many organisms, killing some by disrupting their cell surfaces. The exceptions are the families of Enterobacteriaceae and enterococci, which live well in a medium of bile salts. Lower in the bowel, bile salts are reabsorbed and their effect is diminished.

Normal bowel flora

The normal bowel flora confers **colonization resistance**, a complicated phenomenon depending on competition between microorganisms. Bacteria modify their environment by the production of metabolic products which may alter the local pH or redox potential. Some produce poisonous chemicals such as hydrogen sulphide or volatile fatty acids which can inhibit other organisms. Many of the Enterobacteriaceae produce natural antibiotics, called enterocines, which are harmful to other species, but not to their own. It has also been shown that enterococci inhibit other organisms, such as *Clostridium difficile*. All of these factors combine to favour particular types of organisms in the environments of the bowel.

The protective effect of the normal flora can be reduced by antibiotic action, predisposing a patient to infection by pathogens to which he or she may be exposed. Thus, *Salmonella* infection in exposed people is more likely if they are pretreated with streptomycin or tetracycline.

Immune responses

Immune responses in the bowel are also important. The bowel mucosa is rich in lymphocytes and contains much lymphoid tissue. Cell-mediated immunity can be demonstrated by finding lymphocytes sensitized to the antigens of recently active pathogens. Humoral immunity is provided by secreted immunoglobulin A (IgA). Plasma cells are often plentiful in established mucosal inflammation. The immune responses of the gut are exploited when oral polio vaccine is given; after immunization the mucosa resists invasion by these enteroviruses which are then deprived of their usual means of entry into the body.

Motility

The motility of the gastrointestinal tract assists greatly in clearing pathogens. This is demonstrated when obstruction or stasis alters the local environment. The stagnant stomach becomes colonized by organisms which ferment its contents, quickly losing its acidity. Blind loops or diverticula in the bowel can easily become the foci of abscesses containing bowel flora. Approximately 50% of the weight of faeces is composed of bacteria. The diarrhoea of bowel infections is probably an important mechanism for clearing pathogens from the gut.

Protection against toxins

The microsomal enzymes of the hepatocytes detoxify many drugs and other substances. Toxins absorbed from the gut enter the portal circulation and are intercepted by the liver before they can reach the general circulation. The gut flora naturally generate endotoxin, and the portal venous blood contains significant endotoxin concentrations. However, after passage through the liver, hepatic venous blood contains only negligible amounts. Isolated Kupffer cells can also inactivate endotoxin *in vitro*.

Normal defences of the bowel

1. Gastric acid.
2. Bile salts.
3. Lymphoid tissue.
4. Enterocines.
5. Normal bowel flora.
6. Secretory immunoglobulin A.
7. Motility.
8. Hepatic deactivation of toxins.

General principles of managing gastrointestinal infections

Diagnostic tests in intestinal infections

The most obvious effects of bowel infections are diarrhoea and vomiting.

Stool specimens

Stool specimens are relatively easy to collect, but it is important to include the most liquid part of the stool, which is the most likely location of excreted pathogens. If there is much mucus, some of this should also be obtained (a syringe is a convenient means of collecting liquid or viscous material).

Mucosal specimens

Some parasites and ova are only sparsely excreted and more direct specimens increase the likelihood of positive diagnostic results. Examples include trophozoites of *Entamoeba histolytica* or schistosome ova which may both be more readily detected in scrapings or small biopsies taken from the rectal mucosa at proctoscopy (a small spatula or gloved finger may be used to obtain scrapings). Specimens should be taken from the edge of any ulcer or area of inflammation.

Intestinal fluids

In giardiasis or strongyloidiasis the parasites may not be detectable in stool samples, but are readily demonstrated in duodenal aspirate, biopsy or material obtained by the string test. In the string test the patient is asked to swallow a gelatin capsule containing a weighted length of soft, absorbent string. The free end of the string is fixed at the mouth; the weighted end unwinds from the capsule as it passes through the stomach and into the duodenum. After allowing 30–60 min resting time, the string is withdrawn and the adherent material is examined for parasites.

Vomitus

Vomitus is rarely obtained but it may contain viruses in acute viral gastroenteritis or bacteria in some types of toxic food poisoning. It is worth attempting to collect specimens, especially in patients who do not have coexisting diarrhoea.

Other specimens

Other specimens which may be useful include blood cultures, which are mandatory in patients with fever. Positive blood cultures are seen in a minority of patients with salmonellosis and on occasions in other gastrointestinal infections. Serum may contain enterotoxins, botulinum toxin or antibodies to toxins.

Management of gastrointestinal diseases

Principles of oral rehydration

Few gastrointestinal infections disable all of the mucosal absorptive function of the bowel. In toxin-mediated diarrhoeas there is no damage to absorption; hypersecretion is the problem. In other diarrhoeas mucosal cells may be damaged, but some survive intact and can function sufficiently to absorb fluid and electrolytes. Absorption of sodium and water can be maximized by giving sufficient of each with the optimum amount of glucose to 'drive' the active transport systems. A steady input of this mixture can allow absorption to overtake diarrhoeal fluid loss, even in severely dehydrated patients. At the same time, potassium will flow passively along concentration gradients. This solution can be given until normal hydration is restored, as determined by clinical condition or body weight. Thereafter the sodium intake may be reduced by diluting the solution to approximately half-strength or by alternating it with drinks of water, dilute juice or squash. This is important in small children in whom sodium overload easily occurs, causing peripheral oedema which may be slow to resolve.

Recommended composition of oral rehydration fluid

Sodium:	150–155 mmol/l
Glucose:	200–220 mmol/l
Potassium:	4–5 mmol/l

This solution can be given until normal hydration is restored, as determined by clinical condition or body weight.

Intravenous rehydration

This is indicated for shock, exhaustion precluding oral feeding and failure of oral hydration therapy. The electrolyte solution of choice is half-normal sodium chloride solution (0.45%). This provides adequate sodium, but with less risk of sodium overload than normal saline. Potassium may be added to the solution if required. It is seldom necessary to add any other electrolyte even in children with considerable acidosis; rehydration alone will restore normal homeostasis. A guide to children's fluid and electrolyte requirements is given in Table 8.1.

Infant feeding in diarrhoeal illnesses

Infants who are fed during diarrhoea episodes recover sooner and lose less weight than those who are offered only fluids. Some work in Third World countries suggests

Age	Daily baseline fluid requirement (ml/kg)	Daily sodium (mmol/kg)	Daily potassium (mmol/kg)	Total daily calcium (mmol)
1–2 days	75–100	2.5	2.5	12.5–17.5
Up to 1 year	150	2.5	2.5	12.5–17.5
1–3 years	100	2.5	2.5	20–25
4–6 years	90	2.0	2.0	20–25
7–10 years	70	2.0	2.0	20–25
10–14 years	60	1.5	1.5	30–38

Table 8.1 A guide to fluid and electrolyte requirements.

that rice-water feeds are associated with earlier improvement of diarrhoea than feeds made with plain water.

Secondary acquired lactose intolerance

This is a particular problem in infants and young children. It is commonest after rotavirus infection and enteropathogenic *Escherichia coli* infections, which can cause severe mucosal damage. Lactose absorption depends on lactase enzyme systems in the mucosal brush border, which slowly recover function after mucosal healing. Until this happens, the lactose in milk and other dairy foods cannot be absorbed, so it ferments in the bowel lumen, causing abdominal discomfort, flatulence and acid diarrhoea soon after feeding.

Infants with lactose intolerance can be maintained on lactose-free or low-lactose formulae such as soya-based milks, Pregestimil and some of the low-lactose Galactomin preparations until gaining weight satisfactorily, when normal feeding can be resumed. Older children, including weaning infants, can be given non-dairy solids to replace milk feeds.

Secondary acquired lactose intolerance must be distinguished from the primary acquired alactasia, which occurs particularly in adult oriental and Caribbean people. Other sugar absorption systems are rarely affected by intestinal infection.

Drug treatment of vomiting and diarrhoea

Antiemetic drugs

Antiemetic drugs are often helpful in reducing fluid loss to a level at which oral rehydration can be effective. There are three main types of antiemetic drugs: phenothiazines such as prochlorperazine, metoclopramide and domperidone. Phenothiazines and metoclopramide are slightly sedative at antiemetic doses, and both occasionally cause dopamine-induced dystonic reactions, especially in children and teenagers. Domperidone has dopaminergic actions, but does not cross the blood–brain barrier, and does not cause such noticeable central effects.

Antidiarrhoeal drugs

Antidiarrhoeal drugs are rarely successful in clinical practice. They simply reduce gut motility, allowing fluid faeces to accumulate, but the symptoms resume as soon as the expanded bowel volume is filled. Atropine-like side-effects such as dry mouth and a tendency to urinary retention are a further disadvantage. Antidiarrhoeals can produce severe atropine-like side-effects in small children, and are not therefore recommended for this age group. In general it is better to restore hydration, using oral or intravenous fluids, than to resort to antidiarrhoeal drugs (this is the recommendation of the World Health Organization for treating cholera and childhood diarrhoea).

Laboratory diagnosis of diarrhoea

The diagnosis of infective diarrhoea depends on the identification of the pathogen from faeces by electron microscopy, culture or demonstration of antigens. In the case of viruses, electron microscopy and culture are most often used. As all bacterial bowel pathogens are closely related, diagnosis is usually made by culture and formal identification of individual pathogens. In reference laboratory tests toxins of organisms such as *Clostridium perfringens* can be identified in faeces by polymerase chain techniques.

At least three specimens should be examined. A faecal suspension is inoculated on to a combination of selective media. The more selective the medium, the easier it is to detect the presence of the pathogen sought, but selective media are also inhibitory to the pathogen to some extent, which may mean that low numbers of pathogens will not be detected. A relatively non-selective medium such as MacConkey agar is therefore inoculated in combination with a more selective medium (Fig. 8.1).

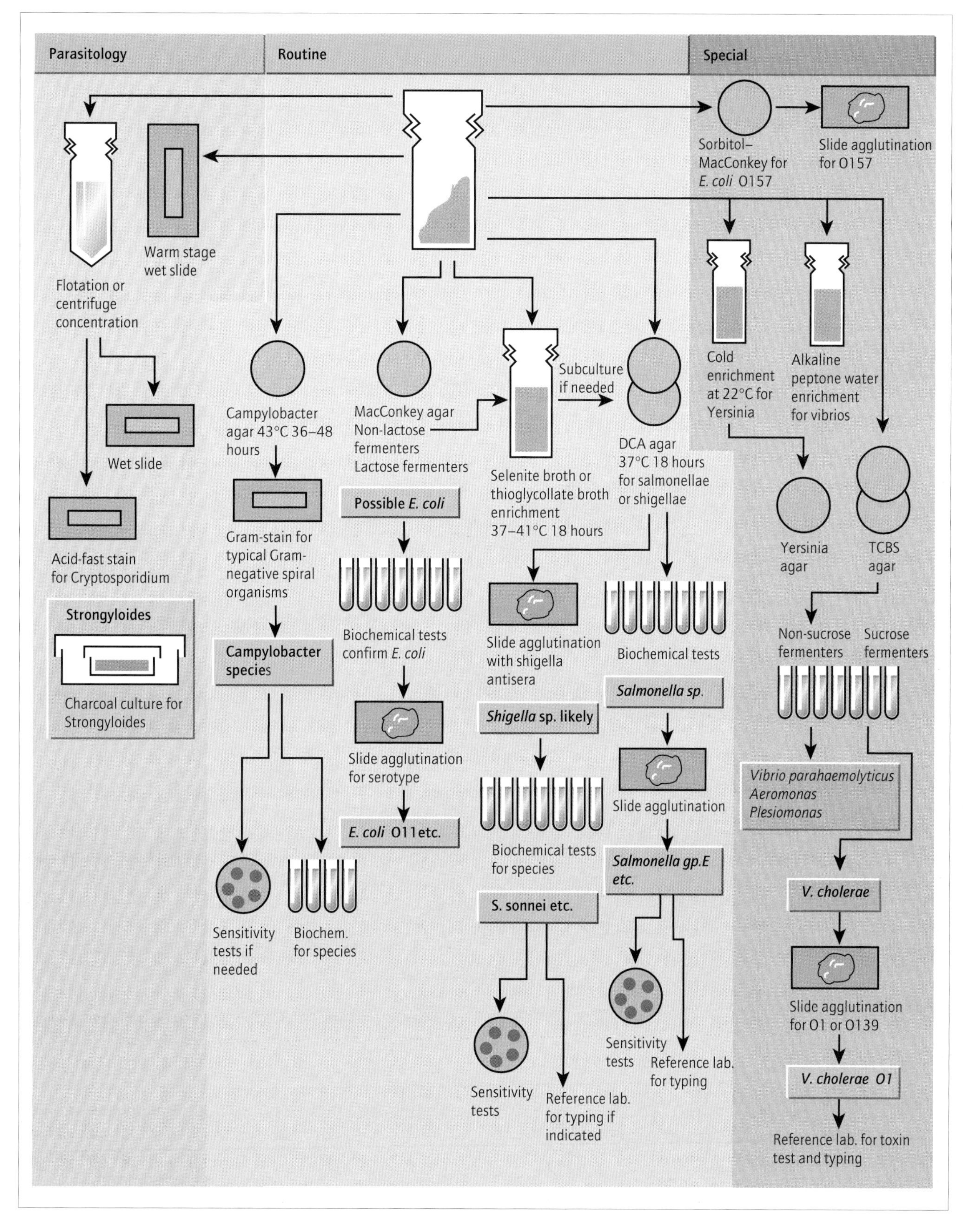

Fig. 8.1 Scheme for the bacteriological investigation of stool specimens. TCBS, thiosulphate–citrate–bile salt–sucrose.

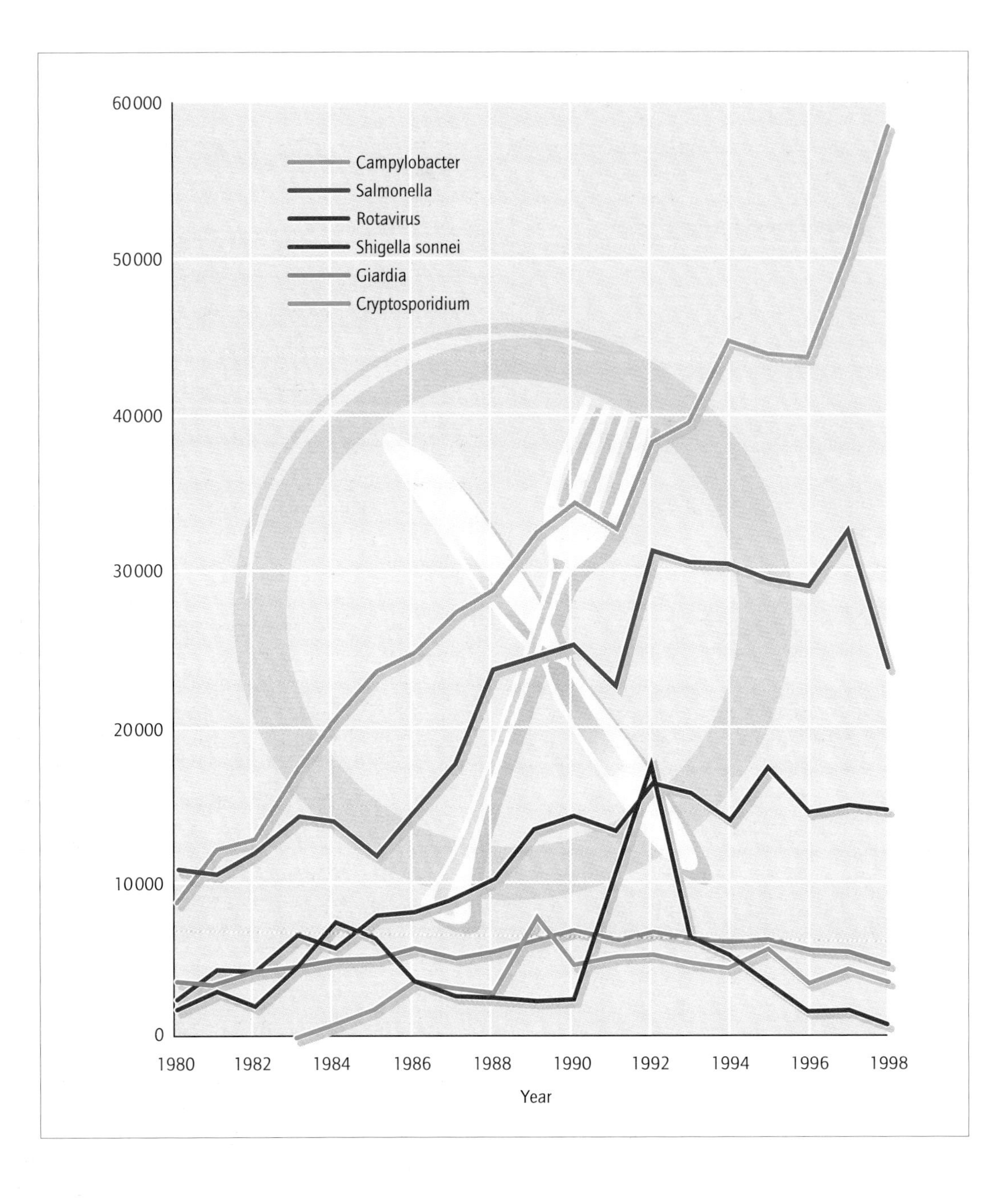

Fig. 8.2 Trends in the incidence of common food-borne infections in England and Wales since 1980.

Managing food-borne disease in the community

In developing countries without safe water supplies, most gastrointestinal infections are water-borne. In developed countries, food is a more important source of infection. It has been estimated, for example, that more than 50% of all cases of infectious gastrointestinal disease in the UK are food-related. The incidence of several food-borne pathogens has increased considerably in recent years (Fig. 8.2).

Ideally, all food produced for human consumption would be free from pathogenic bacteria. In practice, food is frequently contaminated and measures are required to ensure that it is safe at the point of consumption.

A number of measures are aimed at reducing the level of bacterial contamination at source. These include hygienic husbandry and slaughtering of livestock. Such measures are normally the responsibility of central government. Further down the chain of supply, a complex legislative framework exists which controls the processing, storage and preparation of food by retailers. The enforcement of these regulations is the responsibility of local government.

At consumer level, prevention of food poisoning depends on good practices for cooking and storing food and on personal hygiene. Food should be cooked thoroughly and eaten immediately. Cooked and raw food should be stored separately. Perishable items should be kept in properly maintained refrigerators. Food not in a refrigerator should be covered to protect it from flies, rodents and other animals. Food-handlers should wash hands before and after food preparation. Any cuts, abrasions or other skin lesions should be covered before preparing food. A food-handler who is ill (particularly with diarrhoea or vomiting) should not prepare food until fully recovered.

Prevention of food poisoning

1 Safe food production: healthy flocks and herds, avoiding use of sewage to fertilize food crops.

2 Food-manufacturing processes: hygienic slaughtering and meat packing, rodent-free storage of crops, storage at chill or refrigeration temperatures, hygienic packaging, cold chain during distribution.

3 Domestic and commercial food hygiene: adequate refrigeration, avoidance of cross-contamination, usage within spoilage dates, adequate decontamination of food by washing and/or cooking, personal and kitchen hygiene.

Food poisoning is a notifiable disease in the UK. Outbreaks should be promptly investigated to determine the source and implement any necessary control measures. The principles of outbreak investigation are described in Chapter 26.

Viral infections of the intestinal tract

Introduction

Recent large surveys have shown that viral infections are the commonest cause of symptomatic intestinal infections in the Western world. Although rarely severe or fatal, they cause many days of debility and work absence.

ORGANISM LIST

Rotavirus
Adenoviruses (high serotype numbers)
Caliciviruses
Small round structured viruses
Small round viruses
Astroviruses
Coronaviruses
(Hepatitis A and E viruses; see Chapter 9)

Rotavirus gastroenteritis

Epidemiology

Rotaviruses are the most common viral cause of acute gastroenteritis in young children. Up to a million children worldwide die each year from rotavirus diarrhoea, mostly in developing countries; however, the disease is also an important cause of morbidity in developed countries. In the USA approximately a third of all hospital admissions for diarrhoea in children are due to rotavirus infection.

Over 10 000 laboratory-confirmed cases are reported each year in the UK; this probably represents only a small fraction of the total. The peak incidence of infection is between 6 and 24 months. Clinical infection is rare in children over 5 years of age, although subclinical infection is probably common in children and adults. The disease has a highly seasonal pattern; most infections occur during the winter months.

Virology

Rotaviruses are members of the Reoviridae: icosahedral non-enveloped viruses with a segmented double-stranded RNA genome. They are classified by the capsid protein VP6 into seven serogroups (A–G). Most childhood disease is caused by group A strains but group B and C strains are responsible for adult outbreaks. The remainder are mainly animal pathogens. Rotaviruses have a distinctive appearance on electron microscopy, with two shells surrounding a central protein core. The genome codes for proteins VP1–7, and five non-structural proteins (NSP1–4).

Clinical features

The incubation period of about 24 h is followed by a rather abrupt onset of both diarrhoea and vomiting. A mild to moderate fever is common at the onset, but rarely persists.

Vomiting is usually moderate, permitting successful feeding of oral fluids. Prolonged vomiting is rare; if it continues for more than 48 h the diagnosis should be reviewed. The severity of the diarrhoea varies but occasionally causes gross dehydration or even shock. Blood may be seen, especially in infants under the age of 1. This probably reflects the large-scale destruction of mucosal cells, leaving denuded areas in the duodenum and upper ileum. Even the most severely affected cases tend to recover after 2–4 days.

Adults are often immune to the common group A rotaviruses, but a few are susceptible, often suffering transient vomiting as their major symptom. The illness is highly infectious while the vomiting lasts, and susceptible children may be infected by an affected nurse or mother. Group B and C rotaviruses are uncommon, and adults are less likely to be immune to them. They have caused large outbreaks of adult gastroenteritis.

Diagnosis

Electron microscopy of diarrhoea stools will often demonstrate typical rotavirus particles (Fig. 8.3). Human rotaviruses do not grow in cell culture. Enzyme-linked immunosorbent assay (ELISA) tests for rotavirus antigen are available. Latex agglutination is less sensitive but does not require sophisticated equipment. Although polymerase chain reaction (PCR) techniques have been described, their use is confined to research laboratories.

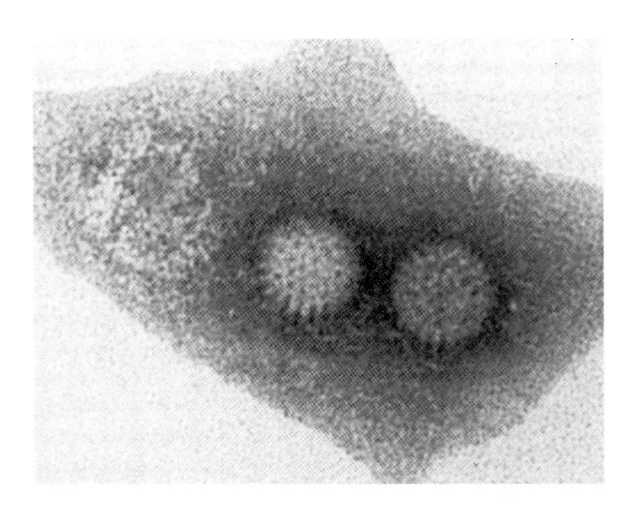

Fig. 8.3 Rotavirus particles demonstrated by negative-stained electron microscopy preparation of a diarrhoea stool. Courtesy of Professor P. D. Griffiths and Ms G. Clewley, Department of Virology, Royal Free Hospital School of Medicine.

Management

Most cases are readily managed with oral rehydration treatment. A minority have persistent diarrhoea, exacerbated by attempts to give milk feeds. This is caused by postgastroenteritis lactose intolerance, and is managed by avoiding milk and yoghurt in the diet, and/or by substituting a low-lactose milk for the child's usual formula.

A few infants have exacerbations of diarrhoea after each feed of oral rehydration fluid, suggesting that there are too few functioning mucosal cells to absorb the glucose in the fluid. This is a transient problem but while it lasts hydration can often be maintained with sugar-free electrolyte solution, such as half-normal saline.

Rare effects of rotavirus infection

Some 1–2% of children admitted with rotavirus infection present with shock and peripheral cyanosis. It is impossible to determine the aetiology of the condition until persisting diarrhoea and resolution of fever point to an intestinal infection. A handful of reports exist of cases whose cerebrospinal fluid contained rotavirus particles during this acute illness.

Rotavirus can cause persisting diarrhoea in immunosuppressed patients, including those with human immunodeficiency virus infection.

Rotavirus vaccines

Orally administered live rotavirus vaccines are currently being developed, using both animal and human virus strains. The protection afforded by these vaccines is variable. A tetravalent vaccine containing reassortant rhesus monkey and human genes was recommended for routine use in infants in the US, although it has recently been with drawn following reports of intussusception in vaccine recipients.

Other viral infections of the gastrointestinal tract

Epidemiology

Small round structured viruses (SRSVs, including Norwalk virus), adenoviruses, astroviruses, caliciviruses and coronaviruses have all been associated with outbreaks of gastroenteritis. Transmission is usually from person to person by the faecal–oral route; however, several food-borne incidents have been reported. Many of these are due to shellfish harvested from sewage-polluted estuaries and eaten raw or without sufficient cooking.

Aerosol spread probably also plays an important role in many of these infections, particularly in winter vomiting disease (see below). All age groups are affected.

Clinical features

These viral infections all have short incubation periods, of 6–36 h. There is more variation of incubation between individuals with the same infection than between infections. In most cases the illness lasts for 1–3 days.

Adenoviruses and caliciviruses tend to affect children, adenoviruses causing mainly diarrhoea and calciviruses mainly vomiting. Adenoviruses have often been identified during surveys of asymptomatic children, but calicivirus usually causes symptomatic infection.

Winter vomiting disease

Calicivirus is a common cause of this disease. The illness has an incubation period of about 1 day followed by several hours of profuse vomiting. There is then a period of 24–36 h of fatigue before recovery is complete. This is an intensely infectious disease which often causes school and family outbreaks, affecting both adults and children.

Viral food poisoning

SRSVs and unstructured viruses are common causes of food-poisoning outbreaks, often associated with shellfish. They are much more often described as causes of adult disease than of childhood infections.

Diagnosis of viral gastroenteritis

An aetiological diagnosis in viral gastroenteritis is most important in the investigation of outbreaks. Diagnosis of individual cases requires a considerable investment of time and resources, and is of less clinical relevance, as it does not influence treatment in most cases. The principal method of diagnosis is electron microscopy, which demonstrates characteristic morphology (Fig. 8.4). ELISA tests for the diagnosis of rotavirus are also available.

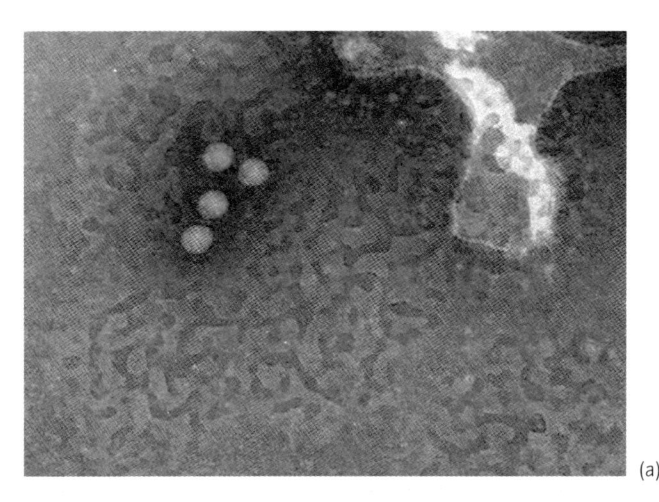

(a)

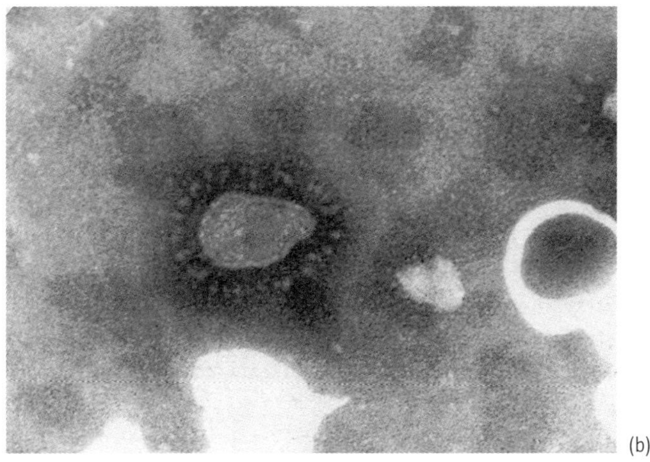

(b)

Fig. 8.4 (a) Small round viruses and (b) a coronavirus in electron microscopic preparation of a diarrhoea stool. Courtesy of Professor P. D. Griffiths and Ms G. Clewley, Department of Virology, Royal Free Hospital School of Medicine.

Bacterial diseases of the gastrointestinal tract

ORGANISM LIST

Common infections
- *Campylobacter* sp.
- *Salmonella* sp.
- *Shigella* sp.
- *Escherichia coli*
- *Clostridium perfringens*
- *Helicobacter pylori*

Uncommon or rare infections
- *Aeromonas* sp.
- *Vibrio parahaemolyticus*
- *Plesiomonas* sp.
- *Yersinia* sp.
- *Clostridium difficile*
- *Vibrio cholerae*

Bacteria which elaborate preformed toxins
- *Staphylococcus aureus*
- *Bacillus cereus*
- *Clostridium botulinum*

Escherichia coli gastroenteritis

Introduction

E. coli are inhabitants of human gut, forming the major part of the facultative anaerobic flora. Their role in human disease has only recently been appreciated due to the difficulties of distinguishing commensal types from those which are behaving as pathogens.

In the 1960s certain serotypes of *E. coli* caused large epidemics of gastroenteritis in infants. Important *E. coli* diseases now include toxin-mediated traveller's diarrhoea (see Chapter 24), and haemorrhagic colitis associated with certain serotypes of verocytotoxin-producing organisms.

Microbiology

E. coli are facultative anaerobes which ferment a wide range of sugars, including lactose, producing acid and gas. They are oxidase-, Voges–Proskauer- and citrate-negative, but produce indole and are methyl red-positive. They are actively motile due to the possession of flagella.

Pathogenesis

The capsular K antigens of *E. coli* facilitate adherence to host cells. Strains possessing a K1 capsule are the most frequent *E. coli* isolated from cases of neonatal meningitis. The mechanism for this is not understood, although it is known that colonization of neonatal rat gut by K1 strains often leads to bacteraemia and meningitis. K88 antigen-bearing strains readily cause scour (diarrhoea) in piglets.

Fimbriae also mediate attachment to mucosal surfaces (first recognized in the contribution of P fimbriae to the pathogenesis of urinary tract infection). Fimbriae play a role in the pathogenesis of enterotoxigenic *E. coli* (ETEC) infections, as toxigenic *E. coli* which cannot adhere to gut mucosa are non-pathogenic. Some fimbrial colonizing factor antigens (CFAs) have been identified, and these include CFA/I, 6–7 nm rigid fimbriae, and the CFA/II family of antigens, 2–3 nm fibrillar fimbriae which are analogous to the K88 antigens of porcine strains. Other similar CFA families have been described in other strains.

Like other Gram-negative bacteria, *E. coli* has a lipopolysaccharide (LPS) antigen with a central lipid A core, an oligosaccharide moiety and polysaccharide chain. The lipid A portion is an endotoxin (see Chapter 1). The polysaccharide chain protects the organism from serum lysis and is the main (O) antigen of the organism. More than 150 O serotypes have been described, many of which are linked to other *E. coli* pathogenicity factors. For example, the most common enteropathogenic *E. coli* (EPEC) are serotypes O26, O55, O111, O114, O119, O125–9 and O142. Strains responsible for the haemolytic–uraemic syndrome (HUS) are usually serotype O157.

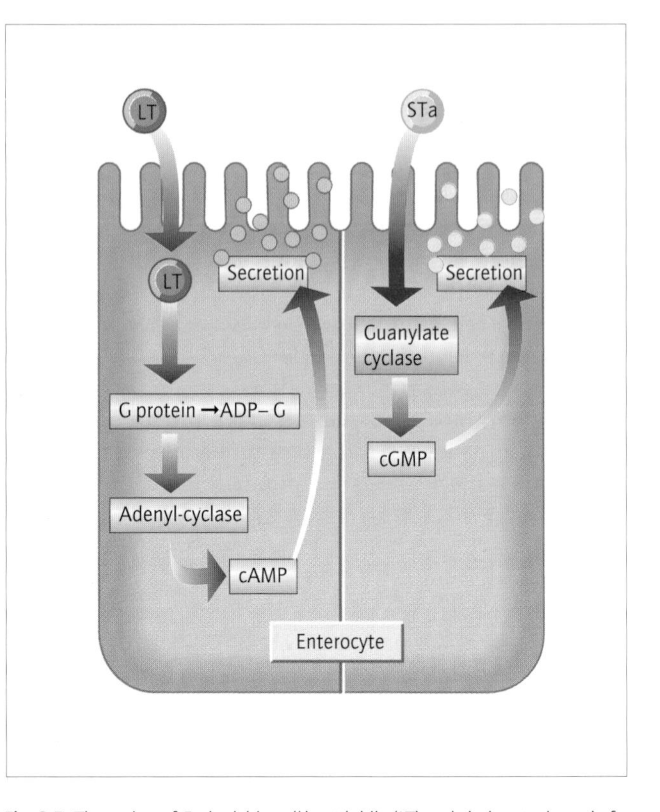

Fig. 8.5 The action of *Escherichia coli* heat-labile (LT) and cholera toxin and of *E. coli* heat-stable toxin (STa).

Enterotoxigenic E. coli *(ETEC)*

Toxins elaborated by *E. coli* are important in the pathogenesis of diarrhoeal and systemic disease. Two main enterotoxins are produced by ETEC, the heat-labile (LT) and the heat-stable (ST) toxins.

The LT toxin consists of polypeptide subunits, A and B. In the native toxin there are five B subunits which mediate attachment to cells via the GD1 receptor, and one A subunit which enters the cell and activates adenylate cyclase. This toxin is biologically and immunologically closely related to cholera toxin (Fig. 8.5).

The ST toxin exists in several forms. It induces hypersecretion by stimulating guanylate cyclase synthesis.

Verocytotoxic E. coli *(VTEC) (or enterohaemorrhagic* E. coli, *EHEC)*

Some *E. coli* strains make a toxin, verocytotoxin (VT) or shiga-like toxin (SLT), which kills the cells in vero-cell monolayers. This toxin exists in two forms; VT1 and VT2. VT1, but not VT2, is neutralized by antibodies to the shiga toxin. Both toxins are important in the pathogenesis of haemorrhagic colitis and HUS. The infective dose of

VTEC is small, probably between 10 and 100 organisms. Large outbreaks can therefore occur when low numbers of organisms exist in food, milk or the environment.

Enteropathogenic E. coli *(EPEC)*

EPEC have been associated with outbreaks of diarrhoea in institutional settings and in infant diarrhoea in developing countries. These strains lack ST and LT, and have no enteroinvasive properties. Despite this, volunteer studies indicate that these organisms are true pathogens. Electron microscopic studies show that EPEC cause attachment-effacing lesions in mucosal cells (destruction of microvilli without evidence of invasion). These strains adhere to Hep-2 cells, an unusual property for *E. coli*. EPEC adherence factor (EAF) is a 94-kDa protein which is encoded on a plasmid. EPEC serotypes without this gene cannot cause disease. Infection with EAF-positive strains stimulates production of EAF antibody, which may be protective.

Enteroaggregative E. coli *(EAEC)*

These are not classic EPEC serotypes and do not possess the EAF plasmid, or elaborate any of the recognized toxins. They adhere to Hep-2 cells in either an aggregative or a diffuse pattern, depending on different factors. It is not yet clear whether the pathogenicity of this organism depends on its enteroadherence property.

Pathogenicity factors for *Escherichia coli*
1. Capsular K antigens (K1).
2. Fimbriae (colonizing factor antigens—CFA/I, CFA/II and others).
3. Lipopolysaccharide O antigens (endotoxin).
4. Enterotoxins (heat-labile and heat-stable).
5. Verocytotoxin (shiga-like toxin).
6. Enteropathogenic *Escherichia coli* adherence factor (EAF).
7. Enteroadherence (mechanism unknown).

Epidemiology

The routes of transmission and epidemiological features of *E. coli* vary considerably between different pathogenic types and in different geographical locations. In conditions of poor hygiene, ETEC is the commonest bacterial cause of diarrhoea in children. Humans are the main source of infection, which is transmitted by contaminated food and water. The disease is uncommon in western Europe and the USA, although ETEC are an important cause of traveller's diarrhoea.

EPEC are usually spread from person to person by the faecal–oral route, although transmission via contaminated baby food also occurs. Most infections occur as outbreaks of infantile enteritis. The incidence of EPEC infections is decreasing in many countries.

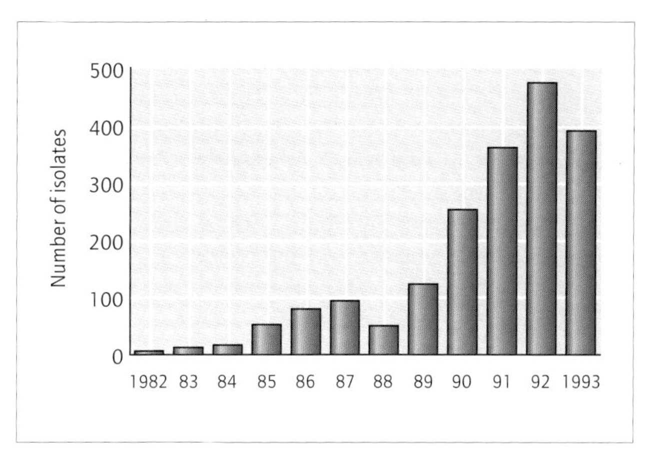

Fig. 8.6 Increasing reports of verocytotoxin-producing *Escherichia coli* infection in England and Wales.

Enteroinvasive *E. coli* (EIEC) infections are also rare in developed countries. Infection is usually food-borne but direct person-to-person spread may also occur. All age groups are affected.

VTEC includes several serogroups of *E. coli*, although by far the commonest is serogroup O157. During the 1980s the recognition of VTEC infection and its clinical sequelae (haemorrhagic colitis and HUS) increased rapidly in the USA and Canada. VTEC is now the commonest cause of acute renal failure in children in Western countries (Fig. 8.6). The infection is spread by contaminated food (notably hamburger meat) and unpasteurized milk. Large outbreaks have occurred in the community, in nursing homes for the elderly and children's day-care centres, and in association with farm visits. The largest outbreak in the UK was associated with contaminated meat sold from a butcher's shop; altogether over 500 people were affected, of whom 18 died.

Clinical features

The incubation period is 1–5 days. The onset is abrupt, with both vomiting and diarrhoea for the first 6–24 h, followed by watery diarrhoea alone. There is often moderate fever at the onset. There is little abdominal pain in uncomplicated gastroenteritis, and the illness is clinically similar to viral gastroenteritis or salmonellosis. A minority of cases exhibit severe disease with rapid dehydration and collapse, or prolonged illness with persisting diarrhoea and sometimes lactose intolerance.

Haemorrhagic colitis

Haemorrhagic colitis affects both children and adults. It begins as an unremarkable diarrhoeal illness, but quickly progresses to a syndrome of bloody diarrhoea and abdominal pain. In spite of the intense illness, fever is not an important feature. Sigmoidoscopy reveals an acutely inflamed colonic mucosa and the condition can be mistaken for acute inflammatory bowel disease. Most cases are self-limiting, recovering in 7–10 days. A handful of cases need prolonged symptomatic support and attempts at specific chemotherapy. In outbreaks, about 10% of children develop HUS (see below), but almost all of the fatalities are in patients over age 65–70.

Haemolytic–uraemic syndrome

HUS is mainly a disease of children, but can also affect adults. The adult form overlaps clinically with the syndrome of thrombotic thrombocytopenic purpura (TTP), but in TTP, thrombosis and platelet depletion are related to functional abnormality of uncleaved von Willebrand's factor, while the pathogenesis of HUS is based on mucosal damage and microangiopathic anaemia, together with renal vascular damage. The important features of HUS are a rising blood urea and creatinine, microangiopathic haemolytic anaemia and thrombocytopenia (Fig. 8.7). Clinical suspicion can be alerted by a raised blood pressure, persistent vomiting or fits. Although most cases of HUS follow a gastrointestinal illness, this may be mild and need not include bloody diarrhoea or abdominal pain.

More than half of clinically apparent HUS cases require haemodialysis, but the outlook is good and almost all VTEC-associated cases recover fully in time.

Diagnosis

Most gastrointestinal illnesses produced by *E. coli* are non-specific, and must be suspected on epidemiological evidence. Outbreaks of traveller's diarrhoea or HUS will alert clinicians and public health specialists.

In HUS the diarrhoeal phase may be over by the time systemic illness is apparent. VTEC becomes undetectable at an early stage and stool cultures are often negative after the fifth or sixth day of illness. The VT gene may be detectable in stools by PCR; serum antibodies to VT often become detectable after 5–7 days.

Laboratory diagnosis

E. coli pose significant problems for diagnosis in the microbiological laboratory, as the pathogen and the normal flora are the same species. Conventional methods of selection cannot therefore be employed.

EPEC are limited to a relatively small number of serotypes. In children under the age of 3 years in whom the diagnosis is suspected, an additional blood or nutrient agar plate is inoculated and several colonies of *E. coli* are 'screened' with polyvalent antisera to the EPEC strains.

VTEC strains are found in several O serotypes, among which O157 predominates. This strain is unusual in not fermenting sorbitol. A modified form of MacConkey's agar, incorporating sorbitol in place of lactose, can be used to identify the majority of these strains. In addition to this, toxin production must be confirmed by immunological or gene-detection techniques. Unselected stool cultures can be tested by DNA hybridization for the presence of the verocytotoxin-producing gene.

Management

In most cases symptomatic treatment is adequate. Specific chemotherapy is indicated in severe or prolonged disease, or in elderly and debilitated patients.

Pathogenic *E. coli* have become resistant to many antimicrobial agents, including broad-spectrum penicillins and cephalosporins, trimethoprim, sulphonamides, chloramphenicol and aminoglycosides. Isolation and sensitivity testing of the *E. coli* are therefore recommended if specific treatment must be contemplated. The agents most likely to be effective are quinolones; oral ciprofloxacin 500 mg twice daily for 3–5 days often succeeds, intravenous ceftriaxone is an alternative.

It is not clear whether antibiotic treatment will influence the development of HUS in VTEC infection. Evidence suggests that some antibiotics stimulate verocytotoxin production, but that quinolones inhibit it. It is very likely that antimotility drugs increase the probability of HUS, possibly by delaying the clearance of VTEC or its toxin from the bowel.

Prevention

Prevention of *E. coli* gastroenteritis is by adequate sanitary and hand-washing facilities. Scrupulous attention to hygiene is important, particularly in nurseries, where infection is common. Hamburger meat should be cooked thoroughly before eating. Travellers to tropical countries should avoid drinking untreated water and eating high-risk foods such as raw vegetables, salad, peeled fruit and undercooked meat (see Chapter 24).

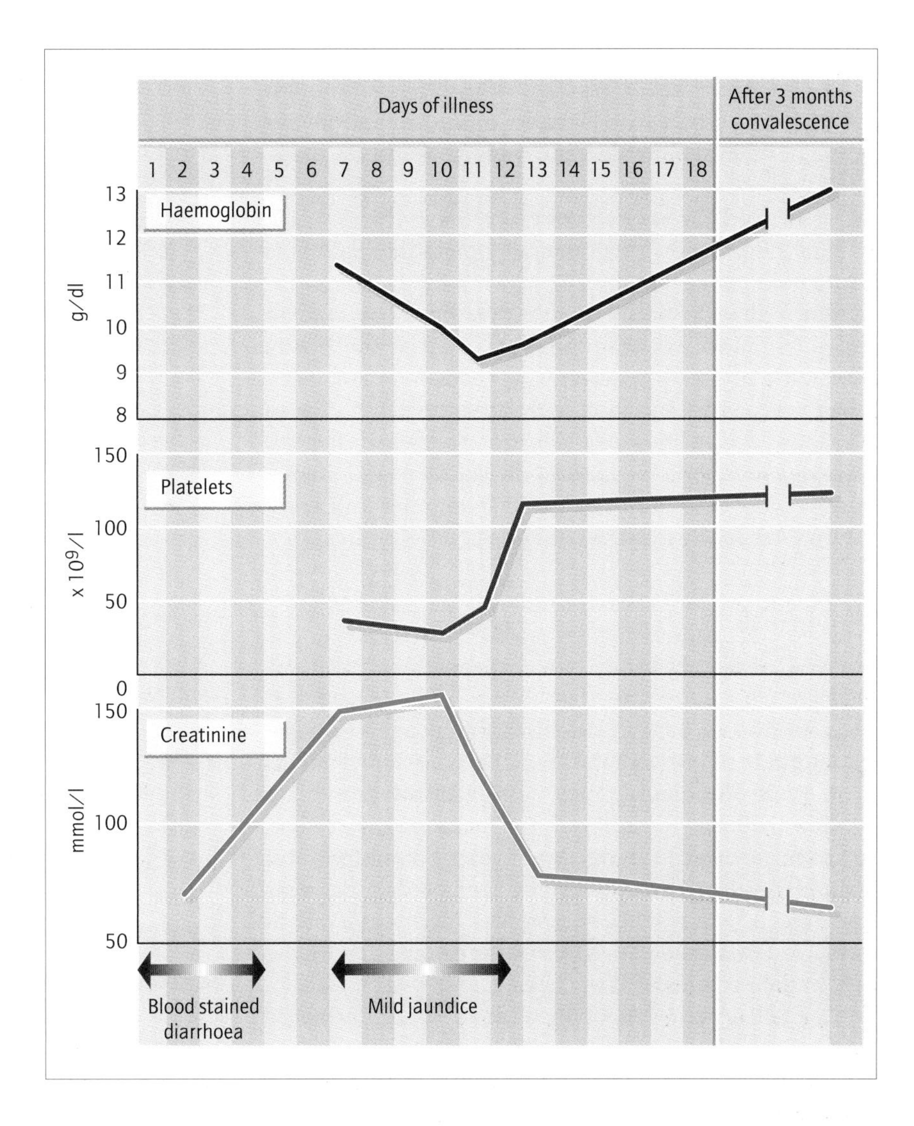

Fig. 8.7 The evolution of haemolytic–uraemic syndrome after an attack of bloody diarrhoea.

Salmonella infections

Introduction

It is important to appreciate the difference between salmonella food poisoning, and typhoid and paratyphoid fevers caused by specific enteric fever salmonellae. The food-poisoning organisms infect both humans and animals, and are biochemically and clinically distinct from 'enteric' salmonellae which are exclusively human pathogens. Enteric fevers are mainly associated with travel, and will be discussed in Chapter 24.

Epidemiology

There are approximately 2200 different serotypes of *Salmonella* that infect animals. Most of these are capable of causing salmonellosis in humans, although only about 200 are reported in any one year in the UK. Currently the most common are *S. enteritidis*, *S. typhimurium* and *S. virchow*.

Poultry is the commonest source of human salmonellosis. Up to 60% of poultry meat may be contaminated with salmonella. Both the shell and occasionally the white of eggs can be contaminated. Other meats such as beef and pork are also well-recognized 'sources' and outbreaks have been caused by contamination of seafood and vegetables. Some *Salmonella* serotypes are particularly associated with one type of food, e.g. *S. enteritidis* phage types 4 and 8 with chicken.

Food-handlers are not usually a source of infection unless they remain at work with diarrhoea. Person-to-person spread by the faecal–oral route sometimes occurs, usually in institutions such as psychogeriatric hospitals and old people's homes.

Many countries experienced an unprecedented rise in salmonellosis during the late 1980s. The commonest epidemic serotype varies between countries, and over a period of years within the same country. In the UK, beef-associated *S. typhimurium* was replaced by poultry- and egg-associated *S. enteritidis* phage type 4 in the 1980s. In mainland Europe the predominant type in the 1990s was *S. enteritidis* phage type 8.

Salmonellosis typically occurs sporadically or in small outbreaks, often within households. Large outbreaks, although more readily detected, are less common. They usually occur in association with large functions such as weddings or in institutions. All age groups are affected; however, morbidity is greatest in children and in the elderly. Most case occur during late summer and early autumn, especially in hot weather.

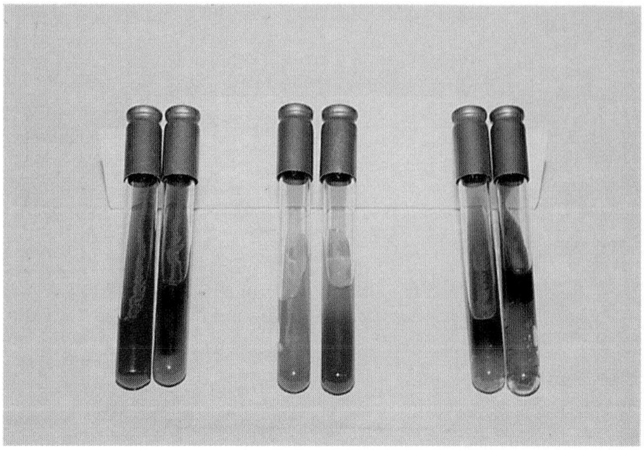

Fig. 8.8 Effect of *Salmonella* growth in triple sugar–iron medium.

Microbiology

Salmonellae belong to the family of Enterobacteriaceae. *Salmonella* is a non-lactose fermenter, produces acid and gas from glucose, metabolizes citrate and produces hydrogen sulphide. These and other biochemical characteristics are used in the identification of salmonella growth in selective media. In triple sugar–iron medium, gas production from glucose causes fractures in the agar column, and hydrogen sulphide blackens the iron-containing agar layer (Fig. 8.8). *S. typhi*, *S. cholerae-suis*, *S. paratyphi* and *S. arizona* can be differentiated from other salmonellae on the basis of their biochemical reactions.

Salmonellae possess lipopolysaccharide (LPS), which is their somatic O antigen, and exists in many types. These, together with the flagellar H antigens, define the serotype which is often given a name, e.g. *S. enteritidis*, *S. typhimurium*, *S. virchow*, etc. The LPS protects the bacterial cell from the bactericidal activity of serum, influences macrophage interactions, decreases susceptibility to host cationic proteins and functions as endotoxin.

The 10 commonest *Salmonella* serotypes in the UK

1 *S. enteritidis.*
2 *S. typhimurium.*
3 *S. virchow.*
4 *S. dublin.*
5 *S. bovis-morbificans.*
6 *S. hadar.*
7 *S. newport.*
8 *S. braenderup.*
9 *S. heidelberg.*
10 *S. montevideo.*

Injection of small amounts of *S. typhi* LPS in human volunteers can reproduce typhoid-like symptoms. *S. typhi* (and also *S. paratyphi* C and some *Citrobacter* spp.) possess the Vi antigen, a polysaccharide capsule consisting of alpha-1,4,2-deoxy-*N*-acetylgalacturonic acid. Its effect is to prevent phagocytosis and to mask the O antigen, reducing the minimum infective dose for organisms possessing this antigen.

Pathogenesis

Salmonella food poisoning is an infection of gut epithelium, which does not extend beyond the basement membrane. The salmonellae first digest the mucosal glycocalyx, and then invade the mucosal cells. Salmonellae which cause systemic disease are thought to be transported through the cells, whereas those strains that cause enteritis remain localized. In intestinal salmonellosis there is excessive fluid secretion from the ileum and jejunum.

Survival within macrophages is an important pathogenic attribute of salmonellae. Intracellular survival is a complex process, controlled by up to 200 genes including the *pagC*, *spv*, and acid response genes.

The Vi virulence antigen is a capsular polysaccharide that appears to be essential for virulence in *S. typhi*. This is genetically controlled by genes found in the *inv* locus. This causes changes in the host cell surface giving it a ruffled appearance following which the bacteria are internalized. The G + C (guanine plus cytosine) content of these genes is different from the rest of the *Salmonella* genome and more similar to that of *Yersinia*, suggesting that this is a pathogenicity island.

Clinical features

The incubation period is 12–36 h. The first symptoms are malaise, nausea, vomiting and often fever. Diarrhoea soon follows and becomes the main feature within 24 h. The diarrhoea is watery and brown, often becoming greenish if it persists. In most cases the fever resolves by the first or second day and the intestinal symptoms improve soon afterwards. In more severe cases illness may persist for many weeks with low-grade fever, continuing diarrhoea and progressive debility. Abdominal pain is not an important feature in most cases.

Elderly patients often fare badly. They have insufficient cardiovascular reserve to withstand sudden fluid loss, and may suffer low-output cardiac failure, myocardial infarction or stroke during acute diarrhoea. They tolerate the necessary rehydration poorly and easily develop pulmonary oedema. Confusion, hypostatic chest infections and the risk of deep-vein thrombosis during immobility are longer-term hazards.

Patients with achlorhydria are especially susceptible to salmonellosis. They are infected by small doses of organisms, and are often severely ill with high fever, intense watery diarrhoea and a high blood urea, out of proportion with the apparent degree of dehydration. They are more likely than other patients to have salmonella bacteraemia.

Salmonella colitis

Salmonella colitis occurs in up to 10% of patients, who then complain of colic and bloody stools. Sigmoidoscopy often confirms the presence of colonic inflammation.

Salmonella bacteraemia

Salmonella bacteraemia and metastatic infections are uncommon. They can occur in association with obvious bowel infection, but are almost equally as common in patients without preceding bowel symptoms. Some salmonellae are more likely than others to produce bacteraemia; *S. dublin* and *S. cholerae-suis* do so in 30–40% of reported cases, but fortunately these are uncommon serotypes.

Sites of tissue damage may become infected by salmonellae, even when bacteraemia has not been apparent. Commonly affected sites are bones and joints, including those of sickle-cell sufferers (Fig. 8.9), and arterial aneurysms. Occasionally a patient presents with a soft-tissue abscess (Fig. 8.10), and laboratory investigation reveals salmonellae in the pus. These infections are presumably blood-borne.

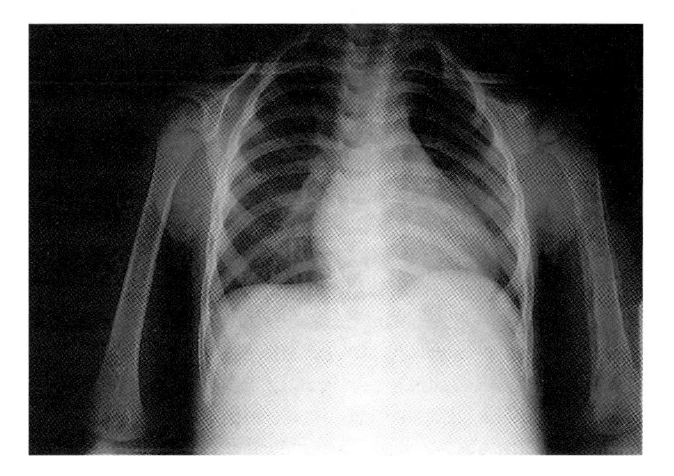

Fig. 8.9 *Salmonella* osteomyelitis in the humerus of a child with sickle-cell disease.

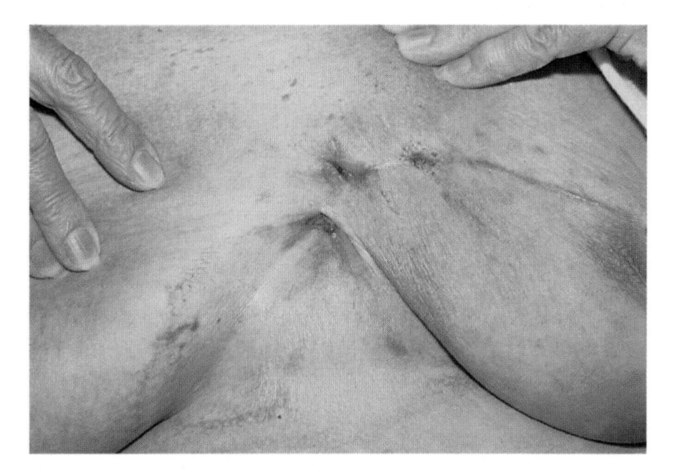

Fig. 8.10 *Salmonella typhimurium* breast abscess: this abscess was unresponsive to treatment with antistaphylococcal agents, but resolved after prolonged co-trimoxazole therapy (there had been no preceding diarrhoeal illness).

Diagnosis of salmonellosis

There are few specific features of salmonellosis, especially in mild cases. It is, however, the commonest cause of persisting diarrhoea without abdominal pain. Recovery of a *Salmonella* from the diarrhoea stool confirms the specific diagnosis. Blood cultures are indicated if fever is high or persists for more than 48 h. A high blood urea, significant dehydration, shock or a history of achlorhydria makes blood cultures mandatory.

Stool cultures should always be carried out in patients presenting with extraintestinal *Salmonella* infections, as the bowel may be the reservoir of infection.

Isolation and identification of salmonellae and shigellae

The selective procedures to identify salmonellae and shigellae are largely the same.

A wide range of selective media have been described for the isolation of salmonellae and shigellae. All contain compounds intended to inhibit the normal faecal flora, but which also inhibit the pathogens to some extent. Thus there is a balance between selection and diagnostic yield. Bile salts select for organisms that inhabit the bowel. More selective still is sodium desoxycholate, found in xylose lysine desoxycholate (XLD) agar, or desoxycholate citrate agar (DCA). Even more inhibitory are media containing bismuth sulphate, such as Wilson and Blair's medium.

Most microbiological laboratories use a combination of selective media; one less inhibitory medium (MacConkey or XLD) for the diagnosis of the more fastidious shigellae, and a more inhibitory medium (DCA or Wilson and Blair) to select for salmonellae.

These media contain indicator systems to aid the selection of colony types for further study. The simplest of these is lactose and neutral red. Organisms that ferment lactose will alter the pH of the medium, changing the colour of the indicator and producing pink/red colonies (Fig. 8.11). As neither salmonellae nor shigellae ferment lactose, these organisms form clear colonies (easily distinguished from the pink colonies of *E. coli* and klebsiellae). In XLD agar more complex changes take place: organisms that ferment xylose include coliforms and salmonellae. This causes a fall in pH, but salmonellae also decarboxylate lysine, causing a counterbalancing rise in pH. Shigellae neither ferment xylose nor decarboxylate lysine. Thus salmonellae and shigellae both appear red (neutral) in this medium, other coliform organisms only produce acid, and have yellow opaque colonies. Ferric ammonium citrate in this medium and in DCA will indicate colonies of hydrogen sulphide producers (salmonellae) which will have a central black dot of iron sulphide.

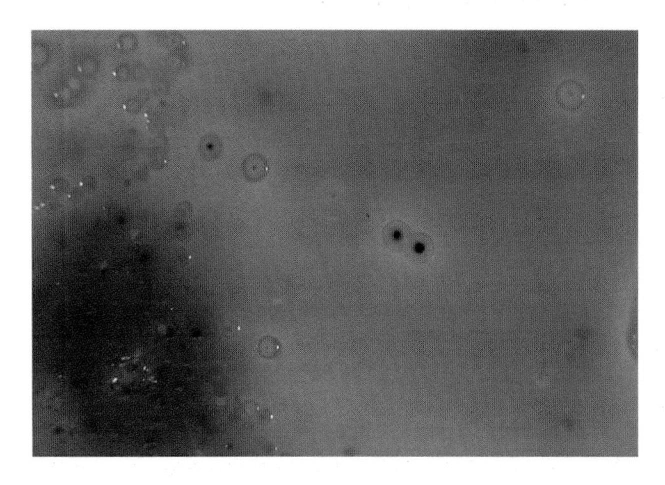

Fig. 8.11 Pink colonies of lactose-fermenting Enterobacteriaceae growing on MacConkey agar.

The next step is the use of biochemical screening tests (including sugar fermentation tests, decarboxylation and dehydrogenation reactions and hydrogen sulphide production) to distinguish salmonella or shigella colonies from *Proteus* colonies, which have similar appearances. These tests utilize combination media such as Kligler iron agar, triple sugar agar or Kohn's tubes. Commercially produced kits (e.g. API ZYM) also detect the characteristic activity of preformed enzymes. Organisms giving characteristic reactions are then subjected to full biochemical and serological identification.

Serological identification of salmonellae and shigellae is unreliable without biochemical confirmation, because of the many antigenic cross-reactions with commensal gut flora.

Further typing of salmonellae is only indicated for the identification and investigation of outbreaks. Typing methods available include sensitivity testing against panels of antimicrobial agents, and phage typing, plasmid typing or comparison of organisms by pulsed-field gel electrophoresis.

Salmonellae that can be phage typed
1 *S. enteritidis.*
2 *S. typhimurium.*
3 *S. virchow.*
4 *S. typhi.*
5 *S. paratyphi.*

Management of salmonellosis

Most mild cases require only oral rehydration. Intravenous fluids may be needed for 12–24 h if dehydration or exhaustion dictates. Even bacteraemic patients may recover uneventfully with non-specific treatment.

Specific antimicrobial chemotherapy is indicated if no spontaneous improvement is evident after 36–48 h rehydration therapy, or if shock is present. It is also indicated in those at special risk. This includes sickle-cell patients, immunosuppressed patients and the elderly, because of the severe complications of salmonellosis that all of these groups may suffer. All patients with positive blood cultures should be treated.

The antimicrobial treatment of choice is limited by the antibiotic resistance that salmonellae have evolved in recent years. Since food-poisoning salmonellae are zoonotic organisms, they have been widely exposed to the antibiotics used in both veterinary and medical practice. They are usually resistant to a wide range of agents, including tetracyclines, sulphonamides, aminoglycosides and broad-spectrum penicillins and cephalosporins. A substantial proportion are resistant to trimethoprim and chloramphenicol. Quinolone agents are most likely to be active against salmonellae.

Recommended regimens include oral ciprofloxacin or ofloxacin. In rare cases of quinolone resistance, trimethoprim, co-trimoxazole or ceftriaxone may be effective. Chloramphenicol is still available for use in cases resistant to other drugs. Short courses of 4 or 5 days' treatment are usually sufficient.

Patients with shock or bacteraemia should be treated with intravenous antibiotics.

Treatment of salmonellosis

1 First choice: ciprofloxacin orally 500 mg twice daily for 3–5 days, *or* norfloxacin or ofloxacin orally 400 mg in the same regimen (contraindicated in patients with epilepsy).

2 Alternatives: trimethoprim orally 200 mg twice daily for 5–7 days, *or* co-trimoxazole orally 960 mg in the same regimen.

3 Invasive salmonellosis: ciprofloxacin or ofloxacin i.v. 200 mg twice daily.

4 Alternative: ceftriaxone 2.0 g daily.

5 Chloramphenicol i.v. 500 mg 6-hourly if other agents are unsuitable.

Sequelae of salmonellae infections

Continuing excretion of salmonellae

This usually stops within 1–4 weeks. Prolonged or permanent excretion is rare. Predisposing factors include gut disorders such as diverticulosis, inflammatory bowel disease or ischaemia, and immunological disorders, including acquired immunodeficiency syndrome (AIDS). Treatment with inappropriate antibiotics, particularly aminoglycosides or ampicillin, may also prolong excretion. Excretion probably occurs less often after treatment with ciprofloxacin (but not after norfloxacin).

Salmonella excretion does not compromise the health of an otherwise fit patient, and is only a minimal cross-infection hazard once diarrhoea has ceased. Excretors who are food, water or dairy workers can, however, initiate food-poisoning outbreaks, and are therefore subject to public health legislation.

Most specialists would declare a patient clear after three consecutively negative, twice-weekly, stool tests.

Metastatic salmonellae infections

This is a particular problem in patients with sickle-cell disease. Metastatic infections should be treated with an antibiotic to which the salmonella is sensitive. Long courses of treatment are often needed to clear the salmonellae from the infected site. Abscesses may need 3 or more weeks' treatment; bone and joint infections may require 6 weeks or longer. Clinical progress and the results of follow-up imaging, or erythrocyte sedimentation rate or C-reactive protein measurements may be used to decide when treatment can be stopped. Resistance to ciprofloxacin has been shown to develop during prolonged treatment; ceftriaxone may be a useful alternative.

Postinfectious disorders

Salmonella bowel infections are occasionally followed by reactive arthritis which can persist for many months (see Chapter 21). This is usually monoarticular, affecting a large joint such as the knee. Symptomatic treatment with non-steroidal anti-inflammatory drugs is helpful, but severe cases may need step-down courses of corticosteroids.

Prevention and control

General principles for the prevention and control of bacterial food poisoning apply to salmonellosis. In the UK salmonella infections in farm animals are notifiable under the Zoonoses Order, and are investigated by the State Veterinary Service.

Shigellosis (bacillary dysentery)

Introduction

Bacillary dysentery is an important worldwide disease. In Western countries the endemic *Shigella* spp. cause self-limiting illnesses which are generally mild. Tropical shigelloses tend to be both more severe and persistent, causing serious morbidity, especially in children. Malnutrition is exacerbated by the prolonged diarrhoea and fever of shigellosis.

Epidemiology

Shigellosis is spread from person to person by the faecal–oral route, by direct contact with faecally contaminated hands, or indirectly from contaminated food, milk or water. Water-borne shigellosis is important in rural tropical areas. Secondary spread within households is common. Outbreaks occur under conditions of crowding and poor sanitation or personal hygiene, for example in prisons, psychiatric institutions and nursery schools. Food-borne outbreaks are less common, but do occur. An incident due to contaminated iceberg lettuce affected several European countries during 1994.

About 3000 cases are reported annually in the UK, of which two-thirds are due to *S. sonnei*. The peak incidence is in children under 5 years of age.

Microbiology

There are four species of *Shigella*: *S. sonnei*, *S. flexneri*, *S. boydii* and *S. dysenteriae*. They share the characteristics of the Enterobacteriaceae and are closely related to *E. coli*. They are relatively inert biochemically and are, for example, non-lactose fermenters, and non-motile. *S. sonnei* and *S. dysenteriae* are biochemically distinct, but *S. boydii* and *S. flexneri* are very similar. Serological characterization is therefore important for identification and typing. *S. dysenteriae* is divided into 10 serotypes on the basis of O antigens, *S. flexneri* into six types and *S. boydii* into 15. *S. sonnei* is serologically homogeneous, so colicin typing is used. For the other species serotyping is sufficient for epidemiological purposes, but phage typing can be performed if necessary. A phage-typing system for *S. sonnei* has recently been developed.

Pathogenesis

Shigellae invade the gut through the M cells which overlie gut lymphoid tissue, as they are unable to gain access through the tight junctions of the epithelial cells. Destruction of M cells in the submucosa triggers the release of cytokines and recruitment of polymorphs. A path for invasion through the tight junctions is opened by the polymorphs. Shigellae are now able to invade the enterocytes from the basement membrane side. Once inside the cytoplasm they spread from cell to cell, a process mediated by expression of the *virG* gene. This gene encodes a 120-kDa protein that polymerizes intracellular actin. The presence of a 220-kB plasmid is thought to be essential for *Shigella* invasiveness, but several genes regulating this process are also found on the chromosome.

S. dysenteriae type 1 expresses a potent exotoxin—the shiga toxin. This enhances local vascular damage. It is structurally related to the shiga-like toxins of *E. coli*. It produces fluid accumulation in the rabbit ileal loop model. Systemic distribution of this toxin results in microangiopathic renal damage and development of the haemolytic–uraemic syndrome. The structure of the toxin is in the form of an A toxin subunit and multimeric B receptor subunit which binds to the Gb3 surface glycolipid.

There is active fluid secretion from the small intestine in shigellosis, but this is much less than in ETEC disease. The clinical features of dysentery are caused by excess fluid loss from the small bowel, a major defect in large-bowel water reabsorption, and large-bowel ulceration and inflammation due to the effects of exotoxins.

Clinical features

After an incubation period of 3 or 4 days there is a prodromal illness of high fever lasting 12–24 h. Non-specific features of this prodrome often occur in children and include meningism, convulsions and confusion. The white cell count is usually high ($12–16 \times 10^9/l$). However in most cases the fever resolves suddenly, and diarrhoea with colic appears, indicating the true diagnosis.

S. sonnei and *S. boydii* tend to cause brown watery diarrhoea in which various amounts of mucus are mixed. Blood may be seen in the stool, but the amount is rarely large. The diarrhoea usually subsides spontaneously after 3–5 days.

S. flexneri and *S. dysenteriae* often cause more severe disease in which irregular fever continues and increasing amounts of mucus and blood appear in successive stools. Distressing colic is exacerbated by attempts to eat or drink. Untreated disease of this kind can lead to dehydration and progressive malnutrition.

Both types of illness can be followed by asymptomatic excretion of the pathogen lasting from days to weeks and occasionally longer. Because of the low infective dose of shigellae, excretors can be hazardous, especially in nurseries and children's institutions.

Diagnosis

The clinical picture usually suggests a diagnosis of dysentery. Bacillary dysentery must be distinguished from amoebic dysentery and inflammatory bowel disease. The mainstay of differential diagnosis is laboratory identification of the pathogen (see p. 160).

Management

Mild shigellosis requires only symptomatic treatment. Antispasmodic agents are helpful in relieving colic. In more severe cases antibiotic treatment can terminate the illness, though it often takes 2 or 3 days for diarrhoea to cease completely. Ciprofloxacin is often successful, and tends to eliminate excretion as well as curing disease.

Prevention and control

The infection can be prevented by safe disposal of faeces and adequate hand-washing. *Shigella* dysentery is notifiable. Patients should avoid handling food until their stools are normal.

Campylobacter infections

Introduction

Campylobacter infections occur as commonly as salmonella infections in most parts of the world, and add significantly to the burden of childhood gastroenteritis in at-risk communities. In developed communities they cause less morbidity than salmonella infections because they rarely produce metastatic or bacteraemic disease. Asymptomatic excretion is uncommon and campylobacters are not transmitted from person to person; however, large food- and water-borne outbreaks have been documented.

Epidemiology

Over 50 000 cases of *Campylobacter* enteritis are reported each year in the UK, and the incidence is increasing.

Most cases are sporadic and associated with the consumption of contaminated poultry, or sometimes the contamination of milk by birds that peck through the tops of doorstep milk bottles. Outbreaks are usually due to consumption of unpasteurized milk or untreated water.

The peak incidence of *Campylobacter* enteritis is during the summer months, about 8 weeks earlier than that of salmonellosis (Fig. 8.12). There is a bimodal age distribution with the greatest incidence in infants and young adults. Infection occurs more frequently in rural areas than in cities.

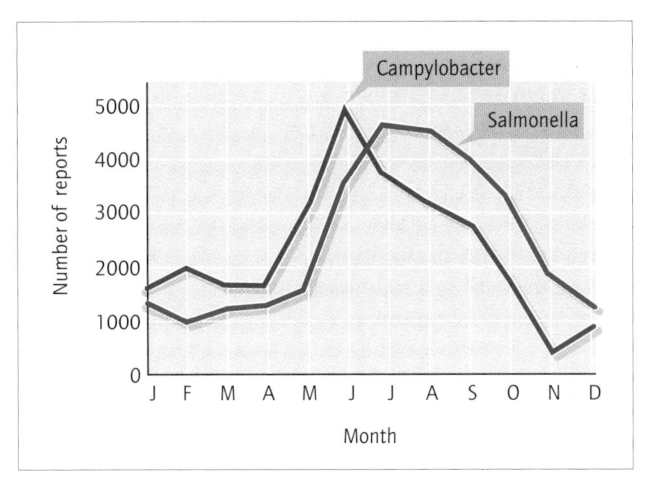

Fig. 8.12 Seasonal incidence of *Campylobacter* and *Salmonella* infections.

Pathogenesis

Campylobacter spp. were first recognized as animal pathogens partly because of difficulties in isolating them from human faeces. They are classified as members of the family Spirillaceae on the basis of their DNA composition. There are four species associated with human enteritis: *C. jejuni*, *C. coli*, *C. fetus* and *C. lari*.

C. jejuni is rapidly killed in the stomach, and the chance of infection is increased if the organism is consumed in milk, which may protect it from gastric acid. The infective dose may be as low as 500 organisms.

Campylobacters possess a cell-wall LPS of low molecular weight similar to that found in *Haemophilus* or *Neisseria* sp., and which contains a lipid A moiety. It is antigenically diverse with many serotypes being described.

Campylobacters possess flagella and are motile, which may play an important role in establishing infection. Convalescent serum contains antibody to flagellin. The surface-exposed flagellar antigens are predominantly strain-specific but species-specific antigens are present.

C. jejuni produces an enterotoxin which causes fluid accumulation in ileal loops, and a cytopathic effect in Chinese hamster ovary cells and Y-1 mouse adrenal cells. Its mode of action is similar to that of cholera toxin and the LT toxin of *E. coli*. Other toxins, including a cytotoxin, may also be implicated in the pathogenesis of *Campylobacter* enteritis. Adherence of *C. jejuni* to the intestine is thought to be mediated through L-fucose receptors.

Pathogenicity factors for *Campylobacter jejuni*
1 Lipid A-containing lipopolysaccharide.
2 Flagellar antigens.
3 Toxins: enterotoxin and cytotoxin.
4 Enterocyte adherence mediated through L-fucose receptors.

Clinical features

The incubation period is usually 3 or 4 days, but can be up to 8 or 9 days. This is followed by approximately 24 h prodromal illness with fever, headache and prostration. The main illness consists of diarrhoea, which is watery and may be bloody. Vomiting may occur at the onset. Abdominal pain is common, and is constant rather than colicky. Abdominal examination often reveals rebound tenderness which may be severe.

Campylobacter infection occasionally causes very severe abdominal pain with little diarrhoea, or only tiny stools containing mostly mucus and blood. Distinction from acute surgical conditions such as acute appendicitis, salpingitis and ectopic pregnancy may therefore be difficult. The abdominal X-ray shows multiple fluid levels in acute bowel infections, and is often unhelpful. X-ray examination should not be omitted, however, as it may show other diagnostic features such as free gas in the peritoneum or, in children, the C-sign of intussusception. Early ultrasound examination may assist diagnosis.

Rare cases of systemic *Campylobacter* infection occur, often affecting children with diseases causing iron overload. The patient is feverish and blood cultures are positive.

Prolonged infection can mimic inflammatory bowel disease, with persistent bloody diarrhoea, abdominal pain and low-grade fever, but stool cultures will indicate the infectious aetiology.

Diagnosis

The diagnosis can rarely be made on clinical findings alone.

Laboratory diagnosis

Special laboratory techniques are required to identify campylobacters in heavily contaminated specimens such as faeces. There are several selective media which permit routine isolation of campylobacters from faeces. The selective agents employed include a mixture of antibiotics including vancomycin, polymyxin B, trimethoprim, amphotericin B and cephalothin. Selection is improved if the cultures are incubated at 43°C. Media that provide effective selection at 37°C are increasingly used to ensure the small number of pathogenic campylobacters that are not thermotolerant can be isolated.

Campylobacters are microaerophilic organisms which grow better at reduced oxygen tension and with an increased concentration of carbon dioxide. Commercially available gas-generating systems make this an easy procedure to carry out.

In most laboratories identification can be confirmed by Gram staining to demonstrate the characteristic 'gull wing' morphology, and by obtaining a positive oxidase reaction. Biochemical testing can be used to differentiate the species, and disc testing of antimicrobial sensitivities can be performed.

Microbiological characteristics of campylobacters

1 Optimum growth at 43°C.
2 Characteristic spiral morphology.
3 Microaerophilic.
4 Oxidase-positive.
5 Resistant to vancomycin, polymyxin B, cephalothin, trimethoprim.
6 Sensitive to erythromycin and ciprofloxacin.

Clinical management

Many mild cases are self-limiting. Specific treatment is indicated in severe or prolonged illness (and a good response will obviate concerns about surgical or inflammatory conditions). The treatment of choice is a short 3- or 4-day course of oral erythromycin. Ciprofloxacin is often effective, but resistance rapidly develops. These drugs can be used parenterally in severe or systemic infections.

Treatment of *Campylobacter* infection

1 First choice: erythromycin orally 250 mg 6-hourly for 3 or 4 days.
2 Alternative: ciprofloxacin orally 500 mg twice daily for 3 days (resistance easily develops).

Prevention and control

Campylobacter enteritis, like salmonellosis, may be prevented by good hygienic practice. Milk-borne spread can be avoided by pasteurization. The most effective long-term measure would be to control infection in poultry, but the existence of many environmental sources of infection makes this difficult.

Infection with 'food-poisoning' vibrios

These members of the family Vibrionaceae, like *V. cholerae*, are all natural residents of water. The range of Vibrionaceae is wide, some living in salt water, some in brackish or fresh estuaries and ponds. None of them produce cholera toxin, but in *V. parahaemolyticus* at least, pathogenicity is probably related to the production of a haemolysin.

Vibrio parahaemolyticus is a natural resident of brackish water, which may be concentrated in the gut of filter-feeding and scavenger shellfish. The most common vehicle of infection is a meal of shrimps or prawns. Most cases occur sporadically, although outbreaks have been reported. The disease is rare in the UK but common in south-east Asia and the USA. Other non-cholera vibrios including non-toxigenic and non-O1 *V cholerae*, *V. alginolyticus*, *V. fluvialis* and *V. mimicus* are occasionally reported. The main symptoms are watery diarrhoea and colicky abdominal pain.

Aeromonas spp. are also members of the Vibrionaceae and are natural inhabitants of water. They can be isolated from diarrhoea stools of some cases of 'food poisoning', and may also be contracted directly from water by fishermen or boatmen. *Aeromonas* infection is relatively common in Japan where fresh-water fish is often eaten raw. The number of infections reported in the UK is small (less than 50 per annum), but increasing. Half of all reported infections are in young adults.

Plesiomonas shigelloides is related to *Aeromonas* and is less certainly associated with food poisoning. Between 30–60 laboratory reports are received each year in England and Wales.

Patient is a 31-year-old woman

	WEEK 1	WEEK 2	WEEK 3	WEEK 4	WEEK 5	WEEK 6
Pains in knees, ankles, back	Diarrhoea and abdominal pain	+++	++	++	+	
ESR (mm/hr)		85	80	80	35	16
Yersinia enterocolitica antibody titre		1:80	1:320	1:320		

Fig. 8.13 Progress of *Yersinia* food poisoning associated with arthritis. ESR, erythrocyte sedimentation rate.

Diagnosis

If the clinician suspects *Vibrio* infection on epidemiological grounds, a specific request should be made for the organisms to be sought (see below). Diagnosis is usually of epidemiological rather than therapeutic value, as the associated illnesses are rarely more than moderately severe, requiring only symptomatic treatment.

Laboratory diagnosis

Vibrionaceae cannot be identified by techniques intended to detect Enterobacteriaceae. Instead, a high-pH selection/indicator medium containing bile salts is used, which inhibits many other bowel flora. The most used medium is thiosulphate–citrate–bile salt–sucrose (TCBS) medium containing a bromothymol blue indicator. Food-poisoning vibrios grow well on this medium. Most are non-sucrose fermenters, producing blue-green colonies, which distinguish them from *V. cholerae*, a sucrose fermenter which produces yellow colonies.

A suitable enrichment medium for vibrios is alkaline peptone water.

Yersinia infections

Yersinia enterocolitica and *Y. pseudotuberculosis* are capable of causing bowel disease. In some countries, for example Belgium, the infection is as common as salmonellosis and *Campylobacter* infections. It is relatively rare in the UK, although the number of reported cases increased more than 10-fold during the 1980s to 580 in 1989.

Transmission occurs from contaminated food, milk and water. Most infections are sporadic, and often follow consumption of raw or undercooked pork, although raw vegetables have also been implicated. A few milk-borne outbreaks have been reported in the USA.

Most infections occur in children under 5 years of age. There is a seasonal pattern, with more infections during autumn and winter.

The associated illness is a mild to moderate gastroenteritis, often with aching abdominal pain. Postinfectious problems, particularly arthritis and erythema nodosum, are common, and can persist for some weeks (Fig. 8.13).

Diagnosis requires special investigations. Culture of *Yersinia* sp. from stool is possible (see below), but diagnosis is often considered late and rising titres of antibodies must then be sought in paired serum samples.

Y. enterocolitica will grow slowly on MacConkey's agar showing pinpoint colonies after 48 h incubation. Specialized *Yersinia* media simplify the isolation of these organisms. Selection is obtained due to the presence of sodium desoxycholate, crystal violet and an antibiotic combination (Irgasan, novobiocin and cefsulodin). *Y. enterocolitica* has translucent colonies with a dark pink centre which may be surrounded by a pale opalescent zone of precipitated bile salts.

Treatment of yersiniosis is possible with oxytetracycline 250–500 mg three or four times daily or doxycycline 100 mg daily, both for 7–10 days.

Y. pseudotuberculosis also causes human disease, but this is more often mesenteric adenitis than diarrhoea. Fever and right lower quadrant abdominal pain must be differentiated from appendicitis, salpingitis or the onset of Crohn's disease. Ultrasound examination will often reveal oedematous terminal ileitis and groups of enlarged mesenteric lymph nodes. Similar findings, even in the absence of diarrhoea, can occur in *Salmonella* and *Campylobacter* infections, demonstrating the overlapping symptom complexes of intestinal infections. It is therefore worthwhile requesting stool cultures in patients with terminal ileitis and mesenteric adenitis, in case specific treatment can be offered.

Helicobacter pylori and acute gastritis

Helicobacter pylori, like the campylobacters, is a member of the family Spirillaceae. It is found in the gastric mucus and surrounding the apices of gastric parietal cells in patients suffering from acute hypochlorhydric gastritis and chronic gastritis. It is the probable cause of about 95% of duodenal ulcers, and a small minority of gastric ulcers. It is found in association with tumours of mucosa-associated lymphoid tissue (MALT), which will often regress if the *Helicobacter* infection is eradicated.

Diseases associated with *Helicobacter pylori*

1 Acute hypochlorhydric gastritis.
2 Chronic gastritis.
3 Duodenal ulcer.
4 Mucosa-associated lymphoid tissue lymphoma.

Pathogenicity

H. pylori leads to acute and chronic inflammation in the gastric antrum. The organism elaborates a potent urease, which produces a local alkaline environment. This enables it to overcome the inhibitory effect of gastric acid and allows it to establish itself in the mucosa. This is thought to stimulate the feedback mechanism for gastric acid, resulting in a high gastrin level and consequent hyperacidity. The CagA phenotype is associated with peptic ulcer disease. The CagA gene product is indirectly related to the production of IL-8 (important in the pathogenesis of inflammation) but more important are the *picA* and *picB* genes that are found associated with the *cagA* genes on a 'pathogenicity island'. The *vacA* gene codes for a toxin that produces vacuolation in tissue culture cells. This gene has a mosaic structure. VacA variation results in differences in the VacA level and the degree of damage in human tissue.

Humans are the probable reservoir of infection, but *H. pylori* has been recovered from untreated water. The origin of the organisms colonizing the stomach is uncertain, but possibilities include person-to-person or food-borne spread. In developed countries the prevalence of *H. pylori* infection is between 20 and 40%; however, up to 95% of duodenal ulcer patients may be colonized. *H. hominis* has recently been identified in the human stomach, and may be another cause of peptic ulcer disease.

Diagnosis

A non-invasive diagnosis can be made using the urea breath test in which the patient swallows ^{13}C or ^{14}C urea orally and samples of exhaled breath are collected over 30 min. Urease in the stomach breaks down the radiolabelled urea and this activity is detected by the presence of rediolabelled CO_2. Biopsy specimens can be examined using rapid urease tests, by histology and by cultivation. A stool antigen test has recently been developed. Bacterial isolation is becoming increasingly important with the rise in antibiotic resistance.

Treatment

The most successful treatment regimen is a combination of amoxycillin, plus clarithromycin or metronidazole, with a proton pump inhibitor such as omeprazole, continued for 1 month. Combined bismuth/clarithromycin preparations may also be used with omeprazole. Resistance to antibiotics is an increasing problem.

Diseases caused by bacterial toxins

ORGANISM LIST

Staphylococcus aureus
Bacillus cereus and other *Bacillus* spp.
Clostridium perfringens, *C. difficile*, *C. botulinum*

Staphylococcal food poisoning

Introduction

Staphylococcus aureus is able to produce a variety of toxins, among which are enterotoxins (A–E). They are preformed toxins produced during replication of the staphylococci in prepared food. Food poisoning occurs when food becomes contaminated by an infected individual, usually from a skin lesion or nasopharyngeal secretions. The organism subsequently multiplies and produces toxin which is absorbed and travels in the circulation to the vomiting centre, causing vomiting by a central effect. It may also cause bowel irritation, leading to mild, transient diarrhoea.

In some countries, including the USA, *S. aureus* is a leading cause of food poisoning. In the UK it is comparatively rare; approximately 100 cases are reported annually.

Clinical features

The incubation period varies from 30 min to 6 h. Malaise and nausea are quickly followed by vomiting, which may be severe and repeated, causing dehydration or even shock. Occasional deaths are reported. The vomiting lasts from 2 to 6 h and is followed by several hours' exhaustion before recovery is complete.

Diagnosis

The illness is similar to winter vomiting disease and other viral infections. Single cases are rarely extensively investigated. The absence of diarrhoea also limits the availability of specimens, though staphylococci may be recovered from vomitus if this is available, and can then be shown to produce enterotoxin.

Family and catering-associated outbreaks are relatively common and may indicate a food source. Remaining food can then be examined for the presence of enterotoxin and/or enterotoxin-producing staphylococci.

Diagnosis of staphylococcal food poisoning

1 Demonstration of toxin-producing *Staphylococcus aureus* in suspected food.
2 Demonstration of toxin-producing *S. aureus* in vomitus.
3 Demonstration of an enterotoxin in suspected food.

Management

Management is symptomatic.

Prevention

Prevention is by hygienic food preparation, particularly covering skin lesions with a waterproof dressing. Food-handlers with extensive staphylococcal skin infection should be excluded from work.

Bacillus cereus food poisoning

Bacillus cereus is an aerobic Gram-positive rod, which forms spores. It is found in soil and dust, and easily contaminates cereals and some beans. It may also occur in animal faeces and occasionally contaminates meat.

Illness occurs following ingestion of contaminated food that has been kept at ambient temperatures after cooking. The disease is relatively uncommon in the UK; fewer than 500 cases are reported annually.

Food-poisoning serotypes produce spores which resist boiling. Boiled foods, particularly rice, may be stored at ambient temperatures for later rewarming (or frying). The spores germinate during storage and highly heat-resistant preformed toxins accumulate in the food. Occasional outbreaks have also been associated with meat and vegetable dishes. The toxins cause vomiting by a local and central effect.

B. cereus of the types found in meat more often produce diarrhoea, mediated by locally acting preformed toxins. Some other *Bacillus* spp., e.g. *B. subtilis*, also occasionally cause food poisoning.

Botulism

Introduction

Botulism is caused by preformed toxins of *Clostridium botulinum*, elaborated during the germination of spores in anaerobic conditions. Different strains of *C. botulinum* produce one of six toxins (A–F). Human disease is usually caused by A or B, derived from soil; occasional cases are caused by type E, derived from estuarine or marine mud.

Home-canned or home-bottled vegetables and salads are the commonest sources of illness, usually causing family outbreaks. In Europe, infection is sometimes due to home-smoked or preserved meats. Commercial canning is highly controlled, but rare community outbreaks have been caused by commercial products. An outbreak of 27 cases occurred in the north-west of England in 1989, caused by canned hazelnut purée contained in yoghurt (Fig. 8.14).

Fig. 8.14 A 'blown' tin of hazelnut purée which had not been properly heat-treated, resulting in the formation of botulinum toxin type b; 27 people developed botulism after eating hazelnut yoghurt prepared from the contents of such a tin. Courtesy of Dr Richard Gilbert, Central Public Health Laboratory.

Clinical features

The toxin causes paralysis by blocking neuromuscular junctions. The incubation period varies from 24–48 h or more, depending on the dose of toxin consumed. Malaise and mild gastrointestinal symptoms may precede neurological features. The onset of neurological disturbance is insidious and subtle, often initially dismissed as 'hysteria'. Paralysis develops from the head downwards, aiding differentiation from the opposite process in Guillain–Barré syndrome. Dry mouth and blurred vision are followed by difficulty with swallowing and speech. Respiratory paralysis and generalized weakness soon follow. Once paralysis is established it can last for several weeks, though eventual recovery is usual, if the patient is supported by ventilation in the meantime. Autonomic dysfunction and smooth-muscle paralysis mean that the blood pressure may be unstable and bowel function will be disturbed.

Diagnosis

Initial diagnosis must depend on clinical findings and a history of recent consumption of suspect food. The cerebrospinal fluid is usually normal. Electromyography shows features of developing denervation (but in the earliest stages may suggest an axonal lesion). *C. botulinum* is not found in the stool, but toxin may be demonstrable in serum. Both the organism and its toxin are demonstrable in left-over food, and sometimes in containers and on utensils. The toxins are destroyed by heating.

Laboratory diagnosis of botulism

1 Demonstration of *Clostridium botulinum* in suspected food.
2 Recovery of *C. botulinum* from unwashed food containers or utensils.
3 Demonstration of *C. botulinum* toxin in food, food containers or patient's serum.

Management

This consists mainly of appropriate intensive care. Polyvalent antiserum is available from regional public health laboratories, and can be given to prevent further fixation of toxin in the nervous system. Antitoxin can also be given prophylactically to others who ate the suspect food.

A minority of cases are mild and can even be self-limiting, but respiratory function should be closely monitored, and measures taken to protect the airway, in case muscular weakness predisposes to aspiration of saliva or stomach contents.

Infant botulism

This is a rare condition with worldwide occurrence. It is cased by rapid replication of *C. botulinum* in the infant bowel. Both the bacteria and high concentrations of the toxin are found in the stool of affected infants. The source of the *C. botulinum* is unknown, but may be direct from soil or derived from food. Some affected infants have consumed honey before their illness.

Affected infants are aged between 6 weeks and 3 months. The onset is insidious, with weak suckling, constipation and shallow respiration. Many infants recover with supportive measures only, including postural drainage, careful feeding or tube feeding and careful observation of respiratory function. Less than half require ventilation. No secondary case has been reported following exposure to the stool of an affected infant.

Clostridium perfringens food poisoning

Clostridium perfringens is an anaerobic Gram-positive rod which inhabits the bowel of many animals, and may contaminate meat during butchery and storage. Its spores survive boiling and will germinate in the anaerobic conditions in stored stews, soups, gravies and large joints of meat. The infection is usually food-borne and occurs when contaminated food such as meat is allowed to stand at room temperature for long periods.

Bacteria multiply in the large bowel, elaborating toxin locally. After 18–36 h incubation the toxin causes watery diarrhoea with colicky abdominal pain which may last for 3–5 days. Outbreaks typically occur in association with large-scale catering with inadequate facilities for cooling and storing food.

C. perfringens can be recovered from the diarrhoea stool and identified as a 'food-poisoning' serotype. It is also possible to demonstrate toxin gene in the stool by PCR.

Clostridium difficile and pseudomembranous colitis

Pseudomembranous colitis is almost always antibiotic-associated, though rare cases occur apparently spontaneously. A milder version of the disease, without severe colitis, is recognized as antibiotic-associated diarrhoea. Antibiotics which most commonly precipitate the condition are those which strongly inhibit normal gut flora, especially faecal streptococci. These include clindamycin, ureidopenicillins and very broad-spectrum cephalosporins. However ampicillin, rifampicin and ciprofloxacin have also been associated with a few cases.

Watery diarrhoea and abdominal pain occur abruptly, usually 3 or 4 days after the commencement of antibiotic treatment. Low-grade fever is common. *C. difficile* and its toxin can both be demonstrated in stool specimens. Colonic biopsies show typical changes, with focal inflammation in lymphoid tissue, expanding and erupting through the muscularis mucosae to emerge in the mucosa. Mucosal destruction and inflammatory exudate produce a thick pseudomembrane. Slow improvement may follow discontinuation of antibiotics, but progressive inflammation and perforation can occur in untreated cases.

The isolation of *C. difficile* from stools does not alone confirm the diagnosis of pseudomembranous colitis, as this organism is carried by up to 30% of hospital patients. The diagnosis is supported by the demonstration of toxin in patients' stools. An ELISA method for the detection of toxins has been developed. A pathological diagnosis can be made by showing typical histological changes on biopsies of affected colonic mucosa (Fig. 8.15).

The treatment of choice is oral vancomycin 250 mg three times daily for a week. Relapse sometimes occurs and can be treated by repeating this course. Metronidazole is often, but not always, effective. Parenteral treatment may be given, with good effect, if oral treatment is impossible.

Non-bacterial toxins and food poisoning

TOXIN LIST

Bean haemagglutinins
Scombrotoxin
Shellfish toxins (diarrhoetic and paralytic)
Ciguatera toxin
Cyanobacterial toxins (blue-green algae)

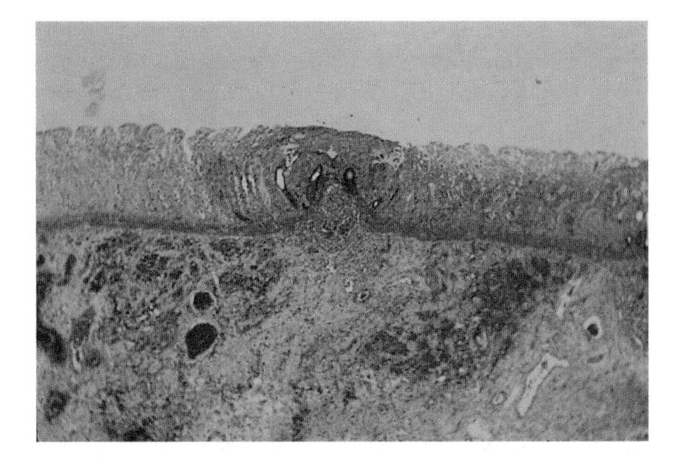

Fig. 8.15 Pseudomembranous colitis: intense inflammation erupts from beneath the muscularis mucosae, producing an accumulation of necrotic debris in place of the damaged mucosa.

Bean haemagglutinins

These haemagglutinins are found in many species of beans. Red kidney beans contain large amounts and are the commonest cause of clinical problems. The toxin is destroyed by vigorous boiling; when the beans are soft they are also safe to eat. Incomplete cooking is particularly likely if beans are insufficiently soaked, or if they are cooked as a casserole ingredient. Canned beans are safe. Illness, consisting of severe vomiting followed by profuse diarrhoea, develops very quickly after consumption of as few as five or six beans. In severe cases the patient has become ill before finishing the meal.

Scombrotoxin

Scombrotoxin is a substance which develops during spoilage of scombroid fish, such as mackerel, tuna and bonito. Sardines have also been implicated in cases of food poisoning. It is thought that the toxin is either histamine or a closely related substance. Even mildly spoiled fish can be toxic and, since the toxin is highly heat-resistant, neither cooking nor canning can make affected fish safe to eat.

Symptoms of intoxication follow 1–4 h after eating fish. They are identical to the symptoms of histamine toxicity, with headache, flushing, urticarial rash and swelling or tingling of the lips and mouth. Spontaneous recovery takes 4–6 h.

Diarrhoetic shellfish poisoning

Diarrhoetic shellfish poisoning is caused by a toxin produced by the shellfish themselves. Some shellfish are well known to be toxic (for instance, the red whelk) but are occasionally mistaken for edible species. The resulting illness is a short-incubation acute diarrhoea whose severity is proportional to the dose of toxin.

Paralytic shellfish poisoning

Neurotoxins are elaborated by dinoflagellates which are low in the food chain of many sea creatures. The dinoflagellates multiply rapidly when the sea is warm and nutrients are plentiful. They sometimes form a visible red or brown 'bloom' (the so-called red tide). Shellfish and other filter feeders consume large quantities of this and the toxin accumulates in their bodies. Humans become poisoned by eating affected clams or crustacea. In tropical waters fish such as groupers can accumulate very large amounts of ciguatera toxin, becoming more poisonous with increasing age and size.

Symptoms of poisoning are caused by interference with neuromuscular junctions. Bradycardia, tingling of the lips and fingers and muscle weakness are common, and can continue for many hours in severe cases. Rare cases of death have been reported in patients who consumed large quantities of shellfish.

Ciguatera poisoning can be life-threatening. It is mainly seen in the Caribbean, where it is well recognized. Local people do not eat large fish. Bradycardia and hypotension can be severe; profound muscle weakness also occurs and can affect respiratory muscles. It may take days or weeks for the symptoms to abate, and persistent compromise of neuromuscular transmission may mean that for many weeks afterwards even a meal of 'safe' fish can precipitate weakness. Intensive support may be required for such severely affected patients. Many less severe cases also occur, with gradual recovery from mild weakness and bradycardia.

Cyanobacterial toxins

These are produced by the blue-green algae that thrive in fresh water when the weather is warm and nitrates or other nutrients are plentiful. An algal 'bloom' is visible in the water. The toxins are extremely irritant, and cause erythema, burning and even blistering of exposed skin and mucosae. Ingestion affects mainly animals, but occasional human cases of diarrhoea, vomiting and abdominal pain are reported. Abnormal liver function tests and sometimes jaundice also occur. Species of cyanobacteria can readily be identified by direct microscopy of affected water.

Poisoning incidents have been reported from many countries, but only rarely in the UK. In 1989 10 young army recruits were affected following canoeing and swimming exercises in a lake containing algal bloom.

Parasitic infections of the gastrointestinal tract

ORGANISM LIST

Protozoa
- *Giardia intestinalis*
- *Cryptosporidium parvum*
- *Cyclospora catayensis*
- *Entamoeba histolytica*

Helminths
- *Enterobius vermicularis*
- *Trichuris trichiura*
- *Ascaris lumbricoides*
- Hookworms
- *Strongyloides stercoralis*
- Tapeworms

Laboratory diagnosis

The simplest investigation is to examine a saline-wet preparation of stool microscopically under the ×10 and ×40 objective. To improve the diagnostic yield, concentration techniques are employed. These include flotation methods such as the zinc sulphate, magnesium sulphate and sucrose flotation techniques or the formol ether centrifugation technique. Flotation methods vary in their ability to concentrate different species, whereas the formol ether technique effectively concentrates helminth ova and protozoal cysts. Individual parasites are identified on the basis of their size (as measured by a microscope eyepiece graticule calibrated against a stage micrometer), and their characteristic morphology. Some examples of protozoan and helminthic parasites are demonstrated in Fig. 8.16.

In patients with acute amoebic dysentery the transit time through the large bowel may be so increased that motile amoebae may be identified in the stool by microscopy. These organisms are very susceptible to cooling and desiccation, and therefore should be processed as soon as possible. Microscopic examination should be performed on a heated stage to preserve trophozoite motility as long as possible. Scrapes from the base of ulcerative lesions demonstrated on proctosigmoidoscopy may show motile trophozoites.

The larvae of *Strongyloides stercoralis* are infrequently seen in the stools. The diagnostic yield is increased if the stool is placed in a Petri dish, mixed with activated charcoal and incubated at room temperature for 5 days inside another dish. Larvae differentiate and migrate into the second dish and can be detected using a plate microscope.

A string test may aid diagnosis of strongyloidiasis and giardiasis, although there are differing reports of its efficiency. The patient swallows a weighted string which passes into the upper small intestine where it remains for approximately 2 h. It is then retrieved and the juice and attached material are concentrated by washing and centrifugation before examination using the ×10 and ×40 objectives. Enzyme immunoassay (EIA) and PCR methods have been described for most of the common protozoan infections, but in practice few of these are available as a routine because most laboratories in developed countries have such a low throughput of specimens that these methods are not cost-effective.

Giardiasis

This diarrhoeal disease is caused by *G. intestinalis*, a flagellate protozoan. Infectious cysts are passed in the faeces of both sick patients and asymptomatic carriers. Cysts

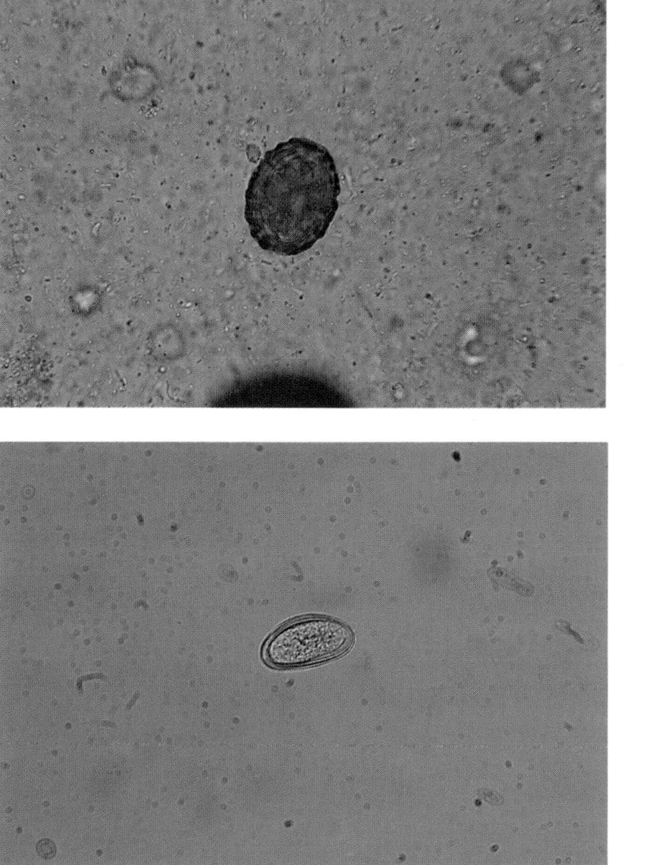

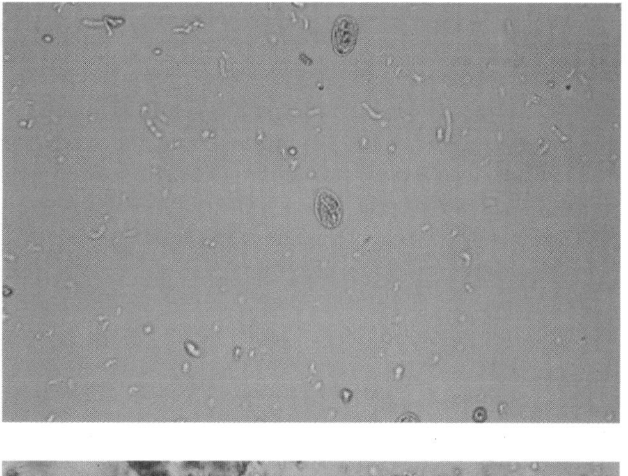

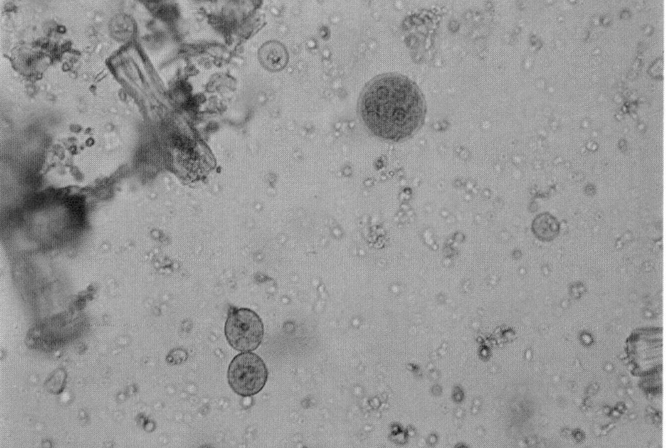

Fig. 8.16 Examples of parasitic infections demonstrated by microscopy of unstained faeces: (a) ovum of *Ascaris lumbricoides*; (b) ovum of *Enterobius vermicularis*; (c) cysts of *Giardia intestinalis*; (d) cyst of *Entamoeba histolytica* containing six nuclei (smaller cyst of non-pathogenic *Endolimax nana* is also seen). (See also Fig. 3.1.)

survive for many days in sewage-contaminated water and infect a new host when the water is consumed. Direct faecal–oral spread is possible but less common. Large outbreaks can occur when water treatment procedures are defective or when breakage of water pipes allows contamination. Dogs may carry the organism and also excrete infectious cysts.

The disease is common in areas of poor sanitation. Children, especially under 5 years, are affected more frequently than adults. The infection also occurs in travellers to countries with inadequately treated water supplies. The number of reports is increasing in the UK, probably due to more widespread examination of faecal samples for cysts.

The effects of infection vary from asymptomatic cyst passage to severe acute diarrhoea. *Giardia* adheres to the mucosa of the jejunum and upper ileum, using a ventral 'sucker'. In heavy infections the whole mucosa may be occluded, leading to malabsorption and steatorrhoea with rapid weight loss. Chronic cases are common; flatulence and passage of greasy, loose stools are particularly noticeable in the mornings and may persist for weeks or months.

Diagnosis is made by demonstrating cysts in the faeces, or by finding trophozoites in string-test, duodenal aspirate or biopsy specimens. Treatment is with metronidazole 400 mg three times daily for 1 week or a single 2.0 g dose of tinidazole. Either treatment may be repeated in the occasional event of relapse.

Cryptosporidiosis

Cryptosporidium parvum is a member of the family Sporozoa. Serotype A has its reservoir in humans, while serotype B is excreted by cattle and can infect other animals, including humans. The infectious dose is extremely low, probably under 50 organisms. Most human cases are caused by serotype B. Human cases are also infectious, and family outbreaks are quite common.

Surface water containing infectious oocysts can contaminate water reservoirs and mains, causing large outbreaks (Fig. 8.17) and sporadic infection. Several outbreaks have been reported in children's nurseries, probably spread by the faecal–oral route. Transmission also occurs between animals and humans by the faecal–oral route. Agricultural and veterinary workers and children who come into close contact with animals are at risk. Immunosuppressed patients suffer persisting disease, once infected.

The organism was first identified in the UK in 1983 and reports have risen steadily as testing for the organism has increased. *Cryptosporidium* is now the fourth most commonly identified cause of gastrointestinal infection in the UK. Seasonal peaks occur in spring and late autumn.

The parasite attaches to the small-bowel mucosa, damaging the brush border of the enterocytes (Fig. 8.18). After an incubation period of 3–6 days there is an acute onset of diarrhoea, often with abdominal pain and colic. The average duration of illness is about 3 weeks.

Diagnosis depends on demonstrating oocysts in the faeces, by acid-fast stains such as modified Ziehl–Nielsen or auramine techniques. No effective specific treatment exists but azithromycin has been reported to ameliorate the symptoms.

Fig. 8.17 Leakage from this drainage system contaminated a public swimming pool, causing a large outbreak of cryptosporidiosis. Courtesy of Dr David Casemore, Rhyl Public Health Laboratory.

Cyclosporiasis

Cyclospora catayensis is a common cause of springtime diarrhoea, discovered in children in Central America. Outbreaks of infection in distant countries have been caused by exported soft fruit, particularly raspberries. Protection of fruit crops from faecal contamination is important, but other sources of contamination have not been ruled out. Diagnosis is by demonstrating the characteristic cysts in the stool.

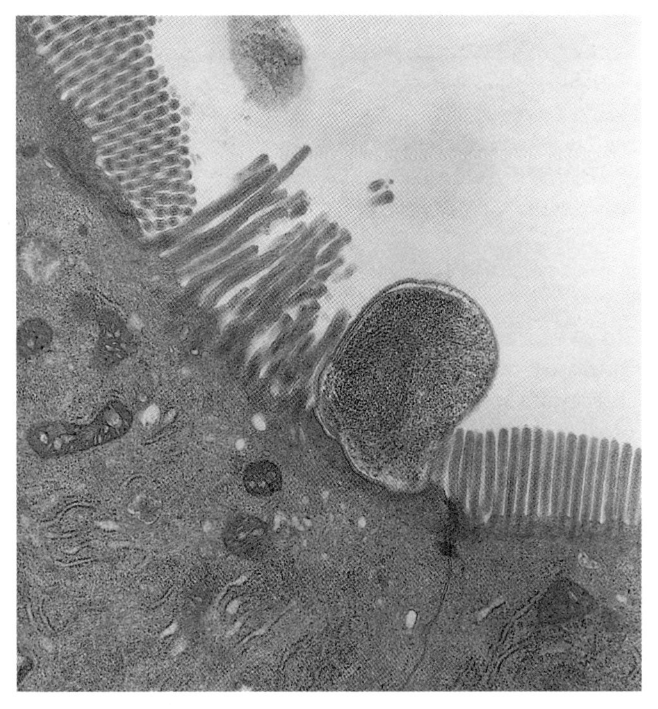

Fig. 8.18 Electron micrograph of a developing trophozoite of *Cryptosporidium parvum*, attached to mucosal cells in the small bowel. The parasite is within a vacuole formed from the host-cell brush border membrane. Courtesy of Dr David Casemore, Rhyl Public Health Laboratory.

Amoebiasis

Introduction

Entamoeba histolytica causes infection and disease in many parts of the world where sanitation is poor. The effects of infection range from asymptomatic excretion of cysts to chronic intestinal infection, sometimes with granuloma formation, to acute amoebic dysentery. Extraintestinal abscesses may also occur, and can threaten life if untreated.

Manifestations of amoebiasis
1 Asymptomatic cyst passage.
2 Chronic intestinal symptoms.
3 Caecal granuloma (amoeboma).
4 Amoebic dysentery.
5 Amoebic liver abscess.
6 Rare other extraintestinal disease.

Epidemiology

Amoebiasis is mainly spread via contaminated food, especially raw vegetables, and water. Infection is common in areas of poor sanitation and personal hygiene. The disease occurs principally in adults and is rare in children below 5 years of age.

Up to 1000 cases are reported annually in the UK. Half of these are known to be acquired overseas, principally in the Indian subcontinent. *E. histolytica* has been differentiated from the non-pathogenic *E. dispar* that is morphologically identical. *E. histolytica* infection makes up approximately 10% of those cases with cysts present in the stool.

Pathogenesis

Cysts of *E. histolytica* are passed in the stools of affected individuals. After ingestion the trophozoites excyst and attach to the mucosa of the large bowel. The amoebic surface membrane carries the glycoprotein galactose *N*-acetyl-D-galactosamine which acts as a lectin and mediates binding to the mucosal cells. Amoebae secrete a pore-forming enzyme, amoebapore, of which there are three closely related types, a, b and c. These all contain six cystine residues in a similar structure to that found in saponins and pulmonary surfactant.

Clinical features

Amoebic dysentery has a variable incubation period, from a few days to 4–6 weeks. Watery diarrhoea then presents and rapidly becomes bloody. A swinging fever develops. Colicky abdominal pain increases and may be replaced by constant pain as disease progresses. The stool is replaced by mucus in which are mixed large quantities of blood. Severe, untreated cases can progress to acute colonic dilatation, with risk of perforation.

Diagnosis

Acute amoebiasis is always a differential diagnosis of bacillary dysentery. Other causes of bloody diarrhoea include VTEC infection, pseudomembranous colitis, ulcerative colitis and Crohn's colitis. Inflammatory bowel disease should be suspected if repeated stool examinations reveal no pathogens. Sigmoidoscopic biopsy is helpful if typical changes of inflammatory bowel disease or pseudomembranous colitis are present. Even in acute amoebiasis it may be difficult to demonstrate amoebae or cysts in the stool; biopsies from ulcer edges will usually show undermined (flask-shaped) ulcers with amoebae packed in their eroded borders.

Laboratory diagnosis

Intestinal amoebiasis is diagnosed by detection of amoebic trophozoites containing red cells. This form is only present in fluid diarrhoea where the specimen must be examined with minimum delay so that motile amoebic trophozoites can be visualized. In specimens which are examined after a delay or in semiformed or formed stools, amoebic cysts should be sought by the formal ether technique. *E. histolytica* can only be differentiated from *E. dispar* by specialized tests such EIA, or culture followed by zymodeme analysis.

Less than half of patients with amoebic dysentery produce antibodies to *E. histolytica* but these antibodies are almost invariably present in patients with invasive amoebiasis. Immunofluorescence antibody and ELISAs have been described. These tend to remain positive after an attack of systemic amoebiasis. A rapid technique using cellulose acetate precipitation is the first to become positive in amoebic liver abscesses and should be used for urgent diagnosis. This positivity is rapidly lost after successful treatment. Species-specific diagnosis has been described detecting antibody to the surface-binding lectin.

Management

This consists of specific treatment, pain relief and the maintenance of hydration, which usually requires intravenous fluid therapy. The treatment of choice is metronidazole 500 mg three times daily intravenously, followed by 800 mg three times daily by mouth when the patient can tolerate oral intake. Treatment should be continued for 5–7 days, depending on the speed of response.

Once cured of the acute disease, many patients continue to excrete infectious cysts. This can be terminated by treatment with an intestinal amoebicide. Diloxanide furoate is the best available choice; the dose is 500 mg every 8 h for 1 week (this may be omitted if the patient is soon to return to an endemic area, as reinfection is inevitable).

Prevention

The disease can be prevented by providing water that has been treated by filtration, and adequate disposal of human faeces. Travellers to tropical countries should avoid drinking water that is not known to have been properly treated and unpeeled fruit and raw vegetables. Known carriers should be given instruction in thorough hand-washing after defecation.

Chronic intestinal amoebiasis

This is common in endemic areas, and causes fluctuating symptoms of abdominal pain and mild diarrhoea. *E. histolytica* preferentially affects the caecum, so right iliac fossa pain and tenderness occur, and must be differentiated from appendicitis. Persisting mucosal inflammation stimulates granuloma (amoeboma) formation. Large amoebomas may produce the same symptoms as caecal carcinomata, and appear the same on barium studies. Stool examination will reveal the amoebic cysts in such cases. The condition responds to treatment with intestinal amoebicides, but surgery may still be needed if the bowel is obstructed.

Amoebic abscess

This may follow either apparent or subclinical bowel infection. The liver is by far the commonest site affected; others include pleura, pericardium and occasionally excoriated perineal or abdominal skin. The localized infection causes intense inflammation and tissue damage which is visible as dark necrosis in advancing skin lesions.

Clinical features

The patient presents with high fever, leucocytosis and a raised erythrocyte sedimentation rate. Many patients have palpable enlargement and tenderness of the liver, but some have a paucity of physical signs. An elevated right diaphragm, small right-sided pleural effusion or slight collapse of the lower lung segment should direct attention to the liver in a case of pyrexia of unknown origin.

Diagnosis

Useful investigations include ultrasound examination and computed tomographic (CT) scanning of the liver. The abscess is usually a single lesion in the right lobe, accessible to diagnostic aspiration and drainage. Left-sided abscesses are rare and difficult to localize on physical examination or ultrasound scan. CT scanning is the investigation of choice for this and other unusual sites.

Amoebae are rarely demonstrable in aspirate, as they are closely adherent to the abscess wall. However, the pus has a typical pinkish (anchovy-sauce) appearance which is diagnostically helpful.

Serological tests are usually positive and permit rapid diagnosis. Culture of the pus should be performed to exclude superadded bacterial infection.

Management

Metronidazole is the treatment of choice. It readily penetrates the abscess wall and can allow rapid healing, often avoiding the need for further aspiration. The dose is 500 mg intravenously or 800 mg orally three times daily for 10–14 days (see Chapter 9).

Helminth infections of the bowel

Introduction

These are endemic wherever there is environmental contamination with human or animal faeces, and some are directly transmitted by the faecal–oral route.

Helminths with direct person-to-person transmission

Some species, *Enterobius vermicularis* (threadworms) and *Trichuris trichiura* (whipworms), inhabit the large bowel and rectum, deriving nutrients from the faeces. They release ova which are detectable by microscopic examination of unstained faeces. Whipworms fix themselves to the bowel wall by their narrow head; their wider body may be seen on proctoscopy, dangling from the mucosa.

Threadworms are freely motile and may even be seen lying in the appendix on histology of surgical specimens. They emerge from the anus, particularly at night, to deposit ova on the perineal skin. This causes irritation which may spread into the vagina, especially in children. Worms can be seen macroscopically as tiny moving threads in the faeces or on the skin. Adults or their ova can be trapped by pressing the sticky side of transparent adhesive tape on to the anal margin at night or just after waking. Microscopy of the tape, stuck to a glass slide, can then be performed to make the diagnosis.

Mebendazole, which is given as a single oral dose of 100 mg, kills the worms and terminates the infection. Threadworms can be paralysed by treatment with piperazine. This is usually given as a mixture of piperazine phosphate and senna in powder form. The senna ensures that worms are passed in the faeces before regaining motility.

Treatment of threadworm infection

1 First choice: mebendazole single oral 100 mg dose (not recommended in pregnancy or for children under 2).

2 Second choice: Pripsen (piperazine 4 mg and sennoside 15.3 mg per sachet). Adult and child over 6 years: one sachet stirred into milk or water and repeated after 14 days (age 3 months–1 year, one-third of a sachet; 1–6 years, two-thirds of a sachet); each sachet contains 7.5 ml of powder.

Ascaris lumbricoides

Epidemiology

Ascariasis is transmitted by ingestion of soil that contains infective eggs, or vegetables contaminated by soil or sewage. In tropical countries, up to 50% of children may be infected. Children who eat soil are most at risk. The peak incidence of infection is between 3 and 8 years of age. Between 500 and 1000 cases are reported each year in the UK, although the incidence is declining.

Clinical features

At the onset of infection ova germinate in the upper gastrointestinal tract and the larvae migrate through the tissues to the lungs. A sufficiently large number of larvae can cause clinically evident lung irritation. Cough, wheezing and mild fever are associated with an eosinophilia; the whole syndrome is called pulmonary eosinophilia. Larvae are coughed from the lungs and swallowed to complete their life cycle in the bowel.

Uncomplicated infection causes no systemic features, except for the occasional passage of a worm *per rectum*. Worms may also be expelled orally if vomiting occurs. They are incidental findings in contrast X-ray studies of the gastrointestinal tract, when they are outlined and their gut is filled by radiopaque medium.

Very heavy worm loads can cause complications, as small ducts, such as the bile and pancreatic ducts, can be entered and obstructed by worms. Closely entwined masses of worms can even obstruct the small bowel (especially when the worms are irritated in the early stages of piperazine intoxication). These conditions may resolve spontaneously, but without treatment there is a risk of recurrence.

Diagnosis

Clinical diagnosis is based on seeing worms in stools, vomitus or contrast X-rays. Eosinophilia without other cause may raise suspicion, especially in the presence of respiratory symptoms.

A laboratory diagnosis is made by demonstrating ova in wet preparations or concentrates of faeces.

Management

The treatment of choice is single-dose mebendazole, which kills the worms without causing hyperactivity. Combinations of piperazine and purgatives are also effective (see text note p. 186), but are less pleasant to take, as piperazine causes significant nausea at effective doses, and may also provoke hyperactivity of the worms. In endemic areas treatment can significantly reduce the worm load, lessening the likelihood of complications and reducing the 'stealing' of nutrients from the bowel (which may be important in an undernourished child).

Prevention

The infection can be prevented by education of children in hand-washing after defecation and safe disposal of human faeces.

Hookworms

Introduction

Hookworm infection is not endemic in westernized countries, but is often present in recently immigrated individuals. Hookworms feed on blood, and heavy worm loads can cause significant anaemia, especially in children and others whose diet is low in iron.

Epidemiology

The infection is transmitted from faecally contaminated soil via infective larvae which penetrate the skin, especially of the feet. All ages are affected, although infection is commonest in children.

The disease is common in the tropics, and rare in temperate climates. About 500 imported cases are reported annually in the UK.

Pathogenesis

Hookworm ova are passed in faeces and infective larvae develop in an extracorporeal soil life cycle. The larvae can attach to and penetrate human skin, from where they migrate through the tissues to the lungs. They then pass to the pharynx and are swallowed, producing intestinal infection. In the intestine they attach to the bowel wall by their grinding mouth parts, damaging mucosa and releasing blood, which is their source of nutrition.

Clinical features

Larval migration in the early stages of infection causes both local skin irritation (ground itch) and pulmonary eosinophilia. Once the intestinal infection is established there are no specific features other than those of iron-deficiency anaemia.

Larva migrans

Larva migrans is a condition of abortive human infection by animal hookworms, usually those of dogs. The larvae penetrate human skin, but remain locally in the tissues until they die. Persisting skin irritation results, often with serpiginous tracks, transient urticarial rashes and respiratory symptoms. See Case 24.1 (p. 460).

Management

The treatment of hookworm infection is mebendazole. Dietary iron supplements may be needed in anaemic children. In rare cases of severe anaemia, transfusion may be given. Cutaneous larva migrans can be treated topically with a paste containing thiabendazole, or with oral albendazole, which penetrates tissues well; irritation persists for a while after treatment.

Prevention

Prevention is by wearing shoes and safe disposal of human faeces.

Strongyloidiasis

Strongyloides stercoralis has a life cycle similar to that of hookworms, except that rhabditiform larvae rather than ova are excreted in the faeces. Many larvae undergo a developmental stage to form filariform larvae in soil before penetrating the skin of a subsequent host; some of them become infectious before being excreted and reinvade the original host by penetrating the bowel wall or perianal skin. Strongyloidiasis can therefore persist lifelong, after the host has left the endemic area.

The disease is common throughout tropical countries, but also occurs in temperate zones. Fewer than 100 cases are reported annually in the UK.

Intermittent migration of larvae causes the syndrome of larva currens. A local urticarial rash (Fig. 8.19) appears several times a week in various sites, usually on the trunk or thighs. Eosinophilia is intermittently present, and larvae may be demonstrated in faeces, duodenal aspirate or string-test material.

Immunosuppressed patients may suffer massive tissue invasion by larvae, without the tell-tale eosinophilia. Sometimes immunosuppression has been caused by increasing doses of prednisolone, given to control 'asthma' which may have been missed pulmonary eosinophilia. The migrating larvae carry bowel bacteria with them, and may enter the lungs (Fig. 8.20) or the meninges, causing secondary coliform infections. The diagnosis must be considered in immunosuppressed patients who originate from the Far East, Africa, Asia or other endemic areas (transplanted organs may also contain larvae and cause infection if the donor had strongyloidiasis).

Treatment consists of albendazole 400 mg twice daily for 3 days, with a second course after 3 weeks if indicated. Thiabendazole daily for 3 days is an alternative, but has a significant incidence of gastrointestinal side-effects. Immunosuppressed patients may need longer treatment, depending on their response.

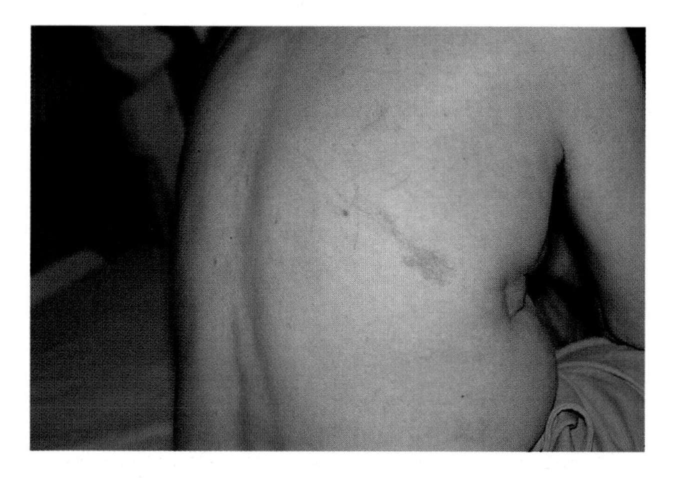

Fig. 8.19 Strongyloidiasis: recurrent urticarial rash of larva currens in a man infected as a prisoner of war in Burma in the 1940s.

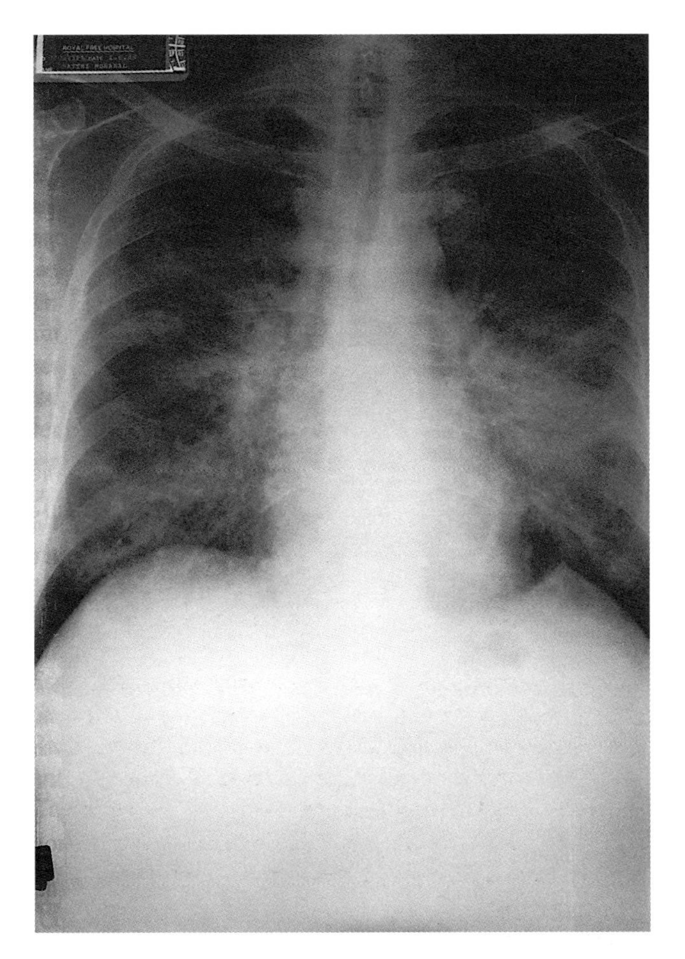

Fig. 8.20 Chest X-ray of an Asian patient who had received increasing doses of prednisolone for 'asthma'; he had *Escherichia coli* bronchopneumonia secondary to massive migration of *Strongyloides stercoralis* larvae.

Treatment of strongyloidiasis

1 First choice: albendazole orally 400 mg twice daily for 3 days. (May be repeated after 3 weeks.)

2 Second choice: thiabendazole orally 25 mg/kg (maximum 1.5 g) 12-hourly for 3 days.

In immunosuppressed patients, albendazole may be given in doses of 400 mg twice daily for up to 4 weeks. These courses may be repeated up to three times with intervals of 14 days.

3 Alternative: Ivermectin 200 µg/kg daily for 2 days.

Toxocariasis

Introduction

This is usually a mild disease, predominantly of children, caused by *Toxocara canis* and *T. cati*, whose hosts, dogs and cats, respectively, excrete eggs in their faeces. After 1–3 weeks' incubation, the eggs become infectious. Children become infected by eating contaminated soil, or raw unwashed vegetables. After ingestion, embryonated eggs hatch in the intestine, larvae penetrate the wall and migrate to the liver and lungs. From the lungs, organisms spread to the abdominal organs (visceral larva migrans) or the eyes (ocular larva migrans).

Clinical features

Clinical features are variable and may include fever, chronic abdominal pain, a generalized rash, pneumonitis, hepatomegaly and focal neurological lesions. Ocular larva migrans may cause endophthalmitis with loss of vision in the affected eye.

Diagnosis and treatment

Diagnostic tests include an ELISA, and demonstration of *Toxocara* larvae by liver biopsy, although this is seldom justified. Anthelminthic therapy is of doubtful value, though albendazole may be helpful.

Trichinellosis

This is a disease caused by a systemically invading roundworm, *Trichinella spiralis*, contracted from eating raw or undercooked pork, horse or bear containing larval cysts. It is rare in the UK.

Symptoms are variable; conjunctivitis, oedema of the eyelids and retinal haemorrhages are characteristic early features. These may be followed by thirst, sweating, chills and remitting fever. Cardiac and neurological complications may occur after 3–6 weeks. There is a marked eosinophilia, which may assist in the diagnosis. Serological tests are available. Albendazole and mebendazole are both used in treatment.

Case 8.1: Difficult recovery from diarrhoea

History: A previously healthy 6-year-old boy became ill with watery diarrhoea and severe abdominal cramps 2 days after a school outing to a 'petting zoo', where he had contact with rabbits, piglets, lambs and kids. Of the 20 other children on the outing, six others also developed diarrhoea. Despite plentiful electrolyte drinks, he became fatigued and the diarrhoea became bloody. On the third day of illness he was admitted to hospital.

Investigations: Haemoglobin 12.2 g/dl, white cell count 11.4×10^9/l (7.8×10^9/l neutrophils), platelets 108×10^9/l, blood urea 4.0 mmol/l, creatinine 65 mmol/l, electrolytes normal.

Management and progress: An intravenous infusion was set up, and rehydration achieved with dextrose–saline. After overnight treatment he was able to take fluids by mouth and the diarrhoea had abated. The infusion was discontinued and feeding gradually recommenced. The following day, he appeared pale and listless, and became unwilling to eat and drink, though he had only three small, loose stools. On the sixth day of illness he vomited once, and appeared minimally jaundiced. Two of the other affected children had now been admitted to hospital with bloody diarrhoea.

Repeated investigations showed haemoglobin 11.2 g/dl, white cell count 10.9×10^9/l and platelets 40×10^9/l, the blood film report described microcytic red blood cells and fragmented red cells (schistocytes), with no platelets seen on the film. Blood urea 7.8 mmol/l, creatinine 152 mmol/l, sodium 130 mmol/l, potassium 4.9 mmol/l, serum bilirubin 27 mmol/l, aspartate transaminase 50 IU/l, alkaline phosphatase 230 IU/l (normal for this age group).

Questions: What blood disorder does the blood count and film indicate?
What is the likely diagnosis?
What infectious organism is probably responsible?
What pathogenicity factor(s) are important in the aetiology?
Is antimicrobial therapy indicated?
What is the prognosis?

Further management and progress: Based on the blood film evidence of microangiopathic haemolytic anaemia and the renal failure, a diagnosis of haemolytic uraemic syndrome (HUS) was made. This syndrome is usually caused by infection with a verocytotoxin-producing *Escherichia coli* (VTEC), which also causes haemorrhagic colitis. Most VTEC isolated from cases of haemorrhagic colitis and HUS are of serotype O157, which is unique among *E. coli* in its inability to metabolize sorbitol. It can be recognized in the laboratory by identifying its sugar metabolism profile and by agglutination testing with specific O157 antiserum. Rare case of colitis and HUS have been related to infections with other *E. coli* serotypes. Two phage-encoded types of verocytotoxin, VT1 and VT2 (which exists in two forms VT2a and VT2b), can be identified serologically, by DNA hybridization tests and by variations in their biological effects. Because of their close homology with the toxin of *Shigella dysenteriae*, the toxins are also called shiga-like toxins (SLT1 and SLT2). The toxins cause death and shedding of the colonic mucosal cells, and the renal endothelial cell damage responsible for HUS. In outbreaks of *E. coli* O157 haemorrhagic colitis, approximately 5% of patients develop HUS.

Stool cultures in this case revealed a heavy growth of *E. coli* O157, which was later shown to be a producer of VT2.

There is no evidence that antimicrobial chemotherapy alters the course of HUS. In most cases, the causative *E. coli* becomes undetectable in the stool by the fourth or fifth day of illness. Furthermore, some antibiotics, such as co-trimoxazole, appear to stimulate increased VT production by the organism, though others, such as ciprofloxacin, may reduce it. Supportive therapy includes control of fluid and electrolyte balance, diuretic therapy for oedema, and blood and platelet infusions as indicated. Although renal support (haemoperfusion or haemodialysis) may be required in up to half of cases, near to 90% of children recover from the renal insult (Fig. 8.7). The 10–15% case fatality in HUS is mainly due to cerebral complications related to oedema, haemorrhage and vasculitis.

Questions: Could it be shown that the ill children were part of a point-source outbreak?
Could the animals have been the source of infection?

Epidemiological investigation: *E. coli* O157 strains were recovered from four of the seven affected children. All produced VT2a. Phage typing showed them all to be phage type 8. Pulsed-field gel electrophoresis patterns were identical in the four isolates.

Faeces was collected from animal pens, and rectal swabs from individual animals at the zoo. Both the lamb and the kid pens produced positive results, and two of the three individual kids sampled were positive. The lamb pen samples and two of the three isolates from kids were of the same type as the isolates from the children.

Haemorrhagic colitis and HUS outbreaks in children are commonly related to animal contact. Lambs and goats are the most often implicated, but calves and adult cattle may also be involved. Consumption of meat, meat products and unpasteurized milk are also sources of infection in children, as in adults.

9 Infections of the Liver

Introduction, 191

Viral infections of the liver, 191
- Hepatitis A, 191
- Hepatitis B, 196
- Delta hepatitis, 199
- Hepatitis C, 200
- Hepatitis E, 201
- Hepatitis G and transfusion-transmitted virus, 201
- Epstein–Barr virus, 202
- Cytomegalovirus, 202
- Yellow fever, 202

Bacterial infections of the liver parenchyma, 203
- Leptospirosis, 203
- Causes of granulomatous hepatitis, 205
- Chronic Q fever (*Coxiella burnetii*), 205
- Brucellosis, 205
- Tuberculous granulomata of the liver, 206

Abscesses of the liver, 206
- Introduction, 206
- Pyogenic liver abscess, 206
- Amoebic liver abscess, 207
- Cholangitis, 209

Parasites of the liver, 210
- Schistosomiasis, 210
- Hydatid disease, 212
- Liver flukes, 213

Introduction

The liver is a complex organ. It receives blood from the intestinal tract via the hepatic portal system, and is sustained by 'systemic' blood via the hepatic artery. The liver cells are the site of detoxification and metabolism of intrinsic and extrinsic substances, including ammonia, complex amines, endotoxin, steroids and many drugs. They are the main site where glycogen is manufactured and are the only important source of blood glucose. They are the only site of synthesis of urea, and of several proteins, including albumin and the clotting factors III, VII, IX and X.

The biliary system is the route of excretion of bilirubin, derived from the metabolism of haem substances. It is also important in the enterohepatic circulation of bile salts. Several drugs are excreted and concentrated in the bile; these include penicillins, cephalosporins and rifampicin. Bile also contains cholesterol, which easily crystallizes, and may form gallstones. Patients with a high turnover of haem excrete high bilirubin loads, and are at risk of pigment stones.

Infections of the liver tend to affect either the cells, producing typical hepatocellular disorder, or the biliary tract, producing cholestasis and features of infection in a hollow organ. Space-occupying lesions in the liver (e.g. abscesses or granulomata) tend to produce the biochemical changes of cholestasis, but rarely cause jaundice.

Viral infections of the liver

ORGANISM LIST

Hepatitis A virus
Hepatitis B virus
Delta virus (hepatitis D)
Hepatitis C virus
Hepatitis E virus
Hepatitis G virus
Transfusion-transmitted virus (TTV)
Epstein–Barr virus
Cytomegalovirus
Yellow fever (and other arboviruses)

Hepatitis A

Introduction

This a common disease worldwide. It is highly infectious and therefore spreads easily among children. Many cases are subclinical; others cause moderate morbidity. Mortality is less than 0.1%, but this means that about 1% of hospital admissions with hepatitis A suffer severe or life-threatening disease.

Epidemiology

In the developing world, where sanitation is poor, most individuals become infected in childhood when the disease is mild or asymptomatic. Immunity is lifelong and outbreaks in adults are uncommon. In more developed

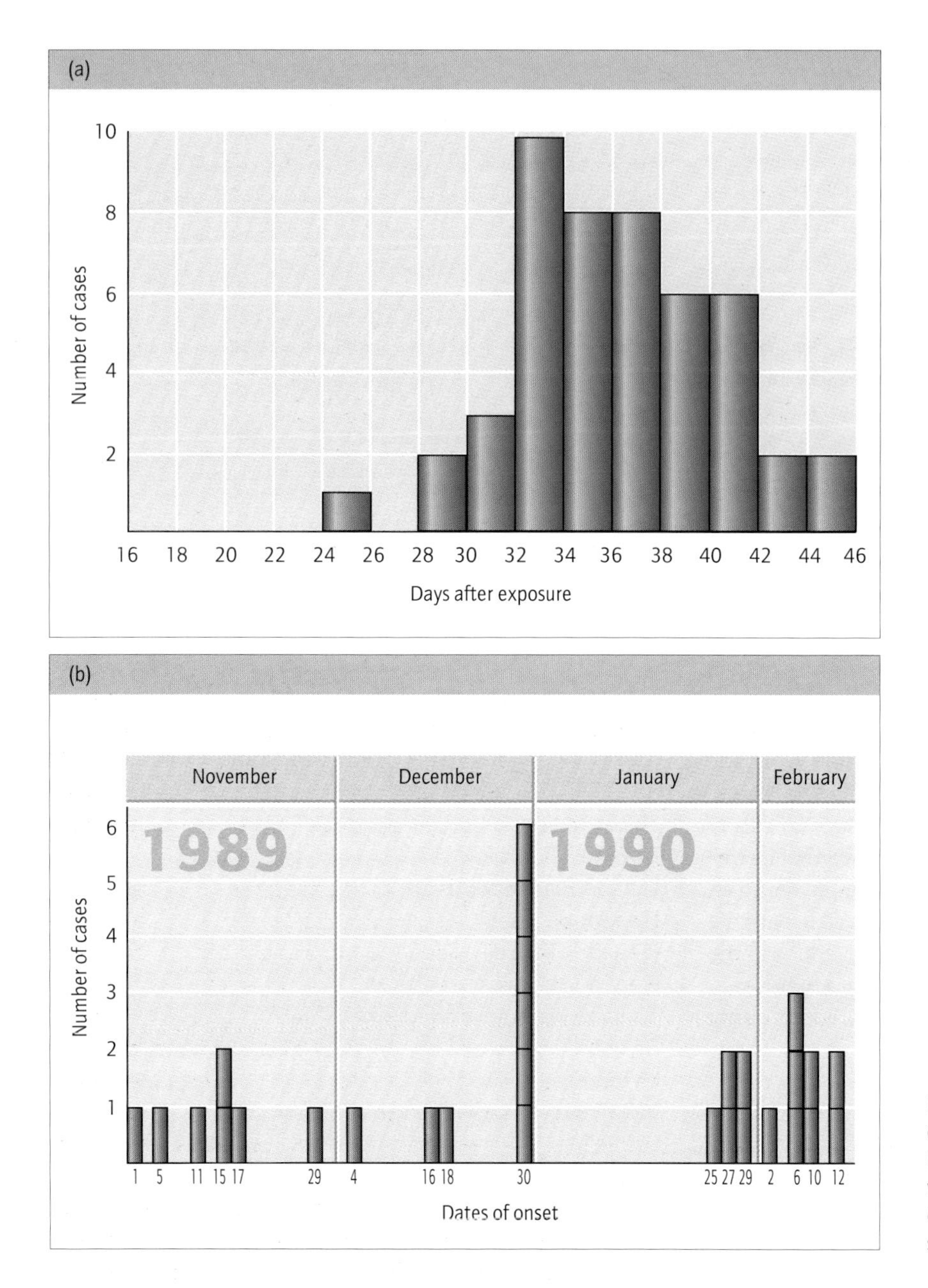

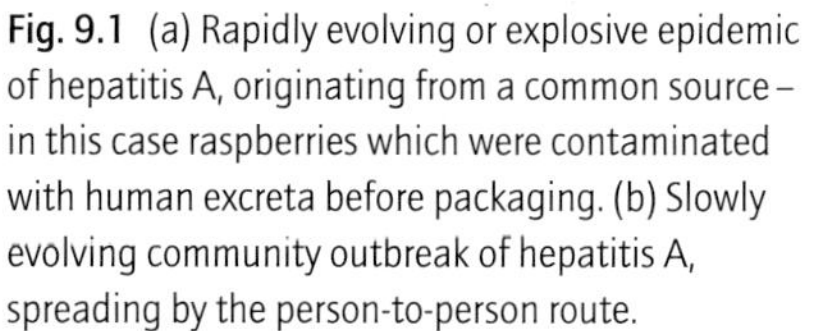

Fig. 9.1 (a) Rapidly evolving or explosive epidemic of hepatitis A, originating from a common source – in this case raspberries which were contaminated with human excreta before packaging. (b) Slowly evolving community outbreak of hepatitis A, spreading by the person-to-person route.

countries, infection in children is less common and many older children and young adults are susceptible. Three distinct patterns of disease then occur: (i) sporadic infections among adult travellers to endemic countries; (ii) epidemics affecting mainly school-age children; and (iii) explosive common-source outbreaks (Fig. 9.1).

Epidemics of hepatitis A in developed countries tend to evolve slowly, last for several months and affect large geographical areas. Spread is from person to person, mainly by the faecal–oral route occurring within households, nurseries, schools and other institutions.

Common-source outbreaks are usually associated with food contaminated by an infected handler, or undercooked molluscan shellfish harvested from contaminated waters. Water-borne outbreaks occur occasionally.

In the UK over 5000 laboratory-confirmed cases of hepatitis A are reported each year. The highest incidence is in children aged 5–9. The proportion of cases acquired abroad may be as high as 20%. The overall incidence is declining, although large epidemics occurred in 1982 and 1990.

Virology

Hepatitis A virus (HAV) is a hepatovirus in the family Picornaviridae. It is a small RNA virus with a single-stranded genome of positive-sense RNA, approximately 7.5 kb in length. A single polypeptide is synthesized and cleaved by viral proteases. The virion is 27 nm in diameter. The virus particle is more stable to heat, detergents and proteases than the other picornaviruses and this may explain the ease with which it is transmitted. The virus is not thought to be cytopathic; hepatic damage is probably related to immunologically generated inflammation and tissue damage. Neutralizing antibodies are directed against groups of closely clustered epitopes on the virus surface.

The virus has been adapted to growth in a wide range of primate cells, after serial passage.

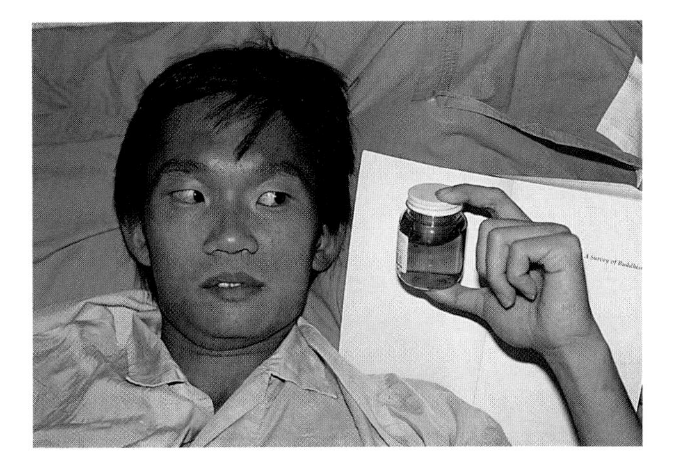

Fig. 9.2 Hepatitis A: typical appearance of jaundice and dark urine in viral hepatitis; the patient is no longer suffering from fever or malaise.

Clinical features

The incubation period is 15–45 days. This is followed by malaise, lassitude, myalgia, arthralgia and variable fever. Features of hepatitis gradually appear after 2–7 days. Many patients complain of nausea or vomiting and a few, especially children, may have loose stools. The urine darkens as bilirubinuria increases, and the stools may be noticeably pale. Jaundice is first seen in the sclerae and later in the skin (Fig. 9.2).

The fever resolves as jaundice develops. At this stage virus excretion ceases, the patient is no longer infectious, and most patients feel better. After 7–10 days the appetite returns and the jaundice begins to resolve.

During the prodrome the white cell count is normal, with a few atypical mononuclear cells. As the prodrome ends, rising plasma transaminases indicate hepatocellular damage, and conjugated bilirubin appears in blood and urine. The early transaminase levels are often 2000–7000 IU but these fall rapidly in most cases, and are often below 100 IU after 7–10 days.

Unlike the other viral hepatitides, hepatitis A commonly causes cholestasis, with a rise in alkaline phosphatase to 250–400 IU during convalescence. Abnormality of bile-salt excretion may then cause itching. Occasionally cholestasis is severe and prolonged, with deepening jaundice and severe pruritis, persisting for months if untreated. This does not indicate prolonged infection; it is a postinfectious event.

Diagnosis

Clinical suspicion depends on a typical history of malaise followed by jaundice, and the absence of risk factors for other types of viral hepatitis. Prolonged vomiting in an otherwise well child might be due to anicteric hepatitis, particularly if the epidemiological history is compatible with hepatitis A. Laboratory investigations should include tests to exclude hepatitis B and C and, where appropriate, other viral hepatitis infections (see pp. 197–200).

Glucose-6-phosphate dehydrogenase deficiency can cause haemolysis and clinical jaundice, often following an acute infection, ingestion of proprietary antipyretic drugs or eating beans (favism). Spherocytosis, sickling and ovalocytosis are all sometimes associated with haemolytic episodes. In travellers, malaria is a common cause of haemolysis, usually accompanied by fever. Haemolytic (acholuric) jaundice is not accompanied by bilirubinuria; urinalysis will therefore suggest this alternative diagnosis.

Persistence of fever in viral hepatitis is unusual once jaundice has developed. In febrile hepatitis a diagnosis of Epstein–Barr virus (EBV) infection, or leptospirosis, may need to be excluded.

Acute cholecystitis can cause jaundice with fever. Pain in the right hypochondrium is a common accompaniment. Gallstones may cause pain and obstructive jaundice. Painless cholestatic jaundice is often a feature of biliary or pancreatic carcinoma. It should be considered in all middle-aged or elderly patients with jaundice. An elevated alkaline phosphatase is a useful warning sign of obstructive pathology.

Some drugs cause hepatocellular damage. Alcohol is the most common, but paracetamol toxicity must not be forgotten. A history of drug treatment or overdose must be sought in jaundiced patients.

Differential diagnosis of viral hepatitis
1 Other infections: Epstein–Barr virus infections, cytomegalovirus, leptospirosis.
2 Infections of the biliary system: acute cholecystitis, acute cholangitis, acute pancreatitis.
3 Obstructive jaundice: gallstones, tumours of the pancreaticobiliary system.
4 Haemolysis, e.g. glucose-6-phosphate dehydrogenase deficiency, malaria.
5 Drug jaundice, e.g. alcohol and paracetamol (hepatocellular), phenothiazines (cholestatic).

Laboratory diagnosis

Immunoglobulin M (IgM) antibodies become detectable when jaundice develops and persist for approximately 3 months, and occasionally for up to 2 years. IgG antibodies persist, and confer lifelong immunity. Specific diagnosis depends on demonstration of specific IgM by antibody-capture enzyme immunoassay (EIA). An EIA to detect specific IgG may be employed to determine immune status before vaccination.

Treatment

Bed rest and simple analgesics ameliorate prodromal symptoms. Antiemetic drugs can safely be given. Antipruritic drugs may be helpful in convalescence, but cholestyramine is more effective for severe pruritus. There is little evidence that bed rest is necessary once the transaminases have shown a significant fall.

Prolonged severe cholestasis after hepatitis A responds to treatment with corticosteroids. Initial dosage with prednisolone 30–40 mg/day can be reduced over 2–4 weeks depending on the rate of fall of the alkaline phosphatase.

Complications

Liver failure

Liver failure is the most common complication. In some cases it is fulminant and early, presenting with altered consciousness before jaundice develops; in others it is subacute and progressive, appearing as an inexorable deterioration in liver function and deepening of jaundice. The prognosis in liver failure is best in patients suffering rapid-onset liver failure and worst in those with insidious onsets.

Clinical indicators of liver failure are persistent vomiting, disturbed behaviour, flapping tremor of the outstretched hands and increasing drowsiness. Poor cerebral function is classically demonstrated by showing the patient's inability to copy a drawing of a five-pointed star, although able to copy a square (constructional apraxia).

The earliest biochemical changes are the disappearance of blood constituents synthesized by the liver. These are glucose, clotting factors and urea (albumin has a longer half-life of 14 days and falls late in liver failure). The transaminase levels may not accurately reflect hepatocellular damage as they inevitably fall when few liver cells remain; however, initial levels above about 7000 IU may alert the clinician to the possibility of failure developing (Fig. 9.3). In established hepatic coma the electroencephalogram shows characteristic 3/s slow-wave patterns.

Treatment. Treatment is based on four main objectives:
1 Maintaining the blood glucose to avoid hypoglycaemic convulsions or cerebral damage. This often requires constant infusion of 10% dextrose solution, and occasionally further glucose supplementation.
2 Minimizing amine production by avoiding excess protein loading (the patient should receive sufficient protein to fulfil daily nitrogen requirements) and emptying the bowel by giving doses of magnesium sulphate. It is no longer considered necessary to add non-absorbable antibiotics to this regimen.
3 Minimizing the likelihood of bleeding while clotting factors are deficient. This is achieved by nursing the patient quietly, avoiding vigorous or traumatic procedures. The risk of bleeding from gastric erosions is reduced by giving H_2-receptor blockers or proton pump inhibitors.
4 Avoiding drugs such as loop diuretics, opiates, major tranquillizers and any drugs whose side-effects include hepatotoxicity.

Short-acting benzodiazepines may be given for sedation. Paracetamol is a useful analgesic, as its toxic metabolite is not produced during liver failure. Transfusions of fresh plasma constitute a protein load, and should be reserved for treating haemorrhagic events (in any case, the clotting factors are quickly consumed and do little to prevent longer-term widespread oozing). Small doses of vitamin K may be given to avoid deficiency during a catabolic state, but large or frequent dosage should be avoided.

Specialist liver units

Specialist liver units offer experienced management, and can arrange liver transplantation when indicated. In hepatitis A, infection is often over by the time liver failure

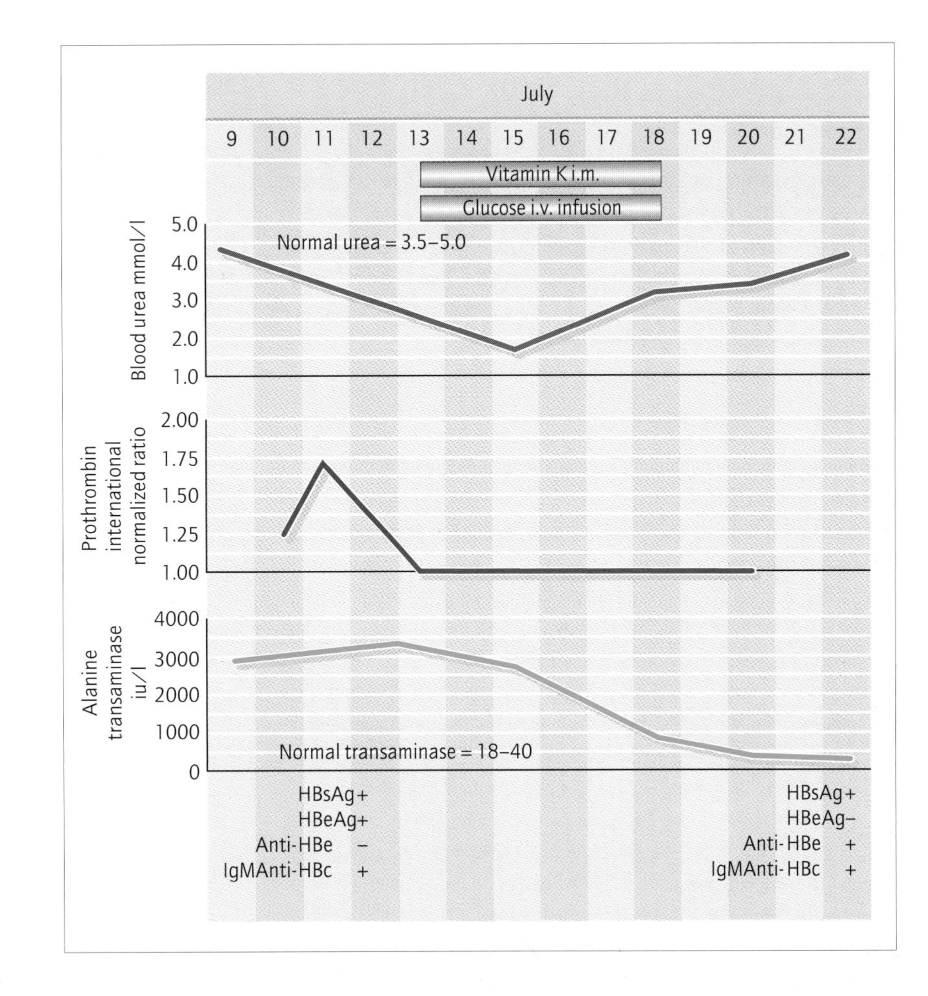

Fig. 9.3 Development of liver failure in a patient with a severe attack of hepatitis B; the patient made a spontaneous recovery with supportive treatment.

is established, so the outlook for a successful transplant is good. Early referral to an appropriate unit gives the best chance of arranging such treatment.

Prevention and control

Hepatitis A is preventable by good sanitation and personal hygiene, which are particularly important in nurseries and schools where spread is likely to occur. The risk of common-source outbreaks can be minimized by proper education of food-handlers and adequate cooking of shellfish.

Non-replicating vaccines against hepatitis A are highly effective, producing lasting immunity 8–10 days after a single dose. They are recommended for travellers to endemic areas, and may be useful in controlling large, continuing outbreaks.

Human normal immunoglobulin (HNIG) is available for postexposure prophylaxis and for travellers making urgent journeys (though vaccine is preferred in most cases, as HNIG is a blood product). An intramuscular injection provides passive protection which lasts for 6–8 weeks. It may be given concurrently with vaccine, at a different site (so-called passive–active immunization) for emergency immunization followed by longer-term immunity. Individuals who travel frequently should be tested for hepatitis A antibody as they may have naturally acquired immunity.

Cases of hepatitis should be isolated until 1–2 days after the onset of jaundice. The disease is notifiable.

Prevention of hepatitis A

1. Safe water supplies.
2. Effective sanitation.
3. Personal hygiene.
4. Safe food-handling.
5. Pre-exposure prophylaxis: human immunoglobulin and inactivated vaccine.

Hepatitis B

Introduction

The importance of hepatitis B lies in its ability to cause prolonged or permanent infectious carriage and persisting liver inflammation, leading eventually to cirrhosis and malignant change in the liver. Vertical and horizontal infection from mother to child can produce successive generations of infectious carriers.

Virology and pathogenesis

Hepatitis B is a small enveloped virus containing partially double-stranded DNA (Fig. 9.4). The DNA contains four major genes which each code for more than one protein. The surface protein gene codes for three polypeptides (including hepatitis B surface antigen, HBsAg) which make up the viral envelope. The C gene codes for the core or capsid protein (hepatitis B core antigen, HBcAg) and the precore protein. The P gene codes for the reverse transcriptase and RNAase, and the X gene for transactivator proteins. HBsAg is the major protein of hepatitis B virus (HBV). The major antigenic determinant of HBsAg is the a antigen; antibodies to this antigen confer protection after infection or vaccination. In addition there are two pairs of mutually exclusive subdeterminants d or y and w or r, giving four phenotypes, adw, adr, ayw and ayr. The frequency of these phenotypes varies in different parts of the world.

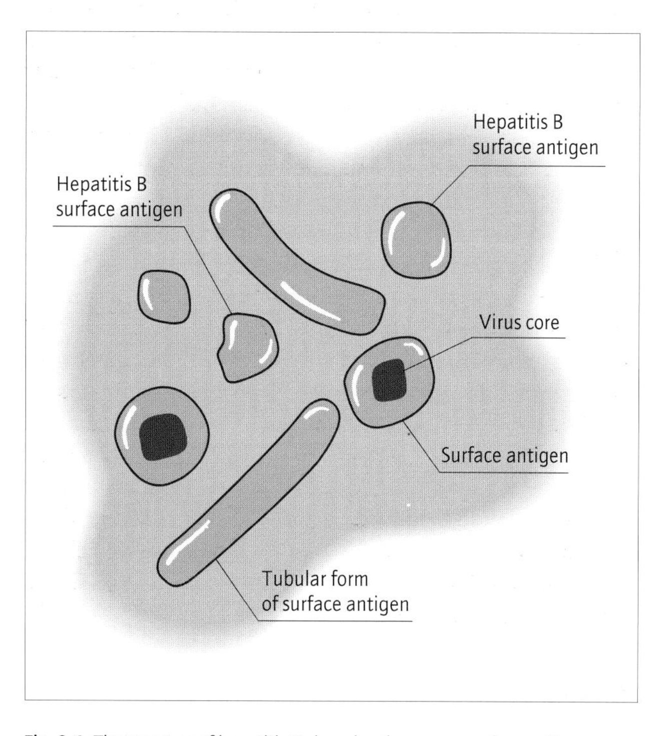

Fig. 9.4 The structure of hepatitis B virus showing excess surface antigen.

HBcAg is the major component of the nucleocapsid. Hepatitis B e antigen (HBeAg) may be generated from this antigen by proteolytic cleavage.

HBV typically causes an acute hepatitis which is characterized by liver cell necrosis and periportal histiocytic infiltration. The necrosis is centrilobular and multifocal. Rarely, massive necrosis occurs associated with considerable disruption of the reticulin framework of the liver. In approximately 10% of patients, persistent infection occurs, leading to chronic hepatitis and, after an average of about 30 years, the development of cirrhosis. Liver carcinoma is an important complication of hepatitis B-associated cirrhosis with an excess risk of 100-fold compared with the general population.

HBV DNA is integrated into host-cell DNA. This insertion is random and may affect important genes near to the site of integration.

Epidemiology

There are around 300 million chronic carriers of the hepatitis B virus worldwide. The incidence of acute disease and prevalence of carriage varies considerably from country to country. In parts of south-east Asia, 10–20% of the population may be carriers, whereas most countries in Europe and North America have carriage rates below 2%.

Where carriage rates are high, acute infection occurs mainly in infants and young children, mostly via intrapartum and horizontal transmission within households. Skin disease and biting arthropods may allow the transfer of body fluids from person to person.

In low-prevalence countries most infections are sporadic and arise in adults through needlestick injuries, shared syringes, bites and scratches, or by sexual contact. Those most at risk include intravenous drug abusers, homosexual men, residents and staff of institutions for the mentally handicapped, surgeons and dentists, laboratory workers and morticians, renal dialysis patients and recipients of unscreened blood and blood products.

Risk groups for hepatitis B in developed countries

1 Intravenous drug abusers.
2 Homosexual men.
3 Sexual contacts of antigen-positive persons.
4 Residents in long-stay homes for mentally handicapped people.
5 Renal dialysis patients.
6 Recipients of multiple blood products.
7 Surgeons, dentists and morticians.
8 Infants of e antigen-positive mothers.

The infection is relatively uncommon in the UK. Approximately 500 confirmed cases are identified each year, mostly in high-risk groups. The prevalence of HBsAg carriage is less than 0.5% overall, although this may be considerably higher in certain ethnic subgroups and in long-stay institutions for mentally handicapped people. A large outbreak occurred in 1984, associated with intravenous drug abuse. Small common-source outbreaks have been reported in association with tattoo parlours and other skin-piercing activities. A large outbreak occurred in 1998, associated with an alternative therapy centre (Fig. 9.5).

Clinical and laboratory features

The incubation period is from 3 to 6 months. Up to 90% of cases are subclinical. Clinical hepatitis B is heralded by a prodromal viral illness and then a period of afebrile jaundice similar to that of hepatitis A. Although it is said that a serum sickness-like prodrome can occur, with mild arthritis and a faint rash, this is rarely clinically helpful. Clinical distinction of acute hepatitis A from hepatitis B is rarely possible but cholestasis, common in convalescent hepatitis A, is rare in hepatitis B.

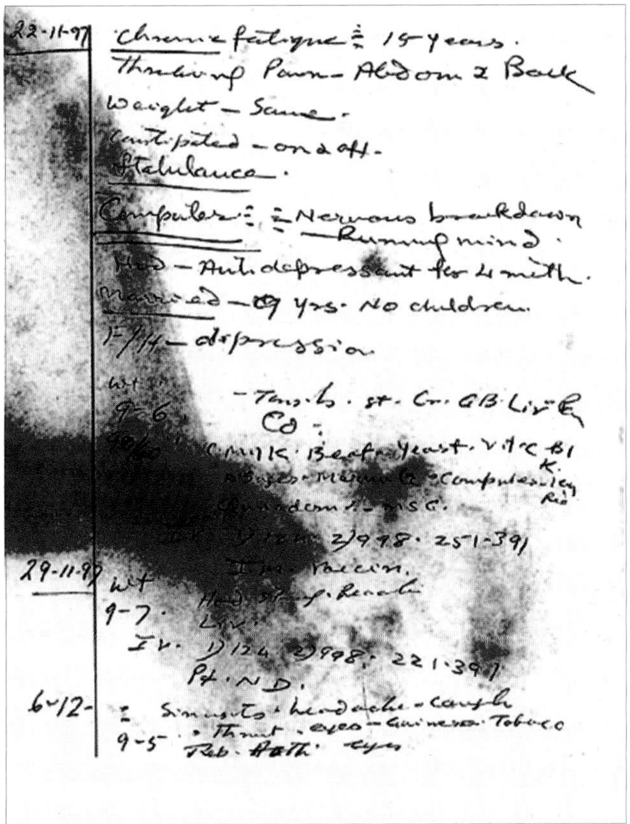

Fig. 9.5 Blood-stained patient records at an alternative therapy centre associated with an outbreak of hepatitis B; a look-back exercise identified 31 cases and one HBe Ag-positive staff member.

During replication of HBV, a large excess of viral HBsAg is produced. This becomes detectable in the serum 2–8 weeks before the clinical illness. As transaminases rise, other viral products are easily detected, including viral DNA polymerase and HBeAg, a blood-borne derivative of core or c antigen. As recovery and convalescence progress, antibodies to the viral proteins appear in the blood, and the proteins themselves gradually disappear. IgM anti-HBc is detectable at the onset of clinical disease. Anti-HBe usually appears in the first 1–3 weeks and HBeAg then becomes undetectable. Anti-HBs appears late in convalescence, after 6 weeks–6 months, and HBsAg then disappears.

This process can cease at any stage, leaving the patient with persisting antigenaemia. Subclinical cases are thought more likely to become carriers. During the state of HBeAg carriage, patients' blood and body fluids contain potentially replicating virus components and are highly infectious. If anti-HBe antibodies fail to develop, the patient's blood remains significantly infectious, even if HBeAg disappears. Patients whose blood is only HBsAg-positive are much less infectious.

If antigens remain detectable after 6 months the patient is considered to be a carrier of hepatitis B antigen. The incidence of carriage is probably in the range of 5–10% of cases. Carriage is more likely in infants and in patients with defective immunity. Many carriers have normal liver function and are well, but those with persisting e antigenaemia, or lack of e antibody, are at risk of chronic persistent hepatitis, chronic active hepatitis, cirrhosis and eventual hepatocellular carcinoma. Female eAg carriers or SAg-positive, anti-e-negative women are also at risk of infecting their infant during or soon after its birth.

Diagnosis

In acute hepatitis B, HBsAg is present in the serum. However, it is also present in long-term carriers of HBV. The diagnosis of acute disease can be confirmed by demonstrating IgM anti-HBc in the serum. This appears 2 weeks after HBsAg (Fig. 9.6) and disappears a few months after uncomplicated infection. Anti-core IgG antibody persists for much longer, perhaps for life. Laboratory diagnosis of acute infections, and establishment of the presence and type of virus carriage, depends on EIA or radioimmunoassay (RIA) to demonstrate antigens and antihepatitis B virus antibodies.

Treatment

Treatment of acute hepatitis B is symptomatic. The patient

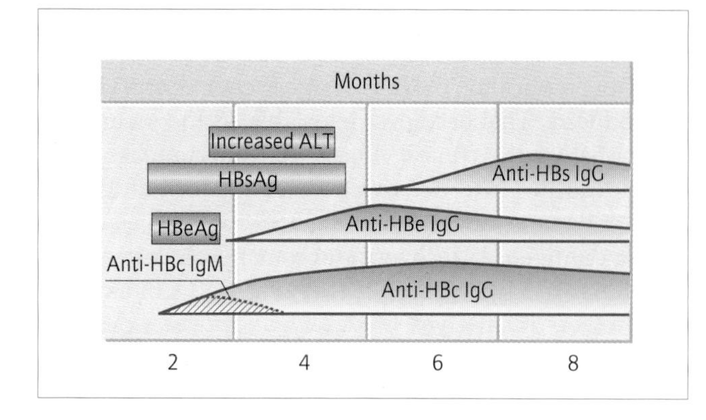

Fig. 9.6 The evolution of hepatitis B: occurrence of hepatitis B virus markers and antibodies in the blood of infected patients. Courtesy of the Communicable Disease Surveillance Centre.

should be followed up to ensure that HBeAg and HBsAg are cleared from the serum, and that anti-HBe is detectable. Patients with antigenaemia persisting after 6 months should be considered for antiviral treatment.

Treatment of chronic hepatitis B

This can reduce infectiousness, liver inflammation and the future likelihood of liver disease. Patients with persisting HBsAg should be evaluated by liver function tests, serology and liver biopsy to determine their infectiousness and exclude other causes of chronic hepatitis (including hepatitis C). The mainstay of treatment is subcutaneous interferon-alpha, 5–10 MU three times weekly for 6 months. If treatment is effective there is often a short illness with abnormal transaminases, occurring up to 6 months after completion of treatment. In different series 25–40% of patients have become HBeAg-negative (see Table 9.1), but few also clear HBsAg. Dosage is limited by influenza-like side-effects. Interferon treatment is not effective in patients who were infected perinatally.

A combination of interferon with the antiviral drugs lamivudine or famciclovir is likely to produce better results, but the optimum regimens have yet to be determined. Lamivudine alone can produce long-term suppression of circulating HBV DNA, with arrest or improvement of liver inflammation, but viral resistance emerges during such monotherapy, and disease activity rebounds on stopping treatment. It is now known that there is a high turnover of infected liver cells in hepatitis B, with the production of millions of viruses daily. By analogy with human immunodeficiency virus (HIV) treatment, it seems likely that multidrug therapy will be required to control viral replication and prevent the emergence of resistant viral clades.

Complications

Liver failure

Liver failure is the most important early complication of acute hepatitis B. Its presentation and management are the same as for hepatitis A. However, liver transplantation does not result in permanent cure because replicating virus may persist in pancreas and other tissue, and will infect the new liver, eventually causing progressive chronic disease despite perioperative prophylaxis with hepatitis B immunoglobulin (HBIG) and/or antiviral drugs.

Rare complications

Rare complications include renal failure, and rash (more common in children, affecting the peripheries early in the illness).

Prevention and control

Immunization

Hepatitis B vaccines contain HBsAg produced from yeast cells using recombinant DNA technology. A course of three injections provides active protection which is probably lifelong in immunocompetent individuals who seroconvert. As approximately 10% of recipients do not produce an antibody response, postvaccination screening should be carried out and a booster dose given if necessary. The proportion of non-responders increases with age. The response may also be reduced if the vaccine is given intradermally or in the buttock.

In the UK, vaccination is indicated for pre-exposure prophylaxis of high-risk groups (see above). In higher-incidence countries universal immunization is offered in infancy or adolescence.

HBIG, (hepatitis B immunoglobin) prepared from the plasma of selected donors provides passive protection against hepatitis B. It is normally used in combination with vaccine to provide passive–active immunity: (i) for infants born to mothers who are HBsAg-positive (particularly those who are also HBeAg-positive and/or anti-HBeAg-negative—see Chapter 14); (ii) following percutaneous or mucosal exposure of non-immune individuals to infected body fluids; and (iii) for sexual contacts of acute cases. Routine antenatal screening for HbsAg is now recommended for all pregnancies in the UK, in order to identify infants for whom prophylaxis will be required.

Immunoglobulin must be given within 1 week of exposure, and preferably within 48 h. A course of vaccine should be started at the same time.

Table 9.1 Factors favouring successful treatment with interferon-alpha, and some cautions.

Hepatitis B	*Hepatitis C*
Short duration of infection	Short duration of infection
No delta or HIV infection	No HIV infection
Higher transaminases	Genotypes 2 and 3 (NOT genotype 1)
Active hepatitis on biopsy	Mild or absent fibrosis on biopsy
Low viral markers in blood (i.e. low HBV DNA)	Low viral markers in blood (i.e. low HCV DNA)
	Younger, female patient
Contraindications to interferon-alpha treatment	*Adverse effects of interferon-alpha*
Pregnancy or risk of pregnancy	Fever and flu-like symptoms (reduced by paracetamol)
Autoimmune diseases	Activation of autoimmune diseases
Active depression or psychosis	Mild depression (rarely, acute psychosis)
Decompensated liver disease	Decompensation of liver failure
Alcohol abuse	Anorexia, fatigue, weight loss
	Leucopenia, thrombocytopenia, anaemia
	Rarely, elevated transaminases
	Reversible alopecia

HBV, hepatitis B virus; HCV, hepatitis C virus.

Other measures

Hepatitis B can also be prevented by ensuring that all blood and blood products are screened and that blood is not collected from donors at risk of infection. All syringes, needles and skin-piercing equipment should be thoroughly sterilized after each use; wherever possible disposable equipment should be used. Carriers of HBsAg may be restricted from certain occupations. The disease is notifiable.

Prevention of hepatitis B

1 Pre-exposure prophylaxis with non-replicating vaccine.
2 Postexposure prophylaxis with specific immunoglobulin and vaccine.
3 Universal antenatal screening.
4 Universal testing of donated blood products.
5 Use of condoms.
6 Use of clean needles and other skin-piercing equipment.
7 Safe disposal of hospital sharps.
8 Use of disposable haemodialysis equipment.

Delta hepatitis

Introduction

Delta hepatitis, or hepatitis D, is caused by a satellite virus which uses HBsAg as its own surface antigen. It therefore only infects antigen-positive patients with either acute hepatitis B, or long-term HBsAg carriage. Simultaneous infection with hepatitis B and D is thought to cause more severe disease than hepatitis B alone.

Virology and pathogenesis

Hepatitis D virus, the delta agent, is an enveloped RNA virus of similar size to hepatitis B virus (38–41 nm compared to 42 nm for hepatitis B). The envelope consists of the same three surface proteins as hepatitis B virus, although the relative amounts are different. The virus core contains only one known protein, the delta antigen. Replication occurs in the nucleus of the cell via a host RNA polymerase. Two complementary strands of RNA are synthesized from the circular genome. One, the antigenome, is an exact complement of the genome. The second, a smaller fragment, is polyadenylated and acts as the messenger RNA for the delta antigen. Infected liver cells become packed with several hundred thousand copies of hepatitis D virus genome.

The delta agent is thought to be directly cytotoxic, although *in vitro* studies are not conclusive. There is evidence that asymptomatic infection is common. The presence of hepatitis B virus is required to provide the envelope proteins, enabling hepatitis D virus to spread from cell to cell, and to express its pathogenic potential. The coexistence of hepatitis D and persistent hepatitis B antigenaemia is associated with an accelerated progression to carcinoma. Delta antigen and IgM antidelta antibody are detectable in the blood during infection.

Epidemiology

Hepatitis D is always associated with a coexistent hepatitis B infection, sharing its routes of transmission. The distributions of the two diseases are therefore similar. Infection occurs endemically among infants and children in areas of high HBsAg prevalence and sporadically among adults exposed to blood and body fluids in low HBsAg prevalence areas.

Prevention and control

Control measures are the same as for hepatitis B, although no specific immunoglobulin or vaccine exists. Controlling the spread of hepatitis B prevents the occurrence of hepatitis D.

Hepatitis C

Introduction

Hepatitis C virus (HCV) is an RNA virus related to the flaviviruses and animal pestiviruses. It is a common cause of blood-borne hepatitis, most cases of which are subclinical. Following the acute infection the liver function tests may fluctuate for many weeks, and, while the majority of cases eventually recover, approximately 40% may have persisting hepatitis with the risk of late cirrhosis or hepatocellular carcinoma. HCV exists in six major genotypes, which have different potential for causing persisting viraemia.

Virology

The existence of HCV was long predicted on epidemiological grounds, as a cause of short-incubation post-transfusion hepatitis. Its existence was proven, and antigens for immunoassay were developed, when viral messenger RNA was purified from the serum of infected patients. Complementary DNA was prepared by reverse transcriptase and inserted into the phage, lambda gt11. Bacteria into which the phage was inserted produced HCV proteins for screening against sera from cases of 'non-A–non-B hepatitis'.

Hepatitis C is a single-stranded RNA virus, with a 9.4-kb positive-sense genome, encoding a single polypeptide. Areas within the major open reading frame code for core and envelope proteins. The RNA shows considerable sequence heterogeneity due to low-fidelity replication, with only 65% homology between the major genotypes.

Epidemiology

Hepatitis C has been found in every country where it has been sought. The disease is transmitted by exposure to blood and body fluids, probably via contaminated blood transfusions. However, spread through contaminated needles and syringes is also important. Sexual transmission probably occurs, but much less readily than with hepatitis B. Vertical transmission from an infected mother is uncommon. The prevalence of anti-HCV antibody in serum is between 1% and 2% in most developed countries. The incidence in the UK is not known at present, but 1 in 1400–2000 blood donors possesses antibodies, and most of these have hepatitis C RNA, demonstrable by polymerase chain reaction (PCR) in their blood.

Clinical features

The incubation period of blood-borne hepatitis C is usually 3 or 4 weeks. Many cases are subclinical; typical illness with malaise followed by jaundice is unusual. Liver failure is extremely rare in immunocompetent individuals.

Diagnosis

The diagnosis of hepatitis C depends on the detection of antibodies to recombinant antigens. The different assays used include antibody capture and antibody competition EIA. Cross-reactions occasionally occur, giving false positive results, for instance during EBV infection, but are uncommon with modern, multiantigen tests. False-positive results may also be obtained in patients who have recently received blood products containing anti-HCV antibodies. A positive test in acute infections can be confirmed by an alternative method, of which the best is viral genome detection by reverse transcriptase PCR. The virus can also be genotyped by sequencing regions of the genome.

In chronic HCV infection, detection of antibodies is not sufficient to make a diagnosis. Demonstration of hepatitis C virus RNA by RT-PCR shows that many seropositive patients are also viraemic.

Treatment

Treatment with subcutaneous interferon-alpha 3 MU three times weekly for 12 months will normalize transaminases in about 50% of patients, with loss of HCV RNA from the blood and improvement in liver inflammation, but the response is sustained in only 15–25% of patients. This response is increased if the antiviral drug tribavirin is added, though given alone it has only a temporary effect. Trials suggest that approximately 45% of patients will gain long-term remission after combined therapy.

This treatment is rigorous, and its benefit for patients with normal transaminases and/or mild histological changes is controversial (see Table 9.1).

Complications

Aplastic anaemia occasionally follows hepatitis C, usually in the convalescent stage. The aplasia is profound and permanent. As bone marrow transplant offers the best hope of cure, blood and platelet transfusions should be avoided as far as possible, so that antihaemocyte antibodies are not induced. Expert haematological advice should be sought. Mixed cryoglobulinaemia, with acrocyanosis and renal failure, is a rare complication of hepatitis C.

Prevention and control

General measures for the prevention of hepatitis B (see above) apply also to hepatitis C. Needle exchange schemes help to reduce new cases among intravenous drug users. There is no vaccine or specific immunoglobulin available. Screening of donated blood has been carried out since 1991.

Hepatitis E

Introduction

This disease behaves rather like hepatitis A and circulates by the faecal–oral route, particularly in north India, Nepal, Bangladesh and north Africa. It is unaccountably severe in pregnancy, causing a mortality of 10–40%, compared with under 1% in non-pregnant patients. Jaundice is often prolonged, lasting 4 or 5 weeks.

Virology

Hepatitis E is a small (32–24 nm) non-enveloped RNA virus, of uncertain classification. It possesses a single-stranded, polyadenylated RNA of approximately 7.5 kb with three open reading frames. Two putative proteins are predicted, one of which is an RNA polymerase. Structural proteins are coded on the second of the three open reading frames. The virus is unstable in storage and has not yet been cultivated artificially.

Epidemiology

The epidemiology of this disease is not well described. Epidemics have been identified in several countries, particularly in South-east Asia and the Indian subcontinent. In these outbreaks, young adults, especially males, are predominantly affected. The source of infection is usually contaminated water; person-to-person spread by the faecal–oral route also occurs.

Diagnosis

Tests for the diagnosis of hepatitis E are in continuing development. An antibody-capture EIA has been developed using recombinant proteins, and is used for demonstrating IgM antihepatitis E virus (HEV) antibodies. Examination of stool by immunoelectron microscopy may reveal the presence of virus-like particles. Methods to detect hepatitis E virus in stools by PCR have been reported.

Clinical features

Clinical HEV infection is similar to hepatitis A with an incubation period averaging 40 days, but has a longer time course. Jaundice may increase for 2–3 weeks after its first appearance, and the average duration of clinical jaundice is 5 or 6 weeks. The infective period is uncertain, but probably coincides with the phase of increasing transaminases. Liver failure can occur, and is more common in pregnant women, who have a high mortality of up to 40%. Persisting cholestasis with high alkaline phosphatase levels has also been observed, but is rare.

Prevention and control

General sanitary measures and personal hygiene are effective in limiting spread. No vaccine or immunoglobulin is available.

Hepatitis G virus and transfusion-transmitted virus

Hepatitis G virus (HGV) is a single-stranded RNA virus of the family Flaviviridae. It is found in the serum of 1–2% of American blood donors, and up to 15% of west Africans, who have an increased likelihood of raised transaminases. To date, there is no evidence linking positive tests for HGV RNA with chronic hepatitis or cirrhosis.

Transfusion-transmitted virus (TTV), like HGV, has been identified in the blood of those receiving multiple donations of blood products. Its role as a pathogen is not yet clear.

Epstein–Barr virus

Approximately 15% of clinically recognized cases of EBV infection have clinical hepatitis. This is usually accompanied by fever, atypical lymphocytosis and positive heterophile antibody tests. The liver function tests initially show moderately elevated transaminases. Cholestasis is common, with alkaline phosphatase levels rising to 400–1000 IU/l, but with few or no clinical consequences. Liver failure is extremely rare (see Chapter 6).

Cytomegalovirus

In developed countries 40% of adults have evidence of immunity to cytomegalovirus; this figure rises to 100% in some developing countries. About 1800 (3%) of the 600 000 children born each year in England and Wales are congenitally infected with cytomegalovirus and about 10% of these have some permanent handicap (see Chapter 17). Infection is almost always asymptomatic in the immunocompetent. On rare occasions fever, mild jaundice, low-grade atypical lymphocytosis and a negative heterophile antibody test are accompanied by seroconversion to cytomegalovirus.

Yellow fever

Introduction

Yellow fever is endemic in parts of the tropics, including Africa, and tropical South America (Fig. 9.7). It is hardly ever imported into European countries because of the International Health Regulations' requirements for immunization of travellers.

The cause is a flavivirus which is transmitted by mosquitoes of the *Aedes* sp.

Epidemiology

Two epidemiologically distinct patterns of disease transmission occur. In Africa, the disease is transmitted from monkeys, the main host, to humans (and from human to human) by the bite of the *Aedes* sp. of mosquito (principally *A. aegypti*). These mosquitoes breed mainly in small water containers such as discarded coconut shells, and inhabit villages and towns. Periodic explosive outbreaks of urban yellow fever occur, often in association with a breakdown in mosquito control measures. Countries between 15°N and 10°S are affected.

In South America, the vector is a forest-dwelling mosquito of the genus *Haemagogus*. The disease occurs endemically among young adult male forest workers. This pattern of transmission is known as sylvan yellow fever. Almost every country in South America has been affected at some time.

The disease has never been reported outside Africa or mainland South America, with the exception of Trinidad.

The incidence of yellow fever has risen sharply in recent years. Between 1986 and 1988, 5395 cases and 3172 deaths were reported to the World Health Organization—the highest in any 3-year period since 1948. Large outbreaks occurred in Peru and Nigeria. Official figures represent only a small fraction of the total cases.

Clinical features

After an incubation period of 6 or 7 days, symptoms appear abruptly, with fever, myalgia and prostration. As

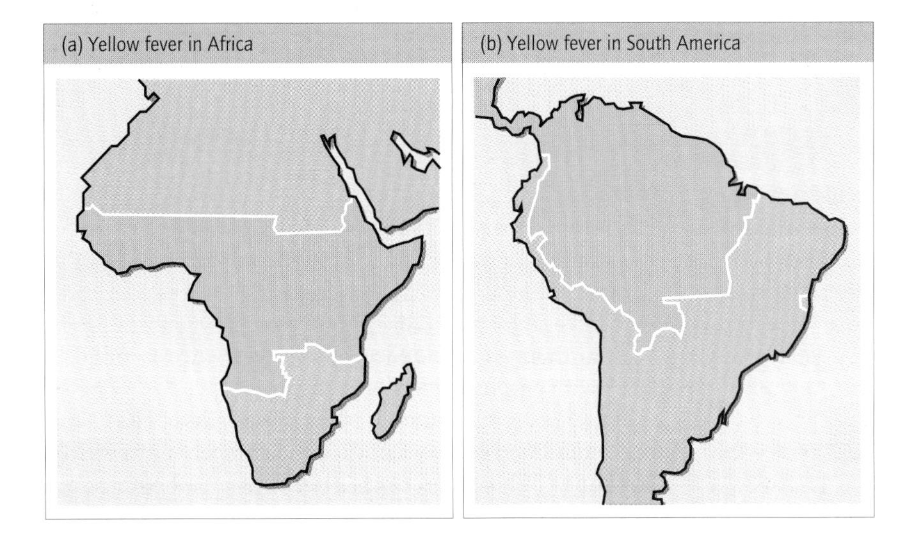

Fig. 9.7 Yellow fever: world map showing affected countries: (a) in Africa; (b) in South America. Courtesy of the World Health Organization.

with other flavivirus infections, a remission of fever and symptoms often occurs after 4 or 5 days. Sometimes the illness ends here, but severe cases progress, with further fever, clinical jaundice, bleeding characterized by haematemesis, and progressive collapse.

Diagnosis

The diagnosis can be suspected in unvaccinated patients who have recently visited endemic areas. Virus can be demonstrated in the blood in the first few days of illness. Serology is the main means of diagnosis, and is carried out by arbovirus reference laboratories.

Prevention and control

Yellow fever vaccine is prepared from the attenuated 17D strain of yellow fever virus. One dose protects for at least 10 years. As with other live vaccines, it may not be given to pregnant women or the immunocompromised. Human immunoglobulin contains no significant yellow fever antibodies and can be given at any time in relation to the vaccine. Several countries in endemic zones of Africa and South America require certification of vaccination.

In endemic areas, the disease is preventable by mass immunization and vector control programmes.

Bacterial infections of the liver parenchyma

ORGANISM LIST

Leptospira spp.
Coxiella burnetii
Brucella spp.
Mycobacteria

It can be seen from this list that bacterial hepatitis is rare. These infections are multisystem diseases in which liver disorder or jaundice is an important or predominant part of the clinical picture. They are important in the differential diagnosis of infectious liver disorders.

Leptospirosis

Introduction

Leptospirosis is a zoonotic disease produced by infection with a variety of *Leptospira* spp. Infections in humans are often due to *Leptospira* of small rodents, cattle, pigs or occasionally dogs. The organisms survive in water, and inoculation of contaminated water infects humans. The disease is important economically among farming communities.

Epidemiology

Transmission occurs when skin or mucosa is exposed to fresh water contaminated by the urine of infected animals or by direct contact with animals. In the UK, the principal animal reservoirs are rats (*L. interrogans* var. *icterohaemorrhagiae*) and cattle (*L. hebdomadis* var. *hardjo*). Approximately 50 cases are reported each year, the majority of which are due to these two serotypes. Two groups are affected: those exposed occupationally, and those infected during recreational exposure. In both groups the disease occurs mostly in men. Farmers, vets and sewer workers are the main occupational risk groups. Recreations associated with infection are swimming, canoeing, windsurfing and fishing.

People at risk of leptospirosis

1 Farmers.
2 Sewage workers.
3 Vets.
4 Water-sport enthusiasts.
5 River fishermen and water bailiffs.

Microbiology

The genus *Leptospira* is divided into two morphologically identical species: *L. interrogans*, which contains all mammalian pathogens, and *L. biflexa*, containing environmental organisms which are not human pathogens. *L. interrogans* is divided into more than 200 serovars on the basis of agglutination reactions. For simplicity, the names of these serovars are shortened, giving the apparent status of a species, so that *L. interrogans* var. *icterohaemorrhagiae* is more commonly known as *L. icterohaemorrhagiae*.

The organisms usually have a preferred mammalian host but can infect a wide range of species, including humans. Thus, *L. icterohaemorrhagiae*, the causative agent of Weil's disease, has its reservoir in the rat, and *L. hebdomadis*, another common human pathogen, occurs in cattle. *L. canicola*'s preferred host is the dog, while in humans it causes canicola fever.

Clinical features

The clinical picture is a variable combination of fever, hepatitis, meningitis, nephritis and rash. In severe cases there may be a marked bleeding tendency. Mild and subclinical infection is common with all types of leptospirosis. While the full severe picture of Weil's disease is commonly seen in *L. icterohaemorrhagiae* infections, it can also occur with other types.

The incubation period is from 1 to 3 weeks, occasionally shorter or longer, and is followed by an illness which may evolve through two or three phases:

1 First, a phase of acute infection lasting up to a week with fever, malaise and (unusually in a bacterial infection) widespread myalgia; at this stage there is bacteraemia and bacteriuria, and a polymorph leucocytosis gradually develops.

2 Second, the appearance of conjunctival suffusion. This may overlap both the preceding and subsequent phase.

3 Finally, the phase of immunopathology with liver, central nervous system and kidney involvement, and the appearance of antibodies in the blood; fever persists throughout the illness (Fig. 9.8).

The meningitis is associated with moderate elevation in the cerebrospinal fluid protein level and lymphocyte count; the glucose level is not altered. Meningitis is more associated with canicola fever, while hepatorenal disturbance is marked in *L. icterohaemorrhagiae* infection.

Bleeding is due mainly to widespread vasculitis, which is the most damaging feature of the disease. While thrombocytopenia is quite common, other features of disseminated intravascular coagulation are rare. A rare but well-recognized feature of severe disease is a widespread nodular pneumonitis.

In practice many patients may be thought to have viral hepatitis or meningitis at presentation. Warning signs are the association of red eyes and fever with hepatocellular liver disturbance, or of hepatocellular disorder with meningitis. Most cases have significant proteinuria and many have microscopic or macroscopic haematuria with or without casts.

The rash is often maculopapular in the early stages, and can be a helpful sign.

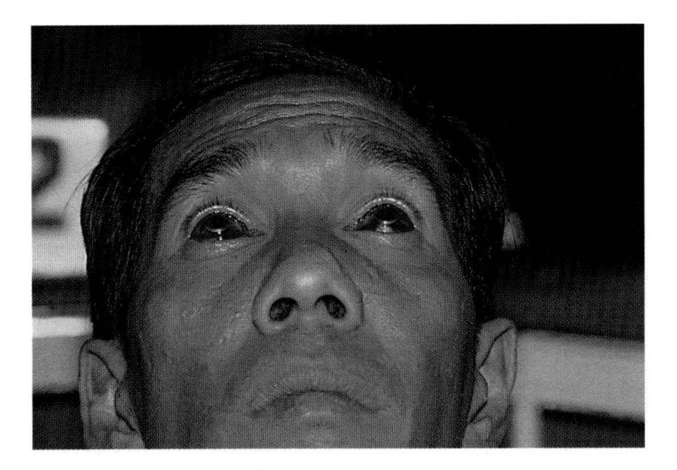

Fig. 9.8 Leptospirosis: jaundice and haemorrhagic conjunctivitis in a feverish farmer. Courtesy of Dr D. Lewis.

Important clinical features of leptospirosis

1 Early myalgia.
2 Hepatitis with fever.
3 Renal impairment.
4 Lymphocytic meningitis.
5 Conjunctivitis.
6 Rash, sometimes haemorrhagic.
7 Thrombocytopenia.
8 Blood, protein and/or bilirubin in the urine.
9 Rare, nodular pneumonitis.

There is a 5–10% mortality in severe cases. Fatalities are due to cardiovascular collapse, often with haemorrhagic myocarditis, and usually associated with renal failure and haemorrhage.

Diagnosis

Clinical diagnosis can be suspected on the basis of epidemiological and clinical features. A useful pointer in flu-like presentations with only fever and myalgia is the finding of protein, blood and/or bilirubin on urine testing.

Laboratory diagnosis

Leptospira can be visualized in the blood of patients who present during the primary phase of the illness, but false-positive results are common. Freshly voided urine can be examined by dark-ground microscopy but this must be done quickly before the organisms die in the acidic environment.

Leptospira can be cultivated from the blood during the first week of illness and from the urine later in the course. The diagnostic yield is often low, as shedding may be intermittent. A specialized medium such as the semisolid agar of Ellinghausen and McCulloch is required for successful culture. *Leptospira* are microaerophilic and grow beneath the surface of the medium. Suspensions of cultures can be examined by dark-ground microscopy to confirm the presence of the organisms, which can be identified by slide agglutination, using panels of antisera.

Diagnosis is usually made by serology. Two alternative techniques are used: the first is the *Leptospira* agglutination technique, in which *Leptospira* of various serogroups (groups of closely related serovars) are agglutinated by patients' serum. More recently, an enzyme-linked immunosorbent assay (ELISA) technique which detects specific IgG and IgM has been described. This makes it possible to confirm the diagnosis on a single serum specimen.

Diagnostic methods for leptospirosis

1 Dark-ground microscopy of blood or urine.
2 Cultures of blood or urine.
3 Serum *Leptospira* agglutination tests.
4 Serum immunoglobulin M enzyme-linked immunosorbent assay.

Treatment

Many cases recover in 10–14 days. Even after the bacteraemic phase, antibiotic treatment can modify the course of severe leptospirosis. The organisms are sensitive to many antibiotics; the treatment of choice is benzylpenicillin. Tetracycline, sulphonamides and erythromycin are all effective, but the effectiveness of cephalosporins is unpredictable.

Treatment of leptospirosis

Benzylpenicillin i.v. 1.2–2.4 g 6-hourly *or* oxytetracycline orally 500 mg 6-hourly, both for 10 days.

Complications

Recovery from leptospirosis is usually complete. Some 1–2% of cases develop late uveitis some weeks after recovery. This can be treated with standard ophthalmological anti-inflammatory preparations.

Prevention and control

The most important preventive measure is to ensure that proper protective clothing is worn during occupational and recreational exposure. Cuts and abrasions should be covered with a waterproof dressing before exposure to river water. Rodent control measures may further reduce the risk of infection on farms. Vaccines which provide some protection against the most common serotypes are available for use in animals. Human vaccines have also been developed, but are associated with a high incidence of adverse reactions and give only limited protection.

Causes of granulomatous hepatitis

Granulomatous hepatitis may be caused by a number of infections, most of which are chronic or subacute. They have to be distinguished from non-infectious causes of parenchymal granuloma, including sarcoidosis, vasculitides, and occasionally drugs such as quinine.

ORGANISM LIST

Coxiella burnetii
Brucella spp.
Mycobacterium spp.
Histoplasma capsulatum

Chronic Q fever (*Coxiella burnetii*)

This can follow acute Q fever, or can commence without previous illness. It usually presents with swingeing fever, abnormal liver function tests with a cholestatic picture, and often clinical jaundice. Half of patients with chronic Q fever also have *Coxiella burnetii* endocarditis (see Chapter 12).

Acute Q fever is a zoonosis which occurs in those exposed to sheep and sheep products. It often presents as atypical pneumonia (see Chapter 7).

Chronic Q fever can be confirmed by demonstrating phase I antibodies to *C. burnetii*. Liver biopsy shows a characteristic histological picture, with many small, non-caseating granulomata which have a peripheral zone of eosinophils (Fig. 9.9).

Tetracycline is the treatment of choice. Two or 3 weeks' dosage is sufficient in uncomplicated cases. If endocarditis is present, long-term treatment is advisable (see Chapter 12).

Brucellosis

Brucellosis is a zoonosis, and is a multisystem disease often with low-grade bacteraemia. There is probably some granulomatous change in the liver of most infected patients. It is rare for liver disorder to be the main feature

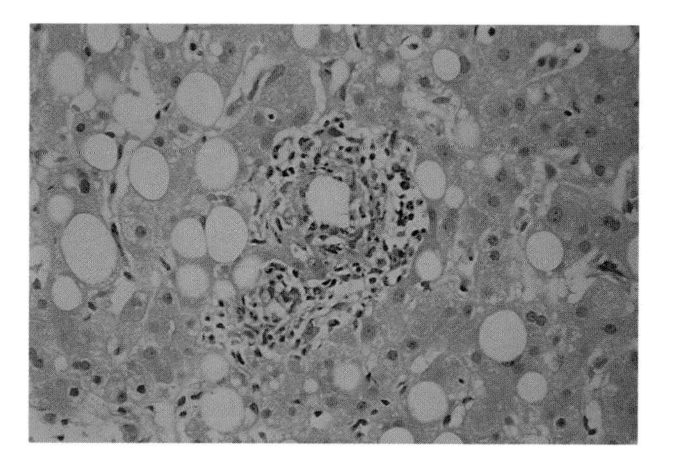

Fig. 9.9 Q fever: liver biopsy showing small granulomata with a peripheral zone of scattered eosinophils.

of brucellosis, but liver granulomata may be discovered during the investigation of pyrexia of unknown origin and should prompt consideration of the diagnosis.

In acute brucellosis blood cultures are often positive, though prolonged culture may be required. Serodiagnostic tests can demonstrate IgG or IgM antibodies to *Brucella* spp.

The treatment of choice is 6 weeks' treatment with tetracycline (doxycycline is useful in a once-daily dose, but some experts favour twice-daily demeclocycline). The addition of rifampicin or streptomycin is recommended to ensure a bactericidal effect (see Chapter 25).

Tuberculous granulomata of the liver

Granulomata are always present in miliary tuberculosis. They are often small and non-caseating, and must be distinguished from the granulomata of brucellosis, sarcoidosis, autoimmune diseases and rare fungal infections. Diagnosis is aided by ensuring that biopsy material is cultured for mycobacteria, fungi and pyogenic organisms.

Serum should also be obtained to permit diagnosis of brucellosis, yersiniosis or autoantibody-positive conditions. The serum angiotensin-converting enzyme level will be elevated if pulmonary sarcoidosis coexists with liver disease, as is often the case.

Abscesses of the liver

Introduction

Liver abscesses are a relatively common cause of pyrexia of unknown origin, and sometimes present without localizing signs. Before the advent of accurate imaging techniques, half of all abscesses were diagnosed at postmortem examination. Nowadays they can be readily demonstrated, and modern antibiotics permit effective treatment.

Pyogenic liver abscess

ORGANISM LIST

Escherichia coli
Klebsiella spp.
Serratia spp.
Other Enterobacteriaceae
Enterococci
Streptococcus 'milleri'
Staphylococcus aureus
Bacteroides spp.
Anaerobic streptococci
Other bowel anaerobes

Many abscesses contain a mixture of two or more organisms. Pyogenic abscesses may be a result of bacteraemia (when other abscesses may also exist, for example, in the brain). They may complicate or follow abdominal sepsis, or they may be apparently spontaneous. They vary from single large lesions to multiple moderate or small ones, and may even be widespread and microscopic.

Clinical features

Patients usually present with high, swingeing fever, marked neutrophilia and a high erythrocyte sedimentation rate. Large abscesses may produce tender enlargement of the liver, even progressing to fluctuant lesions pointing between the ribs. Even without abdominal signs, inflammation below the diaphragm will often produce a small right-sided pleural effusion (Fig. 9.10).

The liver function tests are usually abnormal, showing a moderate elevation of alkaline phosphatase. Clinical jaundice is rare.

Diagnosis

The investigation of choice is imaging. Ultrasound examination will demonstrate most large or moderate-sized lesions. Small lesions, or those whose sonic density is near to that of liver, are best demonstrated by computed tomographic scan. Gallium or magnetic resonance scanning will show moderate to small lesions. Liver biopsy may be necessary to demonstrate multiple small or microscopic lesions.

Microbiological information is obtained from aspiration of the abscess and culture of pus. When the liver lesions complicate bacteraemia, blood cultures are often positive. In bacteraemic cases caused by *Streptococcus 'milleri'*, the likelihood of abscess formation in the brain must be remembered.

Patients who have visited endemic tropical areas may have amoebic liver abscesses (see below). Immunocompromised patients may have abscesses caused by unusual organisms, e.g. *Nocardia* spp. (see Chapter 22).

Management

Treatment should be commenced as soon as blood cultures and pus have been obtained. Initial treatment should be designed to cover the likely range of causative organisms, and can be adjusted later in the light of cultural evidence.

(a)

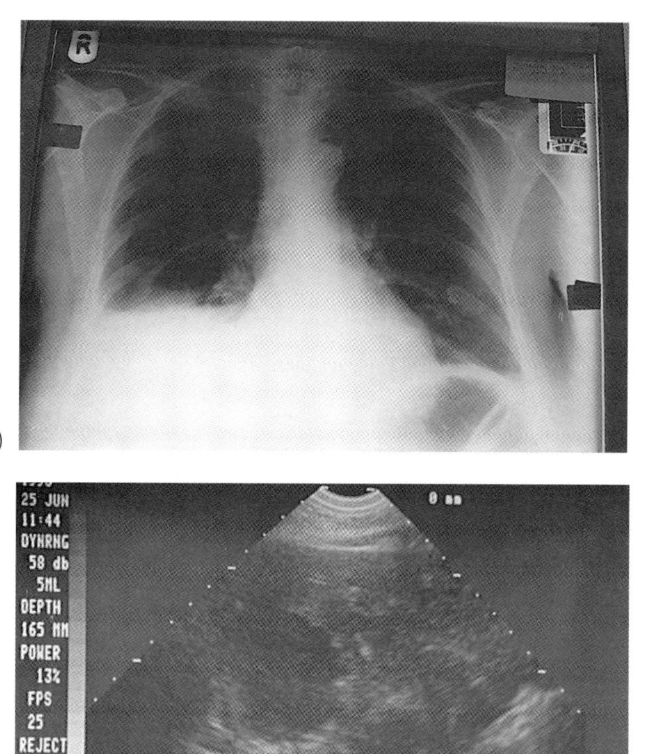

(b)

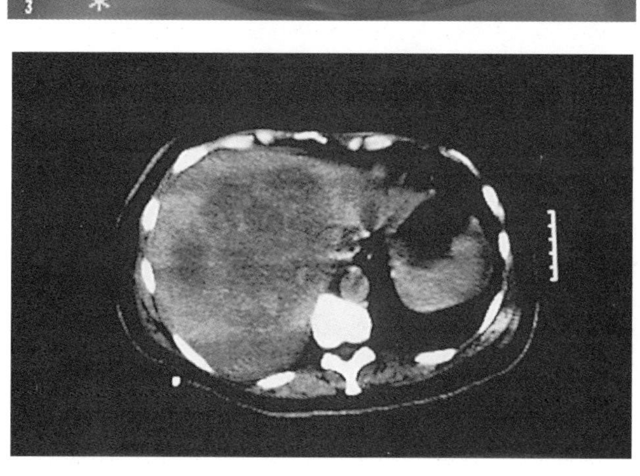

(c)

Fig. 9.10 Liver abscess: (a) pleural reaction at the right lung base; in the absence of significant liver tenderness, the combination of this opacity with neutrophilia and a raised erythrocyte sedimentation rate can lead to the erroneous diagnosis of atypical pneumonia; the huge multilocular abscess is demonstrated by (b) ultrasound and (c) computed tomographic scan.

A third-generation cephalosporin or a broad-spectrum penicillin will be effective against Enterobacteriaceae. Enterococci are also common; the addition of an aminoglycoside will enhance the effectiveness of beta-lactams in this case. Metronidazole may be added to account for anaerobic organisms. If blood-borne infection is suspected, it may be worth including an antistaphylococcal drug in the initial treatment regimen.

Basic treatment for pyogenic liver abscess
Ampicillin i.v. 1 g 6-hourly *or* cefotaxime i.v. 2–3 g 8-hourly; *plus* gentamicin i.v. or i.m. 2–5 mg/kg daily in three divided doses (providing peak blood levels up to 10 mg/l, trough levels less than 2 mg/l); *plus* metronidazole i.v. or rectally 500 mg 8-hourly.
Alternative: meropenem 500 mg–1.0 g 8-hourly.

An adequate response to treatment is indicated by reduction in size of the lesion or lesions, reduction in fever and a falling erythrocyte sedimentation rate.

Drainage of pyogenic abscesses

Drainage of pyogenic abscesses is advisable for moderate or large lesions. Diagnostic aspiration may be sufficient if it is followed by a good response to antimicrobial therapy. Repeated image-guided needle aspiration may speed the healing of large lesions or improve a slow response to treatment.

Persistent reaccumulation of pus requires the insertion of an indwelling drainage catheter and prolongation of chemotherapy, until satisfactory resolution is demonstrated by temperature response and imaging. Some abscesses are inaccessible to needle aspiration, and may require formal surgical evacuation.

Complications

Rupture of a liver abscess into the peritoneal or pleural cavity causes a sudden deterioration in the patient's condition, and may lead to endotoxaemia and collapse. Early treatment should prevent this by controlling the infection and draining abscesses at risk.

Amoebic liver abscess

Introduction

Exposure in endemic areas may lead to amoebic abscess, the commonest complication of *Entamoeba histolytica* infection of the bowel (see Chapter 8). Amoebic abscesses tend to be large and single, usually in the right lobe of the liver (Fig. 9.11).

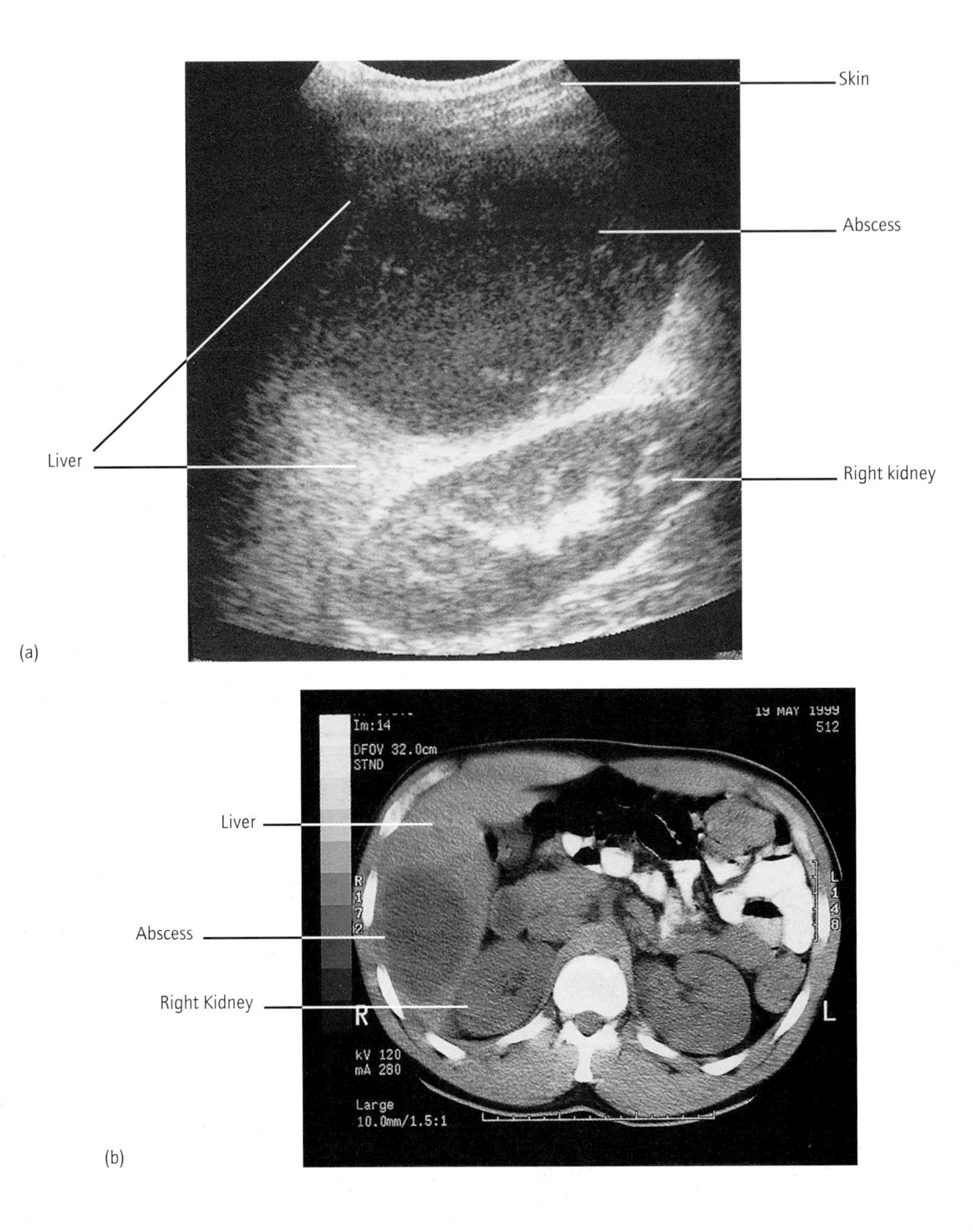

Fig. 9.11 (a) Ultrasound scan of an amoebic abscess. (b) Computed tomographic scan of the same abscess, showing a typical, single, unilocular abscess in the right lobe of the liver. (Cf. the complex, pyogenic abscess in Fig. 9.10.)

Diagnosis

The presentation is identical to that of pyogenic liver abscess. Over half of patients have no bowel symptoms at any time. Diagnostic aspiration is useful, as amoebic abscesses contain brownish-pink, thick (anchovy-sauce) pus rather than greenish or yellow-grey pyogenic pus. Amoebae are rarely seen in the pus, as they are fixed in the wall of the abscess. Bacterial culture of the pus should also be carried out, as pyogenic and amoebic infections occasionally coexist.

The mainstay of diagnosis is serology. Patients with current or recent systemic amoebiasis have positive fluorescent antibody tests (FAT). Residents of Western countries are unlikely to have positive FATs from previous amoebiasis. Residents of endemic areas may have pre-existing FAT responses, but a gel precipitin test is also positive in acute systemic amoebiasis, and quickly becomes negative when the condition is cured. A positive gel precipitin test is strongly suggestive of amoebic abscess.

Treatment

The treatment of choice is metronidazole 500 mg intravenously three times daily for 10 days. Many small and moderate-sized abscesses will heal with medical treatment alone. A few very large lesions, or ones which continue to enlarge after treatment is established, may require aspiration or drainage.

It is unusual for metronidazole to fail, but occasionally an amoebic abscess appears to resist treatment, and the addition of chloroquine 600 mg daily may induce a response. Chloroquine readily penetrates the abscess wall and has an additive effect to that of metronidazole. If the abscess still fails to resolve, the possibility of coexisting bacterial infection must be considered, and broad-spectrum antibiotic treatment may be indicated.

Complications

Rupture of an amoebic abscess

Rupture of an amoebic abscess may release pus into the peritoneum, the pleura or the pericardium causing severe inflammation, shock and collapse. Secondary abscesses themselves may require drainage. Rupture is unusual in right lobe abscesses which are accessible to percutaneous drainage. Abscesses in the left lobe of the liver are more difficult to approach. Surgical drainage should be considered in cases that are unresponsive to medical treatment and/or needle aspiration.

Secondary pyogenic infection

Secondary pyogenic infection is uncommon, but well recognized. It should be considered in all cases of poor response to metronidazole treatment. Serology usually suggests acute amoebic infection, but the pus may not have the typical appearance. Gram-negative rods, enterococci and/or anaerobes may be recovered on culture. Presumptive treatment should be commenced if there is doubt about the aetiology of the abscess (see above). Metronidazole must be given to cover both amoebae and anaerobes.

Prevention and control

See Chapter 8.

Cholangitis

Introduction and epidemiology

Cholangitis is infection of the biliary tree, often associated with surgical disorders of the gallbladder or bile ducts. Typical examples include gallstones, chronic cholecystitis, benign and malignant strictures of the bile duct or of larger hepatic ducts and, occasionally, pancreatitis. It can complicate instrumentation of the common bile duct, for instance endoscopic retrograde cholangiopancreatography, for which antibiotic prophylaxis should always be given.

Pathology

The causative organisms almost always colonize the biliary tree by ascending from the duodenum. *Escherichia coli*, *Klebsiella pneumoniae* and *Serratia* spp. are common pathogens. Enterococci and anaerobes, particularly *Bacteroides fragilis*, are also often found. *Clostridium perfringens* is an occasional cause, which may produce gas in the biliary tree. Debilitated patients and those who have received broad-spectrum antibiotics can occasionally develop candidal cholangitis.

Clinical features

The common presentation is with right hypochondrial pain, fever, often rigors, and a variable degree of jaundice. Laboratory tests show a neutrophil leucocytosis, a raised alkaline phosphatase level and a raised bilirubin.

Many patients have a past history suggestive of biliary or pancreatic disease.

Diagnosis

Contrast imaging of the biliary tree is difficult when the biliary pressure is raised by infection, but ultrasound, isotope or computed tomographic scanning may show gallstones and/or a dilated duct system.

Blood cultures should be obtained, but are not always positive. Bile may be obtained for culture, either by cannulation of the bile duct or by image-directed percutaneous aspiration from dilated intrahepatic ducts. Bile duct cannulation may carry a risk of introducing further organisms, but this should be weighed against the therapeutic advantage of draining the infected bile.

Management

This is usually commenced 'blind', with an appropriate range of antibiotics. Gentamicin or other aminoglycoside, or a broad-spectrum cephalosporin, is appropriate for coliform infections. Ampicillin or amoxycillin, plus an aminoglycoside, will provide effective treatment for enterococcal infection. Metronidazole is the treatment of choice for anaerobes. Imipenem/cilastatin or meropenem provide broad-spectrum action with an antianaerobe effect. The antibiotic regimen can be modified if the response is unsatisfactory or the bacteriological results suggest a need for change.

Surgery or drainage is indicated if purulent bile is trapped or loculated. An early surgical opinion is of value, as cholecystectomy is curative, and avoids the risk of recurrences.

Complications

The most important of these are Gram-negative bacteraemia and shock, abscess formation and erosion of the gallbladder. Clinical vigilance must be maintained, and surgical assistance should be sought if improvement is not maintained.

Colonization of the biliary tree

A patient with one or more strictures in the biliary tree can acquire colonization of the bile without symptoms of infection. At varying intervals there are transient bacteraemias, with sudden fever, rigors and then sweating and defervescence. The episodes may last anything from 15 min to several hours.

It is often difficult to obtain blood cultures at the correct time for a positive result. Some patients become debilitated and lose weight because of frequent fevers, some have clinical jaundice or laboratory signs of cholestasis. The diagnosis depends on suspicion, and imaging should be performed to identify the cause of the condition. Single strictures may be amenable to correction. Occasionally, a condition such as sclerosing cholangitis is responsible. There is then a choice of treating bacteraemias as they arise, or of attempting chemoprophylaxis. Culture of the colonized bile will identify the organism and its sensitivities.

Parasites of the liver

ORGANISM LIST

Schistosome species (*Schistosoma mansoni* most important)
Hydatids, especially *Echinococcus granulosus*
Fasciola hepatica and other flukes

Schistosomiasis

Introduction

Schistosomes are trematodes which derive their name from the large genital cleft in the body of the male, in which the longer, slimmer female lies throughout its adult life. The adult forms live in the visceral veins: *Schistosoma mansoni* in the veins of the small bowel and the portal veins of the liver, *S. japonicum* in the veins of the small bowel and *S. haematobium* in the vesical plexus. The paired schistosomes produce many fertile ova, some of which lodge in the tissues, causing an intense inflammatory response, and some of which penetrate the mucosae and are released in faeces or urine. They 'hatch' on contact with fresh water.

Free ova release motile miracidia which invade fresh-water snails. They are eventually released from the snails as cercariae, which use their motile tails to swim in the water. The cercariae die in a few hours if they are unable to attach to a host's skin and penetrate it, when they shed their tails and become invasive schistosomules. These migrate through the tissues causing an inflammatory reaction and eosinophilia, until they mature and settle as mating pairs in their favoured venous site, producing up to 2000 eggs per day (Fig. 9.12).

Many eggs are excreted from the gut or bladder to complete the life cycle. Some pass through the portal veins, to cause liver inflammation leading to fibrosis and portal hypertension. In the lungs (usually in *S. japonicum* infections) they can cause lung fibrosis and cor pulmonale. Ectopic eggs in the spinal cord or central nervous system can result in transverse myelitis or epilepsy. Eggs in the bladder wall can result in epithelial metaplasia and bladder wall fibrosis, and chronic heavy infection is associated with bladder malignancy.

Epidemiology

It is estimated that 200 million people are infected worldwide. Transmission occurs wherever there is exposure to water inhabited by infected snails. This occurs when fresh-water rivers are used for washing, or in rice-growing areas. The peak prevalence of infection is in children aged 5–15 years.

The reservoir of infection for *S. haematobium* and *S. mansoni* is humans, although domestic animals are also a reservoir for *S. japonicum*. The geographical distribution of disease follows that of the infected snails, which are species-specific. The vector snail for *S. haematobium* occurs throughout the Middle East and Africa, particularly along the Nile valley. *S. mansoni* has a wider distribution in

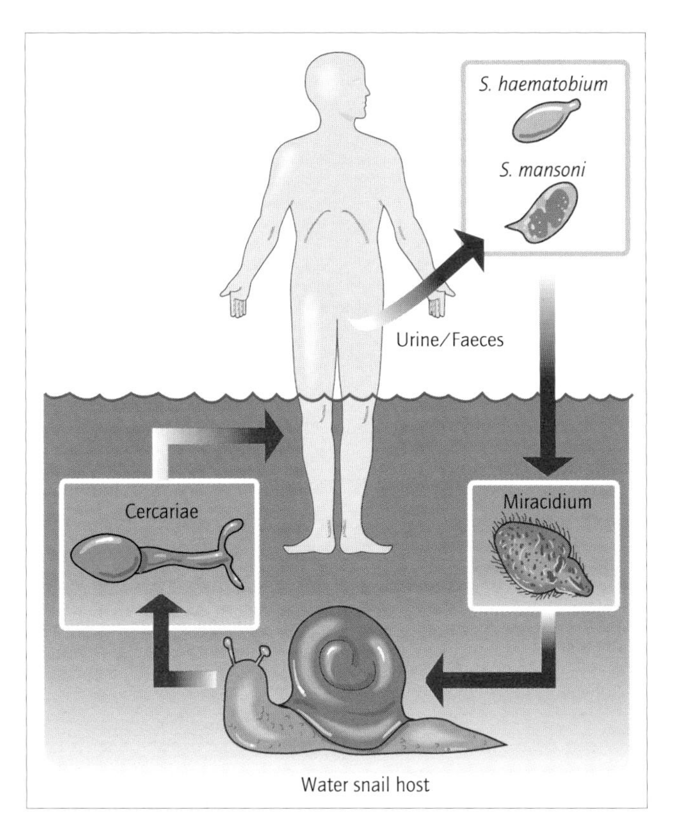

Fig. 9.12 Schistosomiasis: the life cycle of the parasite.

Africa, the Middle East, Brazil, Surinam and the West Indies. *S. japonicum* occurs in the Far East and South-east Asia.

Two patterns of disease are seen: focal (associated with small water sources) and non-focal (associated with large marshy areas, mainly in South-east Asia). In some areas, the disease incidence has increased in recent years due to the construction of reservoirs. Transmission is greatest during the rainy season.

Clinical features and diagnosis

The severity of clinical disease depends on the numbers of invading parasites. Slight infections are rarely symptomatic. More intense invasion may cause a sequence of symptoms and signs, while massive parasite loads can cause severe or even fatal disease. Heavy or repeated infections are an important cause of chronic ill health in endemic areas.

There are several stages of schistosomal infection, as detailed below.

Swimmer's itch

Swimmer's itch results directly from cercarial invasion of the skin. It happens within a few hours of exposure and clears up in 24–48 h. A rash may affect the exposed skin.

Invasive stage

The invasive stage begins after about 24 h as schistosomules migrate through veins and lymphatics to the lungs and viscera. The patient may complain of dry cough and abdominal discomfort. The spleen may be palpable and there is eosinophilia.

Toxaemic phase

A toxaemic phase, often called Katayama fever, follows significant infection after about 15–20 days. A prostrating illness includes fever, lymphadenopathy, splenomegaly, diarrhoea and eosinophilia. Severe cases can be life-threatening, especially in non-immune travellers.

Acute disease

Acute bowel disease begins after 6–8 weeks as ova are deposited in the bowel wall. The main effects are dysentery-like diarrhoea and weight loss, which may persist for as long as 6–12 months.

Chronic disease

The bowel may show fibrotic, polypoid or granulomatous lesions. The liver may become firm and large. Severe cases develop portal fibrosis and hypertension, with splenomegaly, hypersplenism and risk of variceal bleeding (hepatocellular function is usually well preserved).

S. mansoni can cause rarer effects, of which two are important: one is nephrosis, and the other is granulomatous transverse myelitis following ovum deposition in the spinal cord.

S. japonicum tends to cause more intense invasion and more widespread disease. Chronic features include cardiac involvement with cor pulmonale and alterations in the electrocardiogram.

In all types of schistosomiasis, adventitious ova may be deposited in many tissues, including the brain, but overt tissue damage is rare.

Diagnosis

Clinical diagnosis is important up to, and sometimes for 2 or 3 weeks after, the stage of Katayama fever. From then on, eggs are increasingly produced, and antibodies to soluble egg antigen begin to appear, permitting serodiagnosis. Alternatively, detection of circulating 'cathodic' or 'anodic' antigen can be detected.

Ova appear in the stool from 6–8 weeks after infection. If ova are scanty and not detected in stool, they can often be seen in rectal snips (fragments of mucosa teased off the rectal wall with a needle or biopsy forceps, and squashed on a microscope slide).

Management

Praziquantel is the treatment of choice. It is given as a single dose of 40 mg/kg, preferably after an evening meal as this minimizes abdominal discomfort or vomiting. For *S. japonicum* 60 mg/kg is given in two divided doses on the same day. Treatment must be given without delay in suspected neurological cases.

Treatment is not fully effective unless given *after* the deposition of ova has begun. Katayama fever and neurological disease responds poorly. In severe Katayama fever prednisolone is helpful in limiting inflammation and controlling tissue damage. An initial dosage of 40–60 mg/day should be reduced as rapidly as the patient's condition allows.

The faeces can be examined 1 month after treatment, when ova will still be seen, but should by then all be dead. Death is demonstrated by diluting the stool with distilled water. This stimulates any live ova to hatch rapidly, releasing active miracidia.

Prevention and control

The most effective method is to provide safe water and sanitation for communities in infected areas, together with education on avoiding contact with potentially infected water. Snail control is expensive and only likely to suceed in small focal areas of infection. An alternative approach is to reduce egg excretion from the human reservoir by chemotherapy. In high-prevalence areas, it may be necessary to treat the whole population, whereas in low-prevalence areas it is more cost-effective to treat children (who are high egg excreters) and clinically affected adults. Tourists to infected areas should be advised of the risks of swimming in fresh water.

Hydatid disease

Introduction and epidemiology

Hydatid disease is the illness caused by the cystic phase of the small tapeworms *Echinococcus granulosus*, whose primary host is the dog (particularly sheepdogs), and whose cysts affect sheep. The rarer *E. multilocularis* tapeworm affects foxes (but can exist in dogs), and cysts are found in small rodents on which the foxes predate.

E. granulosus is the commonest cause of human disease. Tapeworm eggs passed by infected dogs are often deposited in fields, where they are consumed by their natural secondary host, the sheep. The eggs migrate from the gut to parenchymal organs, usually the liver and occasionally the lung, where they develop into cysts containing several tapeworm scolices. If the cysts are eaten, the scolices are released and attach to the bowel wall, allowing the establishment and maturation of a tapeworm in the dog's gut.

Human disease occurs when tapeworm ova are ingested by humans, often as a result of close contact with a working or pet dog. Cysts then develop, and may enlarge for many years before they become clinically apparent.

Clinical features

E. granulosus infection usually causes a solitary cyst in the liver, or occasionally the lung. The enlarging cyst may produce abdominal pain or swelling, or may obstruct the biliary system, leading to jaundice. Lung cysts can cause partial bronchial obstruction, with repeated chest infections. Rupture of cysts is rare, and more often related to attempted surgery than spontaneous breakdown. Eosinophilia is not seen, unless leakage of cyst contents causes an allergic reaction.

E. multilocularis is more invasive than *E. granulosus*. Its cyst wall cells can invade locally, and also metastasize to distant sites, including the brain.

Diagnosis

Cysts of *E. granulosus* may be visible as thin-walled, fluid-filled structures in plain radiographs, especially of the lungs, and in liver ultrasound scans. Computed tomographic scans may reveal daughter cysts, and on magnetic resonance scans magnified images may demonstrate scolices within them.

ELISA tests, used to detect cyst-associated hydatid antibodies, are highly sensitive. Antigen detection tests can be useful in making an acute diagnosis and, more importantly, in monitoring the response to chemotherapy.

Management

Albendazole is effective in treating *E. granulosus*. It is given in a dose of 400 mg twice daily for 4 weeks, and this may be repeated up to four times, with intervals of 1 or 2 weeks between courses. Liver and renal function should be checked during treatment. As the cysts die, they may become painful and inflamed for a time. Analgesics are then helpful. Corticosteroids are not usually required.

E. multilocularis does not respond fully to albendazole; lifelong suppressive therapy is often necessary.

Surgery without prior chemotherapy is undesirable, as rupture of the cyst may cause an anaphylactic reaction, and can also seed the peritoneum and other tissues with daughter cysts (Fig. 9.13). A common approach is to prepare patients for operation with albendazole for at least 2 weeks. Praziquantel is given peroperatively and for 2 weeks following to act against the protoscolices and reduce the risk of secondary seeding.

It is impractical to excise multiple cysts, or some of those in the brain. Albendazole offers a good chance of disease regression in such cases, and may be alternated with praziquantel 20 mg/kg daily for 1 week. Recent trials suggest that aspiration of cysts can safely be performed after chemotherapy, avoiding the need for major surgery in frail patients or those with multiple cysts.

Liver flukes

Introduction

Liver flukes are trematodes which live in the intrahepatic bile ducts. Their life cycle includes multiple hosts. *Fasciola hepatica* is the only liver fluke endemic in the UK. It has fresh-water snails as its second host. Cercariae migrate on leaving the snail's body, and become attached to vegetation, existing as non-motile metacercariae until eaten by humans or sheep. Wild watercress is the commonest source of human infection. In oriental countries *Clonorchis (Opisthorchis) sinensis* is common. This fluke affects humans, dogs, pigs, cats and even rodents. Its next host is the snail; cercariae then invade fish (especially fresh-water carp) or shrimps which are again eaten by humans and other animals.

Ingested metacercariae excyst in the duodenum, penetrate the bowel wall, cross the peritoneum and eventually migrate through the capsule and parenchyma of the liver to reach the bile ducts.

Clinical features

Acute

The acute presentation is with fever, abdominal pain, liver tenderness and eosinophilia occurring as the parasites migrate and invade the liver. Symptoms vary from mild to severe, depending on the parasite load, and can last for 3 or 4 months. Episodes of obstructive jaundice may occur as flukes settle in the biliary tree.

Chronic

Chronic fascioliasis reflects persisting irritation of bile ducts. Upper quadrant pains, fevers, cholangitis and persisting eosinophilia may occur. Localized stasis and infection can produce liver abscesses. Over a period of years the flukes die, being passed in the faeces or becoming calcified in the liver.

Diagnosis

Eggs are passed in the faeces once adult flukes are established in the liver. They are often scanty and are best demonstrated by concentration techniques. Serological tests are positive early in *Fasciola* infections, but are not helpful in early *Clonorchis* cases.

Fig. 9.13 Hydatid disease: computed tomographic body scan showing: (a) two thin-walled cysts in the right lung (arrows); (b) several cysts in the liver (arrows); and (c) a large cyst in the spleen (arrow). The patient's abdominal pain and the size of the cysts improved dramatically with albendazole therapy.

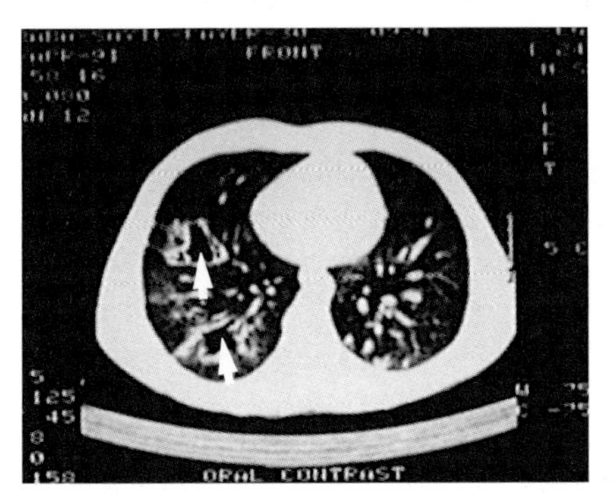

(a)

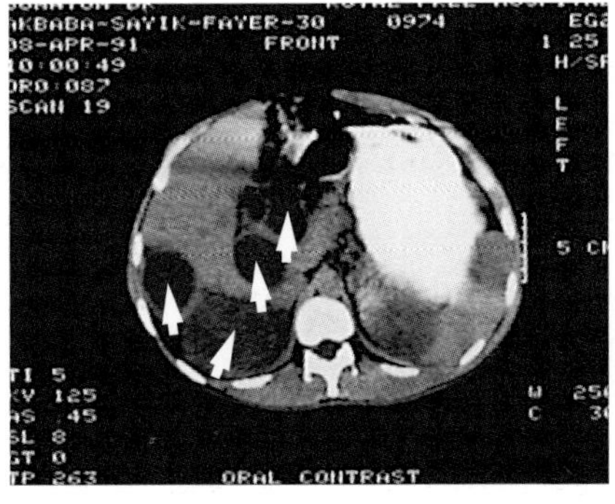

(b)

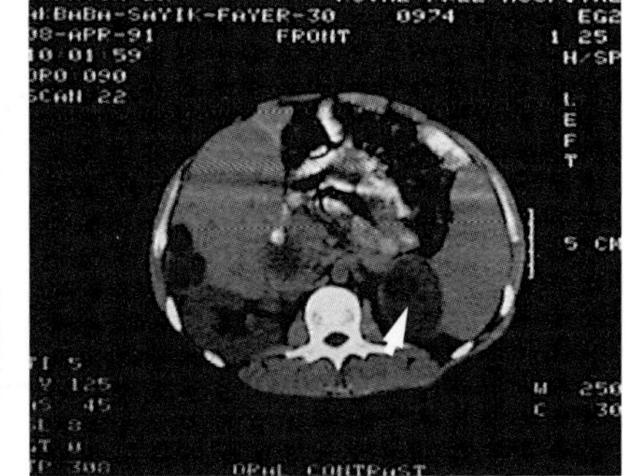

(c)

Management

The treatment of choice is praziquantel 20 mg/kg daily for 3 days. The drug is best taken at night after food to avoid abdominal discomfort and nausea in waking hours. Treatment may need repeating after an interval if egg excretion continues.

Prevention and control

Fascioliasis can be prevented by avoiding consumption of wild watercress, especially if grown on land grazed by sheep. The use of sheep faeces for fertilization of water plants should be avoided. Snail control is technically feasible, but not economically justified.

Case 9.1: Hepatitis B outbreak in an alternative medicine clinic

History and diagnosis: A case of acute hepatitis B was notified in a middle-aged woman whose only risk factor was having received injections at an alternative medicine clinic 4 months previously. The patient's daughter, who had visited the clinic at the same time as her mother for treatment of acne, also developed acute hepatitis B.

Question: How could the consultant in communicable disease control determine whether transmission of hepatitis had occurred in the clinic?

Further enquiries and investigation: The clinic offered a variety of alternative therapies, including autohaemotherapy, in which patients receive an injection of their own blood after it has been mixed with saline solution.

Several possible routes of transmission were identified at the clinic, the most likely being the use of multidose bottles, into which the needles of successive syringes were introduced, for several of the procedures. The immediate control measures were to prevent the use of multidose bottles in the clinic, to give advice on sharps and clinical waste disposal, and to arrange for all clinic staff to be tested for hepatitis B virus (HBV) and HCV.

Question: How can it be determined whether other patients have been at risk?

Further investigation: A 'look-back' investigation was conducted, in which all patients were identified and followed up who had attended the clinic in the 6 months before the first case developed symptoms (i.e. the maximum incubation period for hepatitis B). The look-back period was extended to 1 year when an earlier case of hepatitis C was identified in a clinic attender.

Two of the practitioners in the clinic had evidence of HBV infection and one was HbeAg positive, i.e. highly infectious. Both were required to cease practice by the Health and Safety Executive.

The look-back exercise traced 353 patients who had received autohaemotherapy, of whom 33 subsequently developed acute hepatitis B. Thirty were infected with the same virus strain, suggesting a common exposure at the clinic. This is the largest reported iatrogenic outbreak of hepatitis B in the UK.

10 Infections of the Urinary Tract

Introduction, 215

Pathogenesis, 215

Symptoms, 216

Diagnosis, 216
Three-glass test, 216
Other useful investigations, 216
Bacteriuria and its relationship to infection, 217
Microbiological diagnosis of urinary tract infection, 217

Urethritis and the urethral syndrome, 219
Introduction, 219
Epidemiology, 220
Clinical features, 220
Diagnosis, 220
Management, 220

Cystitis and ascending urinary infections, 221
Introduction, 221
Clinical features, 221
Diagnosis, 221
Management, 221
Urinary tract infections in children, 222
Urinary tract infections in men, 222
Urinary tract infections in pregnancy, 223

Reflux nephropathy (chronic pyelonephritis), 223

Bilharzia, 223

Acute epididymo-orchitis, 223

Bartholin's abscess, 224

Introduction

Urinary tract infections are common in most communities. Both the collecting system of the urinary tract and the kidneys themselves may become infected.

The collecting system is usually bacteriologically sterile, but the urethra opens to the exterior at the perineum, a site rich in potentially pathogenic flora. Protection from ascending infection is derived from: the regular flow of urine, diluting and expelling pathogens; the mucosal defence mechanisms of the urinary tract; the antibacterial properties of the urine itself; and the integrity of the sphincters separating the urethra from the bladder and upper tract. Disturbance of any of these mechanisms predisposes to ascending infection. Examples include the following:

1 Mechanical abnormalities of the urinary tract, which may cause stagnation of urine, with bacterial replication in residual pools of urine.

2 Disruption of the urothelium, allowing the establishment of epithelial infection.

3 Abnormal chemical constituents in urine, e.g. glucose, which may encourage the survival and replication of organisms.

4 Foreign bodies, such as stones, tumours, schistosome eggs and associated granulomata, which may become colonized with pathogens and act as a reservoir of infection.

5 Loss of sphincter function (including that due to indwelling catheters), which destroys an important barrier to ascending colonization of the bladder.

Pregnancy causes dilatation of the collecting system, reduced motility and a large volume of stagnant urine, predisposing to ascending infection.

The kidneys have a complex glandular structure of tubules and blood vessels. They receive about one-third of the cardiac output and are therefore susceptible to blood-borne pathogens, as well as to damage from ascending infections. Previously healthy kidneys may show only a temporary decline in excretory function during bacteraemic disease, but localized infection in the form of renal abscesses can also occur. This is more likely if the kidney has a pre-existing abnormality such as cysts, scarring or infarcts.

Pathogenesis

Uropathic strains of *Escherichia coli* possess pili. This subject is discussed in more detail in Chapter 2. Type I mannose-sensitive pili appear to be important in strains colonizing the bladder. Type P pili favour colonization of the kidney. These pili are coded for by *pap* genes (**p**yelonephritis-**a**ssociated **p**ili). Although P pili are divided into many antigenic types they all bind to the same receptor: α-D-Gal-(1,4)-β-D-Gal (globobiose). Gene expression changes in response to a number of stimuli including temperature and glucose concentration.

Uropathogenic strains also adhere by non-fimbrial mechanisms. Examples include afimbrial adhesins 1

(AFA 1), and the Dr adhesin. Both of these recognize the Dr blood group antigen.

Bacteria sometimes undergo phase variation after attachment to the host's epithelium, changing their expressed antigens to those less readily recognized by phagocytes.

Inflammation is mediated by lipopolysaccharide (endotoxin), which appears to act synergistically with P pili. There is some evidence that uropathogenic strains also produce exotoxins: designated RTX haemolysins.

Symptoms

These are classically divided into symptoms of urethral irritation, bladder irritation or upper tract symptoms originating in the pelvicaliceal system and/or the kidney. Obstruction or severe inflammation of a ureter may also produce typical ureteric pain, felt in the flank and radiating to the lower abdomen and perineum.

Urethral symptoms

Urethral symptoms are burning or stinging at the meatus and in the perineum, with a continuous desire to micturate, causing marked frequency. There is often severe dysuria.

Bladder symptoms

Bladder symptoms include suprapubic aching, often relieved by micturition. Suprapubic tenderness is common. Urethral symptoms often coexist.

Pyelonephritis

Pyelonephritis is represented by loin pain, with tenderness in the renal angle on the affected side. Rigors are common. An obstructed or hydronephrotic kidney may produce a tender palpable or ballottable mass in the upper abdomen.

In clinical practice, symptomatic localization of disease correlates poorly with the true site and extent of infection. Any or all of the classical symptoms may be present in infections of various parts of the urinary tract, and many urinary infections occur without specific symptoms. Spurious symptoms are common, particularly in children. These include failure to thrive, episodic nausea and vomiting, and symptomless fever. Without a high index of suspicion, persisting infection can cause impaired health or, in infants, long-term renal damage. Acute urinary infections may present with watery diarrhoea, spurious meningism or febrile convulsions. Urinary infection can exist in the absence of fever or significant neutrophilia.

Non-specific symptoms of urinary tract infections

1 Failure to thrive.
2 Episodic nausea.
3 Diarrhoea.
4 Meningism.
5 Pyrexia of unknown origin.
6 Febrile convulsions.

Diagnosis

The urinary tract above the urethra is normally bacteriologically sterile, but the urethral meatus and surrounding perineum are colonized with a mixture of skin and bowel flora. In women vaginal flora or pathogens may contaminate the urethra.

Three-glass test

In the three-glass test urination is commenced into the first container; when the flow is established, the mid-stream urine (MSU) is directed into the second container; the last few millilitres (terminal urine) are passed into the third container, making an effort to expel all urine from the bladder and urethra.

The first urine passed contains debris, cells and organisms from the urethra and often contains strands of mucus if urethritis is present. The MSU contains bladder urine. A normal specimen appears clear and transparent, while cloudiness indicates the presence of cells, bacteria or crystals. The final sample contains matter from the clefts of the trigone or from glands adjacent to the pelvic urethra. Mucus strands are often seen in cases with prostatitis. Schistosome ova from the bladder wall are best recovered from terminal urine (Fig. 10.1).

Other useful investigations

A number of screening techniques have been developed for the rapid detection of urinary tract infection. These techniques usually use dipsticks. Glucose can usually be detected in overnight fasting urine in normal subjects. With urinary tract infection, glucose is typically metabolized by bacteria and is undetectable. Inflammation may lead to leakage of blood cells and protein from the urinary epithelium that can be detected by dip-strip tests for blood and protein. Many urinary pathogens catalyse the reduction of nitrate to nitrite and this can also be detected by a dip test.

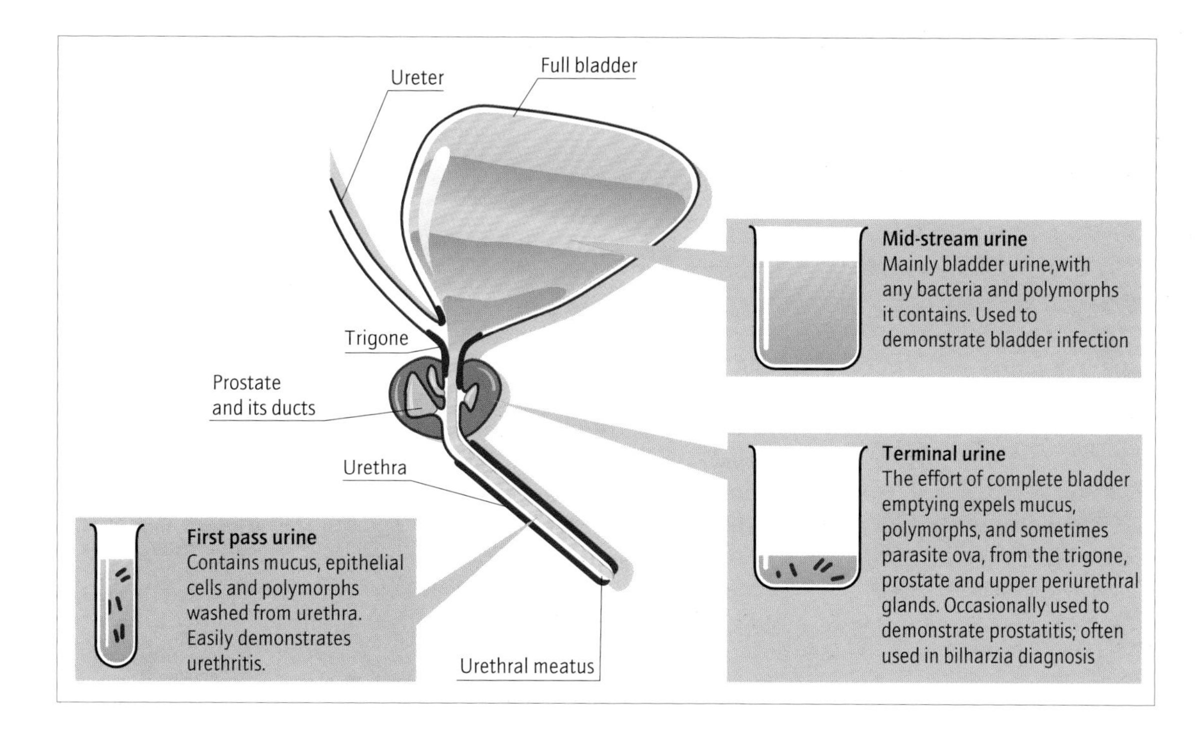

Fig. 10.1 The three-glass procedure for examining urine.

Large numbers of white cells may indicate urinary tract infection: but white cells are a normal part of the urine. Disease is indicated only when they are present in an excess (>10/mm^3). Excess white cells may appear in urine as a result of fever, exercise or contamination from a vaginal discharge. Absence of white cells does not exclude infection as they rapidly lyse in acid urine. Excess white cells in urine can be indicated by test strips that measure the activity of leucocyte esterase. Positive dip-test results for nitrite and leucocyte esterase can provide a rapid presumptive diagnosis of infection. Clinical trials have shown this approach to be both sensitive and specific.

Bacteriuria and its relationship to infection

Although bladder urine is usually sterile, bacterial colonization of the bladder is common in girls and women, and also occurs in boys and men. Because both colonization and asymptomatic infection are frequently encountered, criteria are required to decide whether infection is truly present.

The most helpful test is semiquantitative bacterial culture. Infection is likely to be present if colony counts exceed 10^8/l in women. A concentration of less than 10^7/l suggests that infection is unlikely, while intermediate concentrations require further investigation. The different anatomy of the lower urinary tract in men affects this principle. Infection is considered likely in men and boys when the bacterial count in the MSU is 10^6/l or greater. The purity of the culture is also important as a mixed culture is more likely to indicate a contaminated specimen.

When the significance of bacteriuria is in doubt, urine can be obtained from the bladder by catheterization using an aseptic technique. Suprapubic aspiration of urine is possible in infants, in whom the full bladder is an abdominal organ. Numerical criteria are not applied to these specimens, in which any bacteria are taken to indicate infection.

Microbiological diagnosis of urinary tract infection

Handling the specimen

After collection the specimen should be processed rapidly or refrigerated to prevent multiplication of any contaminating organisms. A rigid view of results must not be taken when interpreting the results of semiquantitative culture; the specimen may have been taken later in the day, so that organisms have not had the opportunity to multiply in the bladder. Early morning urine is more concentrated than specimens taken later. Also, in specimens delayed in transit, organisms which were present in non-significant numbers may have multiplied sufficiently to give a positive result. Despite the potential pitfalls, examination of MSU is the most useful method for investigating urinary tract infection.

The laboratory diagnosis of urinary tract infection is divided into three stages: (i) microscopy; (ii) semiquantitative culture; and (iii) sensitivity testing. Each of these has a contribution to make to clinical diagnosis.

Microscopy

The presence of pyuria (more than 10/mm^3 neutrophils in the urine) is useful evidence of infection when accompanied by bacteriuria In most laboratories, an unspun specimen of urine is examined either in a counting chamber (disposable chambers are commercially available) or by using a microtitration tray and an inverted microscope.

A number of non-infectious conditions also cause pyuria; these include urinary stones, tumours of the urinary tract and reactions to drugs and chemicals such as cyclophosphamide. Pyuria in apparently sterile urine can also be caused by tuberculosis, by rarer conditions such as brucellosis and by chlamydial infections, and when bacterial growth has been suppressed by antibiotic therapy.

Causes of sterile pyuria

1 Urinary stones.
2 Urinary tract tumours.
3 Drug reactions.
4 Tuberculosis.
5 Brucellosis.
6 Chlamydial infections.
7 True urinary tract infection partly suppressed by antibiotics.

Bacteria may be seen readily in unstained uncentrifuged preparations of urine. This is often associated with significant bacteriuria and some laboratories will inoculate a primary sensitivity plate (see below). This approach is justified on the basis that the smallest number of bacteria that can be seen is 10^{4}CFU/ml, which is similar to the significance level.

Microscopy also enables the quality of the specimen to be evaluated. The presence of epithelial cells indicates skin contamination, suggesting that the specimen should be repeated, or the results interpreted with caution.

In addition, white cell casts may indicate renal infection, and red cells or red cell casts may be seen in glomerulonephritis or endocarditis.

Culture

Many methods have been developed for semiquantitative culture of urine. The simplest of these is to inoculate a standard loop of 1 µl. The presence of more than 100 CFU will be the equivalent of >10^{5}CFU/ml (10^8 / l). It is usual to report cultures with 10^{4}CFU/ml as indicating significant infection (see above). A method using filter paper strips to deliver a standard inoculum has been described (Fig. 10.2).

In all culture systems it is important to inhibit the motility of *Proteus* spp. Itself a common pathogen of urine, its swarming growth may obscure other organisms (Fig. 10.3). Bile salts effectively inhibit swarming, and these are incorporated into MacConkey agar. Most urinary pathogens will grow in the presence of bile salts, but some fastidious organisms (see p. 39) are inhibited. Electrolyte-deficient media such as cysteine lactose electrolyte-deficient (CLED) medium prevent swarming and support the growth of more fastidious organisms.

Ascribing appropriate significance to a culture result is important. Most laboratories use a protocol to do this.

The adequacy of the specimen is indicated by the absence of epithelial cells on microscopy with significant numbers of epithelial cells indicating meatal contamination. Excess neutrophils support a diagnosis of infection.

Most important is the number of organisms (see above) and the purity of culture: organisms present in lower numbers may be significant if present in pure culture, whereas larger numbers of enterococci and coliforms present together probably indicate a contaminated specimen and the need for a repeat specimen.

The identity of the organism is also important. Isolation of an organism usually present on the skin but also recognized as a common urinary pathogen may indicate the need for a repeat specimen.

Indicators that bacteriuria is significant

1 Adequate specimen (no epithelial cells).
2 Excess neutrophils.
3 Purity of culture.
4 Number of organisms (appropriate for type of urine specimen).
5 Identity of the organism.

Automated methods

Several automated methods have been described which rapidly identify specimens containing a significant number of organisms for subculture. Growth detection methods are based on changes in optical density, or electrical impedance. Subculture is then performed by conventional techniques.

A semiautomated system uses a series of microtitration trays containing different identification and sensitivity test media. Specimens are inoculated by multipoint inoculator and the plates incubated at 37°C for 18 h. The changes in the media are read by a computer-driven

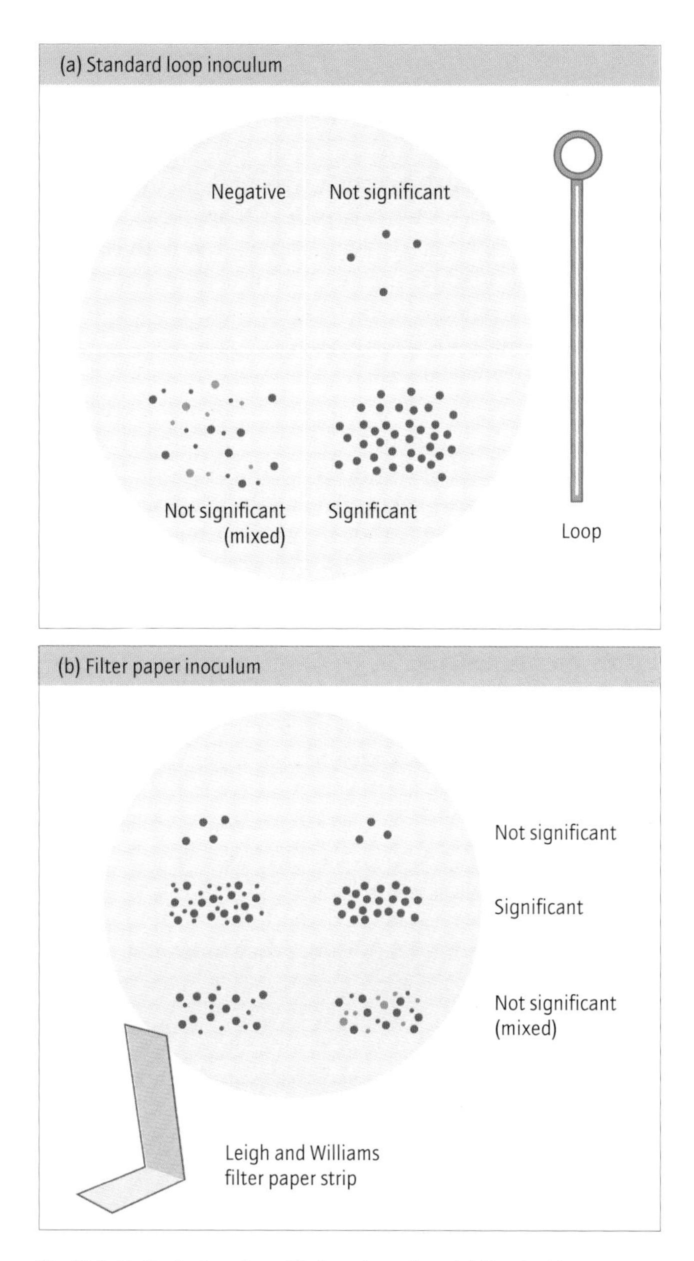

Fig. 10.2 Methods of semiquantitative urine culture. (a) Standard loop inoculum; (b) filter paper inoculum: the filter paper is dipped into the urine and then pressed on to the agar plate.

image analyser. Contaminated cultures are detected by visual inspection of the reactions of test media included in the trays. As isolation, identification and sensitivity testing occur in parallel rather than sequentially, a final result is usually available within 24 h.

Bacterial identification and sensitivity

For most clinical situations, simple or presumptive identification is all that is necessary, and laboratory methods uti-

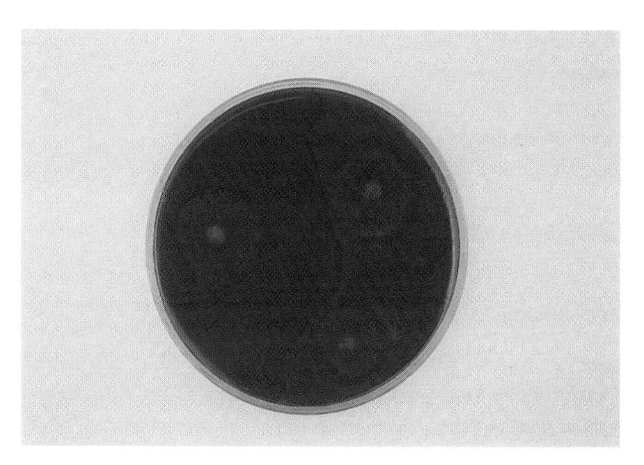

Fig. 10.3 *Proteus* sp. swarming on a blood agar plate after overnight incubation; the same inoculum on a MacConkey's agar plate forms discrete colonies.

lized usually reflect this. Thus, a lactose-fermenting organism capable of indole production may be labelled a coliform. Urease-positive organisms will be reported as *Proteus* sp., and enterococci diagnosed on the basis of their characteristic colonial morphology and hydrolysis of aesculin. Automated and semiautomated methods produce a species or genus diagnosis by use of a computer database.

Sensitivity tests should be performed on all potentially significant isolates. Antibiotics used in the treatment of urinary tract infection include trimethoprim and co-amoxiclav, cephalexin, nalidixic acid, nitrofurantoin, and 4-fluoroquinolones such as ciprofloxacin. In addition, inpatient specimens should be tested against the injectable penicillins (e.g. azlocillin), cephalosporins (e.g. cefotaxime) and the aminoglycosides.

Urethritis and the urethral syndrome

Introduction

Infection of the urethra often occurs as part of a more extensive urinary tract infection. However, the urethra is also often involved in sexually transmitted diseases. In women, it is also often affected by genital infections such as candidiasis and non-specific vaginitis.

ORGANISM LIST

Herpes simplex
Escherichia coli
Staphylococcus saprophyticus (in young women)

Other Enterobacteriaceae
Neisseria gonorrhoeae
Chlamydia trachomatis
Gardnerella vaginale
(Lactobacilli)
(Diphtheroids)
Candida albicans

Epidemiology

This varies according to the infecting organism. Urethritis occurring as part of a general urinary tract infection is commoner in females than males and increases with age. More than two-thirds of urinary tract infections are due to *E. coli*.

Urethritis associated with sexually transmitted disease is usually due to *Chlamydia trachomatis* or *N. gonorrhoeae*, less commonly herpes simplex or *Trichomonas vaginalis* (see also Chapter 15). Males are clinically affected more commonly than females because of the higher preponderance of asymptomatic infections among females and the higher incidence of infection in male homosexuals. The incidence of gonorrhoea and trichomoniasis declined throughout the 1980s, especially among older patients who modified their sexual behaviour in response to the acquired immunodeficiency syndrome (AIDS) epidemic, but infection rates are now increasing, especially in younger age groups. The incidence of both genital herpes and *Chlamydia* infection has been steadily increasing since the early 1980s in both sexes but especially among females for genital herpes.

Clinical features

The main features are frequency and urgency of micturition, with burning dysuria. In men, particularly those with sexually transmitted diseases, there may be a mucoid or mucopurulent urethral discharge, often most noticeable first thing in the morning.

Slightly more than half of all cases are associated with pyuria and a diagnostic growth of bacteria from the MSU. In most cases the pathogen is a coliform, but in women it may be *Staphylococcus saprophyticus*.

Both chlamydial infection and gonorrhoea affect the urethra, and on occasions both conditions coexist. In women with sexually transmitted urethritis there is often an associated vaginal infection and there may be a discharge. The meatus of both the male and female urethra may be affected by genital herpes simplex, with typical vesicular lesions, and painful inguinal lymphadenopathy may occur.

Urethral syndrome

The urethral syndrome in women is the presence of symptoms without significant bacterial growth or demonstrable vulvovaginal infection. About half of sufferers have mid-stream bacteriuria of less than 10^7/l. Surveys show that these women have positive growth in catheter specimens of bladder urine, and probably have low-grade urinary infection which can be treated conventionally. About half of the remainder will have positive investigations for *Chlamydia*.

Among the rest fastidious organisms such as lactobacilli or *Gardnerella* are sometimes found. The clinical significance of these findings is uncertain, but appropriate treatment is sometimes helpful. Finally, *Candida albicans* should be treated if vulval swabs are positive.

There are some non-infectious causes of urethral irritation. In middle-aged women urethral caruncle produces a small, red swelling at the meatus. This responds to surgical treatment. Irritants in the urine may include drugs such as rifampicin, warfarin and cyclophosphamide. Dietary irritants such as cayenne or chilli may also cause urinary symptoms when excreted.

Diagnosis

This depends on a combination of adequate history, appropriate urinary specimens and swab tests for viral, bacterial and chlamydial cultures when indicated.

Management

Many patients can be managed with simple antimicrobial chemotherapy, as for urinary tract infection. *Staphylococcus saprophyticus* usually responds to a 5–7-day course of flucloxacillin. Chlamydial infection responds to tetracycline or erythromycin given in conventional doses for 2 weeks. Other organisms may be treated according to laboratory sensitivity tests.

Relapse or failure to respond should prompt a search for predisposing factors such as vaginal prolapse or atrophic vaginitis, surgical conditions such as caruncle, or carriage of an organism by a sexual partner (even *Candida* infections can recur for this reason).

Reasons that urethral symptoms may fail to respond to antimicrobial therapy

1 There is a predisposing factor (atrophic vaginitis, urethral caruncle, small anterior prolapse).
2 There is chemical irritation from drugs or strong spices excreted in the urine.
3 There is herpes simplex infection.

Cystitis and ascending urinary infections

ORGANISM LIST

Adenoviruses
Escherichia coli
Staphylococcus saprophyticus in young women
Klebsiella pneumoniae
Other coliforms
Proteus mirabilis
Other Proteaceae
Candida albicans

Introduction

Cystitis is infection of the bladder and often involves the upper urinary tract. It is common, but affects the sexes and age groups differently, depending on the prevalence of bacteriuria in each group. Thus, among infants, boys are more commonly affected, particularly those who are uncircumcised. Among children and young adults, females outnumber males by 10 : 1. In the older age groups the occurrence of infection is favoured by prostatism in men, and by incontinence in the frail or disabled of both sexes.

Clinical features

It is impossible to distinguish clinically between bladder infection and that which involves the ureters and renal calyces. Symptoms of frequency, urgency, dysuria and suprapubic discomfort are common. The urine is often cloudy, pale pink or frankly blood-stained. Proteinuria and microscopic haematuria are the rule. Dip tests for nitrite and leucocyte esterase are positive. Spurious or non-specific symptoms are common, especially in young patients.

Symptoms often resolve spontaneously, but this may conceal the presence of continuing low-grade infection. Fever may be slight or absent; neutrophilia is not the rule in simple urinary infections. Continuing infection may be harmful in patients at risk of renal damage or of severe exacerbations of infection (see below).

When infection affects the renal parenchyma there is often pain and tenderness in the renal angle, upper abdomen or loin. Nausea is common; vomiting or loose stools may occur. Abscesses, either renal or perinephric, may form; warning signs are increasing pain, fever and neutrophilia.

Some urinary tract infections are associated with bacteraemia. High fever and rigors are signs of this, and endotoxaemic shock may follow. Bacteraemia is made more likely by foreign bodies in the urinary tract, and often complicates instrumentation or catheterization performed while untreated infection exists.

Diagnosis

A high index of suspicion is needed to detect less obvious cases. Laboratory examination of the urine should be requested whenever infection is a possibility and the blood urea or serum creatinine should be checked. Feverish patients in hospital should always have two blood cultures before chemotherapy is commenced.

Urine examination is sometimes negative when a patient has strongly suggestive signs of urinary infection, or when infection has failed to respond to initial therapy. Microbiological reasons for this have already been discussed. However, pus and organisms from an infected kidney may be trapped behind an obstruction in the ureter. This prevents microbiological diagnosis, and prevents drainage of the loculated infection. A renal excretion scan or intravenous urogram will show reduced excretion from the affected kidney, which may also be swollen or hydronephrotic.

Management

Mild or uncomplicated infections are usually caused by Gram-negative rods. A suitable choice of antibiotic for initial treatment would be nitrofurantoin, trimethoprim, cephalexin or co-amoxiclav. All of these agents are concentrated in the urine, and reach adequate therapeutic levels when given orally. Flucloxacillin is usually effective against *S. saprophyticus*. *Klebsiella* and some other unusual urinary pathogens are uniformly resistant to ampicillin. Very broad-spectrum agents are rarely required, and should only be given when indicated by laboratory test results.

Oral treatment of urinary tract infection

1 First choices: cephalexin 250–500 mg 6-hourly (child 25–50 mg/kg daily in three divided doses); *or* trimethoprim 200 mg 12-hourly (child 2–5 months, 25 mg; 6 months–5 years, 50 mg; 6–12 years, 100 mg, all twice daily; contraindicated in pregnancy and in neonates).

2 Second choices: co-amoxiclav (as amoxycillin) 250 mg 8-hourly (child up to 10 years, 125 mg 8-hourly); dose may be doubled in severe infection *or* amoxycillin 250 mg 6-hourly (if organism sensitive); *or* nitrofurantoin 50 mg 6-hourly with food (child over 3 months, 3 mg/kg daily in four divided doses). Nitrofurantoin has many side-effects, including occasional severe nausea. Co-trimoxazole carries a risk of severe skin reaction or neutropenia, especially in the elderly. It should only be used if there is no satisfactory alternative.

All of the above regimens for 7 days.

Severe infections with fever, neutrophilia or shock should be treated initially with parenteral antibiotics. Broad-spectrum cephalosporins such as cefotaxime or ceftriaxone are useful and have low toxicity. When fever is controlled, treatment may be continued orally, with any effective antibiotic.

The duration of treatment is controversial. Although single-dose amoxycillin treatment is possible, almost half of infecting organisms are now resistant to this drug.

Infection behind an obstruction usually demands surgical treatment. Debris or a ureteric stone can often be removed via a ureteric catheter, but if the obstruction is impassable the kidney must be drained by nephrostomy until the infection is controlled.

Prophylaxis of urinary tract infection

Some individuals suffer repeated urinary infections and often have continuing bacteriuria between attacks. Low-dose antimicrobial chemoprophylaxis will often suppress bacteriuria, improving both health and quality of life. Temporary prophylaxis is indicated while awaiting treatment of a predisposing condition.

Prophylaxis of urinary tract infection

1 First choice: trimethoprim 100 mg at night (child 1–2 mg/kg); *or* nitrofurantoin 50–100 mg at night (child over 3 months, 1 mg/kg).

2 Alternatives: cephalexin 250 mg at night (child 125 mg at night); *or* co-amoxiclav 250 mg at night (child 125 mg at night).

Urinary tract infections in children

Urinary infections in children are often associated with abnormalities of the renal tract. Examples include urethral valves in boys, bladder outflow obstruction, duplex drainage systems and stones (which may themselves be associated with metabolic disorders such as renal tubular acidosis). Many children have ureteric reflux, which allows infection to ascend to the kidneys. This is thought to cause renal scarring. As 20% of all cases of chronic renal failure have evidence of scarring, it is important to detect and treat childhood infections. A consensus view of experts in childhood infections may be summarized as follows.

1 Detect the infection. Urine should be collected and cultured from all feverish children, unless there is an obvious alternative cause.

2 Treat promptly. Antibiotic treatment should be commenced as soon as specimens have been obtained (it can be modified later, if necessary). At this stage all children should have a renal ultrasound examination to outline the kidneys, and a plain abdominal radiograph to detect stones.

3 Maintain prophylactic chemotherapy (see above) and investigate further. Once infection is controlled an excretion scan should be performed to demonstrate excretory function and to outline the collecting system (children over 1 year of age may have an intravenous urogram instead).

All infants, and older children with abnormalities of the other tests, should have a cystourethrogram to search for ureteric reflux. Cystourethrograms may be direct (contrast or isotope is introduced into the bladder) or indirect (the bladder and urethra are imaged using contrast or isotope which has been excreted via the kidneys). Direct studies are necessary to demonstrate minor reflux confined to the lower urethra. Indirect studies are useful for follow-up of major reflux, and avoid the need for catheterization.

Prophylactic antibiotics may be discontinued if bacteriuria has ceased and all investigations are normal. Repeat urine cultures should be performed every 3 or 4 months until the age of 2, following which culture need only be performed if infection is suspected.

Urinary tract infections in men

Urinary infections are relatively common in infant boys, who are four times more likely than girls to have bacteriuria, and in elderly men, when prostatism predisposes to stagnation of residual urine in the bladder. In later childhood and adult life, males are only one-tenth as likely to have bacteriuria as are females.

Men with urinary infections should therefore be investigated, as the likelihood of urinary tract abnormality is high. Imaging of the renal pelvis, calyces and ureters should be performed, to demonstrate deformities or reduplications of the collecting system. If there is evidence of back pressure the possibility of partial urethral obstruction should be considered. This may be due to bladder exit obstruction or to urethral valves persisting into adulthood.

If no pyogenic organisms are found on urine culture, the patient should be investigated for sexually transmitted or other genitourinary infection, even if urethral discharge is not evident.

Urinary tract infections in pregnancy

This is one of the commonest complications of pregnancy. Nearly half of all women who have bacteriuria detected at the first antenatal assessment will develop overt infection. Even asymptomatic bacteriuria should therefore be treated. A further discussion of infections in pregnancy is found in Chapter 17.

Reflux nephropathy (chronic pyelonephritis)

This is thought to be the result of damage to the growing kidney from ascending urinary tract infections in childhood. As in chronic bronchitis a mechanical or anatomical abnormality predisposes the normally sterile organ to bacterial colonization. Recurrent infections then cause further damage and impair function.

Reflux nephropathy, with scarring of the kidney, is associated with a 10% risk of renal failure in later adult life and with a 20% risk of hypertension. Each successive urinary infection causes a dip in renal function which never quite returns to its previous level. About half of all renal scars are present when the first urinary tract infection is diagnosed.

Treatment is directed at controlling infections and preventing recurrences. Prophylactic antimicrobial chemotherapy may have a place. Recurrent infections are often with the same organism, having the same antimicrobial sensitivities. Nitrofurantoin is a useful prophylactic drug, as it is concentrated in the urine while attaining only very low levels in the blood and other systems. Its general side-effects are therefore few (confined to nausea or mild gastrointestinal symptoms). Trimethoprim also produces few side-effects, and its spectrum is such that it causes little significant disturbance of the bowel flora.

Surgery is important in relieving obstruction and removing stones. Its place in correcting ureteric reflux is more controversial, but many would advocate surgery for gross reflux through a patulous ureteric orifice.

In established reflux nephropathy the treatment of hypertension is important, as this can damage the kidneys further, as well as causing hypertensive vascular disease.

Bilharzia

In endemic areas, bilharzia is a common cause of chronic urinary tract disease. The disease occurs throughout Africa and the Middle East, but especially along the Nile valley (see Chapter 9). The adult *Schistosoma haematobium* resides in the venous plexus of the bladder wall. The spiked ova penetrate the mucosa to be excreted in the urine. Haematuria is common in heavy infections which cause severe inflammation. Inflammatory granulomata, fibrosis and calcification distort the bladder and lower ureters, causing repeated infections and progressive reflux nephropathy.

Haematuria and acute urinary symptoms follow invasion of bladder wall venules by schistosomes. Following this, typical ova are excreted. These are best demonstrated in the pellets from centrifuged early morning or terminal urine specimens. Serology is usually positive by this stage.

The treatment of choice for bilharzia is praziquantel. For prevention and control, see Chapter 9.

Acute epididymo-orchitis

ORGANISM LIST

Mumps virus
Coxsackievirus
Escherichia coli
Other coliforms
Neisseria gonorrhoeae (rare)
Chlamydia trachomatis (rare)

Acute epididymo-orchitis usually has a sudden onset. There is pain, swelling and redness of one or both testicles, with or without symptoms of cystitis or urethritis. Fever and other systemic symptoms are common. It must be distinguished from acute torsion of the testis, which requires urgent surgical correction.

There may be obvious features of the underlying systemic disease, e.g. parotitis in mumps or severe muscle pain in coxsackie B infections.

Torsion of the testis, common in fit young men, must be excluded. Testes that undergo torsion often hang horizontally, below the epididymis instead of lying anterior to it. Ultrasound examination is the investigation of choice to detect torsion.

Bacterial pathogens may be demonstrable on MSU examination or by urethral swab. Viral infections are diagnosed initially on clinical grounds and ultimately by cultural and serological methods applicable to the individual viruses.

Treatment is that of the underlying condition. Antimicrobial chemotherapy is appropriate for bacterial infections, and should also be attempted if the aetiology is not obvious. A short course of prednisolone treatment will alleviate oedema and inflammation in viral infections, and will not prolong the natural course of the disease. A well-fitting scrotal support or underpants will reduce discomfort. Anti-inflammatory analgesics are helpful. Stronger analgesics may be required early in the disease.

Chronic or persisting epididymo-orchitis is a rare feature of tuberculosis (particularly of the renal tract), and of brucellosis.

Bartholin's abscess

Bartholin's glands open into the posterior part of the vulval vestibule via a duct on each side. Infection is usually unilateral, causing intense pain, swelling and tenderness deep to the labia on the affected side. Common causative organisms include *Staphylococcus aureus* and *E. coli*, which probably ascend from the perineal skin. Broad-spectrum antibiotics may be effective, but surgical drainage is often required.

On rare occasions the abscess contains chlamydiae or mycobacteria. Culture of the pus obtained at surgery is therefore advisable, to confirm the aetiology.

11 Childhood Infections

Introduction, 225

Measles, 225
- Introduction, 225
- Epidemiology, 226
- Virology, 226
- Clinical features, 226
- Differential diagnosis, 227
- Laboratory diagnosis, 227
- Treatment, 228
- Problems and complications, 228
- Prevention and control, 229

Mumps, 230
- Introduction, 230
- Epidemiology, 230
- Virology, 230
- Clinical features, 230
- Differential diagnosis, 231
- Treatment, 231
- Prevention and control, 231

Rubella, 232
- Introduction, 232
- Epidemiology, 232
- Virology, 232
- Clinical features, 232
- Differential diagnosis, 232
- Laboratory diagnosis, 232
- Complications, 234
- Prevention and control, 234

Chickenpox (varicella), 234
- Introduction, 234
- Epidemiology, 234
- Virology, 235
- Clinical features, 235
- Differential diagnosis, 236
- Laboratory diagnosis, 236
- Treatment, 237
- Problems and complications, 237
- Prevention and control, 238

Human parvovirus B19, 239
- Introduction, 239
- Epidemiology, 239
- Virology, 239
- Clinical features, 239
- Differential diagnosis, 239
- Laboratory diagnosis, 239
- Treatment, 239
- Prevention and control, 240

Human herpesvirus type 6, 240

Human herpesvirus type 7, 240

Kawasaki disease (mucocutaneous lymph-node syndrome), 240
- Introduction, 240
- Clinical features, 240
- Diagnosis, 241
- Treatment, 242
- Complications, 242

Introduction

Many infections, both bacterial and viral, are more common in childhood than in adulthood. Examples include primary herpes simplex, hepatitis A, hand, foot and mouth disease and meningococcal disease. Although they can cause considerable morbidity in individuals, these diseases do not cause fast-moving epidemics and clinically affect only a small minority of children. In contrast, the epidemic diseases of children discussed in this chapter are almost all highly infectious, have a high rate of clinical morbidity, and affect up to 90% of all individuals by the end of childhood.

Nowadays, some childhood infections have been made uncommon by immunization programmes (Fig. 11.1), but a high rate of immunization must be maintained to prevent their resurgence. In epidemics, the few non-immune adults are at risk of infection, and often suffer more severe disease, with more complications, than those infected in childhood.

A few diseases mentioned here are not highly infectious, but are important differential diagnoses of epidemic diseases with rashes. It is convenient to present them in this context.

Measles

Introduction

Measles is a systemic viral infection whose main features are respiratory disease and rash. It is highly infectious among susceptible individuals and almost always produces clinical disease in those infected. In unprotected populations it tends to occur in large epidemics mainly affecting children, but the widespread use of effective immunization programmes has made it uncommon in many parts of the world.

The important impact of measles is threefold:

1 It can be a severe and debilitating illness.

2 Secondary bacterial respiratory disease is common and may be severe.

3 Post-measles encephalitis is life-threatening and can leave severe sequelae.

Measles virus antigen or genome has been demonstrated in bone-associated cells in Paget's disease and otosclerosis. Early reports of virus detection in Crohn's disease were not supported by extensive further research.

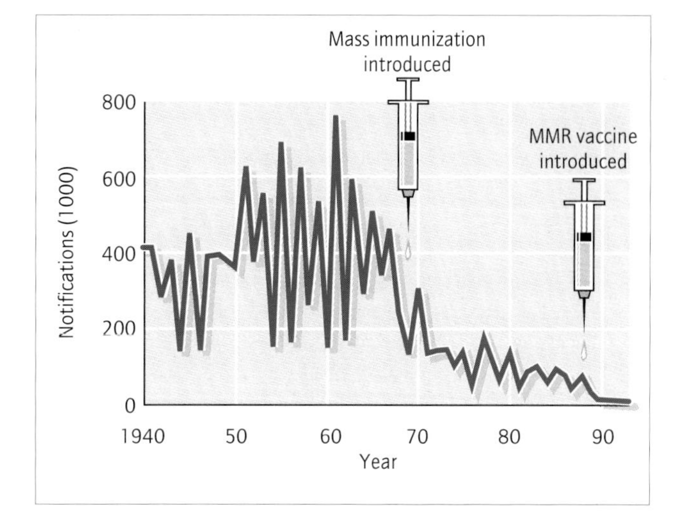

Fig. 11.1 The progress of control of measles after commencement of immunization in the UK.

Epidemiology

The disease is highly infectious with a reproduction rate (see Chapter 1) of 10–18. The attack rate in a susceptible population is usually 95% or greater. Transmission occurs mainly by droplet spread or by direct person-to-person contact, less commonly by air-borne spread or by articles freshly soiled with secretions of the nose and throat. The maximum period of communicability is during the prodromal period and in the first 2 days after the appearance of the rash. The average incubation period is 10 days (range 8–13 days) from exposure to the onset of fever, and 14 days to the onset of rash. Human beings are the only reservoir of infection. Immunity following the disease is usually lifelong.

In the absence of an effective immunization programme, most people are infected in early childhood. The average age at infection is 4 years. Maternal antibody provides protection in young infants; however, this is usually lost by about 6 months of age. An unvaccinated person has very little chance of going through life without becoming infected.

In the UK, large epidemics occurred at regular 2-yearly intervals before the introduction in 1968 of a mass vaccination programme (Fig. 11.1). Initially, the coverage of the vaccine was low (about 50%), and epidemics continued to occur every 2–3 years, although the number of cases dropped by 80%. Vaccination coverage has improved to over 90% in recent years and the epidemic cycle has been broken. Many of the cases reported in recent years are in older unvaccinated children. The case fatality ratio has declined from 1 per 100 in 1940 to 0.02 per 100 in 1989. Deaths from measles are now extremely rare. In epidemic years, the disease has a marked seasonal pattern, with a peak incidence in spring and early summer.

Virology

The measles virus is a paramyxovirus of a single serological type related to canine distemper and rinderpest viruses. The measles virus can only infect primates. Its helical nucleocapsid is made up of a single strand of RNA coated with a protein and associated with an RNA-dependent RNA polymerase. The virus is enveloped and varies in diameter from 120 to 200 nm. The envelope contains three major antigens: (i) the matrix or M protein; (ii) H protein, a glycoprotein responsible for haemagglutination and adsorption of the virus to host cell receptors; and (iii) F protein, a glycoprotein which mediates fusion with the host cell membrane and haemolysis. Unlike other paramyxoviruses, the measles virus does not contain neuraminidase. Among the non-envelope proteins, the large protein (L) interacts with the pyrophosphate protein (P) to form an RNA-dependent RNA polymerase complex.

Like other enveloped viruses, measles is sensitive to ether and is readily inactivated. Despite its high attack rate for the human host it is difficult to culture artificially but it may grow in primary human or simian cells, in which its characteristic cytopathic effect (CPE) is production of multinucleate giant cells. In contrast, vaccine strains produce a spindle-cell CPE.

Clinical features

After 8–13 days' incubation, illness begins with fever and a catarrhal respiratory infection. This is accompanied by extreme irritability and febrile convulsions are common at this stage. There is conjunctival inflammation, running eyes and nose, persistent croupy cough, mucoid sputum and often coarse crepitations on auscultation of the chest. Diarrhoea is common, especially in children.

The buccal mucosa is inflamed and Koplik's spots appear. These are raised white lesions which look like breadcrumbs or grains of salt on a red, inflamed background (Fig. 11.2). They are reliably present for a short time during the prodromal period, on the mucosa of the cheek, adjacent to the upper premolars and molars, but can be much more extensive in some cases.

The rash starts in the hairline and behind the ears on the third or fourth day. It reaches the hips on the next day and then the lower legs. On the face it appears as large, swollen blotches (Fig. 11.3) but elsewhere it is maculo-

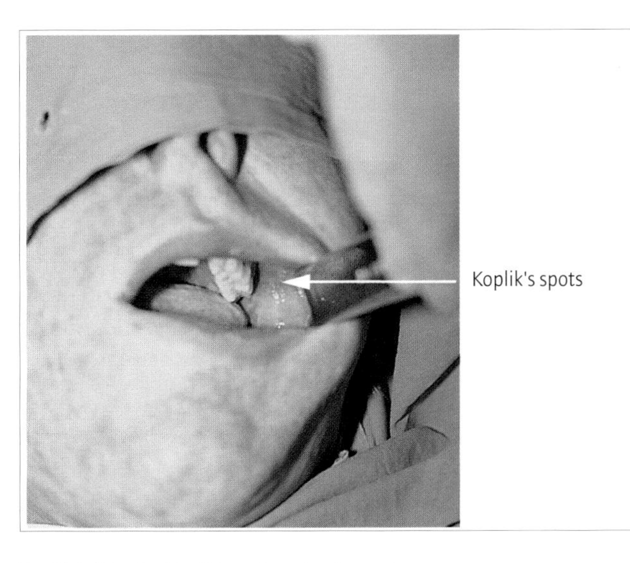

Fig. 11.2 Measles: Koplik's spots during the prodrome.

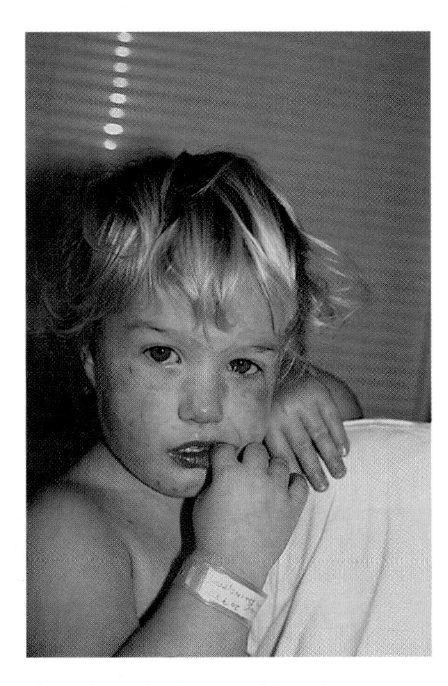

Fig. 11.3 Measles: facies on development of the rash.

papular with largish elements of 1–2 cm in the greatest diameter (Fig. 11.4). It is not pruritic but the skin feels hot and uncomfortable. The Koplik's spots fade as the skin rash evolves.

When the skin rash reaches the lower legs the temperature begins to fall. In the next few days the rash fades to a café-au-lait colour (often called staining); the cough should then subside and the fever resolve. The staining rash fades gradually away within a week or 10 days.

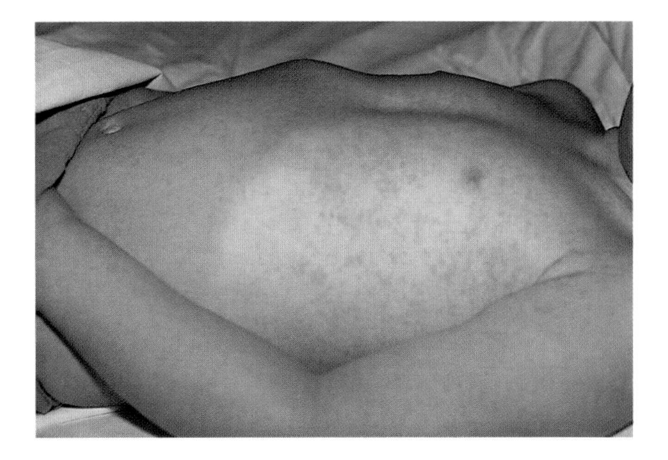

Fig. 11.4 Measles: evolution of the maculopapular rash. By the second day the rash has reached the hips.

Differential diagnosis

In countries where measles is still common, the evolution of the illness is so characteristic that a confident clinical diagnosis is usually possible, even in mild or immunized cases. The absence of prodromal respiratory features or Koplik's spots makes the diagnosis unlikely. The rash of rubella does not spread from the head downwards, and its maculopapular elements are much smaller. Allergic rashes are usually pruritic, unaccompanied by respiratory features and spread in a random way.

Laboratory diagnosis

The laboratory diagnosis of measles is important in countries such as the UK that are close to eliminating the disease, as the natural history and clinical features are increasingly unfamiliar. It is also necessary in patients with deficient cellular immunity in whom the classic clinical features may be absent.

The white cell count is low or normal unless secondary bacterial infection develops. Smears may be made of Koplik's spots and aspirated nasopharyngeal secretions. Demonstration of multinucleate giant cells would support the diagnosis of measles. These preparations may also be examined by an immunofluorescent technique to detect viral antigen. Blood, nasopharyngeal secretions, conjunctival secretions and urine may be taken for culture. Specimens should be transported to the laboratory at 4°C with the minimum of delay due to lability of virus infectivity. Specimens may be inoculated into primary human cells, or continuous-line Vero cells. Isolated virus is identified by the giant-cell CPE, and intranuclear inclusions. Definitive identification can be made by immunofluorescence or virus neutralization.

Complement-fixing, neutralizing and haemolysis-inhibiting antibodies may be sought; they appear with the rash, and reach a peak within 10 days. Thus the first of paired sera should be taken as soon as possible. Immunoglobulin G (IgG) and IgM antibodies may be detected by enzyme-linked immunosorbent assay (ELISA). Antibodies of both classes rise in parallel, but IgM antibodies fall within 3 months, enabling a diagnosis of acute infection to be made. Patients with subacute sclerosing panencephalitis (SSPE) can have persistently elevated IgM concentrations, but this is unlikely to cause diagnostic confusion.

Since the disease has become uncommon, it is usual to confirm the diagnosis by IgM ELISA. Measles IgM can now reliably be detected in a saliva specimen, provided this is collected between 1 and 6 weeks after the onset of symptoms. Reverse transcriptase polymerase chain reaction (RT-PCR) methods are available but usually only used in research studies.

Diagnosis of measles

1 Cytology of Koplik's spots or respiratory mucosal cells (shows multinucleate giant cells).
2 Immunofluorescent staining of cells to demonstrate measles antigen (a rapid diagnostic method).
3 Culture if specimens rapidly transported to the laboratory.
4 Demonstration of measles immunoglobulin M by enzyme-linked immunosorbent assay.
5 Rising titres of various antibodies in paired sera.

Treatment

There is no specific treatment for acute measles. In most cases the risk of severe secondary bacterial disease is greater than that from the viral infection itself, and is responsible for most of the 5–20% fatality rate in third-world outbreaks. Bacterial bronchitis should be easily suspected, sputum and blood cultures should be obtained and antimicrobial therapy should include an antistaphylococcal agent.

Problems and complications

Unusually severe measles

Danger signs include severe respiratory disease when the rash is only beginning, or widespread petechial or haemorrhagic components in the rash. Giant-cell viral pneumonitis, trivial in most cases, can cause severe or fatal respiratory failure.

A nebulized preparation of the antiviral drug tribavirin is obtainable in the UK. Small trials suggest that it may influence the course of myxoviral and paramyxoviral respiratory diseases, though the evidence is not strong enough to make firm recommendations.

A rare effect of measles is keratitis, which usually presents as blurred vision in one or both eyes. Cloudy oedema of the cornea is apparent on slit-lamp examination. It is usual to offer antibiotic drops or ointment to prevent secondary bacterial infection. The disorder is often self-limiting over a period of 3 or 4 weeks.

Measles is a dangerous disease in the immunosuppressed (see Chapter 22).

Secondary bacterial infections

Acute otitis media and lower respiratory infection are the commonest complications of measles. Otitis media occurs during the catarrhal phase of the disease, while bronchitis or bronchopneumonia are indicated by persisting cough and fever when the rash is staining (Fig. 11.5). While *Streptococcus pneumoniae* and *Haemophilus influenzae* are common causes of this, *Staphylococcus aureus* is also a real likelihood, as in influenza.

Mucocutaneous infections occur in debilitated or undernourished children. Impetiginous lesions appear at the angles of the mouth or around the nares. In conditions of poor hygiene, necrotizing, synergistic infection may develop. This is called cancrum oris and is nowadays rarely seen except in the poorest rural communities of the world.

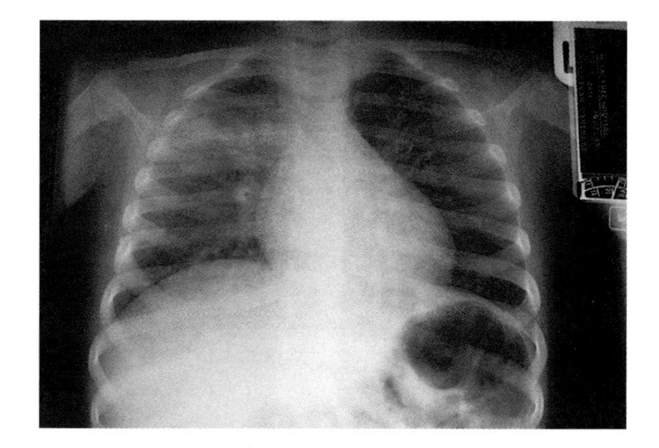

Fig. 11.5 Secondary bacterial bronchopneumonia in measles: this 2-year-old girl developed respiratory distress and high fever as the rash began to stain. The chest X-ray shows extensive nodular pneumonitis as well as a distinct area of consolidation. The patient responded rapidly to intravenous ampicillin plus flucloxacillin, and *Staphylococcus aureus* was recovered from sputum cultures.

Temporary immunosuppression and vitamin A

Measles produces significant immunosuppression with temporary loss of response to skin-test antigens. This predisposes to secondary infections, and in susceptible communities may precipitate or exacerbate tuberculosis. Vitamin A deficiency is common in such communities and exacerbates the immune paresis. A single dose of 10 000 IU of vitamin A in acute measles can reduce mortality by up to 70% in such communities.

Post-measles encephalitis

This complicates about 1 in 1000 infections. It develops in the second week as the acute illness is resolving. The encephalitis is severe and no specific treatment is available. There is no evidence that corticosteroids or other immunosuppressive drugs influence the course, though dexamethasone may be part of the treatment to lower intracranial pressure. Mortality is about 15% and about 50% of survivors have permanent neurological sequelae.

Subacute sclerosing panencephalitis

This is rare, affecting about 1 in 1 000 000 cases, almost always patients who had measles before the age of 2. Neurological disease begins 7–10 years after the acute measles. Clumsiness and poor school performance are followed by progressive spasticity. Myoclonic episodes and salaam-like seizures are common and there is a typical abnormality of the electroencephalogram. The outcome is uniformly fatal.

Measles antigen is plentiful in the brains of SSPE patients, with high antibody titres in cerebrospinal fluid. It appears that virus which lacks surface protein gradually spreads within the syncytium of the brain, without escaping from the cells. Viral antigen can also be demonstrated in the peripheral blood mononuclear cells of sufferers.

Complications of measles

1 Severe, haemorrhagic disease.
2 Measles keratitis.
3 Secondary acute suppurative otitis media.
4 Secondary bacterial bronchopneumonia.
5 Cancrum oris (rare).
6 Postinfectious encephalitis.
7 Subacute sclerosing panencephalitis.

Prevention and control

Measles vaccines are produced from live strains of measles virus which have been attenuated by passage in serial cell culture. In most developed countries, measles vaccine is given to all children at 12–15 months of age. Vaccination before this age is not recommended because of a poor immune response, possibly because of interference by maternal antibody. The vaccine is administered by deep subcutaneous or intramuscular injection, usually as part of a combined measles/mumps/rubella (MMR) preparation. It is contraindicated in children with immune deficiency disorders, the only exception being children with human immunodeficiency virus (HIV) infection, in whom the risks from the disease are greater than those from the vaccine.

A single injection of measles vaccine induces protective antibody in over 95% of recipients. Minor side-effects occur in 5–10% of vaccine recipients. They include fever, which may be accompanied by loss of appetite and a measles-like rash (mini-measles). These signs usually appear between the sixth and tenth day after vaccination. Occasionally, convulsions occur if the fever is high. Simple measures such as tepid sponging, removal of warm clothing and antipyretic therapy reduce the likelihood of postvaccination febrile convulsions. Very rarely (approximately 1 per 10 million doses) SSPE occurs following vaccination. The risk of neurological complications is 10–100 times greater after the disease than after the vaccine.

Very high vaccination rates are required to prevent the spread of measles. It is estimated that more than 95% of the population must be immune to interrupt transmission. Not all children seroconvert following vaccination. Thus, even countries such as Sweden with vaccination rates close to 100% have so far failed to eliminate the disease completely. For this reason many countries recommend a second dose of vaccine at either 4–6 or 11–12 years of age.

Measles is a notifiable disease. All suspected cases must be notified to the local consultant in communicable disease control so that preventive measures may be taken. Children with measles should be kept out of school until they are no longer infectious (5 days after the appearance of the rash). It is particularly important that contact with immunosuppressed children is avoided.

Unvaccinated children who have been in contact with a case of measles may be protected by vaccination, provided that it is given within 3 days of exposure. Immunosuppressed children, for whom vaccination is contraindicated, may be given temporary passive protection by administration of normal human immunoglobulin, which contains significant quantities of measles antibody.

Mumps

Introduction

Mumps is a systemic viral infection commonly regarded as epidemic parotitis. It has many other clinical presentations capable of causing significant morbidity. Fatalities, however, are very rare.

Epidemiology

The disease is moderately infectious, with a reproduction rate of 4–7. Transmission is through droplet spread and direct contact with saliva of an infected person. The maximum period of communicability is in the 2 days before onset of illness. Virus may however be recovered from saliva from 6 days before the onset of parotitis up to 9 days after onset, and from urine up to 14 days after onset. Very mild or subclinical infections are common, and these can also be infectious. The incubation period is 2–3 weeks (average 18 days). Humans are the only reservoir of infection.

In countries that have not yet implemented mass immunization programmes, epidemics occur at 3-yearly intervals, with the peak incidence during winter and spring. The maximum incidence is in children aged 5–9 years. Deaths from mumps are extremely rare; however, approximately 1500 children were admitted to hospital in the UK each year as a result of mumps complications before vaccination was introduced in 1988. The usual reason for admission was meningitis; mumps was the commonest viral cause of meningitis in children. Immunity following natural mumps infection is generally lifelong.

Virology

Mumps virus is a paramyxovirus of one serological type. It contains a single linear strand of RNA which is associated with an RNA-dependent RNA polymerase to form a helical nucleocapsid. The virus is enveloped and is of variable size, between 120 and 200 nm. It is ether sensitive, and is destroyed by heating to 56°C for 20 min, and by ultraviolet irradiation. The envelope is approximately 10 nm thick and consists of three layers: the outer layer contains glycoproteins with haemagglutinin, neuraminidase and cell fusion activity. The middle layer is made up of host cell membrane, and the inner layer contains non-glycosylated viral structural proteins.

Six major proteins are encoded: the nucleocapsid protein (NP), the phosphoprotein (P) and the large protein (L) are associated with the ribonucleoprotein complex. The envelope consists of matrix protein (M), and two glycoproteins that mediate fusion (F) and haemagglutination (HN).

Complement-fixing antibodies recognize the NP antigen. Antibodies to S antigen appear soon after infection and decline over the next few months. Anti-V antibodies rise more slowly, peaking 2–4 weeks after the beginning of the infection, and persist for years. Neutralizing antibodies develop during convalescence and persist for years.

Clinical features

The features of mumps can appear in any order. Parotitis is the most common, occurring in over 70% of cases. It is usually bilateral but the onset may be asymmetrical and other salivary glands may be involved. Parotid tenderness and pain on salivation precede swelling by 2–4 days. After increasing for about 3 days the inflammation subsides over 7–10 days.

Orchitis is common in adult men but rare before puberty. It varies greatly in severity, but some men suffer extreme swelling and pain of the testicle, which may take up to a month to resolve completely. Severe inflammation can result in testicular atrophy; but bilateral atrophy is very uncommon. Mumps orchitis is therefore a rare cause of infertility.

Meningitis and meningoencephalitis affect around 15% of cases, but are often so mild as to be overlooked. Clinically significant cases are often admitted to hospital.

Diagnostic problems arise when mumps presents as isolated meningitis, pancreatitis (occasionally with acute diabetes) or orchitis. It should always figure in the differential diagnosis of these conditions, but not necessarily be first on the list.

Rarer manifestations of mumps include mastitis (which can affect both sexes and all age groups), cochlear infection with hearing impairment, oophoritis in women and arthritis (which often appears in the second week). Myocarditis can occur and may be involved in the rare fatalities in adult cases.

Clinical features of mumps

1 Parotitis.
2 Meningitis and meningoencephalitis.
3 Orchitis.
4 Pancreatitis.
5 Oophoritis.
6 Cochlear inflammation.
7 Arthritis.
8 Mastitis.
9 Myocarditis.

Differential diagnosis

The diagnosis is often made clinically. Parotitis alone must be distinguished from acute cervical lymphadenitis. In parotid swelling the angle of the jaw is enclosed in the swollen gland, whereas it is superficial to cervical lymph nodes. The serum amylase is often raised in parotitis. The white blood cell count is not always a useful feature, as the severe inflammatory effects of mumps can be accompanied by neutrophilia. In cases of doubt there is no harm in giving antibiotics (to cover *Staphylococcus aureus* and *Streptococcus pyogenes*) while awaiting laboratory diagnosis. The other differential diagnosis is pyogenic parotitis, which may also cause a raised amylase. This is usually unilateral; the gland may be fluctuant and any abscess will tend to point below the lobe of the ear, where the cartilage joins the external meatus. Surgical drainage is often required. Sarcoidosis can present with persisting bilateral parotitis.

Isolated orchitis must be distinguished from testicular torsion or pyogenic epididymo-orchitis. Ultrasound imaging is the best investigation for distinguishing these.

The white cell count and differential are usually normal but a leucopenia and relative lymphocytosis may be present. A leucocytosis may be found in patients with meningitis, orchitis or pancreatitis.

Saliva, cerebrospinal fluid or urine may be collected for culture. Specimens are cultured in primary monkey kidney cell or human embryonic kidney cells. Cultured virus may be identified by haemadsorption, neutralization or fluorescent antibody techniques. An RT-PCR technique is also available for diagnosis

Paired serum samples should be taken for serological investigation. A number of techniques have been described, including haemagglutinin inhibition (HI), virus neutralization, complement fixation (CF) and ELISA. Virus neutralization is the best technique for establishing immunity to mumps but it is too cumbersome for routine diagnostic use.

HI and CF tests are simple to perform but lack sensitivity. A fourfold rise in CF antibodies would confirm a diagnosis of mumps. An elevated titre to S antigen with a low or high titre to V antigen in an acute-phase specimen would indicate recent mumps infection. IgM antibodies may be detected by ELISA and titres reach a peak 1 week after the onset of symptoms, remaining elevated for approximately 6 weeks. An IgG ELISA may be used to detect local production of antibodies in the cerebrospinal fluid of patients with mumps encephalitis or meningitis.

Treatment

No specific treatment is available. Analgesics and bed rest are helpful in severe cases. A short course of corticosteroids may reduce pain and swelling in severe orchitis.

Prevention and control

Mumps vaccine is a live, attenuated preparation, usually administered as part of combined MMR vaccine. Two mumps vaccine strains (Urabe Am9 and Jeryl Lynn) have been widely used in immunization programmes. Mass vaccination of children aged 12–15 months was introduced in the UK for the first time in 1988 and has greatly reduced the incidence of the disease. A sustained coverage level of 85% or greater is required to prevent transmission of mumps and to eliminate the disease.

If coverage levels are low (less then 70%), the total number of cases may decline; however, the average age at which infection occurs will increase. This may lead to a net increase in cases among adolescents and adults. Complications occur more frequently in these older age groups; thus a vaccine programme with low coverage may have a negative effect, despite a decline in the total number of cases. It is therefore important to ensure high vaccine coverage levels in young children.

The vaccine is highly effective. More than 95% of recipients develop immunity after a single dose which is usually lifelong. As with other live vaccines, immunodeficiency is a contraindication (see Chapter 26). Minor side-reactions include fever and parotitis (mini-mumps). These occur in up to 5% of recipients, and usually appear in the third week after vaccination. A mild, self-limiting meningoencephalitis occurs after approximately 1 per 11 000 vaccinations with the Urabe Am9 strain, which is not therefore used in British immunization programmes. The incidence of meningitis following the Jeryl Lynn strain is not known precisely, but is considerably lower than that for Urabe Am9. These complications must be compared to the risk of clinical meningoencephalitis following natural infection, estimated to occur in 1 per 200 cases.

Mumps is a notifiable disease. Children with the disease should be kept out of school until 5 days after the onset of parotitis. Vaccination of exposed contacts is of no value, although it may be used during outbreaks to protect those who have not previously been vaccinated.

Rubella

Introduction

Rubella is a systemic viral infection with many features, including a rash. Although highly infectious, it often produces subclinical or trivial disease. It is important because even subclinical viraemia can infect the developing fetus, causing severe tissue damage and progressive developmental defects (see Chapter 17).

Epidemiology

The disease is moderately infectious, although less so than measles or varicella. The reproduction rate is 7–8. Transmission is by droplet spread or direct person-to-person contact. The period of communicability lasts from about 1 week before to at least 4 days after the onset of rash. The incubation period from exposure to onset of fever is 2–3 weeks. Humans are the only reservoir of infection.

In the absence of mass immunization, rubella epidemics occur approximately every 6 years. Children are predominantly affected (the average age at infection is 8 years); however, cases may also occur in adolescents and adults. During rubella epidemics, up to 5% of susceptible pregnant women may catch the disease, resulting in subsequent epidemics of congenital rubella syndrome and rubella-associated terminations of pregnancy.

Virology

Rubella is caused by rubivirus, an alphavirus member of the family Togaviridae. Unlike many other members it does not need an arthropod vector for transmission. It is immunologically distinct and differs serologically from the other alphaviruses. It is an icosahedral enveloped virus 60 nm in diameter. The central core containing the nucleocapsid is 30 nm in diameter. It consists of a single-strand RNA and a nucleocapsid protein (protein C) in a helical form. Two viral glycoproteins, E1 and E2, are associated with the envelope. Haemagglutinin activity is associated with the viral envelope. The virus is readily inactivated by ether, trypsin, ultraviolet light, heat and extremes of pH. It will grow in Vero cells, without cytopathic effect.

IgG and IgM antibodies begin to rise as the rash appears, reaching a peak after 7 to 14 days illness. IgM antibody concentrations fall to low levels within 1 month. HI and CF antibodies rise in the first week and remain elevated for a prolonged period. Passive haemagglutinating antibody begins to rise after 1 month and remains elevated for years.

Clinical features

The average incubation period of 17–18 days is followed by a mild sore throat and mild conjunctivitis, often just a gritty feeling in the eyes. Fever is rarely high and the rash appears on the second or third day. It consists of fine macules; papules are unusual, petechiae rare. The macules coalesce to a generalized 'blush' in 1 or 2 days, and this fades without desquamation in 3–5 days. Lymphadenopathy commonly affects the neck, and suboccipital nodes may be large and painful.

Arthralgia is common in young adults. It affects the small joints of the hands and feet and occasionally large joints. It can last some weeks, so non-steroidal anti-inflammatory agents may be needed until the discomfort gradually subsides.

Differential diagnosis

Clinical diagnosis is difficult, because many patients lack the rash. A rubelliform rash is common in parvovirus infection (in which arthralgia is also common), enterovirus infections, mild allergic rashes and sometimes mild scarlet fever or toxic shock syndrome.

Rubella in pregnancy or exposure of a susceptible pregnant woman can be a disaster. Readiness to suspect rubella and prompt laboratory diagnosis are extremely important in this context. It is also important to reconfirm immunity in an exposed pregnant woman.

Laboratory diagnosis

The diagnosis of rubella is usually made by serological techniques. Viral culture should be attempted when strain characterization is required, when vaccine-related infection is suspected, or in complicated neonatal cases. In the absence of CPE, viral growth is detected by challenging infected cells with enterovirus. Control monolayers are destroyed but infected cells resist enteroviral infection and remain intact. Cultured virus is identified by neutralization of virus infectivity with polyclonal rabbit immunoglobulin. Virus shedding from the pharynx may be scanty, but urine is a useful specimen for viral culture.

HI is a widely used technique which detects both IgM and IgG antibodies by their ability to inhibit the agglutination of chick red cells by rubella antigen. Titres of antibody correlate well with the degree of immunity. The diagnosis of acute rubella can be established if a fourfold rise in HI titre is detected, or specific IgM is detected. ELISA and radioimmunoassay are the simplest methods for detecting specific rubella IgM. Congenital rubella can be diagnosed by the presence of IgM antibody, culture or RT-PCR detection of viral RNA.

Diagnosis of rubella

1 Immunoglobulin M detection by enzyme-linked immunosorbent assay, radioimmunoassay or particle agglutination.
2 Fourfold rise in haemagglutinin inhibition antibodies in paired sera.
3 Viral culture.
4 Demonstration of viral RNA by RT-PCR.

Passive haemagglutination utilizes red cells coated with rubella antigens which are agglutinated in the presence of rubella antibodies. It may be successfully applied for antenatal screening, and the results obtained correlate well with HI. In the single radial haemolysis test the presence of rubella antibody is indicated by complement-mediated lysis of sensitized red cells suspended in an agar gel (Fig. 11.6). It is simple and rapid, facilitating the screening of large numbers of specimens.

Detection of immunity to rubella

1 Single radial haemolysis test.
2 Passive haemagglutination test.

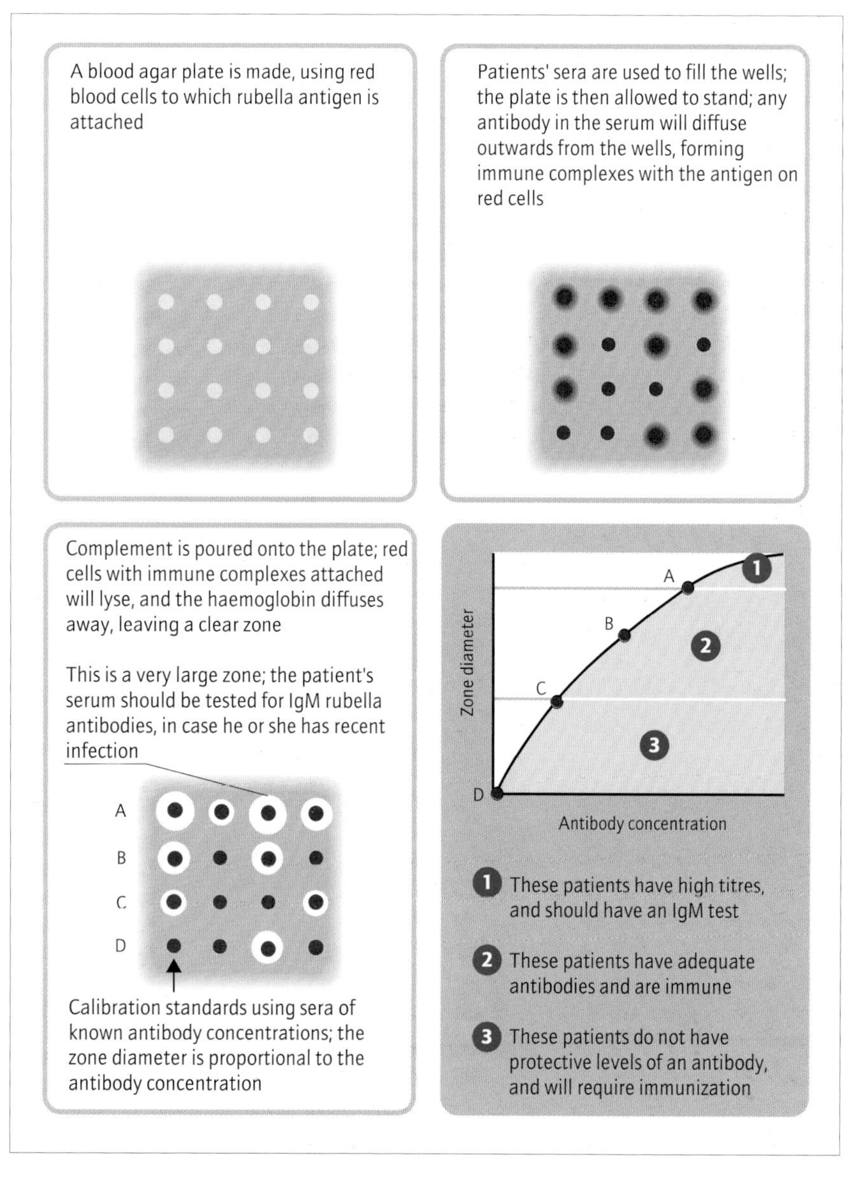

Fig. 11.6 The principle of the radial haemolysis test in screening for immunity to rubella.

Complications

Immune thrombocytopenic purpura

Rubella is one of the commonest infectious precursors of thrombocytopenic purpura. The thrombocytopenia is transient, lasting from 1 to 3 weeks. If mild it needs no treatment, but if necessary it can be treated with a brief course of prednisolone or, if severe, with intravenous immunoglobulin.

Encephalitis

This is clinically evident in about 1 in 5000 cases, occurring soon after the rash. It is variable in severity, and can only be treated symptomatically.

Prevention and control

Rubella vaccine is given as part of a combined measles/mumps/rubella preparation; the aim of the programme is to interrupt rubella transmission and thereby avoid infection in pregnancy (Fig. 11.7; see Chapter 17). Rubella is a notifiable disease. Children should be excluded from school for 5 days after the onset of the rash.

Contact with pregnant women should be avoided. It is particularly important that all healthcare workers who are likely to be in contact with pregnant women are vaccinated against rubella. Pregnant women who have been in contact with a case, particularly during the first trimester, should be tested serologically for susceptibility or evidence of early infection (IgM antibody) and advised accordingly. Laboratory diagnosis of the case should be sought, as the clinical diagnosis of rubella is unreliable.

Chickenpox (varicella)

Introduction

Chickenpox is a systemic viral infection with a characteristic vesicular rash. Infection is almost always symptomatic but is rarely severe except in adults. Rare cases of fetal injury have occurred after infection in pregnancy (see Chapter 17), and the disease is dangerous in infants aged less than 2 weeks and in the immunosuppressed. Patients on moderate and high doses of corticosteroids are at particular risk of severe illness. Active immunization may become widely available in the near future.

Epidemiology

The disease is highly infectious. The most important sources of infection are cases of chickenpox. Patients with herpes zoster are also infectious, and their susceptible contacts can develop chickenpox. Transmission occurs directly by person-to-person contact, by air-borne spread of respiratory secretions or vesicular fluid, and indirectly through articles recently contaminated by discharge from vesicles and mucous membranes. The period of communicability is usually from 1–2 days before, to 6 days after,

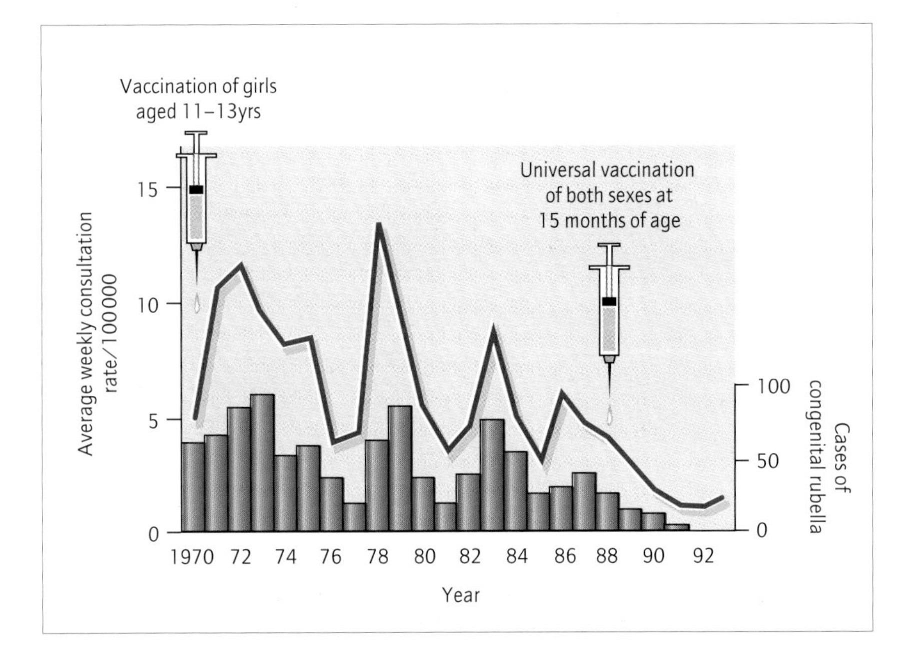

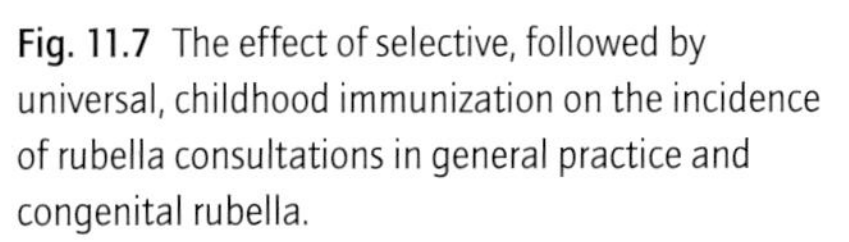
Fig. 11.7 The effect of selective, followed by universal, childhood immunization on the incidence of rubella consultations in general practice and congenital rubella.

the appearance of the first crop of vesicles. This can sometimes be prolonged, particularly in patients with immune deficiency. The incubation period averages 15 days (range 10–25 days). Humans are the only reservoir of infection.

Chickenpox occurs predominantly in young children. Epidemics occur every 1–2 years, usually in winter and early spring. Over 90% of adults have naturally acquired immunity, but in the UK and some other Western countries the incidence of chickenpox in individuals over the age of 14 years is steadily increasing (Fig. 11.8). Shingles occurs mainly in older adults, although children, especially those with immune deficiency, may also develop typical lesions (see Chapter 5).

Virology

Varicella zoster virus (VZV) is a member of the Herpesviridae, and shares morphological and pathological characteristics with the other members. Antigenic variation has not been shown, although strain variation can be demonstrated by restriction endonuclease digestion. The virion consists of a central nucleocapsid core 90–95 nm in diameter and an outer membrane envelope 150–220 nm, from which glycoprotein spikes project. The genome consists of double-stranded DNA of mass 80 MDa, coding for 75 protein antigens. There are five major glycoprotein antigens: gp I–V. Viral infectivity may be neutralized by monoclonal antibodies to gp I, II and III.

VZV can be isolated in continuous cell culture using many simian and human cell lines, but the virus is strongly cell associated and therefore difficult to culture from serum and even from cell-free vesicle fluid. The VZV CPE is characterized by discrete foci of rounded, enlarged cells, and destruction of the monolayer.

Like other herpesviruses, VZV produces lifelong latency, and may reactivate in later life. This reactivation is usually confined to one or two dermatomes served by the dorsal root ganglia, in which the latent virus exists. The mechanism of VZV reactivation remains unknown.

Clinical features

The incubation period is 10–25 days, averaging 15–18 days. Children rarely have a prodromal illness but adults may suffer a few days of fever, headache and myalgia. The appearance of clear vesicles on the head or trunk is commonly the first sign of disease. They develop from small round papules which are transient and rarely noticed (Fig. 11.9). Many of the vesicles are oval with their long axis along the creases of the skin, and are sometimes intensely itchy and inflamed. They evolve to opaque pustules which become umbilicated as they dry to crusts. For several days new lesions appear as the older ones evolve (a process called cropping) but newer lesions are smaller and eventually fail to develop (Fig. 11.10). Lesions appear earliest and most densely on the trunk and face. The hands and feet are relatively spared (Fig. 11.11).

Mucosal lesions can affect the conjunctiva, mouth and perineum. They are very superficial and usually heal without scarring. Oesophageal, gastric and intestinal lesions occasionally cause significant abdominal pain, and synovial lesions can cause joint pain and swelling.

A focal pneumonitis occurs in parallel with the rash. It is rarely clinically significant, but occasionally causes

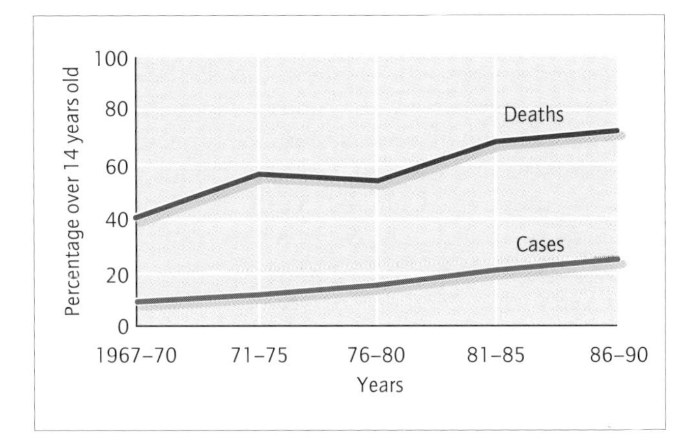

Fig. 11.8 Chickenpox: the increasing incidence in older children and adults in the UK. Courtesy of Dr Elizabeth Miller, Communicable Disease Surveillance Centre.

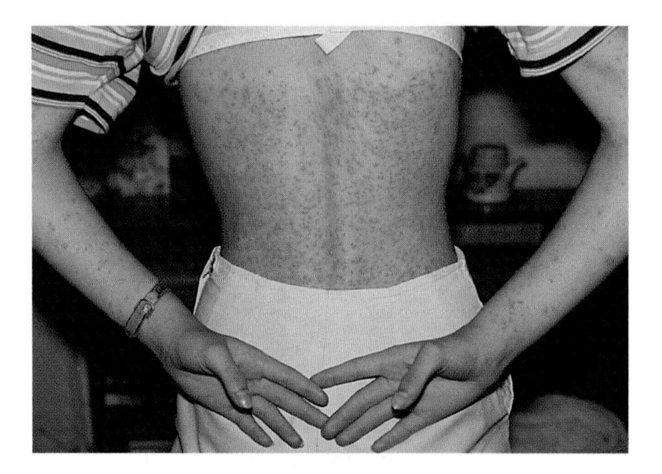

Fig. 11.9 Chickenpox: the very early rash. Many papules are seen, which will all become vesicles in the next few hours.

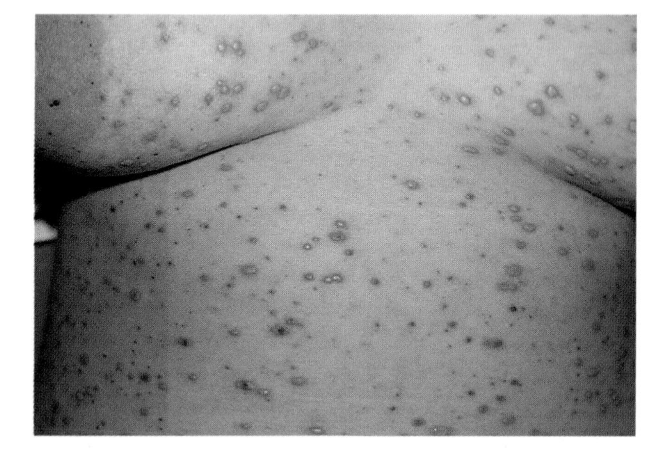

Fig. 11.10 Chickenpox: typical rash showing non-coalescing lesions at all stages of development.

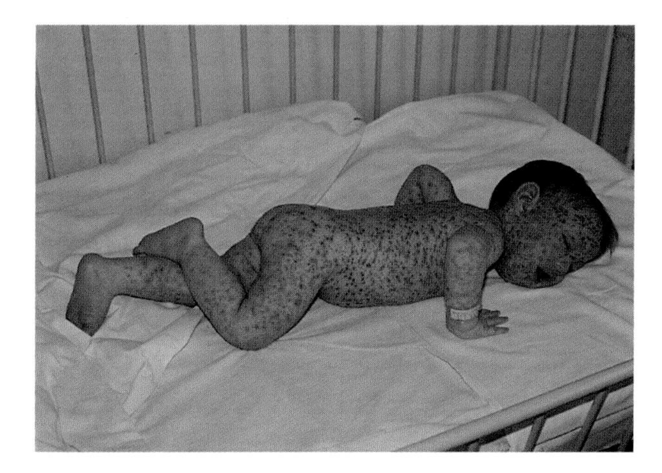

Fig. 11.11 Chickenpox rash: the centripetal distribution spares the hands and feet.

severe respiratory disease with blood-stained expectoration and widespread nodular opacities on chest X-ray (Fig. 11.12). The lung lesions tend to calcify on healing, producing a typical X-ray appearance (chickenpox lung).

Differential diagnosis

The diagnosis of chickenpox and herpes Zoster (see Chapter 5) is usually clinically evident, because of the typical rash. Disseminated herpes simplex, hand, foot and mouth disease, guttate psoriasis and, occasionally, vesiculating allergic eruptions can cause confusion. Rickettsial pox, an arthropod-transmitted disease, is occasionally seen in deprived inner cities in the Americas, and can cause a chickenpox-like rash.

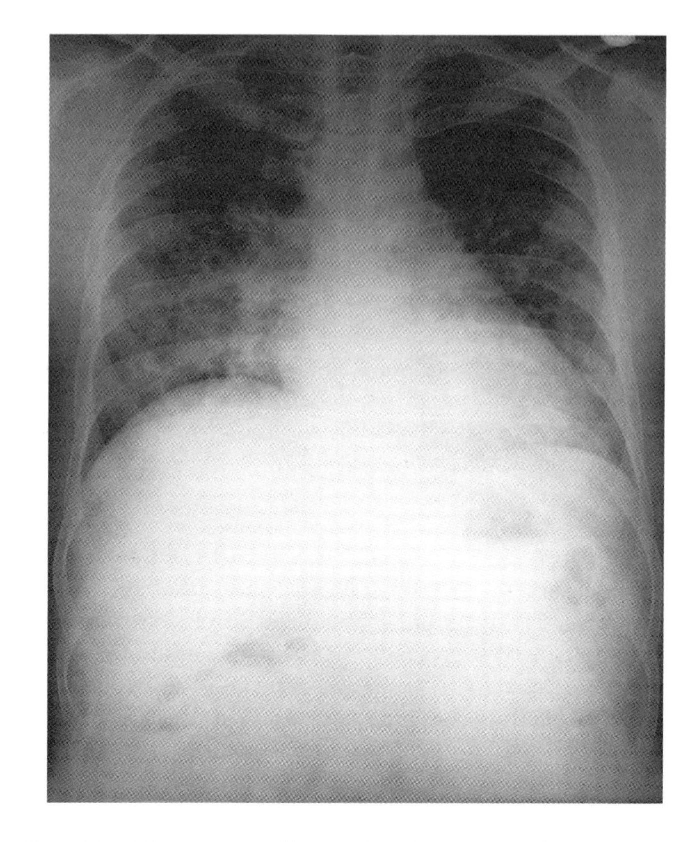

Fig. 11.12 Chickenpox pneumonitis: extensive nodular pneumonitis. Respiratory failure required 5 weeks of assisted ventilation.

Laboratory diagnosis

Specimens which may be collected for laboratory diagnosis include smears of vesicular lesions and tissue biopsy or postmortem material. Multinucleate giant cells can be found in the cellular debris from the base of vesicles and smears of this material can also be examined by electron microscopy, by immunofluorescence and by culture. Immunofluorescence and electron microscopy are more sensitive than viral culture for the diagnosis of VZV infection. Cultured virus is identified by examination under the electron microscope or by immunofluorescence.

The serological diagnosis of VZV infection is complicated by heterotypic antibody responses to herpes simplex virus in patients previously infected with VZV. A fourfold rise in CF antibody titre is diagnostic of acute VZV infection. The CF test is insufficiently sensitive for the determination of past VZV infection. Techniques used for this purpose include ELISA and fluorescent antibody tests. VZV DNA can be detected by PCR.

Diagnosis of chickenpox

1 Electron microscopy of vesicle scrapings.
2 Immunofluorescent staining of vesicle scrapings.
3 Culture of vesicle scrapings.
4 Fourfold rise in complement-fixing antibodies.
5 Varicella zoster virus DNA detection by PCR.

Treatment

Specific treatment is rarely required in children. Adolescents and adults may suffer severe disease and should be treated, ideally starting on the first day of the rash. Oral aciclovir: adult dose 800 mg five times daily for 7 days; child dose 20 mg/kg four times daily for 5 days (under 2 years 200 mg four times daily) reduces the duration of fever and active rash by an average of 1 day, and may prevent serious complications. Immunosuppressed patients should be referred urgently for intravenous therapy. Pruritus may be ameliorated by antihistamines. Fewer lesions will develop, and irritation will be less if the skin is kept cool, with light clothing and frequent cool washes. Mild analgesia may be helpful for painful lesions.

Problems and complications

Severe forms of chickenpox

Widespread tissue damage occurs, with pneumonitis and sometimes intravascular coagulation, renal damage and disturbed liver function. Smokers are at increased risk of pneumonitis. Danger signs are dense and/or haemorrhagic rash, many mucosal lesions, substernal and epigastric pain, probably indicating many mucosal lesions, and reduced arterial oxygen saturation. By the time that abnormal physical signs appear in the chest, respiratory failure is often already established. In the immunocompromised, pneumonitis may be out of proportion with the rash.

Treatment with aciclovir (10 mg/kg 8-hourly by intravenous infusion) may ameliorate the disease, but must be given early for the best effect, as the advancing lesions are not halted until 24–48 h after commencing therapy. Aciclovir must be given by slow infusion over 1 h to avoid nephrotoxicity, which is related to peak blood levels. The blood urea and creatinine levels should be monitored during therapy.

Falling arterial oxygen saturations may lead to the need for assisted ventilation. Renal failure demands modification of the dose of aciclovir, which is excreted via the kidneys, and must often be treated by haemoperfusion or dialysis. The risk of secondary bronchopneumonia is great, and broad-spectrum antibiotic treatment which includes antistaphylococcal activity is highly advisable.

Secondary bacterial infections

Superinfection of skin lesions by *Staphylococcus aureus* or *Streptococcus pyogenes* is common, particularly in children. These infections can be complicated by erysipelas, scarlet fever or toxic shock syndrome (Fig. 11.13). They should be promptly treated with appropriate antibiotics. If skin lesions become abscess-like or necrotic, swabs should be taken and parenteral antibiotics commenced.

Secondary staphylococcal pneumonia is particularly common in adults, especially smokers. Any patient with respiratory symptoms should be treated with an anti-staphylococcal antibiotic as mixed viral and staphylococcal pathology is common. Admission to hospital should be considered. Other secondary bacterial chest infections may occur, and should be treated with co-amoxiclav or a broad-spectrum cephalosporin.

Treatment of chickenpox

1 Symptomatic—keep the skin cool and give antipruritics.
2 Aciclovir orally; adult 800 mg five times daily for 7 days; child 20 mg/kg (maximum 800 mg 6-hourly).
3 *For immunosuppressed or severely ill patients*: aciclovir i.v. (by infusion over 1 h) 10 mg/kg 8-hourly for 5–7 days.
4 Antibiotic treatment of secondary skin or chest infection—include an antistaphylococcal drug.

Post-chickenpox encephalitis and other neurological diseases

Post-chickenpox encephalitis is common, but usually mild and self-limiting. Cerebellar disturbance is rare, with ataxia and nystagmus which develop as the rash heals. Cases with other neurological signs or altered consciousness have a more unpredictable outcome. No specific treatment is indicated. Rare examples of meningitis, destructive retinitis, transverse myelitis or ascending myeloencephalitis are seen, and can occur before, during or after the rash. All are more common in the immunosuppressed. The diagnosis should be sought, as aciclovir is effective in many cases.

Thrombocytopenia

This is not uncommon. It may make the rash appear haemorrhagic, but the concurrent appearance of dependent purpura and microscopic or macroscopic haematuria

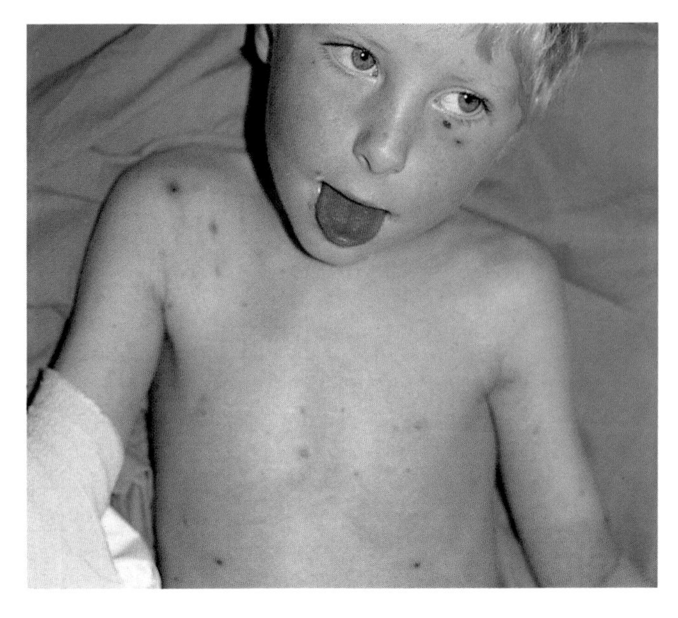

Fig. 11.13 Chickenpox with staphylococcal secondary infection. An abscess has developed in an affected lesion, and the child has an erythematous rash of toxic shock syndrome.

will indicate the diagnosis (Fig. 11.14). The condition is transient and can be managed by a brief course of corticosteroids, with or without platelet transfusion. Intravenous immunoglobulin is also effective.

Prevention and control

Live attenuated varicella vaccines have been developed and shown to be highly effective. To date these have only been used for immunization of non-immune children with leukaemia, or to protect healthcare workers who are in regular contact with immunosuppressed children. Mass vaccination in early childhood has been introduced in the USA, and may become more widely adopted in future, when a combined measles/mumps/rubella/varicella (MMRV) preparation becomes available. Zoster does occur as a complication of vaccination, although less frequently than after the disease.

Varicella zoster immunoglobulin (VZIG) prepared from plasma containing high titres of specific antibody is available for passive immunization. It is indicated for leukaemic and other immunosuppressed contacts of chickenpox or shingles without a definite history of previous chickenpox, for neonates whose mothers develop chickenpox in the period 7 days before to 28 days after delivery, and for non-immune pregnant contacts. There is no evidence that children with asthma or eczema suffer unusually severe disease or complications.

Children with chickenpox should be excluded from school for 1–5 days after the appearance of the first crop of vesicles. Patients are no longer infectious when all of the lesions are dry and scabbed. In hospital, cases of chickenpox and shingles should be isolated, because of the risk to immunosuppressed patients.

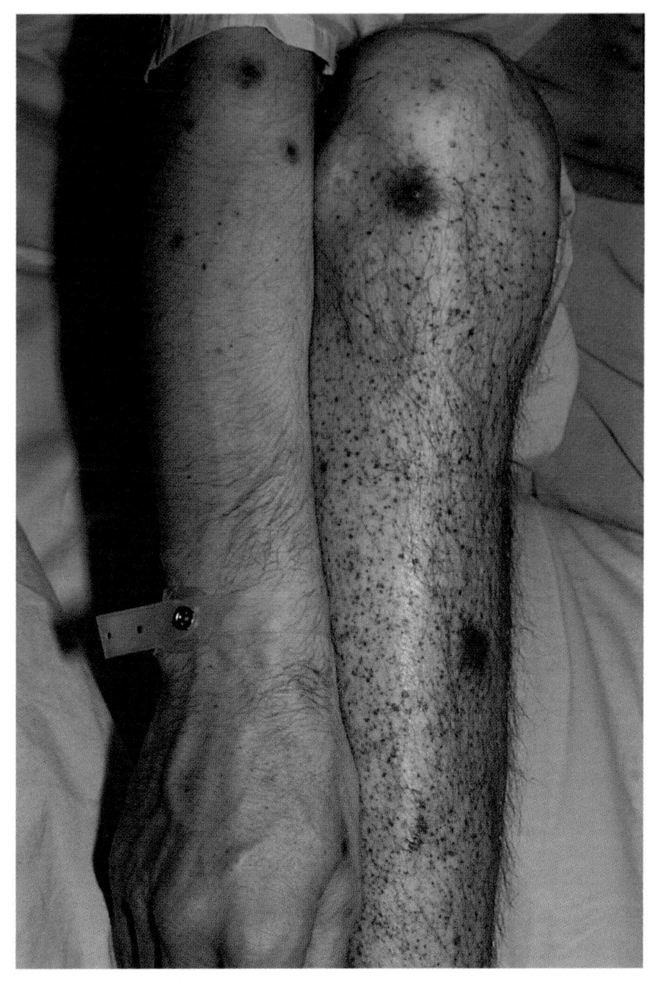

Fig. 11.14 Thrombocytopenia in chickenpox. There has been bleeding into the vesicular lesions and there is a purpuric rash on the leg. The patient had gross haematuria.

Prevention of chickenpox

1 Isolation of patients with fluid in their vesicles.
2 Postexposure varicella zoster immunoglobulin (VZIG) for susceptible:
  (a) immunosuppressed oncology patients;
  (b) immunosuppressed patients taking regular systemic corticosteroids;
  (c) transplant recipients on immunosuppressives;
  (d) non-immune pregnant women; and
  (e) neonates of infected mothers.
3 Live attenuated vaccine under consideration.

Human parvovirus B19

Introduction

Human parvovirus B19 (HPV-19) causes a viraemic disease in which the virus particularly attacks rapidly dividing cells. Epidemics of feverish illness with rash and arthralgia are recognized. Infection of red cell precursors occurs and can cause a transient aplastic crisis in children with inherited haemolytic anaemias. Infection during the second trimester of pregnancy carries a risk of fetal anaemia and hydrops (see Chapter 17). Persistent infection in the immunocompromised can cause prolonged aplastic anaemia (see Chapter 22).

Epidemiology

The disease appears to be highly infectious. Transmission is person to person, by droplet infection from the respiratory tract. Occasionally the infection may be spread through contaminated blood products. Infection is commonest in children between 5 and 14 years. Outbreaks in schools are common, and usually occur in late winter or spring. The incubation period is 7–22 days (average 14 days).

Virology

Parvoviruses are small DNA viruses which infect a wide range of animal species, including humans. A single serotype, B19, infects humans, causing erythema infectiosum, and aplastic crises in patients with haemolytic anaemias.

Parvovirus is a small, dense icosahedral virus which is not enveloped. The genome is made up of a single-stranded linear DNA which may be of positive or negative polarity. Three non-structural proteins are produced, which participate in replication and encapsidation, as well as in regulation of transcription. The two structural proteins are VP1 and VP2. VP2 is the major structural protein.

The virus may be cultivated in tissue culture but this is too insensitive for routine clinical use. The virus is stable to heating at 56°C for more than 1 h. It is resistant to ether and chloroform and survives at room temperature for a prolonged period.

Clinical features

The incubation period of 1–3 weeks is followed by viraemia and often fever. In 2 or 3 days the fever falls and the rash appears. In children the cheeks are often bright red (slapped cheek syndrome). In all age groups the rash appears on the limbs, and less often on the body. It may be rubelliform but is often reticulate. It fades quickly, recurring transiently if the skin is warm. Bizarre petechial or haemorrhagic rashes are occasionally seen, particularly affecting the peripheries, but the patient is rarely severely ill. Painful generalized arthralgia can last for several weeks, improving with a fluctuating course. As in rubella, young women are most affected by the joint symptoms.

A pause in erythropoiesis accompanies the rash (and can also occur in the absence of rash). This is caused by infection of red cell precursors in the bone marrow. Reticulocytes become undetectable in the blood film and there may be transient anaemia, often profound in children with inherited haemolytic anaemias. Erythropoiesis resumes as the infection resolves. Immunosuppressed patients may be unable to clear their infection, leading to continued hypoplastic anaemia.

Differential diagnosis

The acute illness must be distinguished from rubella. This is especially important if the patient or a household contact is pregnant. Clinical distinction is poor, so laboratory tests for both rubella and parvovirus should be performed (both infections can adversely affect pregnancy: see Chapter 17).

Other rash diseases such as scarlet fever and toxic shock syndrome should be suspected if the white cell count is raised. Wide fluctuations in joint symptoms may lead to a suspicion of rheumatic fever.

Laboratory diagnosis

The diagnosis of erythema infectiosum is usually made on the basis of the characteristic clinical features. The difficulties of *in vitro* cultivation of parvovirus mean that serology and genome detection by PCR are the principal methods of diagnosis. Specific IgM and IgG antibody titres rise after the transient viraemia. ELISA for IgG and IgM capture assays have been described.

Parvovirus can be visualized by electron microscopy in the serum of infected patients. Antibodies to parvovirus can be found in up to 50% of the normal population.

Treatment

There is no specific treatment. Non-steroidal anti-inflammatory agents help the arthralgia. Transfusion may be required for aplastic crises. Normal human immunoglobulin contains antibodies to parvovirus, and intravenous immunoglobulin (IVIG) treatment may terminate or ameliorate bone marrow infection in the immunosuppressed. Anecdotal reports exist of improvement in chronic aplasia after IVIG or aciclovir treatment.

Prevention and control

Exclusion of cases from school is of no value, as most susceptible children will already have been exposed by the time the case is diagnosed. If a midwife or healthcare worker has HPV-19 infection, susceptible contacts should be identified, as they can be offered human normal immunoglobulin, which may give a degree of protection from infection.

Human herpesvirus type 6

Human herpesvirus type 6 (HHV-6) was originally thought to be a B-lymphotrophic virus, but is now known to infect many types of cells. Serosurveys suggest that most people are infected early in childhood, subsequently developing antibodies and harbouring latent virus.

Roseola infantum (erythema subitum) is the clinical manifestation of HHV-6 seroconversion. It is an illness of infants in which 2 or 3 days of fever are followed by a widespread morbilliform rash which starts on the trunk. As in other viral infections, leucopenia is usual. Small and large outbreaks have been described, with an incubation period of 10–15 days.

Recent reports suggest that HHV-6 may also be a rare cause of severe hepatitis.

Human herpesvirus type 7

Human herpesvirus type 7 (HHV-7) has been demonstrated in the lesions of pityriasis rosea, and seroconversion to HHV-7 has been associated with this disease.

Kawasaki disease (mucocutaneous lymph-node syndrome)

Introduction

Kawasaki disease is a vasculitic disease of children whose main features are fever, rash, mucocutaneous inflammation and lymphadenopathy. Its cause is unknown, but is likely to be an infection because moderate to large epidemics sometimes occur. Although uncommon, the disease has a significant morbidity and mortality. Cardiac ischaemia due to coronary aneurysms makes Kawasaki disease the commonest cause of heart transplant in children.

Clinical features

The patient is usually a toddler, but rare cases occur in older children and even adults. Illness begins with irregular fever which is often severe. Swelling of the hands and feet is common and a red rash soon appears. There is usually a 'glove-and-stocking' distribution of intense erythema; adjacent areas of the limbs develop roundish slightly raised lesions of various sizes while the rest of the body may be covered with a faint erythema. There is often a shiny or scaly rash on the perineum (Fig. 11.15). Other early features include oral inflammation, fissuring of the

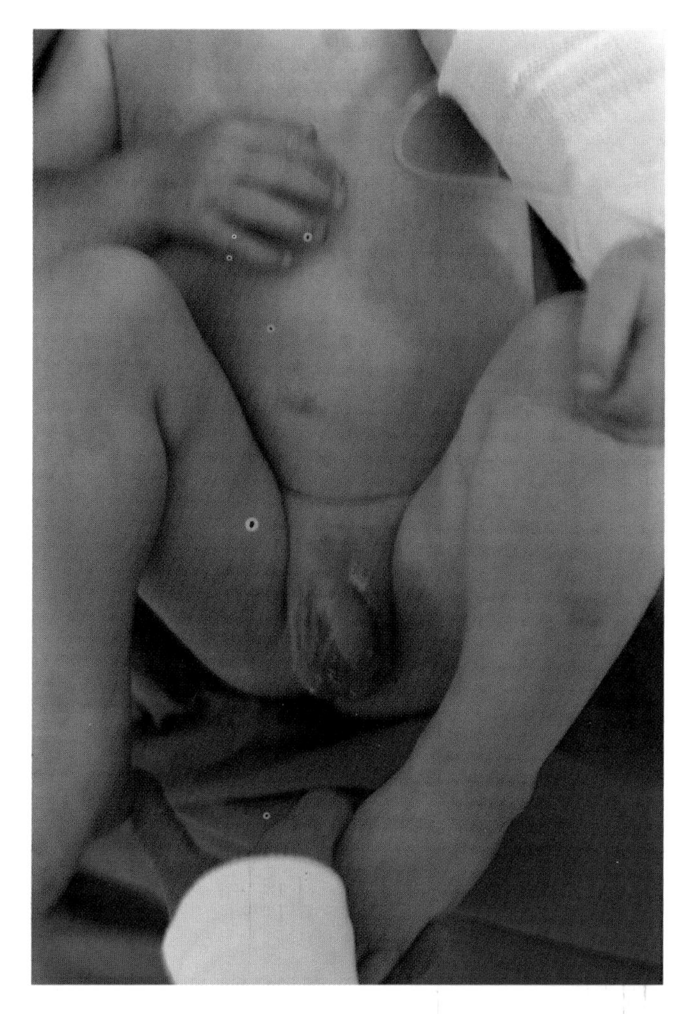

Fig. 11.15 Kawasaki disease: the acute rash affecting the peripheries and the perineum.

lips and conjunctival injection. There is moderate or gross enlargement of cervical lymph nodes.

After 3–6 days the rash on the hands and feet begins to desquamate. The skin from the ends of the digits often breaks away, characteristically in one piece (Fig. 11.16). Elsewhere the rash may persist for many days.

Even after desquamation of the hands the fever persists, and at this stage systemic features become increasingly important. Arteritis predominates and, like polyarteritis nodosa of adults, can cause aneurysm formation. Coronary artery aneurysms can be large and multiple, leading to myocardial ischaemia or infarction.

Mucocoele of the gallbladder, which is often present, varies in severity from simply an ultrasound finding to an acute surgical emergency.

Mild, self-limiting forms of the disease probably occur. Cardiac events without a preceding illness have not, however, been recognized.

Diagnosis

There is no specific diagnostic test, but the evolution of the skin and mucous membrane lesions combined with lymph-node enlargement is quite characteristic. The British Paediatric Surveillance Unit uses a clinical definition.

Clinical definition of Kawasaki syndrome

1 Fever, otherwise unexplained, lasting 5 days or more.
2 Bilateral conjunctival congestion.
3 Generalized erythema of buccal and pharyngeal mucosae.
4 Localized cervical lymph-node enlargement.
5 Polymorphous rash with changes in extremities.
6 Desquamation of hands and feet.

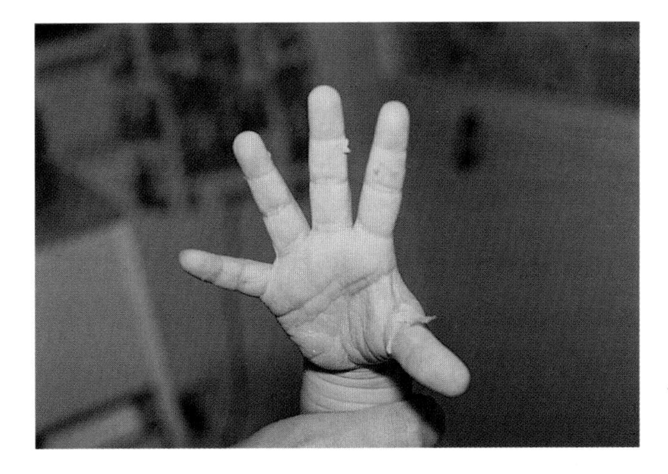

Fig. 11.16 Kawasaki disease: typical desquamation of the fingertips.

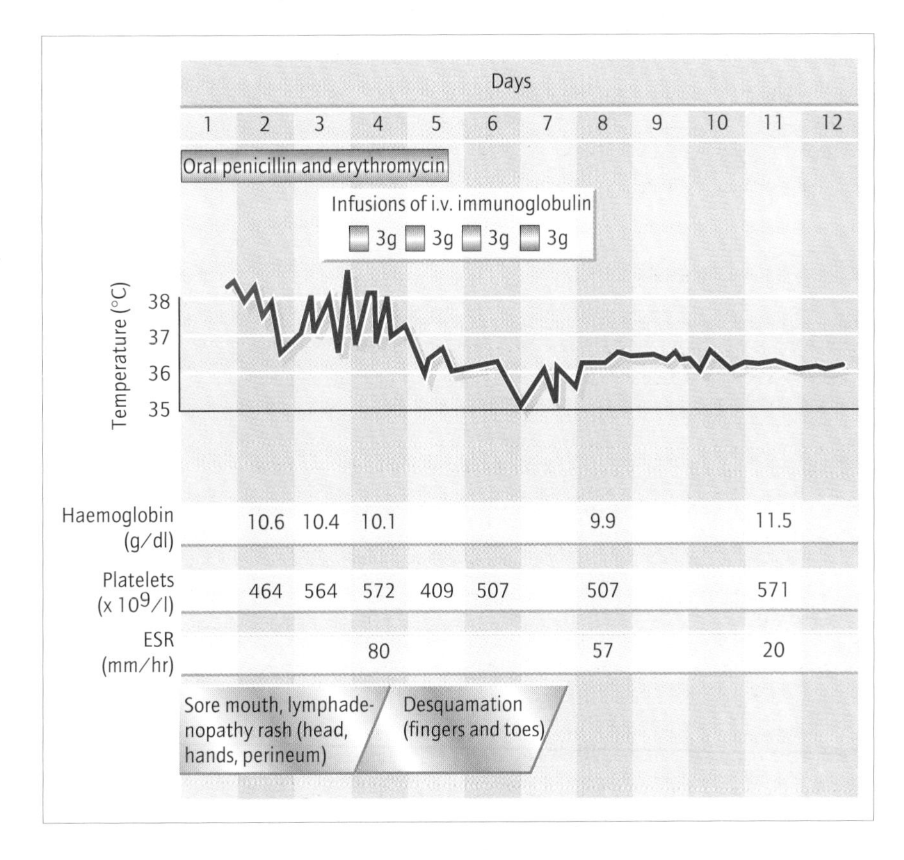

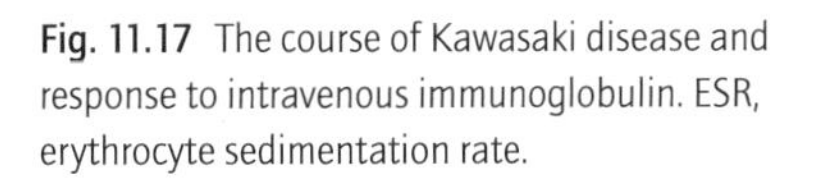

Fig. 11.17 The course of Kawasaki disease and response to intravenous immunoglobulin. ESR, erythrocyte sedimentation rate.

As the illness progresses the erythrocyte sedimentation rate and the platelet count both rise dramatically. There is a moderate leucocytosis. Abnormal liver enzyme levels are common—particularly a rising alkaline phosphatase.

Ultrasound imaging often shows distension of the gallbladder. It is essential to perform echocardiography, both at presentation to detect any existing aneurysm and later to check that coronary artery dilatation is not occurring.

Treatment

Treatment with intravenous immunoglobulin inhibits the development of coronary artery aneurysms. Infusions should be commenced as soon as the diagnosis is evident and should be given daily for 3 or 4 days at a dose of 400 mg/kg. Fever and distress also seem to be reduced by this treatment (Fig. 11.17).

Kawasaki disease is an indication for aspirin treatment, even in children. Full doses of 50 mg/kg daily should be given while the platelet count is raised. Smaller doses, e.g. 10–15 mg/kg daily, may then be used. Most experts would continue aspirin for at least 3 months—longer if the erythrocyte sedimentation rate and platelet count remain raised. Some would add dipyridamole to this regimen, but there is no firm evidence of benefit.

Complications

Mycocardial ischaemia or infarction

This is the most important complication. Repeated echocardiography will give warning of developing aneurysms. Large aneurysms of more than 0.9 mm diameter are dangerous, and paediatric cardiologists may consider surgical intervention if these are seen. Myocardial ischaemia is accompanied by pain, as in adults, and electrocardiogram and cardiac enzyme estimations must be performed. Rarely, infarction has occurred after cessation of active treatment. Severe heart failure following infarction may be an indication for heart transplant.

Mucocoele of the gallbladder

This is a rare cause of abdominal emergency in Kawasaki disease. If peritonism or gross pain and enlargement of the gallbladder occur, surgery may be indicated, but the risk of anaesthesia must be taken into account if the heart is compromised.

12 Infections of the Cardiovascular System

Introduction, 243

Pericarditis, 243
Introduction, 243
Clinical features, 244
Diagnosis, 244
Management, 245
Complications, 245

Myocarditis, 245
Introduction, 245
Pathology and epidemiology, 245
Clinical features, 246
Diagnosis, 246
Management, 246

Infective endocarditis, 246
Introduction, 246
Epidemiology, 246
Pathology, 247
Clinical features, 247
Diagnosis, 248
Management, 249
Complications, 250
Prevention and control, 250

Infective endarteritis, 251

Introduction

The cardiovascular system has three main structural components:

1 The intimal lining of the blood vessels, or the endocardium in the heart, is composed of endothelium and is in contact with the blood. Damage to the endothelium or endocardium disrupts its smoothness and also initiates the mechanisms of platelet adhesiveness, causing a sticky plaque of platelet thrombus to develop. This can both trap organisms and protect them from the immunological mechanisms of the blood. The endothelium has numerous functions, mediated by chemical factors, which initiate or modulate platelet function, coagulation pathways, leucocyte chemotaxis, margination and migration, inflammatory responses and fluid and electrolyte movement.

2 The muscular media of the arteries can occasionally be invaded and damaged by blood-borne organisms. This can cause aneurysms, leading to altered circulation, thrombosis and even occlusive disease. The myocardium is also susceptible to infection; organisms can penetrate from infected heart valves and cause myocardial abscesses; many viruses and some other pathogens can invade the myocardium and cause infective myocarditis which, in severe cases, compromises cardiac function. The toxins of *Corynebacterium diphtheriae* and some other bacteria damage myocardial cells.

3 The pericardium is a mesothelial structure, with the same embryonic origin as the pleura. There is a visceral and a parietal pericardium, with a potential space between. The pericardium can be invaded directly, causing isolated pericarditis, or it can be involved in a pleural or a myocardial inflammatory process. Exudate in the pericardial space can compress the heart, reducing its ability to fill and sometimes leading to dangerous tamponade.

Some infecting organisms target a particular part of the cardiovascular system. Rickettsiae attack endothelial cells and cause endovasculitis, which predisposes to thrombosis, and haemorrhage. Late syphilis causes aortitis which damages the elastic layer of the ascending aorta, with aneurysm formation.

Pericarditis

Introduction

Pericarditis is inflammation of the pericardium. While it is often infective in origin, other causes include reactive pericarditis after myocardial infarct, autoimmune diseases, hypothyroidism, malignant invasion and involvement in the pancarditis of rheumatic fever. Non-infectious pericarditis should therefore be considered in the initial assessment of the patient.

A variety of pathogens can cause pericarditis, either alone or accompanied by myocarditis. Acute pericarditis can be a focal complication of bacteraemic disease, such as staphylococcal or streptococcal septicaemia. Contiguous spread of pus from an empyema or from a liver abscess can cause pericarditis with organisms such as pneumococci, enterococci or *Entamoeba histolytica*. Tuberculosis can cause subacute pericarditis, sometimes with thickening and stiffening of the pericardium (constrictive pericarditis).

ORGANISM LIST

Enteroviruses, especially coxsackieviruses
Influenza viruses
Mycoplasma pneumoniae
Streptococcus pneumoniae
Other Gram-positive cocci
Mycobacterium tuberculosis
Coxiella burnetii

Clinical features

Viral pericarditis

The commonest type of pericarditis is a self-limiting illness with fever, normal or neutropenic white cell count and typical chest pain, accompanied by cardiographic evidence of pericarditis. It is often preceded or accompanied by symptoms of malaise, myalgia or arthralgia, or sore throat, suggesting a viral aetiology. Precordial pain then develops. Usually pleuritic, it often varies with different postures, with swallowing or with the heart beat, and sometimes radiates to the neck, shoulders or back. Auscultation may reveal a pericardial rub, which is often transient and variable, disappearing if effusion separates the pericardial layers.

Suppurative pericarditis

This is usually caused by pyogenic bacterial infection. There is swingeing fever, neutrophilia in the blood and sometimes other features of infection, such as pneumonia or pleural effusion. Pain is often severe, and tamponade may occur, with a falling blood pressure and pulse pressure, and a paradoxical rise in the jugular venous pressure on inspiration.

Diagnosis

Electrocardiography (ECG) shows upward-curved elevated S-T segments in the anterior chest and sometimes other leads. Echocardiography will often demonstrate pericardial thickening, or effusion if present. On chest X-ray, a pericardial effusion shows as an enlarged, globular heart shadow, which may have a double outline—one for the heart itself and one for the border of the distended pericardium (Fig. 12.1).

In viral-type pericarditis, nose swabs, throat swabs, urine and stool should be submitted for virus culture.

An early serum sample should be taken. Immunoglobulin M (IgM) antibodies to enteroviral antigens may be demonstrable in viral pericarditis. Detection of *Mycoplasma*-specific IgM or detection of rising IgG titres

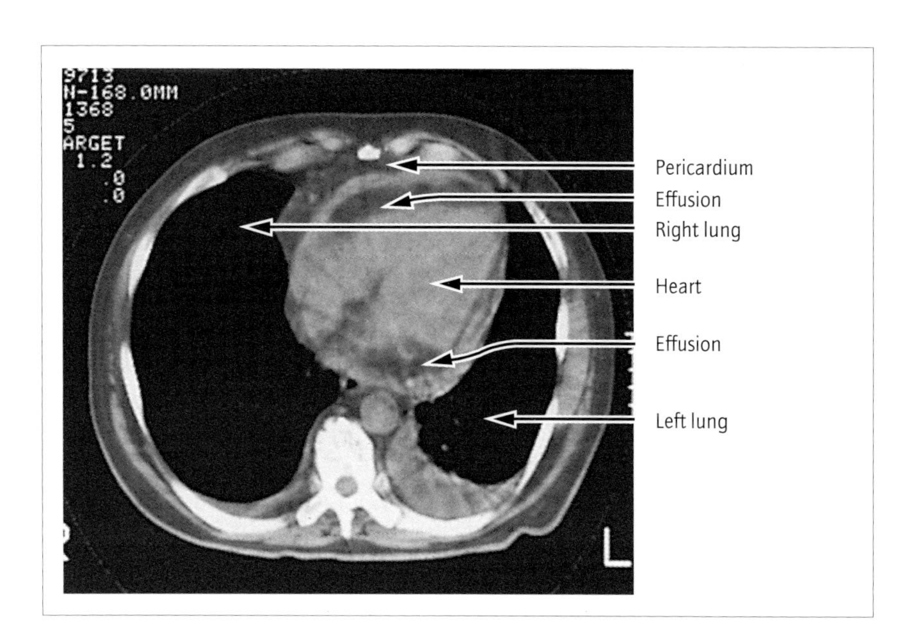

Fig. 12.1 Pericarditis with effusion: computed tomographic scan clearly shows dark fluid between the thickened pericardium and the ventricular muscle.

indicate a diagnosis of *Mycoplasma pneumoniae* infection, and serological test results can also reveal rarer conditions such as Q fever.

In suspected pyogenic infections, blood and sputum cultures should be performed. The chest X-ray should be examined for evidence of pneumonia, abscess or tuberculosis. Tapping a pericardial effusion is not without risk; it should be performed as an aseptic technique with ECG or imaging control. Specimens should be submitted for cytology, bacterial culture, acid-fast staining and mycobacterial culture, as indicated.

Management

Viral pericarditis can be treated symptomatically with bed rest and analgesia. Improvement in fever, pain and ECG signs usually occurs in 2 or 3 days and gradual mobilization can then begin. Dysrrhythmia or heart failure is rare, as are ECG abnormalities other than S-T elevation; these may warn of accompanying myocarditis.

Pyogenic pericarditis must be treated urgently with intravenous antibiotics appropriate to the situation. If there is localized pneumonia, suggesting pneumococcal aetiology, high-dose penicillin is appropriate for sensitive and moderately penicillin-resistant infection. If local strains of *S. pneumoniae* are known to be highly resistant to penicillin, then a broad-spectrum cephalosporin is indicated. Flucloxacillin or another antistaphylococcal drug should also be given if the aetiology of the infection is unknown or a lung abscess suggests possible staphylococcal bacteraemia. In a severely ill patient with few clinical signs, *Streptococcus pyogenes* must be considered; high-dose penicillin is appropriate in this case.

If tuberculosis is suspected, sputum, urine and possibly bronchoalveolar lavage specimens are examined. A tuberculin test should be 'planted' and triple or quadruple therapy then considered. If a decision is made to give antituberculosis treatment, corticosteroids should be given at the same time, to avoid a sudden increase in pericardial effusion or oedema (see Chapter 18).

Complications

Pericardial tamponade is caused by effusion or thickening of the pericardium that interferes with normal filling of the heart. Warning signs are low systolic or pulse pressure, shortness of breath first appearing on exertion, and raised jugular venous pressure, often with a paradoxical rise on inspiration. Tamponade can develop suddenly, and occasionally causes severe heart failure, which can be relieved by drainage. Tapping the pericardium carries a risk of damage to the atrial wall or coronary vessels, and should preferably be carried out by an experienced operator with imaging control (such as computed tomographic (CT) scanning).

In all cases of severe pericarditis, there is a risk of pericardial fibrosis, with slow development of tamponade (constrictive pericarditis). Such cases often present with congestive cardiac failure, and signs of tamponade must be carefully sought as surgical decompression is often curative.

Myocarditis

Introduction

Myocarditis is inflammation of the myocardium. Most cases are probably of infectious aetiology, and the majority of these are viral. Many viral cases are purely myocarditis, but myocarditis can also complicate multisystem infections such as influenza, infectious mononucleosis, mumps, adenovirus or *Mycoplasma* infections. The incidence varies with the seasons and from year to year, as would be expected for an viral infection. A few cases occur as part of a systemic disease such as brucellosis or rickettsial infection, or complicate a bacteraemia.

ORGANISM LIST

Coxsackie B (causes about 50% of enteroviral myocarditis)
Coxsackie A
Echovirus
Influenza A and B viruses
Rubella
Epstein–Barr virus
Cytomegalovirus
Adenovirus
Mumps
Rarities: rabies, hepatitis viruses
Coxiella burnetii
Leptospira spp.
Borrelia spp.
Mycoplasma pneumoniae
Neisseria meningitidis and other pyogenic organisms.

Pathology and epidemiology

The pathogenesis of myocarditis is probably different in different types of infection. In infants and young children, and in adults with systemic disease, the myocarditis is concurrent with the acute infection. This is usually the case with bacterial myocarditis.

In many viral infections in adults, particularly coxsackie B infections, the myocarditis is delayed and occurs when viruses can no longer be isolated from the respiratory tract or bowel. In animals this type of myocarditis can be prevented by disabling cell-mediated immunity. Nevertheless, coxsackievirus RNA has been demonstrated by polymerase chain reaction (PCR) in the myocardium of a high proportion of patients with this type of myocarditis (and also in patients with dilated cardiomyopathy of unknown aetiology).

Clinical features

In adults the disease has a variable presentation. There is usually a history of fever and influenza-like symptoms in the previous 2 weeks, sometimes more recently. Fatigue and exertional dyspnoea are common. Palpitations and precordial pain may be present. Physical examination may reveal fever, tachycardia, dysrhythmia or frank heart failure. Failure of conduction with heart block are common in borelliosis and diphtheria. Symptoms and signs of pericarditis occasionally coexist.

The ECG shows non-specific changes in the T waves, often inversion. There may be prolongation of the P-R or QRS interval, extrasystoles or heart block. S-T elevation is also seen, as in pericarditis, which may also be present. Laboratory tests may reveal elevation of the cardiac enzymes, which can persist for many days, but is not always present. The chest X-ray may show cardiomegaly. In bacterial disease, such as Q fever or leptospirosis, the other systemic signs of the disease are usually present (see Chapters 7 & 9).

Myocarditis is rare in infants, but is often severe or fulminating, with significant cardiac failure accompanying signs of active viral infection. Older children tend to have relatively mild disease.

Diagnosis

The diagnosis is suggested by the relationship of viral symptoms to the development of cardiological abnormalities. Enteroviruses may be recovered in culture from throat or stool cultures. Respiratory viruses can be detected by direct immunofluorescence in nasopharyngeal or throat specimens. PCR-based techniques are becoming available for rapid diagnosis. Endomyocardial biopsy may allow demonstration of virus by culture, immunofluorescence or PCR. Diffuse myocardial inflammation can be demonstrated by magnetic resonance scanning.

Management

Specific antiviral treatment is not yet available for most viral causes of myocarditis, and treatment must therefore be symptomatic. Q fever and other bacterial diseases often respond to appropriate antibiotic treatment. Supportive treatment is important. Bed rest in the early stages may limit both inflammation and the development of heart failure. If heart failure or severe cardiomyopathy develops, increasing antifailure medication, inotropic support or ventricular assistance may be required, and should be supervised by a cardiologist. In severe disease, once the acute viral infection is over, cardiac transplant may be indicated.

Infective endocarditis

Introduction

Infective endocarditis is infection of the lining of the heart, which particularly damages the cusps of valves, but can also affect other intracardiac sites. Platelet thrombi form on the affected sites, and these offer a nidus where pathogens can adhere and replicate. Many thrombi are friable, producing emboli. Endocarditis is important because it can be difficult to diagnose, may require prolonged and closely supervised treatment and is uniformly fatal if untreated.

Epidemiology

Up to the 1950s, the epidemiology of endocarditis was closely linked to the occurrence of rheumatic heart disease. Spontaneous or native valve disease affected the damaged mitral or sometimes aortic valve. Bacteraemia from the teeth or mouth was almost always the origin of the infecting organism, and this was made more likely by the high prevalence of dental caries and gingival disease, together with the high rate of dental interventions that these demanded.

Both rheumatic heart disease and poor oral health have become less common, but the detectable rate of endocarditis has remained at 1000–1500 cases per year. Predisposing factors are now congenital valve conditions, particularly bicuspid aortic valve and floppy mitral valve. Small ventricular septal defects are at risk, because of the large pressure gradient and intense turbulent flow that they cause; vegetations readily form on the downstream side of the orifice. Elderly patients may have small areas of fibrosis or calcification on valves, on chordae tendinae or on the myocardial wall, or occasionally a mural thrombus may become infected. While younger patients are at risk from mouth organisms, older patients may have bacteraemias of genital or bowel origin, and enterococci are relatively common in native valve endocarditis of the elderly.

Intravenous drug abusers may suffer repeated bacteraemias, often with skin organisms such as *Staphylococcus aureus*, but also with pathogens which contaminate their drug preparations. These may include organisms such as *Candida albicans*, derived from lemon juice used as a diluent. Because the source of the contamination is venous, the right side of the heart may be affected. This is otherwise a rare site of endocarditis.

The advent of cardiac surgery and valve replacement has created a new population at risk, either from early infection predominantly with staphylococci from the skin, acquired during surgery, or from later infection of the implanted valve itself. Patients whose artificial valves are at risk of thrombus formation receive long-term anticoagulants, which may reduce the risk of vegetations and endocarditis. These patients tend to be either children undergoing treatment for congenital disorders, or the middle-aged and elderly with acquired valvular damage.

Predispositions to endocarditis

1 Congenital valve disease.
2 Septal defects (usually ventricular).
3 Degenerative valve disease.
4 Rheumatic heart disease.
5 Mural thrombus.
6 Intravenous drug abuse (right-sided infections).
7 Cardiac surgery, including artificial or biological valve implants.

Pathology

ORGANISM LIST

Native valve infection

Young patients	**Elderly patients**
Viridans streptococci	Enterococci
Streptococcus sanguis	*Enterococcus faecalis*
Streptococcus mitior	*Streptococcus bovis*
Streptococcus mutans	*Enterococcus faecium*
Streptococcus mitis	*Streptococcus durans*
Streptococcus 'milleri'	Viridans streptococci
Enterococci	Staphylococci
Staphylococci	*Coxiella burnetii*
Coxiella burnetii	

Prosthetic valve infection
- *Staphylococcus aureus*
- *Staphylococcus epidermidis*
- Viridans streptococci
- Enterococci
- *Coxiella burnetii*

Rare organisms
- Small Gram-negative rods (*Haemophilus*, *Actinobacter*, *Cardiobacter*, *Eikenella*, *Kingella*)
- Enterobacteriaceae
- *Chlamydia psittaci*
- Fungi (e.g. *Candida* spp.)
- Mycobacteria (often 'atypical')

Clinical features

These can be divided into: (i) early manifestations of infection; (ii) embolic events; and (iii) late effects of sepsis and inflammation.

The earliest features are usually fever and heart murmur, with or without associated malaise and fatigue. The onset is often subtle, the fever slight and the murmur easily dismissed as a flow murmur. Characteristically, the murmur changes as vegetations develop and valvular patency alters, but this may not occur for days or sometimes weeks. At this stage there is often a mild leucocytosis, and the erythrocyte sedimentation rate and C-reactive protein are both raised.

Features of embolization may appear after days or weeks. Early emboli are seen in more aggressive endocarditis (for instance, *Staphylococcus aureus* endocarditis), and fungal endocarditis can produce frequent and massive emboli. Endocarditis usually affects the left side of the heart, so emboli are usually systemic. Showers of small emboli may cause petechial skin lesions, episodes of haematuria or splinter haemorrhages in the nails. Large emboli occasionally cause strokes, infarcts of the kidneys or acute arterial occlusions. In rare cases of right-sided endocarditis (which may occur in intravenous drug abusers), showers of pulmonary emboli or pulmonary infarcts are typical. These can occur, paradoxically, in left-sided endocarditis with a septal defect and left-to-right shunt. Emboli may be infected and produce local infective vasculitis (Janeway lesions; Fig. 12.2), infective aneurysms or lung abscesses.

The long-term effects of the infection are those of immunological reaction and tissue damage. Immunological effects include splenomegaly, nephritis and vasculitic lesions of the eyes and skin. The vasculitic lesions occur in the finger pulp or nail margin (Osler's nodes) and in the retina (Roth's spots). Tissue damage affects the valves of

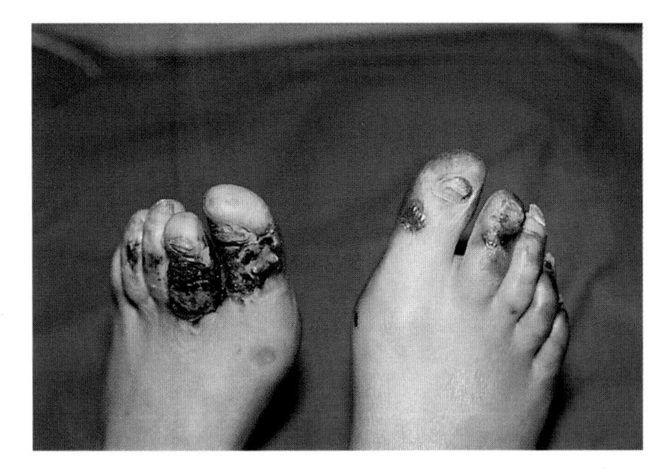

Fig. 12.2 Septic embolic lesions in the feet of a man with acute endocarditis and a huge, friable aortic vegetation.

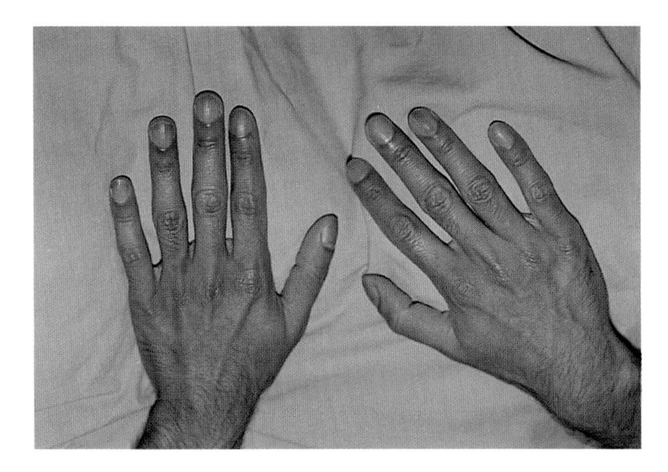

Fig. 12.3 Clubbing of the fingers in endocarditis.

the heart, which may rupture or fragment, causing sudden heart failure. Pus may form in the valve rings or cardiac septum, and occasionally ruptures into the pericardial sac.

Clubbing of the fingers (Fig. 12.3) is caused by an immunological phenomenon that alters the haemodynamics in the capillary loops of the nail beds. In some cases of endocarditis clubbing occurs early, and responds quickly to effective therapy. After longstanding untreated disease, amyloidosis can occur.

Acute (aggressive) endocarditis

Acute endocarditis occurs when highly pathogenic organisms, such as *S. aureus* or *S. pyogenes* infect the heart valves. In contrast to less aggressive infections there is high fever, large vegetations develop quickly, catastrophic valve failure occurs early, and septic emboli or marantic aneurysms are more common. The case fatality rate is higher in aggressive infective endocarditis.

Diagnosis

The diagnosis must be considered in all patients with fever and a known predisposition to endocarditis. However, apparently spontaneous endocarditis can occur in the elderly, with malaise but very little fever. Changing cardiac function or unexplained murmurs should prompt investigation. Longstanding intravenous lines may become colonized and this may lead to endocarditis.

Positive blood cultures, in conjunction with a compatible history of predisposition and clinical features (Table 12.1), will confirm the diagnosis and provide essential information on the bacterial cause and antibiotic sensitivity. Three sets of cultures should be taken (a total of 30–60 ml blood) although there is little additional benefit in taking more than three sets. It is essential that blood cultures are taken before any antibiotics are given. Subacute endocarditis rarely progresses sufficiently quickly to justify introducing antibiotic therapy before waiting for cultures, which should be incubated for 7–10 days.

Care must be taken in interpreting culture results, especially with organisms that are normally resident on the skin. Multiple isolates are required with identical antimicrobial susceptibility to confirm the significance, for example a *S. epidermidis* isolate.

About 10% of patients with endocarditis have negative blood cultures. These should have serological tests for *Coxiella burnetii*, *Chlamydia psittaci* and *Borrelia* spp.

Causes of culture-negative endocarditis

1 *Coxiella burnetii.*
2 *Chlamydia psittaci.*
3 Fastidious streptococci or Gram-negative rods.
4 Mycobacterial endocarditis.
5 Fungal endocarditis (*Aspergillus* spp. are virtually never recovered from blood cultures).

Echocardiography is useful in demonstrating vegetations on affected valves, and sometimes intracardiac abscesses. Transoesophageal echocardiography can demonstrate some aortic valve lesions not visible on standard views. However, a negative echocardiograph does not exclude the diagnosis. Computed tomography or magnetic resonance scans can demonstrate intracardiac

Type of criterion	Description of criterion
Major	**1** *Positive blood culture*, either with an organism recognized as a cause of endocarditis or more than one positive from cultures taken more than 12 hours apart, or three positives from three or four cultures drawn over a period of 1 h **2** *Evidence of endocardial damage*. Echocardiographic evidence of vegetations on valves or supporting structures, on an implant or in the turbulent path of a high-flow jet (as in damaged valve, septal defect or patent ductus), or new valvular regurgitation (not just a changed murmur)
Minor	Known predisposition to endocarditis Temperature of 38 °C or higher Evidence of focal vascultis or emboli Immunological features (Osler's nodes, nephrosis etc.) Single positive blood culture or blood culture result of uncertain significance Serological evidence of infection known to cause endocarditis (such as chlamydial or *Coxiella* infection Echocardiographic finding of uncertain significance
Diagnostic groups of criteria	Two major criteria or one major and three minor criteria or five minor criteria

Table 12.1 Duke University criteria for the diagnosis of infective endocarditis.

abscesses. Contrast ventriculography and angiography may be necessary to define connections between abscesses and the bloodstream, or to check on the patency of the coronary ostia in aortic valve disease.

When surgery is performed, tissue specimens become available for culture and microscopy. Organisms concealed in vegetations may then be recovered, or rarities such as mycobacteria demonstrated by staining and culture.

Management

The mainstay of management is adequate antibiotic treatment. Many alpha-haemolytic streptococci are sensitive to benzylpenicillin, but the *S. 'milleri'* group of organisms and enterococci respond much better if gentamicin is added. Other organisms may need different regimens, depending on the results of sensitivity testing. Chlamydial and mycoplasmal endocarditis should be treated with tetracycline (conveniently given as doxycycline 200 mg daily), with erythromycin as the drug of second choice. *Coxiella burnetii* endocarditis requires tetracycline therapy, but after control of the active infection, treatment with lower doses should probably be continued indefinitely to avoid future recrudescence.

The duration of treatment should usually be 4 weeks. While fully sensitive alpha-haemolytic streptococci can be well treated in this time, less sensitive organisms require more prolonged parenteral treatment with two drugs. Examples of typical regimens are:

1 For penicillin-sensitive streptococci: intravenous benzylpenicillin 2.4 g 6-hourly plus low-dose gentamicin (60–80 mg twice daily) for 2 weeks, followed by either benzylpenicillin alone for 2 weeks or oral amoxycillin 500 mg 8-hourly for 2 weeks.

2 For streptococci less sensitive to penicillin: benzylpenicillin plus low-dose gentamicin for 4–6 weeks.

3 For staphylococcal endocarditis: intravenous flucloxacillin up to 1.5 g 6-hourly plus either fusidic acid 500 mg 8-hourly or rifampicin 300–600 mg twice daily (both of which may be given orally) for 4–6 weeks. For staphylococci resistant to flucloxacillin, vancomycin or teicoplanin may be given (blood levels must be monitored if vancomycin is chosen).

4 For penicillin-allergic patients: vancomycin can be substituted for benzylpenicillin, and can be given alone for the treatment of staphylococcal endocarditis (blood levels must be monitored to avoid nephrotoxicity and ototoxicity). Teicoplanin is also an effective treatment.

5 For children: the doses of commonly used antibiotics are:

(a) Benzylpenicillin i.v. 150–300 mg/kg daily in four to six divided doses.

(b) Gentamicin 2 mg/kg 8-hourly.

(c) Amoxycillin orally 250 mg 8-hourly (may be increased to 750 mg 12-hourly below age 5 years, and 1.5 g 12-hourly over age 5 years).
(d) Vancomycin i.v. infusion 10 mg/kg 6-hourly (with plasma concentration monitoring).
(e) Teicoplanin i.v. 10 mg/kg 12-hourly (may be reduced to 10 mg/kg daily after a good response).

Use of vancomycin or teicoplanin for the treatment of endocarditis

1 Vancomycin i.v. infusion *either* 500 mg over 60–90 min 6-hourly *or* 1.0 g over at least 100 min 12-hourly (peak level 1 h after infusion should not exceed 30 mg/l, trough should not exceed 10 mg/l).
2 Teicoplanin i.v. 400 mg 12-hourly (may be reduced to 400 mg daily after a good response).

A resolving fever and stabilization of cardiac function are signs that infection is resolving. The C-reactive protein level is also a useful monitor of progress, falling to normal as infection is controlled.

During therapy, cardiac function, renal function and C-reactive protein should be monitored twice weekly. The size of vegetations should be followed by echocardiogram.

Complications

Continuing fever

Continuing fever and elevated C-reactive protein are rarely due to inappropriate antibiotic treatment. The usual cause is inaccessibility of the infecting organisms, because they are replicating in necrotic tissue or abscesses. In these circumstances, surgery is often needed to remove pus or devitalized tissue, and to avoid deterioration in haemodynamic function. Accurate preoperative imaging of the infected site, and of the rest of the heart (to assess ventricular function, the extent of any shunts and the integrity of the coronary circulation), greatly assists the surgeon in operating effectively and safely.

Persisting production of emboli

Persisting production of emboli can lead to disabling strokes, myocardial infarction or limb ischaemia. Routine anticoagulation is considered unwise in uncomplicated endocarditis, because of the risk of bleeding from sites of intra-arterial infection. However, when embolization threatens life or function, anticoagulation is indicated to limit the size and friability of vegetations.

Control of the infection is also important in limiting embolization. In some cases, especially of fungal endocarditis, the best chance of arresting the problem is surgery to remove the affected valve and vegetations.

Severe valve dysfunction

Severe valve dysfunction is an unpredicatable event, though it is more likely if infection is uncontrolled. The changing status of vegetations and loss of devitalized tissue can cause sudden haemodynamic changes. Occasionally there is complete valve failure. In left-sided endocarditis this usually causes acute heart failure, with signs of aortic or mitral regurgitation. Emergency surgery and valve replacement is the only reasonable treatment.

Prevention and control

About two-thirds of endocarditis cases affect patients with an identifiable cardiac abnormality, usually with a positive medical history or physical signs of a susceptible lesion. Antibiotic prophylaxis should be offered to at-risk patients undergoing procedures likely to cause bacteraemia. Although there is no firm evidence that prophylaxis prevents cases, the costs of prophylaxis are low, and the benefit of saving even one case is relatively enormous.

The British Society for Antimicrobial Chemotherapy recommendations for the prophylaxis of infective endocarditis are as follows.

1 For standard risk patients (heart valve lesion, septal defect, patent ductus arteriosus or prosthetic valve) for dental extractions, scaling, periodontal surgery without general anaesthesia; for surgery or instrumentation of the upper respiratory tract; for urogenital examination or instrumentation:
(a) amoxycillin 3 g orally 1 h before the procedure; *or*
(b) for penicillin-allergic patients, or those who have received more than a single dose of penicillin in the previous month) clindamycin 600 mg orally 1 h before the procedure.

For both of these regimens, children under five receive a quarter of the adult dose, and children aged 5–10 years receive half of the adult dose.

2 For patients with a previous history of endocarditis: amoxycillin *plus* gentamicin, as for procedures performed under anaesthesia.

3 For the above procedures performed under general anaesthesia in standard-risk patients without prosthetic valves:

(a) amoxycillin 1 g intravenously 1 h before induction and 0.5 g orally 6 h later; *or*
(b) amoxycillin 3 g orally 4 h before and 3 g orally as soon as possible postoperatively; *or*
(c) amoxycillin 3 g orally plus probenecid 1 g orally 4 h before induction (child doses not recommended).

4 For these procedures under general anaesthesia in patients with prosthetic valves or a a previous history of endocarditis:
intravenous amoxycillin 1.0 g *plus* intravenous gentamicin 120 mg at induction, then 0.5 g amoxycillin orally or intravenously 6 hours later (child under 5, amoxycillin a quarter of the adult dose plus gentamicin 2 mg/kg; child 5–10 years, amoxycillin half of the adult dose plus gentamicin 5 mg/kg).

5 For all penicillin-allergic patients having these procedures under general anaesthetic:
(a) vancomycin 1.0 g intravenously over at least 100 min, followed by gentamicin 120 mg intravenously at induction or 15 min before the procedure (child under 10 years, vancomycin 20 mg/kg plus gentamicin 2 mg/kg); *or*
(b) teicoplanin intravenously 400 mg *plus* gentamicin 120 mg at induction or 15 min before the procedure (child under 14 years, teicoplanin 6 mg/kg plus gentamicin 2 mg/kg); *or*
(c) clindamycin intravenously 300 mg over at least 10 min at induction or 15 min before the procedure (child under 5 years, a quarter of the adult dose; child aged 5–10 years, half of the adult dose).

6 For genitourinary procedures:
As for **4** and **5** above, except that clindamycin is not given.

Note: for genitourinary procedures in the presence of colonized or infected urine, the prophylaxis chosen should include an antibiotic effective against the urinary organism.

7 For obstetric, gynaecological and gastrointestinal procedures:
prophylaxis is indicated only for patients with a previous history of endocarditis or with a prosthetic valve: the regimens described for genitourinary procedures are appropriate.

Infective endarteritis

Infective endarteritis can arise by haematogenous spread in any bacteraemic condition, though it is likely that some insult to the endothelium must occur to allow adherence of organisms. The atheromatous lower aorta is particularly at risk, because of its susceptibility to endothelial ulceration, and aneurysm formation with intramural thrombus. In the affected age groups, bacteraemias are often of bowel or urinary tract origin, so *Escherichia coli* or enterococci are common pathogens. Pre-existing aneurysms in other vessels, such as the popliteal or femoral arteries, may also become infected. These are often traumatic or autoimmune in origin, affecting younger people. They may be infected with staphylococci, endocarditis organisms or rarities such as salmonellae.

Infected emboli from endocarditis itself may damage or occlude arteries, causing a metastatic or marantic arteritis which can become aneurysmal.

Clinical indicators of infective endarteritis are not always obvious, but include fever, local pain or enlargement of any aneurysm, raised erythrocyte sedimentation rate and C-reactive protein and episodes of embolization or occlusion in the circulation affected by the disease. Postoperative graft occlusion or leakage of the anastomosis site may be indicators that an infected blood vessel has been operated upon. Imaging of the affected site is helpful in demonstrating vegetations, exuberant thrombus, dissection or leakage from the vessel.

Syphilitic aortitis (see Chapter 15) is a special case in which late infection of the ascending aorta destroys the elastic tissue, causing the development of a potentially massive and fragile aneurysm. This must be treated with care, as the Jarisch–Herxheimer reaction can increase inflammation and friability of the infected tissue.

Case 12.1: PUO in a fit young man

History: A 20-year-old manual worker complained of 2 or 3 weeks history of fevers and fatigue with no other symptoms. He had recently been working at a bridge restoration where rats and nesting pigeons were often found. Six weeks before, he had stayed on his parents' farm in rural Wales, but denied consuming any unpasteurized dairy products. He had always been extremely fit, and had never travelled overseas.

Physical findings: Temperature 37.8°C, pulse 88 bpm, blood pressure 115/70 mmHg, a faint aortic systolic murmur (possibly a 'flow' murmur), some traumatic scratches on the hands, a blister on the right great toe (from his new work boots). Physical examination was otherwise normal, including neurological testing and fundoscopy.

Questions: What immediate investigations and laboratory tests would you request?

Which infectious diagnoses would you consider at this stage?

Laboratory investigations: Haemoglobin 12.3 g/dl, white cell count 11.2×10^9/l (8.7×10^9/l neutrophils), ESR (erythrocyte sedimentation rate) 45 mm/hour, CRP (C-reactive protein) 30 IU/l. Urine dip testing showed + blood but no neutrophil reaction, renal and liver function tests were all normal. The chest X-ray was normal. Blood was drawn for cultures, and serological tests for leptospirosis, brucellosis and *Coxiella burnetii* infection were requested. Serum was saved for possible histoplasmosis investigation, but this was deferred in the absence of skin or chest abnormalities.

Management and progress: After 2 days' observation and bed rest in hospital, his temperature had fallen to 37°C and urine and blood cultures had produced no evidence of growth. He was discharged with early outpatient follow-up arranged. Thirty-six hours later, he was recalled to the ward because four of four blood culture bottles had signalled positive, and Gram-staining showed small Gram-negative rods.

Re-examination showed temperature 38.2°C, pulse 92 bpm, blood pressure 120/55 mmHg, and a $^1/_4$ ejection systolic murmur radiating to the left neck with a faint early diastolic murmur at the left sternal edge. The haemoglobin was 11.2 g/dl, white cell count 13.4×10^9/l. Urine dip test showed ++ blood and + protein.

Questions: What is the likely diagnosis?

Does it fulfil accepted criteria?

What immediate investigation is indicated?

Should antimicrobial therapy be commenced; what regimen would you use?

Further management and progress: The diagnosis of infective endocarditis was made, and fulfilment of the 'Duke University criteria' was complete, with two major criteria documented (see Chapter 14). Treatment was commenced immediately with intravenous amoxycillin plus gentamicin. A Doppler echocardiogram was urgently arranged (presence of diagnostic abnormalities would comprise another diagnostic criterion, but information is also required about cardiac function and the degree of valvular damage at this stage). The result confirmed the presence of vegetations on both leaflets of a bicuspid aortic valve, with restricted outflow and a pressure gradient of 12 mmHg across the valve. Mild regurgitation was demonstrated. Left ventricular volumes and function were normal.

After 4 days' therapy, fever persisted at 38°C, the pulse was 120 bpm and the blood pressure 130/30 mmHg.

The patient complained of shortness of breath and rapid palpitation.

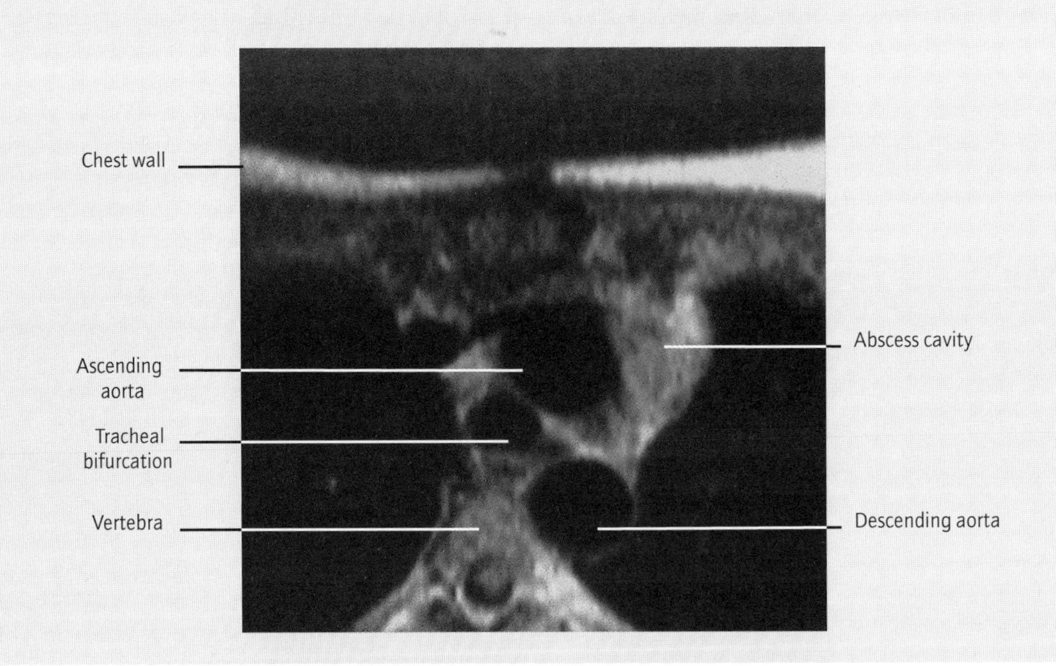

Fig. CS.2 Transaxial NMR image of the upper mediastinum at the level of the tracheal bifurcation, showing a large abscess surrounding the origin of the ascending aorta.

Questions: What is the most likely reason for his failure to respond to therapy?
What urgent investigation is indicated?
What mode of treatment is likely to be needed?

Further management and progress: Laboratory results identified the Gram-negative rod as *Cardiobacterium hominis*, sensitive to amoxycillin plus gentamicin and also to broad-spectrum cephalosporins. Inappropriate antimicrobial chemotherapy was therefore not the cause of the patient's deterioration. In almost every case, the problem is caused by infection inaccessible to the antimicrobial agent, either because of gross vegetations or, more often, to intracardiac abscess formation. Urgent cardiac imaging (in this case, dynamic magnetic resonance imaging, though angiography is often employed) showed a large abscess surrounding the origin of the ascending aorta (Fig. CS.2). Surgery is the only curative option in such cases; urgent abscess clearance and aortic valve replacement was followed by 6 weeks' therapy with intravenous ceftriaxone, and resulted in complete resolution of fever and cardiac abnormalities.

13 Infections of the Central Nervous System

Introduction, 254
Structural and functional considerations, 254
Pathogenesis of CNS infections, 255

Meningitis and meningism, 255
Introduction, 255
Physical signs of meningism, 256
Lumbar puncture, 257
Laboratory diagnosis, 258

Viral meningitis, 260
Introduction, 260
Enterovirus meningitis, 261

Poliomyelitis, 262
Introduction, 262
Epidemiology, 262
Pathogenesis, 263
Clinical features, 263
Diagnosis, 263
Management, 264
Complications, 264
Prevention and control, 264

Herpes simplex type 2 meningitis, 264

Bacterial meningitis, 265
Introduction, 265
Lumbar puncture and CSF changes in purulent meningitis, 265
Meningococcal meningitis, 265
Haemophilus influenzae meningitis, 272
Pneumococcal meningitis, 275
Listerial meningitis, 276
Streptococcus 'milleri' meningitis and ventriculitis, 277

Encephalitis and meningoencephalitis, 278
Introduction, 278
Herpes simplex encephalitis, 278
Arbovirus encephalitides, 280
Rabies, 280
Neuroborreliosis, 281
Spongiform encephalopathies, 281

Cerebral and intracranial abscesses, 282
Cerebral abscess, 282
Other space-occupying lesions of the CNS, 284

Introduction

Structural and functional considerations

The brain and spinal cord form a special compartment of the body. The cerebrospinal fluid (CSF) is different from other body fluids, in particular having a low protein content, including very low immunoglobulin levels. The central nervous system (CNS) has no lymphatics, but depends on migration of lymphocytes to and from the blood circulation.

The layers of the meninges support and protect the brain and spinal cord, and contain the CSF. Together with their blood vessels, choroid plexuses and arachnoid processes, the meninges constitute the blood–brain barrier, a structural and functional barrier between the blood and the CSF. It permits the brain to maintain its own homeostatic environment, with its own acid–base equilibrium. It inhibits the entry of organisms, and of many drugs and toxins.

During an inflammatory process, fast-reacting proteins appear in the CSF, mostly by diffusion from the blood, with the low molecular weight, more diffusible proteins appearing in highest concentrations. Immunoglobulins diffuse into the CSF but can also be locally manufactured once lymphocytes have migrated from the circulation. Both lymphocytes and neutrophils are able to enter the CSF during infection and inflammation.

It is possible to deduce that a specific antibody is being manufactured locally, and not just diffusing into the CSF, by comparing the ratio of its CSF:blood concentrations with the ratio for an antibody unlikely to be involved with acute CNS disease. For instance, if herpes simplex antibody is manufactured locally in a case of herpes simplex encephalitis, its CSF:blood concentration ratio will be higher than that of, say, rubella antibodies. Most virology departments will estimate these types of ratios in suspected viral infections. The relationship between the ratios is an antibody index.

An antibody index

$$\frac{\text{CSF: serum ratio of the antibody in question}}{\text{CSF: serum ratio of reference (e.g. rubella) antibody}}$$

If this index is greater than 2, local production of the antibody in question may be assumed.

When the blood–brain barrier is disrupted by inflammation, proteins and cells readily enter the CSF. This may permit more effective immune responses, and also allows the rapid entry of therapeutic drugs such as penicillins or vancomycin which do not penetrate the intact blood–brain barrier. The disadvantages of inflammation and exudation are the elevation of intracranial pressure, with reduction in cerebral perfusion pressure (the difference between blood pressure and CSF pressure) and reduced cerebral blood flow. Intense inflammation causes protein accumulation and fibrin deposition. This can obstruct the aqueduct, or the foramina of the brain, causing local or general hydrocephalus.

Inflammatory changes can be inhibited by the use of corticosteroids, especially if these are given early, and this may improve the outcome in some diseases.

Rising CSF pressure is also related to changes in the water content of the brain. It is now recognized that increased brain water is not necessarily equivalent to cerebral oedema, as the water may be in other compartments than the interstitium. In meningitis particularly, it is no longer the custom to limit fluid intake as, rather than preventing a rise in intracranial pressure, it reduces blood volume and impairs cerebral perfusion. In encephalitis there may be true cerebral oedema, often demonstrable by computed tomographic (CT) or magnetic resonance (MR) scanning. It may then be possible to reduce oedema with dexamethasone treatment.

Effects of meningeal inflammation

1 Increasing cerebrospinal fluid (CSF) protein and cell count.
2 Increased entry of water-soluble antibiotics.
3 Increased brain water.
4 Increased CSF (= intracranial) pressure.
5 Reduced cerebral perfusion pressure.
6 Risk of CSF obstruction leading to hydrocephalus.

The incidence of deafness following some types of acute meningitis is reduced by early dexamethasone treatment. The reason for this is uncertain; it may be reduction of oedema or inflammation, or possibly protection of the cochlear hair cells from the action of endotoxin.

Pathogenesis of CNS infections

Many infections of the CNS are blood borne; the causative organism colonizes or infects other sites of the body before causing CNS disease. Thus, the viruses of meningitis are often found in the throat and bowel as well as in the CSF; a neonate with cutaneous herpes simplex lesions is at risk of developing encephalitis; in bacterial meningitis, the causative organisms appear first in the nasopharynx, then in the blood and lastly in the CSF.

It is relatively uncommon for the CNS to be infected by the spread of organisms from adjacent structures. However, pneumococcal infection may arise from chronic infection in the paranasal sinuses or the middle ear. This is much more likely if there is a defect in the meninges, usually a tear in the dura mater following head injury, surgery or other damage (in this situation recurrent meningitis may occur). It is thought that amoebic meningitis arises by migration of amoebae through the cribriform plate.

There are rare examples of spread of pathogens along nerve trunks or between segments of the spinal cord. Rabies viruses enter peripheral nerves and migrate to the CNS where they then cause infection. The incubation period is proportional to the length of nerve that must be traversed. Transection of the nerve trunk or amputation of the injured limb will prevent rabies in animals. Herpes zoster viruses can enter the spinal cord via a dorsal root and spread segment by segment, causing an ascending myeloencephalitis.

Modes of pathogenesis of central nervous system infections

1 Infection during viraemic phase of viral infections.
2 Blood-borne spread from local or bacteraemic bacterial infection.
3 Contiguous spread from intracranial infective focus.
4 Entry of bacteria through a defect in the dura.
5 Rare spread through cribriform plate.
6 Rare spread along nerve fibres and connections.

Meningitis and meningism

Introduction

Meningitis is inflammation of the meninges. Meningism is the group of symptoms and signs that accompany the inflammation. The symptoms are headache, neck stiffness, nausea and vomiting and photophobia. The headache is global and usually described as the worst ever experienced. Photophobia is the most variable of the symptoms; in severe cases it may be very marked, but it can occur in a number of other conditions such as tension headache and migraine.

Main features of meningism
1 Headache.
2 Neck and back stiffness.
3 Nausea and vomiting.
4 Photophobia.

Typical signs of meningism are uncommon in infants and small children, who rarely exhibit neck and back stiffness. They usually simply appear hypotonic, though later they may develop opisthotonus. A bulging fontanelle is a useful sign of raised intracranial pressure in an infant. Fever, convulsions and persistent vomiting are important warnings of meningitis in the under 2s, in whom meningism may also be absent.

Warning signs of meningism in infants
1 Bulging fontanelle.
2 Vomiting.
3 Strange high-pitched cry.
4 Convulsions.
5 Opisthotonus.

Meningism can occur without meningitis. It can accompany upper lobe pneumonia, urinary tract infection and high fevers such as the prodromal fever of dysentery. In these conditions CSF examination is normal, and the subsequent investigation and evolution of the disease reveal the true diagnosis. The mechanism of meningism without meningitis is unknown.

Meningism can also occur without significant fever in non-infectious conditions. In subarachnoid haemorrhage, headache and nausea have a very abrupt ('thunderclap') onset and may be accompanied by collapse, loss of consciousness and/or abnormal neurological signs. Malignant infiltration is a cause of meningism in leukaemia and metastatic melanoma. Meningism is a rare adverse reaction to non-steroidal anti-inflammatory drugs, described most often after ibuprofen.

Conditions where meningism can occur without meningitis
1 Small children with high fevers.
2 Upper lobe pneumonias.
3 Acute urinary tract infections.
4 Subarachnoid haemorrhage.
5 Meningeal malignancies.

Physical signs of meningism

The two traditional tests for meningism are: (i) to demonstrate inability to flex the neck and touch the chin to the chest; and (ii) to elicit Kernig's sign (Fig. 13.1). These will easily demonstrate moderate to severe meningism, but mild meningism is sometimes missed. More subtle signs are the tripod sign—the patient is unable to sit up from a supine position without making a tripod with the hands resting on the bed behind him or her; and the inability to curl forwards enough to touch the nose to the knees (Fig. 13.2).

The detection of mild meningism is important, first because it permits early clinical suspicion in developing meningitis, and second because diseases such as tuberculous meningitis may cause only the mildest meningism before major neurological deficit occurs.

Fig. 13.1 Kernig's sign of meningism: the patient's leg cannot be straightened because of hamstring spasm.

(a)

(b)

Fig. 13.2 Meningism can be easily demonstrated: this patient with meningococcal meningitis could not oppose his nose and knees (a), but after recovery (b), the manoeuvre was easily performed.

Lumbar puncture

This is the most rapid diagnostic test for meningitis. It permits the distinction of bacterial meningitis from viral meningitis or meningoencephalitis. It often allows a rapid aetiological diagnosis of infection by Gram stain, polymerase chain reaction (PCR) or slide agglutination test on CSF.

Nevertheless, lumbar puncture occasionally causes herniation of the brain through the foramen magnum after fluid is removed from the spinal theca. This is rare, probably because the duration of inflammation before presentation is short, and has not yet interfered with normal CSF flow. Even if hydrocephalus is present, it is communicating, so that removing CSF from the lumbar theca will simply reduce intracranial pressure. Since fluid is incompressible, only a tiny amount need be removed to reduce intracranial pressure and increase cerebral perfusion pressure and, often, to reduce headache (Fig. 13.3).

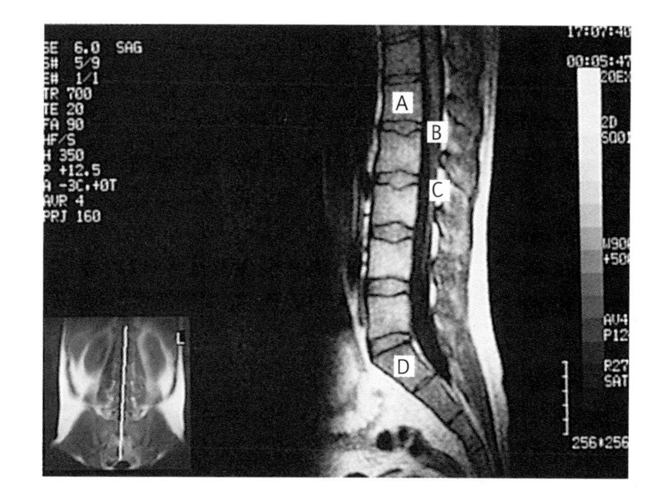

Fig. 13.3 Lateral magnetic resonance image of normal lumbar spine, showing the termination of the spinal cord at the L1/L2 level, leaving a space below filled with cerebrospinal fluid and nerve roots; the optimum track for lumbar puncture is shown by the dotted line. A, body of 1st lumbar vertebra; B, lower extreme of the spinal cord; C, track of the lumbar puncture between spines of the third and fourth lumbar vertebra; D, the sacrum.

The risk of herniation is greater, however, when the history of headache is longer than 3 or 4 days or when a history of cerebellar tonsillar herniation is known. The presence of papilloedema also indicates increased risk. The absence of papilloedema means little, however, as it develops slowly, and is not always present when the CSF pressure is high. If lumbar puncture is contraindicated it is necessary to obtain an emergency CT or MR scan. In most cases, scan results show that lumbar puncture can safely be performed. Otherwise, ventricular puncture may be required.

Indicators of increased risks of lumbar puncture

Reduced cerebral function (Glasgow Coma Scale <13).
Sluggish or dilated pupils.
Prolonged or focal seizures.
Abnormal or decerebrate posturing.
Papilloedema.
Arterial hypertension, bradycardia or abnormal respiratory pattern.
Severe shock.
Coagulopathy.

Unrelieved intracranial hypertension will 'incarcerate' the brain inside the skull, occluding the cerebral circulation and leading to cerebral vein thrombosis or cerebral ischaemia. Anaesthesia, artificial hyperventilation or insertion of a CSF drain may occasionally be needed to avoid this.

It may be possible to deduce the diagnosis and give empirical treatment without performing a lumbar puncture. This is so with childhood meningitis, particularly in the presence of a typical meningococcal rash, when a third-generation cephalosporin is the treatment of choice.

Laboratory diagnosis

Urgent diagnostic and prognostic information can be obtained from CSF examination. However, it is essential that a simultaneous specimen of blood be obtained for glucose estimation so that the result of CSF glucose may be interpreted. Other important investigations include blood cultures, throat swabs and stool specimens (for viral culture).

CSF examination

Three specimens of CSF are collected into numbered universal containers, and one is collected into a blood sugar estimation bottle. From infants, only 10–12 drops of fluid should be taken for each specimen, while from older children and adults 1–2 ml can be collected.

The CSF cell count is performed by examining an unspun, unstained specimen in a counting chamber. The cell types are identified by examination of a Giemsa-stained centrifuged deposit. The presence of neutrophils is associated with bacterial infections such as *Neisseria meningitidis, Streptococcus pneumoniae* or *Haemophilus influenzae*. The presence of lymphocytes is associated with infection with viruses, mycobacteria or *Leptospira*. A lymphocytic or mixed pleiocytosis may be seen in cases of partially treated pyogenic infection or in brain abscess.

Further information can be obtained from biochemical tests. The CSF protein level reflects the degree of meningeal inflammation. It is often two or three times the normal level in viral infections of the CNS, and up to five or 10 times the normal level in bacterial infections. In abscesses and chronic infections, such as tuberculous meningitis, it may reach very high levels. The CSF glucose level is usually normal in viral infections, significantly low in bacterial meningitis, and negligible in advanced tuberculous meningitis. As acute intracranial disorders can raise the blood glucose (often to 8–12 mmol/l), it is essential to compare the CSF glucose with a simultaneous blood glucose estimation.

Normal CSF composition includes:

1 Cells: <6 white blood cells, all lymphocytes; no red blood cells (in an atraumatic tap).

2 Protein: 0.1–0.4 g/l (lumbar theca); 0.07–0.25 g/l (ventricular).

3 Glucose: up to 1.7 mmol/l below blood glucose.

The centrifuged deposit should be stained by Gram's method to demonstrate bacteria. The sensitivity of Gram stain is about 10^4 CFU/ml. This can be improved upon by using acridine orange. Ziehl–Nielsen stain should be employed if tuberculosis is suspected, or indicated by the cell count or biochemical examination. If cryptococcosis is suspected in an immunosuppressed patient, the CSF should be examined by an India ink method which outlines the fungal capsule in sharp relief.

Specimens from the centrifuged deposit are inoculated on to a range of media capable of growing the main pathogens. This usually includes blood and chocolated agar, together with an enrichment broth (such as tryptose broth, intended to amplify the sometimes scanty organisms). All colonies growing after 24 h are identified and sensitivity testing performed. If there is no bacterial growth after 24 h the enrichment broth is subcultured. It is important that all isolates are sent to reference laboratories for confirmation and typing (Fig. 13.4). In England and Wales they should also be reported to the Communicable Disease Surveillance Centre (CDSC; see Chapter 26), to update national records of current types and antimicrobial sensitivities of important pathogens.

The aetiological diagnosis of viral meningitis by CSF culture is often unrewarding. A better diagnostic yield, in enteroviral infections, is obtained from stool culture. Mumps or measles virus may be recovered from urine cultures. In suspected herpetic encephalitis, anti-herpes simplex virus antibody can be measured in the CSF, and the CSF:serum antibody ratio compared with that of rubella or measles (Fig. 13.5). Positive identification of viral agents in CNS disease should also be reported to the CDSC.

Rapid diagnostic techniques

The importance of pyogenic meningitis as a medical emergency has focused research on rapid diagnostic techniques. Some attempted to distinguish bacterial from viral infections, based on the detection of lactate or C-reactive protein. Unfortunately, CSF lactate was difficult to measure reliably, and was not clinically useful. C-reactive protein levels were found to indicate the intensity of inflammation rather than its aetiology.

The common bacteria causing meningitis possess polysaccharide capsules, whose antigens can be detected by rapid methods. These methods include latex agglutination, coagglutination, counterimmunoelectrophoresis and enzyme immunoassay (EIA). Positive results can be

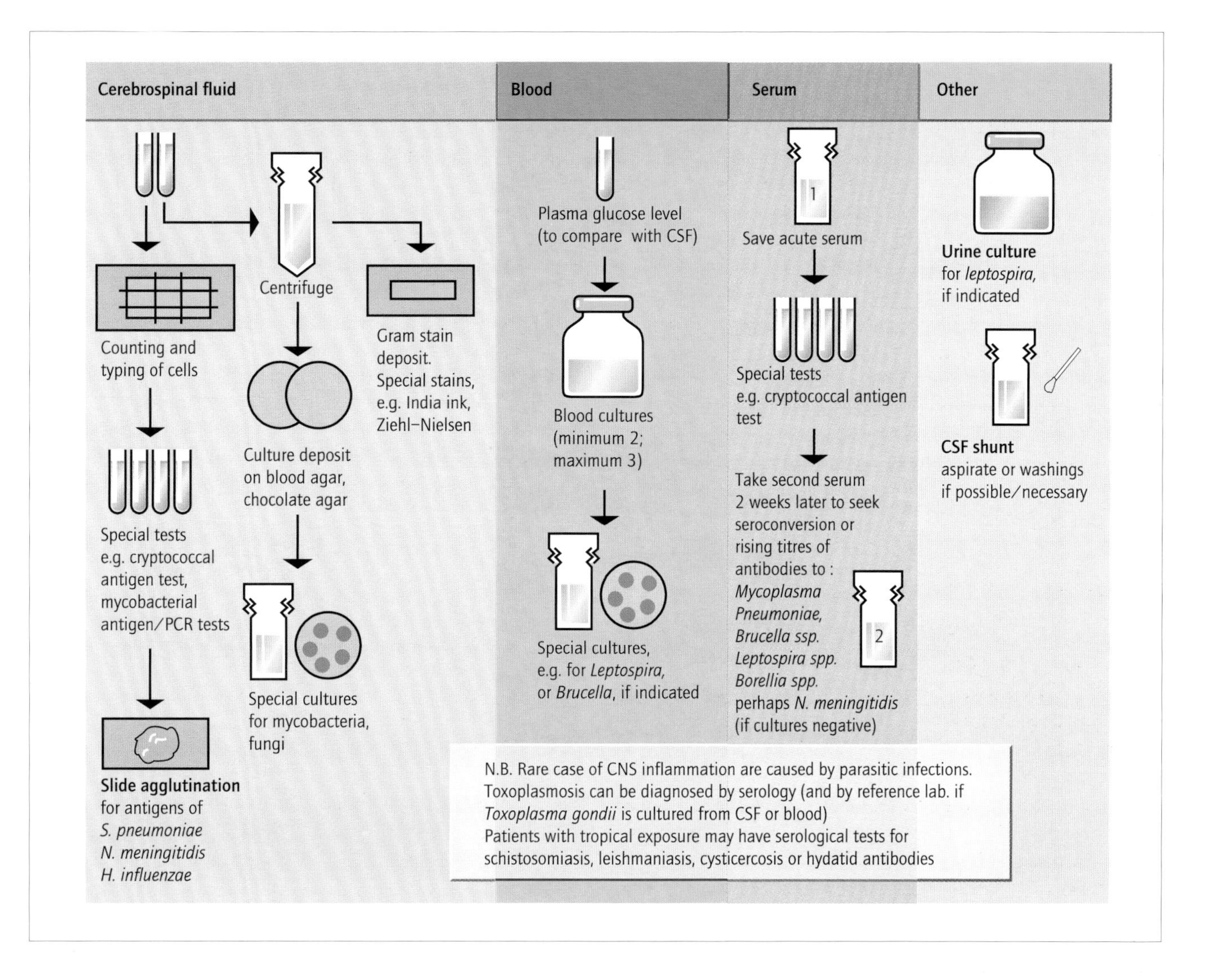

Fig. 13.4 Scheme for the investigation of bacterial meningitis. CNS, central nervous system; CSF, cerebrospinal fluid; PCR, polymerase chain reaction.

obtained after antibiotics have been commenced and cultures become negative. Effective tests are available for *S. pneumoniae*, *N. meningitidis* serogroups A, C and W135 and *H. influenzae*. Satisfactory results are more difficult to obtain with the less immunogenic *N. meningitidis* serogroup B, but recent research indicates that exposing the reagent mixture to ultrasound can improve the sensitivity of agglutination tests sufficiently to obtain a reliable result for serogroup B. The concentration of antigen indicates prognosis in *S. pneumoniae* infection, but studies for other pathogens are awaited.

PCR-based detection methods have had a significant impact on the diagnosis of pyogenic meningitis and are proving especially valuable in the diagnosis of meningococcal disease. PCR has also improved the ability of laboratories to diagnose tuberculous meningitis and has also found an important role in herpes simplex encephalitis and varicella zoster virus menigitis.

Rapid diagnostic techniques that can be performed on cerebrospinal fluid

1 Gram stain of centrifuged deposit.
2 Special stains:
   (a) Ziehl–Nielsen (for acid-fast organisms).
   (b) India ink (for capsulated yeasts).
3 Slide agglutination tests:
   (a) *Neisseria meningitidis* A, C, W135, Y.
   (b) *Streptococcus pneumoniae*.
   (c) *Haemophilus influenzae*.

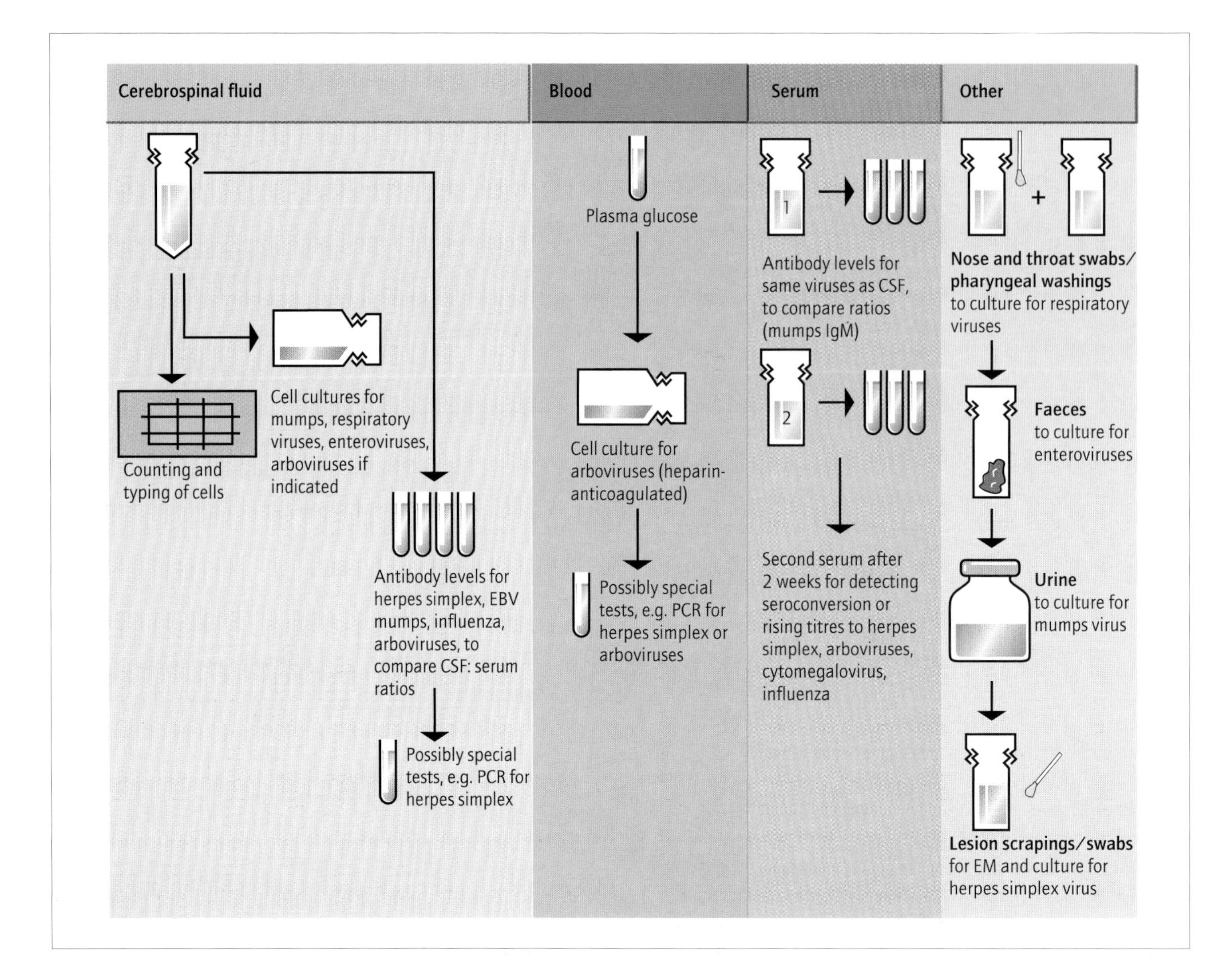

Fig. 13.5 Scheme for virological investigation of central nervous system disease. CSF, cerebrospinal fluid; EBV, Epstein–Barr virus; EM, electron microscopy; IgM, immunoglobulin M; PCR, polymerase chain reaction.

4 Antigen test: *Cryptococcus neoformans*.
5 Polymerase chain reaction DNA detection tests:
 (a) *Mycobacterium tuberculosis*.
 (b) Herpes simplex.
 (c) Varicella zoster virus.
 (d) Enteroviruses.
 (e) *Neisseria meningitidis*.
 (f) *Streptococcus pneumoniae*.

Examination for other pathogens

CSF examination contributes to the investigation of other infective conditions such as syphilis, Lyme disease and trypanosomiasis. In syphilis and Lyme disease serological investigations are important. In suspected cerebral trypanosomiasis, CSF should only be examined after the blood has been cleared of parasites with suramin. Trypanosomes can be concentrated by an anion exchange chromatography. The presence of parasites or of morula cells indicates CSF infection and the need for treatment with arsenical preparations.

Viral meningitis

Introduction

Viral meningitis is common worldwide. Approximately 1000 laboratory-confirmed cases are reported each year in the UK, but this is only a fraction of the total incidence as most infections are mild or inapparent. The illness is usually self-limiting, rarely lasting for more than a week. Its importance lies in the differential diagnosis from tuberculous (see Chapter 18) and acute bacterial meningitis, and from rare diseases such as listeriosis, borreliosis and cryptococcosis. In some places where immunization is still rare, viral meningitis can be caused by poliovirus, and may be followed by paralysis.

ORGANISM LIST

Echovirus
Coxsackievirus
Poliovirus
Mumps virus
Herpes simplex type 2
Varicella zoster virus
Influenza virus type A or B
Arboviruses (usually meningoencephalitis)
A rare effect of many virus infections, such as rubella and Epstein–Barr virus infection.

Enterovirus meningitis

Introduction and epidemiology

Young children are the usual source of enterovirus infection. Faecal virus shedding may persist for several weeks and spread usually occurs as a result of environmental contamination, particularly under conditions of crowding and poor hygiene. Enterovirus meningitis is commonest in children aged 5–14 years, but also occurs in other age groups. Intrafamilial spread is common, and outbreaks have been described in hospital nurseries, boarding schools and other residential institutions.

By far the commonest cause of enteroviral meningitis is echovirus infection. The predominant serotypes are 6, 9, 11, 19 and 30. There are annual epidemics of echovirus infections in the late summer (Fig. 13.6). They cause the so-called summer flu, with fever, sore throat and often headache. A proportion of infected individuals develop meningitis, with or without a preceding or accompanying sore throat. Enteroviruses types 70 and 71 are also important causes of outbreaks.

Coxsackievirus meningitis is caused predominantly by serotypes A9, B4 and B5. Coxsackie A infections affect all age groups; type B disease occurs mainly in infants and preschool children.

Enteroviruses can cause a rare, generalized chronic brain infection in individuals with agammaglobulinaemia.

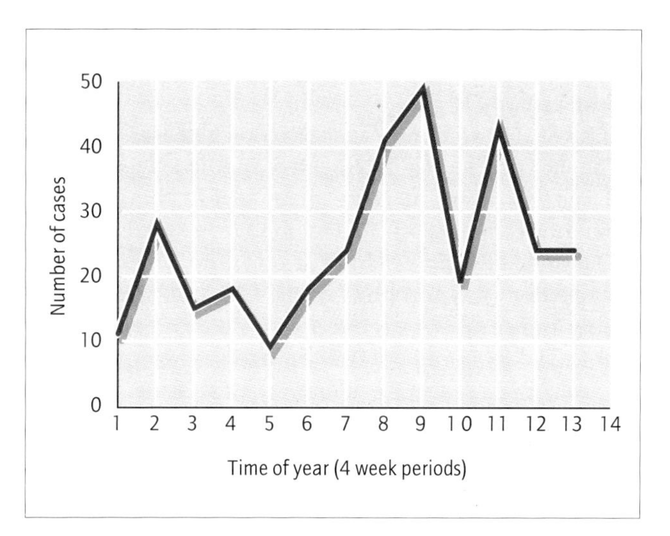

Fig. 13.6 Echovirus epidemic curve: this common infection dominates the epidemic curve for all viral meningitis notifications.

Clinical features

There may be a personal or family history of sore throat in the preceding 5–7 days. The patient then develops increasing headache over 12–36h, usually with nausea and vomiting. There is no alteration in consciousness or neurological function. There is usually meningism, but this varies from minimal to severe and is not always in proportion to the severity of the headache.

Other clinical features are few. There may be cervical lymphadenopathy and, rarely, pharyngitis. The white cell count and differential are almost always normal, as is blood biochemistry.

Diagnosis

The typical clinical and CSF findings during the epidemic period of the year should suggest the diagnosis. The differential diagnoses are rarities; they include tuberculous meningitis, neuroborreliosis and, when the CSF cell count includes neutrophils, intracerebral abscess and partly treated bacterial meningitis. On rare occasions an exacerbation of multiple sclerosis or an autoimmune transverse myelitis causes CSF pleiocytosis, but this is often accompanied by typical neurological abnormalities.

Lumbar puncture and CSF examination should be carried out in patients with other than mild meningism. This is to exclude bacterial meningitis or other diagnoses such as subarachnoid haemorrhage. Cases with minimal meningism may be observed overnight before a decision on lumbar puncture is made, and the investigation often turns out to be unnecessary.

Typical CSF changes in viral meningitis are as follows:
1 Cells: 40–250, all lymphocytes.
2 Protein: 0.45–0.9 g/l.
3 Glucose: 1.0–1.7 mmol/l below blood glucose.

The blood glucose is elevated in diabetic patients, and can be raised in acute brain insult (though viral meningitis is rarely sufficient to do this). It should always be checked at the time of lumbar puncture. There may be a proportion of neutrophils in the CSF cell count if the lumbar puncture is performed early in acute viral meningitis. The presence of 50% or more of lymphocytes, however, is reassuring evidence against acute bacterial meningitis.

A lymphocytosis, or predominantly lymphocytic pleiocytosis, can occur in the CSF in:
1 Tuberculous meningitis.
2 Partly treated bacterial meningitis.
3 Intracranial abscess.
4 Leptospirosis.
5 Lyme borreliosis.
6 Viral encephalitis.
7 Lymphocytic leukaemias.

These conditions should be considered, however briefly, if the presentation or course of viral meningitis is unusual. CSF glucose levels are usually depressed in tuberculous meningitis, an important feature in the differential diagnosis (see Chapter 18).

Enteroviruses are transiently present in the pharynx and the CSF. They are excreted for several days or weeks in the faeces. Enteroviruses and their detection are discussed more fully in Chapter 6. Test systems exist for detecting immunoglobulin M (IgM) antibodies to enteroviruses, but the multiplicity of serotypes makes specific serological diagnosis impractical.

Management

There is currently no specific antiviral treatment for enteroviral diseases, but antiviral drugs that inhibit the attachment and entry of enteroviruses into host cells may soon be available. Symptomatic treatment with bed rest produces early improvement in malaise. Analgesics and antiemetics are helpful. Fever and headache often improve within 48–72 h. Lumbar puncture is often followed by improvement in the headache. Occasionally severe tension headache or cervical spondylosis delays recovery. A non-steroidal anti-inflammatory analgesic will often help, as will a small dose of promethazine for its mild sedative and muscle-relaxing effect.

If there is no improvement in 5–7 days, the differential diagnosis should be reconsidered.

Poliomyelitis

Introduction

Poliomyelitis is an enteroviral infection that causes damage and death of anterior horn cells. This is exceptionally rare in other enteroviral infections. Until recently, polio was the commonest worldwide cause of paralysis and limb-wasting in young age groups. However, no cases have been reported from most Western countries in the last 2 years, and the World Health Organization hopes to begin the process of declaring the disease eradicated in the year 2000.

Epidemiology

Humans are the only known reservoir of infection. In acute infection, virus excretion lasts for up to a week in oropharyngeal secretions and for up to 6 weeks in faeces. Faecal–oral transmission is the major method of spread in conditions of poor hygiene. Where sanitation is good, pharyngeal spread is more important.

There are three distinctive patterns of disease: endemic, epidemic and sporadic. Endemic disease occurred in the prevaccine era and is still seen in some parts of the world with poor vaccine coverage. Virus circulated extensively among infants and young children and most individuals had been infected by 3 years of age. Because a large proportion (over 99%) of polio infections in young children are asymptomatic, the incidence of neurological disease is relatively low when the disease is endemic.

During the 20th century, many temperate countries experienced a shift from endemic to epidemic disease. This occurred as living conditions improved and virus circulation among young children diminished. The effect was to increase the average age at which infection occurred in teenagers and young adults, in whom neurological disease is much commoner; 10% of infections at this age are paralytic. Large outbreaks of paralytic polio occurred in the UK, USA and many other countries during the 1940s and 1950s. These outbreaks were usually due to type 1 poliovirus.

Following the introduction of vaccination in the 1950s, polio was rapidly brought under control. The last outbreak in UK was in 1977 and occurred in unvaccinated itinerant gypsies. Since then there have been only sporadic cases, in unvaccinated travellers to countries where the disease is still common. Some developed countries, e.g. Holland, have continued to experience occasional outbreaks in groups that refuse vaccination on religious grounds. In other countries, outbreaks in highly vaccinated populations have occurred due to vaccine failure. This may occur when reduced-potency vaccine has been used, or due to competitive inhibition of oral poliovaccine in the intestine by other enteric viruses. An unusual outbreak occurred in Finland in 1984 due to an antigenically altered type 3 poliovirus, against which vaccination with inactivated vaccine (routinely used in Finland) was ineffective.

Live polio vaccines carry a minute, but definite, risk that the vaccine virus will mutate to a virulent form. Half of the resulting clinical cases have occurred in vaccine recipients; the remainder affect contacts of recipients.

Pathogenesis

Poliovirus enters the body through the alimentary tract and the oropharynx. This is followed by a viraemic phase, during which the virus enters the meninges and spinal cord. The virus infects and kills anterior horn cells, leading to lower motor neurone degeneration.

Clinical features

The incubation of poliomyelitis varies from 3 to 35 days, but is usually between 1 and 3 weeks. Probably over 90% of cases are inapparent or mild. Sometimes an influenza-like illness, an acute pharyngitis or a mild diarrhoeal disease may be recognized as caused by poliovirus. Otherwise the disease may be divided into non-paralytic and paralytic, depending on the outcome.

Non-paralytic poliomyelitis

Non-paralytic poliomyelitis is simply viral meningitis caused by poliovirus. It is indistinguishable from other enteroviral meningitides, except for the results of viral culture and serology, and it has an identical self-limiting course. It is important as patients excrete the virus in faeces for up to 6 weeks after disease onset.

Paralytic poliomyelitis

Paralytic poliomyelitis begins as a feverish illness, often with myalgia and/or meningism, followed after 2–5 days by the development of lower motor neurone lesions. The muscles affected are often painful in the day or two before paralysis is evident. Trauma, such as vigorous exercise, injection, vaccination or tonsillectomy, makes paralysis of the local muscles more likely. Paralysis is also more likely and more extensive in older age groups.

Paralysis develops over about 48h. Fasciculation is common at the onset, then reflexes and movement are lost. The muscle groups affected are asymmetrical. Upper motor neurone or sensory features are absent. If the upper arms are affected, the diaphragm is at risk; if bulbar weakness occurs, the airway must be guarded.

Clinical spectrum of poliovirus infection

1 Inapparent seroconversion.
2 Non-neurological (usually pharyngeal or diarrhoeal).
3 Non-paralytic (viral meningitis only).
4 Paralytic (anterior horn cell death).

Examination of the CSF reveals a lymphocyte count of up to 500 cells/mm^3, with other changes consistent with viral meningitis. The peripheral white cell count may be raised during the myalgic early stages, but is otherwise normal.

Diagnosis

Specimens from patients with suspected polio infection should be handled with great care, and any laboratories receiving specimens should be warned beforehand, so that inadvertent spread of the virus is avoided during the last stages of disease eradication. Rapid confirmation of the diagnosis is important, and can be made by PCR-based testing of CSF or other specimens. Culture of throat swabs, CSF and faeces should permit identification and typing of the causative virus. Virus persists in the faeces for up to 6 weeks after the onset of paralysis. In later cases, rising titres of antibodies may be detectable in serum, and local synthesis of antibody may be demonstrable by comparing serum and CSF titres.

Typing of the virus may be important in identifying a community outbreak and in distinguishing between wild and vaccine viruses. Poliovirus can be serologically typed into types 1, 2 and 3, and further typed when necessary by RNA studies. Three corresponding attenuated vaccine viruses also exist, which can be distinguished from wild virus by RNA sequencing or PCR.

Diagnosis of poliomyelitis

1 PCR-based detection of poliovirus RNA in CSF or other specimens.
2 Cell culture of stool, throat swab or cerebrospinal fluid (CSF).
3 Serotyping of virus isolates by neutralization tests.
4 RNA studies if vaccine-associated polio is suspected.
5 Demonstration of local antibody production in CSF.
6 Demonstration of rising titres of serum antibodies.

Management

Antiviral therapy, as for other enteroviruses, may soon be available Bed rest is recommended until the fever has resolved, as this may limit the severity of paralysis.

Once the patient is afebrile and stable, physiotherapy should be commenced. The rate and extent of recovery of power are variable, but improvement can continue for up to 2 years. Children are more likely to have a good recovery than are adults.

Complications

Secondary bacterial infection is common: in the chest if the intercostal muscles, glottis or diaphragm are weak, and in the bladder if catheterization is necessary. Appropriate cultures and antibiotic treatment are important in maintaining the patient's health during recovery.

Persisting muscle paralysis may lead to contracture or deformity, especially in growing limbs. Expert orthopaedic advice may be needed to optimize posture and give the best functional result.

Prevention and control

Inactivated poliovaccine (IPV) was introduced in 1956 and was followed by live oral poliovaccine (OPV) in 1962. Since 1967, only OPV has been used in the UK, although IPV is still recommended for immunization of pregnant women, immunosuppressed patients and their household contacts. The standard UK OPV schedule is three doses at monthly intervals starting at 2 months of age, with booster doses before entry to school and at 15 years of age (see Chapter 26).

Both OPV and IPV protect against poliovirus types 1, 2 and 3. OPV has the benefit of being orally administered and provides intestinal as well as humoral immunity, whereas the immunity from IPV is largely humoral. For this reason OPV is the vaccine of choice for control of epidemics, where the aim is to interrupt transmission of circulating virus. Following vaccination with OPV, vaccine virus is excreted in the faeces for up to 6 weeks. OPV can thus spread to, and immunize, close contacts of recently vaccinated individuals. This means that good population immunity can be achieved even where uptake of the vaccine is low.

The main disadvantage of OPV is that it causes paralysis in approximately 1 per 2 million recipients, with a similar risk for non-immune contacts. An average of two vaccine-associated cases of polio are reported per year in the UK. Nevertheless, OPV remains the most widely used vaccine in national immunization programmes. A novel approach used in some countries is a combined OPV/IPV schedule.

Advantages of oral poliovaccine

1 Can be administered orally.
2 Confers mucosal as well as systemic immunity.
3 The vaccine infects and immunizes close contacts.

Disadvantages of oral poliovaccine

1 Vaccine must be kept refrigerated.
2 Three doses are needed.
3 There is a small risk of mutation to virulence, and paralytic disease.
4 The vaccine is infectious within the family and the community.

Although polio remains endemic in many Asian countries, it has been virtually eradicated in most developing countries, notably in the Americas and Europe. The World Health Organization aims to eliminate polio globally by the year 2000.

Polio is a notifiable disease. Cases should be isolated in hospital. In an outbreak, extensive virus circulation usually precedes the first case; a single case of indigenously acquired polio requires vaccination of a wide network of contacts.

Herpes simplex type 2 meningitis

This is almost always associated with primary genital herpes simplex, of which it is a complication. Occasional cases occur in the absence of detectable genital lesions. The meningitis is benign, behaving like an enteroviral infection, but may be treated with aciclovir if the diagnosis is known before resolution occurs.

It is sometimes accompanied by sacral myeloradiculitis with pain in the perineum, buttock or leg, associated with paraesthesiae, and occasionally difficulty of micturition or defaecation. Although the condition is usually self-limiting, it can persist for weeks. It is therefore treated with aciclovir, which should be given intravenously in a dose of 10 mg/kg 8-hourly (see Chapter 15).

Bacterial meningitis

Introduction

This section deals with diseases of children and adults but not of neonates, who have particular types of bacterial meningitis associated with the birth process and early bacterial colonization. Neonatal meningitis is discussed in Chapter 17.

Bacterial meningitis is always a medical emergency because of the high mortality of untreated or late-treated cases. The importance of meningitis is the preventable morbidity which different types cause in various age groups. Prompt treatment can minimize the mortality to 4–8% in childhood types, and to 8–25% in adults. Many cases recovering after early treatment have little or no neurological deficit.

ORGANISM LIST

Neisseria meningitidis
Streptococcus pneumoniae
Listeria monocytogenes
Haemophilus influenzae
Mycobacterium tuberculosis
Streptococcus 'milleri'
Other Gram-positive cocci
Escherichia coli
Rarities:
Leptospira spp.
Borrelia spp.
Brucella spp.

Between 15 and 25% of cases of bacterial meningitis have negative blood and CSF cultures. This is more often because of sampling problems or previous antibiotic treatment than because of an unusual bacterial aetiology.

Rare cases of bacterial meningitis after neurosurgery may be caused by Gram-negative rods such as *Acinetobacter* spp., *Candida* spp. or Enterobacteriaceae.

Lumbar puncture and CSF changes in purulent meningitis

With the exception of tuberculosis, leptospirosis and borreliosis, bacterial meningitis causes neutrophil pleiocytosis in the CSF. All bacterial meningitis causes a raised protein level and low glucose. Endotoxin accumulates in the CSF of patients with Gram-negative bacterial meningitis, but this is usually only measured for research purposes.

Any number of neutrophils in the CSF is abnormal; some patients have fewer than 10/mm^3; others may have 2000 or more. The protein level is usually more elevated in bacterial than in viral meningitis; levels greater than 1.0 g/l are common. A low glucose level is rarely found in viral meningitis, but is usual in bacterial infections.

In bacterial meningitis, the degree of abnormality in the CSF is closely related to the prognosis. In particular, high neutrophil counts, high protein, high lactate and high endotoxin levels are associated with a poor prognosis. Since these levels tend to rise in parallel, the protein level and neutrophil count often give a reasonable indication of prognosis.

Meningococcal meningitis

Introduction

This disease, caused by *Neisseria meningitidis*, is an important cause of morbidity and mortality in children and young adults. It is a bacteraemic disease with a rapid and acute onset. About 80% of cases have signs of meningitis, which often leads to a prompt diagnosis. The minority with bacteraemia alone tend to present with more advanced disease, and have approximately twice the mortality of those with meningitis.

Epidemiology

N. meningitidis is the commonest cause of bacterial meningitis in the UK (Fig. 13.7). Approximately 3000 laboratory-confirmed cases of meningococcal disease are reported annually, 40% of these occurring between January and March. About 60% of the cases are due to serogroup B infections. The epidemiology in the UK has changed in recent years, with a rising incidence since the mid 1990s (Fig. 13.8). The proportion of cases due to serogroup C infections has also increased, and there has been a change in the age distribution with a rise in cases among teenagers. This is in contrast to the 'meningitis belt' of sub-Saharan Africa, where the predominant strains are group A, and sometimes group C, which cause explosive epidemics during the dry season from December to March.

The peak incidence of meningococcal meningitis is at 6 months of age, coinciding with loss of maternal antibody. A second, smaller peak occurs in teenagers and young adults. Many infections are sporadic, but outbreaks sometimes occur in households, schools and military establishments. Group C strains cause outbreaks more frequently than others. During outbreaks, the age distribution shifts towards older children.

Certain conditions predispose to meningococcal infection, notably functional or anatomical asplenia and complement deficiency. There is also some evidence that recent upper respiratory infection (especially influenza) temporarily increases the risk of meningococcal disease.

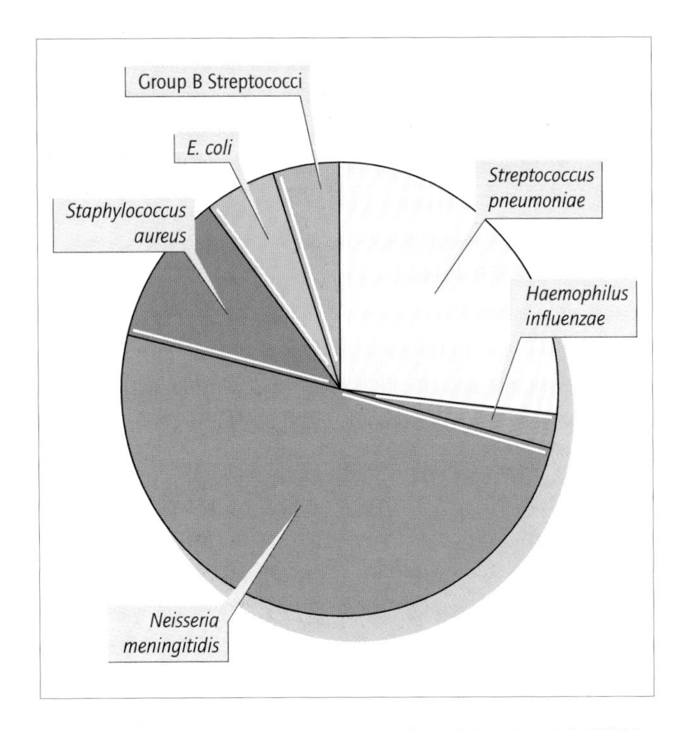

Fig. 13.7 Pie chart of the commonest causes of bacterial meningitis in 1998 in the UK.

Pathogenesis

N. meningitidis are small Gram-negative diplococci. They are carried in the nasopharynx, and cause septicaemia and meningitis in a minority of hosts. They are aerobic, catalase- and oxidase-positive and non-motile, and possess a polysaccharide capsule, which is the main antigen and determines the serotype of the species. The main serogroups of pathogenic organisms are A, B, C and W135 and their prevalence varies with time and geographical location. Up to 50% of the bacterial outer membrane is a lipo-oligosaccharide (LOS), analogous to the lipopolysaccharide of Gram-negative bacteria, in that it contains a lipid A subcomponent. This antigen causes activation of macrophages and release of tumour necrosis factor, a mediator of shock in serious meningococcal septicaemia. The LOS may also assist in invasion of the CNS, by altering the permeability of the blood–brain barrier.

Meningococcal serogroups can be subdivided by serotyping, based on the outer membrane protein (OMP) 2 antigen. This is one of three OMPs which act as porins, controlling the influx of water-soluble molecules through the lipophilic outer membrane. Some OMP types are associated with more serious infections.

Neisseria express either one of two types of pili, which are important in adherence to host epithelium. Class I pili are similar to those produced by *N. gonorrhoeae*. Pilus production can be switched on and off and this may give the organism advantages in colonization and transmission.

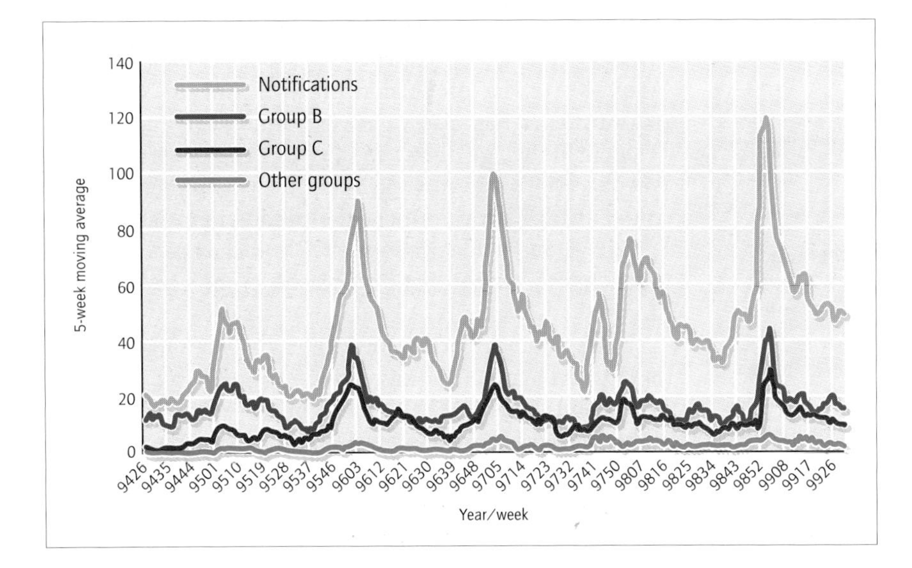

Fig. 13.8 Recent trends in the epidemiology of meningococcal disease in the UK.

Pathogenicity factors in *Neisseria meningitidis*
1 Capsular polysaccharide.
2 Outer membrane lipo-oligosaccharide (endotoxin-like).
3 Outer membrane proteins.
4 Pili.
5 Immunoglobulin A protease.

Antibodies to the capsular polysaccharide are bactericidal and protective, but the response to serogroup B antigen is weak because it cross-reacts with human host tissues. There is a strong immune response to meningococcal iron regulation proteins, which may assist in controlling infection by denying the organism iron. Like many other upper respiratory tract colonists, *N. meningitidis* produces an IgA protease which cleaves secretory IgA.

Complement is important in protection against meningococcal disease. Defects of the alternative complement pathway cause enhanced susceptibility to infection. Rare defects in properdin predispose to catastrophic infection. Immunization permits recruitment of the classical complement pathway and abolishes this predisposition.

Clinical features

The disease has a short incubation period. Most secondary cases are seen 2–5 days after the primary case, although some occur up to 4 weeks later. In most cases the illness develops rapidly, over less than 24 h, but a slower or stuttering onset sometimes occurs and causes diagnostic uncertainty. Fever, malaise and increasing headache are accompanied by nausea and often vomiting. Photophobia may be extreme.

A typical petechial and purpuric rash appears in most patients (Fig. 13.9) sometimes preceded by a measles-like rash or generalized erythema which lasts for only a few hours. The rash, a feature of disseminated intravascular coagulation (DIC), is an important clue to diagnosis, but varies widely in severity and extent. Some cases present with only one or two lesions (Fig. 13.10), and a few have petechiae only in the conjunctiva or other mucosal surface. A careful search for petechiae is warranted, as their presence makes the diagnosis very likely. A few cases present with vasculitic lesions (Fig. 13.11), or extensive, necrotic lesions (purpura fulminans; Fig. 13.12).

Patients with rapidly advancing disease may present with, or quickly develop, features of endotoxaemia such as hypotension, sluggish peripheral circulation and pulmonary oedema. As the disease progresses, drowsiness and confusion are common and coma sometimes develops.

The white cell count is usually raised to $15–20 \times 10^9/l$. The erythrocyte sedimentation rate is low, consistent with DIC, and fibrin degradation products are present in elevated amounts. The prothrombin time or international normalized ratio (INR) is often slightly prolonged, but serious bleeding or very low platelet counts are unusual.

Lumbar puncture reveals purulent CSF with a neutrophilia, raised protein and low sugar levels. Gram-negative diplococci may be seen inside and outside the neutrophils (Fig. 13.13), though they are often undetectable after antibiotic treatment. Blood and CSF cultures are usually positive.

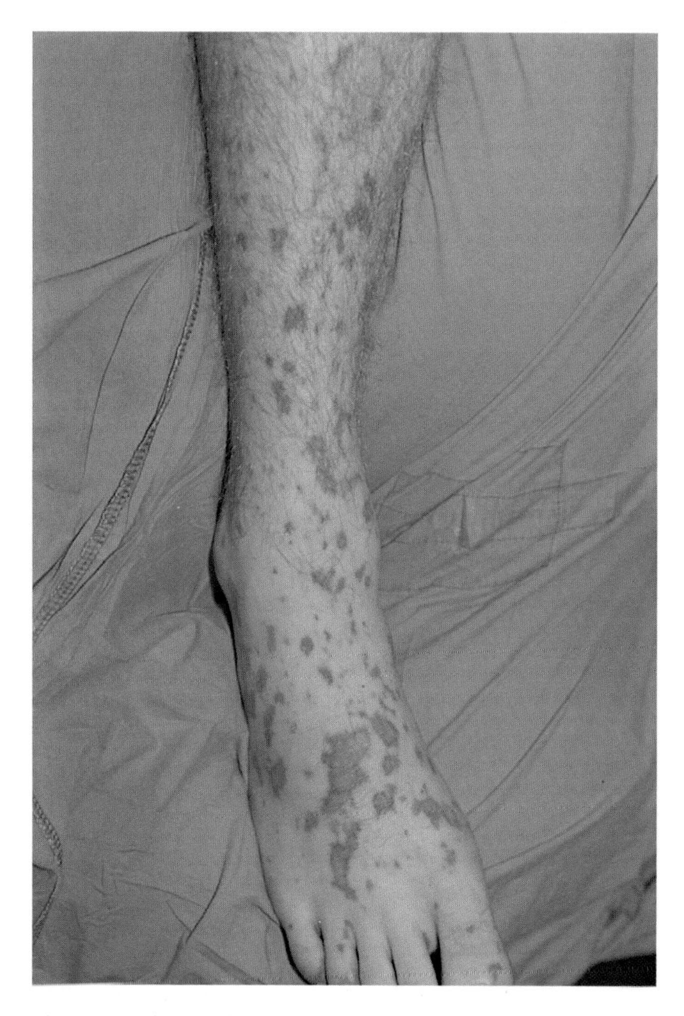

Fig. 13.9 Meningococcal meningitis: acute rash in a teenage boy; the legs, buttocks and elbows are often most affected.

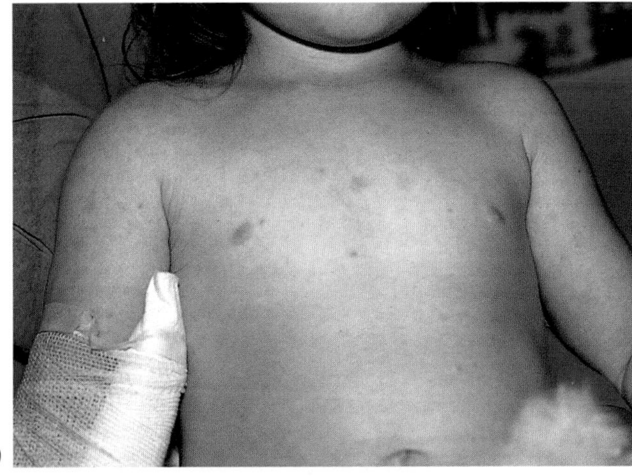

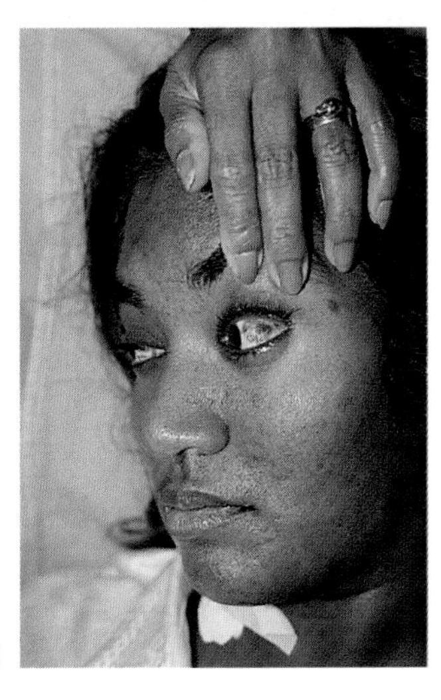

Fig. 13.10 Acute meningococcal meningitis: (a) this 3-year-old girl had only a handful of odd-shaped spots on the trunk and shoulders (arrows); (b) this 20-year-old woman had a conjunctival lesion.

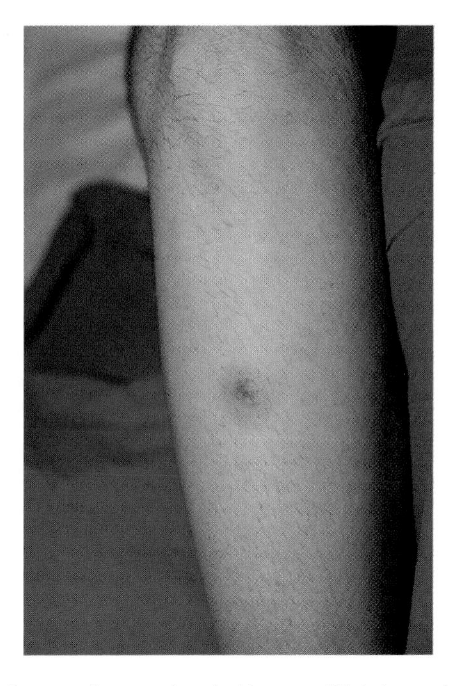

Fig. 13.11 Acute meningococcal meningitis: a vasculitic lesion on the shin; there were three others on the hand and arms.

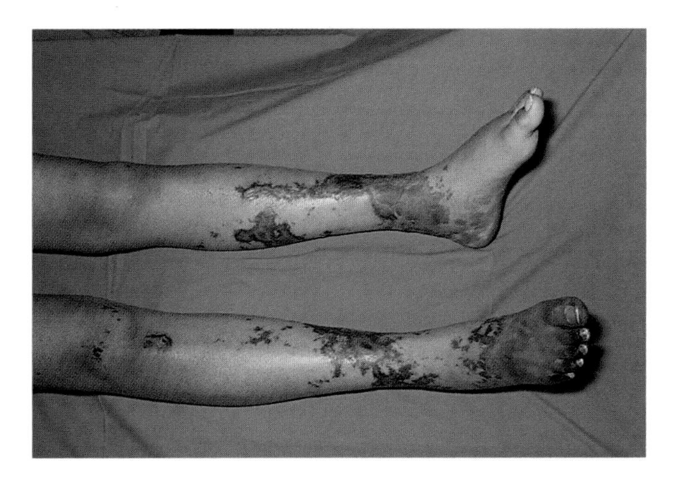

Fig. 13.12 Acute meningococcal meningitis: purpura fulminans.

Unusual presentations

Meningitis in an infant sometimes has a fluctuating onset, with intermittent fever, lassitude, high-pitched crying and perhaps vomiting, separated by periods of apparent normality. A bulging fontanelle or altered responsiveness eventually prompts the performance of diagnostic tests.

Occasionally the disease progresses so fast, typically in a toddler, that the patient presents with sepsis and shock, but without a typical rash. The case fatality rate with this presentation approaches 40%.

Older children and adults may have chronic meningococcal bacteraemia, with fluctuating fever, rash and arthralgias. The flat rash of DIC becomes progressively more papular and autoimmune-like (Fig. 13.14), and is then difficult to distinguish from Henoch–Schönlein disease, which is a common misdiagnosis. Joint swelling and small effusions may appear in the hands, knees or ankles. The white cell count is usually elevated. The erythrocyte sedimentation rate rises as the condition persists, increasing the chance of misdiagnosis of autoimmune or collagen disease. Blood cultures are positive during feverish episodes, though several may need to be taken before the true diagnosis is confirmed. This fluctuating course can persist for weeks, but there is significant risk of progression to acute meningitis or bacteraemia.

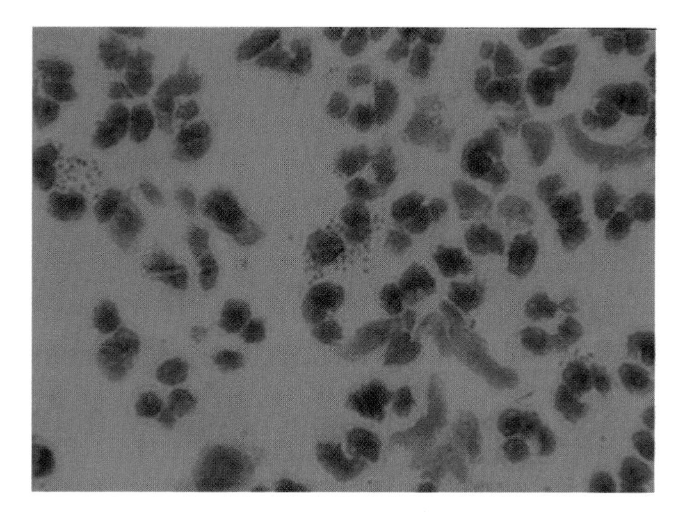

Fig. 13.13 Rapid diagnosis of meningococcal meningitis. Cerebrospinal fluid smear shows neutrophils, and intracellular and extracellular Gram-negative diplococci.

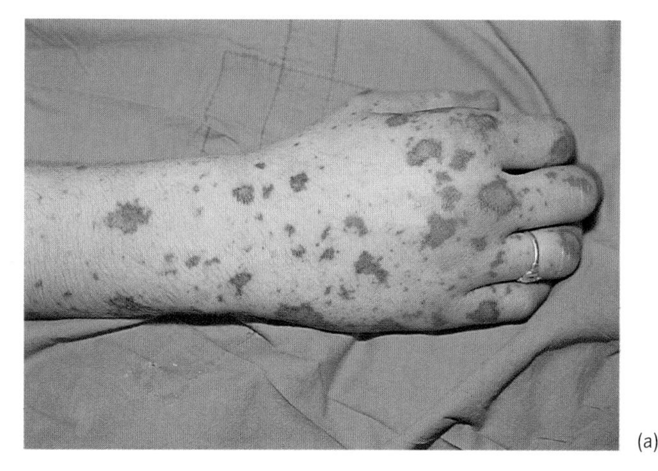

(a)

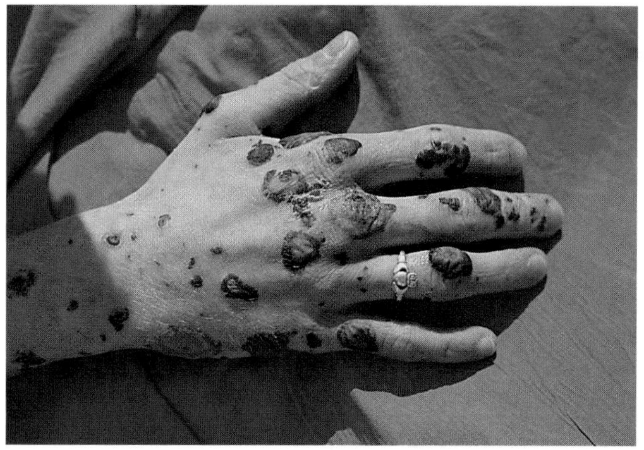

(b)

Fig. 13.14 (a) Early meningococcal rash; (b) the same rash after 1 week's evolution.

Diagnosis

The clinical features are often typical, allowing immediate specific treatment without the need for a confirmatory lumbar puncture. Occasionally, this type of rash is associated with staphylococcal or streptococcal infection, and some meningococcal cases have no rash. Where there is doubt, therefore, lumbar puncture offers the advantage of rapid diagnosis. Blood cultures should always be carried out.

The diagnosis can be confirmed by culture of blood and/or CSF. PCR-based DNA detection allows identification and typing of organisms not identifiable by culture. Late agglutination tests can detect group A or C antigen in CSF. Rising titres of antibodies in acute and convalescent serum will provide retrospective diagnosis. For epidemiological reasons, all *N. meningitides* should be typed by referring cultures in PCR specimens to the reference laboratory.

Management

Almost all meningococci are extremely sensitive to benzylpenicillin. A few have reduced sensitivity, but will still respond to initial high-dose treatment. Attending general practitioners should give a first dose pending hospital referral. This policy saves lives, as the disease can progress alarmingly in the time it takes to reach hospital. For the first 24–48 h, most specialists would give 1.2 g 2-hourly. This is because penicillin is very rapidly excreted, so frequent dosage is required to maintain blood and CSF penicillin levels. Once the fever has resolved, 1.2 g 4-hourly or 2.4 g 6-hourly can be given to complete 5 days' treatment.

Many patients claim to be allergic to penicillin, often meaning that it causes diarrhoea or gastric irritation. The treatment of meningococcal disease is so urgent that only a history of anaphylaxis or angioneurotic oedema should be taken as a contraindication to penicillin. The drug of choice in severe penicillin allergy, for cases of unproven aetiology and for *N. meningitidis* with reduced penicillin sensitivity is cefotaxime or ceftriaxone.

Chloramphenicol is effective against *N. meningitidis*, is well-absorbed and penetrates the CSF well, even in oral dosage. It is little used in Western practice because of its ability to cause agranulocytosis. In developing countries it is inexpensive, and is also available for use as a single- or two-dose course of 'oily injection' (which gives cure rates of about 80%). It is widely used for treating meningitis, though individual reports of resistant organisms are now emerging.

Treatment of meningococcal meningitis

1 *General practitioner treatment*: benzylpenicillin i.v.: infant below 1 year, 300 mg; child 1–6 years, 600 mg; older child or adult, 1.2 g; given immediately.

2 *Hospital treatment*: benzylpenicillin i.v. 1.2 g 2-hourly for 24–48 h (child 1–12 years, 200 mg/kg daily), reducing to 1.2 g 6-hourly to complete 5 days' course (child reduced to 100 mg/kg daily).

3 *Alternative*: cefotaxime i.v. 2.0 g 8-hourly (child 150–200 mg/kg daily) for 5 days; *or* ceftriaxone i.v. 2–4 g daily (child 50–80 mg/kg daily) as a single dose.

4 *In severe allergy to beta-lactam antibiotics*: chloramphenicol i.v. or orally, 2–3 g daily in three or four divided doses (child 50–100 mg/kg daily).

5 *For inexpensive short-course therapy*: chloramphenicol oily injection, one or two doses within 48 h.

Some patients, especially children, suffer convulsions. A single dose of diazepam, intravenously (250 μg/kg) or rectally (500 μg/kg), may be sufficient to control convulsions occurring during the acute fever at presentation. As fever and inflammation are controlled, fits often cease. If fits persist, phenytoin may be given by slow intravenous infusion to a total of 5–15 mg/kg. This can often be followed by oral administration of phenytoin or sodium valproate until the acute illness is controlled, when weaning from the anticonvulsant may be attempted.

Uncomplicated cases become afebrile within 6–12 h of commencing treatment. The headache progressively resolves, skin lesions fade, or heal by scabbing or ulceration and re-epithelialization, depending on their size. Neurological sequelae are rare, but include cranial nerve lesions, visual impairmant and hydrocephalus.

More seriously ill patients with reduced blood pressure and impaired cerebral function will also lose their fever in the first day or two. However, the pathology of the sepsis and accompanying DIC resolves more slowly. There are platelet plugs containing trapped organisms in many small blood vessels. The activated platelets promote inflammation, complement activation and coagulation. There is often little improvement in haemodynamics or cerebral function in the first 24–36 h in such patients. During this time support measures as for Gram-negative bacteraemia and endotoxaemia are often needed (see Chapter 19).

Patients with established pulmonary oedema and severe purpura are gravely ill and require vigorous support. There is no evidence that heparin benefits the DIC once penicillin treatment has been commenced. Corticosteroids, plasmapheresis and extracorporeal membrane oxygenation do not appear to improve the prognosis. Bacterial permeability-increasing factor, which binds endotoxin, appears to confer some benefit in limited trials. Even with the best intensive care, 40% of these patients fail to recover. Many who die have bilateral haemorrhagic necrosis of the adrenal glands (Waterhouse–Friderichsen syndrome).

Complications

Other manifestations of meningococcal infection

Other manifestations of meningococcal infection may precede or complicate the meningitis, or occur alone. This is uncommon, but can cause diagnostic confusion. Many meningococcal infections are fairly trivial, for instance conjunctivitis, pharyngitis or otitis media. Others may be severe, including endophthalmitis, pericarditis, endocarditis, myocarditis and, rarely, septic arthritis. Most will respond to standard antimeningitis treatment.

Necrosis

Necrosis of purpuric and ecchymotic lesions can cause severe pain. Adequate analgesia is important. Some lesions ulcerate and a few leave full-thickness skin defects which later require grafting. Large lesions of fingers or toes can cause loss of digits by gangrene. Nevertheless, secondary infection of these lesions is rare, and healing is often better then expected.

Reactive arthritis

Reactive arthritis is a common late complication, affecting more than 50% of adolescents and young adults after 5–7 days. The fingers or knees are most often affected, usually asymmetrically (Fig. 13.15). Aspiration reveals neutrophils but no organisms. Immune complexes containing meningococcal antigen are demonstrable in the synovium. The condition resolves spontaneously and good symptomatic relief is afforded by non-steroidal anti-inflammatory agents.

Serositis

Serositis with effusion is much rarer than arthritis, but is probably also an immune-complex disorder. A pleural or pericardial effusion may be detected on X-ray, and occasionally requires aspiration. A rapidly reducing course of prednisolone will often terminate persistent or recurrent effusions.

Neurological complications

Neurological sequelae are rare except in the severest cases. Deafness or other cranial nerve lesions are the most likely, and may improve or recover during convalescence. Occa-

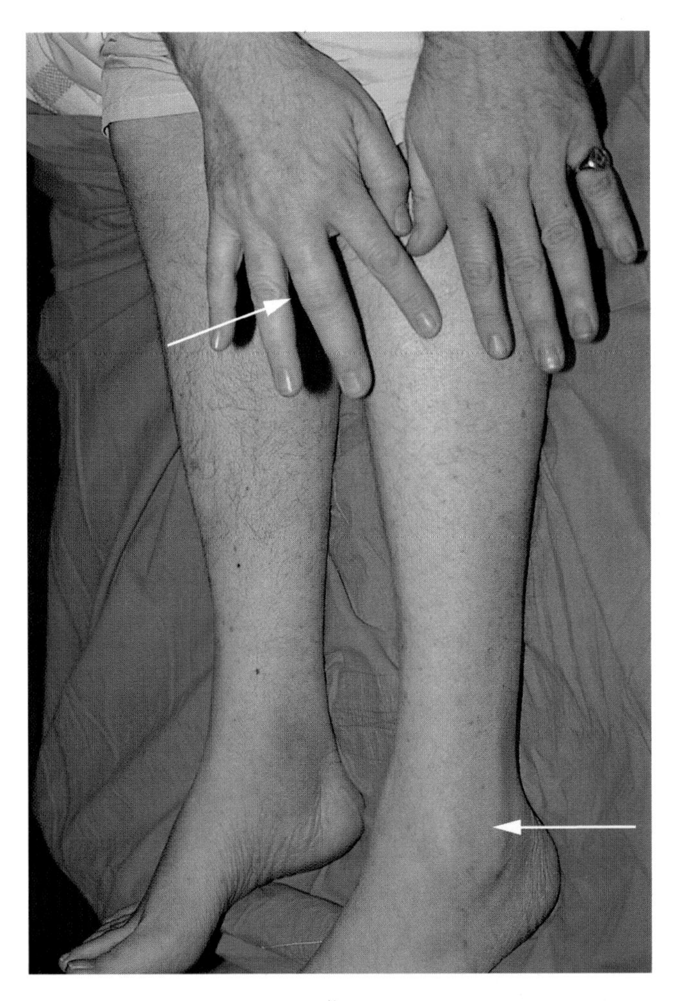

Fig. 13.15 Postmeningococcal synovitis affecting the finger and ankle (arrows).

sionally an apparently recovering patient will collapse with respiratory arrest. This is usually due to acute basilar artery pathology secondary to meningeal or intracerebral purpuric lesions. It is often fatal.

Abscess formation

Abscess formation occurs rarely at the original site of sepsis, such as the middle ear or a paranasal sinus. While this may respond to the treatment of the meningitis, surgical decompression and drainage are often indicated, to remove the persisting infection and to prevent recurrence of the meningitis.

Prevention and control

Meningococcal vaccines are available against groups A, C, Y and W135 strains. Unfortunately, no group B vaccine is available at the present, although several promising candidates are in clinical trials. Vaccines are prepared from purified polysaccharide capsular antigen, and like the early Hib vaccines (see p. 274), they are less effective in young children and provide only temporary T-cell-independent immunity. Vaccination is indicated for those at particular risk from meningococcal disease. These include military recruits, household contacts of cases due to vaccine-preventable strains, asplenic patients and those with certain complement deficiencies. Vaccination has been successfully used to control outbreaks of group C disease in schools and military training camps. Travellers to the African meningitis belt and Haj pilgrims visiting Mecca (where an outbreak occurred in 1987) should also be vaccinated.

Conjugate serogroup C meningococcal vaccines have become available in recent years. These are highly immunogenic and safe in infants from 2 months of age, and induce immunological memory. A conjugate serogroup C vaccine was introduced into the UK immunization schedule in the autumn of 1999.

Chemoprophylaxis should be given to household and mouth-kissing contacts of cases in the 10 days preceding admission. During outbreaks chemoprophylaxis may be more widely extended to include classroom, nursery or other institutional contacts. The regimen is rifampicin 10 mg/kg every 12 h for 2 days, up to a maximum of 600 mg per dose. This differs from the regimen for Hib prophylaxis (see p. 275). Compliance may be poor due to gastrointestinal side-effects. Patients must be informed of the possible adverse effects, which include interference with oral contraception, red colouration of urine, sputum and tears, staining of contact lenses, skin rashes and gastrointestinal reactions. Ciprofloxacin as a single oral dose of 500 mg is a good alternative but is not licensed for this use. Ceftriaxone 1 g intramuscularly may be used in pregnancy. The index case should receive chemoprophylaxis before discharge from hospital to prevent reintroduction of the infecting strain into the household. This is important, but often overlooked.

Prophylaxis of meningococcal disease

1 Rifampicin orally, 600 mg 12-hourly for 2 days (infant 6–12 months, 5 mg/kg 12-hourly; child 1–12 years, 10 mg/kg 12-hourly).
2 Alternative: ciprofloxacin orally, 500 mg, single dose.
3 In pregnancy: ceftriaxone i.m. 1 g, single dose.
4 *Meningococcus* types A plus C polysaccharide vaccine, for school and military outbreaks with these types, 0.5 ml i.m. or deep s.c. (not effective below age of 18 months).
5 Conjugate meningococcus C vaccine for infants.

Swabbing of contacts and cases after chemoprophylaxis is not essential.

A case of meningococcal meningitis, particularly in a school, often causes alarm and may attract media attention. Good communication and public education can minimize this. It is advisable to inform parents when a case occurs in a school, nursery or college. Meningococcal meningitis and septicaemia are both notifiable diseases (see Chapter 26).

Haemophilus influenzae meningitis

Introduction and epidemiology

Haemophilus influenzae meningitis is a rare disease caused by an encapsulated *H. influenzae* of capsular type b (Hib). Childhood immunization is universal in industrialized countries, and antibodies to the organism are naturally acquired by age 4 or 5 in unimmunized populations. Other rare childhood bacteraemic diseases caused by Hib include pneumonia, acute epiglottitis, facial cellulitis and some bone and joint infections. Immunocompromised adults occasionally suffer *H. influenzae* bacteraemias.

A few cases of *H. influenzae* meningitis are a complication of chronic CSF fistulae, many of which occur in adults and over half of which are non-type b infections (Fig. 13.16).

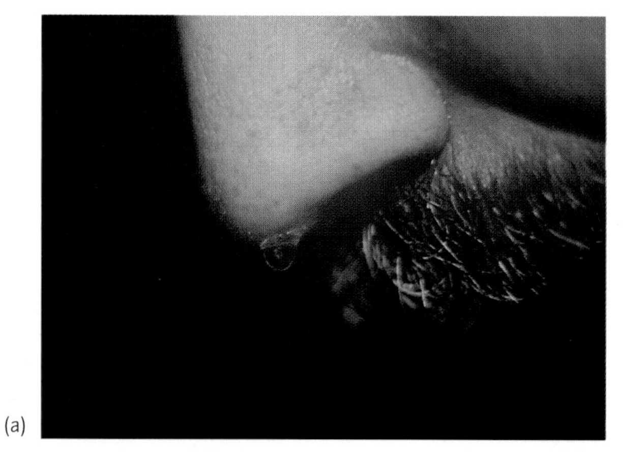

(a)

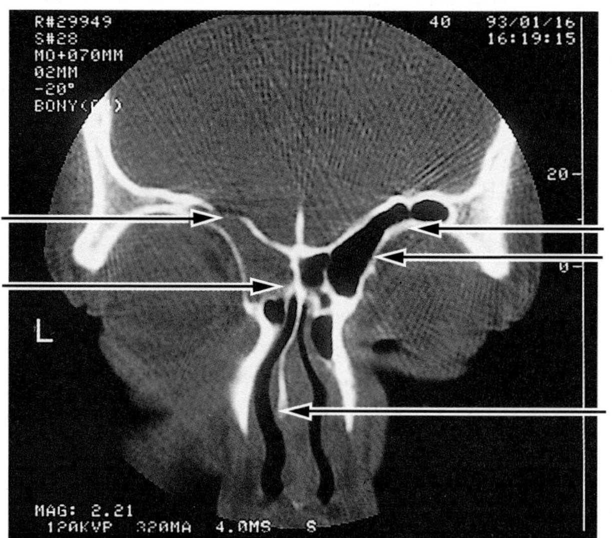

Defect in wall of frontal sinus communicating with subarachnoid space

Left frontal sinus is opacified: it is full of CSF

Right frontal sinus containing air

Roof of orbit

Nasal septum

(b)

Fig. 13.16 Cerebrospinal fluid (CSF) rhinorrhoea in a man who developed *Haemophilus influenzae* type b meningitis many years after a head injury. (a) Clear CSF dripping from the nose; (b) contrast computed tomographic scan showing CSF leaking via a skull defect into the frontal sinus (arrow). The sinus is opaque compared with the air-filled sinus on the right. (Courtesy of Dr W.R.C. Weir and the *Journal of Infection*.)

Microbiology

H. influenzae is a small Gram-negative coccobacillus which is catalase- and oxidase-positive. As it requires nicotinamide adenine dinucleotide phosphate (NADP) and haematin for active growth, it produces little or no growth on unsupplemented blood agar, but it grows well on chocolated blood agar. The organism is aerobic but isolation and growth are favoured by an atmosphere with enhanced carbon dioxide.

Pathogenesis

The main pathogenicity determinant of *H. influenzae* is the capsule. Although the organism is capable of producing one of six different capsular polysaccharides (or none), almost all isolates from clinical disease are serotype b. Susceptibility to Hib infection correlates with the absence of antibodies to the type b capsule. Molecular studies have confirmed the importance of the polyribitol ribosyl phosphate (PRRP) capsule in experimental meningitis. Other pathogenicity determinants include the lipopolysaccharide, outer membrane protein, pilus proteins and the IgA protease.

Pathogenicity factors of *Haemophilus influenzae*
1 Polyribitol ribosyl phosphate (PRP) capsule, especially capsular type b.
2 Cell-wall lipopolysaccharide.
3 Outer membrane protein.
4 Pilus proteins.
5 Immunoglobulin A protease.

In respiratory infections the capsule is not a vital pathogenicity requirement. Infection tends to occur in individuals with bronchiectasis or chronic obstructive airways disease. In this environment the organisms are partly protected from the activities of the host immune system. In addition it appears that bacterial antigens and the host's inflammatory response to them sets up a cycle of damage, colonization and further damage, with progressive loss of respiratory function.

Clinical features

Hib meningitis often has a slower onset than meningococcal meningitis, developing over 3 or 4 days. The condition may fluctuate initially, improving when the temperature is lower, and worsening with feverish episodes.

The white cell count is usually high: $15–20 \times 10^9$/l with a neutrophilia. The blood sugar may also be raised to 8–10 mmol/l in the face of acute cerebral insult.

Diagnosis

Lumbar puncture reveals changes typical of bacterial meningitis. Gram stain will often demonstrate small, Gram-negative rods in the CSF, but these are not easy to identify and are rarely seen if antibiotic treatment has already been given.

If Gram stain and culture are negative, serological evidence of Hib infection may be obtained by performing a latex agglutination test on the CSF, using the Hib PRP antigen. Failing this, blood cultures may prove positive 24–36 h after admission.

Diagnosis of *Haemophilus influenzae* type b (Hib) meningitis
1 Demonstration of small, Gram-negative rods in cerebrospinal fluid (CSF) deposit.
2 Demonstration of Hib antigen in CSF by latex agglutination.
3 Positive CSF culture.
4 Positive blood culture.

Management

The treatment of choice is a broad-spectrum cephalosporin, such as cefotaxime or ceftriaxone, which penetrates excellently into the CSF and is highly bactericidal to Hib. Cefuroxime does not reach such good CSF concentrations and carries a slight risk of late relapse. Antibiotic treatment should be continued for at least 1 week, and probably for 10 days in severe cases.

Dexamethasone makes only a few hours' difference to the duration of fever, and does not alter mortality, but it reduces the incidence of deafness after recovery in some trials. The dose is 8 mg twice daily during the first 3 days of antibiotic treatment.

Treatment of *Haemophilus influenzae* type b meningitis
1 Cefotaxime i.v. 150–200 mg/kg daily in three or four divided doses; *alternative*, ceftriaxone i.v. or i.m. 50 mg/kg daily in a single dose (up to 80 mg/kg can be given i.v.); both for 7 days.
2 Second choice: chloramphenicol i.v. or orally 50–100 mg/kg daily in four divided doses for 10–14 days.

Plus dexamethasone orally, i.m. or i.v. 8 mg twice daily for the first 3 days only.

Other priorities include adequate rehydration and haemodynamic support, control of convulsions (see p. 270) and, on rare occasions, specific measures to control raised intracranial pressure.

Complications

Apart from the immediate problems of treatment, late problems can occur after the initial resolution of fever.

Recurrence of fever

Recurrence of fever is occasionally seen, sometimes with a neurological complication. It probably represents a small subdural area of infection, or cerebral vein thrombosis, which can be demonstrated by enhanced CT or MR scan. It usually responds to continued or increased antibiotic treatment. Tapping of subdural collections is occasionally indicated.

Convulsions

Convulsions may develop during the course of healing. They are rarely caused by frank abscess formation, but a cerebral vein thrombosis, subdural collection, rising intracranial pressure or small area of gliotic scarring may be responsible. A CT or MR scan is indicated to detect these.

Hydrocephalus

Hydrocephalus is rare but can develop at any time after treatment commences. It may present as increasing fits or deteriorating cerebral function with neurological deficit, drowsiness or coma. Imaging is required to make the diagnosis, and neurosurgical advice on temporary or permanent CSF drainage may be necessary.

Neurological impairment

Neurological impairment is rarely gross in those who survive the infection. The commonest problem is reduced hearing, which occurs in 9–15% of survivors. It should always be looked for, as it affects subsequent schooling and social development if untreated. Other cranial nerve defects may occur, including squint, ptosis, reduced visual field or acuity, and impaired balance. These may improve considerably with time, but residual defects do occur. More global neurological problems are rare, but monoparesis, spasticity, learning difficulties and epilepsy are all recognized sequelae of severe childhood meningitis. Intensive follow-up and rehabilitation under the care of a developmental paediatrician is indicated in these cases.

Complications of *Haemophilus influenzae* type b meningitis

1 Impaired hearing in 9–15% of survivors.
2 Convulsions (rarely, persisting epilepsy).
3 Subdural collections of fluid (rarely, subdural empyema).
4 Cerebral vein thrombosis.
5 Rare visual, motor or sensory deficit.
6 Very rare hydrocephalus.

Prevention and control

Polysaccharide Hib vaccines were first produced in the 1970s. These were very effective in older children, but were much less effective in children under 2 years of age, the group most at risk from disease. In addition they did not prime the immune system for subsequent boosting, as the immunity was mainly B-cell mediated.

The immunogenicity of Hib vaccines has now been greatly improved by conjugating the polysaccharide moeity to T-cell-dependent proteins, such as tetanus toxoid or CRM_{197}, a non-toxic derivative of diphtheria toxin. Conjugate vaccines are effective in infants from 2 months of age and elicit an immune response which primes for subsequent boost with natural polysaccharide.

Hib vaccine was introduced in the UK immunization schedule in 1992. Three doses are given at 2, 3 and 4 months of age (see Chapter 26). Booster doses are not recommended. Unvaccinated children over 1 year of age are protected by a single dose; however, immunization is not worthwhile in children over the age of 4. There are no specific contraindications to Hib vaccine. A mild to moderate local reaction occurs in up to 10% of vaccine recipients.

Household contacts of patients with Hib should be given chemoprophylaxis (and vaccine, if unvaccinated) if there is a child under 4 years in the household. The case should also receive prophylaxis before discharge from hospital. Recurrent infections are rare, but have been documented; thus the index case should also receive vaccine. Nursery contacts should be given chemoprophylaxis (and vaccine, if unvaccinated) where two or more cases occur within 120 days of each other.

The regimen for chemoprophylaxis is rifampicin 20 mg/kg daily for 4 days, up to a maximum of 600 mg daily.

Prophylaxis of *Haemophilus influenzae* type b (Hib) meningitis
Rifampicin orally 20 mg/kg daily (maximum 600 mg daily) for 4 days, to *all* household or nursery contacts, when other children below age 4 are present; *plus*
(in unvaccinated children) Hib vaccine 0.5 ml i.m. or deep s.c. as a single dose (infants below 13 months require three injections at intervals of 1 month; the vaccine is not indicated above the age of 4 years).

Haemophilus meningitis is a notifiable disease in the UK (see Chapter 26).

Pneumococcal meningitis

Introduction and epidemiology

Pneumococcal meningitis is the commonest bacterial meningitis in the middle-aged and elderly, but can affect all age groups. While it can be apparently spontaneous it also has an increased incidence in certain susceptible patients.
1 Those with defects of the dura mater, for instance after head injury or surgery (see Fig. 13.16).
2 Those with chronic infections in the skull, for instance chronic suppurative otitis media or sinusitis. Patients with large nasal polyps are susceptible, especially after polypectomy.
3 Individuals who lack a spleen for any reason, for instance in haemolytic diseases, after abdominal trauma or after splenectomy for lymphoma or other malignancies.
4 Alcoholics are at increased risk of all types of pneumococcal disease.

Pneumococcal meningitis is an important disease because of these increased susceptibilities. It also has the highest mortality of the common bacterial meningitides. Many studies show that mortality increases with increasing age.

Clinical features

The incubation period is uncertain, as most cases are intrinsically derived. The onset is often rapid, over 1 or 2 days, but a few cases have a gradual onset, with meningism evolving during the course of an ear or sinus infection.

Most patients have marked meningism by the time of presentation, and impairment of consciousness is common. Severe, watery diarrhoea is a common, non-specific presenting feature, and the patient may be too confused to complain spontaneously of headache. Signs of meningism must be actively sought if such cases are not to be missed.

Other features of severe pneumococcal disease, such as pneumonia or peritonitis, are occasionally present.

The white cell count is usually high: 15–25×10^9/l is common. A low count indicates a poor prognosis. The plasma creatinine is often slightly elevated. Lumbar puncture shows a marked neutrophil pleiocytosis, and Gram-positive diplococci are relatively easy to demonstrate in most cases. Blood cultures are positive within 18–24 h.

Management

Specific treatment for penicillin-sensitive pneumococci is with high-dose intravenous benzylpenicillin. However, penicillin-resistant *S. pneumoniae* (PRSP) are increasingly common, and cefotaxime or ceftriaxone are then the initial treatment of choice.

Unfortunately there are no controlled clinical studies to indicate the optimal course of treatment of patients with PRSP. Pneumonia with PRSP can successfully be treated with higher doses of benzylpenicillin (dosage over 9.4 g daily has been successfully used). The problem is more difficult in cases of meningitis where poor CSF penetration of penicillin makes treatment failure with this agent more likely. Chloramphenicol is often used as an alternative treatment in bacterial meningitis but increasing resistance among pneumococci means that it cannot be relied upon and a high failure rate is reported. As third-generation cephalosporins penetrate the blood–brain barrier they are favoured as the first-line treatment of PRSP meningitis. However, there are now many reports of treatment failure in cases infected with organisms with a cefotaxime minimum inhibitory concentration (MIC) above 2 mg/l. In areas where third-generation cephalosporin resistance is reported (Spain, Portugal, South Africa, parts of the Pacific and increasingly in the Americas) the addition of teicoplanin or vancomycin should be considered as experimental evidence suggests useful synergy. The use of teicoplanin or other agents is currently supported only on the basis of *in vitro* and clinical anecdotal evidence. Patients with PRSP meningitis require careful monitoring which may include a repeat lumbar puncture to document CSF sterility.

Treatment of pneumococcal meningitis
1 Cefotaxime 2 g 8-hourly for 10–14 days; *or* ceftriaxone i.v. 2–4 g daily, single dose for 10–14 days.
2 *For penicillin-sensitive pneumococci*: benzylpenicillin i.v. 1.2 g 2-hourly, then 2.4 g 4–6-hourly for 10–14 days.
3 *For multiply resistant pneumococci*: cefotaxime *plus* teicoplanin i.v. 400 mg 12-hourly for three doses, then 400 mg daily; *or* vancomycin i.v. 500 mg 6-hourly initially, but then to maintain plasma levels at peak not more than 30 mg/l and trough not more than 10 mg/l.

Patients with severe pneumococcal sepsis can develop renal failure, shock and pulmonary oedema. These problems should be treated as in any patient with sepsis (see Chapter 19).

Complications

Abscess formation

An abscess may be the precipitating cause of the meningitis, or may rapidly form during its course. Persistent fever, deteriorating cerebral function, seizures or the appearance of focal signs indicate the need for brain imaging to exclude sepsis in the meninges or adjacent intracranial structures. Surgical drainage is urgently indicated in such cases.

Recurrence

Recurrence is a real possibility after pneumococcal meningitis in a susceptible patient. It is particularly likely in those with dural tears, who may have many attacks. CSF rhinorrhoea or otorrhoea can be confirmed by the presence of tau protein in the fluid, and the dural defect may be demonstrable by contrast injection and CT scan (see p. 272 and Fig. 13.16). Attempts should be made to prevent recurrence by repair of the predisposing condition, by immunization and prophylaxis. Occasionally an attack of meningitis in such patients is caused by an alternative organism, such as *Haemophilus influenzae*.

Other complications

Other complications are similar to those of meningitis in general. They include cranial nerve lesions, early or late hydrocephalus, visual defects or paresis of varying severity. Metastatic pneumococcal sepsis may occur.

Prevention and control

Pneumococcal polysaccharide vaccine contains antigen from 23 capsular types of pneumococci, which account for 90% of the invasive infections in the UK. The efficacy is approximately 70%, but is reduced in children under 2 years and in those with immunological impairment. Vaccination is indicated for those at increased risk (see pp. 144, 273, 412). Revaccination is not generally recommended because of the risk of severe systemic reactions. Asplenic patients and those with nephrotic syndrome, whose antibody levels are likely to decline more rapidly, should, however, be reimmunized after 5–10 years.

Splenectomized patients should be advised that they are at increased risk of pneumococcal disease. In addition to vaccine they should receive long-term penicillin prophylaxis (see Chapter 22).

Improved conjugate vaccines for the prevention of pneumococcal disease will soon be available (see Chapter 7).

Listerial meningitis

Introduction and epidemiology

Listeriosis is best known as an infection of pregnant women and neonates (see Chapter 17). However, 35–40% of recognized cases occur in older children and adults. Approximately half of non-neonatal infections attack patients who are immunosuppressed; the rest affect apparently immunocompetent individuals. Among the immunosuppressed, those at most risk are those on corticosteroid or anticancer therapy, and those with chronic uraemia or immunosuppression following organ transplant. Listeriosis rarely affects patients with acquired immunodeficiency syndrome (AIDS). Most cases of listerial meningitis occur in the summer months.

Listeria monocytogenes is a small, Gram-positive rod which is common in nature. It is found in human and animal faeces, sewage slurry and land on which it is spread, vegetables, soft cheeses, pâtés and some preprepared meat meals which have been kept chilled. The organism is robust, able to grow between 4 and 40°C, and able to survive temperatures of up to 60°C. Minor failures of pasteurization have been followed by large, milk-borne outbreaks. Its ability to replicate at refrigeration temperatures allows it to contaminate chilled food while other bacteria are inhibited.

Pathology

The organism grows well on blood agar and exhibits a narrow band of beta-haemolysis. It is catalase-positive. It is motile by virtue of its flagella, and exhibits a characteristic, end-over-end 'tumbling' motion at 22°C. *Listeria* are able to survive inside macrophages and have frequently been used as a model of intracellular parasitism.

Increasing interest is being shown in listeriolysin as the major pathogenicity determinant. This is one of the family of cysteine-based toxins homologous to streptolysin O and pneumolysin (see p. 143). It appears to act by allowing the organism to escape from phagolysosomes into the cytoplasm where it can then multiply.

Clinical features

Subclinical or mild gastroenteritis-like illness is probably common. The commonest severe manifestation of non-neonatal listeriosis is purulent meningitis. In the immunocompetent it often presents as acute meningitis, and must be differentiated from the commoner types. In about a third of patients it causes either bacteraemia alone or an unusual type of meningoencephalitis with a predominance of brainstem abnormalities.

In the immunosuppressed it may cause meningitis, bacteraemia or occasionally peritonitis; it rarely causes brainstem encephalitis.

In a pregnant woman the infection is often clinically trivial, or merely a colonization of the bowel or genital tract. The only sign of maternal infection may be the birth of an affected infant.

Manifestations of listeriosis

1 In the immunocompetent: acute meningitis; brainstem encephalitis; or, less commonly, bacteraemia.
2 In the immunosuppressed: acute or subacute meningitis; bacteraemia; peritonitis.
3 In pregnant women: mild feverish illness; mild 'gastroenteritis'; asymptomatic colonization of bowel or genital tract.
4 In the neonate: early bacteraemic multisystem disease or later meningitis.

Diagnosis

Diagnosis depends on the results of blood and CSF culture. The disease must be suspected in cases of brainstem encephalitis, where there may be little or no meningism. CT or MR scanning may suggest brainstem and/or meningeal inflammation. On occasions the CSF microscopy and culture are positive while blood cultures remain negative, even without prior antibiotic treatment. It is therefore important to obtain a CSF culture if listerial meningitis is suspected (CSF culture is rarely positive in encephalitic listeriosis).

Management

Management is with intravenous antibiotics. The choice is not easy, as *L. monocytogenes* is sensitive to penicillin, tetracycline and many other agents on laboratory testing but less so, apparently, in clinical use. High-dose amoxycillin may be effective. Some would advocate additional gentamicin, but it is uncertain whether this offers extra benefit. Chloramphenicol is also often effective, and readily enters the inflamed brain. It is probably the treatment of choice in encephalitis. Cephalosporins are poorly active and unreliable in listeriosis.

Treatment of listeriosis

1 Ampicillin i.v. 500 mg 4-hourly or 1 g 6-hourly for 10–14 days; *plus or minus* gentamicin 2–5 mg/kg daily in divided doses (avoid peak concentration above 10 mg/l and trough above 2 mg/l).
2 Alternative: chloramphenicol i.v. or orally 2–3 g daily in three or four divided doses.

Prevention and control

Prevention and control in the community depend largely on food hygiene. Adequate reheating of cook–chill and cook–freeze meals substantially reduces the risk of infection among immunocompromised patients. High-risk foods (especially pâté and soft cheeses) should not be served to such patients. Pregnant women should also avoid these foods. Affected mothers and babies should be nursed in isolation in delivery units.

Streptococcus 'milleri' meningitis and ventriculitis

Streptococcus 'milleri' is the name used for a group of related streptococci including the species *S. anginosis*, *S. constellatus* and *S. intermedius*. They are an uncommon cause of bacteraemia and meningitis. They tend to cause widespread infection with collections of thick pus, and are one of the few causes of simultaneous abscess formation in the brain and the liver. *S. milleri* meningitis often leads to purulent obstruction of the foramina of the brain. The resulting loculated infections produce ventriculitis as well as severe meningitis.

The organisms may be alpha-, beta- or non-haemolytic,

and of Lancefield group A, C, F or G. The commonest type is beta-haemolytic group F. Group A forms typically produce microcolonies on culture. Diagnosis depends on careful assessment of streptococci isolated from blood, CSF or pus cultures.

The importance of the organism is its reduced sensitivity to penicillin, which is, however, more effective than ampicillin. The treatment of choice is either a broad-spectrum cephalosporin or high-dose benzylpenicillin plus standard doses of gentamicin. A search should be made for distant infection and abscesses, particularly in the liver, as drainage may remove an additional source of sepsis.

Encephalitis and meningoencephalitis

Introduction

Encephalitis is inflammation of the brain. It can occur entirely independently of meningitis or the two can coexist, when the condition is called meningoencephalitis.

The clinical features of encephalitis are those of cerebral irritation and dysfunction. The irritation often appears first, initially resembling bad temper or extreme restlessness. Fits may occur at this stage. The personality may change, giving an appearance of distraction or mild confusion. Some patients complain of ataxia or generalized weakness. If the temporal lobes are affected there may be aphasia or loss of short-term memory. On examination the reflexes are often overbrisk and the plantars may be upgoing. Brain swelling may trap cranial nerves, causing focal signs such as ophthalmoplegia or ptosis.

As cerebral dysfunction increases, fits become more likely and drowsiness or coma develops. In extreme cases bulbar function is impaired, the pupil reactions are sluggish or absent, intermittent breathing patterns develop and the outlook is grave.

General features of encephalitis

1 Irritability.
2 Altered personality.
3 Drowsiness.
4 Ataxia.
5 Excessively brisk tendon reflexes.
6 Upgoing plantar responses.
7 Signs of cerebral or brainstem failure (sluggish or absent pupil reflexes, intermittent breathing patterns).
8 Signs of brain swelling (focal neurological signs, papilloedema).

Signs of meningitis may coexist.

Non-infectious causes of cerebral dysfunction can sometimes mimic encephalitis. Acute confusional states can complicate fever, especially in the old and the very young. Certain infections are likely to cause confusion or encephalopathy; legionnaires' disease, typhoid fever and typhus are examples. A non-cerebral source of infection should therefore be excluded in cases of global cerebral disturbance. Drugs may also cause encephalopathy. Opiate analgesics, high doses of phenytoin, indomethacin, cycloserine and some antidepressants are examples.

ORGANISM LIST

Herpes simplex virus type 1
Tick-borne encephalitis virus
Japanese B encephalitis virus
Many mosquito-borne arboviruses
Rabies virus

Encephalitis can be part of many other viral syndromes, or a postviral condition. Acute encephalitis is common and transient in mumps, and is well recognized and more severe in influenza A and B. A severe encephalitis can follow natural measles infection, and a benign form with cerebellar abnormalities is seen after chickenpox (see Chapter 11). These occur 10–14 days after the onset of acute infection, and are assumed to be of immunopathological origin (postinfectious encephalitis). These types of encephalitis are mentioned under their specific disease headings and will not be further discussed here.

Other infections causing fever and encephalopathy

Legionnaires' disease.
Typhoid and paratyphoid fever.
Typhus fevers.
Rare neurobrucellosis.
Rare neuroborreliosis.

Herpes simplex encephalitis

Introduction

Herpes simplex is a common infection of humans, usually affecting the skin or mucosae, but herpes simplex encephalitis is rare. It is important because it attacks previously well individuals, causing a high mortality and much prolonged disability. If recognized early it can be treated with antiviral medication, much improving the prognosis.

Epidemiology

Fewer than 50 laboratory-confirmed cases are reported in the UK each year. Encephalitis is usually caused by type 1 virus, whereas meningitis is usually due to type 2 virus. The disease arises sporadically and affects mainly young and middle-aged adults. Infection in children is rare. Both sexes are equally affected. The case fatality rate may be as high as 70% in untreated cases.

Clinical features

The incubation period is uncertain, as the disease is almost always intrinsic. The onset varies from a rapid onset of fever and 'influenzal' symptoms, quickly followed by neurological abnormalities, fits and coma, to several days of slowly declining cerebral function. In many cases the temporal or frontotemporal regions of the cerebral cortex are affected earliest and most severely. Focal signs may reflect this, with memory loss, dysphasia and unsteady gait in a drowsy or irritable patient. Headache is not a common complaint.

Up to half of cases have a herpes simplex lesion of the lip, skin or eye. There is rarely any other physical sign. The white cell count is normal, except when cerebral necrosis occurs, when there may be a neutrophilia. If there is significant cerebral destruction the aspartate transaminase levels may be raised.

The CSF almost always contains an excess of lymphocytes, though this varies from a few dozen to more than 200. Red blood cells may also be present in the absence of a traumatic tap. The protein level is moderately elevated, and the glucose is normal.

Diagnosis

When focal neurological signs are present, these help to distinguish encephalitis from a confusional state. Imaging is helpful in showing cerebral inflammation and often in demonstrating localization in the temporal lobes. MR is probably most helpful, but CT is also useful. There are typical electroencephalographic abnormalities in many cases. However, not all herpes simplex encephalitis shows localization, and temporal lobe involvement is not pathognomonic of herpes simplex.

Lumbar puncture is carried out as soon as possible after imaging. PCR-based DNA detection techniques usually demonstrate herpes simplex virus DNA in CSF samples, but it is unusual to isolate herpes simplex virus by CSF culture. A wide variety of antigen detection systems have been tried, but none are reliable. Differential herpes simplex and rubella antibody levels may be estimated in blood and CSF, and may suggest intrathecal production of herpes simplex antibodies. It is rarely justifiable to perform brain biopsy, as neither a positive nor a negative result would influence the necessity to attempt specific antiviral therapy.

Diagnosis of herpes simplex encephalitis

1 Demonstration of temporal lobe oedema on brain imaging.
2 Demonstration of herpes simplex virus (HSV) DNA in cerebrospinal fluid by polymerase chain reaction.
3 Demonstration of intrathecal anti-HSV antibody production.
4 Demonstration of encephalitic electroencephalographic changes in the temporal cortex.
5 Demonstration of HSV immunoglobulin M, seroconversion or rising immunoglobulin G titres in serum.

Management

Antiviral therapy with intravenous aciclovir should be commenced urgently. Since herpes simplex is the commonest cause of infective encephalitis in Britain, and antiviral therapy is effective, it is reasonable to give treatment in all cases until the diagnosis is confirmed or refuted. The treatment of choice is intravenous aciclovir 10 mg/kg 8-hourly. There is no evidence that the addition of other antiviral agents or of corticosteroids offers any benefit, though it is reasonable to give dexamethasone if generalized cerebral oedema is demonstrated on imaging.

The response to treatment is not rapid. Fever usually subsides in 2 or 3 days, but cerebral abnormalities improve slowly, and recovery is often incomplete. Treatment is continued for 2 weeks, sometimes longer, but improvement can continue for many weeks after cessation of therapy. Physiotherapy and speech therapy may help the patient regain social and motor skills as quickly and completely as possible.

Arbovirus encephalitides

Although not endemic in the UK, arboviruses are a common cause of encephalitis in both the new and the old world. The arboviruses are a heterogeneous group of organisms found in the alphaviridae, Flaviviridae and Bunyaviridae. They replicate in the cells of vertebrate and invertebrate hosts and some cause encephalitic diseases.

They are transmitted by many different vectors, including mosquitoes, ticks and sandflies (Table 13.1).In the forests of Germany, Austria, Scandinavia and eastern Europe, tick-borne encephalitis is endemic. It is transmitted to humans from its natural rodent hosts in the summer months when ticks are active.

In north India, Pakistan, parts of Indonesia and the Far East, including Japan, Japanese B encephalitis is transmitted by mosquitoes, and causes outbreaks and epidemics with high morbidity.

An outbreak in Malaysia in 1999 was found to have been caused by a newly identified paramyxovirus, designated Nipah virus.

In the USA there are four main types of encephalitis. The commonest is St Louis encephalitis, but eastern equine, western equine and Californian encephalitis also cause significant human disease. All are transmitted from birds and rodents, by mosquitoes, to horses and humans.

Like most arboviral infections, these have incubation periods of 7–12 days. The onset of illness is rapid, with features of viral-type meningitis. Cerebral disturbance develops over the next 1–2 days. The diseases vary in severity and there are many inapparent infections. However, Japanese B encephalitis can have a mortality rate of 7–20%, with up to 30% incidence of intellectual or psychological problems in survivors. Less severe sequelae, such as poor concentration and disorders of balance can follow tick-borne encephalitis. There is no specific treatment, and trials of prednisolone and dexamethasone have failed to show benefit.

Name	Vector	Distribution
Alphaviruses		
Eastern equine encephalitis	Mosquito	Americas
Venezuelan equine encephalomyelitis	Mosquito	Americas
Western equine encephalomyelitis	Mosquito	Americas
Flaviviruses		
Japanese encephalitis	Mosquito	Asia, Pacific islands
Kyanasur forest disease	Tick	India
Louping ill	Tick	UK
Murray Valley encephalitis	Mosquito	Australia, New Guinea
Powassan virus encephalitis	Tick	North America, Russia
St Louis encephalitis	Mosquito	Americas
Tick-borne encephalitis	Tick	Europe, Asia
Bunyaviruses		
Rift Valley fever	Mosquito	Africa
California encephalitis	Mosquito	USA

Table 13.1 Principal human neurological diseases caused by arboviruses.

Arbovirus infections can be diagnosed by PCR-based rapid tests, by culture or serology. Culture diagnosis requires specialized high-containment laboratory facilities and is rarely attempted in clinical practice. Immunofluorescence, EIA and PCR may be used to diagnose specific viral infections. The detection of virus-specific IgM in the blood indicates recent infection.

Mosquito control is an important factor in prevention of the mosquito-borne encephalitides. Safe and effective vaccines exist for tick-borne encephalitis (for which an immune globulin is also available for postexposure prophylaxis) and for Japanese B encephalitis.

Rabies

Pathology and epidemiology

Rabies is caused by neurotropic lyssavirus infection, transmitted from animals to humans by salivary contamination of a bite or open skin lesion. Human infections can result from contact with dogs, wolves, cats, bats, squirrels, skunks, mongooses and occasionally horses or other animals. Rare cases of iatrogenic transmission have occurred in recipients of corneal grafts harvested from patients who have died from 'ascending polyneuropathy'.

The virus enters the peripheral nerves in the infecting lesion and migrates towards the CNS, where it eventually causes an encephalomyelitis. Viruses then pass down the nerves to the salivary and lachrymal glands and the other tissues of the body. The incubation period varies with the length of nerve which the virus must traverse. It is often 3 or 4 weeks for a bite on the face or head, and 60 days or more for a bite on the foot.

Clinical features

Illness is often heralded by irritation or paraesthesia at the site of the originating lesion. Furious rabies, more common in humans, is a basal encephalitis, with fever, altered personality and periodic extreme agitation. Stimulation of the face and mouth, by attempts to drink or by draughts of air, can precipitate spasms of the face and pharynx; the typical picture of hydrophobia. Electrocardiographic signs of myocarditis are common, and death often occurs in 5–7 days, with cardiac failure or arrest. If the history of exposure is not identified, the agitation of the encephalitis can be mistaken for hysteria or 'rabies phobia'.

Paralytic (dumb) rabies presents as a spreading, often ascending paralysis or myelitis. It must be distinguished from Guillain–Barré syndrome, ascending myelitis or poliomyelitis. The disease has a course of 2 or 3 weeks, with death from respiratory complications or progressive encephalopathy.

Diagnosis and management

No tests are available to diagnose rabies before the onset of clinical disease. Rabies virus is a hazard category 4 organism. All virological tests are performed in specialist laboratories (in the UK, the Central Veterinary Laboratories). The diagnosis can be made by identifying rabies genome in CSF, saliva corneal scrapings or urine by PCR techniques. Rabies antigen can be demonstrated in the nerve endings of skin from the neck, or in postmortem brain samples by immunofluorescence. Virus can be cultured in cell cultures, and antibodies demonstrated in serum and CSF. Histological examination of brain tissue shows perivascular inflammation and the characteristic cytoplasmic inclusion bodies known as Negri bodies. These are absent in up to a third of patients, particularly in persons who have received vaccination.

Rabies is always fatal, except in a handful of atypical cases in immunized individuals. Intensive support, high-dose immunoglobulins, antiviral drugs and interferon have all failed to influence the course of the disease.

Pre-exposure immunization is available for those occupationally exposed. Postexposure prophylaxis is highly effective, and has three main components:

1 Washing the wound with any disinfectant will reduce the viral load and greatly reduce the risk of infection. Suturing and debridement may increase the entry of virus into nerves, and should be deferred.

2 Human rabies immunoglobulin (HRIG) in a dose of 20 IU/kg is given as soon as possible; it is infiltrated into the site of every infecting wound, and any remaining is given intramuscularly at a different site. (If wounds are extensive, the HRIG should be diluted to permit treatment of all injured sites.)

3 Immunization with high-potency cell-culture-derived vaccine (in the UK, this is human diploid cell vaccine) is commenced immediately by subcutaneous injection in the deltoid area, with doses on days 0, 3, 7, 14 and 30.

Neuroborreliosis

Among the manifestations of early disseminated Lyme disease is a meningoradiculitis which can occur 2–6 months after the initial infection. Persistent or recurrent lymphocytic meningitis or, rarely, non-specific encephalopathy may also occur. A history of tick bite and/or erythema chronicum migrans, facial or other nerve palsy (see Chapter 25) is present in approximately 70% of cases.

The clinical features are headache, fever, varying degrees of neck and back stiffness, with localized pain and dysfunction of the affected spinal nerve roots, often the lumbosacral ones.

The CSF contains excess lymphocytes, and often a few red cells. The protein is raised and the sugar may be normal or slightly low.

IgM antibodies to *Borrelia burgdorferi* may have disappeared, but IgG antibodies should be present both in serum and CSF. Attempts may be made to recover spirochaetes from CSF culture, but negative results do not exclude the diagnosis.

Treatment is with intravenous cefotaxime or ceftriaxone. However, the optimum duration is unknown; a significant proportion of patients are not cured by 2 weeks' therapy. Doxycycline orally 200 mg daily may also be effective. Effective treatment at this stage prevents further manifestations of infection, such as late arthritis, neuropathy or acrodermatitis.

Spongiform encephalopathies

Spongiform encephalopathies are progressive brain diseases characterized by dementia, movement disorder and terminal paralysis and coma. Typical 'spongiform' histological changes are seen in affected areas of the brain, with vacuolation of damaged nerve cells. Their aetiology is not fully understood, but no nucleic acid is demonstrable in infectious material. Infectivity is associated with an abnormal variant (PrP^{sc}) of a cell surface glycoprotein called prion protein. The agent is extremely resistant to heat and chemical agents, including formalin disinfection; it requires prolonged autoclaving (up to 1 h at 137°C) to sterilize contaminated materials.

Genes for normal prion protein (prp) are found in all animals and susceptibility to infection seems to have a strong genetic basis. It is suspected that infectious prp^{Sc} initiates irreversible conversion of the host's PrP to the abnormal form.

The natural route of infection is probably via the diet. Iatrogenic transmission has followed the use of reusable stereotactic instruments for neurosurgery, human dura mater grafts, biologically derived human growth hormone and human gonadotrophin (manufactured from pituitary glands harvested postmortem). The incubation period for natural infection is probably in the region of a decade or more. prp^{Sc} can be found in lymphoid tissue of the appendix, terminal ileum and tonsil for many months before the onset of clinical disease, following which it accumulates in the brain.

The clinical disease, Creutzfeld–Jakob disease, is a presenile dementia. Rare before the age of 40, its peak onset is in the 60s. Illness begins with clumsiness, ataxia and tremor, and progresses to intellectual and motor impairment, leading to death in 4–24 months. No treatment alters this course.

A new variant of Creutzfeld–Jakob disease (v-CJD) was first diagnosed in the UK in 1995, about 10 years after bovine spongiform encephalopathy (BSE) had appeared in British cattle. v-CJD is characterized by a distinct clinical course and unique neuropathology. There is now adequate evidence to implicate BSE as the source of v-CJD (see case study 26.1). The number of cases is small, but in the absence of a preclinical test, and an unknown incubation period, the size of any future epidemic cannot be predicted at this stage.

Spongiform encephalopathies of humans include:

Creutzfeldt–Jakob disease (CJD): sporadic or iatrogenic; onset age 55–75; mainly dementia/ataxia leading to death after 4–6 months.

Gerstmann–Straussler–Scheinker syndrome: inherited form of CJD; genetic abnormality of PrP; ataxia is marked; survival is longer than with CJD.

Fatal familial insomnia: another inherited disorder of PrP, with malignant insomnia and autonomic dysfunction.

Kuru: now disappearing; ataxic disease with tremor and incoordination; transmitted by cannibalism, only in Papua New Guinea.

Variant CJD (v-CJD): first described in 1995, now accepted as human infection with agent of bovine spongiform encephalopathy, probably diet-acquired; course of about 2 years with about 2 years with paraesthesiae, depression, paralysis, death.

Cerebral and intracranial abscesses

There are many ways in which space-occupying infections of the CNS can occur. Most are derived from bacteraemia which may be transient or never clinically evident. Some are complications of septicaemia. Others follow CNS invasion by contiguous spread from adjacent bone or intracranial sinus. Extradural or subdural collections of pus may compress the brain or spinal cord, or occlude an important blood supply, producing both local and long tract signs without actual CNS invasion.

Infectious lesions of this type often, but not always, produce fever, and are otherwise difficult to distinguish clinically from tumours or localized vascular disease. Imaging is important in making an early diagnosis, and allowing biopsy or aspiration under imaging control. A CT scan will define lesions and their position, and will demonstrate enhancement of the oedema surrounding inflammatory lesions. An MR scan will readily demonstrate oedema, and define fluid in an abscess cavity; enhancement can be used to demonstrate inflammation in the abscess wall.

The importance of early suspicion and diagnosis is paramount as the prognosis is relatively good before impaired consciousness and severe neurological deficit occur, but is definitely bad once these changes are established.

Cerebral abscess

Epidemiology

Traditionally, brain abscess has been associated with suppurative disease of the middle ear and mastoid cavity, and less often with ethmoid or frontal sinus disease. These are still important but, in children, an almost equal number of abscesses are of bacteraemic origin, associated with cyanotic congenital heart disease.

Rarer sources of brain abscesses are bacteraemias in patients with bronchiectasis, or with necrotic and ischaemic bowel lesions. Several reports exist of brain abscess as a rare complication of injection sclerotherapy for oesophageal varices. Bacteraemic disease may be complicated by brain abscess, and listeriosis is occasionally accompanied by brainstem abscess.

ORGANISM LIST

Anaerobic Gram-positive cocci
Prevotella melaninogenicus
Bacteroides fragilis
Fusobacterium spp.
Actinomyces spp.
Aerobic staphylococci and streptococci
Haemophilus aphrophilus
Other *Haemophilus* spp.
Listeria monocytogenes
Facultative Gram-negative rods

The bacteriology of abscesses is often mixed, with more than one anaerobe identified, or a mixture of aerobes and anaerobes.

Clinical features

These are a mixture of headache, reduced consciousness, features of raised intracranial pressure, localizing signs and, rarely, fits (which are more common when the abscess is subdural or cortical). Up to half of patients have papilloedema at presentation, and a similar number have vomiting and altered consciousness. Only about half of patients have fever. Features of raised intracranial pressure are global headache (worse on lying down and often waking the patient at night), increasing somnolence, rising blood pressure and falling pulse.

The headache may develop against the background of pain in the ear or the sinus, and this offers the chance to make an early diagnosis. If the abscess affects the meninges, or ruptures into the CSF, meningism develops, with nausea, vomiting, neck and back stiffness and photophobia.

The localizing signs depend on the site of the abscess. Frontal abscesses cause subtle signs of altered personality, paucity of conversation and apparent depression; parietal lesions cause classic features of hemiparesis and sometimes dysphasia; temporal lesions cause memory defects, sometimes aphasia and altered personality; visual field defects may be a sign of occipital lesions, or ataxia and tremor of cerebellar abscess.

Approximately equal numbers affect the frontal, parietal or temporal lobe. Occipital lobe abscesses occur less often and brainstem abscesses are rare. Multiple lesions are uncommon but may complicate bacteraemic disease, actinomycosis and nocardiosis. Subdural empyema occurs less often than intracerebral abscess, but the organisms involved and the clinical presentations are similar.

Diagnosis

Early imaging is essential to make the diagnosis (Fig. 13.17). In the presence of evolving localizing signs or raised intracranial pressure, it is unsafe to perform lumbar puncture because deformity of the brain may block the flow of CSF and cause herniation of the cerebellum or brainstem through the foramen magnum (coning) if the theca is opened. Lumbar puncture should be deferred until after imaging, even when meningism is present.

Blood cultures should always be set up to identify any coexisting bacteraemia.

When the abscess has been demonstrated, it is a neurosurgical choice whether it should be drained or aspirated at craniotomy, whether needle drainage alone is possible and whether indwelling drainage is required. Pus should be obtained for aerobic and anaerobic culture when drainage is performed.

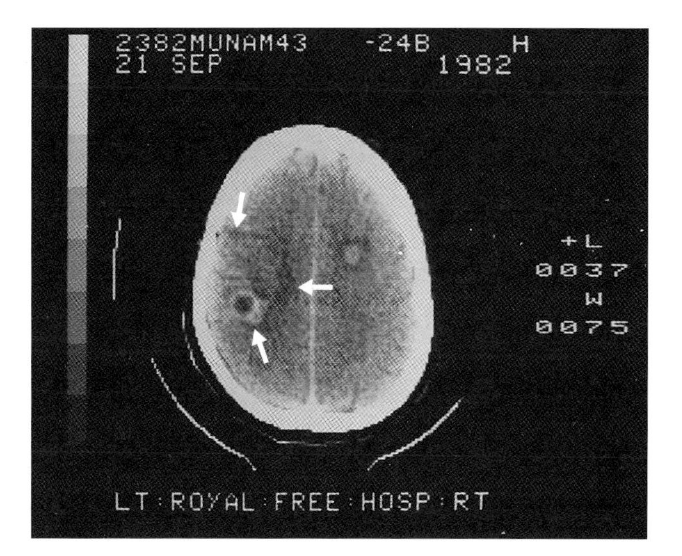

Fig. 13.17 Computed tomographic brain scan, showing abscesses surrounded by oedema (arrows) and zones of bright enhancement (contrast study).

Management

Broad-spectrum antibiotics should be given at diagnosis, to cover the likely range of pathogens. Agents which penetrate the brain well and are effective include cefotaxime, ceftriaxone, chloramphenicol and metronidazole. High doses of the cephalosporin should be given to achieve the best possible tissue levels as early as possible. Some specialists still use penicillin, chloramphenicol and metronidazole, a mixture which has also stood the test of time. However, penicillin penetrates the blood–brain barrier less well than the broad-spectrum cephalosporins. Treatment can be modified if necessary when the results of blood and pus culture are known.

Neurosurgery is an important part of treatment. Single abscesses can be aspirated or drained via burr holes or at craniotomy. In a few cases the abscess cavity may require excision. Small abscesses, especially if multiple, may be treatable with antibiotics alone, but require close follow-up by imaging to ensure adequate resolution.

Treatment of cerebral abscess

1 Cefotaxime i.v. 2–4 g 8-hourly; *or* ceftriaxone i.v. 2–4 g daily; *plus* metronidazole i.v. or rectal 500 mg 8-hourly.

2 Alternative: benzylpenicillin i.v. 2.4 g 4–6-hourly; *plus* chloramphenicol i.v. or oral 2–3 g daily in three or four divided doses; *plus* metronidazole i.v. or rectal 500 mg 8-hourly.

Drainage by needle or burr-hole approach is often indicated.

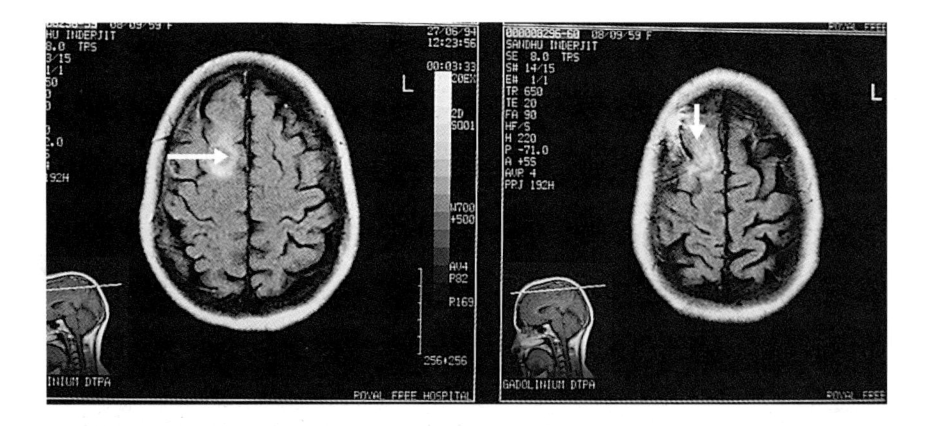

Fig. 13.18 Cerebral cysticercosis: this 44-year-old woman from Kenya had recent-onset epilepsy. The computed tomographic brain scan shows a cystic lesion with enhancing halo in the right frontal lobe (arrows). Serology was positive.

It is important to examine the patient clinically, by X-ray and, if necessary by further imaging, to detect any precipitating lesion. Sinusitis, otitis or, in the spinal column, osteomyelitis are all common and easy to miss if not deliberately sought. The liver and the lung should also be examined for evidence of coexisting abscesses.

Other space-occupying lesions of the CNS

Apart from tumours, tuberculomata and cysticercal lesions can cause chronic space-occupying lesions in the brain. Cerebral toxoplasmosis causes a similar picture in AIDS sufferers.

Tuberculomata appear similar to abscesses on CT and MR scanning. They may be associated with tuberculosis of the meninges, or of another body system. A positive tuberculin test can be helpful in suggesting the diagnosis. Treatment is with triple or quadruple antituberculosis therapy. Corticosteroids are also indicated to avoid early increases in inflammation (see Chapter 18).

Cysticercosis is the condition in which tissue cysts of *Taenia solium* develop in the brain and/or other tissues. It is rare in Western countries where tapeworm infection is extremely rare, but is the commonest cause of epilepsy in individuals from endemic areas. Imaging of the brain shows round cystic lesions, often with surrounding oedema (Fig. 13.18). Cysts with surrounding oedema will often die as a result of the inflammatory immune reaction; calcified cysts are usually dead. Scolices are sometimes visible in living cysts.

The commonest manifestation is epilepsy, but raised intracranial pressure, nausea or vomiting is also often seen. Single cysts may block the flow of CSF, or affect the spinal cord. ELISA tests for cysticercal antibody tests are reliable in indicating the diagnosis. Stool examination for eggs and proglottides is advisable, but is often negative. Treatment with albendazole or praziquantel can ameliorate symptoms and greatly reduce the frequency of seizures.

Treatment of cerebral cysticercosis

1 Albendazole orally 400 mg 12-hourly for 8–10 days.
2 Alternative: praziquantel orally 20 mg/kg 8-hourly for 1 day (consider simultanous prednisolone 40–60 mg, continued for 3–5 days).

Case 13.1: Feverish and lost for words

History: A 56-year-old taxi driver was accompanied to hospital by his wife, who gave a history on his behalf. Four days previously he had developed a fever and malaise, interpreted as a viral infection. Two days previously, he had begun to have difficulty in recalling the names of people and objects, and since then he had become increasingly drowsy and lost his memory of recent events.

Physical examination: This showed a drowsy man, with a Glasgow Coma Score of 13/15, and temperature 38.2°C. He could recall his address and the names of current government officers, but had no recall of the events of the day. He had a severe nominal aphasia. There was no meningism and no focal neurological deficit. The examination was otherwise normal.

Laboratory tests: The haemoglobin was 15.7 g/dl, white cell count 7.8×10^9/l with a normal differential. Renal function, liver function and chest X-ray were normal.

Questions: What investigations would you like to perform next?
What diagnoses would you consider?

Further management and progress: An urgent MR scan was performed, as the neurological features suggested pathology in the left temporoparietal region. This showed no evidence of infarct, space-occupying lesion or abscess. An pale area consistent with localized oedema was demonstrated, affecting a major portion of the left temporal lobe (Fig. CS.3).

Question: Does this suggest an aetiology for his illness?

Diagnosis and management: A feverish illness with central nervous system signs and temporal lobe inflammation is strongly suggestive (but not diagnostic) of herpes simplex encephalitis. Lumbar puncture was performed, and showed a CSF white count of 350 lymphocytes, protein of 0.7 g/l and glucose of 4.5 mmol/l (blood glucose 6.7 mmol/l).

Intravenous aciclovir 10 mg/kg 8-hourly was commenced. Two days later his temperature was normal, and his nominal aphasia had markedly improved. Polymerase chain reaction tests showed the presence of herpes simplex DNA in the CSF. Intravenous aciclovir treatment was continued for a further week. Three months later his only remaining deficit was an unreliable short-term memory, which he overcame by keeping a diary and daily workbook.

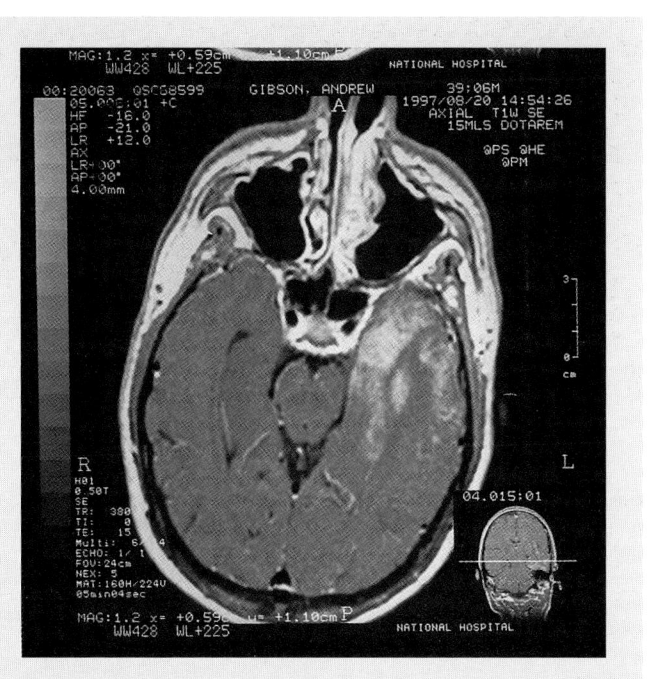

Fig. CS.3 T_2-weighted MR scan of the brain, in which water molecules appear white, showing extensive water accumulation, suggestive of oedema, in the left temporal lobe.

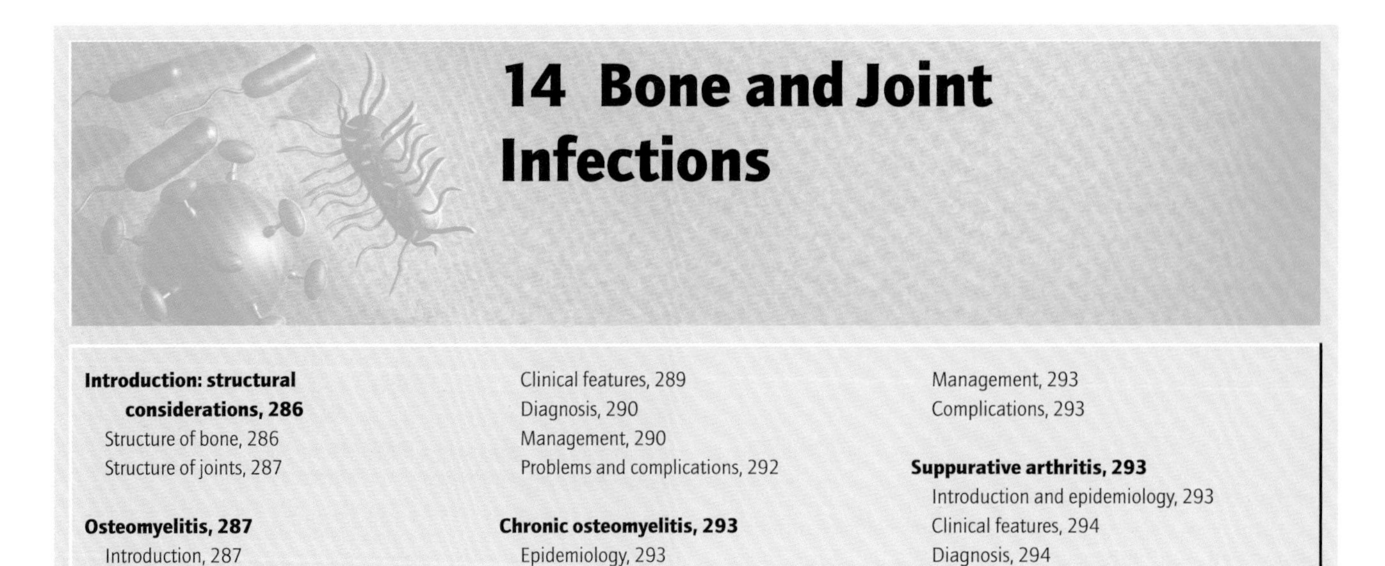

14 Bone and Joint Infections

Introduction: structural considerations, 286
Structure of bone, 286
Structure of joints, 287

Osteomyelitis, 287
Introduction, 287
Epidemiology, 287
Pathology, 287
Clinical features, 289
Diagnosis, 290
Management, 290
Problems and complications, 292

Chronic osteomyelitis, 293
Epidemiology, 293
Clinical features, 293
Diagnosis, 293
Management, 293
Complications, 293

Suppurative arthritis, 293
Introduction and epidemiology, 293
Clinical features, 294
Diagnosis, 294
Management, 295

Introduction: structural considerations

Structure of bone

Bone is a complex connective tissue formed of osteoid material which is hardened by the calcium salt hydroxyapatite. Its structure is best seen in the hard, or cortical, parts of the long bones.

The basic structure is the Haversian system, composed of concentric lamellae of bone surrounding a small blood vessel which runs in a central canal, or Haversian canal. The bone is laid down by cells called osteoblasts, which surround the blood vessel and also form a layer beneath the highly vascular periosteum. The osteoblasts separate the perivascular tissue fluid from the bone tissue fluid, which fills the Haversian system and the Volkmann's system, which forms transverse connections across the bone lamellae. Each Haversian system is bounded by a 'cement line', which separates each set of lamellae, and across which blood vessels do not pass (Fig. 14.1).

If the blood supply of a Haversian system is lost or damaged, the bone lamellae within that cement line will die. The dead tissue may demineralize, fibrose or occasionally be replaced by cartilage. Healing and remodelling can take place if the disease process is controlled.

As bone grows it is remodelled. This is achieved by small, advancing points of bone resorption, performed by groups of cells called osteoclasts or osteocytes. The resorption is followed by ingress of osteoblasts which build a new Haversian system, usually in a different orientation to the first, fragments of which are left supporting the new system (Fig. 14.2).

Each bone is surrounded by a tough, vascular membrane or periosteum. The periosteum is rich in sensory nerves and has a lymphatic supply; both of these are absent from the interior of the bone.

Bone growth in childhood

Bone growth in childhood takes place by the extension of the central, or diaphyseal, part of a bone at a cartilaginous plate, the epiphyseal plate. The growing part of the bone is called the metaphysis (Fig. 14.3). On the other side of the epiphyseal plate is the epiphysis, which is initially composed entirely of cartilage but develops a centre of ossification as the child develops. When the rate of ossification overtakes that of cartilage growth, the epiphysis becomes fully ossified.

Blood vessels do not cross the epiphyseal plate; the epiphysis has its own blood supply, usually derived from a single afferent vessel. In most bones the epiphyseal artery can enter the epiphysis directly, but in the head of the femur it must traverse the joint capsule, as the epiphysis is entirely intracapsular. This artery is more at risk than others of damage caused by pressure or deformity affecting the joint.

In the epiphyseal plate there is a proliferative layer, in which osteoid is manufactured, and a deeper maturation layer. Finally, in the deepest layer, next to bone, is the level at which mineralization, or ossification, occurs. These areas are served by many capillary loops, in which blood flow slows significantly as it passes from the arterial to the venous side. It is thought that this rather stagnant flow may predispose to the deposition of bacteria, and may explain the predisposition of the growing metaphysis to haematogenous osteomyelitis.

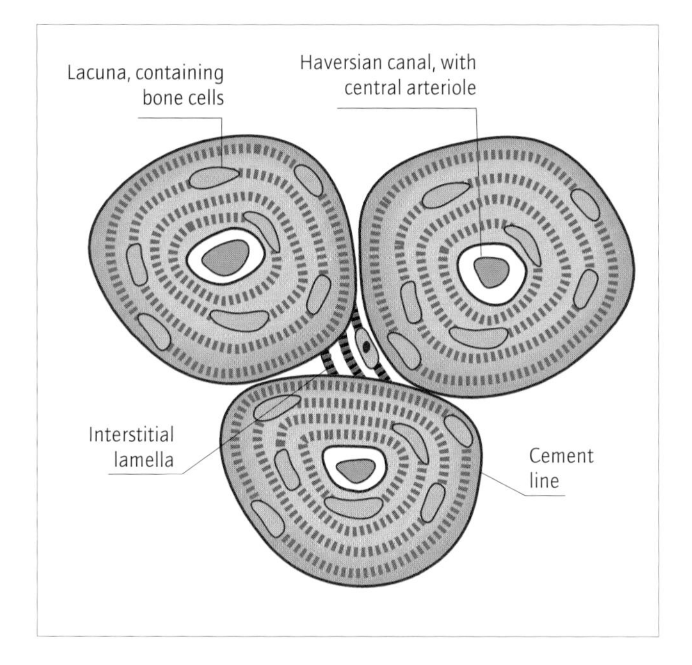

Fig. 14.1 The Haversian system of bone structure.

Structure of joints

Most large joints are synovial joints. The ends of the bones that move over one another are covered with articular cartilage. The bones are held together by a tough, fibrous joint capsule. The capsule is lined with vascular synovial membrane which secretes synovial fluid, and where the capsule joins with the articular cartilage there is a fibrocartilaginous zone overlapped by the vascular edge of the synovial membrane.

The articular cartilage does not grow once adulthood has been reached. It is not bound to the underlying bone by fibrous tissue, and it receives no blood vessels from the bone. It is attached merely by the irregular-shaped, interlocked surfaces of bone and cartilage. The cartilage receives its nutrition by diffusion from the synovial fluid, and since the cartilage has an open, water-saturated structure, this process of diffusion is increased by joint movement. If the cartilage is damaged, it is not replaced, but may repair by a sort of fibrous scar.

Sometimes the joint capsule is attached very near to the end of the articulating bone. The metaphysis is then extracapsular. In the femur, by contrast, the capsule of the hip joint is attached far down the neck, and the femoral neck is therefore intracapsular. This means that infection of the neck of the femur can extend directly into the joint space, or that joint infection may invade directly into the bone.

Osteomyelitis

Introduction

Osteomyelitis is infection of bone. It can arise by haematogenous spread, by extension from an infected joint, by direct invasion as a result of trauma, or by iatrogenic infection following surgery or instrumentation. This section is concerned with acute haematogenous osteomyelitis.

Epidemiology

The commonest type of acute haematogenous osteomyelitis is a disease of children, affecting growing bones. The source of infection is probably a transient bacteraemia, arising from the skin, mouth or respiratory system. The infection usually affects a single long bone (Fig. 14.4), and arises in the metaphysis, possibly because of the rich blood supply and the vulnerable capillary loops (see above).

Adults and elderly people can be affected by a different type of haematogenous osteomyelitis, which develops as a result of a predisposition. An example of this is the ability of bacteria to pass up the valveless venous plexus of the pelvis to the lower spine. Infection, instrumentation or invasive tumour of the lower bowel or urogenital tract can release bacteria into this plexus and cause 'metastatic' Gram-negative spinal osteomyelitis.

Occasionally a haematogenous infection occurs in a bone already damaged by degenerative or malignant disease. This is often an axial bone such as a vertebra or part of the pelvis.

Some intravenous drug users develop pseudomonal osteomyelitis of the spine or pelvis, possibly after injecting into inguinal veins.

Coliform bacteria or salmonellae may escape from an infected bowel and cause distant osteomyelitis, usually of the axial skeleton (Fig. 14.5).

Salmonellae are said to have a predilection for the bones of individuals with sickle-cell disease, possibly because of stagnant bone circulation following sickling crises (Fig. 14.6).

Finally, bone may be infected by extension from an overlying site of soft-tissue infection. An example of this is osteomyelitis underlying a deep pressure sore or a diabetic foot ulcer (Fig. 14.7).

Pathology

Osteomyelitis is caused by pyogenic organisms. Inflammatory oedema and pus therefore form in the infected

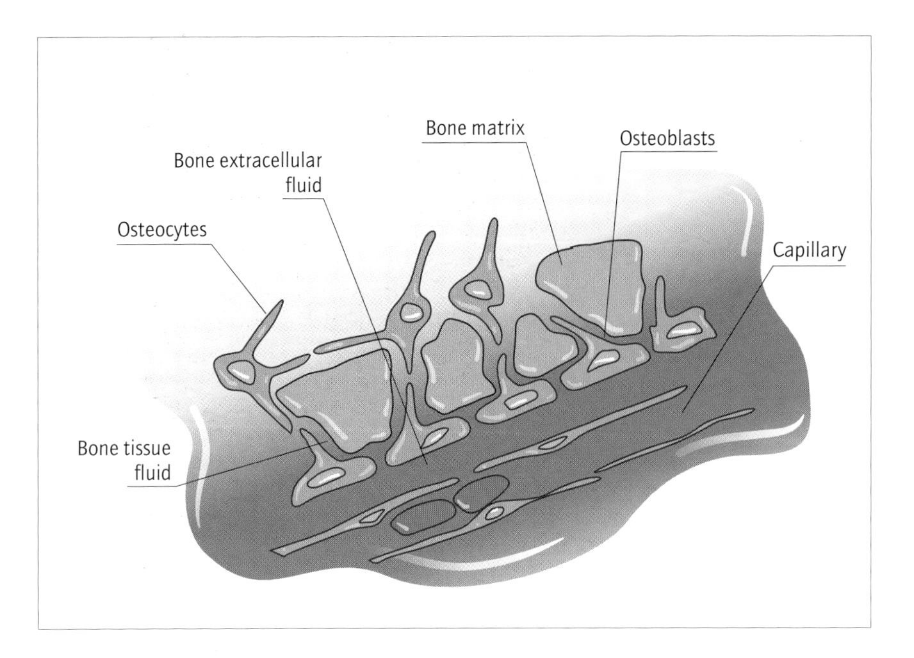

Fig. 14.2 The cells and fluid compartments of bone; antibiotics must enter the bone tissue fluid to treat intraosseous infection effectively.

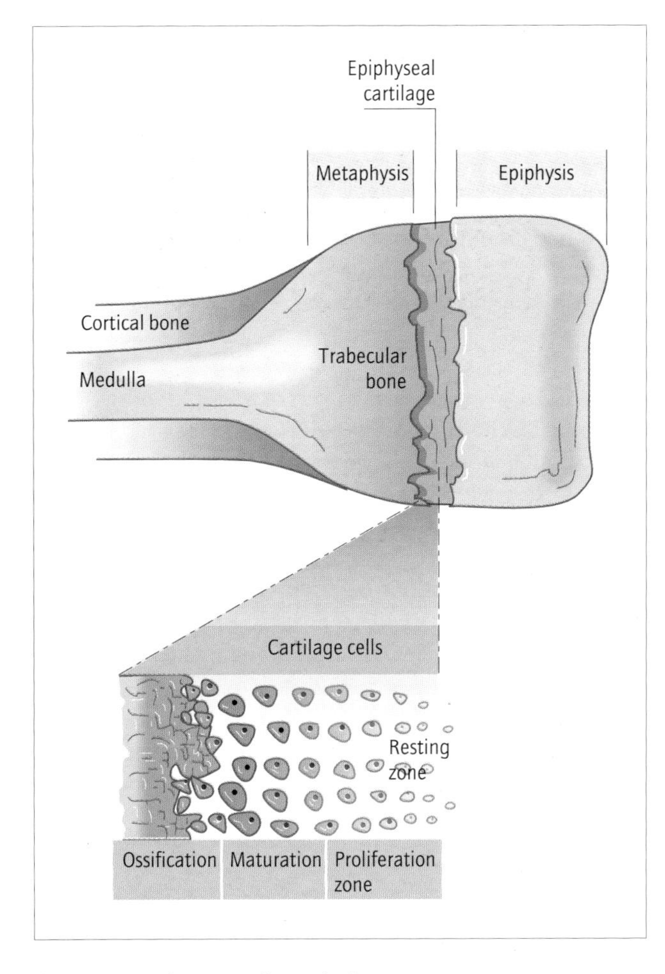

Fig. 14.3 General structure of a growing bone.

bone and track through the Haversian and Volkmann's canals. At the site of infection, bone ischaemia and necrosis occur. When the exudate reaches the periosteal surface of the bone, the periosteum is elevated by the fluid. The osteoblasts underlying the periosteum then begin to generate new bone.

Formation of sequestra and involucra

While the infected area of bone suffers progressive necrosis, it is gradually surrounded by an accumulation of new bone, which encloses it. The old, dead bone is called a sequestrum, and the surrounding live bone is termed an involucrum. These have typical X-ray appearances, with a sclerotic sequestrum surrounded by the more normal-appearing involucrum, from which it is separated by a lucent area (Fig. 14.8). The sequestrum has no blood supply, and so is inaccessible to antibiotics and most immunological processes. It is therefore often a site of persisting infection which prevents the healing of the osteomyelitis.

Brodie's abscess

Sometimes the infected part of a bone is completely replaced by pus. This intraosseous abscess is enclosed in a sclerotic membrane, and becomes surrounded by a sclerotic bony reaction. The infection may become quiescent when it is contained in this way, but there is a high risk of recrudescence and new extension in the future.

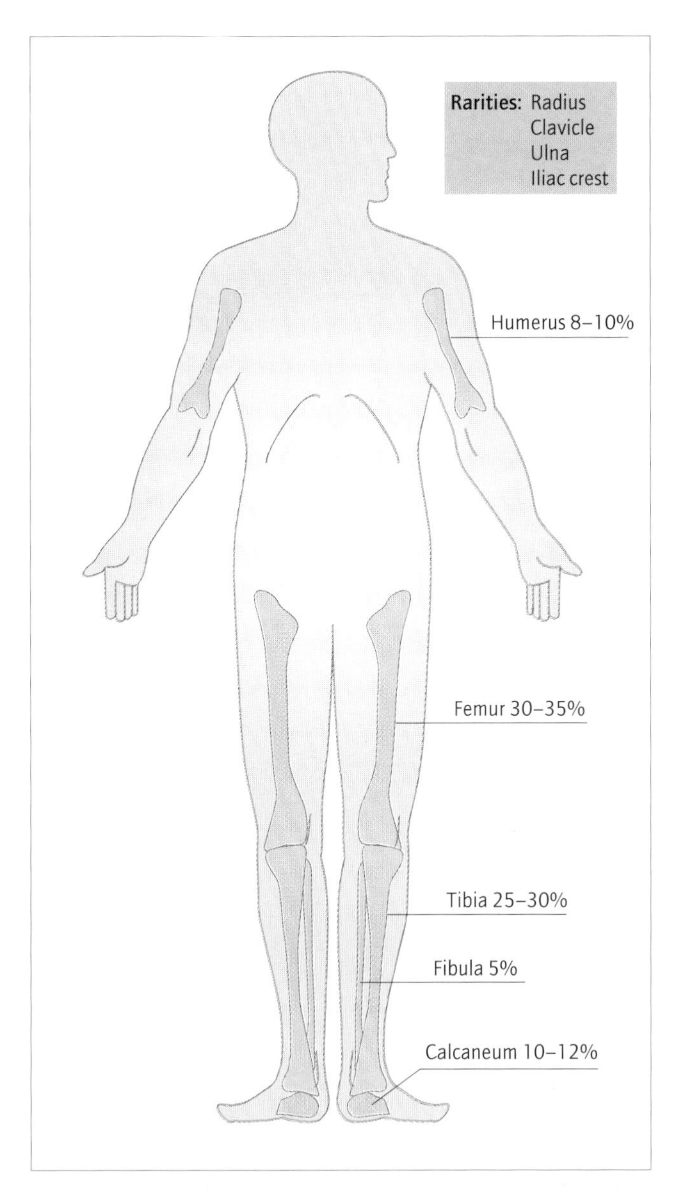

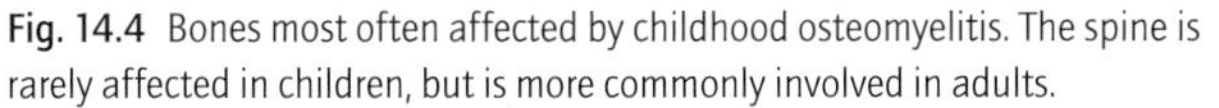
Fig. 14.4 Bones most often affected by childhood osteomyelitis. The spine is rarely affected in children, but is more commonly involved in adults.

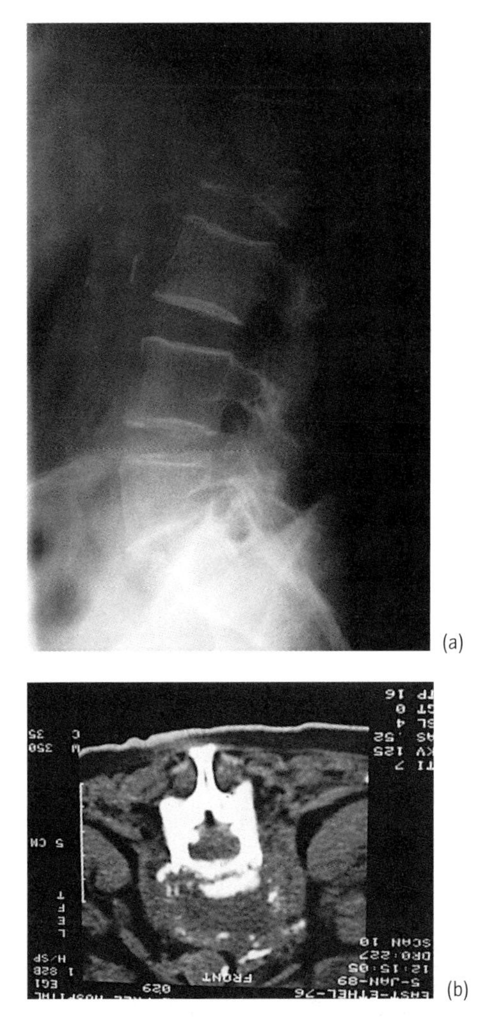

Fig. 14.5 Osteomyelitis of the lumbar spine complicating *Salmonella enteritidis* colitis. (a) X-ray; (b) computed tomographic scan showing destruction of the body of the second lumbar vertebra.

If pus continues to accumulate, it may track through the tissues, causing a local abscess or reaching the skin surface to produce a draining sinus. If the infection is near the joint surface of an intracapsular bone, it may track through the synovial membrane and cause a pyogenic arthritis.

ORGANISM LIST

Staphylococcus aureus (90%)
Streptococcus pyogenes (4%)
Haemophilus influenzae (4%)
Escherichia coli
Proteus spp.
Klebsiella spp.
Pseudomonas spp.
Neisseria meningitidis
Salmonella spp.
Brucella spp.
Mycobacteria
Anaerobic bacteria (anaerobic streptococci, *Bacteroides* spp., *Fusobacterium* spp., etc.)

Clinical features

The onset of osteomyelitis is insidious in most cases. In children a common presentation is with a feverish illness and poorly localized pain. While infection is confined

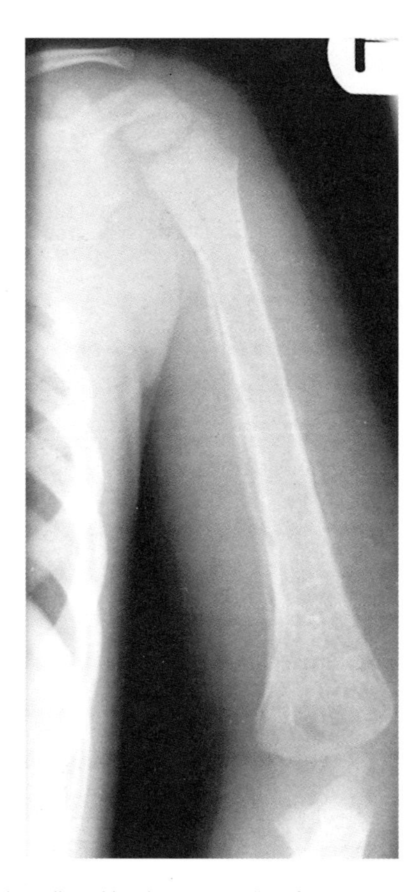

Fig. 14.6 *Salmonella typhimurium* osteomyelitis of the humerus in a child with sickle-cell disease. Note the elevated periosteum.

within the bone, there is little change in the white cell count, and the illness may be passed off as 'flu-like'. Infants and toddlers cannot describe the pain, but instead they stop using the affected limb, displaying a 'pseudoparalysis'.

As infection progresses, the affected bone site becomes surrounded by soft-tissue swelling. Later changes, such as the development of draining sinuses, of deformity of bones or of pathological fractures, are consequences of delayed diagnosis or ineffective treatment.

Diagnosis

X-ray changes are slow to develop, except for soft-tissue shadows. Later on, areas of lucency appear where bone is demineralized or destroyed, and the elevation of periosteum and formation of subperiosteal bone are typical X-ray findings in established disease.

Isotope scans of the skeleton, using technetium or gallium, will show early signs of vascularity or inflammation, respectively. They do not always distinguish between the changes of rheumatic inflammation, infection or tumour circulation.

Magnetic resonance scanning shows extremely early changes of oedema and altered blood flow. Although expensive, it can be an important early diagnostic investigation if others are unhelpful (Fig. 14.9).

Bacteriological diagnosis

Blood cultures are often negative, and pus cannot always be obtained, especially in early disease. Needle aspiration of subperiosteal fluid or pus is sometimes possible. Imaging-directed needle aspiration may yield pus from abnormal bone, or from within or outside the periosteum.

An operative approach can be used to obtain pus, infected bone or adjacent tissue. For a large collection of pus, a sequestrum or a Brodie's abscess, drainage may contribute to treatment, and a bacteriological diagnosis can help the choice of antibiotic management.

Management

Confirming the diagnosis

Confirming the diagnosis is an important step. Blood cultures should be taken, and a careful examination made for skin, chest, throat or ear sepsis, which may provide clues as to the infecting agent. Urine should be cultured, especially in elderly patients.

Imaging studies may show evidence of bone inflammation at a suspicious site. This allows follow-up by clinical and further imaging examination, to confirm healing or to detect the development of pus.

Antimicrobial treatment

Antimicrobial treatment should be commenced as soon as specimens have been obtained. In all age groups an anti-staphylococcal agent should be included. Agents which achieve effective concentrations in bone include fusidic acid, flucloxacillin and clindamycin. In all but the mildest cases, it is advisable to begin with intravenous treatment.

In unvaccinated children up to the age of 5 years, it is advisable to treat empirically for *Haemophilus influenzae* as well as *Staphylococcus aureus*. Cefuroxime is effective against both organisms. Clindamycin is particularly useful in osteomyelitis, being effective against staphylococci and some other Gram-positive cocci, as well as against anaerobes. It penetrates bone very well. Vigilance for pseudomembranous colitis should be maintained when using this drug.

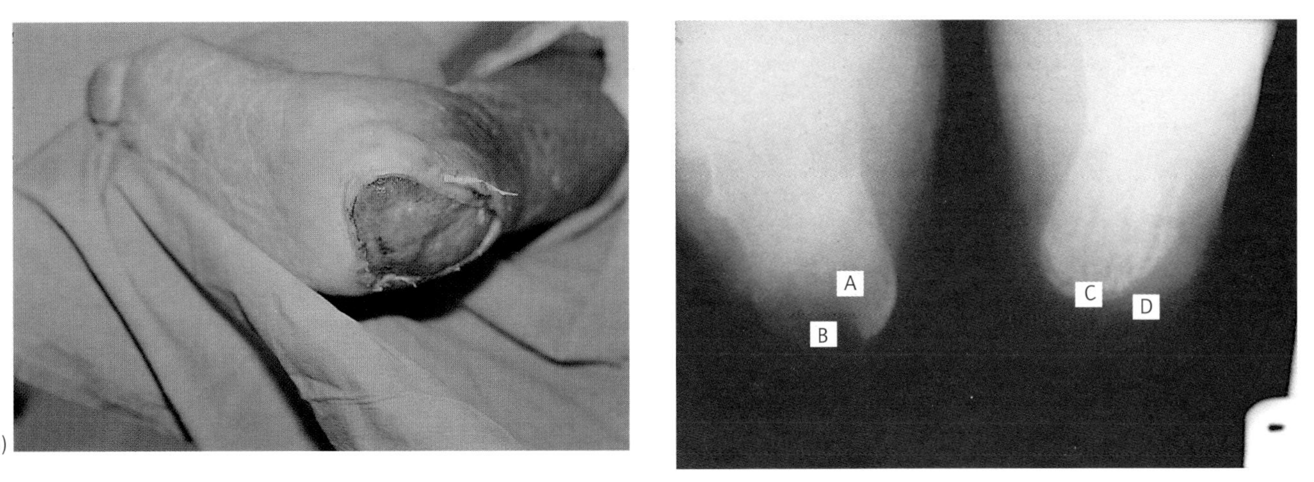

Fig. 14.7 Osteomyelitis of the calcaneum: this immobile, elderly man had a deep heel ulcer colonized with methicillin-resistant *Staphylococcus aureus*; this infection extended to the underlying periosteum and bone. (a) The ulcer; (b) X-ray showing loss of structure of the calcaneum. A, defect in both cortical layer and bone structure; B, absence of soft tissue; C, normal outline of continuous layer of bone cortex; D, soft tissue of the heel.

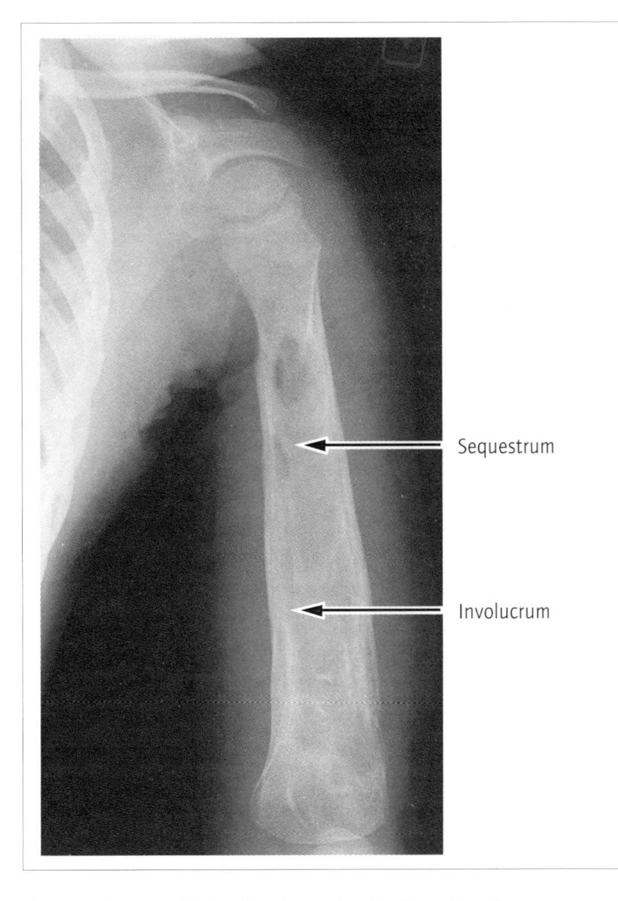

Fig. 14.8 Osteomyelitis in a long bone, showing formation of a sequestrum and involucrum.

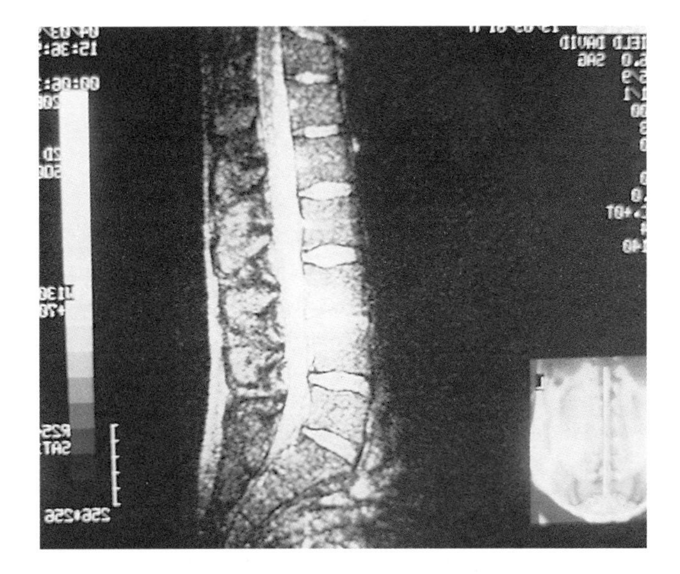

Fig. 14.9 Magnetic resonance scan showing infection of an intervertebral disc (discitis): the patient 'strained' his back while suffering from a severe sore throat. Persisting back pain and fever led to imaging and needle aspiration, revealing a *Fusobacterium necrophorum* infection (see Chapter 7).

In the elderly, especially those with urinary tract disorders, ciprofloxacin is often useful, as it is effective against many potential pathogens, including salmonellae, and has the advantage of being effective when given by mouth. Unfortunately, salmonellae and some other Gram-negative organisms are increasingly resistant to ciprofloxacin; ceftriaxone may then be useful. In osteomyelitis of the diabetic foot, mixed infection is common, often with an anaerobic component. A cephalosporin plus metronidazole, or monotherapy with meropenem may then be used.

Antibiotic treatment of osteomyelitis

1 First choice:

(a) Flucloxacillin i.v. 1–2 g 6-hourly (child under 2 years, 250–500 mg 6-hourly; 2–10 years, 500 mg–1 g 6-hourly *or* cloxacillin i.v. 1 g 6-hourly (child proportionately as for flucloxacillin); *plus*

Fusidic acid orally 750 mg 8-hourly (child up to 1 year, 50 mg/kg daily in three divided doses; 1–5 years, 250 mg 8-hourly; 5–12 years, 500 mg 8-hourly); *or* fusidic acid i.v. adult over 50 kg, 580 mg 8-hourly; adult under 50 kg and child, 6–7 mg/kg 8-hourly; *or*

(b) Clindamycin orally 150–300 mg 6-hourly (child 3–6 mg/kg 6-hourly).

2 For child under 5 years, consider cefuroxime i.v. 60–100 mg/kg daily in three or four divided doses.

3 For elderly people, consider cefuroxime i.v. 750 mg 1.5 g 8-hourly *or* cefotaxime i.v. 2 g 8-hourly plus an anti-staphylococcal drug; *plus or minus* metronidazole 400–500 mg 8-hourly.

4 For methicillin-resistant *Staphylococcus aureus* or enterococci, teicoplanin i.v. 400 mg 12-hourly for three doses, then 400 mg daily (child 10 mg/kg daily, reducing to 6 mg/kg daily after first 2–5 days).

Treatment may be modified in the light of bacteriological information. Difficult Gram-positive infections such as those with enterococci may be treated with teicoplanin. Treatment should be continued until the fever has subsided, the erythrocyte sedimentation rate or C-reactive protein has fallen to the normal range, and healing of the bone is established. In acute, uncomplicated osteomyelitis in children, a 2–3-week course may be sufficient, while more prolonged treatment may be needed for older patients with Gram-negative or mixed infections.

Patients with 'difficult' organisms, such as methicillin-resistant *Staphylococcus aureus* (MRSA), enterococci or mixed facultative and anaerobic infections, are often long-term patients, frail, demented or otherwise debilitated. It is sometimes impossible to maintain continued parenteral treatment in such patients. As long as a steady response is obtained, treatment with combinations of oral drugs such as fusidic acid, rifampicin, trimethoprim and/or metronidazole can be successful.

Surgical treatment

Surgical treatment is usually reserved for those cases requiring release of clinically or radiologically apparent pus. It is rarely justified to operate empirically on a suspected site, as it may not be the exact site of infection. Exceptions are the calcaneum and the sacroiliac area, which are accessible to aspiration.

Evidence of pus formation should be sought by daily examination for fluctuance or localization of inflammation. The erythrocyte sedimentation rate or C-reactive protein can be used to monitor the inflammatory reaction during treatment. When pus is present, drainage will alleviate the infection, reducing fever and allowing healing of the bone.

Surgery may also be indicated when a sequestrum has formed. The orthopaedic specialist will decide whether or when there is adequate new bone formation to compensate for the defect caused by sequestrum removal.

Problems and complications

Failure to respond

Care should be taken that an adequate antibiotic dosage is being used, and that the route of administration is optimal. The causative organism may be unusual (for instance, *Brucella* sp.), and not covered by the chosen antibiotic regimen. Unexpected antibiotic-resistance may be the problem, as with MRSA.

Problems of prolonged antibiotic treatment

Problems may result from the need to use antibiotics for several months in severe infections, though many patients tolerate this surprisingly well. Candidal infections of mucosae or skin folds may be troublesome if broad-spectrum agents must be used; intermittent or concurrent anti-*Candida* medication will often help, if necessary using oral itraconazole or fluconazole (though fluconazole-resistant *Candida* species can emerge, especially in hospital environments).

Prolonged ciprofloxacin treatment can cause anorexia and weight loss. Long-term metronidazole can cause peripheral neuropathy. Prolonged high-dose beta-lactam medication can cause sudden, idiosyncratic agranulocytosis, which quickly resolves if treatment is stopped, but which precludes further use of any beta-lactam because of complete cross-reactogenicity.

Mistaken diagnosis

Bone tumours and cysts, or the effects of trauma, can all cause a painful, warm and immobile limb. Failure to respond to antibiotics should prompt a reconsideration of the diagnosis and a review of imaging studies.

Damage to epiphyseal cartilage

Irreversible damage to epiphyseal cartilage occasionally follows extensive osteomyelitis. This causes progressive distortion and, in children, failure to elongate at the affected site. The least favourable sites for this to occur are the lower femur and the upper tibia, as gait, stature and knee function are all severely affected.

Adjacent suppurative arthritis

Invasion and suppurative arthritis of a neighbouring joint can occur if a bone metaphysis is intracapsular, or partly so. The hip and the shoulder are therefore the joints most affected by this problem.

Chronic osteomyelitis

Epidemiology

The epidemiology of chronic osteomyelitis has changed with the development of powerful antibiotics. Only half of cases, or fewer, are nowadays the result of persisting infection after acute blood-borne osteomyelitis. The remainder of cases follow fractures (sometimes associated with non-union), or complicated surgery.

Bones

The femur and the tibia together account for over 70% of cases in many series. Any bone may be involved, however, including vertebrae, the bones of the foot (especially in diabetics) or the skull (complicating severe otitis externa, for instance).

Bacteria

The bacteria are similar to those of acute disease. Over half of cases are caused by *S. aureus*. Gram-negative infections are slightly less common, and are more often caused by *Pseudomonas* spp. or *Proteus* spp. than by *Escherichia coli*. Mixed Gram-positive and Gram-negative infections occur. Between 10 and 20% of cases are due to anaerobic infection.

Causes of chronic osteomyelitis

1 *Staphylococcus aureus* (> 50%).
2 Anaerobic infections (10–20%).
3 *Pseudomonas* spp.
4 *Proteus* spp.
5 *Escherichia coli*.
6 Mixed Gram-positive and -negative infections.

Clinical features

Clinical features include pain, swelling, deformity and, in the case of fracture or surgery, defective healing or loosening of a prosthesis or pin. Some cases have intermittently or continuously discharging sinuses.

Diagnosis

Diagnosis is made on clinical and radiological grounds, and by identifying a causative organism. Pus or swabs from sinuses are often contaminated by skin commensals or saprophytes, so specimens obtained directly from infected bone or periosteum are preferred.

Management

Appropriate chemotherapy is important, but surgery may also have a major role. Surgical intervention may include removal of affected pins, screws and plates, or of sequestra and other devitalized tissue. Plastic surgery and vascular surgery preocedures may be used to fill tissue defects and optimize blood supply, and bone grafting can restore anatomy and function when infection has been controlled.

Complications

Complications are rare, except for the inconvenience of the lesion and the systemic effects of chronic infection. A rare complication is the occurrence of squamous cell carcinoma in the infected tissue. This is associated with increased pain, progressive bone destruction and metastasis to regional lymph nodes.

Suppurative arthritis

Introduction and epidemiology

Suppurative arthritis is usually a blood-borne infection, arising either from an inapparent bacteraemia or as a complication of a bacteraemic disease. In children the joint infection often arises apparently spontaneously, while in adults pre-existing disease such as rheumatoid arthritis, gout or pseudogout may predispose to septic joint disease, and complicate its diagnosis. In children, but rarely in adults, arthritis of the hip can be an extension from osteomyelitis of the upper femur.

ORGANISM LIST

In children	In adults
Staphylococcus aureus (50%)	*Staphylococcus aureus* (70%)
Streptococcus pyogenes (16%)	Gram-negative rods (15%)
*Haemophilus influenzae**	*Streptococcus pyogenes* (7%)
Gram-negative rods (7%)	*Neisseria gonorrhoeae* (3%)
Anaerobes	*Staphylococcus epidermidis*
Staphylococcus epidermidis	Mixed organisms
	Anaerobes

* Rare where an effective immunization programme is in place.

Rarities
- *Neisseria meningitidis*
- *Borrelia burgdorferi*
- *Salmonella* spp.
- *Brucella* spp.
- *Pasteurella* spp.

Clinical features

The joints most commonly affected are similar in children and adults. The knee is involved approximately twice as commonly as any other joint. The hip, ankle, elbow and wrist are progressively less often affected. In adults the shoulder rivals the hip or ankle in frequency; in children it is about as commonly affected as the wrist.

In children the onset of suppurative arthritis is often abrupt, with fever, pain and swelling of the joint. The main exception occurs when the hip is affected in a small child or infant. The swelling may not be apparent, and the child often complains of abdominal pain. Constipation or abdominal distension may further confuse the diagnosis.

Misleading clinical signs in childhood hip infections
1 Pseudoparalysis of the leg.
2 Abdominal pain.
3 Abdominal distension.
4 Constipation.

In adults the onset is not always rapid. Especially in individuals with pre-existing joint disease, there may be a gradual swelling and slow accumulation of effusion. There may be a clue in the recent medical history, suggesting a source of infection. Recent salmonella gastroenteritis or an episode of urinary retention or urinary tract manipulation may suggest a Gram-negative aetiology.

When the joint infection is part of a bacteraemic disease, there may be other manifestations of the infection on general examination. In staphylococcal septicaemia there may be a cellulitic skin rash, or nodular pneumonitis, with a tendency to abscess formation. In streptococcal infections there may be an erysipelas-like rash on the trunk or arm, a scarlet fever-like rash or a follicular tonsillitis. Painful lymphadenitis tends to accompany severe streptococcal infections. There may be a typical rash in gonococcal bacteraemia (see Chapter 15).

Diagnosis

In children the diagnosis is often clinically obvious, but in adults there is a wide range of differential diagnoses, some of which can coexist with infection. These include trauma-related effusions, rheumatoid arthritis, acute osteoarthritis, acute gout or pseudogout.

Examination of effusion

Examination of effusion is the best rapid diagnostic test. In acute infective arthritis the protein level is high, the glucose level low and there is a neutrophilic pleiocytosis. This picture can also occur in exacerbation of inflammatory joint disease, however, so bacteriological examination is critical. Blood-stained effusions are often of traumatic origin, but infection can occur in the effusion, so culture should always be performed.

Gram-stained preparations will often show typical Gram-positive cocci in staphylococcal and streptococcal infections, and their grouping in masses or chains may indicate the type of bacteria involved. Gram-negative organisms such as *H. influenzae* or coliforms may also be demonstrable, but the small *Haemophilus* organisms in particular are more difficult to see. Most pyogenic organisms will be detectable in culture within 24–36 h, though speciation and sensitivity testing may take a further day. Previous treatment with antibiotics may make culture difficult, slow or impossible. Latex agglutination or polymerase chain reaction (PCR) tests may be helpful in meningococcal or *H. influenzae* infections.

Cultures from other sites

Blood cultures should also be obtained, as should urine cultures and specimens from skin, throat and sputum, when indicated. In patients at risk through known exposure or overseas residence, cultures for mycobacteria should be set up. About 80% of patients with tuberculous arthritis have a strongly positive tuberculin test. Brucellosis is a recognized cause of infective arthritis in African, Middle Eastern and east European countries. It often affects the knee or the spine. A positive IgM ELISA test may indicate recent infection, and is useful if recent antibiotic therapy makes culture unreliable.

Differential diagnosis

Exclusion of differential diagnoses is helpful, and can include examination of joint aspirate for crystals, as well as serological tests for rheumatoid factors and other autoantibodies. On rare occasions arthroscopy and synovial biopsy may be useful in diagnosing typical rheumatoid changes, or in obtaining tissue specimens for culture.

Management

Early eradication of infection minimizes damage to the joint, particularly to the articular cartilage, which is eroded by proteolytic enzymes and other inflammatory mediators. There is general agreement that intravenous antibiotic treatment is justified, to obtain high levels of agent in the synovial fluid and adjacent tissues.

Adults

For adults an antistaphylococcal drug is essential. This may be flucloxacillin or cloxacillin. Fusidic acid (or sometimes rifampicin) should be added, as it penetrates bones and joints well. Teicoplanin is useful for MRSA, and may be used for other staphylococci. Ciprofloxacin is well distributed even when given orally. It is effective against staphylococci and Gram-negative rods, including many salmonellae, but is less effective against streptococci even at high doses. Clindamycin is an option for Gram-positive cocci and anaerobes. Antibiotic treatment is usually continued for 2–4 weeks, depending on the presence of predisposition or complication, and on response. Oral continuation treatment is possible with ciprofloxacin, fusidic acid or rifampicin, but neither of the latter two drugs should be given alone, because of the risk of one-step mutation to resistance by the infecting organism.

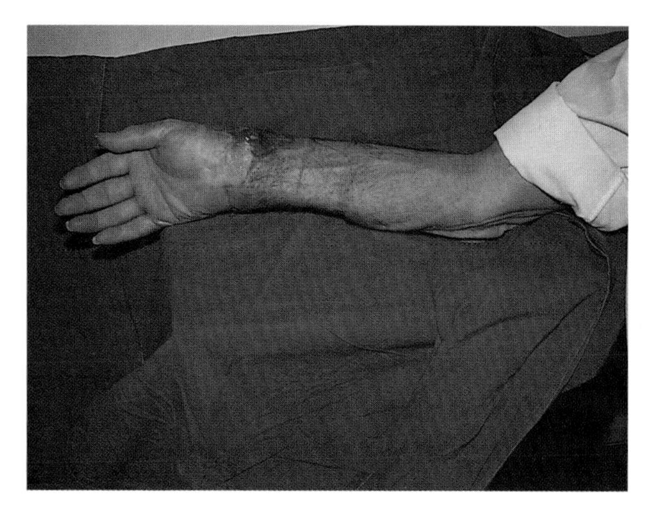

Fig. 14.10 Destructive tuberculous infection of the wrist, with draining sinus. Concurrent pulmonary tuberculosis responded well to triple therapy, but the wrist only healed after debridement and arthrodesis.

Adults with urinary or bowel sepsis

For adults with urinary or bowel sepsis it is advisable to consider the possibility of Gram-negative infection. A broad-spectrum cephalosporin such as cefotaxime is useful in this setting, and can be given with an antistaphylococcal drug.

Children

For children the possibility of *H. influenzae* infection still exists. A broad-spectrum cephalosporin is the antibiotic of choice in this setting, and an additional antistaphylococcal drug can safely be added. Cefuroxime is effective against both *S. aureus* and *H. influenzae*.

Surgical intervention

Surgical intervention is not often needed. When the joint is distended by pus aspiration and irrigation may be performed, and protects the cartilage from inflammatory damage. In severe cases, persistent effusion can be drained by an indwelling suction drain.

In the rare case of extensive joint destruction, it may be possible to carry out arthroplasty or arthrodesis once the infection has been fully controlled (Fig. 14.10).

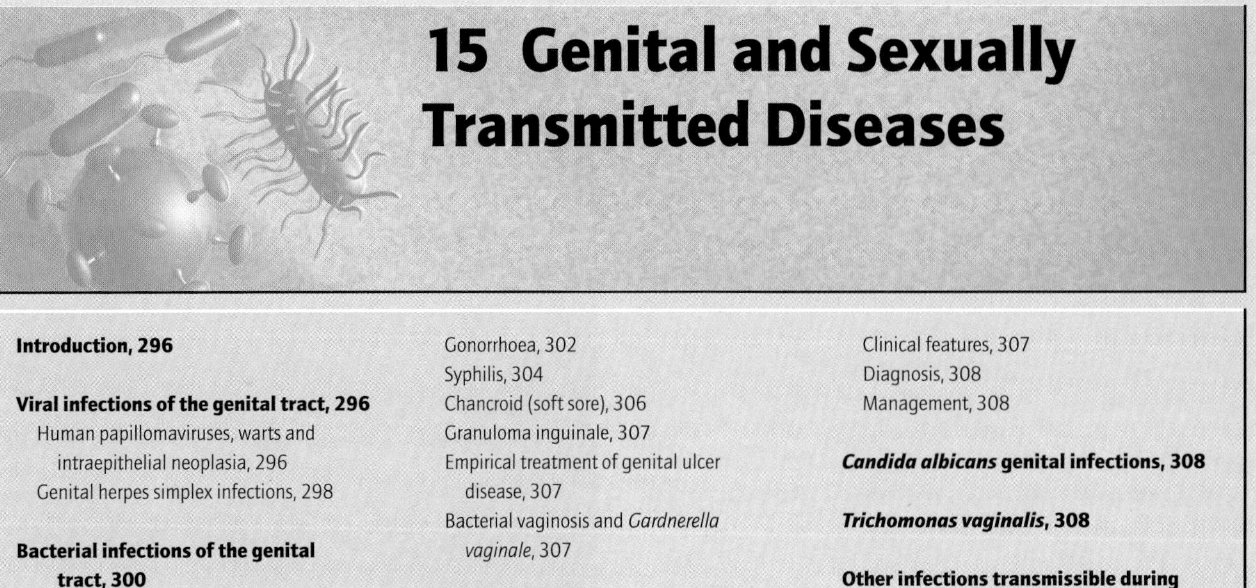

15 Genital and Sexually Transmitted Diseases

Introduction, 296

Viral infections of the genital tract, 296
Human papillomaviruses, warts and intraepithelial neoplasia, 296
Genital herpes simplex infections, 298

Bacterial infections of the genital tract, 300
Chlamydial genital infections, 300
Lymphogranuloma venereum, 301
Gonorrhoea, 302
Syphilis, 304
Chancroid (soft sore), 306
Granuloma inguinale, 307
Empirical treatment of genital ulcer disease, 307
Bacterial vaginosis and *Gardnerella vaginale*, 307

Pelvic inflammatory disease (PID), 307
Introduction, 307
Clinical features, 307
Diagnosis, 308
Management, 308

***Candida albicans* genital infections, 308**

***Trichomonas vaginalis*, 308**

Other infections transmissible during sexual contact, 309

Introduction

The structures of the male and female genital tract differ greatly, but they have their pelvic position and perineal connections in common. Both the vagina and the urethra are lined with squamous epithelium and, in both sexes, the upper genital tract has unique structures which are both germinal and secretory.

In men the genital tract flora reside on the glans of the penis and the urethral meatus. Colonizing and infecting organisms may exist in the normally sterile upper urethra, prostate and epidydymis.

In women the vulva and vagina have a complex flora partly derived from the perineal skin, and partly dictated by the acidic environment of the adult vagina, which is maintained by colonizing lactobacilli. The normal vaginal flora help to inhibit the establishment of *Candida* infections, to which this site is especially vulnerable. The female urethra and cervix are vulnerable to invasion by sexually transmitted pathogens, which may then ascend to cause endometrial and tubal infections, and to affect intra-abdominal organs.

In both sexes the genital tract can act as the portal of entry for systemic or bacteraemic disease. Fastidious organisms are carried and protected in genital secretions, and may be deposited in the consort's genital epithelium, mouth, eye or rectum, depending on the type of sexual activity. It is important to remember these associations when investigating possible sexually transmitted infections.

Trends in incidence of the commonest sexually transmitted diseases are shown in Fig. 15.1. There is a difference in incidence between different ages and sexes. This is because women often become sexually active at an earlier age than heterosexual men, and homosexual men at a later age (Fig. 15.2). The incidence of gonorrhoea and trichomoniasis declined throughout the 1980s, especially among older patients who modified their sexual behaviour in response to the acquired immunodeficiency syndrome (AIDS) epidemic. Since 1989 reported levels of gonorrhoea have increased. In contrast, the incidence of both genital herpes and *Chlamydia* infection has been increasing since the early 1980s in both sexes, but especially among females for genital herpes.

Viral infections of the genital tract

ORGANISM LIST

Papillomaviruses
Herpes simplex viruses

Human papillomaviruses, warts and intraepithelial neoplasia

Introduction

Human papillomaviruses (HPVs) are small DNA viruses, with a circular genome of double-stranded DNA. There are at least 70 types of papillomavirus, loosely associated with types of cutaneous, genital and laryngeal warts. Approximately half of papillomavirus types are only

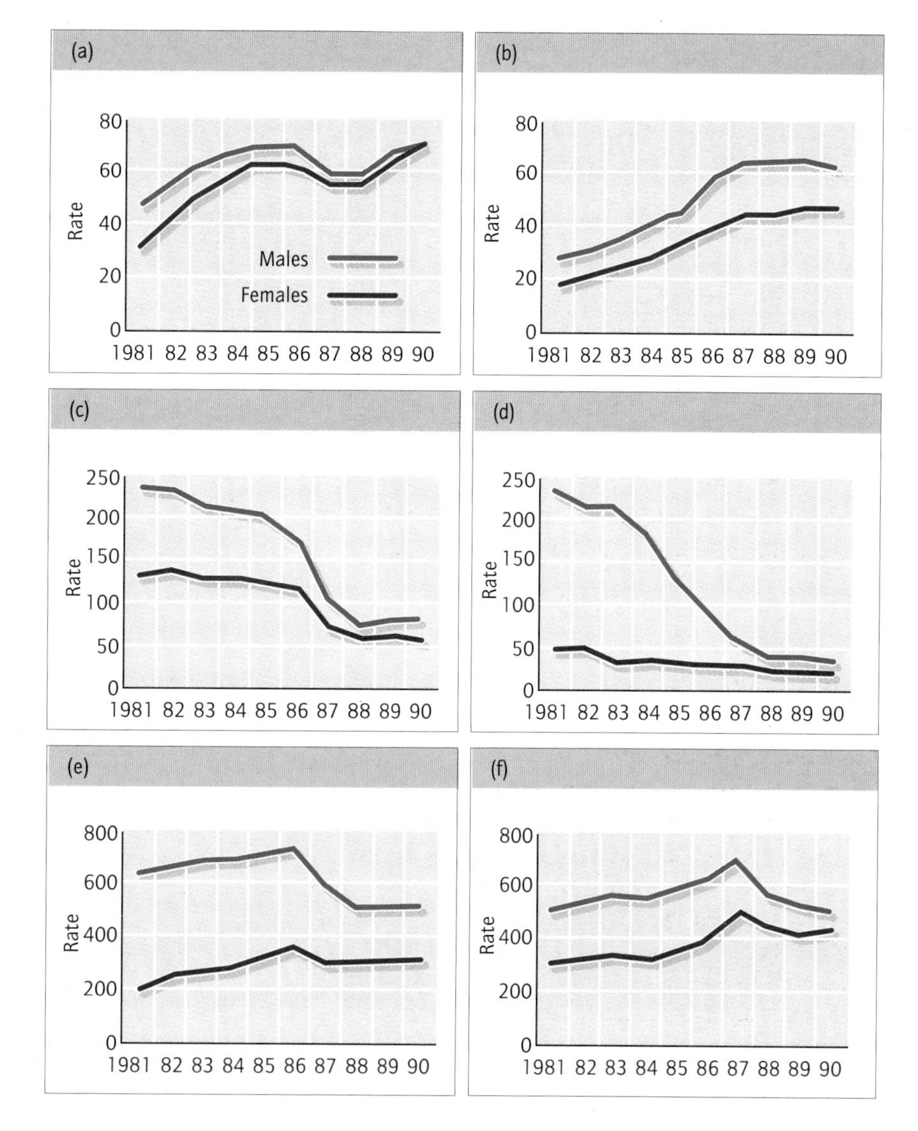

Fig. 15.1 Diagram showing trends in the incidence of common sexually transmitted diseases in the UK. (a) Genital herpes; (b) genital warts; (c) gonorrhoea; (d) infectious syphilis; (e) non-specific genital infection; (f) other conditions not requiring treatment. Courtesy of the Communicable Disease Surveillance Centre.

found in the skin lesions of epidermodysplasia verruciformis.

Genital warts spread by contact and can present as single or scanty lesions on the shaft of the penis, glans, vulva, perineum or anal margin, or occasionally as large areas of moist, shaggy lesions. They are a major problem in pregnancy, when they become very extensive and hypertrophic, and can interfere with vaginal delivery.

The diagnosis is clinically obvious. The best treatment is daily topical podophyllin resin, in the form of a paint. The surrounding skin should be protected with soft paraffin ointment and the paint should be washed off after 6 h, as podophyllin is irritant, and toxic if absorbed. Extensive warts should be treated in successive small areas. Podophyllin is contraindicated in pregnancy but after delivery the warts shrink somewhat, and can be treated conventionally.

Cervical intraepithelial neoplasia and high-risk papillomavirus types

Hybridization studies and polymerase chain amplification have shown the presence of papillomavirus genome in neoplastic cells of the cervix. Typing studies show that types 16 and 18 (and, to a lesser extent, types 31, 33, 35, 39, 45, 51, 52, 56, 58, 59, 66 and 68) predominate in neoplastic cells; other papillomavirus types have similar distributions in malignant and non-malignant cells. This could explain why carcinoma of the cervix is associated with sexual activity at a young age, and with multiple sexual partners.

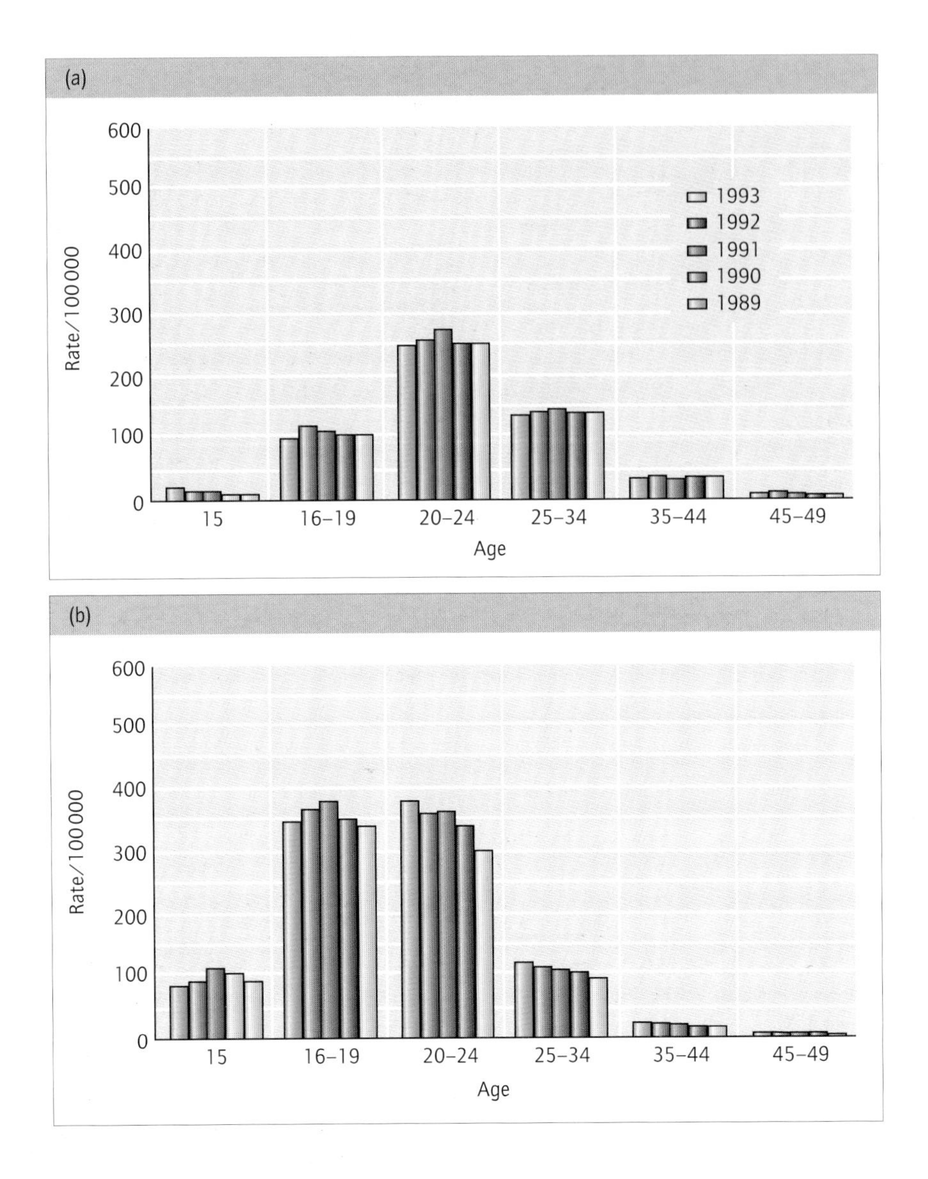

Fig. 15.2 (a) Male and (b) female age- and sex-related incidence of chlamydial genital infection in the UK.

Other intraepithelial neoplasias

Papillomavirus DNA, most commonly HPV types 6 and 11, and sometimes 18, 16 and other types, has been found in vaginal, vulval and penile intraepithelial neoplasias. Whether the virus is a cause or effect of the mucosal change is not clear, nor is it known whether the mild intraepithelial lesions progress to neoplastic disease. There is therefore no firm evidence to connect papillomavirus infections with neoplasias other than cervical carcinoma.

Genital herpes simplex infections

Pathology and epidemiology

Genital herpes simplex is a common infection which spreads by direct contact, usually but not always sexual. It used to be caused mainly by herpes simplex virus type 2 (HSV-2), but in the last two decades type 1 has also become common in genital infections. Both types of infection can be associated with HSV infection in other sites, particularly cold sores and other cutaneous infections. The incidence of both genital herpes and genital warts has increased in recent years, particularly among females. Some of this increase is due to the occurrence of both diseases among patients with human immunodeficiency virus (HIV) infection; it is also likely that ascertainment has improved considerably.

As with other herpesvirus infections, there is a primary and postprimary type of disease, and periods of asymptomatic excretion of virus from the previously affected genital tract. Virus has been demonstrated in the urethra and the vas deferens of asymptomatic men. Antiviral treatment of acute disease prevents neither relapses nor asymptomatic excretion, so that contact tracing is an ineffective means of controlling genital herpesvirus infections.

Clinical features

Primary infection

The incubation period averages 4 or 5 days, but varies very widely. Illness begins with mild fever and malaise, and tender inguinal lymphadenopathy which is usually unilateral and rarely severe. After a day or two, tense superficial vesicles appear and quickly break to become painful ulcers. The commonest site for the vesicles is the coronal sulcus or glans of the penis, the vulva or the anal margin. In severe infections pain may inhibit micturition and cause acute urinary retention. Rare cases of faecal impaction are seen when the perianal area is involved.

Associated neurological problems

Viral meningitis is not uncommon, and may be missed if mild. Lumbar puncture reveals typical cerebrospinal fluid (CSF) changes with lymphocytosis, and HSV-2 can be isolated from the CSF. The meningitis is usually benign, and resolves in a few days.

Radiculitis of the pelvic nerve roots can also occur. There is pain and stiffness of the lower back, sometimes with associated meningism. The neurological lesion is asymmetrical, and may produce paraesthesia or anaesthesia of the buttock, thigh or perineum, with associated disturbance of micturition and defecation. Occasionally, other muscles of the perineum or thigh are affected, with loss of the sartorius or quadriceps tendon reflexes. The course of the disorder is variable, but is often 5–10 days or more. Recovery is gradual, but almost always complete.

Clinical presentations of primary genital herpes

1 Painful inguinal lymphadenopathy.
2 Painful genital ulcers.
3 Lymphocytic meningitis (herpes simplex virus type 2 only).
4 Pelvic radiculitis (herpes simplex virus type 2 only).

Postprimary genital herpes

Postprimary genital herpes is like a cold sore. It may affect the buttock, thigh, perineum or genitalia. The appearance of vesicles is heralded by mild malaise, with local pain or burning. The vesicles last for 2–5 days before healing by drying and epithelialization. Abortive attacks, with the transient appearance of a few papules, also occur. These recurrences can be very frequent, and in women may occur premenstrually for month after month. Although there is little fever or systemic change, women particularly may feel fatigued and suffer radiation of pain to the thighs and back.

Recurrences are unlikely to decline in frequency if they persist for more than a year after the first attack.

Diagnosis

This is often clinically obvious. Herpesvirus particles can be demonstrated by electron microscopy of vesicle fluid or scrapings. Both HSV-1 and HSV-2 grow rapidly in cell cultures, producing typical cytopathic effects in 48–72 h. Seroconversion is demonstrable in primary attacks. Polymerase chain reaction (PCR)-based diagnosis is also available.

Management

Oral medication is sufficient in most cases. Primary and postprimary attacks can be treated with famciclovir 250 mg three times daily, aciclovir 200 mg five times daily or valaciclovir 500 mg twice daily, usually in courses of 5 days. Topical treatment may have a weak local effect, but is ineffective against virus in the lymph nodes and nervous system.

Severe primary infections are best treated with intravenous aciclovir 5 mg/kg 8-hourly, and usually require courses of at least 5 days. Meningoradiculitis may respond better to 10 mg/kg doses.

Frequent recurrences can be prevented by suppressive therapy with oral aciclovir 400 mg twice daily or famciclovir 250 mg twice daily. When recurrences are confined to the premenstrual days, treatment need only be taken at this time.

Treatment of genital herpes

1 Acute attacks: aciclovir orally 200 mg five times daily for 7 days; *or* valaciclovir 500 mg twice daily; *or* famciclovir 250 mg twice daily.
2 Meningoradiculitis: aciclovir i.v. 10 mg/kg 8-hourly for 5–10 days.
3 Suppression of recurrences: aciclovir orally, 400 mg twice daily *or* famciclovir 250 mg twice daily, given continuously, or in the week before menstruation every month.

Complications and cautions

True complications are rare; staphylococcal secondary infection is marked by increasing inflammation and exudation, often of yellowish pus. It responds to treatment with an antistaphylococcal antibiotic such as flucloxacillin orally for 5 or 6 days.

Herpes simplex infection is severe in the neonate and in immunosuppressed patients. In pregnancy it can be transmitted during delivery and (rarely) transplacentally, causing extensive, life-threatening disease in the neonate (see Chapter 17).

HSV infection is often persistent, extensive and invasive in AIDS; extensive lesions lasting more than 1 month comprise an AIDS-defining opportunistic infection.

Aciclovir-resistant organisms sometimes emerge during prolonged suppressive therapy; they lack, or have altered viral thymidine kinase, and are of low pathogenicity, but can cause important disease in highly immunodeficient patients. Foscarnet is an option for treatment of these organisms.

Erythema multiforme accompanies herpes simplex recurrences in a small group of people, among whom tissue types DQW3 and/or DRW53 predominate. HSV DNA can be demonstrated in erythematous skin but not in normal skin. Attacks diminish in severity over a period of 24–36 months, and eventually cease. Suppression of the herpes simplex recurrences prevents the erythema multiforme.

Bacterial infections of the genital tract

ORGANISM LIST

Chlamydia trachomatis (serogroups D–K)
Neisseria gonorrhoeae
Treponema pallidum
Gardnerella vaginale
Actinomyces spp.
Haemophilus ducreyi
Lymphogranuloma venereum (*Chlamydia trachomatis* serogroups L1, L2, L3)
Mycoplasma hominis and *M. genitalis*
Ureaplasma urealyticum

Chlamydial genital infections

Introduction

These infections cause most of the non-specific or non-gonococcal genital infections, and are the commonest genital infections throughout the world. Diagnosis of chlamydial infections has increased steadily over the past 20 years. It has been estimated that up to 50% of cases of non-gonococcal urethritis in the USA are caused by *Chlamydia trachomatis*. Chlamydial genital infections are important not only because of the high transmission rates and high morbidity that they cause, but because of their contribution to infertility due to chronic pelvic infection and their ability to cause significant intrapartum infection of neonates.

Pathology

C. trachomatis is an obligately intracellular bacterium which lacks a cell wall. It is a member of the genus *Chlamydia* which contains two other species, *C. psittaci*, a zoonotic pathogen, and *C. pneumoniae*, a respiratory pathogen of humans (see Chapter 7).

C. trachomatis exists in several serotypes: A, B, Ba and C, which are associated with ocular trachoma (see Chapter 5); D–K, associated with oculogenital infections and neonatal infections (see Chapter 17); and L1, L2 and L3, the causes of lymphogranuloma venereum.

Clinical features

In men the commonest manifestation of infection is urethritis, causing dysuria, urethral and meatal soreness and urethral discharge, which is most noticeable in the morning, before micturition. This may occur alone, or cause persisting symptoms after treatment of gonorrhoea. Ascending infection can cause epididymitis, and acute prostatitis or chronic prostatitis. Chlamydial infection can also cause non-specific proctitis, possibly after anal intercourse.

In women urethritis and cervicitis are common, with symptoms of soreness, dysuria and mucoid discharge, though without significant systemic features. Ascending infection typically causes acute salpingitis, with fever, neutrophilia and lower abdominal pain. It must be distinguished from appendicitis. Wider invasion can occur, producing a picture of perihepatitis (Curtis–Fitz-Hugh syndrome), with high fever, upper abdominal pain and guarding, and abnormal liver function tests. Rare cases of perihepatitis are seen in men, though the route of infection is uncertain.

Many lower genital tract infections are asymptomatic. Pelvic or abdominal disease can therefore occur without preceding genital symptoms. Occasionally, the occurrence of chlamydial infection in a neonate reveals the presence of asymptomatic infection of the parents.

Ocular infection can coexist with genital symptoms in both sexes, usually as a persisting conjunctivitis.

Clinical presentations of chlamydial infections
1 Urethritis.
2 Cervicitis.
3 Proctitis.
4 Conjunctivitis.
5 Salpingitis.
6 Prostatitis.
7 Perihepatitis.
8 Infected neonate.

Diagnosis

Urethritis and cervicitis are often clinically apparent. Diagnostic tests must be performed (see also Chapter 7), as gonorrhoea is the differential diagnosis and may also coexist with chlamydial disease.

Salpingitis does not have the distinctive evolution of appendicitis, and may be accompanied by vaginal discharge, which can be examined microbiologically. Pain and guarding are suprapubic. Swelling, induration and tenderness in one or both fornices are often found on vaginal examination.

Perihepatitis must be differentiated from gallbladder disease and liver abscess. Ultrasound or computed tomographic imaging may help by showing a healthy gallbladder and/or oedema and brightness of the liver capsule.

Laboratory diagnosis of chlamydial genital infections is usually made by antigen detection. Swabs are taken from the urethra, cervix or rectal mucosa. Direct immunofluorescent (IF) or enzyme-linked immunosorbent assay (ELISA) methods are used to detect chlamydial group antigen in the cellular material obtained. The results of IF tests are highly process- and observer-dependent, and are slightly less reliable than ELISA or cultural methods. *C. trachomatis* can be cultured from swab specimens in cultures of cycloheximide- or idoxuridine-pretreated McCoy cells. HeLa cells are also suitable for *Chlamydia* cultures. Positive results are demonstrated by the presence of chlamydial inclusion bodies on Giemsa- or iodine-stained cells, or by antigen demonstration using direct IF or ELISA techniques.

DNA amplification techniques can demonstrate the presence of *Chlamydia*-specific DNA by PCR or the ligase chain reaction, and commercial automated systems are now becoming available to routine microbiological laboratories. Their enhanced sensitivity and specificity means that urine can be used as a specimen, making it easier to screen lower-risk populations.

Serology is not helpful in the diagnosis of chlamydial genital infections.

Management

The usual treatment is a 2-week course of a tetracycline such as daily doxycycline or twice-daily Deteclo. The simplest regimen possible is chosen to encourage compliance. Erythromycin 2 g daily in divided doses is an alternative. Azithromycin in a single dose of 2 g is at least as effective as tetracycline and is effective against *N. gonorrhoeae*, but commonly causes nausea.

Systemic infections require parenteral treatment. Erythromycin is then the treatment of choice.

Treatment of chlamydial infections
1 First choice: doxycycline orally 200 mg on day 1, then 100 mg daily for 2 weeks; *or* Deteclo one tablet 12-hourly for 2 weeks.
2 Alternatives: erythromycin orally 2 g daily in two to four divided doses for 2 weeks; *or* azithromycin orally 2 g single dose.
3 Systemic infections: erythromycin i.v. 500 mg–1 g 6-hourly for 5–7 days.

Surgery for acute salpingitis is avoided if possible, as salpingectomy reduces fertility. Fibrosis of the lumen can still affect tubal function if healing is slow or incomplete.

Role of **Ureaplasma urealyticum** *in non-specific genital infection*

Some non-specific genital infections are not associated with *C. trachomatis*. *U. urealyticum* can be detected by cultural techniques (using media suitable for mycoplasmata; see Chapter 7) in a substantial proportion of these. The importance of this is that *U. urealyticum* is often resistant to tetracycline. It is usually sensitive to erythromycin.

Complications

The most important complication of non-specific genital infection is Reiter's syndrome. This is a postinfectious condition affecting mainly men, and associated with the human leucocyte antigen (HLA) B27 tissue type. It causes prolonged synovitis and connective tissue inflammation (see Chapter 21).

Lymphogranuloma venereum

This is a tropical sexually transmitted disease caused by

the serotypes L1, L2 and L3 of *C. trachomatis*. It has an incubation period of 2–30 days. A small, shallow ulcer or ulcers may appear at the site of inoculation on the genitalia, but most patients present with a slowly enlarging, usually bilateral, lymphadenopathy in the inguinal or femoral region. Suppuration is common, with one or more draining sinuses and slow healing. There are many systemic features, such as fever, meningism, pericarditis, keratitis and skin rashes, all of which slowly resolve. Years later, perineal and inguinal fibrosis may occur, causing rectal strictures in women and genital lymphoedema in men.

Diagnosis can be made by serological testing, or by demonstrating the typical histology of the primary lesion or lymph nodes.

The treatment of choice is tetracycline, which must be given for at least 3 weeks. Treatment of early disease prevents late fibrosis, which is not amenable to antibiotic therapy.

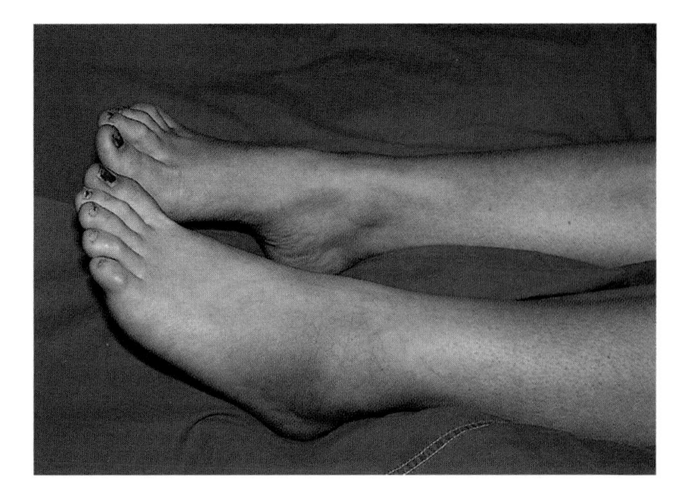

Fig. 15.3 Gonococcal arthritis affecting the ankle 1 week after contact with a new sexual partner.

Gonorrhoea

Introduction and epidemiology

Although less common than non-specific genital infection, gonorrhoea still causes millions of infections worldwide. As well as local infection, it can also cause bacteraemic disease which is often recognized and treated late.

In contrast to viral infections of the genital tract, the incidence of gonorrhoea has fallen in many developed countries in recent years. The decline has been mainly among patients over 25 years of age. In younger age groups, particularly young homosexual males, the incidence has started to increase again after a period of decline during the late 1980s. This trend suggests poor response to prevention initiatives among adolescents.

Clinical features

Gonorrhoea most often causes local genital infection, with urethritis and/or cervicitis, and relatively few systemic effects. Subclinical infection has always been recognized in women, but is increasingly described in men also. The throat or the rectum can be affected, often asymptomatically.

Ascending infection occurs in a minority of cases, causing prostatitis, epididymitis, salpingitis and occasionally perihepatitis.

Gonococcal bacteraemia

Gonococcal bacteraemia tends to present with local inflammatory lesions, rather than a simple fever. The accompanying genital infection may be mild or subclinical.

Arthritis is common, frequently affecting both knees, but the ankle or the wrist are also vulnerable (Fig. 15.3). The patient is feverish; the joint or joints are painful and swollen, and may be surrounded by considerable erythema. There is usually an effusion, which can be large in the knees.

Many patients have a sparse, vasculitic rash, of individual, painful vesicular lesions, with an intensely inflamed halo (Fig. 15.4). These lesions look very like herpes simplex vesicles but are found on extensor surfaces, particularly of the fingers, the elbow or the foot. A few patients have a bruising or haemorrhagic component to the rash, making it similar to a meningococcal rash.

Clinical presentations of gonorrhoea

1 Urethritis.
2 Cervicitis.
3 Proctitis.
4 Pharyngitis.
5 Salpingitis.
6 Prostatitis.
7 Perihepatitis.
8 Septic arthritis.
9 Bacteraemia (often with rash).

Diagnosis

In urethritis and cervicitis Gram stain of the pus, and of

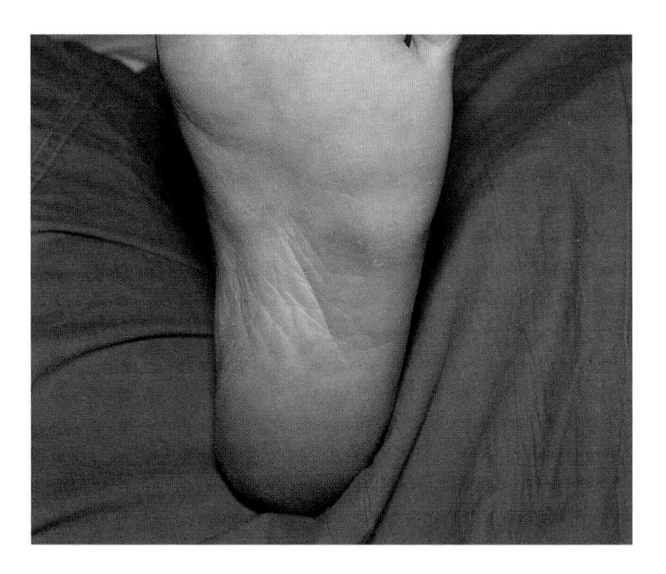

Fig. 15.4 Gonococcal skin lesion: an intensely inflamed, rather vasculitic lesion; the same patient as in Fig. 15.3.

urethral and/or cervical swabs shows many neutrophils and both intracellular and extracellular Gram-negative diplococci. This test is more sensitive in men than in women (97% versus 50%).

Salpingitis usually presents surgically, as a differential diagnosis of appendicitis or 'acute abdomen'. Suprapubic tenderness and guarding, the absence of the usual evolution of appendicitis, adnexal swelling, tenderness and vaginal discharge point to the diagnosis.

Gonococcal bacteraemia should be suspected in any young patient with a fever and arthritis. Aspiration of the affected joint shows many neutrophils, but demonstration of gonococci is difficult until late in the illness. Blood cultures are the main diagnostic test. Gonococci are present in the skin lesions, and can be recovered from vesicle fluid.

When genital swabs are taken, it is important to take rectal and throat swabs also, as one of these may be the only site yielding positive cultures. Specimens obtained by prostatic massage may also be positive.

Neisseria gonorrhoeae is a delicate organism which does not survive for long outside the body. Specimens from the male urethra may be obtained using a platinum loop as many swab materials impair the survival of this organism. Ideally specimens should be examined in a side room by microscopy and inoculated on to culture medium and incubated locally, or transported with the minimum delay to the microbiology laboratory.

Specimens in which *N. gonorrhoeae* is sought are usually heavily contaminated with other bacteria. Antibiotics must therefore be added to the medium to aid selection. The organism is also delicate and fastidious, so the medium must be very nutritious, containing lysed blood and additional growth factors such as yeast extract. The usual medium inoculated is a modified New York City medium or Thayer–Martin medium in parallel with a non-selective medium such as chocolate agar. The plates should be incubated in 5–10% carbon dioxide and increased humidity, and inspected after 24 and 48 h. Suspect colonies are identified by the oxidase test and sugar oxidation tests. Confirmation of identification can be made by serological agglutination tests. The presence of a beta-lactamase enzyme can be detected rapidly using the commercial kits based on a cephalosporin which changes colour when the beta-lactam bond is broken. Susceptibility to penicillin, tetracycline, spectinomycin, ciprofloxacin and a third-generation cephalosporin should be tested. New DNA amplification methods have been described and these can be combined with DNA amplification diagnosis of *Chlamydia* to provide a diagnosis within 24 h.

Several systems of typing exist, based on monoclonal antibodies or auxanograms. They are not in routine clinical use.

Management

To ensure compliance, on-the-spot treatment is offered, as far as possible, for simple genital infections. For penicillin-sensitive organisms this consists of amoxycillin orally 3 g as a single dose, or ampicillin 2 g plus probenecid 1 g (a further dose of the latter treatment is given to women, with an interval of 8–12 h). Penicillin-allergic patients can be given cefuroxime 1.5 g intramuscularly (half the dose in each of two sites) or oral ciprofloxacin 500 mg as a single dose.

Penicillin-resistant gonorrhoea is treated with single-dose ceftriaxone 250 mg intramuscularly. Ciprofloxacin 500 mg orally or spectinomycin 2 g given by deep intramuscular injection are alternatives. A further dose of spectinomycin 1 or 2 g can be given at another site in difficult cases. Ciprofloxacin resistance is an increasing problem, with 30–50% of organisms in Oriental countries now resistant. Azithromycin 2 g orally will treat *N. gonorrhoeae* and coexisting *C. trachomatis* infection.

Salpingitis, perihepatitis and bacteraemia require admission to hospital and treatment with intravenous penicillin (or a cephalosporin if cultures indicate penicillin resistance). A week's course is usually sufficient. Care must be taken that complications of bacteraemia, such as endocarditis, are not overlooked; the C-reactive protein usually remains raised during convalescence if complications exist.

Treatment of gonorrhoea

1 Penicillin-sensitive organisms: ampicillin orally 3 g single dose *or* ampicillin 2 g; *plus* probenecid 1 g orally. Single dose for men, repeated after 12 h for women.
Alternative: cefuroxime i.m. 1.5 g single dose (half the dose in each of two sites).
Systemic disease: benzylpenicillin i.v. 1.2–2.4 g 4–6-hourly for 7 days.

2 Penicillin-resistant organisms: ceftriaxone i.m. 250 mg single dose, *or* ciprofloxacin orally 500 mg single dose, *or* spectinomycin i.m. 2 g single dose, *or* azithromycin orally 2 g single dose.
Systemic disease: cefuroxime i.m. or i.v. 750 mg 6–8-hourly for 1 week.

Complications

The commonest complication of gonorrhoea is reinfection (many infections may occur because of the extreme antigenic variability of gonococcal pilus proteins—the main immunogenic proteins of the organism). Repeated infections, severe infections or those in which treatment is delayed or complicated by penicillin resistance can all lead to tissue damage. In men urethral strictures may need repeated dilatation or even operative intervention. Prostatic damage and chronic inflammation may also follow. In women tubal damage and obstruction can cause infertility and predispose to chronic pelvic inflammation.

Syphilis

Introduction and epidemiology

Syphilis, caused by the spirochaete *Treponema pallidum*, is now relatively uncommon in industrialized countries. It remains important, however, because it is prevalent in some places overseas and in travellers, and remains more common in homosexual men than in other social groups. Early cases must be detected and treated to avoid the problems of congenital infection and of late manifestations, which cause great morbidity and dependence.

The incidence of syphilis has started to rise again in some countries, notably inner cities in the USA, and more recently Russia. A localized outbreak occurred in Bristol in 1997–98. This increase has occurred mainly among intravenous drug abusers and prostitutes.

Clinical features

Primary syphilis

The incubation period can vary from about 10 days to 10 weeks or more. The typical primary lesion or chancre is usually a single, painless ulcer with a border and base of induration. Commonly affected sites are the foreskin, coronal sulcus, vulva, fourchette, uterine cervix or adjacent structures such as the urethra or penile shaft. There is painless enlargement of local lymph nodes. About 5% of chancres are extragenital, affecting the lips, mouth or nipple.

Atypical chancres are common, often taking the form of multiple or painful lesions, easily mistaken for herpes simplex, chancroid or small malignant lesions.

Secondary syphilis

Secondary syphilis develops 6–8 weeks after the primary manifestations, though some patients have no history of chancre. It is a spirochaetaemic disease with fever, rash and generalized lymphadenopathy.

The rash is usually generalized, maculopapular or papular, extending to the palms and soles. It is notorious for variability; serpiginous or discoid lesions can occur, making differential diagnosis difficult. Other cutaneous and mucous membrane lesions are common. There may be superficial oral erosions covered with greyish exudate (mucous patches) serpiginous mouth ulcers (snail-track ulcers) and flat, moist, warty lesions (condylomata lata) on the perineum, especially around the anus. Other systemic manifestations include meningitis, arthritis, arthralgia, mild nephrotic syndrome, patchy alopecia and, in about 5% of cases, iritis or retinitis.

Latent syphilis

Latent syphilis is an asymptomatic state which may persist for years if the early infection is not cured. Slow tissue damage probably occurs throughout this stage, and a few patients have elevated CSF protein levels or mild pleiocytosis. Many patients with latent syphilis will eventually develop late manifestations of the disease.

Late syphilis

Late (tertiary) syphilis can affect many systems of the body. The underlying lesion is the gumma, an indolent, granulomatous lesion which may undergo central mucoid degeneration.

In the nervous system, meningovascular syphilis, tabes or syphilitic paresis (general paralysis of the insane) may occur.

Meningovascular syphilis causes vasculitis and leptomeningitis, particularly at the base of the brain and the upper spinal cord. It can present as meningitis, often with papilloedema, or more often with focal neurological disorder, such as cranial nerve palsy, weakness and wasting of the hands, transverse myelitis or ataxia. The vascular disease can produce strokes or epilepsy. There is a moderate lymphocytosis and a raised protein level in the CSF.

Tabes is now exceptionally rare. It is caused by selective degeneration and demyelination in the posterior columns of the spinal cord and the dorsal nerve roots. It presents with a triad of dysaesthesia and anaesthesia, pains and ataxia. The pains are typically shooting girdle or limb pain, sometimes accompanied by abdominal pain and vomiting (gastric crisis), which may be prolonged or repetitive. Ophthalmoplegia is common. The pupils may be small and fail to react to light, while retaining the reaction to accommodation (Argyll Robertson pupil). Hypoaesthesia leads to severe trophic arthropathy of the legs and feet (Charcot joints).

Syphilitic paresis is a progressive dementia, with or without tabetic features. The CSF contains many lymphocytes and the protein content is high, unlike in Alzheimer's disease or idiopathic psychosis.

In the cardiovascular system aortitis occurs 20–30 years after the original infection. There is intimal thickening and loss of elastic tissue from the root and ascending part of the aorta. The aortic valve ring is dilated, the coronary arteries may be occluded and an aortic aneurysm commonly develops.

In other systems there may be single gummas or gummatous infiltration, causing chronic osteomyelitis or periostitis, nodular liver enlargement or skin lesions which may ulcerate, producing a sticky discharge.

Laboratory diagnosis

Syphilis is a rare but important part of many differential diagnoses, especially of neurological conditions.

In primary syphilis a rapid diagnosis can be made by dark-ground microscopy of exudate from the chancre or of aspirate from enlarged inguinal lymph nodes. The spirochaetes are seen as tightly coiled, 'watch-spring' structures. They also have typical motility, rotating about their long axis, or bending at an angle.

T. pallidum has not been cultivated in artificial medium, so diagnosis is based on serological tests. Enzyme immunoassay (EIA) tests are now the main methods used in many laboratories. These are capable of detecting specific immunogloblin G (IgG) or IgM and are suitable for all of the uses for which traditional tests have been employed. Also, because of their format, they can be automated, allowing large numbers of screening samples to be processed efficiently.

Antibodies may be detected by their interaction with cardiolipin-based reagin antigens, treponemal antigens or *T. pallidum* itself. The test used in most laboratories is the VDRL (Venereal Disease Research Laboratory) test which uses a standardized antigen to perform an agglutination test. In the rapid plasma reagin (RPR) test, carbon particles are incorporated to simplify reading of the test. These tests are subject to biological false-positive results due to cross-reacting antibodies in patients with connective tissue disease, or malaria. However, their value is that they become positive early in the course of infection and become negative after treatment. They are useful therefore for establishing the duration of infection and the treatment status.

Specific treponemal tests use a cultivatable treponeme, the Reiter's treponeme, as the antigen and are therefore less subject to false reactions. Positive results appear later in the course of infection but remain positive for life. The commonest test of this type is the *T. pallidum* haemagglutination assay (TPHA), which is technically simple to perform on large numbers of specimens. Positive results can be titrated and confirmed using a *T. pallidum*-based test.

The most specific test available for routine laboratories is the fluorescent treponemal antibody absorption (FTA-abs) test. *T. pallidum* is bound to a glass slide and patient serum is first absorbed to remove group-reactive antibody, then placed on the slide and incubated. After washing, the binding of specific anti-*T. pallidum* antibody is detected using a fluorescent antihuman immunoglobulin specific for either IgG or IgM. The IgM test is especially useful for detecting acute and congenital infection. The other specific *T. pallidum* test is the *T. pallidum* immobilization (TPI) test which uses live treponemes. The difficulty of providing reagents for this test and its controls means that it is only suitable as a reference technique (Table 15.1).

Management

Penicillin is the treatment of choice for all forms of syphilis. Primary syphilis is curable, with eradication of spirochaetes from the body. It can be treated with daily or twice-daily intramuscular procaine penicillin injections

	VDRL/RPR	TPHA	FTA-abs	IgM (ELISA or FTA)
Congenital infection	+	+	+	+
Primary infection	+	−/+	−/+	+
Untreated secondary infection	+	+	+	+
Treated or late disease	−	+	+/−	−

VDRL, Venereal Disease Research Laboratory; RPR, rapid plasma reagin; TPHA, *Treponema pallidum* haemagglutination assay; FTA-abs, fluorescent treponemal antibody absorption; IgM, immunoglobulin M; ELISA, enzyme-linked immunosorbent assay.

Table 15.1 Results of serological tests for syphilis at different disease stages.

for 10–14 days. Long-acting penicillins are no longer routinely available in the UK. A suitable alternative may be to give amoxycillin in standard doses with probenecid 500 mg 6-hourly to maintain tissue levels. For penicillin-allergic patients, a variety of alternatives exist, including cephalosporins and tetracyclines, but these and the variety of new antibiotics on the market are not well tried in treating syphilis.

Secondary syphilis can be treated in the same way as primary; adequate treatment seems to abolish almost all risk of late manifestations.

Follow-up

The VDRL titre should show significant decline after 6 months and become undetectable after 18–24 months. A persisting positive test is an indication for retreatment. More specific tests remain positive indefinitely. After secondary syphilis or neurological involvement, it is advisable to perform a lumbar puncture after 6 months. If the CSF has not returned to normal, retreatment is indicated.

Treatment

Treatment of late syphilis can provide dramatic improvement in meningovascular syphilis. Other late manifestations can be prevented from progressing or at least slowed significantly, and in some cases worthwhile improvement is achieved by prolonged treatment.

Jarisch–Herxheimer reaction

The Jarisch–Herxheimer reaction is an inflammatory response to spirochaetal antigens. It occurs within hours of the first antibiotic dose, with fever and exacerbation of swelling and inflammation. This can be a serious problem if, for instance, a coronary ostium is further occluded, or if epilepsy or stroke is precipitated. The reaction can be minimized by starting with low doses of penicillin and/or by adding corticosteroids such as prednisolone 30–40 mg daily during the first few days of treatment.

Complications

The most important complications of syphilis are related to infection in pregnancy. Transplacental infection of the fetus is likely, causing abortion, or neonatal and long-term disease (see Chapter 17).

Chancroid (soft sore)

This is a local, ulcerating genital infection caused by *Haemophilus ducreyi*, a Gram-negative rod uncommon in the UK, but widely prevalent in the West Indies, south-east USA, north Africa, the Middle East, China and parts of the Mediterranean. It is therefore often an imported disease. Its importance is that it is often mistaken for syphilis, which may coexist. It can be transmitted by asymptomatic carriers. Small abrasions are especially susceptible to infection with chancroid which, in turn, increases the chance of infection with other sexually transmitted pathogens, including HIV.

After an incubation period of 2–5 days, the typical, soft sore lesion quickly develops. This is a large, irregular, painful ulcer, almost always on the genitalia. The local lymph nodes become enlarged, inflamed and painful, and may suppurate, discharging via the skin and producing a large ulcer crater.

The diagnosis is made by Gram stain and culture of scrapings from lesions or discharge from lymph nodes. The lesion should be thoroughly cleansed, and a cotton wool swab applied vigorously. The swab should be plated on to isolation medium with the minimum of delay. Mueller–Hinton and enriched gonococcal medium with the addition of charcoal has been shown to improve the isolation rate. There is considerable research interest in the rapid diagnosis of chancroid by EIA and other techniques but no method has yet gained acceptance in routine use. The ulcers have a characteristic histological appearance, so biopsy is useful in cases of doubt. Syphilis should always be excluded by dark-ground microscopy and serological testing.

Chancroid responds readily to treatment with sulphonamides, which are ineffective against syphilis. A convenient treatment is a single 2 g dose of sulfametopyrazine (Kelfizine W), which can be repeated after a week if necessary. Side-effects are fewer with 7–10 days' treatment with co-trimoxazole or a short-acting sulphonamide.

Granuloma inguinale

Chancroid is sometimes confused with a condition called granuloma inguinale, an indolent, progressive ulcerating condition confined to the genital skin and subcutaneous tissues. It is a *Calymmatobacterium granulomatis* infection usually seen in tropical climates where poor hygienic conditions prevail. While it may spread by sexual contact, it is only slightly infectious, and is possibly mainly an autoinfection of faecal origin. Tetracycline, erythromycin and co-trimoxazole are all effective treatments.

Empirical treatment of genital ulcer disease

In tropical and developing countries, genital ulcer disease is common, and usually due to granuloma inguinale or chancroid. Genital ulcer disease is an important risk factor in the acquisition of HIV infection. The World Health Organization therefore encourages empirical treatment of genital ulceration whenever possible. Co-trimoxazole is an effective treatment which is readily available and inexpensive in many countries, and a 5–7-day course is usually the treatment of choice.

Bacterial vaginosis and *Gardnerella vaginale*

Bacterial vaginosis is an inflammation of the vagina for which no direct cause is apparent. The vagina depends on a low pH produced by commensal lactobacilli for protection against invading microorganisms. If this is overcome by concurrent infection or antibiotic use, the vagina is colonized by a mixture of organisms, including *Mobiluncus* sp. and *Gardnerella vaginalis*. The mechanisms by which lactobacilli prevent colonization with anaerobic species are poorly understood, but may depend on the bacterial production of lactic acid and hydrogen peroxide. Examination of material obtained by swabbing shows the presence of neutrophils and sometimes of clue cells (squamous epithelial cells covered with adherent bacteria).

Bacterial vaginosis is characterized by variable irritation, soreness and sometimes a slight discharge, which is homogeneous and greenish with a characteristic fishy smell when alkalinized (due to amine production). Bacterial vaginosis in pregnant women is associated with an increased risk of preterm delivery.

Diagnosis is based on the presence of epithelial cells heavily coated with bacteria in the discharge (clue cells), a positive amine test (a characteristic smell that worsens on addition of 10% potassium hydroxide), a pH of greater than 4.5 and a characteristic homogeneous discharge. The presence of at least three out of these four characteristics indicates bacterial vaginosis. Non-specific vaginosis is treated with metronidazole either as a 7-day course of 400–500 mg twice daily or as a single dose of 2 g. Alternatively, topical metronidazole gel or clindamycin cream may be used. Local treatment in pregnant women cures the infection but does not appear to influence neonatal morbidity. Oral clindamycin is associated with a reduction in preterm delivery. Trials of treatment in male partners have shown contradictory results, but may be considered.

Pelvic inflammatory disease (PID)

Introduction

This is a condition of chronic gynaecological symptomatology, which causes morbidity and infertility throughout the world. There is evidence of inflammation in intrapelvic organs, and often in vaginal and cervical swab specimens. Occasionally there is extensive oedema, fibrosis or abscess formation, with much distortion of pelvic structures.

Although *Neisseria gonorrhoeae* or *Chlamydia trachomatis* may be responsible, a single causative organism is often not identified. *Mycoplasma hominis* is an uncommon cause of urinary tract infections, and of bacteraemia or sepsis associated with pelvic disease. It probably contributes to a significant proportion of pelvic infections.

Clinical features

These are pain, malaise and often slight vaginal discharge. The pain may be suprapubic, radiating to the thighs or referred to the lower back. There is often aching discomfort or tenderness on intercourse. Dysmenorrhoea or heavy menstrual periods are common. In severe cases there is chronic fever and/or weight loss.

Pelvic examination shows tenderness and sometimes swelling or induration in the vaginal fornices. Discharge is variable, and may be mucoid or mucopurulent. There may be coexisting anatomical distortion due to varying degrees of prolapse or previous disease of the cervix or fallopian tubes. Intrauterine contraceptive devices rarely predispose to PID.

Diagnosis

Material from vaginal and endocervical swabs often contains moderate numbers of neutrophils. Specific pathogens are rarely recovered, though *Actinomyces* spp. may be associated with intrauterine devices. Tuberculosis should be borne in mind as a differential diagnosis, and appropriate samples should be examined for mycobacteria (see Chapter 18). Endometriosis and malignancy should be excluded. Urinary tract infection should also be sought.

Pelvic ultrasound scan may show fluid in the pouch of Douglas, distortion of the fallopian tubes or inflammatory masses in the pelvis. Cervical cytology and uterine curettage are useful tests, with histological examination and microbiological tests for bacteria, including mycobacteria. Laparoscopy permits direct examination of the pelvic organs, and small biopsies or other sampling may be possible. Actinomycosis can produce granulomatous infiltration of the pelvis, with loss of tissue planes.

Management

This is usually empirical, based on the assumption that the infection is likely to be polymicrobial. The organisms involved may be coliforms, anaerobes and/or a variety of flora from the genital tract. The patient is offered a broad-spectrum treatment, often consisting of metronidazole plus ampicillin, erythromycin, azithromycin or tetracycline. Courses of 2–4 weeks may be successful, but more prolonged treatment may be needed if fibrosis or oedema are extensive.

Removal of any intrauterine device is advisable. Surgical correction of anatomical problems may help to avoid recurrence. In severe cases, removal of an affected tube or ovary or even hysterectomy may offer the best chance of long-term cure. In severe actinomycosis antimicrobial chemotherapy will often reduce the bulk of the infection and restore tissue planes. Prolonged treatment may be needed before surgery can safely be undertaken.

Candida albicans genital infections

Candida albicans is a resident of moist skin and mucosae, which readily infects the squamous epithelium of the vagina and vulva or the glans of the penis. Inflammation and maceration of the affected surface are often accompanied by a soft, white exudate which tends to form plaques or spots. Predispositions include diabetes, antibiotic treatment, hypercalcaemia and disorders of cell-mediated immunity.

The genitalia of adult women are protected from candidal infection by lactic acid and peroxides produced by *Lactobacillus acidophilus*, with which the vagina is heavily colonized. This colonization is lost in pregnancy and after antimicrobial chemotherapy. The inflammatory symptoms of vulvovaginal candidiasis are often accompanied by a voluminous, cheesy vaginal discharge with little odour. Itching and oedema of the mucosa are common.

The diagnosis is often clinically apparent. Microscopy of the exudate shows many budding yeasts, and extensive formation of pseudohyphae, a mark of aggressive candidal growth. Positive cultures are readily obtained.

Superficial infections respond to topical treatment with nitroimidazoles such as clotrimazole or econazole creams. Vaginal candidosis is better treated with pessaries or vaginal cream. Clotrimazole or econazole pessaries are well tolerated in courses of 1–5 days, depending on the preparation used. Single-dose clotrimazole vaginal cream is also effective. Nystatin pessaries are an alternative, but must be used daily for 2–4 weeks.

Persistent or recurrent infections can be treated with oral triazole drugs. Itraconazole or fluconazole can be given as a 1-day course.

Treatment of vaginal candidiasis

1 Clotrimazole vaginal tablets 500 mg one at night as a single dose, *or* econazole pessaries 150 mg one at night as a single dose, *or* clotrimazole 10% vaginal cream 5 g at night as a single dose.

2 Alternative: nystatin pessaries one or two at night for 14–28 days.

3 Systemic treatment: itraconazole orally 200 mg twice daily for 1 day *or* fluconazole, single 150 mg dose.

4 Additional treatment for vulval involvement: clotrimazole 1% cream *or* econazole 1% cream twice daily; *or* nystatin cream to anogenital area three or four times daily.

Trichomonas vaginalis

Trichomonas vaginalis is a flagellate protozoan that thrives in the low oxygen tension of the female lower genital tract. It produces symptoms of soreness and irritating serous or mucoid discharge. It can also colonize the male urethra where it is rarely clinically evident.

Smears of vaginal discharge will often contain motile or

dying protozoa, demonstrable by light microscopy of wet preparations. The organisms can also be transported and cultured in *Trichomonas* medium. Smears can be made from the incubated medium to demonstrate the organisms. This is useful in mild infections when examination of direct vaginal smears may be negative.

It is quite common for cytologists to see *Trichomonas* on Papanicolaou-stained cervical smear preparations. This suggests that even in women, infection may be trivial or inapparent.

Treatment with oral metronidazole 400 mg 8-hourly for 1 week is usually successful. Rare cases of metronidazole resistance have been described. They may respond to single-dose tinidazole 2 g, or to a single 2 g dose of metronidazole.

Other infections transmissible during sexual contact

Sexual contact usually includes prolonged skin-to-skin contact, as well as significant exchange of body fluids, particularly saliva and genital secretions. Minor trauma or small skin lesions can permit exchange of blood, or inoculation of blood on to a partner's mucosae. Extensive anal and rectal contact, or contact with urine droplets increase the likelihood of transmission of several infections not usually spread by sexual contact.

Infections spread by skin contact
- Scabies
- Pubic lice
- Impetigo

Infections spread by body fluids
- Hepatitis B
- Cytomegalovirus
- Epstein–Barr virus

Infections spread by blood
- Hepatitis B
- HIV (see Chapter 16).

Infections spread by faeces
- Shigellosis
- Giardiasis
- Amoebiasis

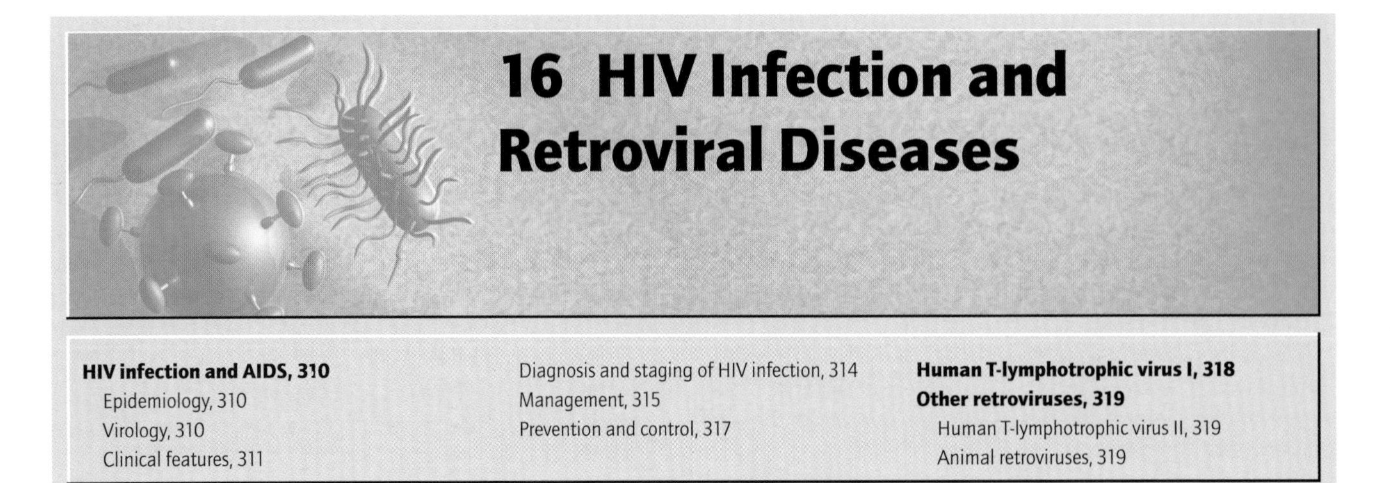

16 HIV Infection and Retroviral Diseases

HIV infection and AIDS, 310
- Epidemiology, 310
- Virology, 310
- Clinical features, 311
- Diagnosis and staging of HIV infection, 314
- Management, 315
- Prevention and control, 317

Human T-lymphotrophic virus I, 318

Other retroviruses, 319
- Human T-lymphotrophic virus II, 319
- Animal retroviruses, 319

ORGANISM LIST

Human immunodeficiency virus 1 (HIV-1)
HIV-2
Human T-lymphotrophic virus 1 (HTLV-I)
HTLV-II
Simian immunodeficiency virus (SIV)
Porcine endogenous retroviruses (PERVs)

HIV infection and AIDS

Introduction

The human immunodeficiency viruses HIV-1 and HIV-2 cause lifelong infection of a number of tissues, but particularly of CD4 helper lymphocytes. Untreated individuals develop profound helper cell deficiency and associated immunodeficiency over a period of several years (acquired immunodeficiency syndrome, AIDS) and die of uncontrolled [illegible] of HIV disease. HIV-1 is distributed worldwide and produces more severe and rapidly progressive disease than the less aggressive HIV-2, which is mostly found in West Africa. HIV is spread by sexual, blood-borne and vertical transmission and is capable of affecting whole communities, profoundly disrupting social structure and commercial activity.

Epidemiology

AIDS was first recognized as an important epidemiological problem in the early 1980s among homosexual men in the USA. The number of cases has risen exponentially, and by the end of 1998, over 47 million people are estimated to have been infected worldwide, nearly 14 million of whom have already died. Two-thirds of AIDS cases have occurred in Africa. It is also estimated that there were 32 million adults infected with HIV at the end of 1998 and that new infections are occurring at a rate of about 16000 a day.

Two major patterns of disease transmission have emerged during the course of the AIDS pandemic. In Africa and some other developing nations, the predominant modes of spread are through heterosexual intercourse and by vertical transmission to infants from infected mothers. In developed countries, the major routes are through sexual intercourse between men and sharing of contaminated drug-injecting equipment. Receipt of contaminated blood and blood products was an important route of spread in the early stages of the epi[illegible]nic, although this has largely stopped since most coun[illegible]s now screen blood donations.

[illegible]ll countries of the world are affected, although the [illegible]est impact of the disease is in Africa (Fig. 16.1). Only [illegible] cases reported have occurred in Europe. In some [illegible]ean countries, for example Switzerland, the inci[illegible]of AIDS may have reached a plateau.

Virology

[illegible] an enveloped RNA virus belonging to the [illegible] subfamily of Retroviridae. HIV-2 is a closely [illegible]in that has been identified from patients in West [illegible]re is considerable homology between HIV-2 [illegible] immunodeficiency virus (SIV). In common [illegible]RNA viruses, HIV has a relatively unstable [illegible]ng rise to numbers of closely related strains or [illegible] of virus, designated a to f. These strains form groups which have different distributions in the world, and slightly different characteristics of infectiousness.

The genetic information of HIV-1 and HIV-2 is found on a single strand of RNA. The genome encodes for a number of proteins essential for viral replication. These include *pol* which encodes reverse transcriptase. This is responsible for generating complementary DNA which is incorporated into the host genome. The *gag* gene encodes a precursor protein which is cleaved to form the core protein

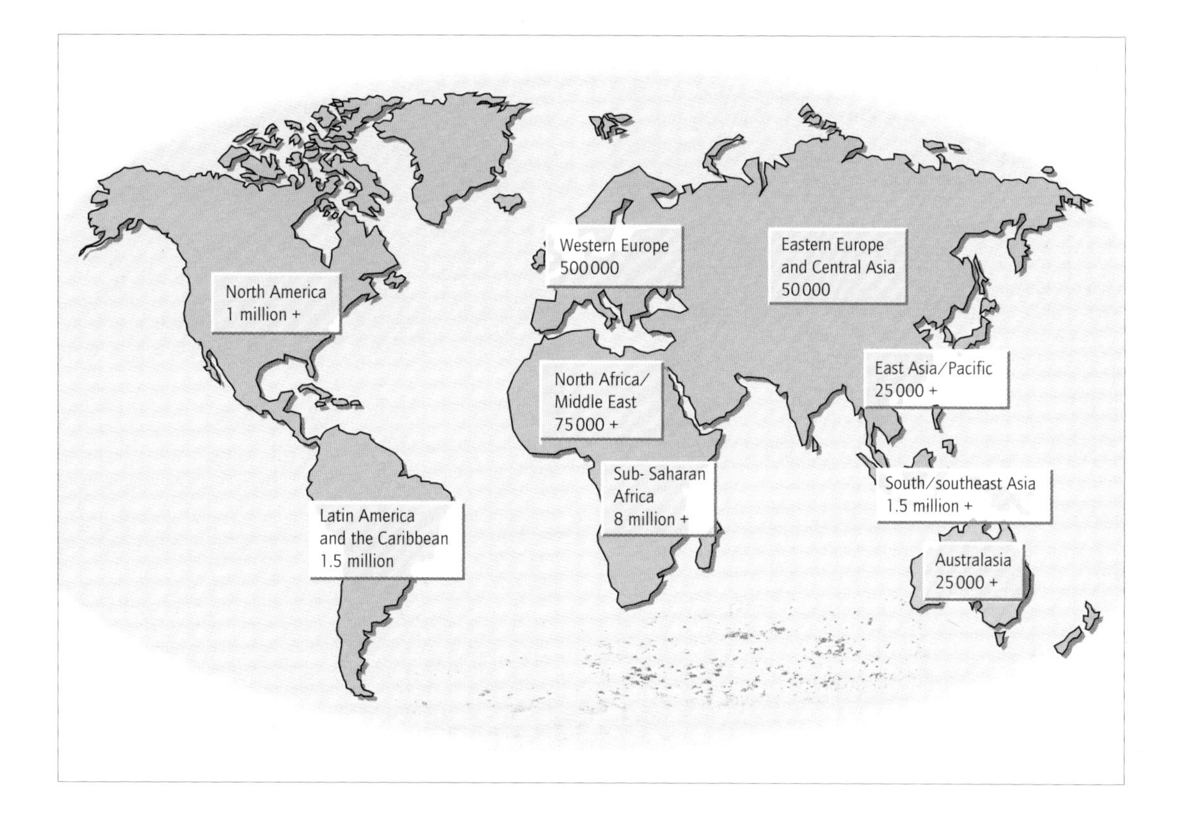

Fig. 16.1 Global distribution of acquired immunodeficiency syndrome (AIDS) cases.

p24 which is detectable in early infection. The *env* gene codes for a 160-kDa glycoprotein which is cleaved to form the envelope glycoprotein responsible for the attachment of HIV to the target cells via the CD4 receptor. Virus is found in the serum in CD4-positive T cells and in tissues in CD4-positive T cells, macrophages and follicular dendritic cells.

Once the virus has entered a target cell its complementary DNA integrates into the host genome and leads to constant production of viruses. It is thought that the immune system maintains a vigorous antiviral response until it is eventually exhausted and depleted. Coinfection with organisms that stimulate T-cell multiplication is thought to hasten the onset of immunodeficiency.

Clinical features

There are three phases to the course of infection with HIV. The first is the phase of seroconversion during which antibodies to HIV become detectable in the serum. The second phase is a period of asymptomatic, latent infection, during which virus replication proceeds and cell-mediated immunity deteriorates until the last, symptomatic phase of the infection is reached. During the symptomatic phase the patient eventually develops serious opportunistic diseases, and is then said to fulfil the definition of AIDS.

Seroconversion illness

Seroconversion usually occurs 6–8 weeks after infection, but incubation can range from 4 to 12 weeks. Studies show that at least 75–80% of infected patients suffer a feverish illness during seroconversion, and about 60% have a rash. Those who seek medical advice may have a mild generalized lymphadenopathy and many have atypical mononuclear cells in the blood, suggestive of acute viral infection. The rash may consist of macular, acne-like or ovoid and papular lesions, characteristically on the upper chest and back (Fig. 16.2). They may appear rather scuffed or abraded, but are not painful or itchy. Oral lesions and genital ulcers are both commonly seen. A few patients present with a lymphocytic meningitis. Most illnesses are mild, but all tend to be prolonged, lasting from 2 to 6 weeks. They resolve without specific treatment. There is a high viraemia during seroconversion, with a profound fall in the CD4 lymphocyte count and significant immunosuppression. Some patients present with an opportunistic infection, such as candidal oesophagitis or *Pneumocystis*

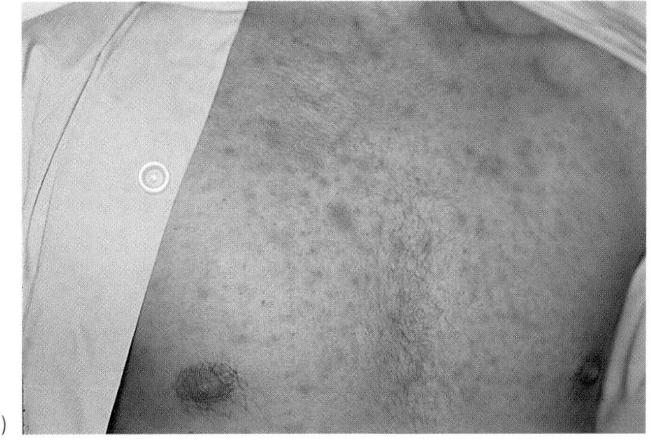

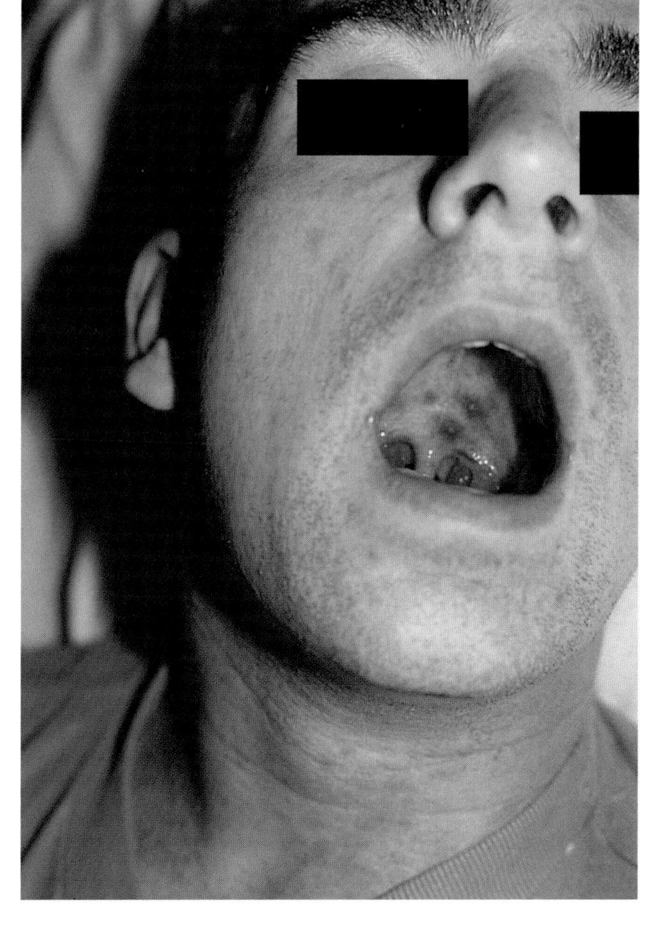

Fig. 16.2 (a) Rash of large, erythematous lesions and (b) oral lesions in human immunodeficiency virus (HIV) seroconversion illness.

carinii pneumonia (PCP), which respond well to specific treatment and do not recur until after the latent phase.

The low CD4 count is mirrored by a raised number of cytotoxic CD8 T lymphocytes, which contribute to termination of the high viraemia. However, the CD4 count is never restored to its original level, and a variable, low viraemia (<1000 genomes/ml) may persist throughout the following latent phase. A severe seroconversion illness and persistent detectable viraemia are thought to predispose to early progression to symptomatic disease.

Latent phase

Untreated, the latent phase of HIV infection lasts from 18 months to 15 years or more, with an average of about 8 years. For much of this time the individual is well, not unduly susceptible to infections and recovers apparently normally from common and seasonal infections. The total CD4 (T4) helper-cell population slowly declines, and CD4 helper function is increasingly impaired. This decline is probably exacerbated by each intercurrent infection.

The viral load in the blood falls to a much lower level than during seroconversion, tending to remain relatively constant at a so-called 'set point', which is often below 1000 genome copies per ml. The lower the set point, the longer the asymptomatic period is likely to last. Some cases have a very prolonged latent period (long-term non-progressors), and are studied to identify good prognostic factors or factors that protect from progression. They tend to have sustained, strong antiviral CD8 lymphocyte reponses, and some lack the coligand chemokine CC5, which aids the entry of HIV into CD4-positive cells. Their virus does not evolve from non-syncytium-inducing (NSI) to syncytium-inducing (SI) forms, typical of isolates from late, symptomatic cases. However, even these patients have an inexorable rate of progression to increasing viraemia, immunosuppression and symptomatic disease.

Symptomatic HIV infection

An increase in the loss of CD4 cells and increasing viraemia heralds the end of the latent phase. Before they become truly symptomatic many patients develop generalized lymphadenopathy. Rubbery, mobile nodes from 1 cm upwards in diameter are easily palpable. They have a characteristic histological appearance of reactive histiocytosis. The lymphadenopathy persists through the symptomatic stages.

Certain infections may occur at about this time. Herpes zoster is common and can be severe, recurrent or multidermatomal, sometimes even leading to scarring of the skin or eye. Prompt treatment with aciclovir is therefore indicated. Bacterial infections also occur, including pneumococcal and *Salmonella* infections. Pneumococcal pneumonia is often bacteraemic, but with a misleadingly low white cell count, with mild or absent neutrophilia. It responds to vigorous antimicrobial treatment (see Chapter 7). *Salmonella* infections may also be bacteraemic and there may be repeated recurrences of both diarrhoea and of bacteraemia despite adequate treatment.

Skin and mucosal infections become increasingly common; these include seborrhoeic dermatitis, molluscum contagiosum, relapsing herpes simplex infections, and oral and genital candidiasis.

Full-blown AIDS

Full-blown AIDS does not coincide exactly with any absolute level of CD4 cell count or circulating viral load, but is likely to develop when the CD4 cell count falls to $0.4–0.2 \times 10^9$/l. Many AIDS patients have counts far below this, and may have undetectable counts for months before their death. The viral load in the blood is often above 10 000 genome copies per ml.

Many cases present with opportunistic conditions so characteristic of cell-mediated immunodeficiency that AIDS can be definitively diagnosed, even without confirmation of HIV infection. Others require the additional confirmation of HIV antibody and/or antigen testing. In some clinically evident opportunistic diseases a clinical diagnosis alone fulfils the criteria for a presumptive diagnosis of AIDS (Table 16.1).

Some AIDS-defining conditions are malignancies, such as Kaposi's sarcoma, lymphomas or invasive carcinoma of the cervix. These may have an infectious aetiology, for instance human herpesvirus type 8 (HHV-8) in Kaposi's sarcoma and Epstein–Barr virus (EBV) in some lymphomas, and represent a defect of immune surveillance of potentially oncogenic infections.

AIDS in infants and children

There are several differences between adults and children in both the presentation of AIDS and the apparent degree of CD4 cell depletion. The CD4 cell count is higher in infants and young children than in adults. Adult reference values are not reliable indicators either of CD4 cell loss or of immunosuppression in children.

Adults possess humoral immunity to a wide range of pathogens encountered before the onset of HIV infection. Young children do not have this advantage. Furthermore, in those under 2 years of age, T-cell-independent antigens are ineffective immunogens. Children are therefore at risk from bacterial infections to which they cannot mount an effective immune response without CD4 helper cells. The occurrence of repeated pneumonias, gastroenteritis, and skin and upper respiratory infections is a common presentation of childhood AIDS.

Some viral infections, instead of being terminated or maintained in a latent state, become persistent, relapsing or progressive. Measles, chickenpox and especially EBV infection can cause severe disease and persisting fever. EBV infection cannot be terminated and results in lymphocytic infiltrative disease, particularly lymphocytic interstitial pneumonitis (LIP), which is progressive and fatal.

Disease
Bacterial infections; multiple or recurrent (child < 13 years)
Candidiasis (trachea, bronchi or lungs)
Candidiasis (oesophagus)
Cervical carcinoma (invasive)
Coccidioidomycosis (disseminated or extrapulmonary)
Cryptococcosis (extrapulmonary)
Cytomegalovirus retinitis
Encephalopathy (milestone loss with no other cause in children)
Herpes simplex; ulcers for > 1 month, or bronchial, lung or oesophageal lesions
Histoplasmosis (disseminated or extrapulmonary)
Isosporiasis (diarrhoea > 1 month)
Kaposi's sarcoma
Lymphoid interstitial pnemonia (child < 13 years)
Lymphoma (Burkitt's, immunoblastic, cerebral)
Mycobacteriosis (disseminated, including extrapulmonary TB)
Mycobacteriosis: (pulmonary TB)
Pneumocystis carinii pneumonia
Pneumonia (recurrent within 1 year)
Progressive multifocal encephalopathy
Salmonella (non-typhoid) septicaemia, or recurrent
Toxoplasmosis (cerebral: onset > 1 month old)
Wasting syndrome: weight loss (> 10% of baseline with > 30 days' fever or diarrhoea; no other cause)

TB, tuberculosis.

Table 16.1 Conditions defining a clinical diagnosis of acquired immunodeficiency syndrome (AIDS). In the USA a total CD4 lymphocyte count below 0.2×10^9/l is also considered diagnostic.

Attempts are made to protect children from these problems by immunization, prophylaxis and early treatment of feverish illnesses. Even measles vaccine should be given to HIV-infected children, whether they have HIV-related symptoms or not, because the risk from measles is much greater than that from the live attenuated vaccine.

Diagnosis and staging of HIV infection

The future healthcare commitment associated with a positive diagnosis means that diagnostic tests must be not only sensitive but also highly specific. All positive results must be subject to strict verification procedures. The main tests employed in the diagnosis of HIV are antibody tests. A wide range of different formats have been used, including antibody capture enzyme immunosorbent assay (EIA), competitive EIA and rapid agglutination tests. These may incorporate HIV-1 and HIV-2 peptides or recombinant antigens. Indeterminate and positive results must be checked by alternative methods. Discrepant results and suspected HIV-2 infections should be confirmed by a reference laboratory.

Polymerase chain reaction (PCR) techniques can demonstrate viral RNA in the blood in adults, and in neonates possessing antibody which may be of maternal origin. Enzyme-linked immunosorbent assay (ELISA) tests for antibodies to HIV antigens are very sensitive and become positive soon after PCR-based tests, with antibody to the p24 antigen appearing first.

In the Western blot test, HIV antigens are separated by sodium dodecyl sulphate-polyacrylamide gel electrophoresis (SDS-PAGE) and blotted on to nitrocellulose. The patient's antibodies bind to the different antigens. A positive result is indicated if two or more of the bands p24, gp41 and gp120/160 are positive. Sequential appearance of antibody bands on successive Western blots in a patient with a positive PCR test allows distinction between early and late seroconversion illness, and is useful in research on therapy.

For a short period (the window period) immediately after infection, patients may test HIV antibody-negative. For patients seeking to check their serological status after a recent exposure, a negative result cannot be confirmed until a further negative result is obtained 3 months after the last exposure to the virus. PCR-based tests for viral RNA may permit an earlier confirmation of infection in this case.

Diagnosis of HIV infection

1 Screening test for HIV gp41 antibodies.
2 Confirmatory test using a different technique.
3 Test for p24 antibodies in suspected early disease.
4 PCR-based test for HIV viral RNA (useful in suspected perinatal transmission when the infant may have passive antibody in the blood, and for early detection of recent infection).

The progress of HIV disease can now be closely monitored by a combination of viral load measurement, CD4 lymphocyte counts and the patient's clinical status (Fig. 16.3). These measurements are used as a guide to the initiation of treatment, a check on the effectiveness of the

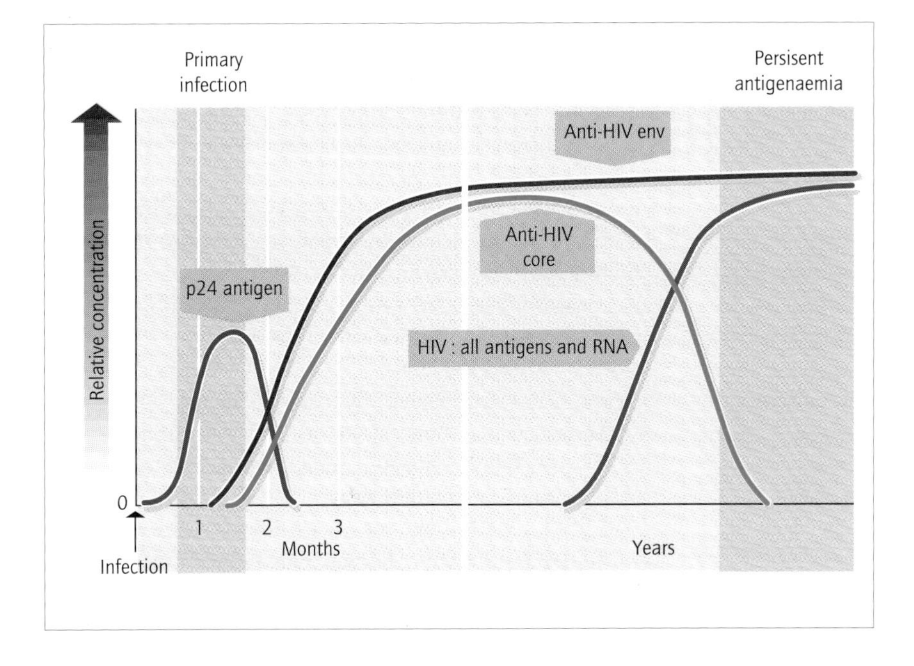

Fig. 16.3 The progression of antigen (Ag) and antibody appearance in the blood of a person infected with human immunodeficiency virus (HIV).

regimen chosen and a warning of developing drug resistance. Most patients have these tests approximately 6-monthly when their disease is stable and more often after the initiation of treatment or treatment changes.

Management

Antiretroviral drugs

Nucleoside-related reverse transcriptase inhibitors (NRTIs)

These include zidovudine (AZT, ZDV), zalcitabine (DDC), lamivudine (3TC), didanosine (DDI), stavudine (D4T) and abacavir (ABC). Resistance to a single NRTI used as monotherapy emerges within weeks or months, especially to lamivudine, and cross-resistance is common.

Adverse effects include nausea (often diminishing with continued therapy), vomiting, diarrhoea, and a varying incidence of neuropathy, pancreatitis, hepatic disorder and reduced red or white blood cell counts. Hypersensitivity reactions are common with abacavir. Lactic acidosis is a rare, late reaction to NRTIS, caused by defective mitochondrial DNA synthesis.

Viral protease inhibitors (PIs)

These include saquinavir (best tolerated), ritonavir, indinavir (most potent) and nelfinavir. All, particularly ritonavir, inhibit the cytochrome P450 system, causing adverse reactions with many drugs. Rifampicin, an enzyme inducer, seriously reduces blood levels of PIs. Rifabutin dosage must be halved if given with PIs, and its ocular toxicity is enhanced by ritonavir. Adverse interactions may also occur with azole antifungal drugs, some antiarrhythmic drugs and some antihistamines. Indinavir is relatively insoluble, so that extra fluid intake is necessary to prevent crystalluria and renal tubular damage.

Long-term use of PIs is associated with raised cholesterol and triglyceride levels, decreased glucose tolerance, hyperuricaemia and lipodystrophy with typical 'buffalo hump'. Statin drugs can ameliorate some of these problems.

Non-nucleoside reverse transcriptase inhibitors (NNRTIs)

Nevirapine is the longest-used NNRTI. Rash limits its use in about 20% of patients. Efavirenz and delavirdine have also been developed, and are licensed for use in some countries. These are effective drugs, but one-step viral mutation to resistance rapidly occurs if they are used alone.

Principles of antiretroviral therapy (highly active antiretroviral therapy, HAART)

The aims of antiretroviral therapy are: to minimize viral replication (and therefore the rate of viral mutation to resistance); to reduce infectiousness; to prevent the progress of immunosuppression; and where possible to allow reconstitution of lymphocyte counts and function.

Aims of HAART

To minimize viral replication.
To minimize the emergence of resistance to antivirals.
To reduce infectiousness.
To halt the progress of immunological damage.
To permit reconstitution of immune function.

In most situations, previously untreated patients are likely to respond to most types of drugs, but all will possess small populations of virus with resistance to one or more types of drug, and these must be reduced, rather than selected for, by antiviral therapy. In some countries, especially where treatment modalities and options are limited, up to 10% of patients have primary resistance to one type of drug. Multidrug therapy with good treatment compliance and close follow-up of response is therefore important.

In all settings where resources allow, multidrug therapy with at least three drugs is the ideal. Two of the drugs are usually NRTIs; effective and well-tolerated combinations include zidovudine plus lamivudine and zidovudine plus DDI. Stavudine plus either lamivudine or DDI is also useful. On the other hand, zidovudine plus stavudine appears to be a slightly antagonistic combination and DDI plus zalcitabine is rather toxic.

The third drug may be a PI or an NNRTI. The addition of a PI usually obtains a rapid fall in viral load, by 1.5–2.0 logs in about 1 month. Using an NNRTI as the third drug obtains a similar fall, but often over 10–12 weeks. After a further period of treatment many patients have a fall in viral load to levels below 50–500 genome copies per ml, as virus loads decrease in less accessible 'sentinel' sites such as the genital tract and central nervous system. The longest-surviving infected cells are the 'memory' T cells, with a lifespan of many months.

Continuation of therapy then allows gradual increase in CD4 cell counts, with reconstitution of immune function, though the repertoire of antigen specificity and the completeness of T-cell function probably never approaches normal. Nevertheless, immune and inflammatory responses increase, initially manifested as the appearance of increased lymphadenopathy in mycobacterial infections, or increased ocular inflammation in retinal cytomegalovirus disease. While primary prophylaxis may be discontinued after commencing HAART, secondary prophylaxis and treatment for opportunistic disease must therefore be continued until CD4 counts are in the range of $0.4–0.5 \times 10^9$/ml. Abolition of cytomegalovirus viraemia and *Mycobacterium avium-intracellulare* (MAI) infection usually occurs at this stage, and the risk of *Pneumocystis carinii* pneumonia is negligible. The use of HAART has resulted in a fivefold fall in the incidence of full-blown AIDS in Europe between 1993 and 1998.

Emergence of resistance to therapy

RNA viruses have an unstable genome because they lack any RNA 'proof-reading' mechanism during replication. Thus, there are many genetic variants present in a virus population ('quasi-species', or 'clades'). The genetic variants may be as successful as the parent virus, or less so if they replicate or infect less well. Some mutations yield a replicative advantage in the face of immunological or therapeutic pressure. Approximately 10 billion HIV virus particles are produced daily, undergoing one mutation in 9200 nucleotides. Thus, every possible single drug-resistant mutant is likely to be generated every day, but multiple changes become increasingly less likely. In practice, resistant virus can be detected in patients not previously exposed to treatment, but almost all will be suppressed by multidrug therapy. A single mutation can confer resistance to lamivudine and certain NNRTIs. When these drugs are given in partially suppressive regimens resistant quasi-species will emerge within weeks. For drugs such as zidovudine and certain protease inhibitors, three or more mutations must occur before resistance is established. Treatment with agents that require multiple mutations to resistance should therefore be preferred. Mutants that are no longer detectable will reappear rapidly if the affected drug is restarted.

Even with multidrug therapy effectiveness is eventually limited by the development of viral resistance, and a change (switch) to alternative drugs is indicated, but a long duration of symptom-free life and avoidance of opportunistic disease is possible. Prolonged therapy is exacting for the patient because of demanding dosing regimens, the need to schedule drugs relative to meals, to drink extra fluids and to avoid comedication. Multidrug formulations are increasingly available to minimize these problems.

Examples of antiretroviral regimens

For initiating treatment: zidovudine plus lamivudine plus EITHER indinavir OR nevirapine.

Alternative: stavudine plus didanosine plus EITHER ritonavir OR nevirapine.

Switch therapy: saquinavir plus ritonavir plus EITHER a nucleoside analogue OR an NNRTI.

Salvage of late increase in viral load: nelfinavir plus indinavir plus nevirapine or efavirenz.

UK and USA recommendation for prophylaxis after transcutaneous exposure: zidovudine plus lamivudine plus indinavir for 4 weeks.

When to commence treatment and prophylaxis for HIV

For seroconversion illness: the high viraemia and infectiousness at this stage make treatment theoretically desirable, even though the condition is self-limiting. It is thought that about 70% of sexually transmitted infections result from contact with a seroconverter. Adverse effects and difficult compliance make it unlikely that lifelong treatment will be practicable. Trials of different treatment regimens and durations are under way.

For established HIV disease: opinions differ slightly over the balance between viral suppression and the long-term problems of therapy. Most experts would begin triple therapy in asymptomatic patients at viral loads between 50 000 and 100 000 copies per ml or at CD4 cell counts of $0.3–0.5 \times 10^9$/l. Patients developing HIV-related symptoms would be treated regardless of CD4 count or viral load.

Switch of therapy: this is offered if primary therapy produces unsatisfactory results, for drug intolerance or adverse effect, and for relapse of previously controlled infection. Many experts regard an increase of 1–1.5 logs of viral load or a rise to 5–10 000 genomes per ml from an undetectable level to indicate a switch. Compliance must always be reviewed before a decision to switch therapy.

Prophylaxis after percutaneous exposure: the risk of HIV infection after percutaneous exposure to infected blood is about 0.3%, increased five- to 16-fold in deep injuries, with hollow needles and with high 'donor' viral loads. Zidovudine monotherapy offers about 80% protection (as does nevirapine in chimpanzees exposed to HIV-1). Individuals with recognized percutaneous exposure to infection are nowadays offered triple therapy with zidovudine plus lamivudine plus indinavir for a period of 4 weeks. This should commence as soon as possible after the event, preferably within an hour or two. Pretreatment blood samples should be taken for HIV testing, and consent should be obtained whenever possible for testing of the 'donor' for HIV and other blood-borne infections. Further tests should be performed after treatment, with the final test 6 months postexposure. The availability of occupational counselling, consultation 'hotlines' and prepre-pared treatment packs are extremely useful. Support is required to minimize anxiety and to respond to nausea and other adverse events.

Prophylaxis for pregnancy in an HIV-positive woman: perinatal transmission from an HIV-positive woman can be minimized by safe delivery techniques, and antiretroviral therapy for the woman and her newborn (see Chapter 17).

Prophylaxis after sexual exposure to HIV: there is no consensus on the efficacy or practicality of prophylaxis after sexual exposure to HIV.

Eradication

With the advent of HAART it had been hoped that eradication of HIV infection might be possible. Decay of HIV-infected cells probably occurs in three phases: the first phase from HIV infected memory T-cells; the second phase from virus trapped in the macrophages and follicular dendritic cells; the third phase is less well characterized and possibly related to non-replicating CD4 cells or neural cells. It is not known what period is necessary before this third pool of infected cells has been eradicated. Thus, for the present, eradication of HIV infection remains a distant goal.

Prevention and treatment of opportunistic diseases

Prophylaxis and treatment of opportunistic conditions becomes important as immune competence declines. Patients with HIV infection should be vigorously investigated and promptly treated for opportunistic infections, to maintain their health, and to prevent further damage to the immune system (for instance, CMV infection is known to accelerate CD4 cell loss). As the lack of inflammatory reaction makes the presentation of infection insidious and atypical, investigations such as bronchoalveolar lavage or brush biopsy are necessary to test for *Pneumocystis*, mycobacterial, herpesvirus and fungal infections, all of which may present similarly, or coexist (Fig. 16.4). Skin or lymph-node biopsy is also useful in the diagnosis of poxvirus, mycobacterial and fungal infections, and in distinguishing between lymphoma, Kaposi's sarcoma

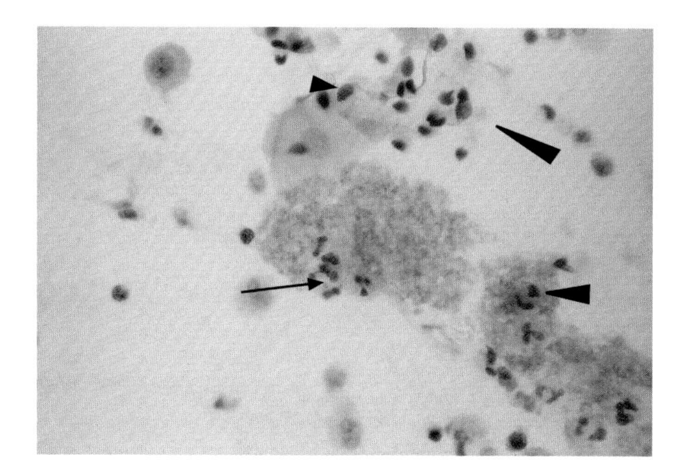

Fig. 16.4 Material from bronchoalveolar lavage, examined with a Papanicolaou-type stain, showing dark blue, oval bodies of *Pneumocystis carinii* (arrowheads) and a pale blue foamy macrophage; a group of small, Gram-positive diplococci is also present (arrow), probably *S. pneumoniae*.

and bacillary angiomatosis (due to *Bartonella henselae* infection).

Prophylaxis is usually given after a first episode of the condition (secondary prophylaxis). Primary prophylaxis is rarely offered as it might encourage infection with resistant organisms in, for instance, mucosal candidiasis. However, PCP is so common and serious that patients are offered primary prophylaxis when their CD4 count is consistently below $0.3\times10^{9}/l$. In countries where BCG immunization is not used, skin test-positive individuals may be offered isoniazid prophylaxis when their CD4 count falls below $0.4\times10^{9}/l$ (as there is a 10% per year incidence of tuberculosis in such cases).

Common regimens of prophylaxis include co-trimoxazole or nebulized pentamidine for PCP, aciclovir for herpes simplex and ganciclovir to suppress cytomegalovirus retinitis (Table 16.2). Children often receive regular intravenous immunoglobulin to prevent recurrent pyogenic infections. Rare infections have occurred with aciclovir-resistant herpes simplex and fluconazole-resistant fungi.

Prevention and control

The most effective control measure is the use of condoms. Despite extensive health education campaigns, this has failed to control the spread of HIV infection, particularly among heterosexuals who tend not to perceive themselves at risk. Treatment of genital ulcer disease may reduce the risk of HIV acquisition. Other preventive measures include screening of blood donations, needle exchange schemes for intravenous drug abusers and antenatal screening, which is now routinely recommended in pregnancy for women in the UK. Individuals in high-risk cate-

Opportunistic infection	Prophylaxis	Treatment
Pneumocystis carinii pneumonia	**1** Co-trimoxazole orally 960 mg 12-hourly (rash may compel use of alternative) **2** Pentamidine isethionate by nebulized inhaler: 300 mg every 2 weeks or 600 mg every 4 weeks	**1** Co-trimoxazole 120 mg/kg daily in divided doses for 14 days (may be given i.v. 1.44 g 12-hourly) **2** Atovaquone orally 750 mg 8-hourly for 21 days **3** Pentamidine isethionate i.v. 4 mg/kg daily for 14 days **4** Pentamidine isethionate by nebulizer 150 mg daily for 21 days
Cytomegalovirus retinitis (suppression)	**1** Ganciclovir by i.v. infusion 5 mg/kg daily or 6 mg/kg on 5 days per week (check blood count frequently) **2** Oral ganciclovir 1 g 8-hourly (2 g 8-hourly after a recrudescence) **3** Oral valganciclovir	**1** Ganciclovir by i.v. infusion (over 1 h) 5 mg/kg 12-hourly for 14–21 days **2** Foscarnet by i.v. infusion 20 mg/kg over 30 min, then 20–200 mg/kg daily (according to renal function) for 14–21 days (check blood) count, liver and renal function)
Cryptococcal meningitis	Fluconazole orally 100–200 mg daily	**1** Amphotericin up to 1 mg/kg daily by i.v. infusion (depending on renal function, plasma potassium, blood count, febrile reaction) for 8–12 weeks **2** Liposomal amphotericin up to 6 mg/kg daily if amphotericin fails or is intolerable **3** Fluconazole i.v. or orally 400 mg, then 200–400 mg daily for 8–12 weeks
Toxoplasmic encephalitis	Pyrimethamine-sulphadoxine one tablet weekly	Sulphadiazine 1.0 g three times daily plus pyrimethamine 50 mg daily for several weeks with repeated CT scans to confirm response
Mucocutaneous *Candida*	Not usually given due to risk of resistant yeasts emerging	**1** Amphotericin orally (suspension) 200 mg 6-hourly **2** Fluconazole 50 mg daily orally for 7–30 days Itraconazole 200 mg daily orally for 15 days
Mycobacterium avium-intracellulare	**1** Rifabutin 300 mg/day orally **2** Azithromycin 1200 mg weekly	See Chapters 18 and 22
Herpes simplex (mucocutaneous)	Aciclovir 200 mg 6-hourly orally	Aciclovir orally 200–400 mg five times daily; i.v. 5 mg/kg 8-hourly (reduced in renal impairment: check blood urea)
Molluscum contagiosum		Cidofovir 10% ointment topically

Table 16.2 Prophylaxis and treatment of some opportunistic conditions in acquired immunodeficiency syndrome.

gories should be encouraged to be tested for HIV, so that they know their diagnosis, and can participate in prevention and control measures.

Strategies for the prevention of human immunodeficiency virus (HIV) infection
1 Safe sex: condom use.
2 Screening of blood products.
3 Needle exchange schemes.
4 Antenatal screening.
5 Voluntary testing of those in high-risk categories.
6 Vigorous control and treatment of genital ulcer diseases.

Human T-lymphotropic virus I

Human T-lymphotrophic virus I (HTLV-I) is an oncovirus member of the family Retroviridae which infects T lymphocytes. Infection spreads by blood transfusion, vertically during breast-feeding (with an 18–30% risk of transmission), and to a much lesser extent by the sexual and perinatal routes.

HTLV-I is most common in southern Japan, where the seroprevalence is up to 16%, and in the Caribbean, where it ranges from about 2.5 to 6%. The seroprevalence in native Europeans is probably below 0.5%.

Seroconversion is probably symptomless, but two serious diseases can follow: (i) HTLV-associated myelopathy (HAM, or tropical spastic paraparesis), after an average interval of about 4 years (lifetime risk approximately 0.25%); and (ii) acute T-lymphoblastic leukaemia, after an average interval of about 30 years (lifetime risk 2–5%). Infected children may suffer repeated attacks of infective dermatitis which respond only temporarily to antibiotics. Uveitis is relatively common in infected individuals.

Prevention depends on safe blood transfusions, avoidance of breast-feeding by infected mothers and, to a lesser extent, on the use of condoms. No completely satisfactory screening test is available for transfused blood, but the best options are utilized in various countries, and family screening of affected individuals permits effective identification and counselling of cases.

Other retroviruses

Human T-lymphotropic virus II

This is a closely homologous retrovirus which has been identified in T lymphocytes less commonly than HTLV-I. It has been recovered from the lymphocytes in a case of hairy-cell leukaemia, but an aetiological association has not been established.

Animal retroviruses

A number of retroviruses have been identified in animals. Some, such as simian immunodeficiency virus (SIV) and feline immunodeficiency virus (FIV), can cause immunological disorders in infected animals. SIV has caused asymptomatic seroconversion in a few monkey handlers. Some retroviruses of pigs (porcine endogenous retroviruses, PERVs) can infect human cells, and are a potential problem in the development of xenotransplantation. They have not so far been identified in human recipients of pig islet cells or other limited 'transplants'.

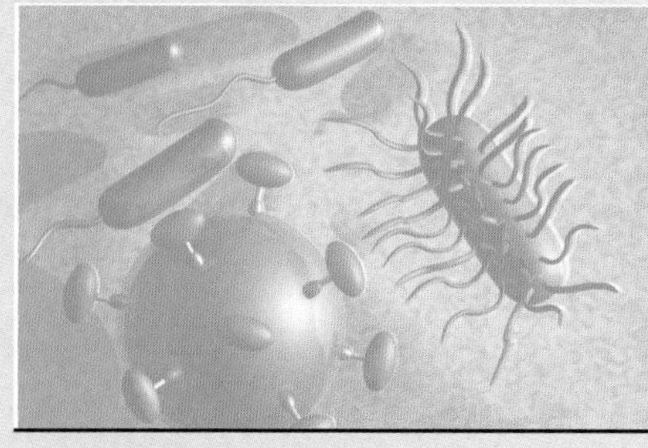

17 Congenital and Perinatal Infections

Introduction, 320

General effects of transplacental and intrapartum infection, 320
Transplacental infection, 320
Congenital infections, 320
Intrapartum and perinatal infections, 321

Transplacental, intrapartum and postnatal infections, 322
Congenital rubella, 322
Congenital and neonatal cytomegalovirus infections, 324
Congenital parvovirus infection, 325
Congenital and intrapartum herpes simplex infections, 326
Varicella embryopathy and neonatal varicella, 326
Congenital HIV infection, 327
Gonococcal neonatal ophthalmia, 328
Congenital and neonatal listeriosis, 328
Congenital syphilis, 330
Congenital toxoplasmosis, 330

Infections that spread only by the intrapartum and perinatal routes, 331
Neonatal and perinatal hepatitis B, 331
Human T-lymphotropic virus type I infection in infancy, 332
Neonatal and infant chlamydial infections, 332
Neonatal bacteraemias, 333
Neonatal staphylococcal infections, 334

Maternal infections related to childbirth, 334
Introduction, 334
Puerperal fever, 335
Breast abscess during lactation, 335
Rare zoonoses in pregnancy, 336

Introduction

The fetus and neonate can be exposed to infection in a variety of ways (Fig. 17.1). Some maternal infections cause viraemia or bacteraemia during pregnancy; there is then a chance that organisms will cross from the maternal to the fetal circulation, causing transplacental infection. The fetus may die of the infection, may recover *in utero* and be born with or without long-term sequelae, or may have a continuing infection at birth (so-called congenital infection).

Even if a blood-borne pathogen does not cross the placenta, the baby may be infected by contact with the mother's blood during birth. Alternatively, a pregnant woman's genital tract may be colonized or infected by a transmissible pathogen at the time of delivery. The neonate may become infected during birth, by contact with infectious genital secretions. The signs of these infections usually appear in the first 2–6 weeks of life (or, rarely, after several years as in the case of syphilis or toxoplasmosis). If the membranes rupture prematurely, organisms can ascend from the maternal genital tract, colonizing the amniotic sac and sometimes causing amnionitis. The infant is then exposed to a kind of intrapartum infection a few days before delivery.

In the neonatal stage, the infant has close contact with its mother, ingesting her breast milk and having intimate face-to-face contact. Older children may eat food from their mother's spoon or have food chewed for them by her; she may clean a dropped comforter or dummy by licking it. Infants occasionally ingest maternal blood if breast-feeding from cracked nipples. The extent of such contacts varies in different cultures. Infection can thus be transmitted from mother to child via saliva, blood or milk in early infancy. Occasionally, infections are transmitted via intimate contact with other family members.

General effects of transplacental and intrapartum infection

Transplacental infection

Transplacental infection causes intrauterine infection of the fetus, but not necessarily any long-term effect. After intrauterine infection with varicella or parvovirus, most fetuses have recovered completely by the time of delivery, and few either die *in utero* or have permanent sequelae.

Congenital infections

Children born with congenital infections often have a multisystem disease, as is the case with congenital rubella. Hepatitis, pneumonitis, meningoencephalitis and blood disorders are common, thrombocytopenic purpura is often seen, and the infant often excretes large amounts of the causative pathogen. Many of these acute problems will resolve if the infant survives, but permanent tissue damage may also be present.

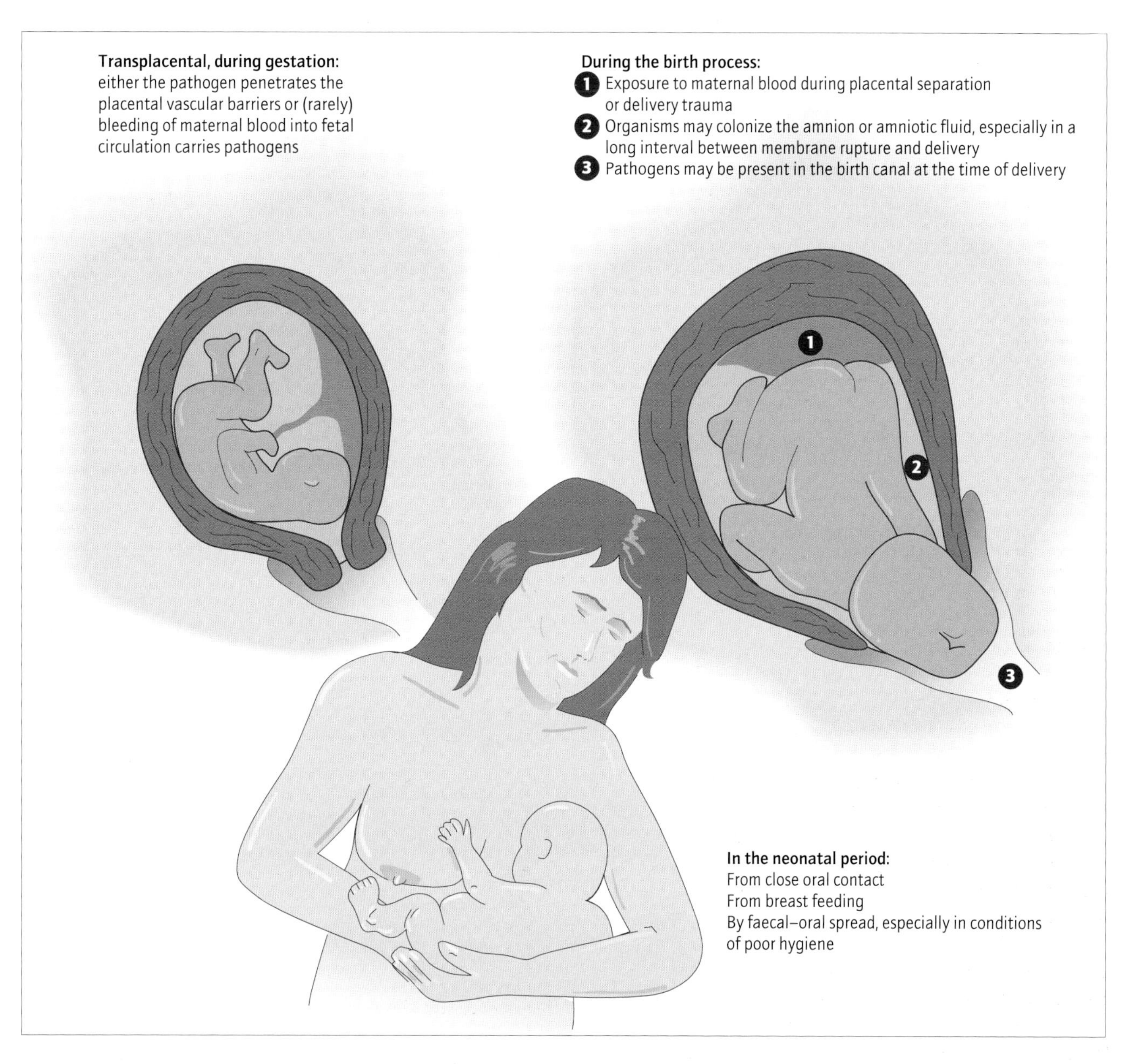

Fig. 17.1 The three common routes of mother-to-child transmission of infections.

The brain, the heart and the inner ear are the tissues most often permanently affected, causing microcephaly, epilepsy, heart murmurs and deafness which can be progressive. Retinopathy is also common, but may not affect sight severely. Isolated nerve deafness is probably the commonest result of intrauterine infection.

Intrapartum and perinatal infections

Intrapartum and perinatal infections rarely cause severe or multisystem disease; indeed with a few exceptions they are often inapparent. Because they take time to develop there is an opportunity for prophylaxis or early treatment, which is particularly important in those few infections which can cause delayed effects or complications.

Outcomes of intrauterine and intrapartum infection

1 Intrauterine infection and recovery.
2 Intrauterine infection and fetal death or stillbirth.
3 Born infected but no disease develops.
4 Born infected, disease develops later.
5 Born infected with active disease.
6 Born with permanent or progressive tissue damage.

Transplacental, intrapartum and postnatal infections

ORGANISM LIST

Transplacental
- Rubella virus
- Cytomegalovirus
- Human parvovirus B19
- Herpes simplex virus
- Human immunodeficiency virus
- Varicella zoster virus
- Vaccinia virus
- *Listeria monocytogenes*
- *Treponema pallidum*
- *Toxoplasma gondii*

Intrapartum
- Cytomegalovirus
- Herpes simplex virus
- Hepatitis B virus
- Human immunodeficiency virus
- *Escherichia coli*
- Group B *Streptococcus*
- *Chlamydia trachomatis*
- *Neisseria gonorrhoeae*
- *Listeria monocytogenes*

Postpartum
- Cytomegalovirus
- Hepatitis B virus
- Varicella zoster virus
- Human T-lymphotropic virus I
- Human immunodeficiency virus
- Herpes simplex virus
- Enteroviruses (e.g. echovirus type 11)

It will be seen that many of these pathogens can infect the fetus and infant by more than one route. Therefore, for each organism mentioned, the routes of transmission and the different clinical disorders produced will be discussed.

Congenital rubella

Introduction and epidemiology

The risk of congenital rubella depends on the stage of pregnancy at which maternal infection occurs. If it occurs during the first trimester, between 80 and 90% of pregnancies carried to term will result in congenital rubella. The risk approaches 100% for infections in the first month. It declines to around 50% during the second trimester and approaches zero after 30 weeks' gestation.

Before the introduction of rubella vaccine, outbreaks of congenital rubella occurred every 5–7 years, following the natural epidemic cycle of rubella (Fig. 17.2; see also Chapter 11). In a typical epidemic year, over 1000 infections in pregnant women were reported in the UK. Approximately 90% of these pregnancies were terminated; however, in countries where termination of pregnancy is not readily available, the impact of these epidemics is much greater.

Following the introduction of rubella vaccination, congenital rubella is now rare, and has almost been eliminated in many developed countries. There is a residual risk among immigrant women, mainly of Asian origin, who have not received rubella vaccination.

Permanent or progressive effects of congenital rubella infection

1 Cardiac: patent ductus arteriosus with or without pulmonary stenosis.
2 Cataracts.
3 Dysplasias of the retina or uveal tract.
4 Increasingly delayed motor and sensory development.
5 Nerve deafness.

Clinical features

Rubella in pregnancy approximately doubles the risk of fetal death. Among survivors the effect varies from severe infection and multiple permanent defects to an apparently normal, but infected, neonate. The severity and nature of the clinical problems depend on the stage of pregnancy at which infection occurred.

Reversible effects

Reversible effects of active infection are often present at birth, causing hepatitis and jaundice, haemolysis and thrombocytopenic purpura, and often a low-grade meningoencephalitis. Some cases have metaphyseal dysplasia with patchy mineralization of the ends of the long bones. Most affected infants are of low birth weight and fail to attain their expected developmental milestones. There is a high mortality in severely affected infants, but in survivors these reversible effects resolve in the first 2–6 months of life.

Permanent effects

After infections in the first month of pregnancy, over half of those born will have multiple defects. Patent ductus arteriosus with or without pulmonary stenosis is the commonest. Deafness may be profound, or only detectable by audiometry. Dense cataracts are common after infection in the first month; smaller or faint opacities may occur after second-month infections. Retinal pigment dysplasia is often evident on ophthalmoscopy. Most infants have a severe brain syndrome inhibiting motor, sensory and intellectual development. This can result in a severely disabled child, further impaired by defective vision and hearing.

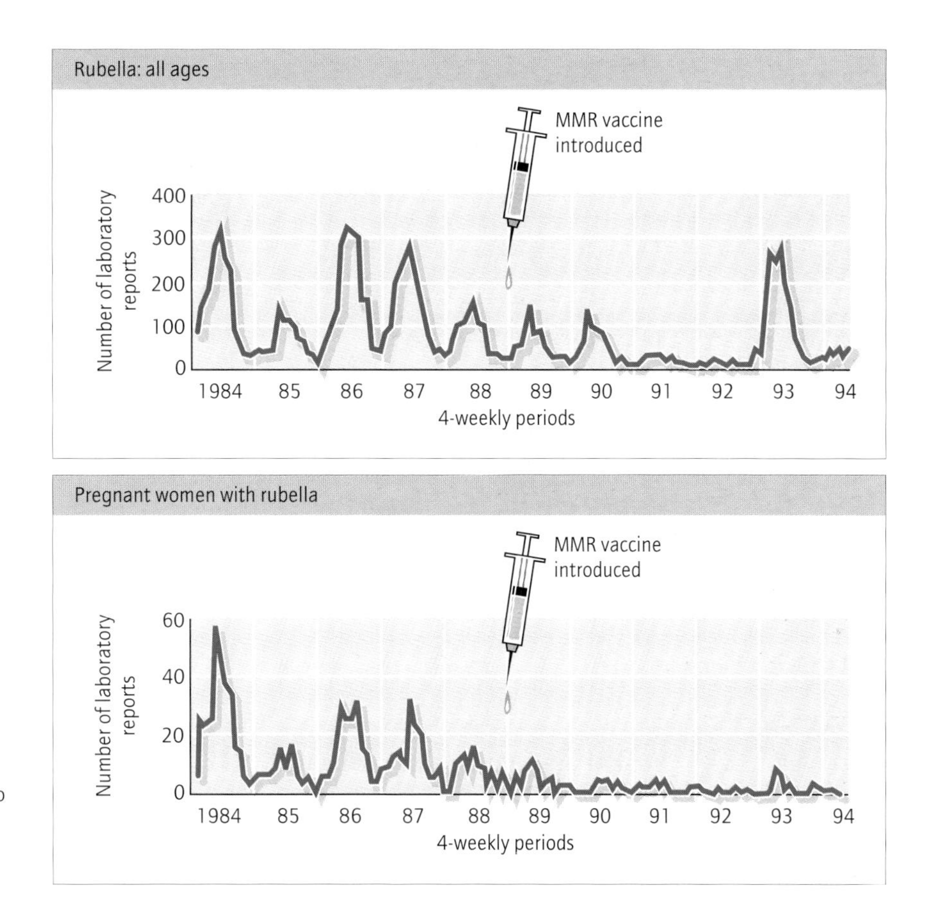

Fig. 17.2 Epidemics of rubella, and their relation to rubella infections in pregnancy; the numbers decline after the introduction of mass immunization.

Hearing defects are the commonest effect of infection later in pregnancy, and may worsen during the early years of life.

Diagnosis

Exposure during pregnancy should be confirmed, if possible, by serological testing of the contact, as clinical diagnosis is extremely unreliable (see Chapter 11). A positive immunoglobulin M (IgM) antibody test confirms recent rubella and indicates that the contact has been infectious. The serological status of the exposed pregnant woman should have been ascertained at her booking appointment. Immunity to rubella does not absolutely protect from reinfection, however, as just under 10% of congenital rubella cases in England are shown to follow maternal reinfection during pregnancy.

The pregnant woman who was seronegative at booking, or whose immune status is unknown, should have an immediate test for IgM rubella antibodies. If negative, the test should be repeated after 2 weeks, or sooner if a feverish illness develops. A final test may be performed after a further week or 10 days. A woman who previously possessed antibodies may not develop an IgM response, but should be followed up to detect a secondary rise in IgG antibody levels. The aim is early detection of rubella infection, so that a risk assessment can be made, and termination of pregnancy can be offered if appropriate.

A rubella-infected neonate will have a positive IgM rubella antibody test, which persists until the third month of life. The absence of IgG antibodies excludes the diagnosis of congenital rubella, as neither mother nor child can have been infected.

Laboratory diagnosis of rubella

1 In pregnant woman: serum immunoglobulin G (IgG) at booking, to check immune status; IgM after suspected exposure or illness (repeat after 2 weeks).

2 In fetus: culture of rubella virus from amniotic fluid; fetal blood sampling for serology (only worthwhile after 20 weeks).

3 In neonate: positive IgM (absence of IgG excludes the diagnosis).

Management

Pregnant women with possible rubella infection should be isolated from other antenatal patients.

Most women opt for termination of pregnancy if infected during the first trimester; in later pregnancy there is a balance between the likely fetal damage and the desirability of termination.

Human normal immunoglobulin does not offer reliable postexposure prophylaxis.

The rubella-affected neonate remains highly infectious for several months, and should be isolated from pregnant women. Supportive treatment during the active infectious phase often includes blood or platelet transfusions, sometimes with treatment for immune thrombocytopenia.

The extent of permanent defects should be ascertained by detailed examination. Hearing tests are difficult to perform in infants; observational tests can be complemented with audiometry when the child is old enough. Cataracts can be treated surgically during the first year of life, allowing normal visual development. Cardiac abnormalities often require correction in the early months. Those with severe physical and intellectual disabilities need lifelong support.

Prevention

Rubella vaccine is a live attenuated vaccine produced from the RA 27/3 strain of virus. A single dose elicits protective antibodies in over 95% of recipients. The duration of protection is not yet known, as the vaccine was first produced less than 30 years ago; however, long-term follow-up shows that vaccine-induced antibodies wane at a similar rate as those acquired from natural infection. This suggests that, for most individuals, protection will be lifelong.

Immunization of all infants early in life (usually as combined measles/mumps/rubella (MMR) vaccine) aims to provide indirect protection, by interruption of rubella transmission among children. Success of this strategy depends upon achieving high coverage (greater than 90%). This approach is succeeding in the majority of countries.

Prevention of congenital rubella

1 Universal immunization in childhood.
2 Antenatal screening for immunity.
3 Postpartum immunization of susceptibles.
4 Isolation of cases in antenatal units.
5 Active attention to diagnosis in pregnancy.
6 Counselling in pregnancy.

Live rubella vaccine is contraindicated in immune suppression and in early pregnancy because of theoretical damage to the fetus. However, follow-up of several hundred babies whose mothers were accidentally vaccinated in early pregnancy has shown no increased risk of congenital abnormalities. Termination of pregnancy following inadvertent rubella vaccination is not therefore indicated.

Congenital and neonatal cytomegalovirus infections

Epidemiology

Cytomegalovirus is the commonest cause of congenital infection in the UK, occurring in 3 per 1000 live births. Although most babies have no symptoms at birth, about 10% of these asymptomatic babies will subsequently develop deafness and neurological impairment. In the USA, an estimated 2500 infants a year are born with symptoms.

Clinical features

Congenital infection

Congenital infection is apparent at birth in about 10% of cases. It presents with prematurity, low birth weight, hepatomegaly, splenomegaly, thrombocytopenia and prolonged jaundice. About 25% of clinically affected neonates have cerebral irritability, fits or abnormal muscle tone or movement. Pneumonitis is uncommon, and ventilation is rarely required.

Permanent defects occur in about half of all symptomatically affected infants. Microcephaly and sensorineural deafness are the commonest problems, and often coexist. Microcephaly improves or disappears with growth in a third to a half of those affected. Deafness is a solitary finding in about 10% of cases. Other problems include cerebral calcification, hemiplegia, diplegia or quadriplegia, and psychomotor retardation. Choroidoretinitis and myopathy have also been reported.

Permanent effects of congenital cytomegalovirus infection

1 Microcephaly.
2 Nerve deafness.
3 Cerebral calcification.
4 Upper motor neurone disorders.
5 Psychomotor retardation.
6 Choroidoretinitis (rare).
7 Myopathy (rare).

The prognosis for later childhood development is poorest in those who have neurological defects at birth. Microcephaly alone does not confer such a poor prognosis as hard neurological signs. Children without neurological defects at birth have a good prognosis, but it is not known whether deafness or other defects may become apparent in late childhood or adulthood.

Intrapartum and perinatally acquired infection

Intrapartum and perinatally acquired infection is often inapparent, although fever, poor growth, pneumonia or late-onset jaundice occasionally occur. It is common for an infant to contract cytomegalovirus infection from its seropositive mother; the majority of all seroconversions occur before school age.

Diagnosis

Infection in pregnancy

Cytomegalovirus infection in pregnancy is rarely apparent, and is often diagnosed by finding congenital infection in the neonate. In the rare situation of a symptomatic primary infection in pregnancy, seroconversion would be demonstrable. Demonstration of IgM antibodies is not completely reliable, as these can appear in postprimary infections, which rarely affect the fetus.

Diagnosis in the neonate

The diagnosis of congenital infection in the neonate depends on demonstrating IgM antibodies in serum, or cytomegalovirus excretion during the first 20 days of life. Urine culture and throat swabs are the best sources of virus isolation. Cytomegalovirus DNA can be demonstrated in these specimens and in blood by polymerase chain reaction (PCR) techniques.

Intrapartum and perinatal infection

Intrapartum and perinatal infection produces positive IgM and cultures after 20 days. IgM and cytomegalovirus excretion up to 20 days of age indicates intrauterine infection; after 20 days it indicates intrapartum or perinatal infection. Infants who are not tested before 20 days of age cannot have a certain microbiological diagnosis of congenital infection because infection in early infancy is extremely common.

Management

In pregnancy

If new infection is diagnosed during pregnancy, counselling is based on the knowledge that only 10% of children have any detectable disorder at birth, and that a proportion of these will recover. There is no specific treatment.

Neonatal

The infected neonate needs no active treatment if it appears normal; supportive treatment is needed for temporary acute problems. Future follow-up to detect possible sensorineural deafness is important. Those who present with neurological disorders require additional follow-up with regular hearing and neurological assessments, and physiotherapy, rehabilitation and/or educational support as appropriate.

Prevention

Opportunities for prevention are limited. Blood transfusion and organ donation from cytomegalovirus-seropositive donors to seronegative recipients should be avoided. Since cytomegalovirus in adults is often asymptomatic, screening in pregnancy would require repeated blood sampling. It has been suggested that susceptible pregnant women can reduce their risk by avoiding contact with the urine and saliva of young children, although this is of unproven benefit. No vaccine is available.

Occupational exposure to young children has not been shown to carry additional risk of cytomegalovirus infection.

Congenital parvovirus infection

This is probably quite common during parvovirus B19 epidemics, when children and adults may develop slapped cheek syndrome or arthralgia and rash (see Chapter 11). It is important to undertake diagnostic tests in a woman with these symptoms, or with fever during a parvovirus epidemic, and to exclude rubella, which parvovirus may closely resemble.

Women who possess IgG antibodies to parvovirus B19 are immune, and not at risk of infection. Fetal infection will occur in about half of maternal infections during pregnancy. In almost all cases both mother and fetus recover uneventfully and a normal infant is born. Termination of pregnancy is not indicated for maternal parvovirus B19 infection.

Infection in the first 20 weeks of pregnancy can damage the fetal red blood cells. Rapid fetal growth in these weeks must be paralleled by equally rapid production of red blood cells if the fetus is not to become anaemic. Human parvovirus B19 can only infect cells during the S phase of mitosis, and intense infection of the actively dividing red cell precursors causes temporary aplasia, which can be severe enough to produce fetal anaemia and hydrops. Fetal death occurs in about 9% of pregnancies affected in the first 20 weeks, and hydrops in about 3% affected between the 9th and 20th week. Hydrops is detectable by serial fetal ultrasound, and severely affected cases may be offered intrauterine blood transfusion.

Congenital and intrapartum herpes simplex infections

Primary herpes simplex infections may be accompanied by viraemia and, if this occurs in pregnancy, transplacental infection can result.

Congenital infection

Infants born with congenital infection tend to have severe disease with a high mortality. The commonest features are pneumonitis, meningoencephalitis with fits and neurological signs, hepatosplenomegaly and cytopenias. A minority of infants have herpetic lesions of the skin or mucosae.

Diagnosis of congenital infection is important, as treatment reduces mortality from 80–90% to 10–15%. Virus particles can be seen on electron microscopy of throat swabs, bronchial secretions and scrapings from lesions. The virus grows well in cell culture, producing early and diagnostic cytopathic effects. Herpes simplex DNA is detectable by PCR in these specimens and in cerebrospinal fluid. Intravenous aciclovir, 10 mg/kg 8-hourly is effective and well tolerated.

Intrapartum infection

Intrapartum infection occurs when a mother is excreting herpes simplex virus at the time of delivery. In this case the infant develops skin, conjunctival, oral or genital lesions within a few days of birth. The condition should be treated vigorously, as half of cases will develop disseminated disease. Intravenous aciclovir is the treatment of choice.

Varicella embryopathy and neonatal varicella

Varicella embryopathy

Varicella embryopathy is a rare deformity following varicella infection, usually during the 13th to 20th week of pregnancy. About 12% of adults are susceptible to varicella, so maternal infection during pregnancy is not uncommon. Large surveys suggest that varicella embryopathy occurs in about 2% of infected pregnancies.

The usual deformity is cicatricial contracture of a limb, with hypoplasia and reddened scars suggestive of old, zoster-like lesions. If the head is affected, microcephaly and unilateral microphthalmia may occur. There is no evidence that immunoglobulin or aciclovir treatment affects the likelihood of embryopathy after the mother has developed chickenpox.

If a woman with no history of chickenpox is exposed to infection during pregnancy, she should have her antivaricella antibody titres estimated. If IgG antibodies are present, no further action need be taken. If she is not immune, the woman may be offered postexposure prophylaxis with varicellar zoster immunoglobulin (VZIG). This can prevent or ameliorate infection if given within 10 days of exposure.

Neonatal varicella

Neonatal varicella is a life-threatening, disseminated disease, with a mortality of up to 40%. It occurs in neonates born to women with no immunity to varicella. The neonate has no protective maternal antibodies. It can acquire infection from its mother if she develops chickenpox within 1 week before or after delivery: if before delivery, she cannot develop antibodies in time to confer transplacental protection; and if after delivery, the infant may contract infection in the neonatal period.

The neonate should be offered postexposure prophylaxis with VZIG. If the mother has had chickenpox for less than 8 days, this should be given as soon after birth as possible. Immunoglobulin given to the pregnant mother will not protect the infant. If the mother's disease appears within 28 days after birth, VZIG should be given to the infant as soon as possible, preferably within 48 h. The infant should be followed up, and treated with aciclovir if varicella develops despite VZIG prophylaxis.

Maternal varicella

Clinical varicella in a pregnant woman is often no more severe than in the non-pregnant woman. As in other adults there is a risk of pneumonitis, which is greater in smokers; also the feverish illness can precipitate early or premature labour. Many experts therefore recommend offering oral aciclovir treatment to avoid possible severe or complicated disease. In severe or complicated disease, intravenous aciclovir and appropriate antibiotics should be used as indicated. Aciclovir is not licensed for use in pregnancy, but increasing experience so far shows no evidence of adverse effects. The little evidence available suggests that it may reduce the risk of fetal infection.

Aciclovir should be used when clinically indicated to treat maternal disease (see Chapter 11).

Congenital HIV infection

Epidemiology

Congenital human immunodeficiency virus (HIV) infection is rare in the UK, where the prevalence of infection in the antenatal population is low (0.4% in London, 0.01% outside London). Between 1982 and 1994 only 127 cases of congenital acquired immunodeficiency syndrome (AIDS) were reported. By contrast, in some parts of Africa the prevalence reaches 25% and HIV-1 is the most important cause of congenital infection. The risk of vertical transmission also varies—from 17% in Europe to almost 30% in Africa. The prevalence of HIV infection in the UK is continuing to rise in the antenatal population; thus congenital infections will become more common in future.

These figures are well established for HIV-1 infection. It is likely that they differ for the less common HIV-2 infection, which appears to be less easily transmissible. There are insufficient data from which to deduce the probable risk.

Clinical features

A variable percentage of infants born to HIV-positive mothers are premature or of low birth weight. This may be related to the mother's health and does not inevitably mean that the child is infected.

Some infected infants begin to have severe infections and fail to thrive in the first 12–18 months of life, while others remain persistently antibody-positive for several years without any adverse effect. The reasons for these different presentations are uncertain.

Diagnosis

All infants of infected mothers will have HIV antibodies in the blood at birth. In the majority these will be transplacentally acquired, and will decline in titre, disappearing between 6 weeks and 6 months of age. HIV antigen, if present, is diagnostic of true HIV infection but may be absent for the initial weeks or years of life. The most reliable diagnostic test is the presence of HIV RNA, detectable by sensitive PCR techniques.

Management

In pregnancy

HIV infection during pregnancy should be treated as indicated with conventional multidrug therapy; the only possible adverse effect on the fetus which has been demonstrated is a rare syndrome of mitochondrial dysfunction which has affected a small number of neonates. Some other drugs, including sulphonamides, some antifungal drugs and systemic pentamidine are contraindicated in pregnancy; others, such as rifampicin, are not recommended but are not known to produce adverse effects. Care should therefore be taken with drugs given prophylactically or electively, though life-saving treatment should be given as in the non-pregnant.

Before and after delivery

Zidovudine treatment of mother and infant reduces the risk of vertical transmission of HIV-1 by 60–70%. The optimum regimen includes maternal therapy for the last 11–12 weeks of pregnancy and during labour. The infant is usually treated for 6 weeks after birth, but there is evidence that much shorter regimens are equally effective. Shortening or omission of the maternal phase of treatment reduces effectiveness, but still affords some benefit.

Recent evidence suggests that nevirapine given during labour, and to the neonate at birth, is safe, and at least as effective as zidovudine in preventing vertical transmission of HIV.

If a woman is already taking multidrug therapy during pregnancy, this should be continued, and the viral load monitored. A switch of therapy may be used for the last trimester, to aim for a maximum antiviral effect. In this case, the infant should receive zidovudine, unless the mother's virus is known to be resistant to it, when an alternative may be offered (or treatment omitted if the mother has an undetectable viral load).

Other methods for reducing vertical transmission include the avoidance of breast-feeding, and elective Caesarean section in which a bloodless field is obtained before the fetal sac is opened. In Western countries, these precautions can prevent the majority of mother-to-infant transmissions. The World Health Organization continues to recommend breast-feeding by HIV-positive mothers in developing countries, where the protective effect of breast milk against opportunistic infections outweighs the small additional risk of HIV transmission.

Inhibition of maternofetal transmission of HIV

For the mother: from the 14th week of pregnancy, zidovudine 500 mg daily orally; then, during labour, zidovudine 2 mg/kg by i.v. infusion during the first hour, then 1 mg/kg hourly until the umbilical cord is clamped.

For the infant: from before 12 h of age, zidovudine 2 mg/kg 6-hourly orally, continued for 6 weeks.

The HIV-infected infant

The HIV-infected infant may need no specific treatment for months or years. The normal programme of infant immunization should be followed, including measles vaccine, even in symptomatic children, as the risk from childhood infection is far greater than that from the vaccines. However, inactivated polio vaccine (IPV) should be used, not the live, oral vaccine (OPV).

There is no reason to exclude HIV-positive children from nursery or school, though they should be aware of the potential infectiousness of their blood. Cuts and grazes should be cleaned with water, soap or mild disinfectant and covered with a simple dressing. Spillages of blood or body fluids should be absorbed with wadding or absorbent granules, and the area cleaned with soapy solution or a mild household disinfectant/cleanser; the wadding or granules may be placed in a plastic bag for disposal (commercial spillage kits are convenient for this purpose).

Gonococcal neonatal ophthalmia

Neonatal ophthalmia is a severe conjunctivitis occurring within the first 48 h of life. *Neisseria gonorrhoeae* is an uncommon cause, as most cases are due to *Chlamydia trachomatis* infection. Gram stain of the purulent conjunctival discharge shows many Gram-negative diplococci, confirming the diagnosis. Chloramphenicol eye ointment, applied three times daily for 7 days, is the treatment of choice.

If a mother is known to be infected with *N. gonorrhoeae* at the time of delivery, gonococcal ophthalmia can be prevented by treating the infant with an effective antibiotic. Benzylpenicillin is effective for sensitive gonococci; cefuroxime or cefotaxime is recommended for penicillin-resistant gonococci.

As with chlamydial neonatal infections, both parents should be offered investigation and treatment for gonorrhoea and other sexually transmitted diseases.

Congenital and neonatal listeriosis

Introduction

Both congenital and neonatal listeriosis are rare in the UK. Fewer than 100 cases a year are reported. Outbreaks occur from time to time, due to consumption of contaminated foods such as soft cheese or pâté (Fig. 17.3).

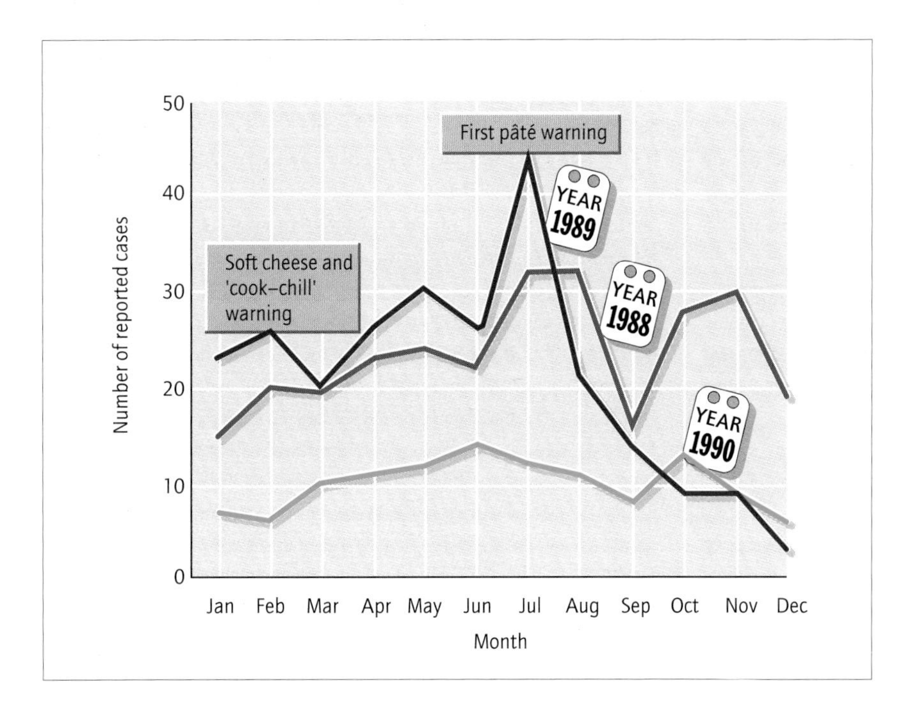

Fig. 17.3 Recent trends in mother/infant listerial infections in the UK: effect of health education and food hygiene measures following epidemics in the 1980s. Courtesy of Dr Jim McLauchlin, Central Public Health Laboratory.

Clinical features

Congenital listeriosis

Congenital listeriosis is the result of transplacental infection with *Listeria monocytogenes* during maternal bacteraemia. The bacteraemic infection in the mother is usually trivial or inapparent. Features such as transient fever, diarrhoea, backache and pruritus are described, but these are common in pregnancy and rarely lead to specific diagnosis. The first sign of a problem is usually recognition of the disease in the offspring. An infected mother–baby pair may become the focus of nosocomial transmission in a neonatal unit.

Listeriosis in early pregnancy often results in fetal death. Otherwise, affected infants may be born prematurely, or at term, with severe disease. Characteristically this is a bacteraemic multisystem disease with granulomatous infiltration of parenchymal organs. There is often a purplish, nodular rash on the lower body and legs. Hepatosplenomegaly is common, and the other features of congenital infection are seen, including meningoencephalitis, thrombocytopenia and variable pneumonitis.

Neonatal listeriosis

Neonatal listeriosis follows intrapartum exposure to maternal birth passages colonized by *L. monocytogenes*. The infant becomes ill within the first 2 weeks of life, usually with meningitis and bacteraemia.

Diagnosis

In congenital listeriosis the presence of an unusual rash may suggest the diagnosis. Disease presenting in the neonatal period must be distinguished from the more common *Escherichia coli* or group B streptococcal meningitis and bacteraemia. The recent occurrence of a case or cases in a delivery unit should increase the index of suspicion.

Blood and/or cerebrospinal fluid cultures will provide the definitive diagnosis. In neonates with fever or sepsis, meconium, urine and gastric aspirate should also be submitted for culture at the time of birth. Products of conception may yield positive culture after abortion. If the mother is feverish, blood cultures should also be obtained from her. Positive cultures may also be obtained from placental tissue or lochia.

Serological tests may demonstrate listerial antibodies, but these often originate from remote infections or infections by strains of low pathogenicity. Serology is unreliable in making a diagnosis of recent or current listeriosis.

Diagnosis of neonatal and congenital listeriosis

1 Culture of meconium, urine and gastric aspirate at birth.
2 Culture of blood and cerebrospinal fluid at birth or in neonatal fever or meningitis.
3 Culture of products of conception (placental tissue or lochia).

Management

Affected mothers and their infants should be managed in isolation.

L. monocytogenes is sensitive *in vitro* to many antibiotics, including penicillin, ampicillin, tetracyclines, aminoglycosides, imipenem, trimethoprim and co-trimoxazole. It is also sensitive to rifampicin, ciprofloxacin and vancomycin, which are either contraindicated in pregnancy or the neonate, or penetrate the cerebrospinal fluid poorly. Most cephalosporins are ineffective or only weakly effective.

The response to treatment is not always as satisfactory as expected from laboratory tests. Improvement may be transient, and relapses often occur. This is common with penicillin alone, and can also occur during treatment with other drugs. Some recommend high-dose ampicillin or amoxycillin; others favour the combination of ampicillin plus gentamicin. The usual penicillin plus gentamicin treatment for neonatal meningitis or septicaemia is also likely to cure, or at least ameliorate, neonatal listeriosis. If a good response is obtained, 2–3 weeks' treatment may be sufficient, but 4 or even 6 weeks may be wise to avoid relapse after initial slow response.

When the response to other treatments is suboptimal, chloramphenicol may be effective. It should not be used in combination with other drugs, as there is the possibility of antagonism, and courses longer than 2 or 3 weeks carry a high risk of agranulocytosis. If used in the neonate, whose liver detoxifies the drug inefficiently, it must be given initially in a divided dose of 25 mg/kg per day, and plasma concentrations must be monitored. The therapeutic range is 15–25 mg/l; levels above these may cause the grey baby syndrome.

Treatment of congenital listeriosis

1 First choice: ampicillin i.v. 50–100 mg/kg daily in four divided doses; *plus* gentamicin i.v.—up to 2 weeks, 3 mg/kg 12-hourly; over 2 weeks, 2 mg/kg 8-hourly (may be given intrathecally 1–2 mg daily). Plasma concentrations should be measured to obtain peak (1 h) concentration not above 10 mg/l and trough not above 2 mg/l.
2 Alternative: benzylpenicillin i.v. 50 mg/kg daily 8-hourly; *plus* gentamicin i.v. as above.
3 Second choice: chloramphenicol i.v. 25 mg/kg daily in four divided doses. Plasma concentrations should be measured to obtain range of 15–20 mg/l.

Prevention and control

Pregnant women should be discouraged from eating foods likely to contain high counts of *L. monocytogenes*. These include pâtés and soft, ripened cheese such as Brie and Camembert. Other food that may contain significant bacterial counts include cook–chilled meals and ready-to-eat poultry: pregnant women should reheat these types of food before consumption. They should also avoid contact with potentially contaminated material such as aborted animal fetuses on farms.

Congenital syphilis

This is now extremely rare. Untreated syphilis in a pregnant woman often causes fetal death and early abortion. However, succeeding pregnancies are more likely to survive to term. The infant may have signs of disease at or soon after birth. If untreated, later manifestations can occur, appearing in childhood, the teens and even in adulthood.

The affected baby is often feverish and has features similar to those of secondary syphilis: rash, condylomata, and mucosal fissures and inflammation. Osteochondritis may cause pain. Half those affected even mildly have persistent rhinitis ('snuffles').

Early diagnosis is best made by dark-ground microscopy of material from mucosal or skin lesions. Serology may demonstrate maternal antibodies; only antibodies persisting after 3–6 months of age indicate true infection. The fluorescent treponemal antibody-absorption (FTA-ABS) test and IgM enzyme-linked immunosorbent assay (ELISA) are unreliable in the neonate.

Late manifestations tend to appear between the ages of 12 and 20 years. Neurological disorders such as nerve deafness, optic atrophy or pareitic neurosyphilis respond poorly to treatment. Interstitial keratitis often causes damaging corneal opacity, and synovitis of the knees (Clutton's joints) commonly accompanies this protracted inflammatory condition. Other features include bossing of the frontal bones, chronic periostitis of the tibias (sabre tibia), notching of the incisors (Hutchinson's teeth), 'mulberry' deformity of the first permanent molar and a high arched palate.

The treatment of choice is benzylpenicillin. Neonates can be treated with 300 mg 6-hourly for 10 days, and a good response should be expected. Neurological disease may respond slowly to high-dose treatment, and courses of several weeks may be justified. Interstitial keratitis and Clutton's joints also respond slowly. There seems to be a hypersensitivity component to the keratitis, which can be suppressed by topical corticosteroids; this treatment should be continued until spontaneous remission occurs.

Congenital toxoplasmosis

In the UK toxoplasmosis is a rare infection in pregnancy, affecting between 1 in 1000 and 1 in 2000 pregnancies. It is much more common in France and other continental countries, where attack rates of 2–6% of pregnancies are reported. The great majority of maternal infections are subclinical.

Infection in early pregnancy less often affects the fetus than infection in the third trimester. Overall, transplacental infection occurs in about a third of affected pregnancies, and has only been reported in primary infection. The result of infection depends on the stage during which infection occurs. The greatest risk of significant disease following toxoplasmosis is from infections occurring between the second and sixth month. Most later infections cause only seropositivity in the neonate.

Severely affected neonates may be stillborn or may die soon after birth. Others have cerebral calcification, cerebral palsy or epilepsy. Choroidoretinitis is usual, but may not be evident until some months after birth. This may be the only feature in mildly affected infants.

Clinical features of congenital toxoplasmosis

1 Intrauterine death or stillbirth.
2 Cerebral calcification.
3 Cerebral palsy.
4 Epilepsy.
5 Early or late choroidoretinitis.

Diagnosis of toxoplasmosis in the pregnant mother is confirmed by the presence of IgM antibodies or by seroconversion. IgM antibodies may also be demonstrated in affected neonates or in cord blood. A measure of the avidity of the antibodies may also be determined to give an indication of the duration of the infection. High titres of specific IgM in the infant's cerebrospinal fluid are strongly suggestive of congenital infection. *Toxoplasma gondii* can be detected in the products of conception or from the infant's cerebrospinal fluid by PCR by culture in a continuous cell line or by mouse inoculation.

Intervention is possible if the maternal infection is recognized. Management is expectant in third-trimester infections. Termination may be considered for infections in the second to sixth month of pregnancy. Treatment with spiramycin, a macrolide antibiotic, is thought to reduce the incidence of transplacental infection by reducing maternal parasitaemia. However, fetal infections still occur after spiramycin treatment, and the risk of clinical disease in an infected fetus is unchanged by spiramycin treatment. Fetal blood spiramycin concentrations are only 50% of those in the mother, so cannot cure an already infected fetus. It is not known whether newer macrolides and azolides are more effective.

The most effective regimen is pyrimethamine and sulphadiazine, but pyrimethamine is teratogenic, especially if given in the first trimester, and can produce bone marrow suppression. The combination of pyrimethamine and sulphadiazine, with folinic acid supplements, can be used in later pregnancy when there is definite evidence of fetal infection. It is given as 4-week courses alternating with spiramycin for the remainder of the pregnancy.

Diagnosis of *Toxoplasma* infection in mother–infant pairs

1 In the pregnant woman: immunoglobulin M (IgM) antibodies or seroconversion in other antibodies (indication to perform fetal sampling).

2 In the fetus: culture of amniotic fluid; culture of fetal blood samples; serodiagnosis on fetal blood samples (from 20 weeks).

3 In the neonate: IgM antibody in blood; positive culture of blood, cerebrospinal fluid, placenta or products of conception.

Women who have toxoplasmosis are advised to wait until the IgM antibodies have disappeared before attempting to conceive. A small minority have persisting IgM antibodies; after a year it is unlikely that they have persisting parasitaemia, and further delay is not usually recommended.

Treatment of mothers and infants with toxoplasmosis

1 For the infected pregnant woman: spiramycin 3 g/day in divided doses for 3 weeks (perform fetal sampling afterwards).

2 If fetal infection is confirmed in the first 6 months, and termination is not contemplated: sulphadiazine 3 g daily in three or four divided doses *plus* pyrimethamine, single dose, 50 mg daily *plus* folinic acid supplements for 4 weeks; *alternating with* spiramycin 3 g daily in divided doses for 4 weeks. Continue alternating therapy until delivery.

3 For the infected neonate: sulphadiazine 50–100 mg/kg daily in two divided doses *plus* pyrimethamine, single doses 1 mg/kg daily *plus* folinic acid supplement for 3 weeks; *alternating with* spiramycin 100 mg/kg daily in two divided doses for 3 weeks. Continue alternating therapy for 1 year.

Infections that spread only by the intrapartum and perinatal routes

Neonatal and perinatal hepatitis B

Epidemiology

In the UK, the prevalence of hepatitis B surface antigen (HBsAg) carriage in the general population is less than 0.5%. Both neonatal and perinatal infection are very uncommon. The risk of infection is between 10 and 25% for infants born to mothers who are hepatitis B e antigen-(HBeAg-)negative, but rises to 90% for those born to e antigen-positive mothers.

In communities originating in some parts of the world, notably Asia and some Mediterranean areas, the prevalence of infection in the antenatal population exceeds 10%, and neonatal and perinatal infections are very common.

Clinical features

Infants who have been infected by hepatitis B at birth rarely suffer significant illness; they simply develop the antigen and antibody sequence characteristic of acute hepatitis B infection. In the great majority (70–90%) healing is arrested at the stage of hepatitis B e and s antigenaemia (HBeAg and HBsAg), leaving the child at risk in later life of cirrhotic or malignant complications. Female children will also be at high risk of passing infection to their own infants.

Children in whom infection at birth is avoided or prevented may still be exposed during intimate contact with an antigen-positive mother or other family member during nursing and early childhood. Childhood hepatitis B infection is usually mild but may be icteric and in exceptional cases can be life-threatening or fatal. The risk of persisting e-antigen positivity after the acute infection is greatest when seroconversion occurs below 6 months of age; after the age of 9 months the risk is no higher than in older age groups. Women who are HBsAg-positive and HBeAg-negative, but who lack hepatitis Be-antibody (i.e. who have no HBe markers) are also likely to infect their newborn.

Diagnosis

All pregnant women are screened to identify those with hepatitis B antigenaemia. The populations most at risk are Mediterranean, Far Eastern, Asian and Caribbean. Effective postexposure prophylaxis for the newborn is possible because the infant's infection is contracted at the time of birth, and almost never *in utero*.

Management

The labour and delivery of a hepatitis B-antigenaemic mother should take place in a single-occupancy suite, as her blood and body fluids present an infection hazard to other mothers and infants.

As soon as possible after birth the neonate should receive appropriate immunization (Table 17.1). In infants at high risk, hepatitis B immune globulin (HBIG) prolongs the incubation period of the infection and allows time for active immunization with hepatitis B vaccine. The first dose of vaccine should be given at the same time as HBIG, but in a different intramuscular site. Neonates and infants mount a good immune response to the vaccine. Passive–active immunization reduces the incidence of permanent antigenaemia to about 5% in infants exposed to perinatal infection.

Human T-lymphotropic virus type I infection in infancy

Human T-lymphotropic virus type I (HTLV-I) is a retrovirus which infects human T lymphocytes. Seropositivity for this virus is associated with the occurrence of T-cell leukaemia and with a rare acquired neurological disease called tropical spastic paraparesis. The populations in which HTLV-I is prevalent include negroes of African and Caribbean origin, and Far Eastern populations, particularly from southern Japan.

The main route of transmission from mother to infant is via breast milk (which contains significant numbers of lymphocytes). Infants of infected mothers are many times more likely to acquire infection if breast-fed than if they are given artificial feeds. Avoidance of breast-feeding, if possible, offers important protection to the child.

Serological tests for HTLV-I are currently unreliable, as there are many false positives for each true positive result. Only Western blotting or tests dependent on identifying seropositivity to recombinant *env* and *gag* gene products offer dependable results, and these are too cumbersome and expensive to use as screening tests. It is hoped that screening for at-risk populations will soon be available.

Neonatal and infant chlamydial infections

Introduction

Chlamydial infections are among the commonest sexually transmissible diseases now encountered throughout the world. They have superseded gonorrhoea and syphilis, which are more readily detected and treated before and during pregnancy. Many pregnant women shed increasing numbers of chlamydiae from the genital mucosa as term approaches, so many neonates are exposed to intrapartum infection.

Clinical features

Chlamydial neonatal ophthalmia

Chlamydial ophthalmia is a severe conjunctivitis which appears within 3 or 4 days of birth. It may appear first in one eye, but usually becomes bilateral. There is oedema of the lids, closing of the eye and a purulent exudate which runs from the palpebral fissure. Parting the lids reveals a swollen, bulging conjunctiva.

Neonatal chlamydial pneumonitis

Neonatal chlamydial pneumonitis is a moderately severe, persistent infection which is often, but not always, preceded by ophthalmia. It presents at 3–6 weeks of age with tachypnoea and a repetitive, staccato cough. The chest X-ray is often abnormal, showing streaky or radiating perihilar opacities with or without areas of consolidation and air bronchograms. Untreated, it can last for many weeks, producing debility, failure to thrive and significant respiratory impairment.

Hepatitis B status of mother	Baby receives HBIG	Baby receives vaccine
HBsAg-positive and HBeAg-positive	Yes: 200 IU immediately	Yes: three doses plus booster
HBsAg-positive, no e-markers (or not determined)	Yes: 200 IU immediately	Yes: three doses plus booster
Mother had acute hepatitis B in pregnancy	Yes: 200 IU immediately	Yes: three doses plus booster
HBsAg-positive and anti-HBe-positive	No	Yes: three doses plus booster

Neonate's dose of hepatitis B vaccine is half of adult dose given at birth, then at 1 month and again at 2 months; a booster is given at 1 year.

Table 17.1 Recommendations for prophylaxis of hepatitis B in infants of antigen-positive mothers (UK National Screening Committee, 1998). (See Chapter 9).

Diagnosis

Differential diagnosis

Differential diagnosis of neonatal ophthalmia is important, as both parents may be colonized by the causative organism, and should be offered appropriate examination and treatment. An identical ophthalmia is caused less often by *Neisseria gonorrhoeae*. Staphylococcal or pneumococcal infections also occasionally cause purulent neonatal conjunctivitis.

Specimens should be obtained for bacteriological examination, and should include swabs in appropriate transport medium for isolation of gonococci. Scrapings from the inflamed conjunctiva can be examined for typical chlamydial inclusions in the epithelial cells by direct immunofluorescence for elementary bodies, or cultured in McCoy cells for isolation of *Chlamydia trachomatis*, demonstrable by ELISA for the presence of antigen or by DNA amplification techniques. Serology is unhelpful in the diagnosis of mucosal infections. However, chlamydial pneumonitis can be confirmed by the diagnostic presence of IgM antibodies to *C. trachomatis* in the infant's blood.

Management

Chlamydial neonatal ophthalmia

For chlamydial ophthalmia the treatment of choice is tetracycline eye ointment applied three times daily for 10–14 days. Chloramphenicol drops or ointment are not recommended as relapse can follow apparently effective treatment. Systemic treatment should also be given with oral erythromycin 125 mg 6-hourly for 10 days, as there is a high probability of coexisting early chlamydial pneumonitis.

Mild 'sticky eye' without conjunctival redness or positive cultures is often non-infectious, caused by sticky secretions from an engorged lacrimal sac, and relieved by twice-daily massage of the sac to empty it.

Chlamydial pneumonitis

For chlamydial pneumonitis the treatment of choice is erythromycin orally in a dose of 125 mg 6-hourly for 10–14 days.

Prevention

This depends on detection and treatment of maternal (and paternal) infection before the birth of the infant.

Neonatal bacteraemias

Epidemiology

The epidemiology of neonatal bacteraemia is markedly different from that in other age groups. Approximately 60% of infections are caused by *Escherichia coli* and group B streptococci. The remainder are due to other Gram-negative bacteria, staphylococci, *Listeria monocytogenes, Neisseria meningitidis, Haemophilus influenzae* and *Streptococcus pneumoniae* (Fig. 17.4).

Neonatal infection is closely related to colonization of the maternal genital tract. Bacteraemia tends to occur in the first week of life. Meningitis often occurs later, at 2–3 weeks. It is more likely in cases where there is a long interval between rupture of the membranes and delivery, which allows colonization of the amniotic sac with maternal genital flora.

Clinical features

In the first few days of life there are few specific clinical features of bacteraemia. There is often a fever. The infant is listless, with poor or absent muscle tone (floppy baby) and is uninterested in feeding. The neutrophil count may rise, though in very small babies this is not always a reliable finding.

Meningitis also presents non-specifically, with features of bacteraemia. There may also be vomiting, drowsiness, convulsions or a strange, high-pitched cry.

These presentations should be taken seriously, especially if convulsions occur in the absence of a biochemical cause. All cases of doubt should be investigated. Blood cultures, urine cultures and cerebrospinal fluid examination should all be performed. Gastric aspirate, meconium and urine may also be cultured in the first days of life. Treatment for bacteraemia and meningitis should be commenced while results are awaited.

Management

Intravenous antimicrobial chemotherapy should be given, with a spectrum broad enough to cover both *E. coli* and group B streptococci.

Broad-spectrum cephalosporins penetrate the cerebrospinal fluid readily, have low toxicity and do not require blood-level monitoring. Drugs such as cefotaxime or ceftazidime are convenient and effective. Treatment should be continued for about a week, with high doses for the first 2–3 days. The choice of antibiotic can be modified if indicated when the results of cultures are available.

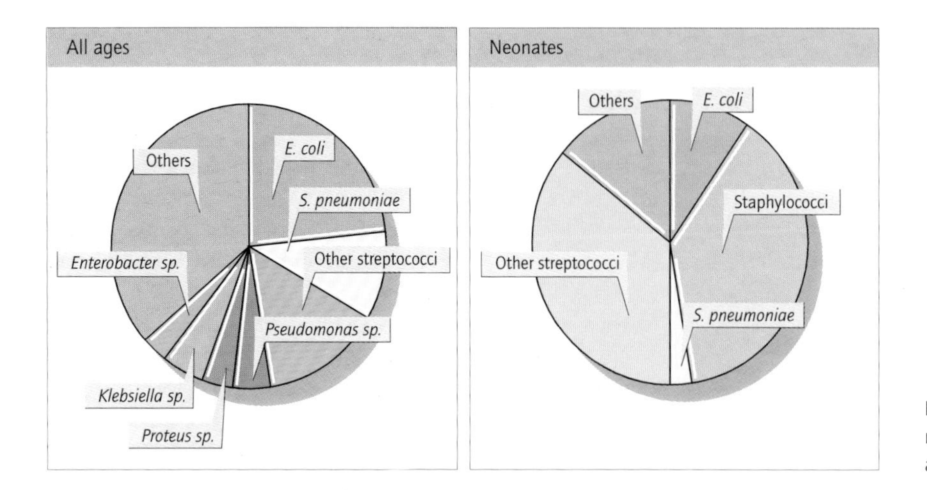

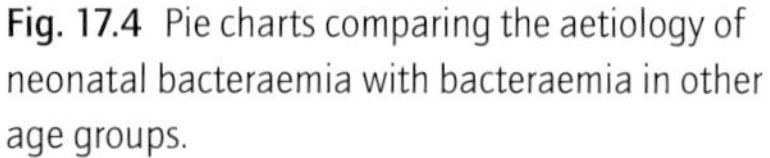
Fig. 17.4 Pie charts comparing the aetiology of neonatal bacteraemia with bacteraemia in other age groups.

Prevention of neonatal sepsis

This depends largely on hygienic and expeditious delivery of an infant. Delayed delivery (longer than 48 h after rupture of the membranes) should be avoided. Attempts to screen mothers for group B streptococcal colonization immediately before delivery have not proved reliable in identifying all cases at risk.

Neonatal staphylococcal infections

The commonest neonatal infection with *Staphylococcus aureus* is umbilical infection. The cut end of the umbilical cord easily accepts colonization and infection, leading to an impetiginous or weeping lesion. Spread to surrounding skin and subcutaneous tissues can occur.

Any area of broken skin may also become infected, as may the conjunctiva.

Bullous impetigo (Lyell's syndrome)

Although *S. aureus* bacteraemia is rare, the effects of staphylococcal toxins are more common. Staphylococcal pyrogenic exotoxins damage the layers of the epidermis allowing the surface layers to separate and form fragile blisters and bullae. The surface of erythematous, non-blistered skin can be rubbed off with gentle shearing stress (Nikolsky's sign). The resulting bare patches closely resemble scalds, and the condition is often called the scalded skin syndrome.

The lesions are heavily colonized with staphylococci, which can be recovered by culture from swabs. Staphylococcal skin lesions are highly infectious. The infant and mother should be separated from others in the neonatal nursery. Most cases will recover readily on treatment with oral flucloxacillin; severe infections need parenteral treatment. Lyell's syndrome can present with extensive exfoliation. This needs vigorous treatment with antibiotics; the infant requires warmth, humidity, ample fluid replacement and strict hygiene to avoid superinfection of the bare areas.

Control of infection measures should always include hand-washing after each contact with affected individuals. Chlorhexidine is often applied topically to the umbilical stump in the first day or two of life, but this does not preclude the need for strict hygiene generally in the nursery.

Maternal infections related to childbirth

Introduction

Around the time of delivery a woman may have increased susceptibility to conditions such as urinary tract infections because of her changed anatomy and physiology. She may also be more severely affected than others by certain infections such as genital warts or candidiasis because of altered physiology during pregnancy. Any acute feverish illness can cause premature labour, particularly in the third, and sometimes the first trimester.

There are other susceptibilities in pregnancy and the puerperium because of the presence of tissues or body functions that do not occur at any other time. The placenta is unique to pregnancy; it can be invaded and damaged by some pathogens causing severe maternal morbidity and putting the pregnancy at risk. This is important in some unusual infections including brucellosis, enzootic ovine abortion and, in exposed populations, falciparum malaria, which is dangerous to both mother and fetus. After parturition the placental bed provides a portal of entry for pathogens, which can invade the blood, causing puerperal fever. Finally the breasts become highly vascular and filled with secreted milk. Engorgement and stasis are common, and occasionally lead to the formation of breast abscesses.

Puerperal fever

ORGANISM LIST

Escherichia coli
Streptococcus pyogenes
S. pneumoniae
Staphylococcus aureus
Other coliforms
Mycoplasma hominis
Clostridium perfringens
Bacteroides fragilis

Introduction and epidemiology

Puerperal fever is defined as any significant feverish illness occurring within 14 days of childbirth, miscarriage or termination of pregnancy. It is a severe, usually bacteraemic infection caused by entry of pathogens either through the bare placental bed or through traumatic lesions of the cervix, vagina or perineum.

Clinical features

Most cases begin within 4–7 days of delivery. Fever may be the only sign, but there may be back pain, offensive lochia, faintness or signs of sepsis with or without shock. Disseminated intravascular coagulation occurs early, especially if there are retained products of conception.

There may be clinical features particular to the causative organism. Erysipelas-like lesions on the trunk or arms or a scarlatiniform rash may occur in streptococcal infection. The rash and shock of toxic shock can accompany staphylococcal cases. Intravascular haemolysis, jaundice and even crepitation of vulval tissues may indicate clostridial infection.

Management

Fever in the early puerperium should be taken seriously. If it is not obviously due to a local infection of the skin, breast or urinary tract, investigations should be carried out and initial treatment begun without delay.

Cultures of blood, urine, lochia and genital swabs should be obtained. The possibility of retained products of conception should be considered; ultrasound examination may help to exclude this. Any retained products should be evacuated (by an experienced operator, as the uterus may be oedematous or friable). Evacuated products should be submitted for microbiological examination.

Initial antibiotic treatment should be active against Gram-negative rods, *Streptococcus pyogenes* and anaerobic organisms. A reasonable regimen would be a mixture of an aminoglycoside, high doses of benzylpenicillin and metronidazole. A broad-spectrum cephalosporin could be substituted for the aminoglycoside and penicillin, but large doses are needed for an adequate antistreptococcal effect. Treatment may be modified when microbiological information becomes available. Some patients need intensive support, including treatment for shock, renal failure, adult respiratory distress and disseminated intravascular coagulation (see Chapter 19).

Prevention and control

Delivery should be as hygienic and atraumatic as possible. The placenta should be delivered without undue delay, and retained products should be promptly evacuated. Similar attention should be afforded to premature deliveries, stillbirths and terminations of pregnancy.

When a mother has a severe infection, her infant should be observed closely in the neonatal period as it is at increased risk of bacteraemia or meningitis caused by the same organisms.

Breast abscess during lactation

This is a distressing and painful condition, usually caused by a *Staphylococcus aureus* infection. Bacteria probably ascend via the milk ducts, and replicate in an area of stagnation. The abscess often develops soon after breastfeeding commences when the milk flow is relatively intermittent, and difficulties with cracked or infected nipples are commonest. Avoiding breast engorgement and nipple trauma reduces the risk of breast abscess.

A typical hot, tender lesion is palpable in the affected breast, and may point towards the surface as a red fluctuant area. 'Blind' treatment with moderate oral doses of flucloxacillin is justified in early cases, and may permit early resolution without significant interruption of breastfeeding. The antibiotics are secreted in milk, but are safe for the infant, though rare cases of penicillin rash can occur. In more severe cases pain may prevent continued feeding; aspiration or drainage of fluctuant lesions may afford relief, and antibiotic treatment will clear residual infection. Lactation can be maintained by gently expressing milk from the affected side and allowing the infant to feed from the other.

Antibiotics during lactation

Few antibiotics are contraindicated during lactation. Precautions apply to sulphonamides, which are excreted in small amounts in breast milk, and could exacerbate kernicterus, or cause skin rashes or rare haemolysis in glucose-6-phosphate dehydrogenase deficiency; and chloramphenicol, which may cause neutropenia in the infant. Isoniazid is also excreted in small amounts; if it is given, both mother and baby should receive pyridoxine supplements and be monitored for possible adverse effects.

Antimicrobials safe to use in lactation

These include penicillins, cephalosporins, aminoglycosides, erythromycin, standard doses of metronidazole (but large single doses are not recommended), aciclovir, rifampicin, ethambutol, pyrazinamide and chloroquine (which is not excreted sufficiently for a prophylactic effect in the baby).

Antimicrobials best avoided in lactation

These include tetracyclines, ciprofloxacin, vancomycin, high-dose sulphonamides, imipenem and ganciclovir.

Rare zoonoses in pregnancy

These are brucellosis, Q fever and enzootic ovine abortion (a *Chlamydia psittaci* disease of sheep). All cause abortion in their primary bovine or ovine hosts, and severe infection with placental damage and abortion in pregnant women. However, if both mother and fetus survive, the baby is not permanently harmed.

Farmers, farmers' wives or women working with animals are the population at risk. Q fever has also spread from cats during the birth of kittens. The diagnosis may be suspected on epidemiological grounds. Urgent serodiagnosis should be sought (see Chapter 7).

Early treatment is essential, and should not harm the fetus. Q fever and enzootic ovine abortion may be treated with intravenous erythromycin, to which chloramphenicol can be added if a prompt response is not obtained. Q fever may also respond to monotherapy with chloramphenicol. Ciprofloxacin has been used in maternal Q fever, but has not been shown to abolish placental infection. It is not recommended in pregnancy, but its use may be justified in this serious situation.

Brucellosis is difficult to treat, as tetracycline, by far the best choice, may harm the fetus. The risk of co-trimoxazole plus rifampicin is small, and is justified in this situation. Treatment should be continued for at least a month, and retreatment with tetracycline should be given to the mother after delivery to avoid the high risk of relapse after co-trimoxazole treatment (see Chapter 25).

18 Tuberculosis and Other Mycobacterial Diseases

Introduction, 337

Epidemiology, 337

Microbiology, 338
Classification, 339

Pathogenesis, 339
Primary tuberculosis, 340
Postprimary tuberculosis, 340
Miliary or disseminated tuberculosis, 341
Other types of mycobacterial infection, 341

Primary tuberculosis, 341
Primary pulmonary tuberculosis, 341
Endobronchial tuberculosis, 341
Pleural tuberculosis, 342
Tuberculous pericarditis, 342
Lymph-node tuberculosis, 342
Tuberculous peritonitis, 342

Diagnosis of tuberculosis, 344
Tuberculin tests, 344
Histology and cytology as diagnostic tools, 346
Microbiological diagnosis, 346

Postprimary tuberculosis, 349
Introduction, 349
Pulmonary tuberculosis, 349
Tuberculous meningitis, 351
Renal tuberculosis, 352
Tuberculous epididymitis, 352
Tuberculosis of the female pelvis, 352
Tuberculosis of bones and joints, 353

Miliary or disseminated tuberculosis, 353

Treatment of tuberculosis, 354
Introduction, 354
Drug-resistant tuberculosis, 354
Multidrug-resistant tuberculosis (MDRTB), 355
Mycobactericidal drugs used in tuberculosis (first-line drugs), 355
Additional and second-line drugs, 356

Non-tuberculous mycobacterioses, 357
Introduction, 357
Mycobacterium chelonei, 357
Mycobacterium fortuitum, 357
Mycobacterium marinum, 357
Mycobacterium ulcerans, 357

Prevention and control of tuberculosis, 357
Introduction, 357
Bacillus Calmette–Guérin vaccine, 357
Management of the case and close contacts, 358
Screening of immigrants, 358

Leprosy, 358
Introduction, 358
Clinical features and grading of disease, 359
Diagnosis, 359
Treatment, 359
Complications, 360
Prevention and control, 360

Introduction

Tuberculosis is a granulomatous disease caused by infection with some species of mycobacteria. Depending on the portal of entry of the infection, and on the degree of haematogenous spread, many organs and systems of the body can be affected. The commonest sites of infection are the lungs or the lymph nodes. Other important sites are the bowel, the peritoneum, the meninges, the kidneys, the bones and joints and the skin.

The disease is important because it is common in many countries (Fig. 18.1), the pulmonary form is highly infectious, and tuberculosis can cause severe morbidity and mortality in people of all ages. Drug-resistant tuberculosis is a rapidly increasing problem worldwide.

Epidemiology

Tuberculosis is one of the commonest infections of humans. It is estimated that 1.7 billion people are infected worldwide, that there are 20 million active cases and 3.3 million deaths a year. Most of these cases are in the developing world, although the incidence in many developed countries is now starting to rise after almost a century of steady decline (Table 18.1). There are a number of possible reasons for this recent increase: a deterioration in living conditions, particularly in inner cities and among refugees; spread of the human immunodeficiency virus (HIV); intravenous drug abuse; and worsening standards of health care. In developed countries, the disease mainly affects two groups: elderly, predominantly male, 'skid row' populations, and recent immigrants from developing countries. Healthcare workers, teachers and veterinary workers also have an increased risk of disease compared to the general population. Outbreaks occur in conditions of crowding, such as shelters for the homeless, hospitals and prisons. In 1990, the incidence in the UK was 10.3 per 100 000 population, slightly lower than the average for western Europe (14.0 per 100 000).

The prevalence of infection detected by tuberculin testing rises with age. In the UK, tuberculin sensitivity among school children aged 10–13 years, who are

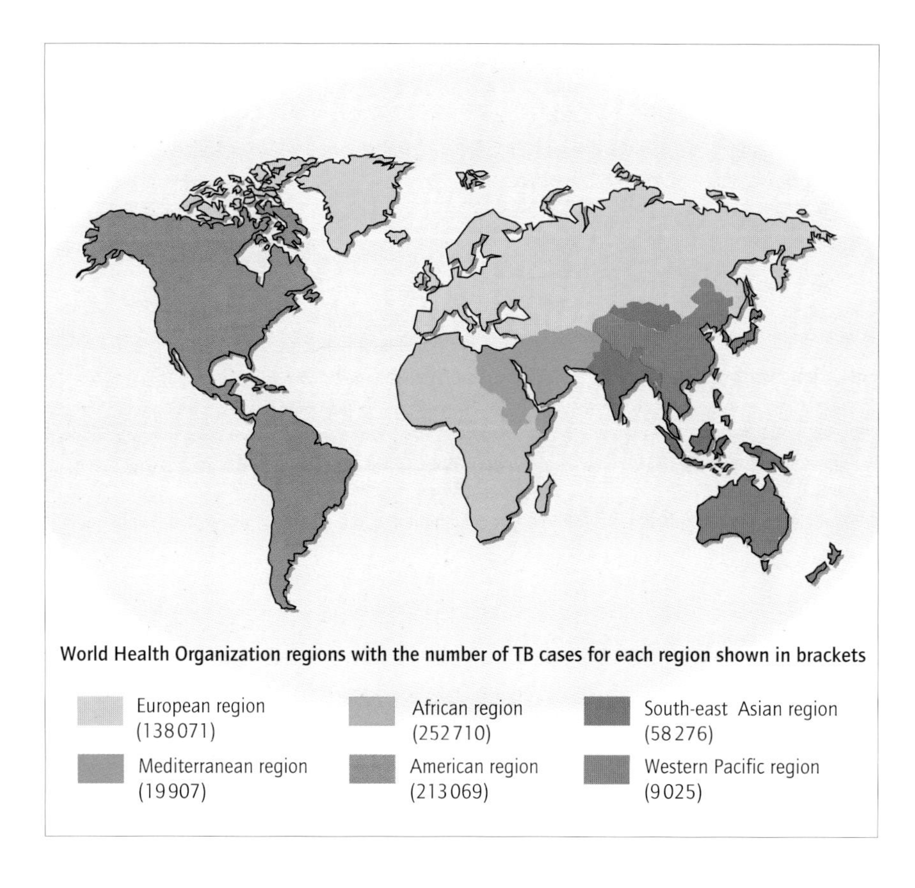

Fig. 18.1 Incidence of tuberculosis (TB) reported in different regions of the world in 1993. Courtesy of the World Health Organization.

Country	Incidence per 100000	Current trend
Austria	18.3	Up
Belgium	14.8	Down
Denmark	6.5	Up
Finland	15.5	Down
France	15.0	Down
Germany	18.4	Down
Ireland	17.9	Up
Italy	7.3	Up
Netherlands	9.2	Up
Norway	8.5	Up
Spain	23.1	Down
Sweden	6.5	Stable
Switzerland	16.5	Up
United Kingdom	10.5	Stable

Table 18.1 Tuberculosis: recent trends in western Europe (from Raviglione *et al.* (1993) *Bulletin of the World Health Organization* **71**, 297–306, with permission).

routinely tested prior to bacillus Calmette–Guérin (BCG) administration, is between 1 and 2%.

Most infections in developed countries are caused by *Mycobacterium tuberculosis*. Infection due to *M. bovis* is, however, still a problem in countries where the disease has not been controlled in cattle, and milk is consumed raw.

The rise of tuberculosis among HIV-infected individuals is often due to reactivation rather than primary infection. This phenomenon has mainly been observed in areas where the prevalence of HIV infection is high, such as parts of Africa and some US cities. Another problem in these areas is the emergence of strains of *M. tuberculosis* resistant to several antituberculosis drugs.

Microbiology

There are more than 85 species of mycobacteria, and new species continue to be proposed. Most mycobacteria are environmental organisms which rarely, if ever, cause human disease. Their natural habitat is soil, and fresh and estuarine water. Only a minority of species are obligate human pathogens. The remaining human diseases arise when host defences are reduced by physical or immunological defect.

Classification

Obligate pathogens

This important group includes those organisms whose isolation or detection implies that disease is present and requires specific chemotherapy. It includes the organisms of the *M. tuberculosis* complex: *M. tuberculosis, M. africanum, M. bovis*, BCG, and also *M. leprae*, which is in a group of its own.

Pulmonary opportunists

M. kansasii and *M. xenopi* infection are found most commonly in patients who have anatomical abnormalities of the lower respiratory tract. Chronic obstructive airways disease and bronchiectasis are common predisposing conditions. They have, however, rarely been isolated from patients with acquired immunodeficiency syndrome (AIDS). Infection is usually pulmonary, resembling indolent progressive pulmonary tuberculosis.

The most commonly isolated mycobacteria in the UK

1 *Mycobacterium tuberculosis.*
2 *M. kansasii.*
3 *M. avium-intercellulare.*
4 *M. bovis.*
5 *M. xenopi.*
6 *M. chelonei.*
7 *M. fortuitum.*

Skin pathogens

This group includes *M. marinum*, a cause of chronic granulomatous infection and ulceration of the skin, and sometimes subcutaneous tissue. It is found in old, concrete-lined swimming pools, rivers and fish tanks or aquaria. Infection occurs when the organism enters through broken or macerated skin. *M. ulcerans* is the causative organism of buruli (tropical) ulcers. This is a chronic destructive ulcer, often of the foot or leg, which erodes the skin and the subcutaneous tissues, including bone. It is thought to be due to inoculation of the organism by sharp vegetation, though it has never been isolated from the environment in endemic areas. *M. tuberculosis* can also act as a skin pathogen, causing a chronic scarring infection of the skin, principally on the face. The disease is called lupus vulgaris because of the wolf-like facies of sufferers in the preantibiotic period.

Opportunist pathogens

In the past, opportunist infections with mycobacteria were uncommon. Now they are commonly seen in HIV-infected patients, and occur in transplant patients, those treated for lymphomas and those with inherited cellular immunodeficiencies.

AIDS-related opportunists

Organisms of the *M. avium-intracellulare* (MAI) complex often affect AIDS patients in the latter part of the disease process. The infection is acquired through the gastrointestinal tract but as immunity wanes, invasion occurs and the organism can be isolated from the blood and other tissues. Infection often takes the form of cervical adenopathy. Infection with *M. avium-intracellulare* is rare in immune-competent patients but has been associated with outbreaks of pulmonary infection in a hospital environment. Several 'new' mycobacteria have been isolated from patients with HIV infection including *M. genevense, M. haemophilum, M. cookei* and *M hiberniae.* These are uncommon infecting agents even in HIV-infected patients.

Rapid growers

These organisms include *M. fortuitum* and the *M. chelonei* complex, and differ from other mycobacteria in the speed of their growth. Isolates grow on Löwenstein–Jensen and other selective media in 2–48 h. The organisms have low virulence but can be a problem in some circumstances. *M. chelonei* has caused systemic infections in neutropenic patients. Both *M. chelonei* and *M. fortuitum* can cause abscesses in patients who have been injected with 'sterile' fluids contaminated by this organism (for example, at injection sites in diabetics whose needles or multidose insulin containers have become contaminated). *M. fortuitum* is a rare cause of opportunistic bone and joint infections.

Pathogenesis

This has proved difficult to research; however, a number of important pathogenic attributes of mycobacteria can be identified. These mechanisms are gradually being elucidated by the use of molecular cytology and immunology techniques.

In contrast to many bacteria, the cell wall of mycobacteria is very lipid rich. Up to 40% of the dry weight of the organism is made up of lipid, and several lipid antigens have been identified as potential pathogenicity determinants.

Handling of mycobacterial antigens by the immune system

Once ingested by antigen-presenting cells, mycobacterial antigens are processed and presented in the classic way, in association with human leucocyte antigen (HLA) class II molecules. T lymphocytes bearing receptors which can recognize this complex bind to the macrophage, causing the release of interleukin 2 (IL-2), and the activation of T helper cells. IL-2 stimulates both antigen-specific T cells and antigen non-specific N cells. All of these cell types are capable of producing gamma-interferon, which has an important role in activating bactericidal mechanisms in macrophages.

Basis of the wasting effect of mycobacterial infections

The macrophage releases several cytokines while reacting to mycobacteria. These include tumour necrosis factor-alpha (TNF-alpha), IL-3 and granulocyte–macrophage colony-stimulating factor (GM-CSF), which probably have an important role in inducing symptoms such as fever and wasting. Mycobacterial lipoarabinomannan is a strong inducer of TNF.

Inhibition of cell-mediated immune responses

Macrophage activation may be inhibited by mycobacterial lipid antigens such as lipoarabinomannan, and capsule-like materials such as the phenolic glycolipid of *M. leprae*. These are thought to act by scavenging reactive oxygen intermediates and interfering with the efficiency of the microbicidal oxidative system. Protein antigens of mycobacteria, such as catalase and superoxide dismutase, may also act in this way.

Mycobacterial antigens may interfere with the activation of T-cell responses. Lipoarabinomannan has been implicated in blocking the stimulatory effect of gamma-interferon by inhibiting its interaction with macrophage surface molecules such as protein kinase C. Lymphocytes and macrophages which are recruited to the site of infection, but cannot destroy the mycobacteria, contribute to the progressive formation of granulomas.

Mechanisms of survival within macrophages

A key characteristic of the pathogenic mycobacteria is the ability to survive inside macrophages. Mycobacterial ingestion is mediated via the CR1 and CR3 receptors which do not stimulate microbicidal oxidative responses. Mycobacteria are able to survive inside macrophages by inhibiting phagosomal–lysosomal fusion or by escaping from the phagosome into the cytoplasm, in both cases avoiding the effects of lysozymes.

Heat-shock proteins are produced by phagocytosed mycobacteria, and this may also be an adaptive mechanism to survival inside the macrophage.

Possible pathogenicity factors of *Mycobacterium tuberculosis*

1 Lipoarabinomannan (induces tumour necrosis factor and scavenges oxidizing molecules).
2 Catalase (scavenges oxidizing molecules).
3 Superoxide dismutase (scavenges oxidizing molecules).
4 Inhibitors of phagosomal–lysosomal fusion.
5 Heat-shock proteins.

Primary tuberculosis

This is the result of primary infection in a non-immune host. There is usually a small focus of inflammation, with a few mycobacteria surrounded by a dense granuloma. When this occurs in the lung it is called a Ghon focus. Regional lymph nodes are often enlarged, and the combination of primary granuloma and enlarged nodes is called a primary complex.

Primary tuberculosis is often clinically silent; indeed, naturally acquired immunity often follows healing of an inapparent primary complex. Healed lesions slowly calcify, leaving an irregular, X-ray-dense shadow which is easily identified.

Postprimary tuberculosis

This is a reinfection or reactivation illness in a person previously sensitized to mycobacterial antigens. It can follow primary infection or immunization after a few weeks to many years. The resulting immune reaction causes the formation of exuberant granulomata, often with central, cheesy necrosis, called caseation. In pulmonary infection, necrotic tissue is coughed away, leaving cavities. In solid organs or soft tissues, the caseous material resembles pus, and may discharge. The abscess-like lesion is indurated rather than hot and inflamed, and is therefore called a cold abscess.

In postprimary tuberculosis the immune system is suppressed and mycobacteria survive within macrophages. Numerous organisms are released from caseating tissues. Postprimary, cavitating pulmonary tuberculosis is highly infectious to susceptible individuals because of the large numbers of mycobacteria released during coughing.

Miliary or disseminated tuberculosis

This type of disease occurs when mycobacteria are disseminated via the bloodstream. It may follow rupture of an active granuloma into a blood vessel. Many organs are affected. The severity of disease varies from mild fever and malaise to severe, debilitating illness. Macroscopically detectable granulomata are visible in the organs as small white nodules rather like millet seeds (the origin of the expression miliary tuberculosis).

Other types of mycobacterial infection

In addition to the pathogens of tuberculosis in humans, environmental (atypical) mycobacteria are occasionally pathogenic. They tend to enter the soft tissues via small skin lesions or injection sites, causing a granuloma of the skin or a subcutaneous cold abscess.

ORGANISM LIST

Human organisms
- *Mycobacterium tuberculosis*
- *M. africanum*
- *M. bovis*
- BCG

Other mycobacteria capable of causing typical tuberculosis
- *M. kansasii*
- *M. avium-intracellulare complex*
- *M. xenopi*

Mycobacteria that can cause cold abscesses
- *M. fortuitum*
- *M. chelonei*
- *M. scrofulaceum*
- BCG (if mistakenly injected subcutaneously)

Mycobacteria associated with skin lesions
- *M. marinum*
- *M. ulcerans*

Organism of leprosy
- *M. leprae*

Primary tuberculosis

Introduction

Primary tuberculosis characteristically affects certain organs and systems. The commonest affected organ is the lung, but the pleura, lymph nodes, peritoneum, pericardium and meninges are other sites where primary disease occurs.

Clinical features

The clinical presentation is a combination of the common general features of tuberculosis and of signs and symptoms related to the affected site. There are fever, night sweats, anorexia and weight loss. There is no predictable change in the blood count, the liver function tests or other blood biochemistry. The erythrocyte sedimentation rate or C-reactive protein is often raised.

Erythema nodosum

Erythema nodosum sometimes accompanies primary tuberculosis, and is occasionally the first sign of the disease. Other associations with erythema nodosum, including sarcoidosis or streptococcal infections, are important differential diagnoses (see Chapter 5), but tuberculosis should always be actively excluded.

In rare cases, erythema nodosum develops early in the treatment of tuberculosis. This is similar to the situation in leprosy, where the condition is well recognized as a complication of treatment, occurring with a surge of cell-mediated immune reactivity.

Affected site

The clinical features relating to the affected site are often helpful in suggesting the diagnosis, because of both the typical symptoms and the typical type of lesion produced.

Primary pulmonary tuberculosis

Primary pulmonary tuberculosis, when it is clinically evident, often causes a persistent, dry cough. The site of the infection is usually the periphery of the midzone of one lung. The primary focus is too small to produce abnormal signs on physical examination, but is visible on chest X-ray as a fluffy, opacity with a diameter of 1–2 cm. The mediastinal lymph nodes on the affected side may be enlarged and, together with the primary focus, form the primary complex (Fig. 18.2).

Endobronchial tuberculosis

Endobronchial tuberculosis sometimes occurs as a primary infection in children. Granulomata develop in the bronchial mucosa of one of the larger airways, causing

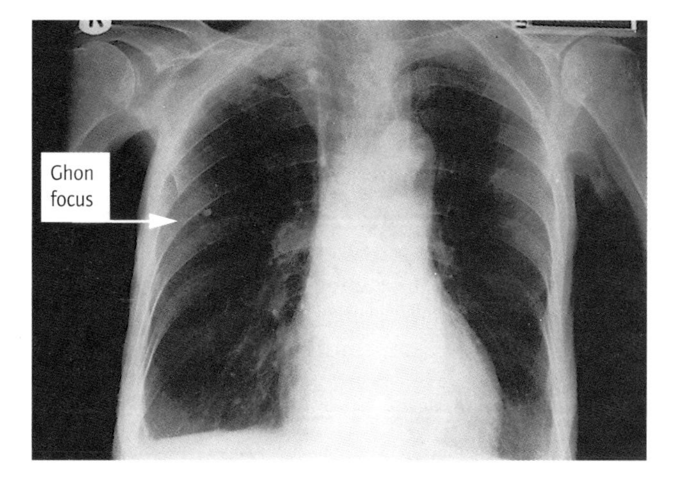

Fig. 18.2 Primary tuberculosis: Ghon focus on chest X-ray. The density of the lesion suggests healing and calcification.

partial obstruction. There is a persistent, wheezy cough and often a fixed wheeze on auscultation over the affected airway. The narrowed section of bronchus may be visible on chest X-ray, and can be demonstrated on computed tomographic (CT) or magnetic resonance (MR) scanning.

Pleural tuberculosis

Pleural tuberculosis is commonest in young adults. Pleuritic pain is the characteristic clinical feature, and a pleural rub is occasionally heard. Pleural effusion may cause shortness of breath and typical physical signs on chest examination; the pleural rub disappears as the pleural surfaces are separated by effusion. The chest X-ray shows a typical opacity with an upcurved surface (Figs 18.3 & 18.4).

Tuberculous pericarditis

The features of tuberculosis are combined with those of pericarditis (see Chapter 12; Fig. 18.5). In late cases the patient may present with heart failure due to tamponade, or with constrictive pericarditis.

Lymph-node tuberculosis

Lymph-node tuberculosis usually affects the cervical or mediastinal nodes. In children, caseating or suppurating lymphadenitis may be due to *M. avium-intracellulare* or, less commonly, *M. scrofulaceum*. The swelling may develop slowly or alarmingly suddenly, when lymphoma must be urgently excluded. Cold abscess formation is

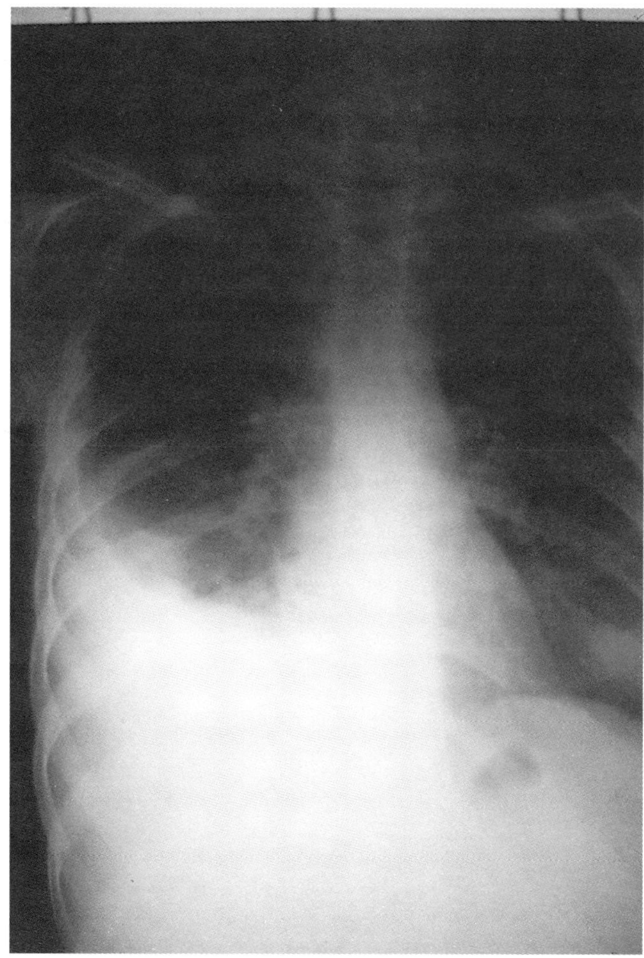

Fig. 18.3 Primary tuberculosis: pleural effusion in an Indian teenager with malaise, night sweats, weight loss and low-grade fever.

nowadays rare, as early diagnosis and treatment usually prevent it (Fig. 18.6).

Mediastinal lymph-node swelling may cause cough, due to extrinsic tracheal irritation. Dull, central chest pain is common. Physical signs are few, unless compression of an airway produces a fixed wheeze. The chest X-ray often shows a nodular shadow widening the mediastinum, with or without narrowing of a central airway (Fig. 18.7).

On rare occasions a swollen lymph node ruptures into an airway, protruding into it and obstructing it. Complete tracheal obstruction may be fatal, but urgent bronchoscopy with bypass or resection is sometimes possible.

Tuberculous peritonitis

Tuberculous peritonitis is most common in Asians. It presents with abdominal discomfort and distension.

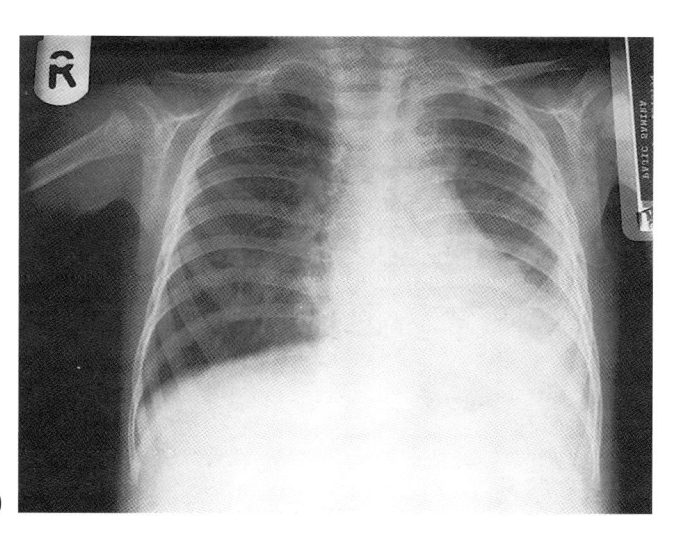

(a)

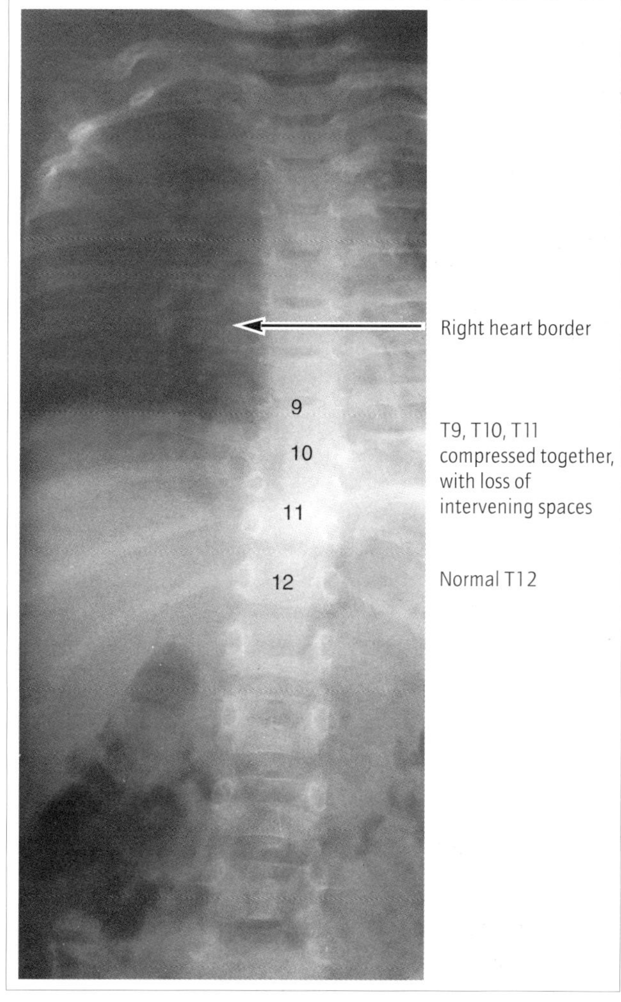

(b)

Fig. 18.4 This Bosnian refugee was born in internment and did not receive bacillus Calmette–Guérin (BCG) vaccination. At the age of 18 months she had persistent cough, fever and back pain, with evidence of: (a) left-sided pleural disease on chest X-ray; and (b) spinal osteomyelitis, with loss of two disc spaces and vertebral volume in the lower thoracic spine.

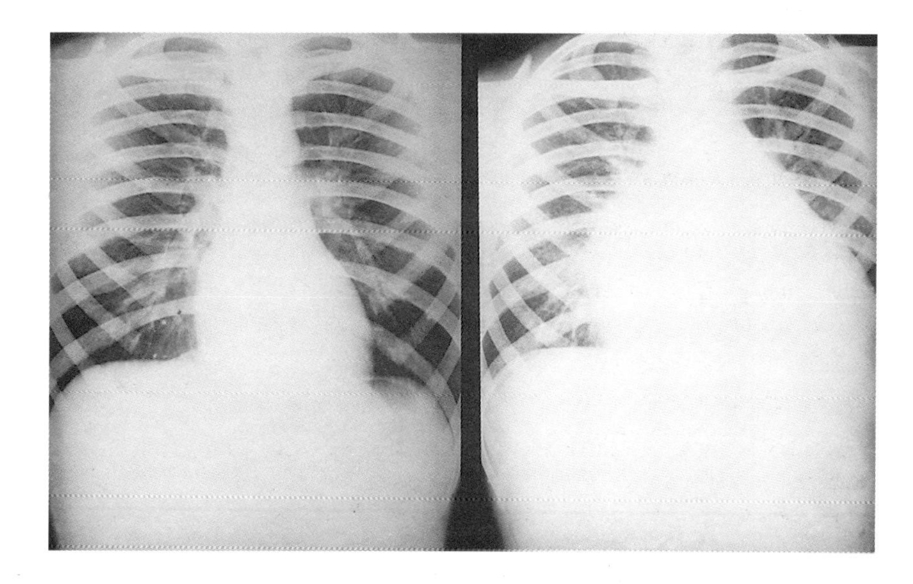

Fig. 18.5 Tuberculous pericarditis in a 57-year-old Pakistani businessman: he presented with several days' increasing chest pain, anorexia and fever. Echocardiography revealed the small pericardial effusion, which rapidly enlarged over the next 10 days.

Minor ascites is common, showing as separation of bowel loops on X-ray or as a fluid collection in the pelvis or peritoneal reflections on imaging. Massive ascites is extremely rare.

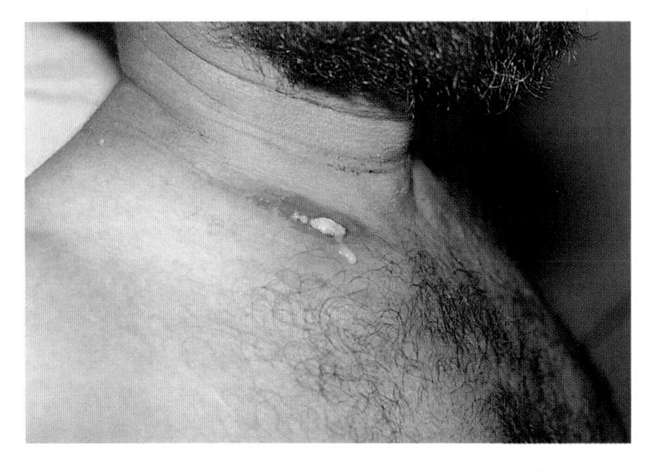

Fig. 18.6 Tuberculous lymphadenitis: this 42-year-old man had typical symptoms and a swollen cervical lymph node for 4 months. Chemotherapy did not prevent the formation of a draining sinus.

Diagnosis of tuberculosis

The diagnosis of tuberculosis is suggested by the typical general features, and localizing symptoms and signs. Confirmation of the diagnosis is not always easy, as there are relatively few mycobacteria in the primary lesion, and they are encased in a dense granuloma. Few cases of primary pulmonary tuberculosis have a positive sputum examination. Pleural, pericardial or ascitic fluids rarely yield positive smears or cultures. Only about 1 in 5 cases of tuberculous meningitis have positive cerebrospinal fluid (CSF) bacteriology. Bacteriological examination should always be performed, as it is diagnostic when positive. It is not reliable in excluding primary tuberculosis.

Tuberculin tests

The tuberculin test is an intradermal test for cell-mediated hypersensitivity to tuberculoprotein. Three methods of testing are available: (i) the Heaf test, which is widely used for screening (Fig. 18.8), for instance before BGC

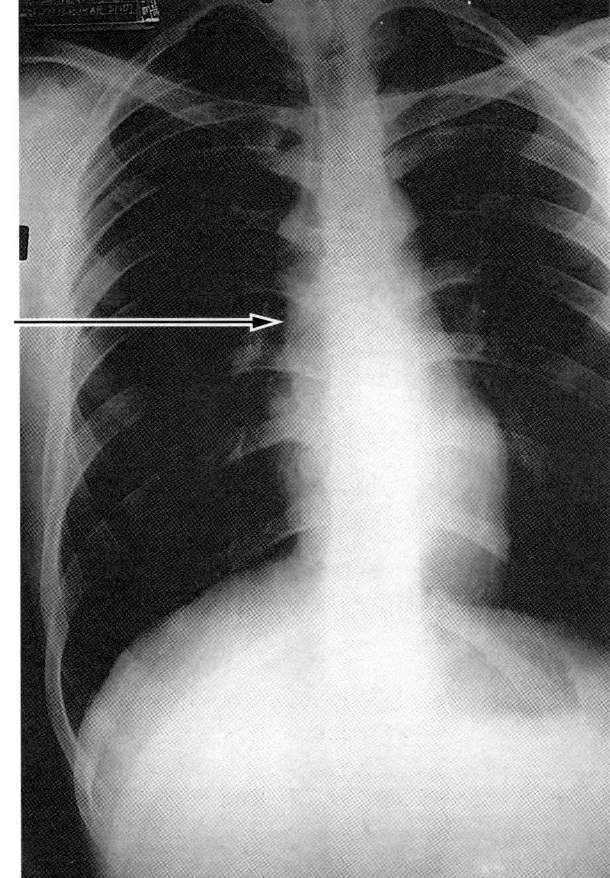

Fig. 18.7 Tuberculous lymphadenitis: this teenager presented with typical symptoms of tuberculosis, plus a cough and substernal chest pain. The swollen gland was compressing the trachea.

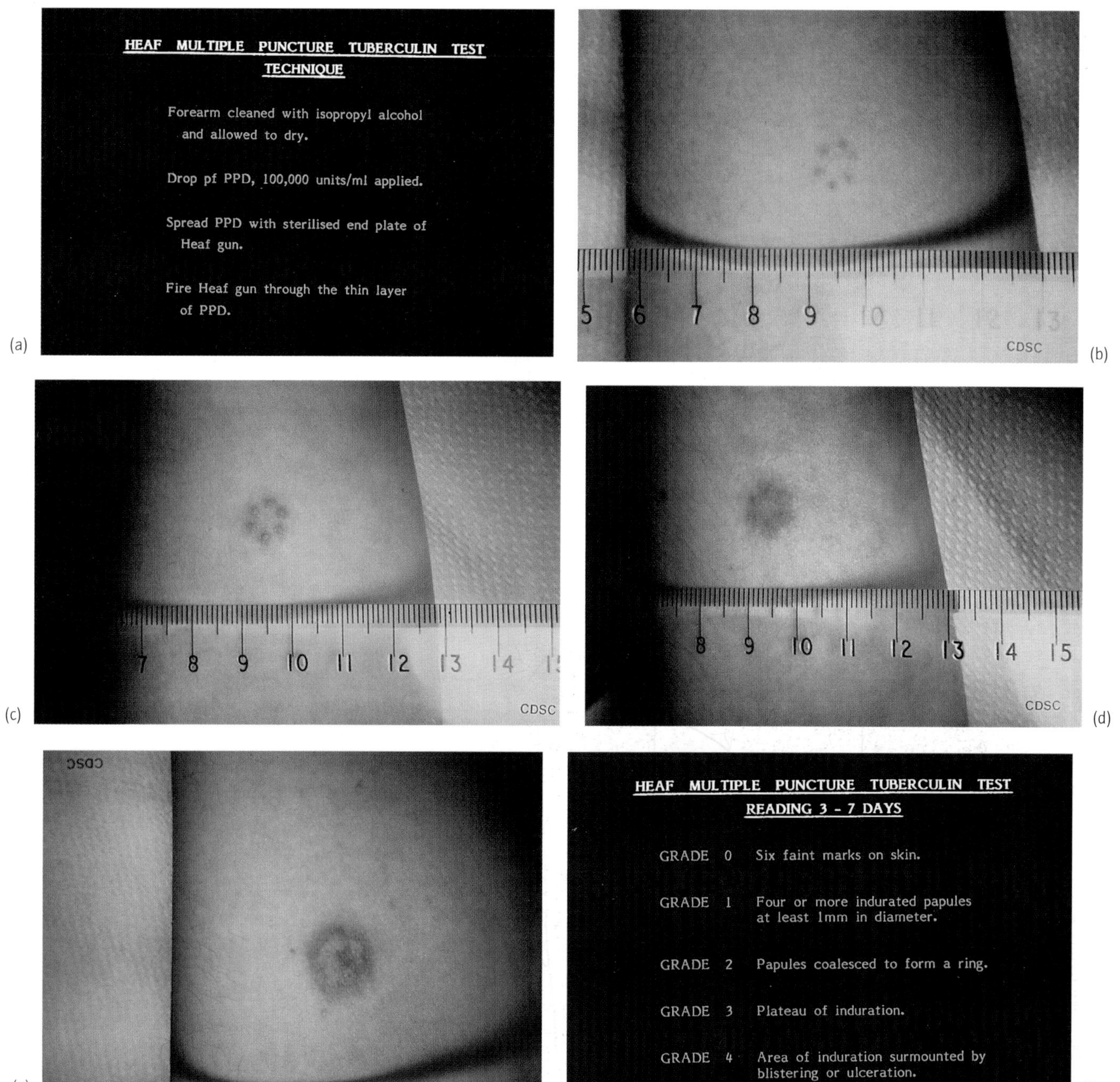

Fig. 18.8 Heaf test method and grading of results. (a) Summary of technique; (b) grade 0; (c) grade 1; (d) grade 2–3; (e) grade 4; and (f) description of grades.

immunization and in which concentrated purified protein derivative (PPD) is inoculated into the epidermis, using an automated Heaf gun; (ii) the tine test, which is similar to the Heaf test (this test is difficult to perform reproducibly, and is rarely used in Britain); and (iii) the Mantoux test, which is particularly applicable to individual cases, and can be performed in different dilutions.

The tuberculin test is negative in those who have never been infected with tuberculosis. It becomes positive 3–5 weeks after infection. A strongly positive tuberculin test is good evidence of active tuberculosis unless the diagnosis is disproved. A grade three or four Heaf or tine test, a strongly positive (greater than 15 mm) 1 : 1000 (10 tuberculin unit) Mantoux test, or a positive (greater than 6 mm induration) 1:10000 (1 tuberculin unit) Mantoux test can all be taken as indicating tuberculosis (Fig. 18.9).

Strongly positive tuberculin tests can be found in individuals repeatedly exposed to tuberculoprotein or mycobacteria. Nurses and doctors in chest or infectious diseases departments, overseas aid workers or previously immunized individuals recently exposed to a case of tuberculosis may show such reactions. If the positive individual is well and has a normal chest X-ray, treatment is probably not indicated, unless the positivity is related to a recent family contact (see section on prevention and control, below).

The tuberculin test may be misleadingly negative if performed too soon in the course of the disease, as it may not have had time to convert to a positive response. This can occur in erythema nodosum or tuberculous meningitis, which can both present soon after exposure to infection. It is worth repeating the test after 1 or 2 more weeks in cases where the disease is strongly suspected.

A debilitated patient or one with overwhelming tuberculous infection may have a falsely negative tuberculin test because of suppression of cell-mediated responses. This is not such a common problem in primary tuberculosis, which is rarely severely debilitating, as in postprimary disease (see below). Improvement in the patient's general condition or 7–10 days' antituberculosis treatment often restores the ability to mount a response.

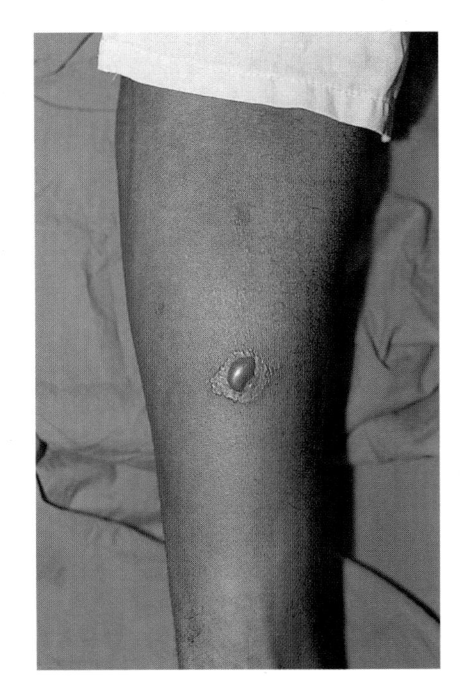

Fig. 18.9 Strongly positive Mantoux test: this patient had a short history and high fever; the main reaction is 18 mm in diameter. (0.1 ml of fluid containing 1 or 10 units of purified protein derivative is injected intradermally: the diameter of the resulting induration is measured 48–72 h later.)

Histology and cytology as diagnostic tools

Infected tissue such as pleura or lymph node is often accessible to needle biopsy or fine-needle aspiration, which may permit demonstration of acid-fast bacilli (Fig. 18.10). Excision biopsy of cervical lymph nodes is a straightforward procedure, which contributes to cure.

Histological examination of biopsy material shows granulomata. In the absence of caseation, a firm diagnosis of tuberculosis cannot be made, and sarcoidosis or other granulomatous conditions must be excluded by other means. Tissue sections can be stained to demonstrate acid–alcohol-fast bacilli, which are diagnostic if present. Culture of unfixed tissue should always be carried out, as this increases diagnostic yield by about 50%. It is important not to place all of the tissue obtained into formalin or other fixative, as this prevents its use for culture.

Microbiological diagnosis

Specimens

Sputum is the most important specimen for the diagnosis of pulmonary tuberculosis. Early morning specimens are preferable. In some patients excretion of bacilli may be intermittent or scanty; other patients find it impossible to produce a satisfactory specimen. In these circumstances bronchoalveolar lavage is indicated, but morning aspiration of the gastric contents (to recover swallowed organisms) is a low-cost alternative that may yield a positive diagnosis. Unfortunately, microscopic examination of

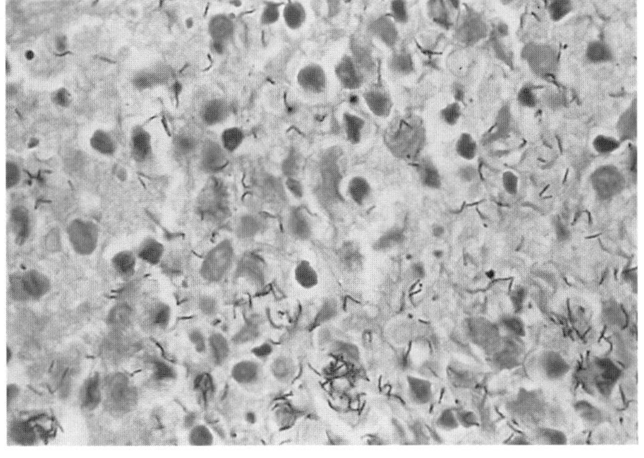

Fig. 18.10 Ziehl–Nielsen-stained material obtained from a caseating mediastinal lymph node. Many acid–alcohol-fast bacteria are seen, with the typical 'cording' or clustering appearance of *Mycobacterium tuberculosis*.

gastric aspirate can give false positive results, so these specimens are only suitable for culture.

Cerebrospinal fluid is required for the diagnosis of meningitis. Other naturally sterile fluids may be examined, but the diagnostic yield is often disappointing: culture and histological examination of a pleural biopsy increases this yield in pleural tuberculosis. Early morning urine specimens can be used for the diagnosis of renal tuberculosis. Twenty-four-hour collections are less useful because of frequent contamination with other bacteria. Pus may be submitted for acid-fast staining and mycobacterial culture if tuberculosis is suspected. Blood culture has become more important in the diagnosis of mycobacterial infections following the HIV epidemic. *Mycobacterium avium-intracellulare* infection can readily be demonstrated in blood culture. In these patients faecal smear and culture can also yield a positive diagnosis, as intestinal infection is present before dissemination takes place.

In the laboratory, specimens which are normally sterile can be processed without decontamination, on non-selective media, while those which possess a normal bacterial flora require decontamination before inoculation. Cerebrospinal fluid, pus and blood do not require decontamination but sputum and faeces do. Urine, which may sometimes be contaminated, should be examined by Gram stain and, if non-mycobacterial organisms are seen, is subject to a decontamination procedure (see below).

Specimens that may be examined for mycobacterial infection

Body fluids

1 Sputum (D,S,C).
2 Gastric aspirate (C).
3 Effusion fluids (S,C).
4 Early morning urine specimens (S,C).
5 Cerebrospinal fluid (S,C).
6 Pus (D,S,C).
7 Blood (C).
8 Faeces (D,S,C).
9 Fine-needle lymph-node aspirate (S,C).
10 Bronchoalveolar lavage (BAL) specimen (S,C).

Tissues

1 Lymph-node biopsy (H,S,C).
2 Liver biopsy (H,S,C).
3 Pleural biopsy (H,S,C).
4 Uterine curettings (H,S,C).
5 Biopsy of affected skin (H,S,C).
6 Bone marrow (H, S, C).

(Key: C, culture; D, decontamination; H, histology; S, smear and acid-fast staining.)

Microscopy

The examination of sputum smears is central to the diagnosis of pulmonary tuberculosis. Early morning samples are collected, and if there is delay in processing these should be refrigerated to prevent overgrowth of other bacteria. Since sputum is contaminated with mouth flora, it must be decontaminated before it can be used for culture. Several specimens from the same patient can be pooled and decontaminated using 4% sodium hydroxide. This acts by killing the other bacteria present in the specimen. This effect is not absolutely specific, so care is necessary if false negatives are not to result.

The treated specimen is centrifuged, and the deposit used for microscopical examination and inoculation of culture medium (see below). Slides can be stained in two ways: by a modification of the Ziehl–Nielsen method, in which hot carbol-fuchsin is used to stain the mycobacteria; or the phenol auramine technique, which uses a fluorescent dye easily visible under ultraviolet illumination. Auramine staining enables large numbers of specimens to be screened quickly using a UV microscope. All positive slides are then overstained by the Ziehl–Nielsen method so that the identity of the fluorescent objects can be confirmed. This technique is particularly suited to laboratories with a large throughput. In smaller laboratories, or where an ultraviolet microscope is not available, Ziehl–Nielsen is the method of choice.

The lower limit of detection by stained smear is approximately 10^4 CFU/ml. Excretion of bacilli can be intermittent and, thus, a negative result does not exclude the diagnosis. The importance of microscopical diagnosis of tuberculosis cannot be overemphasized. Not only does it provide a rapid and relatively sensitive diagnostic technique, but it also identifies patients who are excreting large numbers of organisms and are therefore most infectious.

Culture of mycobacteria

All but a few species of mycobacteria are slow-growing. This means that other bacteria present in the specimen would rapidly overgrow if not adequately suppressed. Specimens from sterile sites can be inoculated directly on to isolation medium but those from sites with a normal bacterial flora (e.g. sputum) require decontamination. Growth of contaminating species is further inhibited by the incorporation of dyes, for instance malachite green in Löwenstein–Jensen medium, or antibiotics, as in Kirchner's selective medium. Culture must be performed in screw-capped containers to prevent the release of infectious organisms, and to prevent desiccation of the culture medium (Fig. 18.11).

Fig. 18.11 Culture of mycobacteria on Löwenstein–Jensen medium. Results at 3 weeks show the typical, breadcrumb-like growth of *Mycobacterium tuberculosis* in the middle bottle (left is *M. fortuitum*, right is *M. kansasii*).

Many different media are used in the isolation of mycobacteria. All contain a source of fatty acids; fresh eggs in Löwenstein–Jensen, or purified oleic acid in Middlebrook's medium. Liquid media increase the diagnostic yield because they permit inoculation of a larger amount of specimen. The optimal approach is to use a combination of solid and liquid media. A positive diagnosis is made when colonies grow on solid medium, or liquid medium becomes cloudy. Colonies of *Mycobacterium tuberculosis* on solid media are usually rough and a buff colour, and the smear often shows cording (see below), but these characteristics are not sufficiently typical to allow a presumptive diagnosis.

Automated methods

More recently, the diagnosis of mycobacterial infection has been improved by the introduction of radiometric detection of mycobacterial growth. This technique utilizes a Middlebrook broth which incorporates ^{14}C palmitate. This is metabolized by the organism to produce $^{14}CO_2$, which is detected by the machine. More recent innovations have overcome the need for radioactivity and metabolic changes induced by growth are detected by changes in a fluorogenic substrate in the medium. The most modern systems also allow continuous monitoring of the culture bottles. This has reduced the time taken to detect a positive isolate from about 21 days to approximately 10 days.

Identification

Organisms isolated in this way must be shown to be mycobacteria on the basis of acid-fast staining by Ziehl–Nielsen's method. Acid-fast bacteria are reported as '*Mycobacterium* sp. isolated'. The organisms are then subcultured on to identification and sensitivity media. Identification is often performed in two stages—screening and definitive. The first stage uses screening tests which will identify different members of the *M. tuberculosis* group by their microscopic morphology (cording), ability to grow on medium incorporating paranitrobenzoic acid, and their pyrazinamide sensitivity or resistance.

Definitive diagnosis is based on biochemical tests, ability to grow at various temperatures, and ability to produce coloured pigments in the presence or absence of light.

Molecular techniques

Mycobacterial DNA can be detected in clinical specimens, using the polymerase chain reaction (PCR). This technique has a similar sensitivity to that of culture, but is much quicker, giving a result in 24–48 h.

Typing mycobacteria

Several techniques are now available to type *M. tuberculosis* using analysis of restriction fragment length polymorphisms with several different probes. These methods are very valuable, not only in worldwide epidemiological studies, but in episodes of transmission within the hospital, laboratory or community.

Sensitivity testing

Resistance determinants in mycobacteria are not located on transmissible plasmids, phages or transposons, as in other bacteria, but arise by spontaneous mutation. Mutation to rifampicin resistance occurs once in 10^8 cell divisions. In a patient with pulmonary tuberculosis there are approximately 10^{13} mycobacteria, making resistance inevitable. In combination chemotherapy the likelihood of resistance developing to all of the agents in a combination is multiplied (i.e. 10^8 rifampicin $\times 10^6$ ethambutol and 10^6 pyrazinamide $= 10^{20}$, i.e. the risk is once in 10^{20} divisions). Generally, if 1–10% of a patient's mycobacterial population is resistant to an antimicrobial, laboratory tests will indicate resistance to that drug. By using combinations of antituberculosis agents, the risk that an organism resistant to all of the prescribed agents will develop is very low. If patients do not receive appropriate medication, it is possible that mycobacteria will be exposed to single agents, increasing the risk of induction of resistance.

Resistance is described as primary if it arises in patients never previously treated for tuberculosis. It implies that there has been transmission of a resistant strain in the community. Secondary resistance arises during therapy and indicates poor compliance or an inadequate prescription. Multiple drug resistance is defined as the presence of at least rifampicin and isoniazid resistance.

Resistance to rifampicin is caused by mutations in a particular area (resistance hot-spot) of the beta-subunit of the RNA polymerase gene *rpoB*. Resistance to isoniazid can arise via deletion of or mutation in the catalase *katG* gene, or mutation in the *inhA* or *aphC* genes. Studies of isoniazid-resistant populations of *M. tuberculosis* suggest that there are other mechanisms still to be discovered. Although the mechanism of action of pyrazinamide is uncertain, resistance is probably associated with mutation in the pyrazinamidase gene *pncA*. Resistance mechanisms have also been determined for streptomycin, fluoroquinolones and ethambutol.

In countries where antituberculosis therapy is closely controlled, wild strains are usually sensitive to all agents. Where there is little or no control, patients may take only one drug (or one effective drug) for long periods of time, and resistance is more common. A high cost of medical care, poor advice and the choice of inadequate regimens contribute to this. Extreme poverty may deny treatment to many. Also, patients must understand the need to continue therapy after they return to apparently normal health. Where drugs are provided free by government schemes, they may be sold by patients to supplement inadequate incomes.

Testing for antimicrobial resistance

The resistance ratio method

This is the method most commonly employed in UK reference centres. It uses Löwenstein–Jensen medium into which are incorporated antituberculosis drugs in differing dilutions. Essentially, a minimum inhibitory concentration (MIC) is obtained, of the test organism and of a range of simultaneously tested control strains. An organism with an MIC below that of the controls, by a factor of at least four, is considered sensitive.

The proportion method

In the proportion method a culture is inoculated on to plates containing test antibiotic or no antibiotic. A strain is considered resistant to a test antibiotic if the proportion of bacteria growing in the presence of drug, compared to the non-drug control, is greater than 1%.

Growth detection methods

All of the current automated mycobacterial culture methods can be used to perform rapid susceptibility testing, giving results in about 10 days. The rate of growth of the test organism is compared in the presence and absence of the antituberculosis agent.

Postprimary tuberculosis

Introduction

Postprimary tuberculosis produces expanding granulomata which readily caseate, releasing numerous mycobacteria. The tissues affected are different from those involved in primary disease. The commonest site affected is the apex of the lung. Less common are kidneys, bones and joints, the male and female genital tracts and the bowel. In contrast to primary tuberculosis, spontaneous remission is uncommon. Progressive tissue destruction occurs, causing increasing debility and, eventually, death.

Pulmonary tuberculosis

Introduction and epidemiology

Pulmonary tuberculosis is one of the most important epidemic respiratory infections worldwide. Approximately 75% of all infections due to *M. tuberculosis* present as pulmonary tuberculosis.

Pathology

The usual cause is *M. tuberculosis* but, as discussed above, other mycobacteria can cause pulmonary disease, often in people with debility or immunosuppression. Pulmonary tuberculosis may exist alone or coexist with postprimary tuberculosis in other sites.

Clinical features

The typical quartet of fever, night sweats, anorexia and weight loss is usually prominent. Cough is usual and there is often sputum, which may appear purulent but is rarely copious. A few patients have episodes of blood-streaking in the sputum and the occasional case expectorates fresh blood or blood clot. A few patients have no sputum, and children rarely expectorate their sputum, as it is immediately swallowed.

Physical examination may be surprisingly uninformative. Most patients with pulmonary tuberculosis have few abnormal signs. Crepitations over the affected apex or dullness to percussion and perhaps bronchial breathing are the commonest findings.

The chest X-ray often shows typical changes which are virtually diagnostic. A ragged-edged opacity is seen in the apex of the lung, occasionally of both lungs. The opacity contains small and large lucencies which represent cavities. Often one or two particularly large and thick-walled cavities are present (Fig. 18.12).

The erythrocyte sedimentation rate and C-reactive protein may be elevated, but are not always abnormal. As debility and weight loss progress, the blood albumin falls and the anaemia of chronic inflammation often develops. On rare occasions bizarre haematological abnormalities occur, including thrombocythaemia, leukaemoid reactions or pancytopenia. The pathogenesis of these abnormalities is poorly understood.

Diagnosis

The diagnosis is often evident from the history and the chest X-ray appearances. Confirmation is obtainable from examination of sputum, brochoalveolar lavage or early morning gastric aspirate. In early disease and in non-debilitated patients the tuberculin test is often strongly positive. However, in debilitated patients and those with advanced disease, cell-mediated responses are depressed. The test is then negative but usually becomes strongly positive within a week or two of commencing effective therapy.

While 70–75% of all reported tuberculosis cases have pulmonary disease, of those with non-pulmonary disease about 30% also have lung lesions. Thus, approximately 10% of pulmonary tuberculosis cases have coexisting non-pulmonary disease.

Pulmonary tuberculosis often responds readily to appropriate chemotherapy, but disease in other sites may be much slower to improve. For instance, a patient with pulmonary and articular disease may have an improving chest X-ray and a deteriorating joint (see Fig. 14.10). Similarly, a patient with pulmonary and renal involvement may have completely healed lung lesions, but persisting low-grade infection in a kidney. It is therefore important to examine cases of pulmonary tuberculosis for evidence of extrapulmonary and disseminated disease. At the least, a series of urine examinations and a full physical examination should be carried out.

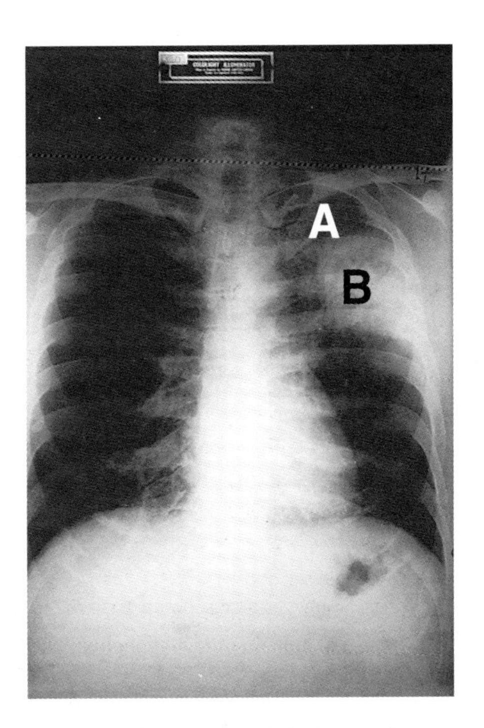

Fig. 18.12 Postprimary pulmonary tuberculosis. Sputum microscopy and culture were both positive. A, apical opacity; B, large thick-walled cavity.

Complications

Before treatment

Infected caseous material or sputum may overflow into other parts of the bronchial tree, causing tuberculous bronchopneumonia. This causes multinodular segmental or lobar opacities on chest X-ray and significant impairment of respiratory function.

Occasionally a cavity will erode a pulmonary blood vessel, causing haemoptysis. The appearance of blood-streaked sputum or fresh blood clots is alarming. Bed rest, lying on the side of the lesion to minimize drainage into the opposite lung, and a check of the platelet count are advisable. Many cases resolve, especially if anti-tuberculosis treatment is in progress. A few cases have severe bleeding, requiring transfusion, or even emergency lobectomy, but catastrophic and fatal haemorrhage is very rare.

Patients with severe, extensive pulmonary disease sometimes develop tuberculous laryngitis, with harsh cough and hoarseness. Such patients are extremely infectious. However, they usually respond rapidly to antimicrobial chemotherapy, though some scarring and hoarseness may remain after healing.

After treatment

After the start of treatment, there is a sudden release of immune function from the suppressive effect of the disease. This is most intense in patients who are severely debilitated or cachectic when treatment begins. The resulting surge of inflammation can cause high, swingeing fever, anaemia, a rapid rise in the erythrocyte sedimentation rate and a rapidly falling albumin, leading to hypotension and peripheral oedema. Such patients may need circulatory support; whole blood is better in such cases than plasma substitutes. Oral prednisolone or intravenous hydrocortisone will reduce the inflammatory responses. The dose should be tailored to achieve stability of blood pressure and body weight, and then gradually tailed off as the patient gains condition over 3–6 weeks. Attention to nutrition, with adequate protein, vitamin and mineral intake, assists in recovery.

In patients with both pulmonary and extrapulmonary disease, care should be taken to ensure that both have been adequately treated before chemotherapy is discontinued (see above).

Tuberculous meningitis

Introduction

At least 200 cases per year of tuberculous meningitis are reported in the UK. Meningitis is a relatively common manifestation of primary tuberculosis in children and young adults, in whom it can develop within 3–4 weeks of exposure. In adults and the elderly it is more often associated with extensive tuberculosis elsewhere in the body. It is a diagnostic opportunistic condition in AIDS, and is particularly seen in cases originating in Africa.

The disease is important because the diagnosis is often difficult, and early CSF changes are similar to those of viral meningitis. Delayed diagnosis or treatment can lead to severe complications which may be irreversible.

Clinical features

The disease can present as fever of unknown origin, personality disorder or meningitis of slow onset, or with a neurological complication. Headache may not be prominent, but mild or minimal meningism is often demonstrable by careful examination. Fever is almost always significant. Older patients may have signs of pulmonary tuberculosis, or of disease elsewhere. Not all patients complain of the classic quartet of fever, night sweats, anorexia and weight loss.

Haematological and biochemical examination of the blood is often normal, though tuberculosis can produce unpredictable changes in the white cell count. If the meningitis is part of miliary or systemic disease, the liver function tests may be mildly abnormal. The erythrocyte sedimentation rate is not predictably altered.

The CSF appears clear in tuberculous meningitis, even though it contains excess white cells. From 10 to 2000 lymphocytes per cubic millimetre may be found, sometimes more. The CSF protein is elevated, sometimes so much that it forms a 'spidery' white clot if the fluid is allowed to stand. The glucose is low, and may be undetectable in advanced cases.

Diagnosis

The most important factor in diagnosis is suspicion. Examination of the CSF is mandatory. In many cases imaging of the brain is indicated before lumbar puncture. This may show oedema or hyperaemia of the meninges, but is often normal in early cases.

Mild CSF changes, with slightly raised protein, a few excess lymphocytes and a slightly low glucose level, are difficult to distinguish from the changes of viral meningitis. A careful search should therefore be made for acid-fast bacilli, and cultures for mycobacteria should be set up. However, only 20–25% of cases have organisms identified in the CSF. Polymerase chain amplification of mycobacterial DNA in CSF samples is very important in making an early diagnosis and has been shown to be more sensitive than culture in some studies.

A strongly positive tuberculin test is good evidence of the diagnosis. A negative test is unhelpful (see above). Childhood tuberculous meningitis sometimes presents before the tuberculin test has converted to positivity. It is always worthwhile repeating the test after 1–2 weeks, or after a few days' trial of therapy when suppression of cell-mediated immunity has ceased.

Management

Treatment should be commenced without awaiting bacteriological confirmation. It is usual to commence quadruple therapy, including isoniazid, rifampicin, pyrazinamide and ethambutol. Streptomycin does not cross the blood–brain barrier, and is not a first-line drug in tuberculous meningitis.

The granulomatous inflammation of tuberculosis easily causes vascular and neurological lesions, especially if fibrosis occurs. Corticosteroids should therefore be given for the first 2–4 weeks of treatment. A typical dose would be prednisolone 40 mg daily for the first 7–14 days, and then tailing off as fever resolves. Corticosteroids may also aid improvement if neurological abnormalities are present when treatment is started. As long as effective antituberculosis treatment is given concurrently they have no adverse effect on the tuberculosis.

Trials of short-course chemotherapy have not been completed for tuberculous meningitis. Treatment is usually continued for at least a year, depending on the speed and completeness of response.

Complications

Tuberculomata

Space-occupying granulomata may already exist when the patient presents. As treatment improves cell-mediated immunity, they may enlarge and cause neurological signs. Treatment with dexamethasone or prednisolone often reduces inflammation and swelling, and should be given until the tuberculomata have responded to chemotherapy. On rare occasions they fail to respond, and require neurosurgical intervention.

Focal lesions

Focal spinal cord or brain lesions can be caused directly by granulomata or by granulomatous compression of blood vessels. Cranial nerve lesions, paraparesis and cauda equina syndromes are the commonest problems. Cauda equina lesions are often due to multiple granulomata, and tend to present with a mixture of leg weakness and bladder dysfunction.

Hydrocephalus

Hydrocephalus may occur as a result of extensive fibrosis or as part of a poorly controlled granulomatous process. In either case, corticosteroids offer hope of prevention and improvement. Temporary or permanent CSF drainage may be needed, depending on the response to further therapy.

Renal tuberculosis

Introduction

Tuberculosis of the renal tract is assumed to be blood-borne in origin. The kidney is almost always involved; granulomatous lesions of the collecting system, the ureters, the vas deferens and epididymis occur when mycobacteria descend from the kidney.

Initially, a destructive granuloma of the medulla affects one or more pyramids. Swelling deforms the adjacent renal calyx, and destruction of the pyramid may cause a bulbous enlargement of the calyx. If the disease is untreated, the kidney is slowly replaced by a tuberculous abscess. If the ureter becomes blocked by seedling granulomata, back pressure and hydronephrosis complicate the infectious process.

Clinical features

The clinical findings are initially minimal. Dull flank pain on the affected side is common in established disease. Low-grade or swinging fever is also common. Many cases have classic symptoms of urinary tract infection and some have superimposed bacterial infections, as deformity and partial ureteric obstruction predispose to bacterial colonization. Some cases come to light because they present with a complicating epididymitis.

A significant proportion of patients with renal tuberculosis have coexisting pulmonary disease. At least a chest X-ray should therefore be carried out, and sputum, bronchoalveolar lavage or gastric aspirates examined.

Diagnosis

The diagnosis is based on the demonstration of typical renal lesions and culture of mycobacteria from urine. The specimen of choice is the whole of an early morning voiding of urine, in which acid-fast bacilli will have been concentrated overnight.

Imaging of the kidneys may show typical distortion of the calyces, expansion of a kidney by the inflammatory process, or areas of calcification related to partial healing. It is advisable to perform imaging which can show ureteric obstruction, as early relief of obstruction will avoid effective loss of a kidney.

Tuberculous epididymitis

This presents as a progressive, indurated swelling of the affected epididymis. The inflammation is subacute, with dusky red swelling and relatively little pain. Both sides may be affected, often asymmetrically. The diagnosis is usually made by demonstrating the associated renal infection. If this proves difficult, biopsy with histology and culture will confirm the aetiology and exclude malignancy.

Tuberculosis of the female pelvis

Tuberculosis of the female genital tract usually causes subacute or chronic salpingitis. The mucosa of the tube is affected, and granulomata develop in the endometrium, many of them being shed during the menses. Progressive distortion and eventual occlusion of the tubes lead to infertility, which is often the only manifestation of the disease.

Clinical features are few in most cases. They include minimal vaginal discharge, mild to moderate suprapubic discomfort, or dyspareunia in the presence of a large cold abscess. Occasionally, the condition presents subacutely with suprapubic pain and a moderate fever.

The diagnosis may be suggested by an ultrasound appearance of tubal swelling. Laparoscopy permits inspection and biopsy of the tubes. Premenstrual endometrium is obtained by dilatation and curettage for histological and microbiological examination. Typical granulomata or acid-fast bacilli may be seen in the tissue. Mycobacteria can often be recovered by appropriate cultural methods.

Tuberculosis of bones and joints

Introduction

Tuberculosis of bones and joints originates via blood-borne infection. The joints most often affected are the spine and the hip. The knee and the wrist are less common sites, and involvement of other joints is uncommon. Synovitis without joint involvement is occasionally seen, especially affecting the extensor tendons of the wrist and hand.

The granulomatous process begins in the cartilage of the joint, and spreads by a process of caseation and destruction into adjacent bone. The capsule of a joint may rupture, allowing a cold abscess to extend, sometimes erupting at the skin as a sinus. An example of this is the appearance of an abscess in the groin resulting from infection in the lumbar spine, with pus tracking along the psoas from its spinal origin (see Fig. 20.6). Destruction of bone surfaces and cortex can lead to severe deformity of the joint. The classic example of this is Pott's disease of the spine, in which infection originates in a disc and spreads to the two adjacent vertebrae (see Fig. 18.4b). Collapse of the vertebral bodies produces an angular kyphosis at the level of the infection.

Clinical presentation

The clinical presentation is often with fever and pain. Soft-tissue swelling of the synovium and capsule follows, and effusion develops. Clinical examination will reveal these signs in limb joints, but in the spine and sometimes the hip, MR imaging may be required to demonstrate soft-tissue changes and features of inflammation. X-ray changes occur late, and considerable disease must exist before they are detectable. The X-ray changes are indistinguishable from those of pyogenic chronic osteomyelitis.

Diagnosis

Diagnosis depends on suspicion and specific tests. Joint effusions usually have a high protein content and a predominantly lymphocytic pleiocytosis. Synovial swellings may produce an exudate containing soft masses of inflammatory material, which look similar to melon seeds.

Demonstration of acid-fast bacilli or culture of mycobacteria may be possible from aspirated effusion or synovial biopsy. In advanced disease, bone biopsy may yield positive results.

Problems in treatment

There is often a considerable delay between the start of treatment and the resolution of inflammation and pain. Sinus formation and bone destruction may progress for some weeks after treatment is commenced.

Surgical drainage and debridement may contribute to cure in extensive disease (see Fig. 14.10). There is little evidence that bed rest or splinting of joints affects the rate of healing or degree of final deformity. However, both may be useful in limiting pain in the early stages of treatment.

Miliary or disseminated tuberculosis

In miliary tuberculosis blood-borne dissemination of mycobacteria gives rise to small granulomata in many organs. They are visible radiographically in the chest X-ray and pathologically on the surfaces and cut sections of the solid organs. In rare cases they can be seen in the retina on fundoscopy.

The diagnosis of miliary tuberculosis may be suggested by the chest X-ray appearances (Fig. 18.13). There is rarely significant sputum production, and sputum examination is usually negative. There is often a modest elevation of liver alkaline phosphatase levels in the blood, because of the many space-occupying granulomata in the liver. The white cell count and erythrocyte sedimentation rate are not predictably abnormal.

Liver biopsy material may show small granulomata. These are not always caseating, and acid-fast bacilli are not always demonstrable on Ziehl–Nielsen or auramine staining. Some of the material should be saved unfixed for culture, as this may yield a diagnostic growth of mycobacteria, allowing speciation and sensitivity testing.

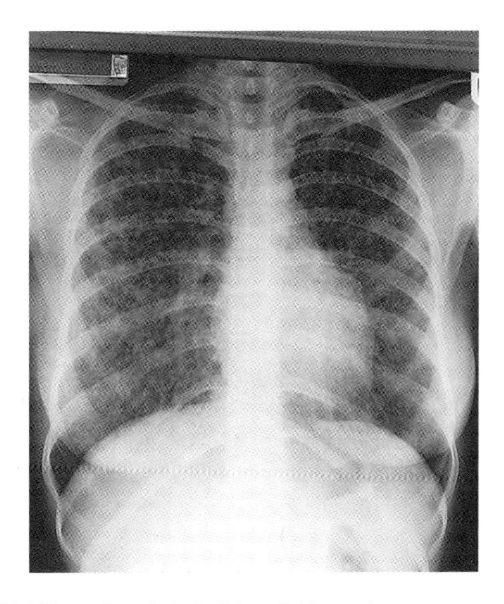

Fig. 18.13 Miliary tuberculosis. Both lung fields contain numerous small, ill-defined, round opacities (the solid organs are similarly affected).

Disseminated tuberculosis may present simply as a fever, often with weight loss, with or without an abnormal erythrocyte sedimentation rate, white cell count or liver function tests. Biopsy of reticuloendothelial tissue does not reveal granulomata. The tuberculin test may be strongly positive, but is negative in about 40% of cases. This condition, known as cryptogenic miliary tuberculosis, must be suspected on epidemiological grounds or by exclusion.

Treatment of tuberculosis

Introduction

The mainstay of treatment in tuberculosis is effective multidrug antimicrobial chemotherapy. In the great majority of cases this will produce cure of the disease with a negligible chance of relapse. There are two phases of treatment.

Intensive phase (2 months)

For the first 2 months of treatment, three first-line drugs are given, of which two should be rifampicin and isoniazid, and the third should be pyrazinamide. Ethambutol may be added as a fourth drug if there is no contraindication. This regimen aims to deliver at least two or three effective drugs pending the results of sensitivity testing. It greatly reduces the load of mycobacteria by using drugs with a range of actions to attack rapidly replicating intracellular and extracellular mycobacteria. It also initiates therapy against more slowly metabolizing organisms.

Continuation phase (4–7 months)

In this phase, rifampicin and isoniazid are continued and the additional drugs are stopped. In most Western countries the sensitivities of the patient's mycobacterium will be known before the beginning of the continuation phase. Provided that pyrazinamide has been used throughout the intensive phase, and the organism is sensitive to it, the continuation phase should be continued for 4 months. If the third drug was ethambutol, or the organism was pyrazinamide resistant, continuation therapy should be for 7 months. Most patients therefore receive a 6- or 9-month course of treatment, and for a sensitive organism this is curative in close to 99% of cases. Continuing follow-up after appropriate treatment of fully sensitive tuberculosis is therefore rarely indicated.

Approximately 5% of patients receiving rifampicin plus isoniazid have to stop one of these drugs because of unwanted effects. If unwanted effects or resistance of the organism prevent the use of rifampicin or isoniazid, then other drugs must be substituted. Rifampicin plus ethambutol can be given for 9 months, with regular monitoring for ethambutol-related ocular toxicity (see below). If rifampicin cannot be given, treatment with three alternative drugs must usually be continued for 18 months.

The drugs must be taken as a continuous course, to avoid the emergence of resistance, with its risk of failed therapy or relapse. In many countries this is encouraged by directly observed therapy (DOT), using rifampicin-containing short-course regimens (DOTS; directly observed, short-course).

'Intermittent' regimens

Several different intermittent dosing regimens have been tested. These have used drugs daily for 2 months and then two or three times weekly for 4 months, and drugs three times weekly for the whole treatment period. All of these regimens appear equally effective.

Drug-resistant tuberculosis

This is tuberculosis resistant to either rifampicin or isonazid. The rates of these resistances vary in different countries; in indigenous cases in Britain, isoniazid resistance occurs in about 6% of cases, and rifampicin resistance in about 2%.

Multidrug-resistant tuberculosis (MDRTB)

This is tuberculosis caused by *M. tuberculosis* resistant to rifampicin and isoniazid, with or without resistance to other antimycobacterial drugs. Approximately 2% of previously untreated patients in Britain have MDRTB, but much higher rates exist in some other countries. The risk of MDRTB is increased in the following circumstances:

1 A past history of TB, especially if incompletely or intermittently treated.
2 Birth or exposure in higher-risk areas (e.g. Asia, Africa, southern and eastern Europe, former USSR, Latin America).
3 Contact of a person with MDRTB.
4 Disease unresponsive to first-line therapy: fever continuing after 2 weeks, sputum smear positive after 2 months or sputum culture positive after 3 months.

MDRTB is not more infectious than sensitive disease, but is more difficult to treat and more likely to relapse. Patients with positive sputum smears should avoid close association with new contacts, especially immunosuppressed individuals. When in hospital they should be nursed in a negative-pressure isolation room. For initial therapy, ideally at least five drugs should be given of which three have not been taken before; an optimum regimen of three or four drugs can be chosen for continuation when sensitivities have been confirmed. At least 18–24 months treatment will be needed.

Support and encouragement is needed if these demanding drug regimens are to be followed continuously for the required duration. DOT is particularly helpful in this situation. Further resistances may emerge during therapy; regular sputum cultures should therefore be monitored throughout therapy and for at least 5 years after stopping drugs. Some patients never achieve permanently negative sputum cultures, but may stabilize their clinical condition. They must be educated about the risks their organism presents to their contacts and warned of the danger to immunosuppressed contacts.

Mycobactericidal drugs used in tuberculosis (first-line drugs)

Rifampicin

Daily dose: 600 mg (450 mg if < 50 kg), child 10 mg/kg; intermittent dose: 600–900 mg three times weekly, child 15 mg/kg three times weekly.

This is the most important drug in the treatment of tuberculosis, as it kills slowly replicating mycobacteria throughout the course of treatment. Its inclusion therefore permits short-course chemotherapy of 6–9 months' duration. It is well absorbed by mouth and widely distributed in the body. It penetrates moderately well into the CSF.

The unwanted effects of rifampicin include gastrointestinal distress and a red-orange discoloration of urine, tears and other body fluids. Soft contact lenses will also become discoloured. Urticarial rashes or itching are rare effects.

The most important adverse effect is hepatocellular liver damage. There is always a temporary elevation of transaminases in the blood when antituberculosis treatment is commenced, but this should peak by 3–4 weeks, and slowly subside thereafter. Levels should not climb to more than three or four times the upper limit of normal. In patients with pre-existing hepatic impairment the rifampicin dose should not exceed 8 mg/kg daily.

In some cases of miliary tuberculosis, an inflammatory reaction in early treatment produces liver enzyme abnormalities unrelated to drug toxicity. It is worth temporarily withdrawing antimicrobial drugs or giving a few days' course of prednisolone. When liver function tests have returned to normal specific treatment may be cautiously reintroduced or corticosteroids stopped.

Rifampicin occasionally causes a viral-like syndrome of fever, myalgia and anorexia, particularly in patients on intermittent or interrupted therapy. This does not improve as therapy continues, and the drug must usually be withdrawn.

Rifampicin is a powerful inducer of hepatic enzymes. It increases the metabolism of the contraceptive pill, sulphonylureas and corticosteroids. Patients using oral contraception may be advised to take two pills daily instead of one, but barrier methods are probably more reliable during rifampicin therapy.

Rifabutin is a closely related drug, useful in some infections due to 'atypical' mycobacteria. Up to 30% of patients develop uveitis during long-term therapy. This may be controllable with anti-inflammatory treatment.

Isoniazid

Daily dose: 300 mg, child 5 mg/kg; intermittent dose: for adult or child, 15 mg/kg three times weekly.

This drug is highly effective in killing rapidly replicating mycobacteria, allowing early dramatic reduction in mycobacterial load. It is given with rifampicin for the whole duration of chemotherapy, and contributes to the effectiveness of short-course treatments. It penetrates well into the CSF.

Unwanted effects include nausea and vomiting, hypersensitivity rashes and occasional cerebral disturbance or convulsions.

It can produce hepatocellular damage, especially in middle-aged and elderly patients.

Peripheral neuritis is an important side-effect which occurs in slow acetylators of the drug. The neuritis is painful and disabling, improving only slowly and often incompletely on withdrawal of isoniazid. This side-effect can be completely avoided by giving pyridoxine supplements to patients taking isoniazid. The usual pyridoxine dosage is 10 mg/day.

Ethambutol

Daily dose: for adult and child, 15 mg/kg; intermittent dose: for adult and child, 30 mg/kg three times weekly or 45 mg/kg twice weekly.

This is a slightly less powerful drug than rifampicin and isoniazid. It is useful in combination with the two main drugs for the intensive phase of therapy. It can also be used as one of two drugs for continuation therapy, but the duration of treatment must be longer than with rifampicin plus isoniazid continuation therapy.

Ethambutol can cause optic neuritis, loss of red-green colour discrimination and visual impairment. These effects are particularly likely in the elderly and in patients with renal impairment. Prolonged or high dosage also carries the risk of ocular effects. The patient notices loss of acuity and colour appreciation, both of which are reversible if the drug is promptly discontinued. Regular enquiry for visual symptoms and testing of colour vision before and during treatment are advisable. The drug should not be given to preschool children or others who cannot understand and report visual defects.

Pyrazinamide

Daily dose: 2.0 g (1.5 g if < 50 kg); intermittent dose three times weekly: 2.5 g (2.0 g if < 50 kg), child 50 mg/kg; twice weekly: 3.5 g (3.0 g if < 50 kg), child 75 mg/kg.

This is a bactericidal drug which is well absorbed and enters the CSF particularly well. It is highly effective against replicating intracellular organisms, but not against slowly metabolizing organisms later in the course of treatment. It is a useful addition to initial therapy, especially in the treatment of tuberculous meningitis, but is less useful after the first 2–3 months.

Pyrazinamide can be hepatotoxic, and also produces rashes, including urticaria. On occasions it can precipitate acute gout.

M. bovis is resistant to pyrazinamide.

Streptomycin

This is an effective drug whose usefulness is limited by ototoxicity, vertigo and nephrotoxicity, as well as the necessity for intramuscular administration. It is most useful as part of the intensive phase of treatment. The usual dose of 1 g daily should be reduced in those over age 40, in small patients and in those with renal impairment. Streptomycin can be given three times weekly once the sputum is smear-negative. It is highly advisable to check pre- and postdose streptomycin levels at 2-weekly intervals.

The total dose given should not exceed 100 g, above which toxicity becomes much more likely.

Additional and second-line drugs

Amikacin

This aminoglycoside may be effective when streptomycin is not. It must be given intramuscularly, and has the same side-effects as other aminoglycosides, so should not be given in combination with them. Like streptomycin, it can be given in a daily or a three times weekly regimen.

Capreomycin

This aminoglycoside is only used in tuberculosis. The principles of its use are the same as for amikacin. As well as ototoxicity and vertigo, it can cause renal impairment, hepatotoxicity and skin reactions.

Prothionamide

This is a bacteriostatic drug, which is useful in resistant *M. tuberculosis* infections. It penetrates well into the CSF. Side-effects are mainly those of gastrointestinal distress, but there is a long list of rare side-effects, including rashes, blood dyscrasias and liver dysfunction.

Clarithromycin

This macrolide drug is particularly useful in treating *M. avium-intracellulare* infections, but may also be effective against *M. tuberculosis*. It is a broad-spectrum antibiotic, which can predispose to *Candida* infections. Its use is limited by nausea in some patients.

Ciprofloxacin

This broad-spectrum antibiotic has proved effective against several types of mycobacteria, and it is well tolerated by most patients.

Para-aminosalicylic acid (PAS)

This is a mycobacteriostatic drug, well absorbed by mouth, but which must be given in voluminous doses, usually as numerous gelatin capsules three or four times daily. Its main problem is its frequent association with gastritis and dyspepsia, but it can also cause hepatotoxicity and occasional rashes.

Cycloserine

This is a rather toxic mycobacteriostatic drug, which can be used in combination with other antimycobacterial agents. It has unpleasant side-effects, including headache, dizziness, depression, agitation, convulsions and allergic rashes. It tends to be a drug of last choice.

Non-tuberculous mycobacterioses

Introduction

These conditions are caused by accidental acquisition of environmental *Mycobacterium* species, commonly resulting in skin infections or inoculation abscesses. More extensive disease, or mycobacteraemic disease, can occur in immunosuppressed individuals (see Chapter 22).

Mycobacterium chelonei

This organism is found in water, and can inhabit inadequately cleaned and chlorinated hydrotherapy pools It is most commonly seen in cold abscesses at injection sites, particularly in insulin-dependent diabetics, who have many injections. Outbreaks of bacteraemia have occurred in immunosuppressed patients.

The infection can be treated with amikacin. Attempts at excision or drainage often result in extension or recurrence. Additional or alternative drugs include clarithromycin, cefoxitin, doxycycline and clofazimine.

Mycobacterium fortuitum

This organism also causes inoculation abscesses. It is a rare cause of bone and joint disease, especially in the diabetic foot. It is often sensitive to a wide range of antimicrobial agents, including amikacin, co-trimoxazole, clarithromycin, cefoxitin and ciprofloxacin.

Mycobacterium marinum

This organism is found in river and pond water, and also affects domestic fish tanks. It infects epidermal abrasions and typically causes 'swimming pool granuloma', an indolent, granulomatous lesion on the dorsum of the hand or the finger. Untreated lesions sometimes spread to involve subcutaneous tissue, including fascia and tendons.

It is sensitive to a number of drugs, including rifampicin, ethambutol, ciprofloxacin, co-trimoxazole, streptomycin and doxycycline.

Mycobacterium ulcerans

This is the causative organism of tropical ulcer. It is an environmental organism, probably inoculated into the skin of the leg or foot by spiky vegetation. The resulting lesion is an expanding, undermining ulcer whose true extent is much larger than the visibly broken skin. Most cases respond slowly to multidrug treatment, but debridement and wound hygiene are important in managing extensive disease.

Prevention and control of tuberculosis

Introduction

The most effective interventions for the control of tuberculosis are those that improve living conditions. The decline in tuberculosis that occurred during the latter half of the 19th century and the early part of the 20th century preceded other control measures and is attributed to improvements in housing conditions, nutrition and social deprivation (see Chapter 26).

Reductions in infection due to *M. bovis* have been achieved in many countries by a combination of testing and treating cattle and pasteurization of milk.

Bacillus Calmette–Guérin vaccine

BCG vaccine contains a live attenuated strain derived from *M. bovis*. It is given as a single intradermal dose. The vaccine is contraindicated in patients with immunosuppression, including asymptomatic HIV-positive individuals. A local reaction develops at the immunization site within 2–6 weeks, beginning as a small papule which increases in size; it may ulcerate and gradually heals, leaving a small scar.

Estimates of protection by BCG have varied in different field trials, from zero in one study in India, to 90%. Many studies, including earlier studies in the UK, have shown protection of approximately 70%, lasting for at least 20 years. Some of the variable results may be caused by a gradual change in the genome of BCG over the years, a phenomenon which has only recently been recognized.

Policies for the use of BCG vary considerably between countries. In developing countries, where infection in young children is common, the vaccine is routinely administered at birth. Many developed countries only offer the vaccine to groups at particular risk of tuberculosis. These high-risk groups may include contacts of cases with respiratory tuberculosis, healthcare workers, teachers and immigrants from developing countries. In the USA, the number of groups recommended for BCG is very limited and control rests on case detection and contact tracing (see below). In the UK BCG is given routinely to all tuberculin-negative school children at 10–13 years of age.

Tuberculin testing

BCG vaccine can safely be given without prior tuberculin sensitivity testing to infants up to 3 months of age. In older infants, children and adults, a tuberculin test should be performed first.

A negative tuberculin test indicates that the individual has not previously been infected or received BCG vaccination; such individuals can be given BCG. A weakly positive test indicates past infection or previous vaccination and BCG is not required. A strongly positive reaction may indicate active disease; such individuals should be referred for further investigation.

Management of the case and close contacts

Most patients with pulmonary tuberculosis can be treated at home and need not be separated from other household members, provided chemoprophylaxis is given to young children in the household (see below). Where hospital admission is required because of severe disease or for social reasons, the patient should be nursed in a single room until no longer infectious. With modern antimicrobial therapy and fully sensitive organisms, this is usually achieved within 2 weeks, even though some bacilli may still be seen in sputum smears. Patients who are sputum-negative or with non-pulmonary disease can be nursed in a general ward.

Close contacts of sputum-positive cases should be tuberculin tested and have a chest X-ray. Close contacts are defined as household members, and classroom contacts (if the index case is a teacher or a school child). Where the tuberculin test is negative, it should be repeated 2–3 months after the last exposure to determine whether tuberculin conversion has occurred. Chemoprophylaxis is indicated for the following contacts.

1 Children under 16 with a strongly positive tuberculin test, irrespective of BCG vaccination status.

2 Adults with a strongly positive tuberculin test and no previous BCG vaccination.

3 Adults who have tuberculin converted.

4 Young (<35 years) Asian adults with a strongly positive tuberculin test.

Isoniazid is the drug of choice for chemoprophylaxis. It is usually given for 6 months. Isoniazid plus rifampicin can be given for 3 months. Longer-term or continuing chemoprophylaxis may be indicated for HIV-positive contacts.

BCG vaccine should be given to unvaccinated contacts under 35 years of age who remain tuberculin-negative.

Drug-resistant tuberculosis

There have recently been large outbreaks of drug-resistant tuberculosis in refugees, and 'down and outs', and some smaller hospital outbreaks among immunosuppressed patients.

Control of the situation is gained by energetic case finding, adequate isolation of cases in hospital (see Chapter 23), close supervision of drug taking (directly observed therapy in the patient's own environment) and diligent contact tracing and follow-up.

Screening of immigrants

Screening is indicated for all immigrants from countries where tuberculosis is common, such as the Indian subcontinent. The aims are to detect active disease, identify infected individuals who may require chemoprophylaxis, and identify unvaccinated individuals who may require BCG.

Leprosy

Introduction

Leprosy is an indolent disease, mainly affecting the skin, nerves and mucosa, but also capable of infecting the eye, muscles and testicles. It is caused by *M. leprae*, which has low infectivity and is extremely slow-growing, with approximately one replication per fortnight. It has never been cultured in artificial media, but will grow slowly in some animals.

Spread is by close contact, usually among families and particularly from individuals with extensive multibacillary mucosal lesions.

Its importance is that it produces peripheral nerve lesions, leading to paralysis and anaesthesia of limbs, trophic ulcers and Charcot joints. In untreated cases there is no means of preventing these effects, which can produce devastating deformities, including autoamputation.

Clinical features and grading of disease

The clinical features depend on whether the sufferer is sensitized to the organism. However, the first presentation is often rather mild, and is called indeterminate leprosy. An ill-defined area of skin gradually loses some of its pigmentation, and becomes hypoaesthetic. The face or hand is often affected, and the patient often presents because of a cut or burn of the site, precipitated by diminished sensation

Tuberculoid disease

Tuberculoid (TT) disease occurs in those who mount a strong cell-mediated response to the infection. It is characterized by one or two areas of hypopigmented skin, and localized, asymmetrical inflammation and thickening of peripheral nerves. Nerve thickening may be visible or palpable in the ulnar nerve at the elbow, the accessory nerve in the posterior triangle of the neck or the peroneal nerve at the knee. Biopsy of affected skin shows typical granulomata, with palisades of epithelioid cells, surrounding collections of Langhans-type giant cells and active lymphocytes. Bacteria are rarely, if ever, seen in these lesions, which are termed paucibacillary.

Lepromatous disease

Lepromatous (LL) disease is associated with an absent cell-mediated response. There is intense oedema of the affected tissues, which contain no granulomata and no lymphocytes. Only undifferentiated macrophages are seen, and these are packed with acid-fast bacilli. There is mycobacteraemia, with spread to distant areas of the skin, nerves, nasal and pharyngeal mucosa, eye, muscles, testicles and reticuloendothelial tissues such as the spleen, liver and bone marrow (especially that of the phalanges). This is multibacillary disease.

Disease is expressed as multiple, often symmetrical lesions. Skin lesions are nodular, and not hypoaesthetic because the small nerves are oedematous rather than inflamed. The nasal mucosa is affected early, and the discharge from it is highly infectious. Facial and lip swelling, often with a collapsed bridge of the nose, causes a typical (leonine) facies. Nerve trunks are swollen, the Schwann cells become packed with bacilli and proliferate, adding to physiological and anatomical disruption. There is gradual loss of sensation, starting with small fibre functions and eventually affecting all modalities. Involvement of the eye can cause conjunctivitis or keratitis; muscle involvement causes weakness of small muscles, including the smooth muscles of the skin and superficial blood vessels; bone involvement leads to loss of alveolar bone from the jaw, of the nasal septum and of phalangeal joint surfaces.

Borderline leprosy

Borderline (BB) leprosy is an intermediate form in which lesions contain a lymphocytic infiltrate and macrophages evolve to epithelioid cells, but no giant cells are seen and bacilli survive within the epithelioid cells. There is a continuous spectrum of intermediate forms between tuberculoid, the intermediate borderline and the extreme lepromatous disease (TT–BT–BB–BL–LL). These are distinguished by histological features, by the degree of localization of skin, nerve and other lesions, and by the number of bacilli (on a scale of 1–6) detectable in lesions.

Borderline disease is unstable, and a small fluctuation in immune response or bacterial activity can cause a shift either way along the spectrum.

Diagnosis

The three aspects of diagnosis comprise:

1 Physical examination for typical skin lesions, anaesthesia in lesions or nerve distributions, and nerve thickening.

2 Biopsy of atypical or indeterminate skin lesions (this is also helpful for staging the disease).

3 Examination of split-skin smears for acid-fast bacilli (a small cut is made through the epidermis, without drawing blood, and the blade is then used to scrape a little tissue fluid from the exposed dermal tissue and transfer it to a microscope slide, where it is allowed to dry before staining). Nose-blowings are useful specimens, containing many bacilli in lepromatous disease. PCR-based diagnostic methods have been described.

Treatment

Multidrug treatment is now recommended worldwide. The first-line drugs are rifampicin, dapsone and clofazimine. Rifampicin reduces the bacillary load in a few days, while dapsone kills residual organisms. Clofazimine is as effective as dapsone, and also has an intrinsic anti-inflammatory effect, which lessens the likelihood of reactive conditions complicating treatment. Occasionally prothionamide, minocycline, ofloxacin or clarithromycin are used when other drugs are contraindicated.

The optimal regimen depends on the bacillary load, and for this purpose, leprosy cases are divided into paucibacillary (indeterminate, TT and BT cases), and multibacillary (BB, BL and LL cases—those with multiple skin or nerve lesions, and those with bacilli in their split-skin smears).

Paucibacillary cases are given rifampicin 600 mg monthly plus dapsone 100 mg daily for 6 months.

Multibacillary cases are given rifampicin 600 mg and clofazimine 300 mg monthly plus dapsone 100 mg and clofazimine 50 mg daily for 2 years.

Clofazimine has the disadvantage that it causes first reddening of the skin, viscera and body fluids, and eventually a grey skin colour. The three-drug regimen is adequate, even for dapsone-resistant organisms, and prevents the emergence of further resistance. Prothionamide is a useful second-line drug, used if clofazimine is refused or if adverse reactions occur to other drugs.

Patients are followed up 6-monthly after completing treatment. Paucibacillary patients are examined for new lesions until disease free for 2 years. Multibacillary patients also have split-skin smears, and treatment is continued if these are positive. Follow-up of these patients continues for 5 years without disease.

Complications

Several types of immunological change can occur in leprosy; all are more common after the start of treatment:

1 Changes in the staging of the disease, either downgrading towards the lepromatous end of the spectrum or upgrading towards the tuberculoid end.

2 Type 1 reactions, which represent increasing sensitivity to bacterial antigens, are associated with swelling of both skin and nerve lesions, with a risk of rapidly developing paralysis or anaesthesia (especially in the relatively unstable BB stage).

3 Type 2 reactions, affecting many LL and some BL stages, with the development of nodular, inflamed lesions of erythema nodosum leprosum (ENL) on the face and limbs, and sometimes inflammation of the uveal tract, fingers, peripheral nerves and testicles.

In all cases multiple drug treatment should be continued. Mild type 1 reactions which do not threaten nerve function will often respond to standard doses of aspirin, or to chloroquine 150 mg three times daily. These drugs can be combined for their additive effect. Type 2 reactions respond well to thalidomide in diminishing doses from 400 mg daily to 50 mg daily (but great care must be taken to avoid using this in a pregnant woman). The lowest dose can be given for some months if necessary. Severe reactions of either type may threaten nerve or eye function, or cause severe inflammation and fever. These reactions are treated additionally with prednisolone 40–80 mg daily, with a slow reduction in the dose.

Many patients need physiotherapy, orthopaedic shoes or supportive limb braces to overcome established nerve lesions, and to protect anaesthetic extremities from injury. In some, tendon transplants may improve hand or wrist function. Emergency surgery to open nerve sheaths or surrounding fascia can limit damage caused by severe inflammatory swelling.

Prevention and control

This is based on case finding, often by local health workers, on treating infectious cases who are the reservoir of infection; and on educating the public about the curability of the disease and the needlessness of suffering nerve lesions and disfigurement.

19 Bacteraemia and Sepsis

Introduction, 361

Sepsis and sepsis syndromes, 361
The acute (or adult) respiratory distress syndrome (ARDS), 362
The pathogenesis of sepsis syndromes, 362

Epidemiology of bacteraemia and sepsis, 364
Staphylococci and streptococci, 364
Enterobacteriaceae, 364
Anaerobic and aerobic opportunists, 364
Community-acquired bacteraemias, 364

Defences of the blood, 365
Phagocytes, 365
Alternative complement pathway, 365
Iron binding, 365
Spleen, 365
Antibodies, 365

Pathological accompaniments of bacteraemia, 367
Pathology at the site of origin, 367
Effects of exotoxins, 368
Effects of inflammatory responses, 368
Metastatic infections, 368

Diagnosis of bacteraemia, 368
Clinical diagnosis, 368
Laboratory diagnosis, 369

Management of bacteraemia and sepsis, 369
Antibiotic treatment, 369
General management of bacteraemic patients: systems and life support, 369

Staphylococcal septicaemia, 374
Introduction and epidemiology, 374
Pathology, 374
Clinical features, 374
Diagnosis, 374
Management, 374

Streptococcal septicaemia, 375
Introduction and epidemiology, 375
Pathology, 375
Clinical features, 375
Diagnosis, 376
Management, 376
Complications, 376

Gram-negative septicaemia, 376
Introduction and epidemiology, 376
Clinical features, 376
Diagnosis, 377
Management, 377
Complications, 378

Follow-up of patients with bacteraemia, 378

Introduction

Bacteraemia is the condition in which bacteria circulate in the blood. Analogous pathology can occur with blood-borne fungi (e.g. *Candida* spp.), viruses (e.g. dengue) or parasites (e.g. *Plasmodium falciparum*). Such infections are not always associated with fever or illness. Repeated cultures of blood show that bacteria are detectable for 12–48 h after mild tissue trauma, such as dental extractions, and endoscopic examination of the bladder or biliary tree. Transient, self-limiting bacteraemia is also recognized in *Salmonella* and listerial infections.

Conditions where transient bacteraemia is common, and usually clinically inapparent
1 Dental treatment.
2 Endoscopic procedures.
3 Urinary tract infections.
4 Bowel infections.

These bacteraemias and their damaging effects are controlled and terminated by the natural bactericidal and anti-inflammatory properties of the blood cells and vascular endothelium (see below). When the physiological balance between the effects of infection and the body's defensive systems is overcome, systemic effects of infection develop, and cause the features of sepsis.

Sepsis and sepsis syndromes

The definitions of sepsis and sepsis syndromes have been unified considerably by the recommendations of both European and American consensus groups. They define steps in a continuum of increasingly severe illness (Table 19.1). Sepsis is usually associated with infection, but the same sequence of events can follow non-infectious insults, such as trauma, burns, haemorrhage, pancreatitis and injury by chemical toxins. In the absence of infection some experts call the response the systemic inflammatory response syndrome (SIRS), and the severe form is often called the multiorgan dysfunction syndrome (MODS).

Clinical disorder	Definition
Sepsis	Clinical evidence of infection plus evidence of a systemic response manifested by two or more of: Temperature > 38°C or < 36°C Heart rate > 90 beats/min Respiratory rate > 20 breaths/min or $P\text{CO}_2$ < 4.3 kPa (<32 mmHg) White cell count > 12 × 10^9/l or < 4 × 10^9/l (or > 10% immature neutrophils)
Severe sepsis	Sepsis associated with organ dysfunction: Hypotension Lactic acidosis Oliguria Confusion Hepatic dysfunction
Septic shock	Severe sepsis with hypotension despite adequate fluid resuscitation (Refractory septic shock is present if hypotension persists for > 1 h despite fluid resuscitation and pharmacological support)
Hypotension	Defined as: a systolic blood pressure of < 90 mmHg or a fall of > 40 mmHg from baseline in the absence of other causes of hypotension

Table 19.1 Definitions of sepsis-related conditions.

The acute (or adult) respiratory distress syndrome (ARDS)

The lung is particularly susceptible to damage in sepsis, and can suffer a syndrome of non-hydrostatic pulmonary oedema which may cause respiratory failure and threaten life. Its usual definition is:

1 A condition with acute onset (usually 2–4 days after the initiating insult).

2 Bilateral interstitial infiltrates seen on chest X-ray.

3 An arterial $P\text{O}_2$ of <8 kPa (regardless of PEEP, positive end-expiratory pressure).

4 A pulmonary artery occlusion pressure of <18 mmHg.

At the onset of ARDS, the excess interstitial fluid is a pure exudate, but this may progress to inflammatory alveolar exudate (containing neutrophils) and then to a fibroproliferative picture. Mild cases can resolve at an early stage.

The pathogenesis of sepsis syndromes

This is a complex interaction of the immune response, priming and activation of neutrophils, and endothelial activation. Many contributory factors are now known, but the mechanisms of their responses are poorly understood.

Initiating factors

The most familiar of these is endotoxin (lipopolysaccharide, LPS). It is a component of Gram-negative bacterial cell walls, which is taken in and immunologically presented by macrophages, which become activated. In the blood it is bound by lipopolysaccharide-binding protein (LBP), and may then be cleared, or react with the CD14 ligand on macrophages, again activating them. Another protein, called bacterial permeability-increasing protein (BPI) is highly homologous with LBP, and also binds and inactivates LPS. Antigens other than LPS can probably also interact with macrophage CD14. Superantigens can bind T lymphocytes with macrophages which are not presenting processed antigen, through the macrophage V-beta receptor (see Chapter 1).

The role of macrophages

Activated macrophages secrete tumour necrosis factor-alpha (TNF-alpha) and interleukin 1 (IL-1), leading to many responses such as activation of T cells, endothelial activation and 'priming' of neutrophils, which makes them rigid and adhesive, immobilizing them in the microcirculation. Activated macrophages also secrete IL-8, a chemokine powerfully attractive to neutrophils; it causes primed neutrophils to release superoxide radicals and lysosymes, which are intensely inflammatory and further damage the endothelium to which they are bound (Fig. 19.1).

The role of TNF-alpha

TNF-alpha alone can induce a sepsis response, and is probably a final signal in many of the processes of sepsis. As well as its immunological effects, it causes upregulation of intercellular adhesion molecules (ICAMs) which react with leucocyte beta-2 integrins to allow endothelial attachment of neutrophils and macrophages.

The role of neutrophils

Endothelium-bound neutrophils which have been primed and activated produce many substances which seriously damage the underlying endothelium. Agranulocytic individuals are not susceptible to severe sepsis syndromes or to ARDS. Early evidence suggests that antibody preparations binding neutrophil elastase can protect against ARDS or reduce its severity.

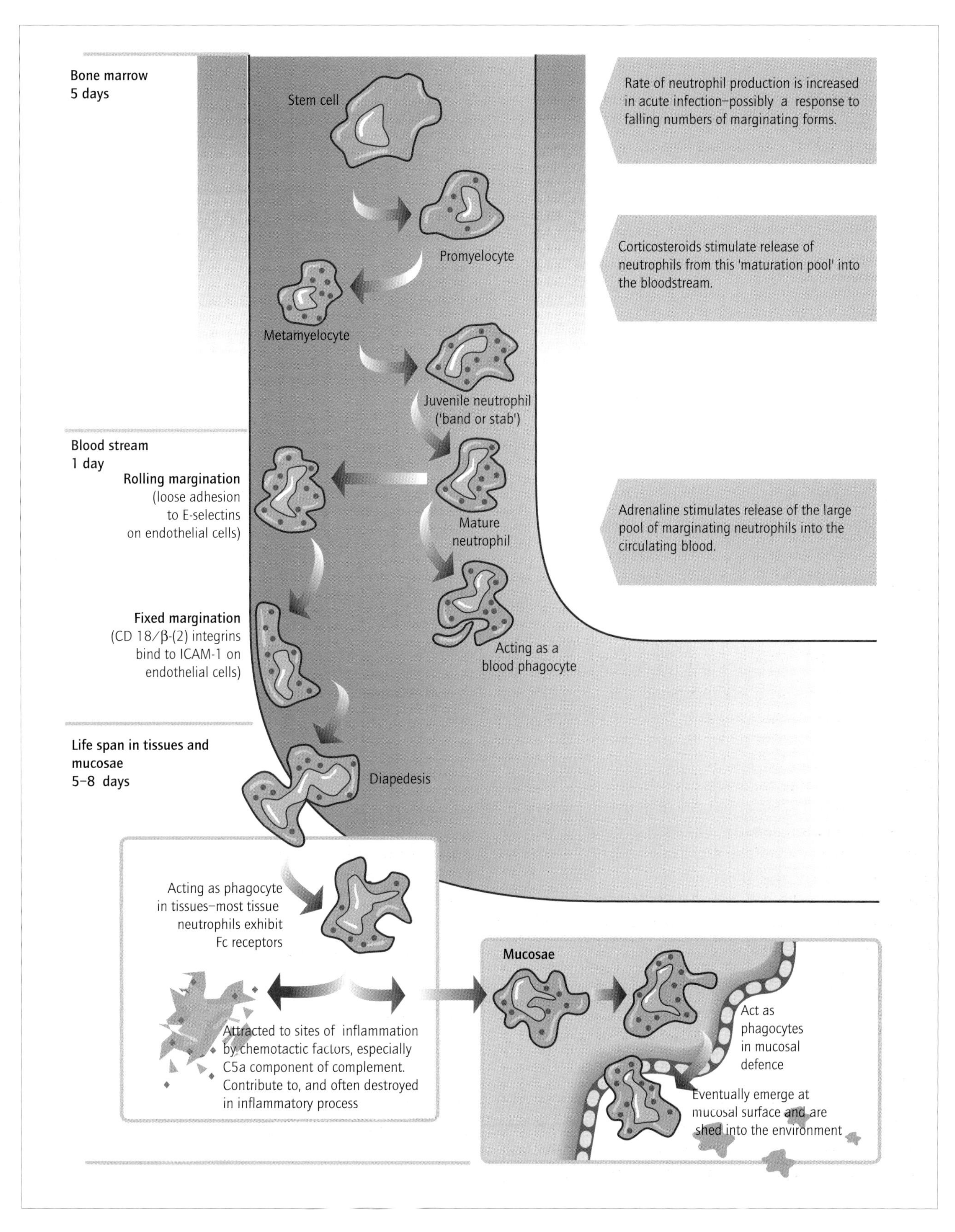

Fig. 19.1 The life cycle of the neutrophil.

The range of abnormalities in sepsis

Endothelial damage leads to extravasation of fluid, with tissue oedema and an early fall in blood volume and albumin levels. Activated endothelium generates nitric oxide, a powerful vasodilator which causes a loss of peripheral resistance not fully compensated by increased cardiac output. These factors predispose to hypotension and shock. Poor tissue perfusion and poor lung function lead to widespread hypoxia (and hypoxic macrophages secrete more IL-8, recruiting more neutrophils).

Disseminated intravascular coagulation

Endothelial damage activates the coagulation cascade, causing fibrinogen consumption, fibrin deposition, activation and consumption of platelets and eventually disseminated intravascular coagulation (DIC). This exacerbates hypoxia, further damages endothelium and brings the risk of haemorrhage.

Epidemiology of bacteraemia and sepsis

Bacteraemia is common. Approximately 30 000 blood isolates are reported from laboratories in the UK each year, and approximately 1% of hospital patients have bacteraemia on admission. Over 10% of hospital patients have features of sepsis, and sepsis affects over 50% of intensive care patients. This probably represents only a small fraction of the true incidence, as many cases are not diagnosed and not all laboratories make regular reports. The infections reported are biased towards more seriously ill patients and may not be representative. Nevertheless, it is clear that many bacteraemias are acquired by people in hospital, who are already ill. Approximately 60% of reported bacteraemias fall into this category. Neonates and the elderly are at greatest risk.

A wide range of pathogens cause bacteraemia. The distribution of the most important organisms is shown in Fig. 19.2. They may be considered under four general categories: (i) staphylococci and streptococci; (ii) Enterobacteriaceae; (iii) anaerobic and aerobic opportunists; and (iv) community-acquired bacteraemias.

Staphylococci and streptococci

These are both community- and hospital-acquired. The epidemiology of staphylococcal bacteraemia is described later in this chapter (p. 374). Over 7000 streptococcal bacteraemias are reported each year, of which half are due to *Streptococcus pneumoniae*. Pneumococcal bacteraemia

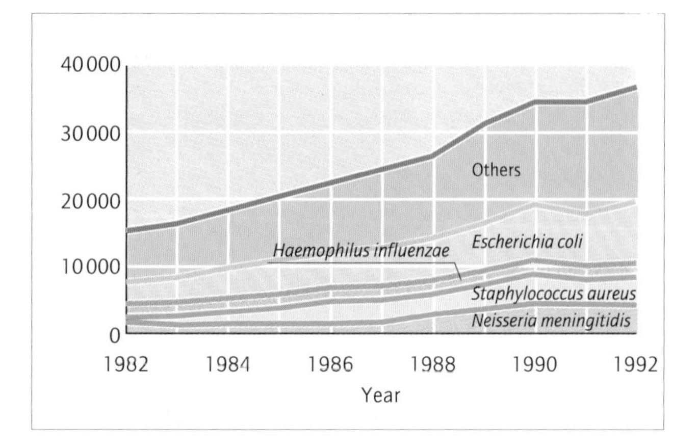

Fig. 19.2 Trends in the occurrence of bacteraemias in the UK.

occurs mainly in the elderly and is often accompanied by pneumonia. It also affects patients of all ages with risk factors such as sickle-cell disease, asplenia, chronic renal, cardiac, liver or lung disease and diabetes mellitus. It is associated with an average mortality of greater than 20%.

Enterobacteriaceae

Enterobacteriaceae include *Escherichia coli*, *Klebsiella*, *Citrobacter*, *Enterobacter*, *Proteus* and *Salmonella* spp. These gastrointestinal organisms are often hospital acquired. The great majority (between 7000 and 8000 per year) are due to *E. coli* and occur in elderly, often surgical, patients. *Klebsiella* infections commonly complicate antibiotic treatment.

Anaerobic and aerobic opportunists

Anaerobic opportunists include *Bacteroides*, *Clostridium* and anaerobic streptococci, and aerobic opportunists include *Acinetobacter*, *Aeromonas*, *Pseudomonas* and *Serratia*. These are also mainly hospital acquired, affecting elderly, postoperative, debilitated or immunosuppressed patients.

Community-acquired bacteraemias

These include *Neisseria meningitidis*, *Haemophilus* spp. and *Listeria monocytogenes*. Meningitis is often an accompanying feature. Meningococcal bacteraemia occurs mainly in children under the age of 4. A second, smaller peak of meningococcal disease occurs in young teenagers (see Chapter 13). Asplenic patients and those with complement deficiencies are at increased risk of meningococcal disease. *H. influenzae* disease is now rare as a result of childhood immunization. Listerial bacteraemia particularly affects adult immunocompromised patients and pregnant women. Infection in pregnancy often results in stillbirth or neonatal bacteraemia (see Chapter 17).

Defences of the blood

Phagocytes

Phagocytes are mobile cells which ingest particulate matter, including bacteria. The main phagocytes of the blood are the neutrophils, but cells of the monocyte–macrophage line are also phagocytic and play a vital part in antigen presentation in the immune system. Pathogens are ingested into vesicles called phagosomes. Within the neutrophil these fuse with lysosomes, forming phagolysosomes into which are liberated proteolytic enzymes and highly oxidative agents, including free radicals, which poison and destroy the bacteria. Neutrophils are particularly protective against staphylococci, yeasts and pseudomonads.

Neutrophils readily leave the bloodstream by rolling along the endothelial capillary wall, then adhering to endothelial cells, and finally passing between the cells into the tissues (Fig. 19.1). Many neutrophils are destroyed in the inflammatory process; others emerge on the surfaces of epithelia and are shed from the body.

Alternative complement pathway

The alternative complement pathway is a soluble defence mechanism. The third component of complement, C3, adheres to bacterial surfaces, where it is slowly broken down to C3a and C3b. In the presence of factor B, active C3bBb is formed, and is stable in the presence of properdin. The 'membrane attack' complex can then be formed at the cell surface, disrupting the membrane and destroying the bacteria (Fig. 19.3).

The alternative complement pathway does not depend on the presence of antibody, and therefore provides a non-specific means of destroying bacteria. It is particularly important in defending against Gram-negative cocci, and people with defects in this pathway are at increased risk of gonococcal and meningococcal bacteraemia. These bacteraemias are not more severe in complement-deficient patients, but in rare individuals who lack properdin they are aggressive and fulminating.

Iron binding

Iron binding depends on both specific and non-specific iron-binding proteins. Bacteria replicate inefficiently, and have reduced capacity to produce toxins when they lack iron. Iron-binding proteins such as ferritin increase greatly in infection and compete with bacteria for the iron they require.

Spleen

The spleen aids removal of bacteria from the blood. It not only provides conditions for phagocytosis in its sinusoids, but it removes engorged phagocytes from the bloodstream and promotes their destruction by other cells. People who lack a functioning spleen are at increased risk of pneumococcal bacteraemic diseases.

Antibodies

Antibodies circulate in the blood and can adhere to or agglutinate bacteria. Bound antibody or antigen–antibody immune complexes can activate the classical complement pathway (Fig. 19.4). This is much faster than the alternative pathway, rapidly destroying bacteria and generating large amounts of chemotactic factors to summon phagocytes and cytotoxic cells. Phagocytes bear Fc receptors which will bind to antibody on bacteria. This process, called opsonization, promotes rapid phagocytosis of the organism. As well as directly disabling organisms, therefore, antibodies also enhance or recruit other defensive mechanisms of the blood. Defects of the alternative complement pathway are negated if the individual has antibodies to an organism, as the classical complement cascade can be rapidly activated.

The disadvantage of antibodies is that they take time to develop. Unless the patient is already immune, the non-specific bactericidal mechanisms must 'buy time' while an antibody response is mounted.

Bactericidal properties of the blood

1 Phagocytes.
2 Alternative complement pathway activity.
3 Iron binding (deprives bacteria of iron).
4 Specific antibody.
5 Classical complement pathway activity.
6 Removal of capsulated bacteria by the spleen.

The natural defences of the blood are overcome when bacteria enter the circulation faster than they are removed,

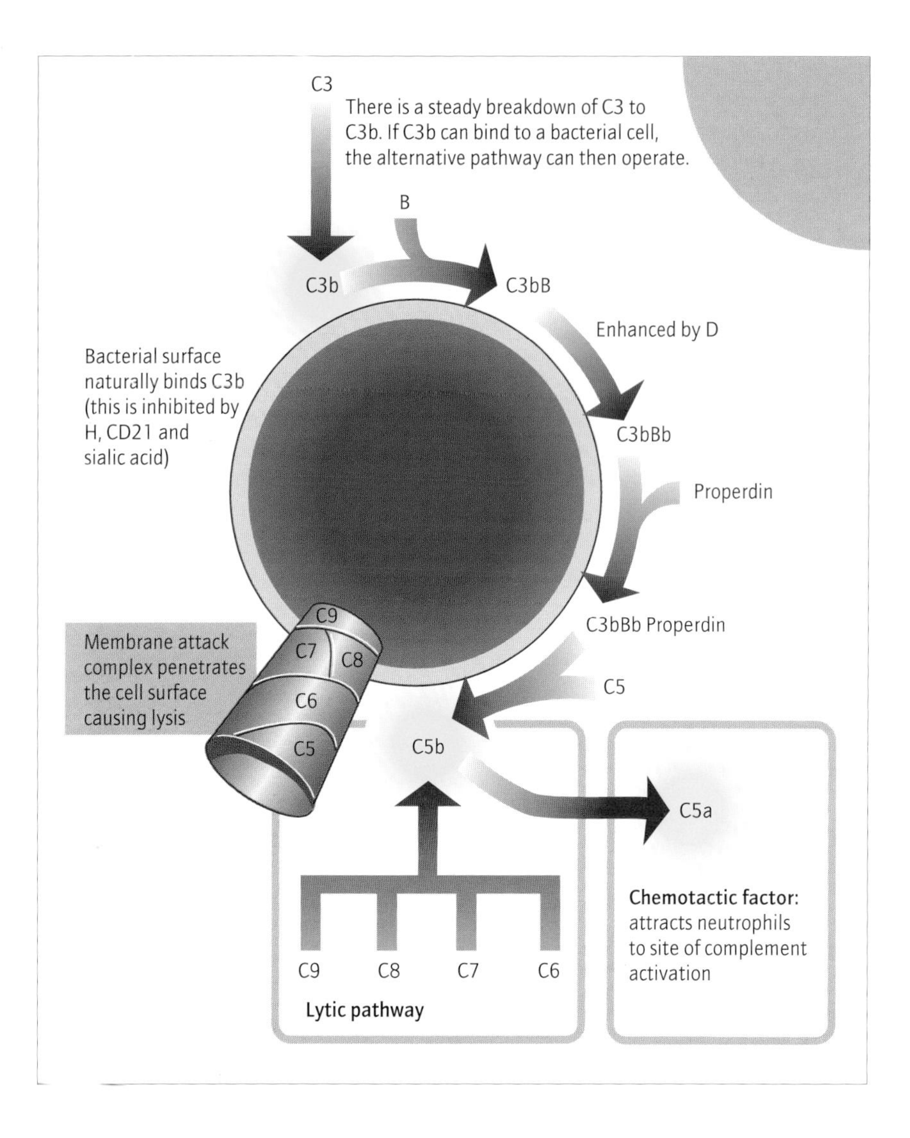

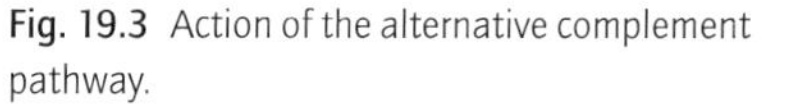

Fig. 19.3 Action of the alternative complement pathway.

or when the bacteria replicate in the blood. There are three ways in which bacteria invade the bloodstream.

1 *Escape from a site of natural occurrence* is a common means of entry. Typical sites of origin include the bowel, urogenital system or skin. When these tissues are damaged by trauma, inflammation or malignancy, natural barriers to the passage of bacteria are disrupted. Manipulation, examination under anaesthesia or surgery all increase the likelihood that bacteria will be released into the bloodstream.

Some pathogens cause bacteraemia by colonizing surface mucosae, invading the epithelium and then escaping into the blood. This is the probable pathogenesis of pneumococcal, meningococcal and *H. influenzae* bacteraemic diseases.

2 *Release from a site of deep sepsis* is also common. Abscess cavities are lined with granulation tissue, which is full of tiny, fragile blood vessels. These are easily invaded and entered by pathogens. The same is true of tissue surrounding a devitalized or necrotic area.

3 *Inoculation of bacteria* can occur by bite, scratch, trauma or 'needlestick'. Some bacteria, e.g. *Streptococcus pyogenes*, are so pathogenic and so successful at evading the defences of blood and tissues, that a trivial injury can lead to overwhelming infection. Animal bites may inoculate organisms such as *Pasteurella multocida*, flea bites transmit plague, and inoculation accidents in hunters are an important means of transmission of tularaemia. Bacteria may replicate at the site of the inoculum or in a draining lymph node, from where bloodstream invasion proceeds. Bacteria can also gain direct access to the circulation when intravenous prosthetic devices become infected by skin organisms including *S. aureus*, *S. epidermidis* and *Corynebacterium jeikeium*.

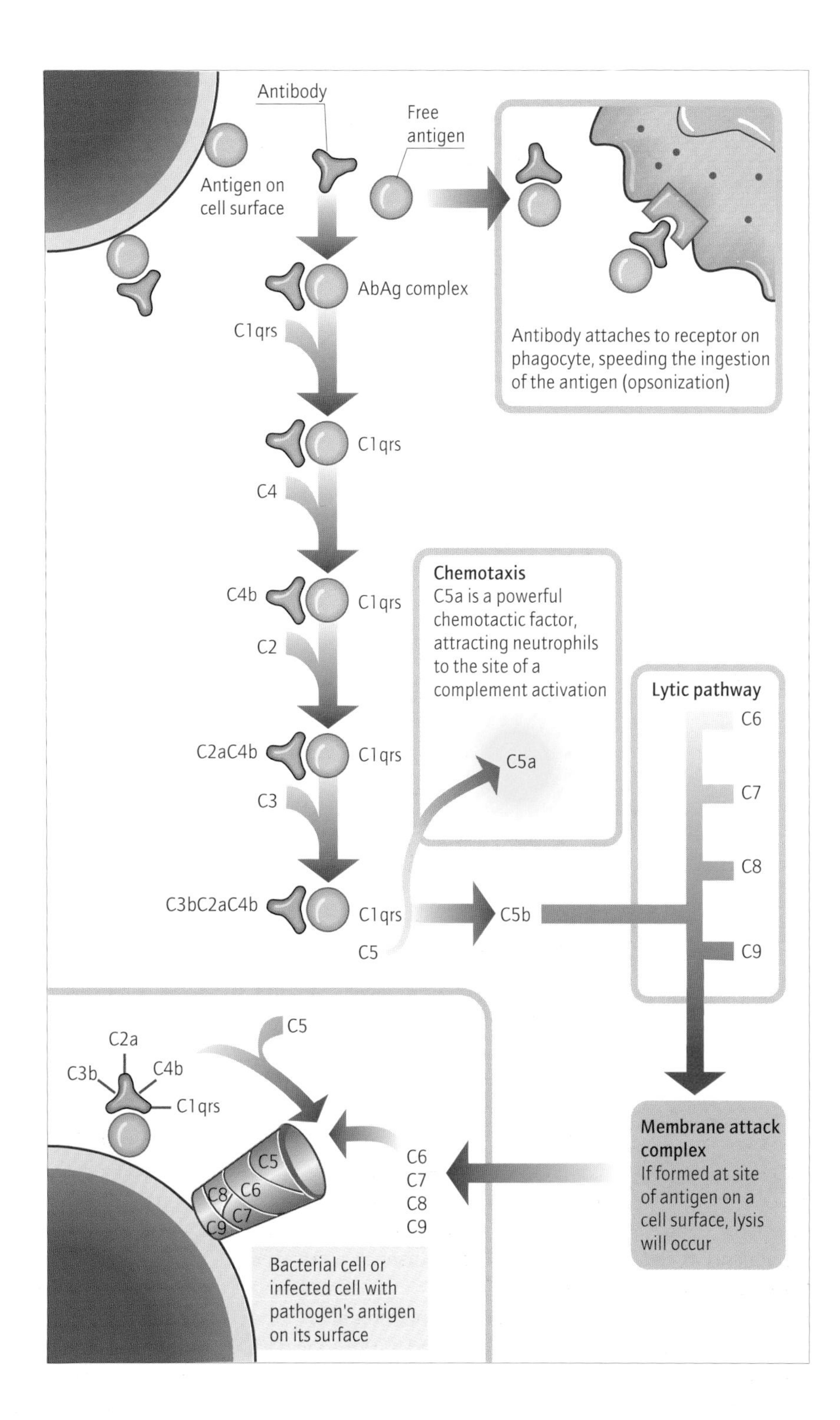

Fig. 19.4 Action of the classical complement pathway.

Pathological accompaniments of bacteraemia

Pathology at the site of origin

Pathology at the site of origin is sometimes obscured by the effects of the bacteraemia itself. An asymptomatic gallstone or biliary stricture can cause overwhelming Gram-negative sepsis which will defy treatment until the causative local infection is removed. Renal or hepatic abscesses, infection behind a ureteric stricture, a small undrained empyema or an infected intracranial sinus are all capable of maintaining a bacteraemia, unless the locu-

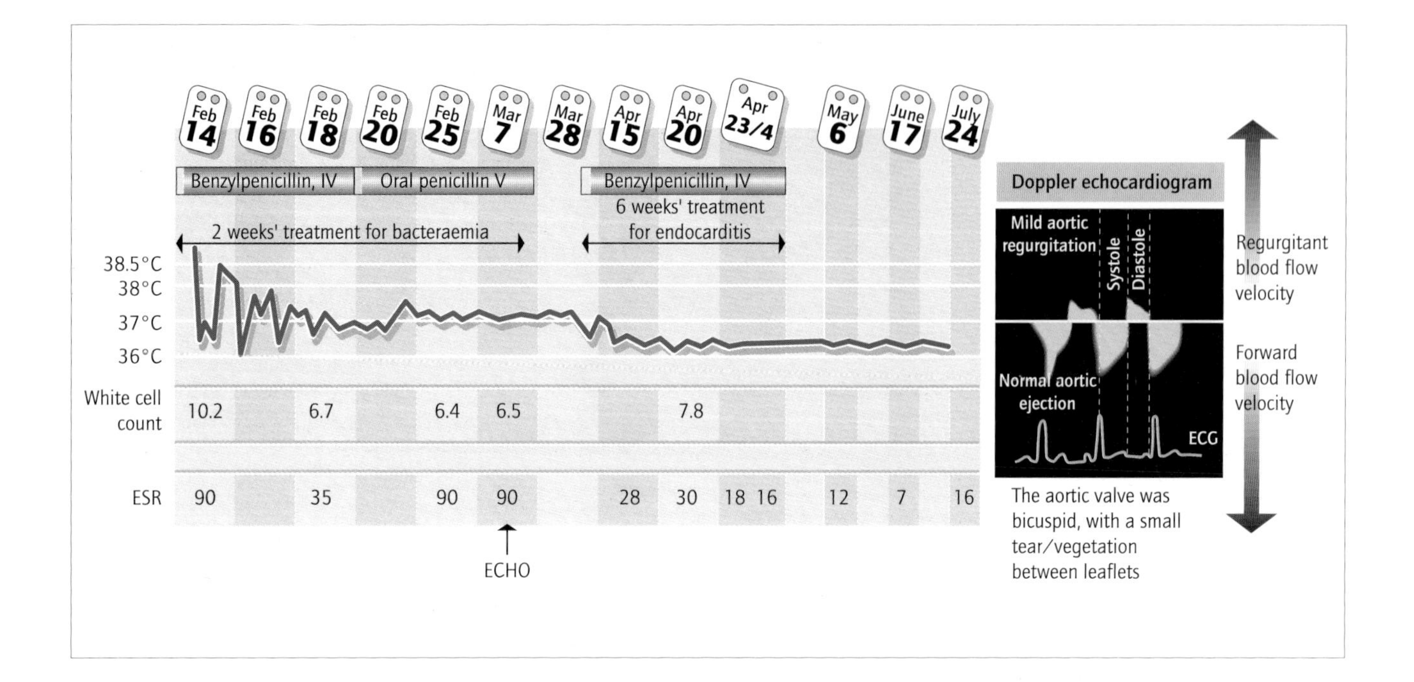

Fig. 19.5 Endocarditis complicating bacteraemic disease: this 27-year old presented with septic arthritis of the ankle and was found to have *Streptococcus pyogenes* bacteraemia; after completion of treatment at 2 weeks, the erythrocyte sedimentation rate (ESR) rose and a systolic murmur led to the diagnosis and treatment of endocarditis. ECG, electrocardiogram; ECHO, echoencephalogram.

lated infection is drained. These predisposing conditions should always be actively sought.

The heart must always be examined to exclude endocarditis. Apparent cure of bacteraemic disease after antibiotic therapy can be followed by recrudescence and cardiac damage because of inadequately treated endocarditis (Fig. 19.5).

Effects of exotoxins

Effects of exotoxins are widespread. Staphylococcal and streptococcal exotoxins can both cause rashes, and staphylococcal enterotoxins cause diarrhoea. Streptococcal toxins include leucocidins, haemolysins and hyaluronidase. Both Gram-positive and Gram-negative bacteria can produce toxins which cause tissue necrosis. *Clostridium perfringens* bacteraemia is sometimes associated with severe, toxin-mediated haemolysis.

Effects of inflammatory responses

Effects of inflammatory responses contribute to fever and rigors. Complement activation, kinin activation and release of toxic agents from phagocytes are highly damaging, and add 'secondary mediators' to the classical pathways of fever production (see Chapter 1). They promote endothelial damage, platelet activation, vascular shunting and poor ventilation–perfusion matching.

Metastatic infections

Metastatic infections occur when tissues are seeded with bacteria from the bloodstream. Common sites affected include the lungs, the bones, the kidneys and the susceptible endocardium. In some bacteraemias, e.g. *Streptococcus 'milleri'*, the brain is also at risk. Meningitis, arthritis and empyema are relatively frequent complications of bacteraemias caused by Gram-positive cocci.

Some bacteraemias are trivial or silent, but present with complications. *Salmonella* bacteraemias may not be accompanied by bowel symptoms, but result in osteomyelitis in sickle-positive individuals, or soft-tissue abscess in others. Staphylococcal joint infection is probably the result of silent bacteraemia, as is streptococcal pericarditis or pneumococcal peritonitis.

Diagnosis of bacteraemia

Clinical diagnosis

Suspicion of a bacteraemia can sometimes be aroused by the presence of characteristic physical signs. Typical rashes may accompany streptococcal, staphylococcal, meningococcal and gonococcal bacteraemias. Large-joint arthritis commonly accompanies gonococcal bacteraemia. Lobar pneumonia is often present in pneumococcal bacteraemia. Nodular lung opacities or lung abscesses may be seen in staphylococcal bacteraemia. Disease or obstruction of the urinary tract, biliary tree or bowel often leads to Gram-negative and/or enterococcal bacteraemia, often with a contribution from the anaerobic *Bacteroides fragilis*. Recent instrumentation or surgery makes bacteraemia more likely.

In contrast, early hypotension is much more common in Gram-negative bacteraemias, particularly those caused by Enterobacteriaceae. Shock is usually absent until the late stages of Gram-positive bacteraemias. An important exception to this is the early presence of shock in *Clostridium perfringens* bacteraemia.

Laboratory diagnosis

Blood cultures should immediately be obtained, and incubated without delay. A minimum of two sets of cultures is essential; more than three do not significantly increase the likelihood of a positive diagnosis. Other specimens which should be cultured include urine, and sputum if obtainable. Specimens such as pus, urinary catheter urine, wound swabs, drain aspirate, throat swabs, joint aspirate, cerebrospinal fluid and diarrhoea stool may be cultured in appropriate circumstances.

Blood cultures are most likely to be positive if collected when the fever is rising. This is less relevant where continuous bacteraemia is the rule, as in endocarditis or cannula-related sepsis. The yield from blood culture increases with the volume cultured. However, with high volumes the dilutional effect of the culture medium is reduced, and the bacteriocidal properties of the blood may prevent a positive culture being obtained. A 1 in 10 dilution is usually optimal. In neonates and small children a positive culture can be obtained with a smaller volume of blood, as the degree of bacteraemia is usually higher.

The blood culture media should be sufficiently nutritious to support the growth of aerobic and anaerobic pathogens. More than one medium is required and microbiology laboratories provide blood culture 'sets'. These vary from laboratory to laboratory but usually consist of one bottle suitable for the growth of aerobic, capnophilic and facultative anaerobic organisms and a second culture containing added reducing agents to aid the isolation of anaerobic species such as *Bacteroides*. Blood culture bottles can be subcultured within 6 h for a rapid result and the supernatant examined by Gram stain or tests for bacterial antigens such as pneumococcal capsular polysaccharides and streptococcal group antigen to yield a rapid presumptive diagnosis. Many laboratories now use automated blood culture systems which detect the presence of bacterial growth by radiometric, infrared or colorimetric detection of carbon dioxide production or changes in electrical impedance of the medium. These systems are continuously monitored, allowing rapid detection of positive cultures without the need for routine subculture, saving laboratory time and materials, and speeding positive diagnosis.

Management of bacteraemia and sepsis

Antibiotic treatment

The mainstay of treatment is control of the underlying infection. A rational choice of empirical treatment may be possible if the history and examination point to a particular pathogen or group of pathogens (Tables 19.2–19.5).

General management of bacteraemic patients: systems and life support

The infected patient with sepsis needs support until antimicrobial therapy has time to act. The ambitions of supportive treatment are ideally to support life and prevent further deterioration by:

1 Maintaining the blood pressure.
2 Maintaining the cardiac output.
3 Counteracting the effects of low peripheral resistance.
4 Maintaining blood and tissue oxygenation.
5 Minimizing the effects of endotoxin and other damaging agents.

Supporting the circulation

The initial step is to obtain an adequate intravascular fluid volume. There is controversy as to whether crystalloid fluids (such as saline solutions) or colloids (such as albumin or polygels) are most effective. Crystalloids are widely distributed in the interstitial fluid as well as intravascularly; they provide filterable solute for the renal tubules and can replenish lost intracellular fluid, but they are also quickly excreted. Colloids tend to remain in the vascular space; they improve blood pressure and organ perfusion but are not filtered through glomeruli. Different colloid substances are broken down in the circulation at different rates, and the longer-lasting ones cannot be removed if the patient becomes overhydrated or develops pulmonary oedema.

Current practice is to give crystalloids first for an

Organism	May originate from	Antimicrobial treatment
Staphylococcus aureus	Skin, nasopharynx (50% of all cases). Intravenous devices or pacemaker systems (recently implanted) (25% of all cases). Bone and joint disease or surgery or prosthesis (10% of all cases). Genitourinary disease or surgery (2% of all cases)	Cloxacillin or flucloxacillin *Alternatives or additions*: fusidic acid*, rifampicin*, teicoplanin, vancomycin, clindamycin*†, ciprofloxacin
Streptococcus pneumoniae	Lung infection (50% of all cases). Intracranial or meningeal infection (10% of all cases). Patients with altered immunity, splenectomy or alcoholism (up to 10% of all cases). Rarely, intrapartum infection or peritonitis	Benzylpenicillin or erythromycin *Alternatives and for penicillin-resistant organisms*: third-generation cephalosporins, e.g. cefotaxime or ceftriaxone *For multiply resistant organisms*: teicoplanin or vancomycin
Coagulase-negative staphylococci	Intravascular devices including pacing systems, haemodialysis and parenteral nutrition systems (over 50% of all cases). Peritoneal dialysis systems. Intracerebral shunts and valves connected with the right atrium. Intracardiac prostheses	Depending on results of laboratory testing, but start with cloxacillin or flucloxacillin plus one alternative, as for *S. aureus* *Other possibilities*: erythromycin, trimethoprim, ciprofloxacin
Group A streptococci	Skin and soft tissues (most cases). Rarely, postpartum infections, genitourinary disease or surgery, bone and joint infection, pharyngitis	Benzylpenicillin, third-generation cephalosporin or erythromycin *Alternative*: clindamycin
Group B streptococci	Neonatal infection (over 50% of all cases) Genitourinary infection or surgery, including termination of pregnancy	Benzylpenicillin *plus* gentamicin *Alternatives*: third-generation cephalosporins
Group C and G streptococci	Skin and soft tissue infection	Benzylpenicillin or third-generation cephalosporin
Streptococci 'milleri'	Meningitis, ventriculitis, brain or liver abscess	As for group B streptococci

* May be given orally, but beware of emerging resistance.
† Beware of pseudomembranous colitis.

Table 19.2 Sources of coccal bacteraemias, and suggested empirical treatments.

immediate effect, and then to follow up with colloid. The most useful monitors of circulatory function are the blood pressure, the central venous pressures and the urine output. The blood pressure need not be completely normalized; systolic pressures of 50–60 mmHg are adequate if cerebral function, urine output and tissue perfusion are satisfactory. A right atrial pressure of 5–10 mmHg is usually adequate; higher pressures are not often helpful and may contribute to congestive cardiac failure and poor tissue perfusion. The pulmonary capillary wedge pressure may be measured, using a flotation balloon catheter. If it is higher than the right atrial pressure, the left ventricle is not adequately moving blood from the lungs, and systemic perfusion is probably poor.

The difference between the body's core temperature and skin temperature gives an idea of skin, and therefore tissue, perfusion. If the skin temperature (usually measured at the foot) is more than 2°C lower than the core (measured rectally or at the eardrum), this suggests inadequate tissue perfusion. Equal or nearly equal temperatures indicate vasodilatation and increased cardiac output.

Drugs used to support the circulation

These are only effective if there is sufficient intravascular fluid volume, indicated by adequate central venous pressure.

DOPA agonists. These increase myocardial contractility via $beta_1$-adrenergic receptors. Dopamine is routinely used in infusions of up to 4 μg/kg/min, as this has an inotropic effect on the heart and a vasodilator effect on the kidney and general circulation. Higher doses cause vasoconstriction and tachycardia, and may contribute to heart failure. Dobutamine may be added in infusions of 2.5–10 μg/kg/min for extra inotropic effect. Dopexamine also

Organism	May originate from	Antimicrobial treatment
Escherichia coli (the cause of over 50% of all Gram-negative bacteraemias and about 25% of all bacteraemias)	Urinary tract colonization or infection, especially after instrumentation. Intestinal disease or surgery. Pancreaticobiliary disease, instrumentation or surgery	Ampicillin, amoxycillin or third-generation cephalosporin, e.g. cefotaxime or ceftriaxone *Alternatives*: gentamicin or other aminoglycoside (20–30% risk of resistance to ampicillin) Plasmid-mediated resistance to antibiotics is easily acquired in hospital, and has also (rarely) been associated with community-acquired infections
Klebsiello sp.: *K. pneumoniae* subsp. *pneumoniae* subsp. *aerogenes* subsp. *ozaenae* *K. oxytoca*	*K. pneumoniae* subsp. *pneumoniae* causes about 50% of all *Klebsiella* bacteraemias. Origins are as for *E. coli* plus rare cases of suppurative lobar pneumonia	Resistant to ampicillin Hospital-acquired strains often resistant to first- and second-generation cephalosporins and gentamicin *May be sensitive to*: third-generation cephalosporins, ureidopenicillins, e.g. piperacillin, carbapenems, e.g. imipenem/cilastatin or meropenem, tobramycin or amikacin
Enterobacter sp. and *Serratia* sp.	As for *E. coli*. Increasingly common in intensive care units and immunosuppressed patients	Third-generation cephalosporins Gentamicin and other aminoglycosides Ureidopenicillins Carbapenems
Proteus sp. (a member of the family Proteaceae which also includes Morganella and Providencia)	Genitourinary and pelvic disease, surgery and instrumentation. Renal or urinary tract stones (to which *Proteus* predisposes by metabolizing urea and forming alkaline ammonia)	As for *Enterobacter*
Salmonella sp.: including *S. typhi* and *S. paratyphi* sp.	Food- and water-borne intestinal infection	Ciprofloxacin *Alternatives*: ceftriaxone, chloramphenicol (if sensitive) (see also Chapter 8)

Table 19.3 Sources of Gram-negative bacteraemias and suggested empirical treatments.

increases renal perfusion by acting on peripheral dopamine receptors. It is commenced at 500 ng/min and increased in 0.5–1.0-μg steps at 15-min intervals up to 6 μg/min for an optimal effect.

Adrenoceptor agonists. As well as the DOPA agonists, both alpha (noradrenaline, phenylephrine and methoxamine) and beta or mixed (adrenaline metaraminol and isoprenaline) adrenoceptor agonists are used for a broader-spectrum inotropic and chronotropic effect. The alpha adrenoceptor agonists cause a degree of vasoconstriction and are useful for maintaining blood pressure in severe vasodilatation.

Phosphodiesterase inhibitors. These (milrinone and enoximone) increase cardiac output in the presence of raised filling pressures.

Vasoconstrictors. These (angiotensin, vasopressin and their analogues) raise blood pressure, but at the expense of intense vasoconstriction which may compromise tissue perfusion.

Although blood pressure, cardiac output and peripheral vascular resistance can be manipulated by using combinations of these drugs, the evidence that tissue perfusion or oxygenation is improved by these effects is lacking. As each patient has a different severity of disease and different combination of problems, controlled trials of treatment outcomes are difficult to perform.

Managing hypoxia and ARDS

Increasing the percentage of inspired oxygen will increase the arterial oxygen saturation, and this can be monitored by pulse oximetry or arterial blood-gas estimations. Removal of carbon dioxide is not increased, however, and care must be taken not to allow the patient to become severely hypercapnic as this impairs cardiac function, increases acidosis and raises intracranial pressure.

Positive-pressure ventilation will overcome decreased lung compliance and may also expel fluid from the

Organism	May originate from	Antimicrobial treatment
Pseudomonas sp. and *Burkholderia* sp.	Immunocompromised patients (25% of all cases). Intensive care unit cases. Biliary tract disease, surgery or instrumentation	Gentamicin, tobramycin or amikacin. Azlocillin or piperacillin. Cefotaxime or ceftazidime. Aztreonam. Imipenem or meropenem. (Often a combination of aminoglycoside with another drug is used, as resistance is common in hospital-acquired infection)
Acinetobacter sp.	Immunocompromised patients (33% of all cases). Intravenous devices. Cardiac prostheses. Rarely, neurosurgical procedures	Third-generation cephalosporins, e.g. cefotaxime, ceftriaxone. Imipenem or meropenem *Alternatives*: gentamicin or other aminoglycoside if sensitive. Piperacillin if sensitive
Bacteroides sp. (especially *B. fragilis*)	Abdominopelvic disease or surgery. Liver abscess. Rarely, postpartum	Metronidazole *Alternatives*: clindamycin, meropenem
Clostridium sp. (including *C. perfringens*)	As for *Bacteroides*, but also biliary tract sepsis, perineal or lower limb trauma, soil-contaminated or necrotic wounds	Metronidazole *Alternative*: benzylpenicillin
Anaerobic cocci (including *Peptococcus* and *Peptostreptococcus*)	Postpartum infections. Abdominopelvic disease or sepsis. Rarely, neonatal bacteraemia	Metronidazole, benzylpenicillin or ampicillin *Alternative*: clindamycin
Enterococci (including *Enterococcus faecalis*, *E. faecium* and *Streptococcus durans*)	Intravascular cannulae, especially in intensive care unit patients (25% of all cases). Pancreaticobiliary disease, surgery or instrumentation. Abdominopelvic disease or surgery	Ampicillim plus gentamicin or vancomycin *Alternatives*: vancomycin plus or minus gentamicin. Imipenem/cilastatin, meropenem (not *E. faecium*)

Table 19.4 Sources of 'hospital' bacteraemias and suggested empirical treatments.

oedematous lungs. Recent experience suggests that low-pressure/low tidal volume ventilation aids alveolar patency and minimizes barotrauma to the lungs. If hypoxia persists, ventilation in the prone position may optimize ventilation–perfusion matching and raise Pa_{O_2}. Positive end-expiratory pressure or continuous positive airways pressure may afford a better improvement. They have the disadvantage of reducing venous return and cardiac output, so often cause a requirement for increased cardiovascular support.

Moderate doses of corticosteroids may be beneficial in improving lung function and preventing progression from the inflammatory to the organizing and fibrotic stage of ARDS. Moderately long courses of treatment may be required.

There is little evidence of benefit from extracorporeal membrane oxygenation (ECMO), but it may support a patient with a short-term reversible oxygenation deficit. Neither inhaled vasodilators such as nitric oxide or prostacyclin, nor surfactant have proved beneficial.

Antagonists of the mediators of sepsis

Plasma exchange to remove toxins and inflammatory mediators has been trialled on several occasions in both Gram-positive and Gram-negative infections. Benefit has not been demonstrated. It is now generally accepted that high-dose corticosteroids have no demonstrable benefit.

Direct inhibition of endotoxin, TNF-alpha or IL-1 is theoretically possible, using monoclonal antibodies. Animal studies show that this is effective if the antibodies are given before signs of sepsis are established, but this is impractical in human medicine, and controlled trials have shown either no benefit or slight disadvantage in antibody treatments. It is now recognized that beta adrenoceptor agonists have anti-inflammatory properties; they can reduce the production of TNF-alpha, IL-1 and IL-6, while those with alpha agonist action tend to have the opposite effect. A discriminating choice of adrenoceptor agonist may be more important in this respect than in direct haemodynamic support.

Nutrition

Patients with bacteraemic diseases have very high calorie

Antimicrobial agent	Adverse reactions and problems	Action to avoid adverse reactions and problems
Benzylpenicillin	Skin rashes	Avoid penicillins; cephalosporins may be given cautiously
Flucloxacillin	Cholestasis	Not usually intolerable; resolves on discontinuing drug
Ampicillin	Anaphylaxis (rare)	Avoid all penicillins. Avoid other beta-lactams where possible
	Agranulocytosis (rare)	Check white cell count during prolonged, high-dose treatment. Will occur with all beta-lactams – avoid
Erythromycin	Nausea	Substitute another drug (clarithromycin or non-macrolide drug)
Cephalosporins	Skin rashes	Avoid cephalosporins
	Thrombocytopenia (rare)	Check platelet count during high-dose treatment. Oral cephalosporins may be given cautiously in future
	Defective coagulation (rare)	
Aminoglycosides	Impaired renal function	Renally excreted; reduce dose in renal impairment; always monitor peak and trough blood levels. Avoid co-administration of frusemide (also ototoxic)
	Ototoxicity	
	Exacerbation of myaesthenia gravis	Avoid
Clindamycin (a lincoside)	Antibiotic-associated diarrhoea and pseudomembranous colitis	Discontinue drug if diarrhoea occurs. Persisting diarrhoea with fever and abdominal pain, may be treated with oral vancomycin (see p. 180)
Vancomycin (a glycopeptide)	Hypersensitivity reaction with vasodilatation, collapse and possible renal failure	*Never give by rapid injection.* Always infuse dose over at least 90 min for each 500 mg
	Ototoxicity	Reduce dose in renal impairment, monitor peak and trough blood levels. Use with *caution* if combined with aminoglycoside
	Nephrotoxicity	
Teicoplanin (a glycopeptide)	As for vancomycin, but does not cause hypersensitivity reaction. Less toxic	*May* be given by intravenous or intramuscular injection. No requirement for blood-level monitoring
Ciprofloxacin (a 4-quinolone)	Skin rashes	Discontinue or substitute an alternative drug
	Diarrhoea	Avoid
	Precipitation or exacerbation of epilepsy	
	Enhances effect of warfarin, aminophylline and sulphonyl ureas	Avoid or use with caution and frequent monitoring

Table 19.5 Adverse effects of antimicrobial agents used in treating bacteraemias.

requirements, and rapid turnover of macro- and micronutrients. Controlled trials show that nutritional support improves the outcome of treatment for bacteraemia and sepsis. Enteral feeding is physiological and prevents atrophy of the intestinal mucosa. Nasogastric intubation is unpleasant and increases management requirements in intubated patients, so endoscopically placed gastrostomy catheters are increasingly used when long-term enteral nutrition is needed.

Patients with reduced bowel function cannot absorb enteral nutrients and must be fed parenterally. This is usually done via a dedicated lumen of a multilumen right atrial catheter. A gradual change is made to enteral feeding as soon as improving bowel function allows.

General management of bacteraemic patients

1 Optimize intravascular and tissue fluid volumes using crystalloid and colloid infusions, and measure arterial and central venous pressures.

2 Support circulatory function, using inotropic and/or vasoactive agents.

3 Optimize oxygen delivery to the tissues by giving oxygen with or without positive-pressure ventilation.

4 Maintain adequate nutrition, using enteral or parenteral feeding techniques.

Staphylococcal septicaemia

Introduction and epidemiology

Staphylococcus aureus is the commonest cause of Gram-positive septicaemia. Half of all cases are community-derived and affect previously healthy people. Other cases are hospital-acquired, related to intravenous devices, pacemakers and surgical wounds, or affect intravenous drug abusers, in whom *S. aureus* is the commonest blood pathogen. It is important because it affects all age groups, has an insidious onset, and the diagnosis is often delayed because sufferers do not initially appear severely ill. It carries a high mortality (25–70%) in the elderly.

Pathology

Many patients with staphylococcal bacteraemia have no obvious predisposing factor, and the origin of the bacteraemia is not understood. There appears to be no tendency for a particular strain of *Staphylococcus* to predominate.

Phagocytosis is important in defence against staphylococci. Children with cystic fibrosis have poor phagocyte function, and suffer severe staphylococcal chest infections but rarely septicaemia. Patients whose phagocytes cannot mount a bactericidal respiratory burst, as in chronic granulomatous disease, are at risk of repeated staphylococcal infections.

Clinical features

The onset of illness is often slow, with increasing fever and malaise. Helpful physical signs include pain in a large joint (usually without effusion), severe periarticular tenderness of all joints, or numerous small pustules on the skin. Patients usually have high, swinging fever, but are haemodynamically normal. Rare cases have coexisting toxic shock syndrome (see Chapter 5).

The white cell count is often misleadingly normal, but the proportion of neutrophils is usually near 90%. After some days a conventional leucocytosis develops. The platelet count may be slightly reduced. Renal function is moderately impaired; proteinuria and microscopic haematuria are common. In half or more of cases the chest X-ray is abnormal, showing nodular opacities, abscesses, consolidation, effusion or pleural abscess.

In untreated cases, progressive renal failure is common, because of numerous microabscesses in the kidneys. Disseminated intravascular coagulation is also common and can cause loss of digits. Metastatic infections are a real danger and include aggressive endocarditis, pneumonia, suppurative arthritis, soft-tissue abscesses and abscesses of the kidney, liver or brain.

Diagnosis

Even in patients who have received oral antibiotics, blood cultures are usually positive in all bottles within 18–24 h.

Management

The antimicrobial treatment of choice is intravenous flucloxacillin, in divided doses totalling at least 6 g/day. Flucloxacillin does not reach bactericidal levels in all tissue sites. Either fusidic acid or rifampicin are therefore added. Both are well distributed in the body, but rifampicin penetrates the cerebrospinal fluid better than fusidic acid and may be preferable in the presence of meningitis or endocarditis. Fusidic acid is irritant to peripheral veins but both drugs are well absorbed orally or intragastrically. Neither fusidic acid nor rifampicin should be given alone, because bacteria can make a one-step mutation within 4 or 5 days to become completely resistant to them.

Most *S. aureus* are sensitive to aminoglycosides on laboratory testing. However, these have relatively poor penetration into abscesses, and often give disappointing results when used as antistaphylococcal monotherapy.

In patients who have skin allergies to penicillins, cephradine or cefuroxime may be substituted, but should not be given to patients who have anaphylactic reactions to penicillin. In such difficult situations, ciprofloxacin, teicoplanin or clindamycin may be useful.

Vancomycin is the treatment of choice for organisms resistant to flucloxacillin and other antistaphylococcal drugs (MRSA, methicillin-resistant *S. aureus*). It must be given as an infusion over 90 min to avoid the severe hypersensitivity reaction of vasodilatation (red man syndrome) and acute hypotension. It also avoids high peak levels, which predispose to nephrotoxicity and ototoxicity. Peak and trough levels at the end of infusion and immediately before the next dose should not exceed 30 and 10 mg/l, respectively. The dose should be reduced in renal impairment. Aminoglycosides and loop diuretics enhance vancomycin toxicity. Teicoplanin may be an effective and less toxic alternative, but some staphylococci have increased minimum inhibitory concentration (MIC) and minimum bacteriocidal concentration (MBC) values for this drug.

Important antistaphylococcal drugs
Flucloxacillin i.v. 1–2 g 6-hourly (child under 2 years, 250–500 mg 6-hourly; 2–10 years, 500 mg–1 g 6-hourly); *or*
Cloxacillin i.v. 500 mg–1 g 4–6-hourly (child under 2 years, 125–250 mg 6-hourly; 2–10 years, 250–500 mg 6-hourly); *plus*
Fusidic acid orally 750 mg 8-hourly (child under 1 year, 50 mg/kg daily; 1–5 years, 750 mg daily; 5–12 years, 1.5 g daily, all in three divided doses) or i.v. 580 mg 8-hourly; *or*
Rifampicin orally or i.v. 300–600 mg 12-hourly (child under 3 months, 5 mg/kg 12-hourly; 3 months–12 years, 10 mg/kg 12-hourly to maximum 600 mg/day).

Alternative drugs
1 Cephradine i.v. 1–2 g 6-hourly (child 50–100 mg/kg daily in four divided doses).
2 Cefuroxime i.v. 750 mg–1.5 g 6–8-hourly (child 60–100 mg/kg daily in three divided doses).
3 Trimethoprim by slow i.v. injection or infusion 150–250 mg 12-hourly (child 6–9 mg/kg daily in two or three divided doses).
4 Vancomycin i.v. infusion over 90 min, 500 mg 6-hourly (neonate up to 1 week, 15 mg/kg followed by 10 mg/kg 12-hourly; 1–4 weeks, 15 mg/kg followed by 10 mg/kg 8-hourly; child over 1 month, 10 mg/kg 6-hourly); plasma levels should peak (1 h after dose) not above 30 mg/l, trough not above 10 mg/l.
5 Teicoplanin i.v. 400 mg 12-hourly for three doses, then 400 mg daily—may be reduced to 200 mg daily after good response (child over 2 months, 10 mg/kg daily—may be reduced to 5 mg/kg daily after good response).
6 Clindamycin orally 300–600 mg 6-hourly—may be increased to 900 mg 6-hourly in life-threatening disease (child over 1 month, 30–40 mg/kg daily—*minimum* 300 mg daily at any age); discontinue immediately if diarrhoea or colitis develops.

Coagulase-negative staphylococci

Coagulase-negative staphylococci are common causes of bacteraemia in neonatal intensive care, particularly associated with cardiothoracic surgery and the use of indwelling intravenous devices. Many are resistant to flucloxacillin. They must therefore be treated as methicillin-resistant staphylococci, with teicoplanin, vancomycin, gentamicin, fusidic acid, clindamycin or trimethoprim, depending ultimately on the results of sensitivity testing.

The duration of treatment will depend on response. Uncomplicated cases can be adequately treated with 7–10 days' antibiotics. Cases with multiple abscesses or loculated infections need longer courses. Drainage may be required for empyema, large lung abscesses or secondary bone infection. Endocarditis needs long-term treatment and follow-up (see Chapter 12).

Streptococcal septicaemia

Introduction and epidemiology

Streptococcus pyogenes is a commensal of the skin and upper respiratory tract. It can also colonize the female genital tract following examination or manipulation during pregnancy and delivery.

Many cases of streptococcal septicaemia originate from skin or throat infections, which may not themselves be severe. Examples include small cuts or abrasions, which may have healed by the time bacteraemia is recognized. Streptococcal puerperal fever may follow normal or assisted delivery, Caesarean section or other gynaecological procedures (see Chapter 17).

Although a rarer cause of bacteraemia than either *Staphylococcus aureus*, meningococci or pneumococci, *Streptococcus pyogenes* is a dangerous organism; the mortality of treated streptococcal septicaemia is 25–30% in most Western countries.

Pathology

No specific predisposition to streptococcal bacteraemia is recognized. There are many strains of *S. pyogenes*, some more virulent than others. Different strains predominate for several years at a time in slow epidemic fluctuations. M1 types have produced serious disease in the UK in the 1990s. These strains have shown a significant incidence of resistance to erythromycin.

Clinical features

The presenting complaint is usually of severe feverish illness with a short history of 1 or 2 days. A preceding event may be recognized 3–5 days previously, but mild sore throats or skin lesions may have healed in this time. A sinister warning feature is swelling and severe pain in lymph nodes draining the original site. This may be seen in the axilla, or in the inguinal or tonsillar nodes. Spreading erythema, abscess formation or necrosis and sloughing of the nodes is occasionally seen. Septic arthritis or abscess of the parotid gland may occur. Puerperal fever may be marked by pelvic pain and offensive lochia.

Occasionally areas of erysipelas, cellulitis or a scarlet fever-like rash develop. The appearance of erysipelas, other than in the typical sites on the face or leg, is strongly suggestive of bacteraemia. Necrotizing fasciitis is a rare feature of *S. pyogenes* bacteraemia. Many patients lack any helpful physical sign.

Non-specific features of severe illness may be prominent. These include watery diarrhoea, persistent vomiting, meningism or pain in the back or thighs.

The white cell count shows a modest leucocytosis of 12 or $13 \times 10^9/l$. Renal and liver function is usually maintained, as is the blood pressure, until the preterminal stages. In later or untreated cases, an ill-defined pneumonia or a pleural effusion may occur. Shock and profound hypoxia can develop very quickly and are extremely difficult to reverse.

Diagnosis

Initial suspicion must often be based on clinical clues. Appropriate specimens from throat, skin or other lesions may produce a growth of beta-haemolytic streptococci within 24 h. Blood cultures are not reliably positive for up to 3 days in some cases, as *S. pyogenes* has a slow initial growth curve.

Management

Benzylpenicillin is the treatment of choice, except in the presence of streptococcal necrotizing fasciitis. Clindamycin is highly effective in low redox potentials and is indicated in this case.

A suitable dose of penicillin is 1.2 g 2-hourly. It is notoriously difficult to eradicate streptococcal bacteraemia, so the dose should not be reduced too soon after an apparent response. High dosage for 4–5 days may be needed before intermittent recurrences of fever and toxaemia are controlled. A further 7 days or more of conventional dosage is then advisable.

Tissue damage caused by streptococcal toxins can progress for many hours after the start of treatment. Erythemas may appear; inspired oxygen requirements may increase; patients may need a period of positive-pressure ventilation or inotropic support. Close monitoring should therefore be continued until a definite response is established.

Few drugs are as effective as penicillin. Broad-spectrum cephalosporins are not reliable alternatives, nor is ciprofloxacin. Erythromycin is ineffective against a proportion of virulent strains. Poor progress after an initial response may be better managed by increased penicillin dosage (limited only by nausea, drowsiness or other penicillin toxicity) than a change of drug. A change of therapy should be made only for compelling reasons, and clindamycin may be the best alternative.

Complications

A common complication of treatment is a mild febrile reaction to the high penicillin dosage. So long as the patient's physical condition is improving, this is not an indication to change the treatment.

The classic poststreptococcal conditions of rheumatic fever, erythema nodosum and erythema multiforme are rarely seen, but scarlet fever may occur. When it does, it is often severe, with pleural effusion, ascites, renal impairment and hypotension (streptococcal toxic shock syndrome).

Tissue necrosis occasionally occurs in affected skin or lymph nodes even when the bacteraemia is relatively easily controlled.

Gram-negative septicaemia

Introduction and epidemiology

Gram-negative septicaemia almost always originates from abdominal or pelvic pathology. In young age groups appendicitis, trauma or gastroduodenal surgery are the commonest predispositions. In middle age, gallbladder or biliary disease and bowel surgery are more common. In the elderly, or those such as multiple sclerosis sufferers, the urine is often colonized with organisms and is an important source of bacteraemia. *Escherichia coli* is by far the commonest cause of Gram-negative bacteraemias.

Clinical features

While many Gram-negative bacteraemias are continuous, some are intermittent, causing brief episodes of fever once or twice daily or weekly (Fig. 19.6). In either case, rigors, tachycardia and a lowered blood pressure accompany the fever. In continuous bacteraemias the blood pressure can be sufficiently low to compromise cerebral and renal perfusion, while in the intermittent type, it usually normalizes when the bacteraemia ceases after 30–90 min.

A significant prolongation of the prothrombin time tends to occur early in Gram-negative bacteraemias, and elevation of fibrin degradation products indicates DIC. The white cell count usually shows a neutrophilia, but

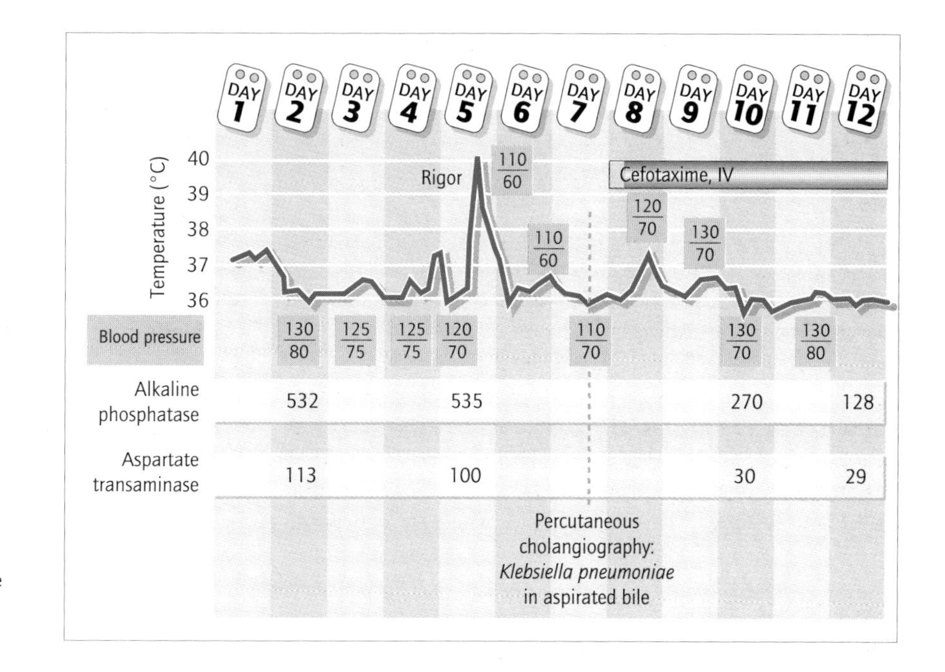

Fig. 19.6 Intermittent Gram-negative bacteraemia: this patient's gallbladder was colonized with *Klebsiella pneumoniae*; his fevers were accompanied by rigors and hypotension; the fever and liver function improved with antibiotic treatment.

in overwhelming sepsis there may be a low white cell count.

Diagnosis

Blood cultures are essential, and should be combined with attempts to obtain specimens from infected sites of origin of the bacteraemias. Imaging of the pancreaticobiliary system or renal drainage system is often helpful. Disease of the bowel may reveal itself by causing pain or bowel dysfunction when ulcer disease, malignancy or diverticulitis are the origin of the infection. Patients with colonized urine, however, often deny urinary symptoms.

Management

Immediate management should include empirical antimicrobial chemotherapy and a search for any underlying condition. Intensive supportive measures are also often required. The choice of antimicrobial chemotherapy needs to include initial cover for Enterobacteriaceae, enterococci and anaerobes. Combinations of drugs are recommended. A broad-spectrum cephalosporin is effective against many coliforms, and can be combined with metronidazole for its antianaerobic effect. Meropenem is effective against both Gram-negative rods and anaerobes, and has some action against enterococci. If enterococcal sepsis is likely, amoxycillin plus gentamicin is indicated, and will also be effective against many Gram-negative rods, but requires the addition of metronidazole if an antianaerobic action is also required.

Typical regimens for treating Gram-negative bacteraemia

Cefotaxime i.v. 1–2 g 8-hourly (neonate, 100–200 mg/kg daily; child, 150–200 mg/kg daily, both in two to four divided doses); *or*

Ampicillin i.v. 500 mg–1 g 4–6-hourly (child under 10 years, 50–100 mg/kg daily in four to six divided doses); *plus*

Gentamicin by slow i.v. injection 2–5 mg/kg daily in three divided doses (child under 2 weeks, 3 mg/kg 12-hourly; 2 weeks–12 years, 2 mg/kg 8-hourly). Plasma concentrations should peak (1 h after dose) not above 10 mg/l, trough not above 2 mg/l; *plus*

Metronidazole rectally 1 g 8-hourly or i.v. 500 mg 8-hourly (child, any route, 7.5 mg/kg 8-hourly).

Alternatives

Meropenem i.v. 1 g 8-hourly (2 g 8-hourly for meningitis) (child 3 months–12 years, 10–20 mg/kg 8-hourly).

Imipenem (with cilastatin) i.v. 2–4 g daily in three or four divided doses (child over 3 months, 60 mg/kg daily in four divided doses—maximum 2 g/day).

For pseudomonal infections

Gentamicin or another aminoglycoside may be given *with* (but not mixed in the same syringe or infusion) azlocillin i.v. 2–5 g 8-hourly (neonate, 100 mg/kg 12-hourly; 7 days–1 year, 100 mg/kg 8-hourly; 1–14 years, 75 mg/kg 8-hourly); *or*

Piperacillin i.v. all ages, 200–300 mg/kg daily in four to six divided doses.

The antibiotic spectrum may be narrowed, or the least toxic drugs selected, when the results of culture are available.

Complications

The most important complication is failure to respond to appropriate antibiotic treatment. The usual reason for this is loculated infection, in which the organisms are inaccessible to the antibiotic.

Follow-up of patients with bacteraemia

Patients responding to treatment of their bacteraemia should be reviewed during convalescence as recurrence sometimes occurs. A particular danger is a persisting undetected predisposition (such as a carcinoma of the colon or a biliary stricture) or unrecognized complicating endocarditis. Examination should always be performed to detect abdominal tenderness, an enlarged liver or gall-bladder or developing heart murmurs. It is also useful to review the erythrocyte sedimentation rate or C-reactive protein, as a rise in either can give early warning of continuing infection.

20 Pyrexia of Unknown Origin

Introduction: causes of pyrexia of unknown origin, 379

Initial assessment, 380
Epidemiological database, 380
Evolution of the feverish condition, 380

Physical examination, 381
Localized bone or joint pain, 381
Soft systolic murmurs, 382
Subtle skin rashes, 382
Mild meningism, 382
Mild localized abdominal tenderness, 382
Chest X-ray, 382

Initial laboratory investigations, 383
Granulocyte counts, 383
Lymphocyte count, 384
Platelet count, 384
Red cells, 384
Stained blood film, 384
Non-specific tests for inflammation, 384
Blood biochemical tests, 384
Urine biochemical tests, 384

Initial microbiological investigations, 385
Microscopy of easily obtained specimens, 385
Initial cultures, 385
Initial serological investigations, 386

Interpretation of initial findings, 386

Serological tests (and probes) for infections, 386

Tests for connective tissue and granulomatous diseases, 387
Elevated ESR, 387
Autoantibodies, 387

Tissue diagnosis (imaging and biopsy), 388
Ultrasound scans, 388
Echocardiography, 388
Computed tomographic scans, 388
Isotope scans, 388
Magnetic resonance scans, 388
X-ray techniques, 389
Laparoscopy, 390
Laparotomy and thoracotomy, 390
Making the most of biopsy material, 390

Trials of therapy, 391
Risks of trials of therapy, 391
Uses of trials of therapy, 391

When no improvement can be obtained in PUO, 392

Introduction: causes of pyrexia of unknown origin

A pyrexia of unknown origin (PUO) is a raised body temperature for which there is no obvious cause. Fever is defined as a body temperature reaching at least 37.8–38°C. Many trivial and self-limiting infections cause short-lasting fevers. These include many viral infections whose formal diagnosis would be tedious and unproductive. They are excluded from the list of conditions to be investigated in PUO by making a more detailed definition: pyrexia of unknown origin is a fever which has continued or frequently recurred for a period of at least 3 weeks, and for which no cause has been found after one or more medical consultations or assessments.

Because fever is a response to cytokine signals, via endothelial activation, infection is not the only type of disease which can cause fever. Non-infectious causes of PUO are common and important. The causes of fever can be described under six main headings:
1 Infections (45–55% of cases).
2 Malignancies (12–20% of cases).
3 Connective tissue disorders (10–15% of cases).
4 Hypersensitivity disorders.
5 Rare metabolic conditions.
6 Factitious fever (fever induced deliberately by the patient).

A low-grade fever is common in embolic or thrombotic conditions, such as pulmonary embolism, probably because of endothelial damage and activation.

Important infectious causes of pyrexia of unknown origin
1 Tuberculosis (usually non-pulmonary, miliary or cryptic).
2 Sepsis (especially in a hollow organ) or abscess, e.g. dental abscess.
3 Imported diseases (such as typhoid fever or brucellosis).
4 Infective endocarditis.
5 Infections with a long time course (such as infectious mononucleosis, toxoplasmosis or Q fever).

Important malignant causes of pyrexia of unknown origin
1 Lymphomata (Hodgkin's and non-Hodgkin's).

Continued

2 Leukaemias (especially monocytic or myelomonocytic types).
3 Histiocytoses (usually in small children).
4 Renal adenocarcinoma.
5 Primary hepatic carcinoma.
6 Rare, atrial myxoma.

Connective tissue disorders which may present as pyrexia of unknown origin
1 Sarcoidosis (and rare granulomatous diseases).
2 Polyarteritis nodosa.
3 Systemic lupus erythematosus.
4 Wegener's granulomatosis.
5 Juvenile rheumatoid arthritis.
6 Dermatomyositis.

Common causes of hypersensitivity with fever

Drugs
1 Sulphonamides (including Salazopyrin).
2 Penicillins.
3 Rifampicin.
4 Isoniazid.
5 Streptomycin.
6 Phenytoin.
7 Methyldopa.

Environmental factors
1 Fungus-infected hay (farmer's lung).
2 Bird proteins (pigeon-fancier's lung).

Rare metabolic causes of pyrexia of unknown origin
1 Porphyrias (acute intermittent or mixed types).
2 Familial relapsing polyserositis (familial Mediterranean fever).
3 Some cases of vasoactive intestinal polypeptide-producing tumour (VIPoma) and glucagonoma.

Most causes of PUO are not exotic diseases; they are common diseases presenting without their usual symptoms and physical signs.

Initial assessment

Assessment of a patient with PUO must include a search for both infectious and non-infectious conditions. Screening procedures work more efficiently if the population to be screened is preselected to include a high proportion of likely positives. The physician can choose initial and follow-up investigations in a structured way by using an epidemiological and clinical database to indicate which tests are likely to be useful. Otherwise an almost infinite range of tests could be performed, each with a different sensitivity and specificity, presenting the investigator with a hugely complex task when interpreting results.

Epidemiological database

While this applies particularly to infection, it also includes the history of exposure, predisposition and protection for other types of febrile disease. A patient may have been exposed to infection by known contact with other cases, by travel, food, water, occupation or recreation, or by association with animals, including farm animals or pets.

Exposure to allergens should also be sought; this could be iatrogenic exposure to antibiotics or other drugs, or to environmental agents at work, home or play. These might include bird proteins (as in pigeon-fancier's lung), organic dusts such as cotton or contaminated hay (byssinosis and farmer's lung) or industrial dusts and vapours, including vinyl chloride monomer or beryllium, which can both cause inflammatory or granulomatous lung disease.

Predisposition may be indicated by a family history either of such rare disorders as relapsing serositis, or to Reiter's syndrome and connective tissue diseases, including systemic lupus erythematosus and rheumatoid arthritis.

The patient may have a history of exposure to carcinogenic agents, such as radiation, including intensive radiotherapy. A history of sustained immunosuppressive therapy, for instance with cyclosporin, also indicates an increased likelihood of some malignant diseases because of impaired 'immune surveillance'.

Protection or resistance may be the result of natural immunity following previous infection, or it can be induced by immunization. Temporary resistance can also be obtained by the use of chemoprophylaxis, as for malaria, or passive immunization, as for hepatitis A. While none of these confers absolute protection from a condition, they reduce the likelihood of a particular disease and allow the investigator to choose priorities in the differential diagnosis of the fever.

Evolution of the feverish condition

Although the current complaint may be fever, this might have been preceded by symptoms either of an earlier stage of the disease, or by a recent precipitating condition. Such a history can point to appropriate diagnostic tests at an early stage in investigation.

The severity or behaviour of the fever itself is not often helpful. The tertian fever of malaria is an exception, occurring when the disease is well established. Similarly, the undulant fever of chronic brucellosis, the escalating fever of early typhoid and the relapsing fever of *Borrelia recurrentis* infection can be diagnostically helpful (Fig. 20.1), but they do not always occur in their classic form.

Developing viral infections are often marked by prostration, myalgia, arthralgia and shivering attacks. Transient diarrhoea, constipation, sore throat or cough could hint at the systemic site of the problem. Bacterial infections may similarly produce transient localizing symptoms. Abscesses and loculated sepsis are often accompanied by intermittent bacteraemias, indicated by rigors: severe shaking chills which make speech and other movement difficult (Fig. 20.2). In postinfectious conditions and connective tissue diseases transient rashes can occur and recur. They may be visible only when the temperature is highest (Fig. 20.3) or when the skin has been warmed by bathing.

If fever is due to a postinfectious disorder, the precipitating infection probably occurred 10–14 days earlier. Typically it would be a sore throat or a viral-type infection with respiratory symptoms or a rash. Next most likely would be a gastrointestinal complaint.

Physical examination

Sometimes subtle physical signs can be helpful in making a diagnosis; they should always be sought, and acted upon when found.

Localized bone or joint pain

Bone or joint pain can be extremely mild at the onset of skeletal infections, appearing as discomfort, stiffness or, in children, reluctance to move the affected part. When they affect the leg or lumbar spine, they are often more obvious on standing or walking. Such symptoms precede X-ray changes by many days or weeks, and should therefore be further investigated by other imaging techniques (see Chapter 14).

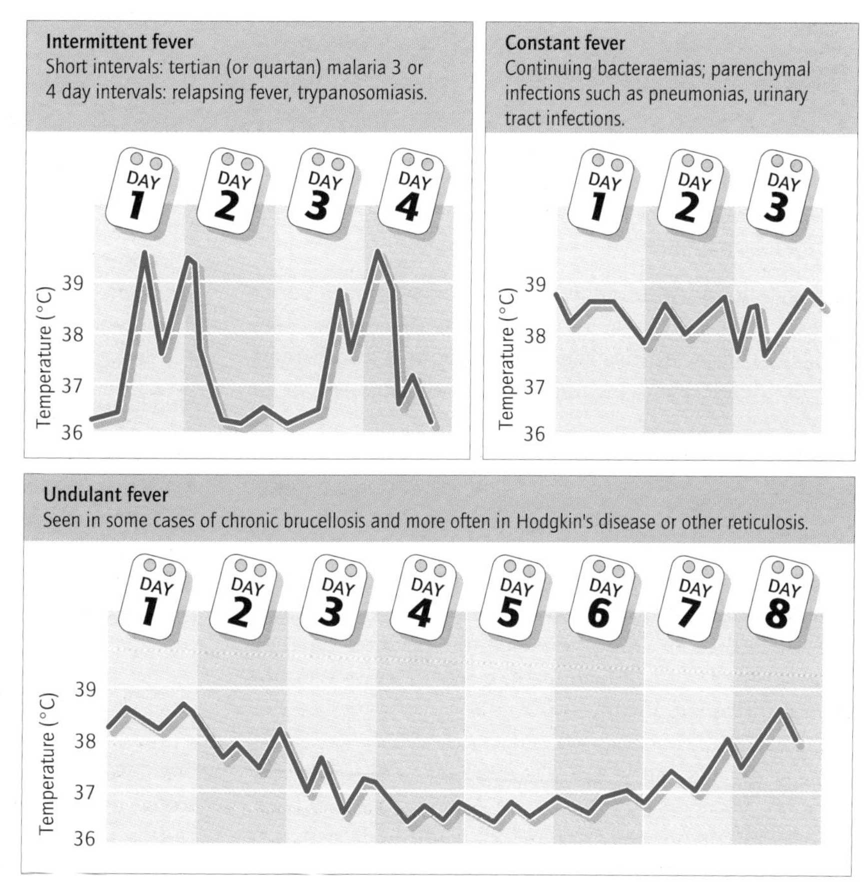

Fig. 20.1 Some classic patterns of fever. (a) Intermittent, constant and undulant fever. (b) Swingeing fever with periods of remission, high swingeing fever and a normal temperature chart.

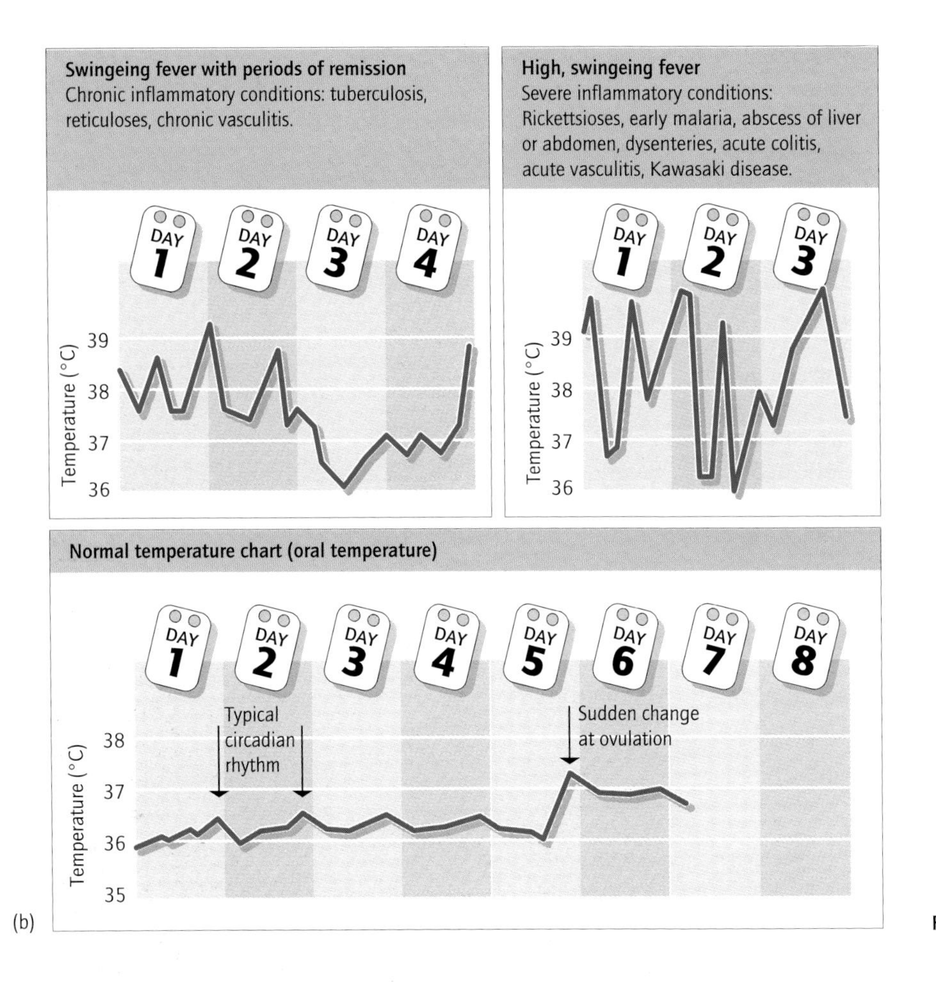

Fig. 20.1 *Continued*

Soft systolic murmurs

Soft systolic murmurs may be flow murmurs in feverish patients, but may also be signs of endocarditis, pericarditis or myocarditis. They should be reassessed regularly to detect changes, and further investigated at an early stage.

Subtle skin rashes

Subtle skin rashes may be the only sign of embolic or vasculitic phenomena. Showers of petechiae, Osler's nodes or simply small, vasculitic lesions of the digits can be signs of endocarditis or immune vasculitis. Splinter haemorrhages, small retinal haemorrhages and cytoid bodies have a similar significance.

Mild meningism

Mild meningism is easy to pass off as 'difficult' or 'uncooperative' behaviour but is a warning sign of central nervous system disease such as tuberculous meningitis. It is important to suspect such disease before more definite (and often more irreversible) signs develop. If there is doubt, early investigation is essential.

Mild localized abdominal tenderness

Mild localized abdominal tenderness is also easily dismissed. Some patients may be excessively sensitive, or trying to help by pointing up every possible sign, but it is difficult to describe vague visceral discomfort. Trivial abdominal signs may indicate peritoneal or subphrenic pathology. If vague signs persist, they should be followed up.

Chest X-ray

The chest X-ray is so important in assessing PUO that initial examination is not complete without it. Extensive pulmonary consolidation or cavitation can exist with few or absent physical signs (Fig. 20.4). Granulomas, infiltra-

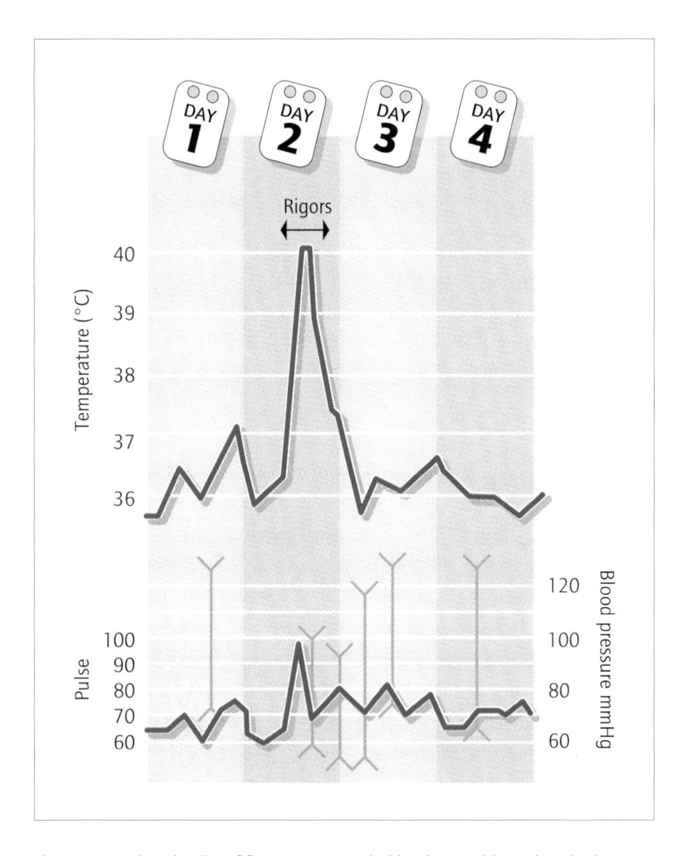

Fig. 20.2 Isolated spike of fever accompanied by rigors. This patient had a *Klebsiella pneumoniae* infection of the obstructed biliary tree, with intermittent bacteraemias.

tions, small pleural effusions and mediastinal swellings are other important abnormalities immediately detectable by X-ray.

Initial laboratory investigations

Granulocyte counts

High neutrophil counts often occur in pneumococcal and *Haemophilus influenzae* infections, in which the white cell count may reach $15–25 \times 10^9$/l. Intermediate elevations of around $12–16 \times 10^9$/l are common in *Streptococcus pyogenes* infections and in the presence of abscesses. While most bacterial infections are accompanied by neutrophilia, important exceptions include early *Staphylococcus aureus* infections, even bacteraemias. However, the disease may still be suspected if the differential count is performed, because of the high proportion of neutrophils, often 90–95%. Typically, enteric fevers and brucellosis induce neutropenia but in early infection there may be a neutrophilia as inflammation begins.

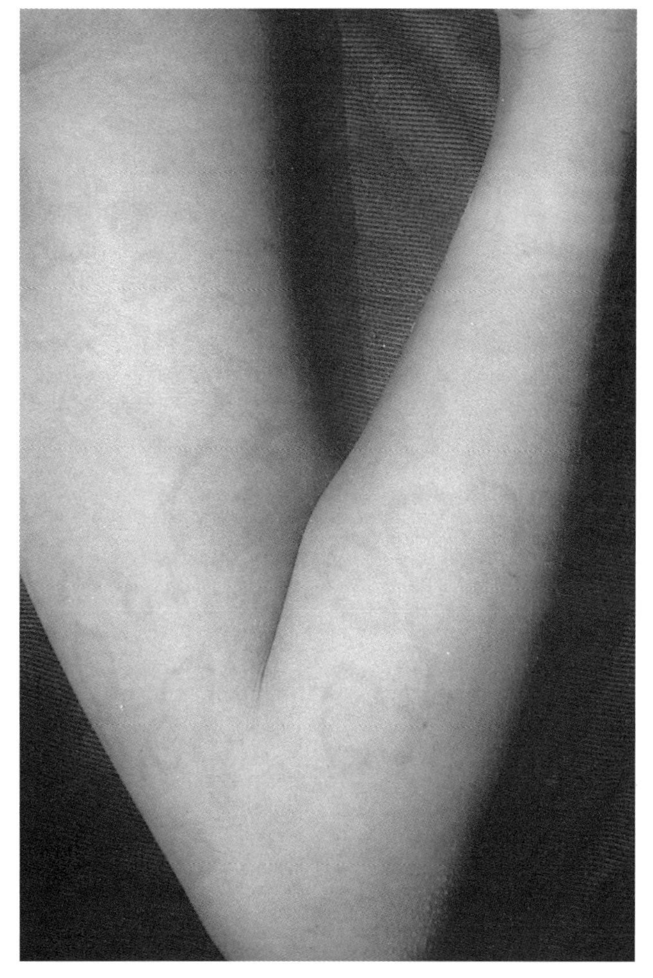

Fig. 20.3 Postinfectious rash: erythema marginatum appearing during an episode of fever.

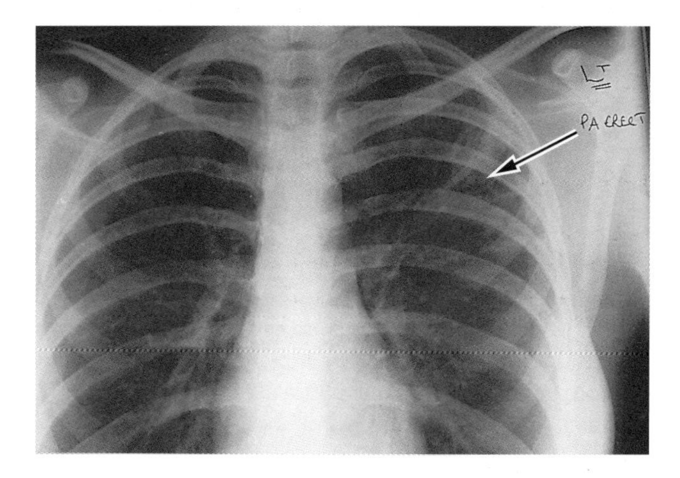

Fig. 20.4 Chest X-ray of a midwife returning from work in Africa with persisting low-grade fever and cough. The chest was clinically clear, but this cavitating opacity (arrow) prompted investigation and treatment for tuberculosis.

Eosinophilia may indicate an invasive helminth infection. It is a helpful sign in early and late schistosomiasis, filariasis, strongyloidiasis and liver fluke infection.

Lymphocyte count

The lymphocyte count is usually normal or low in viral infections, with the exception of the mononucleoses. The proportion of lymphocytes may appear raised because of mild neutropenia, possibly caused by the toxic effects of interferon. Scanty atypical or activated mononuclear cells are often seen in acute viral infections such as hepatitis A or rubella, but these do not approach the numbers seen in the mononucleoses. Pertussis toxin causes marked lymphocytosis ($15–25 \times 10^9/l$), which is an important diagnostic feature of pertussis.

Platelet count

The platelet count is rarely altered. Exceptions are severe Gram-negative sepsis in which intravascular coagulation may deplete platelet numbers, and malaria in which a similar mechanism operates. A reduced platelet count is seen in established dengue fever, and is usual in dengue haemorrhagic fever. A high platelet count is usual in established Kawasaki disease and other severe vasculitides, and is a rare finding in disseminated tuberculosis.

Red cells

Apart from obvious parasitization, distortion and granule formation in malaria, abnormalities of red cell production or survival can alter the blood film. In parvovirus B19 infection, erythropoiesis is arrested by viral invasion of red cell progenitors. This causes a fall in the reticulocyte count. Conversely, the reticulocyte count will rise, causing an increase in the mean corpuscular volume, if haemolysis occurs, as in *Mycoplasma pneumoniae* infections.

Stained blood film

The stained blood film can be diagnostic if it contains tropical parasites such as microfilaria or trypanosomes. It may be worth examining films taken in special circumstances, e.g. at night or after a dose of filaricide, to increase the likelihood of detecting parasitaemia.

Non-specific tests for inflammation

Plasma viscosity and the erythrocyte sedimentation rate
(see Chapter 1)

The erythrocyte sedimentation rate (ESR) is moderately raised (e.g. 35–50 mm/h) in acute infections and other inflammatory conditions. It is often markedly raised (e.g. over 70 mm/h) in the presence of persisting abscesses, in some pneumonias, such as legionnaires' disease, and in severe connective tissue diseases and hypersensitivity reactions.

C-reactive protein

C-reactive protein is an acute-phase protein which responds rapidly to changing levels of inflammation. It is elevated to an unpredictable extent in bacterial infections, viral infections, parasitic diseases and non-infectious conditions, and may be high when the ESR is near normal. Its best use is in monitoring the response of inflammation to treatment (as in the management of endocarditis).

Blood biochemical tests

Aspartate transaminase and alkaline phosphatase levels

Slight elevations of transaminases are common in acute viral infections, but rarely to levels above 60–100 IU/l. Higher levels point to specific liver inflammation or occasionally to severe and widespread tissue damage. Elevated alkaline phosphatase levels are usually of liver origin. They can indicate space-occupying lesions in the liver, but do not distinguish many small lesions, such as granulomata, from larger lesions, such as abscesses or tumours.

Other enzymes

Blood amylase levels may be high in pancreatitis or inflammation of the salivary glands. Creatine kinase elevations are seen in myositis, including myocarditis, and in toxic shock syndromes.

Urine biochemical tests

Urine biochemical tests for protein, blood, nitrite and excess neutrophil lactate dehydrogenase (LDH) may give early indication of infection (see Chapter 2). Products of haemolysis may be present in severe malaria, and bilirubin may be pesent when jaundice is too mild for clinical detection. Screening tests for porphyria are carried out on urine samples (preferably taken during episodes of fever).

Initial microbiological investigations

Microscopy of easily obtained specimens

This is a rapid diagnostic procedure which is easy to overlook. As well as a search for neutrophils and pathogens in urine (Fig. 20.5) and stools (see Chapter 2), investigation of aspirates from abscesses and enlarged organs such as lymph nodes and the spleen can aid early diagnosis. Splenic aspirate can reveal Leishman–Donovan bodies in visceral leishmaniasis, and fine-needle aspirate of lymph nodes may be stained to demonstrate granulomas, malignant cells or acid–alcohol-fast bacilli. Bronchial aspirate can be examined by various stains or polymerase chain reaction (PCR) techniques to reveal fungal, herpesvirus or respiratory virus infections. Actinomycosis can be diagnosed by the Gram-stained appearance of 'sulphur granules' from pus.

Initial cultures (see also Chapter 2)

Blood cultures

Blood cultures are essential, and should be obtained before antimicrobial treatment whenever possible. At least two sets should be taken, at different times and preferably when the temperature is rising or has just risen, as fever is induced when pathogens are released into the blood.

Other specimens for bacterial culture

Urine culture is also important, as both acute and chronic urinary infections can exist without localizing symptoms or signs. The early morning urine (EMU) can be cultured for mycobacteria. Stool cultures may be positive in enteric fevers when blood cultures have failed for any reason. It may be important to obtain cerebrospinal fluid (usually

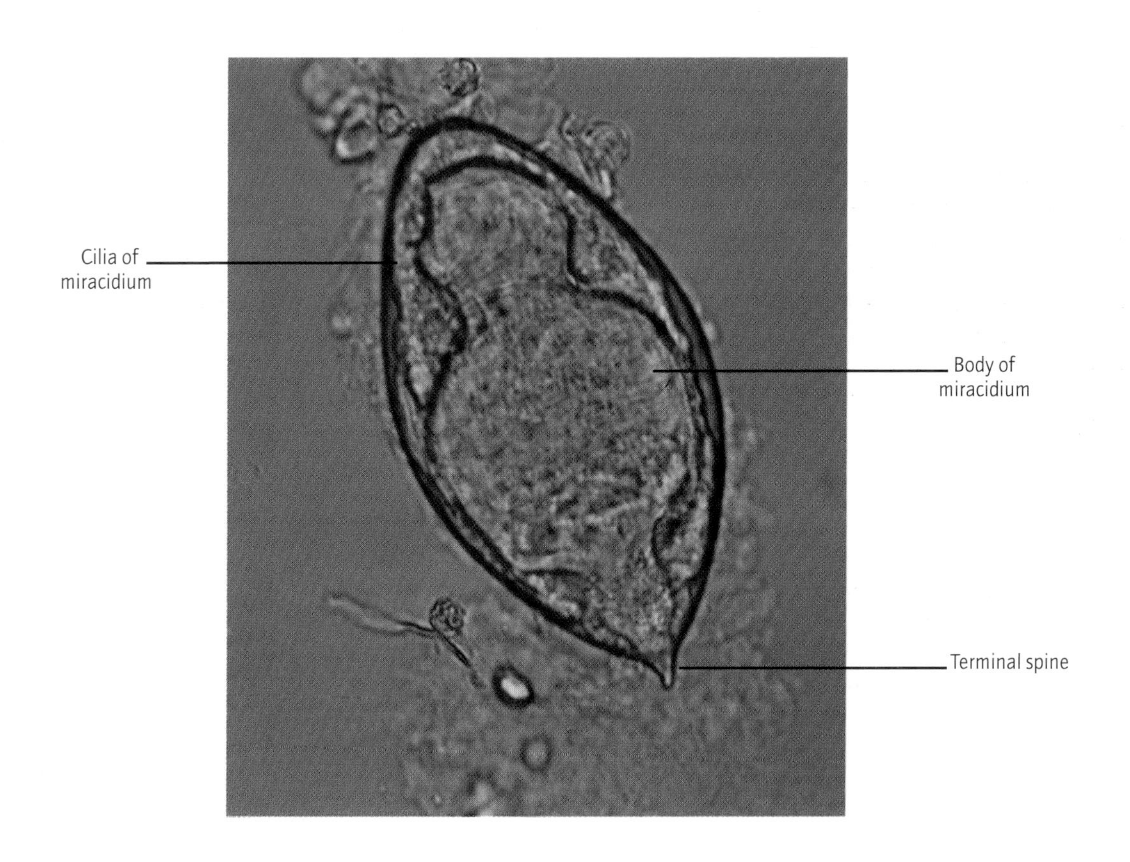

Fig. 20.5 Ovum of *Schistosoma haematobium* in the urine of an engineer with episodic haematuria: he had been working on a river estuary in Africa.

after brain imaging) if meningeal or neurological signs are present.

Pus, discharge or vesicle fluid must always be collected for culture. It is more helpful to the microbiologist if a significant volume can be obtained, rather than just a smear on a swab. To preserve fastidious anaerobes, pus may be inoculated into blood culture media.

If readily obtained specimens are not available, or the patient cannot, for instance, expectorate, specimens should be obtained by specific techniques, such as fine-needle aspiration or bronchoalveolar lavage.

Cultures for virological investigation

Cultures for virological investigation are equally important, especially in the immunosuppressed, and should not be omitted. For respiratory viruses, nose and throat swabs, nasopharyngeal aspirate, bronchial aspirate and lavage specimens may be processed. For enteroviruses, throat swabs, stool and (if appropriate) cerebrospinal fluid may be examined. Urine can be cultured for cytomegalovirus and paramyxoviruses. Respiratory specimens, vesicle scrapings and fluid are appropriate for herpesvirus cultures.

Initial serological investigations

These need not be elaborate, unless the clinical situation demands it, but serum must always be obtained as early as possible, for comparison with later specimens. The need for paired sera, taken 10–14 days apart, means that some serological tests must be deferred. Many tests, however, are useful screening tests and can provide early results. A positive monospot test is virtually diagnostic of Epstein–Barr virus infection and can solve the diagnosis in patients lacking a significant mononucleosis. As well as immunoglobulin M (IgM) and other antibody tests for infectious agents, autoantibody tests for connective tissue diseases are useful early tests.

Tuberculin test

The tuberculin test has poor specificity, but is helpful if strongly positive. Tuberculosis is common among PUO patients, so a tuberculin test should be performed as soon as possible. A strongly positive Mantoux test, or a Heaf reaction of grade three or more, indicates active tuberculosis unless proved otherwise.

Initial work-up plan for pyrexia of unknown origin

1 Full history.
2 Physical examination.
3 Chest X-ray.
4 Blood count, differential count and morphology.
5 Erythrocyte sedimentation rate (and/or C-reactive protein).
6 Liver and renal function tests.
7 Dipstick urine tests.
8 Microscopy of blood film and urine (cerebrospinal fluid (CSF), stool and lesion fluid if indicated).
9 Culture of blood, urine and respiratory specimens (CSF, stool, pus and lesion fluid if indicated).
10 Acid-fast stain and tuberculosis cultures (sputum, gastric aspirates, early morning urine and CSF) if indicated.
11 Heterophile antibody detection test.
12 Tuberculin test.
13 Save baseline serum.

Interpretation of initial findings

The results of initial tests will become available during the first week of investigation. Many diagnoses will be made from these results, including bacteraemias, culture-positive infective endocarditis, pneumonias, urinary tract infections, pulmonary tuberculosis, Epstein–Barr virus infection and common parasitic diseases. Some malignancies such as leukaemias will also be revealed.

While results are accumulating, the physical examination should be reviewed a number of times, both to check on doubtful signs and to detect new or changing ones. A cardiac murmur can suddenly become obvious, finger clubbing can develop or lymph nodes enlarge.

Further investigation is indicated if no diagnosis is apparent or if initial assessment suggests a line of investigation. In some cases the patient's condition is poor, and a wider range of tests must be completed without delay. Three main types of investigation are possible:

1 Further serological tests for infections.
2 Extensive tests for connective tissue or immunological diseases.
3 Tissue diagnosis (imaging and/or biopsy and culture).

Serological tests (and probes) for infections

These tests can be used to follow on from initially negative microscopy and culture tests, and to investigate a wider range of possible infectious conditions. The choice of investigations will depend on the results of epidemiological and clinical investigations already carried out. Examples of useful clinical indicators include:

1 Anaemia

Haemolytic: consider malaria, *Mycoplasma pneumoniae* infection, *Escherichia coli* O157 infection with haemolytic uraemic syndrome, clostridial sepsis and rarities such as thrombotic thrombocytopenic purpura (in human immunodeficiency virus (HIV) infection), bartonellosis or babesiosis.

Of chronic disease: consider tuberculosis.

With haemophagocytosis: consider tuberculosis, brucellosis, leishmaniasis, cytomegalovirus infection (and immunosuppression).

2 Eosinophilia: consider any tropical helminth infection, schistosomiasis or liver fluke, as well as the rare trichinellosis (also occurs in Europe).

3 Hepatocellular disorder: consider viral hepatitis, leptospirosis, rickettsioses, Epstein–Barr virus infection, primary cytomegalovirus infection and rarities such as bartonellosis and viral haemorrhagic fevers.

4 Cholestatic changes: consider granulomatous conditions such as tuberculosis, brucellosis, Q fever and rare histoplasmosis or cryptococcosis; also consider single or multiple liver abscesses, or cystic conditions and rare schistosomiasis and liver flukes.

5 Sterile pyuria: consider tuberculosis, brucellosis, or rare *Chlamydia trachomatis* or *Mycoplasma hominis* infection.

6 Persistent cough or respiratory symptoms: consider tuberculosis, enteric fever, tularaemia, respiratory viral infections, particularly paramyxoviruses, adenoviruses and rare hantavirus pulmonary syndrome, as well as pulmonary eosinophilia due to migrating parasites including *Strongyloides*.

7 Persisting meningism, encephalopathy or fits: consider tuberculosis, borreliosis, any arboviral encephalitis, cysticercosis and rare listeriosis, syphilis or neurobrucellosis.

8 After travel abroad: consider zoonoses such as brucellosis, Q fever, tularaemia, hantavirus infections, leptospirosis, schistosomiasis, also the rare melioidosis and malarial tropical splenomegaly.

Serological and probe tests for infection in pyrexia of unknown origin

1 Antigen detection (blood antigen for hepatitis B or D, cerebrospinal fluid (CSF) antigen for cryptococcosis, tissue antigen for cytomegalovirus, blood probes for hepatitis C RNA and viral haemorrhagic fever RNAs, CSF probes for herpes simplex virus DNA and mycobacterial DNA.

2 Single-serum antibody tests (immunoglobulin M for acute viral infections, borreliosis, toxoplasmosis or malarial tropical splenomegaly, gel precipitin test for amoebiasis, diagnostic titres of *Legionella* antibodies).

3 Paired-serum antibody tests (for atypical pneumonias, leptospirosis, yersiniosis).

Some patients take longer than the traditional 10–14 days to produce diagnostic titres or rises of antibody levels. A serological diagnosis cannot be made before 6–9 weeks in some cases of legionellosis, borreliosis and leptospirosis. Even the antistreptolysin O titre (ASOT) may take 3 or 4 weeks to reach diagnostic levels. It is always worth taking a late serum specimen if a serological diagnosis has not been evident in the first month of fever.

Tests for connective tissue and granulomatous diseases

Connective tissue and granulomatous diseases are good mimics of infection. Some, such as systemic lupus erythematosus, are particularly like viral infections, presenting with fever, neutropenia and sometimes rash. Others may cause neutrophilia, for example polyarteritis nodosa and Wegener's granulomatosis. These further mimic bacterial disease by producing focal inflammatory lesions in the respiratory system. The granulomatous disease sarcoidosis must be distinguished from tuberculosis, especially when it presents without its typical pulmonary features.

Elevated ESR

An elevated ESR is a prominent finding in these conditions. While a high ESR is not exclusive to connective tissue diseases, it is unwise to dismiss such a diagnosis unless the finding can be otherwise explained.

Autoantibodies

Autoantibodies are detectable in many connective tissue diseases, and may be diagnostic. Sarcoidosis is not associated with autoantibodies but if there is lung involvement the serum angiotensin-converting enzyme (SACE) level is elevated, occasionally even when pulmonary function and chest X-ray appearances are normal.

Serological tests for connective tissue disorders
1 Rheumatoid factor.
2 Antinuclear factor, double-stranded-DNA antibodies.
3 Other specific antibodies (smooth-muscle, mitochondrial, thyroid, etc.).
4 ANCAs (antineutrophil cytoplasmic antibodies: p-ANCA and c-ANCA).
5 SACE (serum angiotensin-converting enzyme; elevated in pulmonary sarcoidosis).

Tissue diagnosis (imaging and biopsy)

The availability of accurate imaging techniques has greatly simplified the investigation of PUO. They can be used to demonstrate the anatomy of tissues and organs, to test their function and to detect inflammation within them. Imaging can identify lesions suitable for biopsy, and assist in guiding the biopsy needle.

Ultrasound scans

Ultrasound scans are non-invasive and relatively inexpensive. The technique depends on showing differences in the sonic density of tissues and will delineate the anatomy of organs, and lesions within organs, especially if these are outlined by thin fatty planes. It is particularly useful for demonstrating enlargement of the liver, gallbladder, spleen or kidneys, and for detecting abscesses, cysts or space-occupying lesions. It can demonstrate soft-tissue swelling and periosteal elevation in early osteomyelitis. It can also define pelvic lesions, particularly in and around the female genital tract.

Its limitation is its requirement for expert operation and interpretation, as the anatomical definition is not perfect and the ultrasonic beam makes shadows which can be confused with lesions. Nevertheless, it is regularly used in investigation of PUO cases, and is often diagnostic.

Echocardiography

Echocardiography is ultrasonic imaging of the heart which can show a two-dimensional picture of the beating heart and its valves. Doppler echocardiography demonstrates the direction, velocity and turbulence of blood flow. These techniques can show vegetations on heart valves, abscesses of the valve rings and septum, pericardial effusions and even dilatations of the coronary arteries. The only limitation of echocardiography is its occasional inability to demonstrate small lesions, giving rise to false negative results.

Computed tomographic scans

Computed tomographic (CT) scans are high-definition computed tomograms derived from axial X-rays of the body. They accurately reveal the anatomy of organs and can demonstrate lesions of 0.5 cm or less. Contrast media can demonstrate increased blood supply to inflamed lesions by enhancing the radiodensity of affected tissue. CT-guided biopsy and aspiration can be performed (Fig. 20.6).

Isotope scans

Isotope scans involve the intravenous injection of radioisotopes which will become concentrated in abnormal tissues. Subsequent scanning with an appropriate detection system, usually a gamma camera, will produce a picture demonstrating the affected area. The most commonly used isotopes are technetium, which demonstrates increased blood flow to inflamed tissues, and gallium, which accumulates in areas rich in inflammatory mediators. Technetium is used for bone scans, which can demonstrate inflammatory lesions in bone long before X-rays can show altered bone anatomy. Gallium can reveal abscesses or foci of inflammation in many organs and tissues (Fig. 20.7).

Isotope scans can also be used to demonstrate the relationship between structure and excretory function in the kidneys and liver. Appropriate isotope scans can show whether the isotope is cleared from the organ and concentrated in the excreted fluid, and can then use the concentrated isotope to make an image of the ureter or the bile ducts.

Magnetic resonance scans

Magnetic resonance (MR) scans detect the density of magnetic atoms in the tissues. The most abundant magnetic atom in the body is hydrogen, present in body water. MR scans are therefore good for showing vascularity of tissues, but can also demonstrate subtle variations of the water content. Oedema is easily seen, so inflammation is detectable by MR imaging, even before there is a change in radiodensity (i.e. in X-ray or computed tomographic appearance). So-called 'dynamic' MR scanning techniques can image moving organs such as the heart, and MR techniques can also demonstrate blood flow. It is possible to enhance MR images with

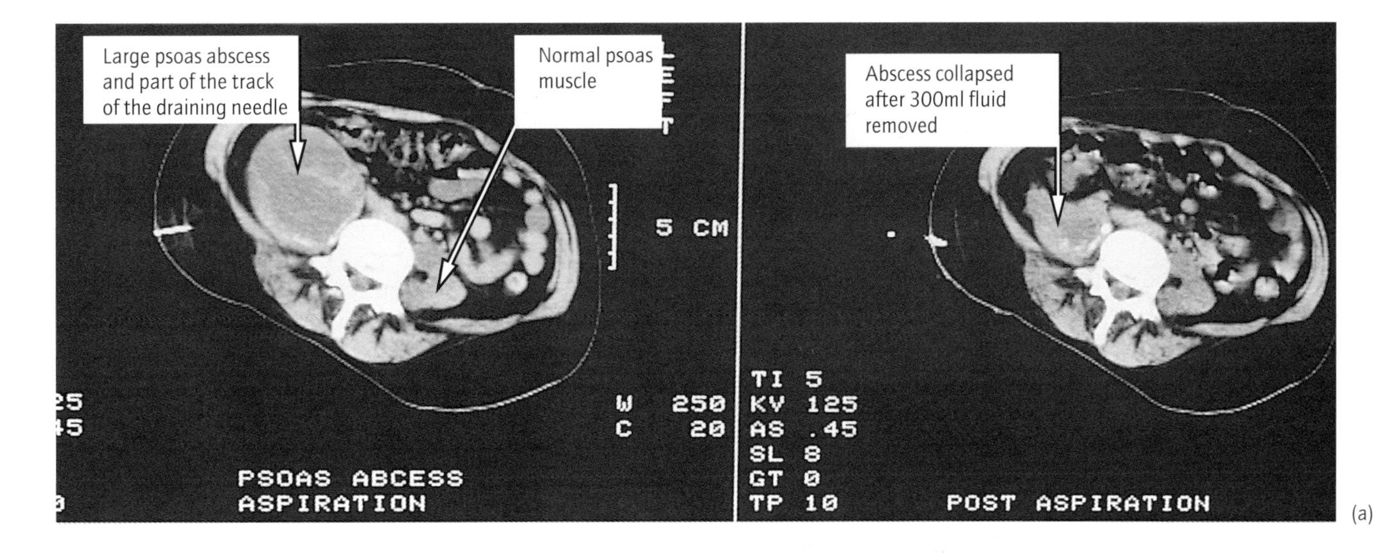

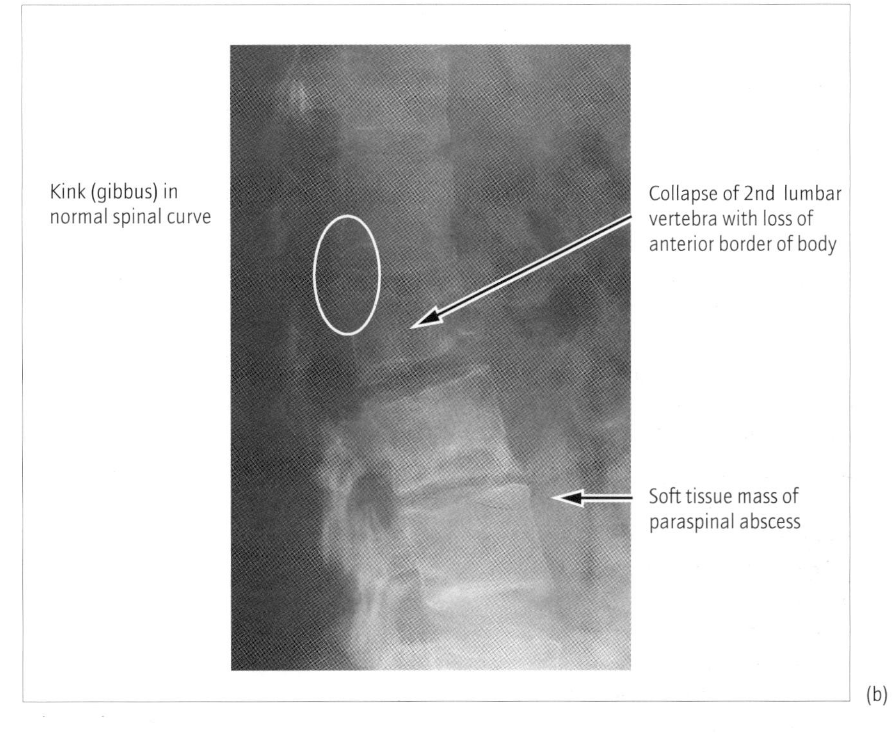

Fig. 20.6 (a) Computed tomographic-guided aspiration of pus from a large psoas abscess: as well as confirming tuberculosis this was a therapeutic procedure – 300 ml of infected material was removed; (b) X-ray appearance of the same abscess.

media containing magnetic atoms, some of which are radioisotopes, but radioisotopes are potentially damaging to some tissues, while magnetic fields are not. A disadvantage of magnetic resonance scans is that the scanners' enormously powerful magnets can disturb moveable magnetic implants, such as pacemakers, haemostatic clips or metal stents. Patients whose implants may have magnetic properties cannot undergo MR scans.

X-ray techniques

X-ray techniques are still important in the investigation of PUO. This is particularly true in bowel disease, where scanning images may not be able to distinguish between the fluid-filled lumen of the bowel and an abscess in the folds of the peritoneum. Contrast studies can demonstrate mucosal disease (Fig. 20.8), perforations and fistulae of the bowel and of other hollow organs.

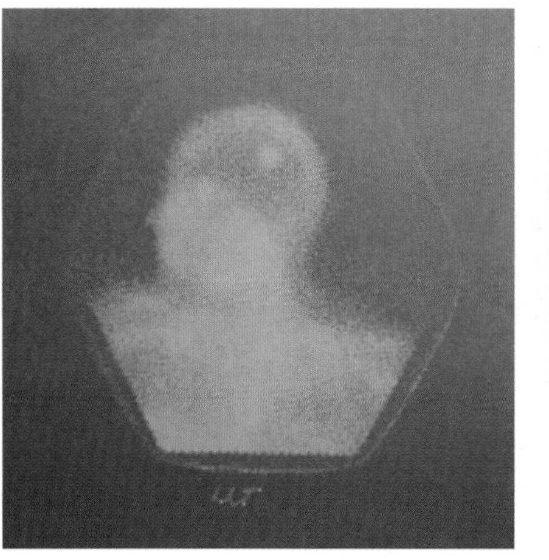

(a)

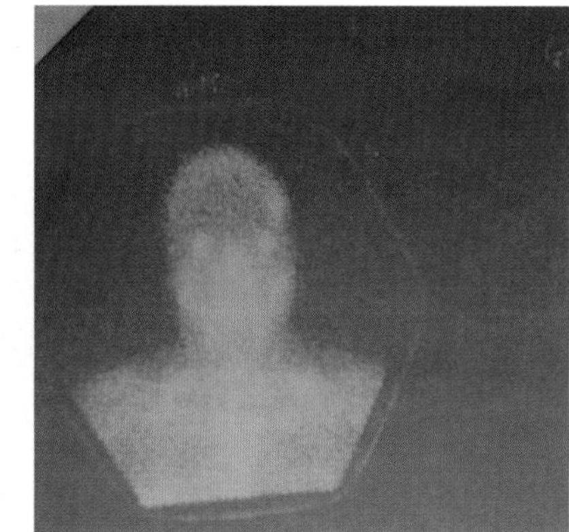

(b)

Fig. 20.7 (a,b) Gallium scan showing concentration of activity in the left parietal bone: this teenager with fever, anaemia and weight loss had a localized lymphoma.

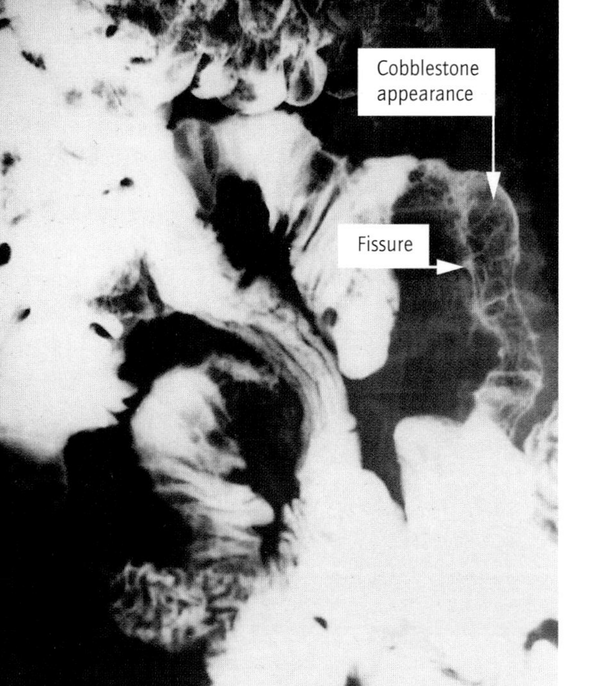

Fig. 20.8 Barium meal examination showing cobblestone change and fissuring in the terminal ileum: this Indian patient had ileal tuberculosis.

Laparoscopy

Laparoscopy allows direct examination of organs in the abdomen or pelvis. It is useful for detecting abnormalities of the abdominal lymph nodes, the liver and the female genital tract. Biopsies can be obtained under laparoscopic control. Other procedures for examining tissue planes or the lumen of the organs include mediastinoscopy, bronchoscopy, and endoscopy of the bowel, biliary tract and urinary system.

Laparotomy and thoracotomy

Laparotomy is rarely necessary in the investigation of PUO. It can, however, be helpful when lesions are demonstrable by imaging of the liver, lung or lymph nodes but biopsy has proved impossible or unproductive. A minilaparotomy or minithoracotomy can then be performed, to obtain an open biopsy. This may permit distinction between, for instance, tuberculosis, sarcoidosis and reticulosis.

Making the most of biopsy material

It is usual to fix biopsy material in formalin for histological examination, but portions should also be retained in sterile water or saline for bacteriological and virological examination, or for special staining, for instance to demonstrate tumour antigens, immune complexes or enzyme activities. Bone marrow, or aspirates from abscesses and cysts, can be cultured like blood cultures or directly inoculated on to culture media. Solid tissues must be treated in the laboratory before inoculation. Smears may be prepared for later virological or bacteriological study. Specimens for viral culture may be refrigerated, but should not be frozen. Although cultures cannot be performed on fixed specimens and serology is rarely possible, DNA can survive fixation and may still be detectable by hybridization techniques.

Imaging and biopsy procedures in the diagnosis of pyrexia of unknown origin

1 Ultrasound scans (especially abdominal and pelvic; B, C).
2 Computed tomographic scan of the head or body (B, C).
3 Magnetic resonance scan of the head or body (B, C).
4 Gallium scan (whole body).
5 Technetium bone scan.
6 Laparoscopy (B, C).
7 Bronchoscopy (B, C).
8 Minilaparotomy and minithoracotomy (B, C).
9 Liver biopsy (C).
10 Lymph-node biopsy (C).
11 Bone marrow biopsy (C).
12 Temporal artery biopsy.
13 Skin lesion biopsy (C).

B indicates that biopsy may be performed; C indicates that standard and tuberculosis culture should be performed.

Trials of therapy

Risks of trials of therapy

Ideally, trials of therapy should be done late or not at all in the investigation of PUO. This is because the drugs used can make further investigation difficult or impossible. Antimicrobial agents can also cause hypersensitivity reactions, which might exacerbate fever or cause problems such as rash, blood disorder or organ failure. They may also prevent recovery of organisms by culture while failing to treat the disease adequately. This is common in enteric fevers, listeriosis, brucellosis and *Streptococcus pyogenes* infections. Antimicrobial therapy alone may fail to resolve abscesses or loculated infections, which must still be identified and drained to cure the condition. Standard therapy for tuberculosis includes broad-spectrum agents such as rifampicin or streptomycin, which will also interfere with culture of pyogenic organisms. Tetracyclines and co-trimoxazole may partly inhibit malarial parasites or other protozoa, masking parasitic infections.

Finally, corticosteroids can inhibit the immune responses which make serodiagnosis possible, and can abolish a positive tuberculin test response. Corticosteroids can sometimes mask and allow untreated infection to advance, leading to complications such as perforation of hollow organs or dissemination of infection.

Risks of trials of therapy

1 Reduced usefulness of diagnostic cultures.
2 Modification of infection without cure.
3 Adverse reaction to therapy complicating the illness.
4 Corticosteroids may reduce the usefulness of immunological tests.
5 Corticosteroids may permit progressive infection with reduced signs of inflammation.

Uses of trials of therapy

In spite of the risks, there are situations where trials of therapy are useful. The most important of these are cases in which a diagnosis is strongly suspected after investigation, but cannot be proved. If the suspected diagnosis is a specific infection, treatment may be attempted, using the narrowest possible antimicrobial spectrum. Obtaining the expected response is then supportive of the presumed diagnosis.

Care must be taken not to treat another condition inadvertently; for instance, trial of antituberculosis treatment should be performed using isoniazid (INAH), pyrazinamide, ethambutol, etc., rather than drugs also active against pyogenic organisms. Patients undergoing a trial of corticosteroids should be examined frequently, and X-rays or imaging results should be reviewed, in case infection is potentiated by immunosuppression. Failure to improve on reducing doses of steroids should cause review of the presumed diagnosis.

When the patient's condition is critical a 'blind' trial of therapy may be unavoidable. All possible specimens should first be obtained for examination. The physician must then make a best-guess decision on the treatment or treatments to try. If possible, treatments should be introduced sequentially but this is not always feasible. Once started, treatment should not be stopped before it has had a chance to produce results; while many pyogenic infections improve promptly, enteric fevers may take up to a week and tuberculosis as much as a month to respond by improvement of fever.

Reasons to attempt trials of therapy

1 To gain further evidence for a suspected diagnosis by obtaining the expected response to therapy.
2 In an emergency when the patient's condition is unsatisfactory.

When treatment is apparently effective it is difficult to decide how long it should be continued. This will depend on the type of condition suspected and on any laboratory data available. The temperature chart gives the earliest indication of success. Other helpful data include improvement in X-ray abnormalities, or normalization of the C-reactive protein, which is a good indicator of response in endocarditis, as is the ESR in tuberculosis. Successful treatment of tuberculosis may improve cell-mediated immunity, causing the tuberculin test to become strongly positive and confirming the diagnosis. Treatment can then be continued for the appropriate 6 or 9 months. In connective tissue disorders the ESR is a useful guide to response. Non-specific indicators of response include improving serum albumin levels and regained body weight.

When no improvement can be obtained in PUO

This occurs in about 5% of most published series. It is most common in cases with mild or subacute fevers. About half of these cases eventually recover and most of the others remain feverish but do not deteriorate. The commonest cause of prolonged high fever, defying diagnosis, is lymphoma.

Investigation of PUO is therefore well worthwhile, usually leading to diagnosis and cure, and rarely ending in failure.

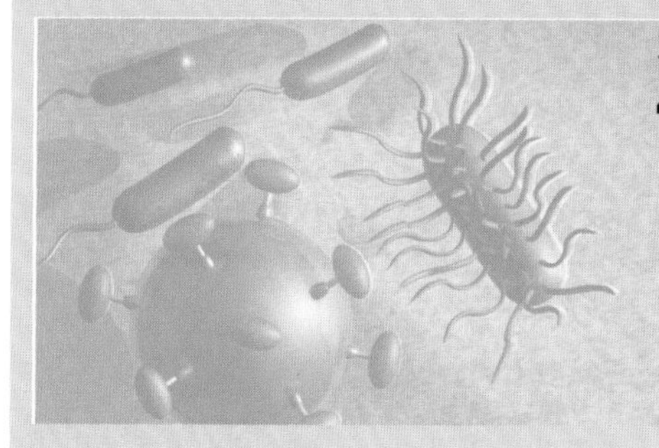

21 Postinfectious Disorders

Introduction, 393
Normal features of convalescence, 393

Features of postinfectious disorders, 394
Possible pathogenesis of postinfectious conditions, 394

Erythema multiforme, 396
Epidemiology, 396
Pathology, 396
Clinical features, 396
Management, 397
Complications, 397

Guillain–Barré syndrome (ascending polyneuritis), 397
Epidemiology, 397
Pathology, 397
Clinical features, 397
Management, 398

Henoch–Schönlein disease, 398
Epidemiology, 398
Clinical features, 398
Management, 398

Poststreptococcal glomerulonephritis, 398
Epidemiology, 398
Clinical features, 399
Diagnosis, 399
Management, 399

Reye's syndrome, 400
Introduction and epidemiology, 400
Clinical features, 400
Diagnosis, 400
Management, 400

Rheumatic fever, 400
Introduction and epidemiology, 400
Clinical features, 400
Diagnosis, 401
Management, 401
Prevention, 401

Reiter's syndrome, 402
Introduction and epidemiology, 402
Clinical features, 402
Management, 403

Thrombotic thrombocytopenic purpura (TTP), 403
Introduction and pathology, 403
Management, 403

Introduction

Most people expect to make a steady improvement after suffering an acute infection. This process of convalescence is complex, and depends on a number of overlapping physiological events.

Normal features of convalescence

Suppression of active infection

In many cases, as in the cure of tonsillitis, meningitis or endocarditis, this is probably achieved by eradication of the causative pathogen by the action of antibiotic, antibody, phagocytosis or immune reaction. In other cases the infected cells are shed, as in influenza, or destroyed by a cytotoxic immune response, as in hepatitis A.

In some diseases the pathogen is suppressed rather than destroyed. In toxoplasmosis, tissue cysts survive and reactivation is prevented by a continuing immune reaction. Viruses of the herpes group remain latent unless the immune system is impaired sufficiently for recrudescent infections to occur. It is now suspected that some viral pathogens are suppressed by selecting inactive mutants from the range of mutations which occur during rapid replication. Hepatitis e antigenaemia may cease when an e-antigen-negative mutant is selected, while e-antigen-positive viral products are destroyed by anti-e antibodies.

Methods of terminating active infection
1 Destruction of pathogens and their antigens.
2 Shedding or destruction of virus-infected cells.
3 Making pathogens inaccessible to the immune system (e.g. in pseudocysts).
4 Selecting antigenically inert mutants of the pathogen.
5 Establishment of latency.

Return of immune responses to resting states

As the amount of microbial antigen falls, antigen presentation will lessen and the stimulus to further immunological activation will subside. However, many activated lymphocytes have been generated, including helper T cells, cytotoxic T cells and clonally proliferating B and T cells, as well as suppressor T cells. It is now known that many activated T cells are programmed to self-destruct by an enzymic process (called apoptosis) that limits their lifetime. The presence of increased levels of interleukin-2 while infection persists tends to delay this destruction and maintain immune activation.

As antigen levels decline, the production of immune complexes slows down, slowing the activation of the classical complement pathway and the coagulation cascade followed by a reduction of inflammatory reactions in the tissues.

Events favouring reduced immune activity and inflammation

1 Falling levels of antigen and immune complexes.
2 Reduced antigen presentation.
3 Reduced cytokine production.
4 Reduced immune complex-mediated complement activation.
5 Apoptosis of activated T cells.

Repair of tissue damage

The catabolic state of acute infection must be reversed to allow protein construction and repair of damaged tissue. This in turn permits the restoration of normal bodily functions; for instance, absorption of nutrients after bowel infection or glucose metabolism after severe liver infection. The time needed to complete this process can be surprisingly long; for instance, there is mild but measurable hypoxaemia for several weeks after acute bronchiolitis in infants. During this period the metabolic rate is increased, with concurrent tachycardia and often a feeling of easy fatiguability.

Features of postinfectious disorders

In postinfectious disorders the process of convalescence is interrupted, sometimes for a considerable time. This is often because of an inflammatory condition which arises in different tissue from that affected by the original infection. In some cases the postinfectious condition overlaps the acute infection in time, as when erythema nodosum complicates primary tuberculosis. In other instances it follows the acute infection with little or no interval, as when erythema multiforme complicates episodes of herpes simplex. Finally the postinfectious condition can follow the acute infection with an interval of 2 weeks or more; this is seen in poststreptococcal nephritis and post-measles encephalitis (Table 21.1).

Possible pathogenesis of postinfectious conditions

There are several mechanisms by which postinfectious conditions may arise. However, although theoretical considerations can suggest explanations, the true pathogenesis of most conditions is poorly understood.

Persisting low-grade infection

Persisting low-grade infection is a possible mechanism, especially in postinfectious encephalitis, which may be histologically indistinguishable from acute infectious encephalitis. Early evidence for persistence of infection came when measles virus was recovered from the brain tissue of cases of subacute sclerosing panencephalitis (SSPE). Immunoglobulin M (IgM) antibodies can persist for weeks or months in some infections such as hepatitis A, infectious mononucleosis and toxoplasmosis, which can all be followed by a debilitating and prolonged convalescence. More recent work on erythema multiforme associated with herpes simplex has used polymerase chain amplification to demonstrate herpes simplex virus DNA in epidermal lesions.

Molecular mimicry

Molecular mimicry also offers a good explanation for some conditions. In the Guillain–Barré syndrome there is evidence of immunological attack against components of myelin. Experimental demyelination of nerve roots very similar to that seen in Guillain–Barré syndrome can be produced in animals by antimyelin antibodies, and about 25% of patients have antiganglioside antibodies. Nephritogenic strains of *Streptococcus pyogenes* produce a unique antigen which cross-reacts with glomerular structures.

Immune complex disease

Immune complex disease has been postulated as a cause particularly of postinfectious arthritis. In the synovitis which often follows meningococcal disease, meningococcal antigen has been demonstrated in biopsies of affected synovium. Similarly, in Reiter's syndrome, chlamydial antigen has been found in synovium.

Persisting inappropriate immune reaction

Persisting inappropriate immune reaction is probably a mechanism in some of the cytopenias. It is known that haemolysis in *Mycoplasma* pneumonia and Epstein–Barr infections is related to the inappropriate production of anti-I and anti-i anti-red cell antibodies, respectively. Thrombocytopenia is likely to be caused by a similar mechanism; indeed, antiplatelet antibodies can be demonstrated in some cases.

A condition like systemic lupus erythematosus occa-

Disorder	Common associations	Rare associations
Aplastic anaemia		Non-A/non-B hepatitis
Arthritis	Rubella *Meningococcus* *Yersinia*	*Salmonella* *Shigella* *Campylobacter* Mumps
Encephalitis	Varicella Measles Influenza Mumps	Rubella Yellow fever vaccine
Erythema multiforme	Herpes simplex	*Mycoplasma pneumoniae*
Erythema nodosum	Tuberculosis Leprosy	*Yersinia*
Glomerulonephritis	*Streptococcus pyogenes*	Hepatitis B Mumps
Guillain–Barré	*Campylobacter* Cytomegalovirus	Hepatitis A or B Respiratory viruses
Haemolysis	*Mycoplasma pneumoniae*	Epstein–Barr virus Syphilis
Haemophagocytic syndrome		Epstein–Barr virus Cytomegalovirus
Reiter's syndrome	Chlamydial genital infections	*Shigella*
Reye's syndrome	Varicella Influenza	Aspirin as a cofactor
Rheumatic fever	*Streptococcus pyogenes*	
Serositis	*Meningococcus*	
Thrombocytopenia	Rubella Mumps Varicella	Epstein–Barr virus Tuberculosis
—		
Thrombotic thrombocytopenic purpura (TTP)		

Table 21.1 Some postinfectious disorders and associated pathogens. HIV infection; other infections.

sionally occurs after infectious mononucleosis. This is associated with positive anti-DNA antibodies, joint pains and raised erythrocyte sedimentation rate (ESR), but has a limited duration of weeks or months. Acquired thrombotic thrombocytopenic purpura (TTP) is caused by antibody-mediated inhibition of the cleavage of von Willebrand factor to its active form.

Failure of termination of immune response

Failure of termination of the normal acute immune response may be responsible for severe conditions such as the haemophagocytic syndrome. Apparently uncontrolled proliferation and activity of macrophages and other immune cells produces infiltration of the liver, spleen and bone marrow which is hard to distinguish from a malignant histiocytosis. Immunosuppression predisposes to this condition.

Susceptibility of the patient

Special susceptibility of the patient may also influence the occurrence of postinfectious conditions. It is known that non-secretors of blood group substances are more likely than others to develop rheumatic fever. Reiter's syndrome is almost exclusive to patients with the human leucocyte antigen (HLA) B27 tissue type, but non-Reiter's arthropathies have a weaker association with this tissue type. A total of 88% of patients who have recurrent erythema multiforme after herpes simplex episodes have tissue type DQw3, and 71% have DRw53.

Possible mechanisms of postinfectious disease
1 Persistent low-grade infection.
2 Molecular mimicry.
3 Immune complex disease.
4 Persisting inappropriate immune activity.
5 Failure to terminate the normal acute immune response.
6 Special immunological susceptibility of the patient.

Erythema multiforme

Epidemiology

This skin disorder affects mainly teenagers and young adults. It has been described after streptococcal infections, *Mycoplasma pneumoniae* infections and after exposure to drugs such as sulphonamides and long-acting penicillins. The authors have seen it as an accompaniment of Epstein–Barr virus infection in toddlers. Rarer causes include barbiturates, diphenoxylate and traumas such as radiotherapy.

Repeated episodes of erythema multiforme occur in some patients who have cold sores. The skin rash often follows 2 or 3 days after the onset of the mucocutaneous lesions, and lasts up to a week. The severity of the rash is related to the extent of the herpetic lesions on each occasion.

Pathology

The skin lesions are associated with oedema and necrosis of epidermis with dilatation and inflammatory infiltration of the subepidermal blood vessels. The oedematous epidermis may become raised, forming bullae. In herpes simplex-associated disease, viral antigen and viral genome can both be demonstrated in the keratinocytes of the affected skin, but are not present in normal skin.

Clinical features

Illness begins abruptly, with fever and developing rash.

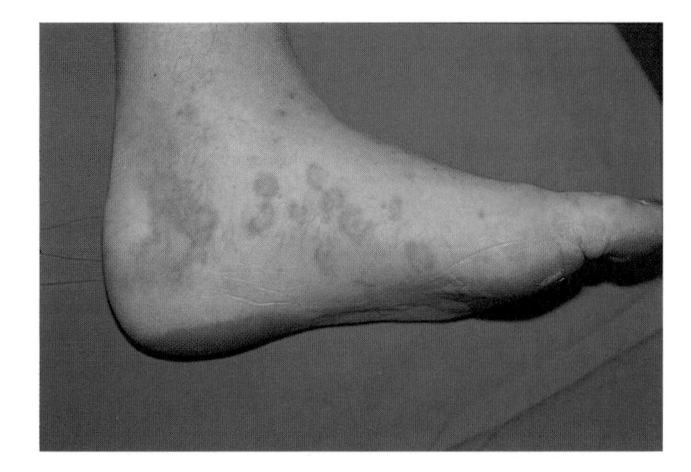

Fig. 21.1 Target lesions of erythema multiforme.

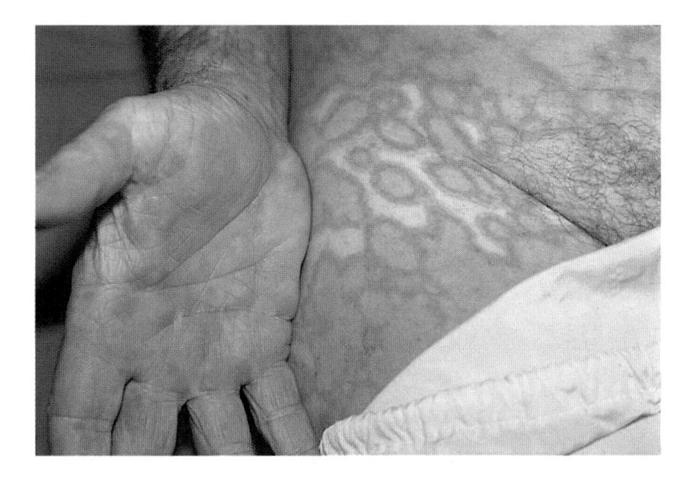

Fig. 21.2 Erythema multiforme: showing the rash, which also involves the palms.

Typically, many of the lesions are target- or iris-shaped with an erythematous halo surrounding an oedematous or dusky centre (Fig. 21.1). The rash is densest on the extremities, including the palms and soles (Fig. 21.2), but is also exaggerated in areas exposed to sunlight or trauma. Untreated, the lesions last for 2–6 weeks, some of them coming and going all the time. There may be arthralgia, especially of the ankles. There is no characteristic change in the white cell count. The ESR is usually high: 70 mm/h or more.

Stevens–Johnson syndrome

Stevens–Johnson syndrome is the term used when erythema multiforme affects mucosae as well as skin. The disease is usually severe, with many skin bullae and severe, sometimes haemorrhagic inflammation of the oral,

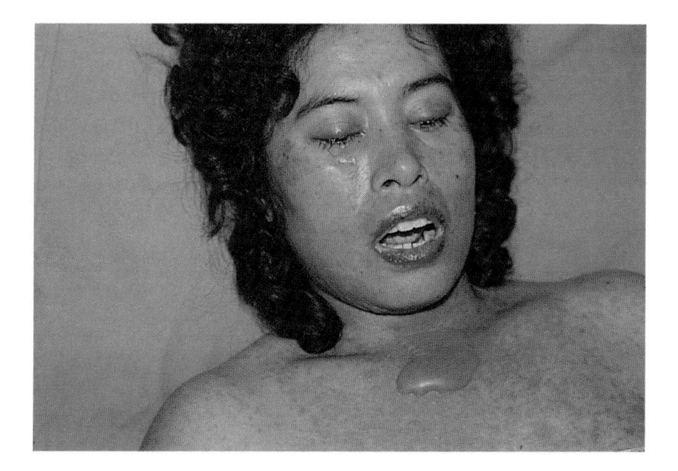

Fig. 21.3 Stevens–Johnson syndrome: severe involvement of the oral and conjunctival mucosae and epidermal necrolysis.

conjunctival and genital mucosae. There is danger of widespread epidermal necrolysis, in which the keratinized layer of the epidermis becomes loose, and can be moved over the lower epidermis, or rubbed off it (Nikolsky's sign; Fig. 21.3).

Management

The rash will often resolve progressively if the cause can be identified and removed. Severe cases, particularly of Stevens–Johnson syndrome, are often treated with corticosteroids, though there is little firm evidence of benefit. The ESR can be used as an objective measure of improvement.

Patients with epidermal necrolysis should be nursed in a warm room and kept well hydrated, as heat and moisture are rapidly lost from the exposed lower epidermis. There is risk of secondary infection, particularly with *Staphylococcus aureus*, which may cause bacteraemia. If a patient develops fever, tachycardia or shock, blood cultures should be obtained and early treatment commenced.

Complications

These are few, once the patient has passed the severe part of the illness. The inflammation is superficial, and the epidermal layers are replaced, with only an occasional atrophic scar. After severe mucosal ulceration, adhesions may develop, for instance between the lip and gum, or in the conjunctival sac. These may require division and split-skin grafting.

Guillain–Barré syndrome (ascending polyneuritis)

Epidemiology

This is a condition in which there is demyelination of the nerve roots. It often follows a viral-type respiratory infection; 25% follow campylobacter enteritis or other gastrointestinal illness, with an interval of 1–3 weeks. About 10% follow cytomegalovirus infections, and occasional cases are associated with infectious mononucleosis, *Mycoplasma pneumoniae* or legionnaires' disease.

The disease affects all age groups, but tends to be more prolonged in middle-aged individuals and the elderly.

Pathology

Inflammation and demyelination in the spinal and cranial nerve roots are presumed to be of immunological origin, but the responsible antigen has not been identified. If inflammation is sufficiently severe, axonal degeneration in nerve trunks may follow.

Clinical features

The first symptoms are usually paraesthesiae followed by weakness in the feet and lower legs. At this stage there may be no objective signs, and the patient may be thought hysterical. The disease progresses up the spinal levels, with weakness of the trunk and arms, and finally facial paralysis. Some patients describe prodromal aching or pain in the muscles. As the paralysis intensifies, tendon reflexes are lost and the patient becomes immobile. There may be respiratory failure because of intercostal and diaphragmatic paralysis. Early facial paralysis warns of sudden respiratory failure. Sensory features include loss of light touch and joint position sense. There are no upper motor neurone signs; the plantar responses, when they can be elicited, are downgoing. There is little involvement of sphincters, and this is transient if it occurs.

Abnormal laboratory findings are confined to the cerebrospinal fluid (CSF). Early in the course of the illness there is an excess of lymphocytes. The CSF protein level gradually rises, and may reach levels of several grams per litre.

Paralysis may cease to progress at any stage, but rarely takes longer than 2 or 3 weeks to stabilize. Some cases progress to complete paralysis within 2–3 days. After stabilization there is progressive recovery. This may be rapid, taking 2 or 3 weeks until the patient can walk independently. In very elderly patients, or when significant nerve fibre degeneration has occurred, it can take 3–12 months or more, and the recovery of muscle strength may be compromised by disuse atrophy and a degree of denervation.

There is a 10–20% mortality, largely from the side-effects of respiratory intensive care. Of those who recover, 90% return to full motor activity; the remainder have variable residual weakness.

Electrophysiological investigation in established disease shows marked slowing of conduction velocities and diminished evoked muscle and sensory nerve action potentials. These are scattered and mild in the first few days, and may be difficult to demonstrate.

Management

Monitoring of the vital capacity and the arterial oxygen saturation are important, as ventilation is often required for those with respiratory paralysis. The paralysed patient also needs skin care and attention to pressure areas.

Plasmapheresis has resulted in remission of prolonged paralysis, but may require supplementation with immunosuppressive therapy for a variable time to prevent relapse. Intravenous immunoglobulin is a safer and more convenient means of obtaining remission, and is probably as effective as plasmapheresis.

Henoch–Schönlein disease

Epidemiology

This is a vasculitic condition which presents with fever, purpuric rash, arthralgia, synovitis, abdominal pain and glomerulonephritis. There is no unique precipitating condition, but there has often been a preceding respiratory infection and some sufferers have evidence of recent streptococcal disease. Children and young adults are most often affected, though cases occasionally occur in the middle-aged.

Clinical features

Rash

The rash affects mainly the extensor surfaces of the feet, ankles and legs, and commonly affects the buttocks. The lesions are raised and of variable size. A variable proportion are purpuric, and those on the feet may be haemorrhagic and bullous (Fig. 21.4). Capillaries in affected skin contain IgA immune complexes and inflammatory infiltrate.

Synovitis

Synovitis most often affects the ankles and knees, but any joint can be involved. Tendon sheaths around the ankle, and sometimes the wrist, are often swollen and painful.

Abdominal pain

The abdominal pain is caused by swelling and haemorrhagic lesions of the bowel. Severe cases may have vomiting and abdominal rigidity. Blood may be seen in the stools. Nodular mucosal swelling occasionally predisposes to intussusception.

Nephritis

The nephritis is a focal glomerulitis pathologically identical to IgA nephropathy. Haematuria and proteinuria are common. Renal failure can occur, with hypertension and oedema.

Main features of Henoch–Schönlein disease

1 Rash on the legs and/or buttocks.
2 Joint pain.
3 Abdominal pain.
4 Proteinuria (and often microscopic haematuria).
5 Reduced renal function.

Management

Management is usually symptomatic, as the disorder is usually self-limiting and benign. Non-steroidal anti-inflammatory agents often help pain, especially of the joints, but should be avoided in renal impairment. Corticosteroids do not alter the course of the disease. Most children who develop nephritis will recover, but about a quarter are at risk of deteriorating renal function later in life.

Problems in the course of the disease may be due to bowel haemorrhage or intracerebral haemorrhage. A minority of patients require dialysis for renal failure and a few do not recover.

Poststreptococcal glomerulonephritis

Epidemiology

This disorder affects children of age 1–5 years, and is rare in other age groups. It can follow *Streptococcus pyogenes* infection of the skin or throat. It usually occurs 10–14 days

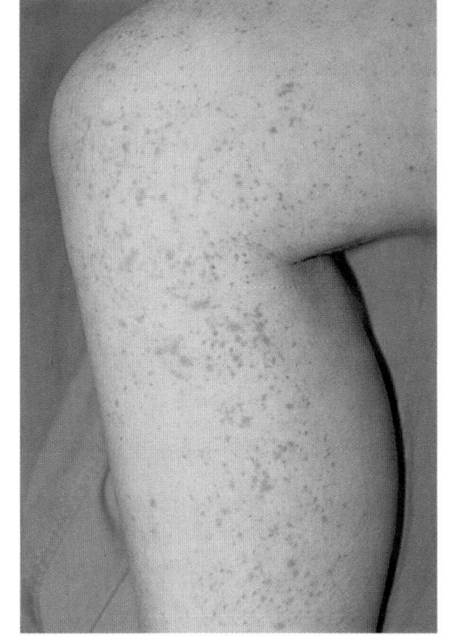

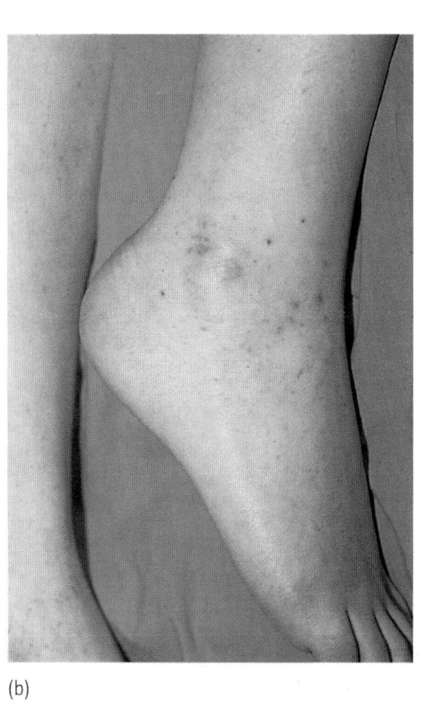

Fig. 21.4 (a,b) Henoch–Schönlein disease: characteristic rash with raised and haemorrhagic elements.

after the initiating infection, but can appear after as long as 4 weeks.

Clinical features

The onset is abrupt, with fever, malaise and often loin pain. There is puffiness of the feet and face. Haematuria and oliguria are often noticed by the mother.

Physical findings are of mild loin tenderness, oedema and hypertension. There is mild to moderate haematuria and proteinuria; significant protein loss is rare. Urine microscopy shows epithelial casts which contain red blood cells, confirming the glomerular origin of the problem.

Main features of poststreptococcal nephritis

1 Fever and loin pain.
2 Facial and dependent oedema.
3 Haematuria and oliguria.
4 Impaired renal function.
5 It follows 10–21 days after streptococcal infection.

Diagnosis

Transient microscopical haematuria is common in acute streptococcal infections and does not indicate glomerulonephritis. Glomerulonephritis can occur in a variety of other infections, including viral hepatitis, infectious mononucleosis, atypical pneumonias and childhood viral infections. Drugs, including penicillins and cephalosporins, occasionally cause idiosyncratic nephritis. Care should be taken to show that the glomerulonephritis is truly associated with streptococcal infection, especially if it is severe or prolonged.

Evidence of recent streptococcal infection is provided either by persisting positive culture from throat or skin, or by significantly elevated antistreptolysin O titre (ASOT) or other antibody titres (see Chapter 6).

Renal biopsy is rarely carried out as the natural course of the disease is short, with diuresis and improvement in creatinine levels within 7–10 days. Histology shows swollen glomeruli with endothelial proliferation and neutrophil infiltration. Irregular deposits of immune complexes can be demonstrated on the glomerular basement membrane.

Management

The throat or skin infection should be treated with penicillin or a suitable alternative (see Chapter 6). The glomerulonephritis almost always resolves without specific treatment. Diuretics may be indicated to reduce oedema; a few patients require protein and potassium restriction. It is rare for dialysis to be indicated. Once the kidneys have recovered, renal function should be expected to remain normal.

Reye's syndrome

Introduction and epidemiology

This is a rare childhood disease consisting of encephalopathy, cerebral oedema and fatty infiltration of the liver, with a characteristic picture of liver dysfunction. It is associated with transient but severe damage to mitochondrial structure and function. Disrupted and necrotic mitochondria are demonstrable on liver biopsy in affected cases.

The syndrome often follows infection with varicella or influenza B, and occasionally echovirus infections, atypical pneumonias and adenovirus infections. There is a strong association between Reye's syndrome and aspirin treatment for the preceding feverish illness, and a steady reduction in the incidence of Reye's syndrome followed the discontinuation of aspirin use in children. Aspirin is therefore not recommended as an antipyretic for children below the age of 12 years.

Clinical features

Illness begins a few days after the onset of a viral-type infection. There is increasing drowsiness, vomiting and often convulsions. Hepatomegaly is common, but jaundice is not often seen. There is biochemical evidence of liver dysfunction, with a raised blood ammonia and low blood urea. The blood glucose may fall, leading to hypoglycaemic convulsions. There is sometimes significant impairment of blood clotting.

The most important aspect of the disease is raised intracranial pressure and reduction of cerebral perfusion, which can cause permanent cerebral damage.

Main features of Reye's syndrome

1 Drowsiness and vomiting.
2 Raised intracranial pressure.
3 Hepatomegaly.
4 Biochemical signs of liver failure.
5 It occurs within 1 week of respiratory or gastrointestinal illness.

Diagnosis

Diagnosis is not always easy, as there is a clinical overlap with several inborn errors of fatty acid oxidation and urea synthesis. These possibilities should be investigated, particularly in children below the age of 15 months. Expert clinical advice is helpful in this respect.

Management

Management is directed at maintaining adequate cerebral perfusion pressure, with the aid of intracerebral pressure monitoring and paediatric intensive care facilities. Admission to such facilities is a matter of urgency. The disease is self-limiting; recovery occurs slowly as the mitochondria recover from the temporary insult. If cerebral damage is prevented, full recovery can be achieved.

Rheumatic fever

Introduction and epidemiology

Rheumatic fever is a multisystem inflammatory disease which follows throat infection with a wide range of *S. pyogenes* serotypes. It does not occur after streptococcal skin infections. Most affected individuals are between the ages of 5 and 16 years, though a few may be up to age 25 or 30. After this age, the susceptibility to rheumatic fever seems to be lost. Rheumatic fever has been rare in Western countries for some decades, but in the 1980s large epidemics occurred in North America. This may be because of the reappearance of epidemic types of *S. pyogenes* with an antigenic structure mimicking that of cardiac and other tissues. The streptococci recovered from North American cases produced mucoid colonies, and most were of M type 18.

The throat infection preceding rheumatic fever is variable in severity. Only half of the North American cases had throat symptoms sufficient to warrant a medical consultation, and only a quarter of all cases received antibiotic treatment.

Clinical features

There is an interval of 2–3 weeks between the acute infection and the onset of rheumatic fever. The various manifestations of the disease have been classified into major and minor features, or criteria, to aid clinical diagnosis (Table 21.2).

The commonest presentation is with fever and flitting or migratory polyarthritis.

Arthritis

The arthritis is acute and painful, with synovitis and often effusion. It affects large joints, particularly the knees and ankles, but the joints of the arm can also be involved. Typically, one joint is affected for a few days, but then improves rapidly while another becomes inflamed (flitting arthritis).

Major manifestations	Minor manifestations
Arthritis (70%)	Fever
Carditis (50%)	Previous episode(s)
Chorea (20%)	Raised ESR or CRP
Rash (12%)	ECG abnormalities
Nodules (5%)	Arthralgia

The percentages shown are for recent North American cases. CRP, C-reactive protein; ECG, electrocardiogram; ESR, erythrocyte sedimentation rate.

Table 21.2 Major and minor manifestations of rheumatic fever.

Carditis

The carditis is a pancarditis which can cause heart failure. About 10% of cases have evidence of this at presentation.

There is acute pericarditis, often with a degree of purulent exudate.

The myocardium is inflamed, and may contain typical Aschoff bodies. These are inflammatory bodies consisting of degenerative epithelioid central areas surrounded by haloes of inflammatory cells. As a postmortem finding these are pathognomonic of rheumatic fever. The myocardial inflammation is often reflected by cardiographic changes, such as prolongation of the P-R interval, axis changes or other conduction changes. The commonest findings on examination of the heart are a soft first heart sound, a third heart sound or a short systolic or diastolic (Carey–Coombs) murmur.

Endocarditis

The endocarditis is usually macroscopic, with verrucous vegetations on the affected valves. The mitral valve is most often affected, the aortic valve less so (boys are more likely than girls to have aortic disease). The right heart valves, especially the pulmonary, are rarely affected. The most common valve lesion is mitral regurgitation, but vegetations can cause stenotic features. When heart failure occurs, it is often related to severe valvular disease, but exacerbated by myocardial inflammation. Echocardiography is useful to demonstrate the valvular lesions and to evaluate left ventricular function.

Chorea

Chorea is often seen in older children and girls, and on occasions it is the sole major manifestation of the disease. The abnormal movements may be mild and partly disguised, making the patient appear restless, or they may be more gross and obtrusive. They can develop late in the illness, when other manifestations are subsiding, and can last for weeks or months.

Rash

The rash is classically erythema marginatum (see Fig. 20.3), a rash of raised, serpiginous lesions which comes and goes rapidly, waxing and waning with the fever from hour to hour. More often there is an urticarial, multiform or even erythema nodosum-like rash, which is similarly changeable.

Nodules

Nodules are usually less than 1 cm in diameter. They are subcutaneous and often found over tendons on the extensor surfaces of the arms. They are most often seen in patients with severe carditis.

Diagnosis

This is based on clinical assessment using the major and minor criteria, and on demonstrating evidence of recent streptococcal infection (see Chapter 6). Two major criteria are sufficient to make the diagnosis; if only one is found, one minor criterion plus evidence of streptococcal infection should be demonstrated.

The presence of Aschoff bodies in biopsy material such as skin nodules may be helpful, but the inflammatory changes in tissue other than myocardium are rarely specific.

Management

The treatment of choice is aspirin, but other non-steroidal anti-inflammatory agents may be used, and are preferred in younger children, if sufficiently effective. Corticosteroids are not effective.

Bed rest and/or diuretic treatment may be indicated if there is heart failure. Fatalities from acute rheumatic fever are rare, but severe congestive heart failure or pericardial effusion can sometimes be life-threatening.

Prevention

This is important, as a first attack is often followed by recurrences each time a streptococcus invades the throat. With each succeeding attack there is fibrosis of the valve rings, shortening of the chordae tendinae and increasing deformity of heart valves, leading to valve failure and increasing the risk of infective endocarditis.

Phenoxymethylpenicillin 250 mg twice daily is safe and effective prophylaxis. Erythromycin or a first-generation cephalosporin may be given to penicillin-allergic patients. Prophylaxis is usually continued until the individual leaves school. For those working in schools or hospitals it may be continued for longer, but the risk of recurrence is small after the age of 30.

Reiter's syndrome

Introduction and epidemiology

Reiter's syndrome is a multisystem disorder which affects individuals with tissue type HLA B27. In its commonest form it follows non-specific genital infection (usually chlamydial) in men. It can also affect patients of either sex after an attack of bacillary dysentery.

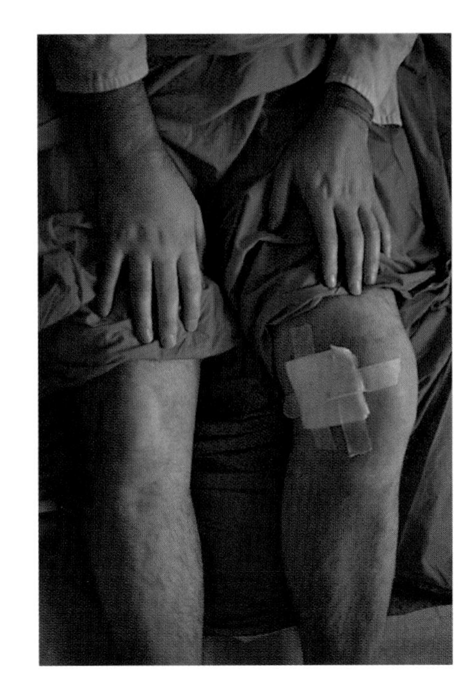

Fig. 21.5 Reiter's syndrome: synovitis of the left knee and tendinitis of the dorsum of the right hand.

Clinical features

Illness begins 10–14 days after the acute precipitating condition. The syndrome consists of synovitis, sacroileitis, conjunctivitis, mucocutaneous rashes and aortitis. Fever is common. There may be a polymorph leucocytosis in the peripheral blood. The ESR is usually raised to 70 mm/h or more and the C-reactive protein is also very high. Chlamydial antigens have been demonstrated in the affected synovium, but this and tissue typing are expensive and not necessary for an adequate diagnosis in most cases.

Synovitis with associated conjunctivitis

This is the commonest presentation. The synovitis typically affects tendon sheaths, especially of the hand and foot, but also of the elbow and shoulder. There is also inflammation of both large and small joints (Fig. 21.5). The knee, ankle and shoulder are often affected. Sacroileitis is prominent, causing low back pain, local tenderness and discomfort on 'springing' the pelvis. Sacroileitis is often radiographically demonstrable, showing as soft-tissue swelling and perisacral sclerosis. As it does not occur in rheumatic fever, this helps to differentiate between that condition and Reiter's syndrome with aortic valve involvement.

Circinate balanitis

Circinate balanitis is the typical mucocutaneous rash. This is often a red, roughly circular slightly weeping lesion with scaly margins, affecting the periurethral part of the glans penis. Non-specific papular or scaly rashes may also occur on the glans and at the angles of the mouth.

Aortitis

Aortitis is not always clinically evident. Sometimes a soft systolic murmur is audible, but in rare, severe cases aortic incompetence may occur and lead to sudden heart failure.

Keratoderma blennorrhagica

Keratoderma blennorrhagica is a thickened, scaling rash of the palms and soles, in which flat vesicles are often seen. It is characteristic of late Reiter's syndrome. It often presents after the earlier features have responded somewhat to treatment.

Uveitis and cardiac dysrhythmias

Uveitis and cardiac dysrhythmias are rarer manifestations.

Main features of Reiter's syndrome
1. Synovitis.
2. Sacroileitis.
3. Conjunctivitis.
4. Aortitis.
5. Circinate balanitis.
6. Late keratoderma blennorrhagica.

Management

This is symptomatic. The initiating illness should be treated if still active, but the Reiter's syndrome is probably related more to antigen already fixed in the affected tissues than to continuing active disease.

Non-steroidal anti-inflammatory drugs are the mainstay of treatment. Systemic corticosteroid treatment is avoided, if possible, but corticosteroid eye drops may be necessary to control uveitis. Rare cases of aortic valve disruption may need to be treated surgically, but control of inflammation is important to provide a good basis for any implanted valve structure.

The disease has a prolonged and relapsing course, often requiring several weeks or months of treatment with anti-inflammatory agents.

Thrombotic thrombocytopenic purpura (TTP)

Introduction and pathology

Thrombotic thrombocytopenic purpura is a multisystem syndrome characterized by a rapid onset of purpura or bleeding with a low platelet count, microangiopathic anaemia, often renal failure and variable neurological disturbances often leading to severe encephalopathy and coma. Pathological investigations show a mixture of endothelial damage, and activation of platelets with microthrombi in the vasculature of many organs. Although TTP was originally thought analogous to the toxic endothelial damage of haemolytic–uraemic syndrome (HUS; see Chapter 8), it is now known that acquired TTP is associated with dysfunctional von Willebrand factor: native, ultralarge von Willebrand factor is not cleaved to its functional form, because the metalloprotease enzyme required to perform this cleavage is damaged by an acquired autoantibody.

TTP is often associated with systemic lupus erythematosus, other collagen disorders, or the use of cytotoxic agents such as cyclosporin or mitomycin. It has now been described as a complication of HIV infection, and is probably related to the polyclonal B-cell activation which occurs in this infection resulting in the production of unusual autoantibodies.

Management

Successful management depends on early recognition of the disease, permitting treatment before irreversible cerebral or other damage is established. Fresh, or fresh frozen plasma can reconstitute the deficient enzyme, and is more effective if combined with plasmapheresis to remove the abnormal antibody. Serum should be collected from the patient before plasmapheresis, so that tests for HIV infection and other predisposing conditions can be carried out.

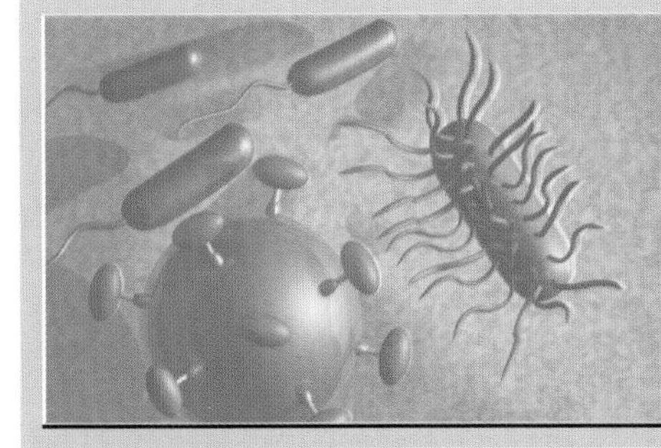

22 Infections in Immunocompromised Patients

Introduction, 404

Classification of infections in immunocompromised patients, 404

Neutropenia, 405
- Epidemiology, 405
- Prevention of infection during neutropenia, 406
- Treatment of fever in neutropenic patients, 406

T-cell deficiency, 408
- Opportunistic pathogens in T-cell deficiencies, 408
- Diagnosis of opportunistic infection, 410

Hypogammaglobulinaemia, 411
- Congenital hypogammaglobulinaemia, 411
- Functional hypogammaglobulinaemia, 411
- Opportunistic infections, 412

Complement deficiency, 412

Splenectomy, 412
- Preventing infections in asplenic patients, 412

Infections in transplant patients, 413
- Organ transplant patients, 413
- Antirejection therapy, 413
- Importance of cytomegalovirus in post-transplant patients, 413
- Bone marrow transplant patients, 415

Introduction

Earlier chapters have described how organisms can overcome host defences and cause infection. This chapter discusses the infections that occur when host defences are reduced by disease, medical treatment or inherited disorders.

Each or all of the components of the immune system can be compromised. Differing patterns and degrees of immune compromise result in different patterns of infection as different opportunities are opened to invading organisms. Immunocompromised patients may be infected by organisms with little pathogenicity, that are usually incapable of causing primary infection, for example, organisms that form part of their bacterial flora or are derived from the environment. The interpretation of microbiological culture results is therefore more difficult in these cases. Also, pathogens which affect non-compromised patients infect the immunocompromised, often causing severe or persisting disease. The absence of an immune response may modify the clinical picture; typical clinical features may not occur (for instance, the typical rash in varicella or measles).

The recognition and treatment of infection in immunocompromised patients is very important, as any delay can allow the establishment of potentially fatal disease. The classic features of high fever, sepsis and shock are often absent, as the development of these depends on immune functions. Hospital admission, extensive investigation and vigorous therapy are usually indicated.

Classification of infections in immunocompromised patients

Immunodeficiencies are classified into primary/congenital immunodeficiency (including defects in B cells, T cells, complement and phagocytes) and secondary or acquired immunodeficiency resulting from malignancy or immunosuppressive therapy.

Immune deficiencies can also be classified into seven groups:

1 Disorders of the innate immune system, e.g. phagocytosis.
2 Neutropenia and neutrophil dysfunction, e.g. congenital granulomatous disease (CGD).
3 T-cell deficit.
4 Hypogammaglobulinaemia.
5 Complement deficiencies.
6 Splenectomy.
7 Broad-spectrum immunodeficiency related to haematological or other malignancy, chemotherapy or immunosuppression for transplants.

This is a simplification, as deficiency in one component of the system leads to imbalance and failure of other components (for example, T cells are the main component of cell-mediated immunity, but T-cell help is also essential for the optimal activity of the humoral response).

Disorders of innate immunity are often found in hospital practice, when, for example, treatment breaches the physical barriers to microbial invasion (see Chapter 23). Infections associated with acquired immunodeficiency

Immune deficit	Caused by	Bacterial infections	Other infections
Complement	Congenital	*Neisseria* spp. *Streptococcus pneumoniae*	
Spleen	Surgery, trauma, sickle-cell anaemia (functional)	*S. pneumoniae* *Haemophilus influenzae* (type b)	*Plasmodium* spp. *Babesia* spp.
Gamma-globulin	Congenital, multiple myeloma, CLL, AIDS	*S. pneumoniae* *H. influenzae* (non-capsulate)	*Pneumocystis carinii* *Giardia intestinalis* *Cryptosporidium parvum*
Neutrophils	Chemotherapy of leukaemia and bone marrow transplantation dysfunction, e.g. CGD	Enterobacteriaceae Oral streptococci *Pseudomonas aeruginosa* *Enterococcus* spp.	*Candida* spp. *Aspergillus* spp.
T cells	Marrow and other transplantation, AIDS, cancer chemotherapy, lymphoma, steroids	*Listeria monocytogenes* *Mycobacterium tuberculosis* *M. avium-intracellulare* *Salmonella* spp. *Rhodococcus equi*	*P. carinii* *Toxoplasma gondii* *Cryptosporidium parvum* *Leishmania* spp. Herpesvirus CMV Varicella-zoster virus *Cryptococcus neoformans* *Histoplasma* spp. and other systemic yeast infections

AIDS, acquired immunodeficiency syndrome; CGD, congenital granulomatous disease; CLL, chronic lymphocytic leukaemia; CMV, cytomegalovirus.

Table 22.1 Common deficits in immune function and the infections with which they are associated.

syndrome (AIDS) are discussed in Chapter 13. The other immunodeficiencies and the common infections associated with them are summarized in Table 22.1.

With the exception of some congenital abnormalities, immune deficits are rarely single. Patients undergoing chemotherapy for leukaemia or bone marrow transplantation are primarily neutropenic, but also have impaired cell-mediated immunity, predisposing, for example, to cytomegalovirus (CMV) infection. Treatment involves the use of many skin-piercing cannulae, predisposing to *Staphylococcus epidermidis* and other skin-derived infections. Fungal infections are also common in these patients, facilitated by the combination of neutropenia and decreased cell-mediated immunity.

Neutropenia

This can be an adverse reaction to treatment with a number of drugs, but also accompanies acute leukaemia or its treatment. The risk of infection increases significantly once the neutrophil count falls below $0.5 \times 10^9/l$, and is proportional to the period of neutropenia. More than half of all patients suffering an episode of neutropenia will develop an infection. The mortality from these infections is high if not promptly treated.

Epidemiology

The epidemiology of infection in neutropenic patients is complex. Not only are patients and their underlying diseases diverse, but treatment protocols are constantly evolving and this results in a changing pattern of infection, both between patients and at different stages of treatment in the same patient.

Bacterial infections

Bacteraemia occurs in 20–30% of neutropenic patients. The principal bacteria implicated in these patients are Gram-negative rods, but Gram-positive cocci are also important. The frequency with which these organisms cause infection changes with developments in treatment and antimicrobial prophylaxis. Hospital patients are susceptible to colonization with resistant organisms because of both the administration of antibiotics and the effects of serious underlying disease. Hospital organisms may be transmitted on the hands of medical and nursing attendants or ingested in food, notably washed vegetables.

Enterobacteriaceae and *Pseudomonas* spp. are the most commonly isolated Gram-negative pathogens. They are usually derived from the patient's own intestinal flora, gaining access to the circulation when the rapidly multiplying intestinal epithelium is damaged by antineoplastic agents or X-irradiation. Improvements in drug treatment of these pathogens have reduced both the incidence and mortality associated with them. Although they remain the commonest Gram-negative bacilli isolated, the prophylactic use of broad-spectrum 4-fluoroquinolones has diminished the incidence of these infections. A decreasing incidence of *Klebsiella* spp., *Serratia* spp. and *Enterobacter* spp. has also been noted in many centres.

Infections with Gram-positive organisms have become more common in neutropenic patients, rising from around 30% of bacterial infections in the mid-1960s to 60% in the mid-1990s. In most centres, the main organisms reported are *Staphylococcus epidermidis*, the oral streptococci *Streptococcus mitis* and *S. oralis*, *Enterococcus* spp., *Staphylococcus aureus* and *Corynebacterium jeikeium*. Methicillin-resistant *S. aureus* (MRSA) strains are an increasing problem in hospitals where these organisms have established themselves.The almost universal use of long-term intravascular access devices such as Hickman and Portacath catheters favours infection with organisms derived from the skin. Infection with oral streptococci may follow mucositis secondary to chemotherapy.

Although uncommon, enterococcal infections are becoming more important because of the use of 4-fluoroquinolones for prophylaxis and cephalosporins for treatment of fevers; enterococci are resistant to both of these agents. Aminoglycoside- and vancomycin-resistant enterococci (VRE) are an increasing problem in intensive care medicine.

Bacteria associated with infections in neutropenia
1 Enterobacteriaceae (*Klebsiella*, *Serratia* and *Enterobacter* spp.).
2 Skin-derived organisms: *Staphylococcus epidermidis*, *S. aureus*, including methicillin-resistant *S. aureus*, and corynebacteria, including *Corynebacterium jeikeium*.
3 Other streptococci including oral streptococci and enterococci.
4 *Pseudomonas* spp.
5 *Bacillus cereus*.
6 Other coagulase-negative staphylococci.

Fungal infections

Patients with prolonged neutropenia are at risk of invasive fungal infection. The frequent use of potent antibacterials encourages colonization by *Candida albicans*, the most commonly isolated organism. Infection with yeasts such as *C. dublini* and *C. krusei* which are naturally resistant to antifungal prophylaxis can cause a difficult problem. Infections with *Candida glabrata* and *C. parapsilosis* have been associated with intravenous cannulae.

Aspergillus spp. are the most common filamentous fungi causing invasive disease, although infections with *Fusarium* spp. and *Pseudoallerchia boydii* have rarely been reported. *Aspergillus fumigatus* is the most commonly isolated species, but *A. flavus* is increasing and *A. niger* is also reported. Infection is acquired by inhalation. *Aspergillus* spores are normally present in the air, and more are released during building work. They pose a grave threat to severely neutropenic patients causing a progressive pneumonia associated with a high mortality.

Fungi associated with infections in neutropenia
1 *Candida albicans*.
2 *Candida parapsilosis*.
3 *Candida dublini*.
4 *Candida krusei*.
5 *Candida glabrata*.
6 *Aspergillus fumigatus*.
7 Other *Aspergillus* spp.

Prevention of infection during neutropenia

Protective isolation (see Chapter 23) is intended to reduce the risk of infection in neutropenic patients. The patient is nursed in a side room, and is given sterile water and food with a low microorganism content. Attendant staff should observe hand-washing routines and may use sterile gloves when working with patients. *Aspergillus* infection can be reduced if the air entering the patient's room has been high-efficiency particulate air (HEPA)-filtered. This regimen is designed to prevent the replacement of the normal flora with potential pathogens or resistant organisms.

As the major infecting agents in neutropenic patients come from their own indigenous flora, various suppressive antimicrobial regimens have been used to prevent infection. Non-absorbable antibiotics have been given orally to reduce the numbers of bacteria in the bowel. A combination of framycetin (later replaced with neomycin), colistin and nystatin ('FRACON'), or of gentamicin, vancomycin and nystatin have both been used. The efficacy of these regimens has never been clearly established, but they do appear to select for aminoglycoside-resistant organisms, notably strains of *Klebsiella* spp.

Current antimicrobial prophylaxis is based on the concept of colonization resistance, which suggests that the obligate anaerobic flora prevent colonization by facultative Gram-negative organisms, by competing for intestinal attachment sites and nutrients, and by production of bacteriocins and toxic free fatty acids. Antibiotics should therefore be targeted to facultative organisms, sparing the protective obligate anaerobic flora. Co-trimoxazole prophylaxis had some success, but patients still became colonized with resistant organisms. This has been prevented by the addition of colistin to the regimen. Side-effects such as skin rashes and neutropenia are common.

The 4-fluoroquinolones norfloxacin, ofloxacin and ciprofloxacin are used for prophylaxis in some centres. Quinolone prophylaxis is usually combined with agents active against fungi (see below). Such regimens are superior to neomycin–colistin and co-trimoxazole–colistin, both in preventing infection with Gram-negative bacilli and in selection of resistant organisms. Ofloxacin and ciprofloxacin may be more effective than norfloxacin, which is less well absorbed. This suggests that these agents have a role both in suppressing Gram-negative bacilli in the gut and in inhibiting the early stages of infection. The quinolones are less active against Gram-positive infections, which are currently increasing.

Antifungal prophylaxis

Oral nystatin alone or in combination with oral amphotericin B is a useful non-absorbed prophylaxis, as it reduces fungal colonization of the mouth and gut. More recently, fluconazole, which is well absorbed orally, has been successful in preventing yeast infections, but it has no useful activity against *Aspergillus* spp. or some yeasts such as *C. krusei* and *C. glabrata*. Itraconazole has useful activity against *Aspergillus* spp. Patients on prophylaxis with fluconazole may become infected with *Candida* spp. that are naturally resistant to it.

Prevention of infection in neutropenic patients

1 Protective isolation.
2 Ward hygiene, including steam-pressed linen (to kill spores).
3 Filtered air supplies.
4 Suppressive antimicrobial regimens.
5 Antifungal prophylaxis.

Treatment of fever in neutropenic patients

Fever in neutropenic patients is assumed to indicate infection unless proved otherwise. Untreated bacterial infections progress rapidly; more than half of patients with *Pseudomonas* bacteraemia will die within 24 h unless appropriate treatment is prescribed.

Many trials of empirical therapy have been performed in search of a best treatment. Consideration of local organisms and their resistance pattern is important. Aminoglycosides alone are inadequate therapy in neutropenic patients and are therefore usually combined with a beta-lactam antibiotic. The importance of *Pseudomonas* as a pathogen means that the regimen should have optimal activity against this agent. A ureidopenicillin, such as azlocillin or piperacillin, plus amikacin, has obtained response rates above 60%—greater if Gram-negative bacteria were isolated. A combination of ceftazidime and amikacin has been a popular regimen but, following a large multicentre trial, the European Organization for Research and Treatment of Cancer (EORTC) proposed that a carbapenem, such as meropenem or imipenem, can be used as initial therapy. It may be necessary to include amphotericin or a glycopeptide to deal with fungal infection, *Corynebacterium jeikeium*, MRSA or *S. epidermidis*. Empirical treatment is later modified, according to the results of culture or response to initial therapy.

In this environment of empirical, broad-spectrum treatment, several organisms are major opportunistic pathogens because of their innate resistance to many antimicrobials. These include MRSA, *S. aureus*, *Stenotrophomonas maltophilia* and *Enterococcus faecium*. Glycopeptide therapy may be necessary if Gram-positive organisms are suspected, and failure to respond to antibacterial agents may indicate the need for antifungal therapy with amphotericin B. Liposomal preparations enable patients to tolerate this agent.

Treatment of fever in neutropenic patients (IDSA recommendations)

Initial therapy

Either meropenem *or* ceftazidime; *plus or minus* vancomycin.
Alternative: aminoglycoside *plus* piperacillin *or* azlocillin.

After reassessment at 3 days

If patient is now afebrile and low risk (cultures negative; neutrophils $>100\times10^9$/l): change to oral therapy with a

quinolone or beta-lactam for a further 4 days, or until neutrophils are approaching 500×10^9/l.
If patient remains febrile: take further cultures, investigate further, and change to alternative empirical regimen.

Reassessment at 5–7 days
If patient is afebrile and low risk: continue treatment for 5–7 days more, then stop and review.
If fever continues and neutrophils are >500×10^9*/l*: stop treatment and review.
If fever continues and neutrophils are <500×10^9*l*: add amphotericin B and continue existing regimen for 2 weeks after fever subsides and there are no signs of infection.

The outcome of therapy is related to the degree and duration of neutropenia. The cytokines GM-CSF (granulocyte–macrophage colony-stimulating factor) and G-CSF can be used to stimulate neutrophil production. This significantly shortens the period of neutropenia, reducing the number of infections and the days of fever.

Patients with chronic granulomatous disease are nowadays treated with interferon, which restores the defect in granulocyte function.

Fungal infection, especially with *Aspergillus* spp., can be a difficult problem for immunosuppressed patients. Liposomal amphotericin B is tolerated in much larger doses than other forms of the drug, and amphotericin B colloidal dispersion may have similar tolerability. Surgery can be useful for pulmonary *Aspergillus* infection; infected segments of lung can be excised after recovery of neutrophil numbers, to remove a locus of infection.

T-cell deficiency

T-cell deficiency is increasingly common, with increasing use of corticosteroids, cyclosporin and other immunosuppressive agents in the chemotherapy of malignancies, and in transplantation medicine. The human immunodeficiency virus (HIV) epidemic has also contributed to the number of patients with T-cell deficiency, but the recent introduction of HAART (highly active antiretroviral therapy) has enabled physicians to reverse the deficiency and has seen patients develop immune reconstitution syndromes. This is discussed in more detail in Chapter 16. Congenital T-cell deficiencies are rare and often associated with T-cell dysfunction or combined with a hypogammaglobulinaemia.

Opportunistic pathogens in T-cell deficiencies

The main pathogens are those that, in the human host, have an intracellular location. Among viruses, the naturally latent herpesviruses are common opportunistic pathogens. Mycobacteria and *Listeria* are well-recognized intracellular bacterial pathogens. Fungal and parasitic infections are also problems for patients with these immunodeficiencies.

Viral infections

Many viral infections occur in the course of leukaemia treatment, or transplantation immunosuppression (see below), often caused by the herpesviruses: herpes simplex, cytomegalovirus (CMV) and varicella zoster virus. Many of these infections can be prevented by prophylactic aciclovir. Unfortunately, a few herpes simplex and CMV isolates are now resistant to aciclovir and ganciclovir, respectively. Hepatitis B, adenovirus, papillomavirus, polyomavirus and Epstein–Barr virus (EBV) may also cause clinical problems. Chickenpox infection, or herpes zoster, caused by varicella zoster virus, can be life-threatening: patients with no immunity to chickenpox can be protected after exposure by varicella zoster immunoglobulin (VZIG).

In children, measles can be life-threatening, complicated by giant-cell pneumonia and encephalitis. Predisposed children who have been exposed to measles virus should be protected passively with human normal immunoglobulin (HNIG). Live attenuated vaccines are contraindicated in children with immunodeficiencies. Because of the risk of intrafamilial transmission, live oral polio vaccine is also contraindicated in the siblings of immunodeficient children (but MMR vaccine can be given, as transmission of these vaccine viruses is exceptionally rare).

Bacterial infections

The principal bacteria associated with T-cell deficiency are mycobacteria, including *Mycobacterium tuberculosis*, *M. kansasii*, *M. avium-intracellulare* and *M. chelonei*. *M. tuberculosis* and *M. kansasii* are primarily respiratory pathogens, and often cause typical granulomatous lung disease, but in severe T-cell deficiency they can cause disseminated or miliary disease. *M. avium-intracellulare* is acquired by the gastrointestinal or respiratory route, and may cause bowel infection, lung infection or disseminated disease. *M. chelonei* is often acquired by inoculation and then causes local abscesses, but it has also been found to cause bacteraemia and fever of unknown origin. *Listeria monocytogenes* is an important cause of meningitis, and occasionally of peritonitis or bacteraemia. Often the only clue to the listerial aetiology is a history of immunosuppression.

Treatment of mycobacterial infections in T-cell deficiency
1 *Mycobacterium tuberculosis* and *M. kansasii*: treat as routine with triple or quadruple therapy, guided by antibiotic sensitivity testing (see Chapter 18).
2 *M. avium-intercellulare*: useful drugs include rifabutin orally 450–600 mg/kg daily; ethambutol orally 15 mg/kg daily; clarithromycin orally 250–500 mg 12-hourly for 2–4 weeks, then 250 mg 12-hourly; and amikacin i.m. or i.v. 7.5 mg/kg 12-hourly.
3 *M. chelonei*: co-trimoxazole 960 mg 12-hourly, reducing to 480 mg 12-hourly after 2–4 weeks.

Fungal infections

In contrast with the effect of neutropenia, superficial fungal infections are uncommon in patients with deficient T-cell function, but deep-seated or disseminated infection can be caused by organisms such as *Histoplasma capsulatum*. Cryptococcal infections, although increasing, are usually found only among patients on high doses of immunosuppressive drugs or with HIV infection. Patients with HIV disease often suffer from severe oropharyngeal and oesophageal candidiasis. *Pneumocystis carinii* pneumonia (PCP) is the commonest infection in AIDS patients (see Chapter 15). Before the HIV epidemic, it was usually reported in patients with leukaemia (particularly lymphoblastic leukaemia), congenital T-cell defects and corticosteroid therapy.

Treatment of systemic yeast infections in the immunosuppressed
1 Amphotericin by i.v. infusion starting with 250 μg/kg daily and increasing over 3–4 days to 1 mg/kg daily, adjust dose to minimize fever, nausea, hypokalaemia, renal impairment (which may be ameliorated by giving prednisolone before each dose) and rare neurological or haematological effects; some experts stop at a total dose of 1.0 g, others are guided by clinical cure.
Alternative: Liposomal amphotericin (must be made up in warmed solution) by i.v. infusion; tolerated in doses of 3–4 mg/kg daily; indicated for severe infections or when standard preparation is not tolerated.
2 For histoplasmosis or sensitive cryptococcal meningitis: fluconazole orally or i.v. 400 mg daily reducing after response to 200 mg daily; treat until clinical features are abolished (at least 6–8 weeks for cryptococcal meningitis).

Prophylaxis of PCP

Prophylaxis initiated before immunosuppression can prevent the development of disease in patients at risk of *P. carinii* infection. It is particularly valuable when immunosuppression is related to bone marrow or organ transplantation. It can also be used in patients with AIDS, in whom PCP is uncommon until the T-cell count falls below 200×10^9/l. Monitoring of CD4 counts can indicate the appropriate time for institution of prophylaxis. Oral co-trimoxazole 480–960 mg daily is the prophylaxis of choice, but many patients are unable to tolerate this regimen. Alternatives include substitution of the sulphamethoxazole with dapsone, which provides equivalent protection with a lower incidence of adverse events and has the useful side-effect of decreasing the likelihood of reactivation of toxoplasmosis. Another effective prophylactic is aerosolized pentamidine 150 mg every 2 weeks or 300 mg 4-weekly, although this is less effective in children or in those who have already had an episode of pneumocystis infection.

Treatment of PCP

Co-trimoxazole orally or intravenously, 120 mg/kg per day in divided doses, is the treatment of choice. Pentamidine is an alternative, indicated for patients with a history of adverse reaction to co-trimoxazole. It is potentially very toxic and can cause hypotension during or immediately after intravenous administration. The dose is 4 mg/kg daily for at least 14 days; inhaled pentamidine 600 mg daily may be effective, and avoids severe systemic side-effects. Other regimens, used when patients fail to respond to the primary therapies, include atovaquone, a ubiquinone which interferes with parasite cytochrome metabolism, and trimetrexate, an antifolate drug related to methotrexate, which is given with folinic acid.

Parasitic infections

Toxoplasma gondii infection

Most *Toxoplasma gondii* infections in immunocompromised hosts result from reactivation of quiescent bradyzoite cysts. Rare, primary infections can cause fulminating meningitis or encephalitis. They usually present as multiple space-occupying lesions of the central nervous system. As with PCP, HIV infection is now the commonest predisposition. Hodgkin's disease, cardiac transplantation and acute leukaemia are non-HIV conditions associated with *Toxoplasma* infection.

Treatment and prophylaxis of toxoplasmosis

The treatment of choice is a combination of pyrimethamine 50 mg daily and sulphadiazine 4 g daily in divided doses. This is well absorbed orally and crosses the blood–brain barrier, though sulphadiazine can be given intravenously if necessary. Patients with cerebral toxoplasmosis usually respond by lysis of fever within 48 h. Failure to respond should prompt a search for an alternative diagnosis. In patients with persisting immunocompromise, suppressive treatment is often indicated after successful therapy. Dapsone is useful for this purpose. A number of salvage treatments have been devised for cerebral toxoplasmosis, including clindamycin plus pyrimethamine, or trimetrexate plus leucovorin.

Cryptosporidiosis

Cryptosporidium parvum infection is very difficult to manage in immunosuppressed patients, who suffer continuing profuse, watery diarrhoea with abdominal pain. There is no tendency to natural resolution, as in immunocompetent children and adults.

C. parvum is naturally resistant to a wide range of disinfectants and antibiotics. A small proportion of patients appear to gain some benefit from spiramycin and others from paromomycin. Recent reports suggest that azithromycin in daily dosage produces a significant benefit. At present, treatment is mainly symptomatic, with antidiarrhoeal agents and antispasmodic drugs. Severely immunocompromised patients are advised to boil all drinking water, to reduce their risk of infection.

Isopora belli infection

Isopora belli is a coccidian parasite which, unlike *Cryptosporidium*, is susceptible to antimicrobial therapy. Patients should be treated with co-trimoxazole. Alternatives include metronidazole, furazolidine, quinacrine, nitrofurantoin and newer macrolide antibiotics. Patients may be maintained on suppressive doses of co-trimoxazole or a weekly dose of sulphadoxine–pyrimethamine (Fansidar).

Strongyloides stercoralis hyperinfection

Strongyloides stercoralis, a nematode infection, is acquired by direct penetration of infective larvae through intact skin followed by invasion of the small bowel by adults which produce further larvae, perpetuating the infection (see Chapter 8). Infection can remain largely asymptomatic for more than 40 years. When cellular immunity is reduced, uncontrolled multiplication of the parasite can develop. This is known as the hyperinfection syndrome, which may be complicated by Gram-negative septicaemia, pneumonia or meningitis, as larvae deposit bacteria in the tissues.

Management of strongyloidiasis

Treatment with albendazole 400 mg daily for 3 days results in eradication in up to 80% of immunocompetent patients, but courses of 400 mg 12-hourly for up to 4 weeks may be needed in the immunosuppressed. Ivermectin is a potentially useful alternative. The complication of Gram-negative septicaemia or meningitis which may arise during uncontrolled hyperinfection syndrome should be treated vigorously with third-generation cephalosporins such as cefotaxime.

Microsporidia have recently been recognized as important pathogens of patients with T-cell deficiency, particularly in AIDS.

Microsporidian species causing infection in immunosuppressed patients

1 *Encephalitozoon cuniculi.*
2 *Enterocytozoon beneusii.*
3 *Encephalitozoon intestinalis.*
4 *Nosema connori.*
5 *Vittaforma corneum.*
6 *Nosema* spp.

Diagnosis of opportunistic infection

Because of the diversity of potential infectious agents and the need for urgent therapy, rapid, accurate diagnosis is important in all immunosuppressed patients.

Blood cultures

All patients should have at least two blood cultures taken from different sites. If these are taken via an indwelling intravenous cannula, parallel cultures should be taken from a peripheral vein, to check whether any growth originated only from the catheter or also from the bloodstream. Blood should also be drawn for mycobacterial culture by conventional liquid medium, or in an automated detection system (which gives the advantage of earlier positive results; see Chapter 18). In patients potentially exposed to the systemic mycoses, cryptococcal antigen tests may be performed and cultures should be made to media designed to support fungal growth (in a containment level 3 laboratory: see Chapter 23).

Bronchoalveolar lavage

Pulmonary lesions should be investigated by bronchoalveolar lavage (see Chapter 7). The washings are examined by Gram, Ziehl–Nielsen and silver methenamine methods for the diagnosis of bacteria, mycobacterial pathogens and PCP. Herpesviruses may be detected by electron microscopy, and by rapid CMV culture methods. PCR and other DNA amplification techniques are now available for several diagnoses (e.g. CMV, HSV, VZV and *M. tuberculosis*). Urinary antigen tests for *Legionella* infection can be performed.

Cultures from susceptible sites

Whenever fever occurs, sites such as long lines, peripheral cannulae, ventricular shunts and drains, tracheostomy sites and urinary catheters must be reviewed as potential origins of infection. Skin wounds, ulcers or multiple papules should also be investigated. Regular sets of cultures for surveillance or screening are taken in some settings, particularly haematology services. The intention is to give early warning of colonization with potentially dangerous or resistant opportunistic pathogens. The results of recent surveillance cultures may indicate which empirical therapy may succeed when fever occurs, but the cause of the fever is not always derived from superficial sites that can be routinely screened.

Early imaging in suspected localized infections

Cerebral toxoplasmosis can be suggested by the clinical picture of focal neurological deficit, and appearance of a ring-enhancing lesion on computed tomographic scanning. Serology is not diagnostic, as only immunoglobulin G antibodies are usually detectable, indicating either past infection or reactivation. Definitive diagnosis depends on brain biopsy, but this is usually only undertaken if the patient fails to respond to antitoxoplasma therapy. Nocardiosis can produce cerebral and lung abscess. The rarer, subacute cryptococcosis can cause lung, skin and bone abscesses.

Strongyloides should not be forgotten

The diagnosis of strongyloidiasis should be attempted for all patients with a history of exposure before immunocompromising therapy is commenced (this includes locally immunosuppressive therapy for inflammatory bowel disease, which may also allow undiagnosed infection with *E. histolytica* to multiply out of control). It can be made by serology and/or jejunal sampling by string test. Diagnosis of amoebiasis is made by examining stools and serology (see Chapter 8).

Diagnosis of opportunistic infection

1 Blood culture.
2 Bone marrow culture.
3 Broncholaveolar lavage.
4 Cerebrospinal fluid culture.
5 Brain scan or other imaging.
6 Surveillance cultures: mouth swabs, sputum, urine, sites of indwelling cannulae, areas of skin inflammation.
7 (Tests for strongyloidiasis or amoebiasis if indicated.)

Hypogammaglobulinaemia

Congenital hypogammaglobulinaemia

Congenital hypogammaglobulinaemia has two forms: (i) X-linked agammaglobulinaemia, in which patients become susceptible to infection after the first 6 months of life when maternal antibody is lost; and (ii) common variable immunodeficiency, which can occur at any age, most commonly the third decade.

Infants aged 3–7 months always show a dip in gammaglobulin levels, as maternal antibodies are lost, and generation of the child's own antibody levels takes over. Infants may temporarily have gammaglobulin levels well below the accepted lower limit, but are not susceptible to opportunistic infections.

Functional hypogammaglobulinaemia

Functional hypogammaglobulinaemia develops in patients with multiple myeloma, due to arrested B-cell maturation. It also occurs in patients with chronic parasitic infections such as leishmaniasis and trypanosomiasis, due to polyclonal B-cell activation and overproduction of low-affinity, non-specific antibody at the expense of specific high-affinity antibody.

Patients with deficiency in T-cell function are also susceptible to pyogenic, particularly pneumococcal and other respiratory, infections due to the loss of T-cell help in antibody production.

Opportunistic infections

The main impact of hypo- or agammaglobulinaemia is on the respiratory and gastrointestinal tracts, with resulting failure to thrive. Patients suffer recurrent and chronic respiratory infections with *Streptococcus pneumoniae* and non-capsulate *Haemophilus influenzae*. *Mycoplasma pneumoniae* and chlamydial pneumonias are also more common and persistent in affected adults. In the intestinal tract, infections with *Campylobacter*, *Giardia* and *Cryptosporidium* may be more persistent than in normal subjects. Rare cases of progressive enteroviral infection with meningoencephalitis occur in agammaglobulinaemic patients. Enteroviruses are demonstrable by culture and PCR techniques in brain, cerebrospinal fluid, muscle and other tissues.

Recurrent suppurative lung infections lead inevitably to bronchiectasis if immunoglobulin is not replaced regularly. Intravenous immunoglobulin is readily available and effective in preventing this. It is therefore indicated in infection-prone patients with congenital agammaglobulinaemia. The role of immunoglobulin in myeloma and other malignant diseases is less clear.

Complement deficiency

Hereditary complement deficiencies are rare but give rise to recurrent pyogenic infections, depending on which component of the complement pathway is deficient. Deficiency in the later components of the complement cascade, C7–C9, results in reduced ability to generate the membrane attack complex and achieve lysis of Gram-negative bacteria. The clinical consequence is recurrent infection with Gram-negative cocci, usually *Neisseria meningitidis*.

Kindreds with deficiency of components of the alternative complement pathway suffer more frequent and more serious *S. pneumoniae* infections, including meningitis. This is due to the importance of the alternative complement pathway in opsonizing pneumococcal cell wall components for clearance (see p. 142). Defects in this pathway increase the likelihood, but not the severity, of meningococcal and gonococcal bacteraemias. Rare individuals with properdin deficiency can develop devastating meningococcal infection. Immunization against the organisms carrying high risk will reduce the likelihood of severe infections in patients with alternative pathway defects, by allowing the recruitment of the classical complement cascade (see Chapter 1).

Deficiencies of individual complement components have been described affecting each of the components of the cascade. Acquired complement deficiency occurs in immune disorders such as systemic lupus erythematosus.

Splenectomy

Following splenectomy there is a continuing risk of serious sepsis, with an incidence of approximately 0.5–1.0% per year. The risk varies with age and is particularly high in infants and children. Risk also varies with the indication for splenectomy; high mortality is associated with splenectomy for lymphoma and thalassaemia. The increased risk diminishes with time after splenectomy, but is never eliminated. Patients with sickle-cell disease have functional asplenia and suffer similar susceptibility to sepsis.

The most important infecting organism for splenectomized patients is *S. pneumoniae*, which causes approximately two-thirds of infections in most series. Other important bacteria are *H. influenzae* and *Escherichia coli*. Malaria may follow a fulminant course in patients with splenectomy. Splenectomy is an important predisposition to rare *Capnocytophagia canimorsis* infections, which usually follow dog bites.

Preventing infections in asplenic patients

Prophylactic vaccination with 23-valent capsular polysaccharide pneumococcal vaccine should be offered 2 weeks before an elective splenectomy. Antibody responses are reduced in magnitude and duration after splenectomy, so vaccination should be repeated 3–6 years later. Local reactions are common on revaccination. Improved responses may be obtained with conjugate vaccines, currently under trial. *Haemophilus influenzae* type b (Hib) conjugate vaccine should also be given.

Antimicrobial prophylaxis with penicillin V should also be prescribed but its value may decrease as the prevalence of penicillin-resistant pneumococci increases. Patient education is important in encouraging patients to consult their physician quickly at the onset of a fever, or possibly to use an antibiotic regimen prescribed for early treatment without consultation. In reality many patients who undergo splenectomy are lost to follow-up, occasionally with fatal consequences.

Prevention of infection in splenectomized patients
1 Immunization against *Streptococcus pneumoniae* and *Haemophilus influenzae*, preferably before splenectomy.
2 Antimicrobial prophylaxis.
3 Patient education and information.
4 Use of alerting card or bracelet.

Evolution of infection risks in transplant patients
1 Risk derived from original disease (e.g. renal failure or haematological malignancy).
2 Risk of hospital admission and surgery.
3 Risk of disease contracted from transplanted tissue (e.g. toxoplasmosis).
4 Early risk of opportunistic infections during strong immunosuppression.
5 Later risk of opportunistic infections due to chronic suppression of cell-mediated immunity.

Infections in transplant patients

Organ transplant patients

Organ transplant patients have usually received their transplants because of underlying disease of the organ concerned. This may be non-infectious, as in terminal renal failure or ischaemic heart disease. Some transplants, however, are performed for diseases caused by severe or persisting infection; examples include liver transplants for acute or chronic liver failure caused by viral hepatitis, and heart transplant for myocarditis of infective origin. Fortunately, the original infection has often been terminated by the same immune and inflammatory response that damaged the affected organ. In a few cases, the infecting agent is not entirely removed with the infected organ, and recrudescence of local or generalized infection can occur. Thus hepatitis B virus can exist in the pancreas, and possibly other tissues, and will eventually reinfect a transplanted liver.

Antirejection therapy

Antirejection therapy produces major suppression of cell-mediated immune responses, designed to disable the effects of cytotoxic and natural killer cells on the transplanted organ. In the early post-transplant period, aggressive treatment with corticosteroids, azathioprine and cyclosporin is given. Corticosteroids and azathioprine are broad-spectrum immunosuppressants, which affect both B- and T-cell function, and also depress phagocytes and eosinophils. Cyclosporin has a narrower-spectrum but strong effect against cell-mediated immune responses. As the transplant stabilizes and the early risk of acute rejection passes, the degree and spectrum of immunosuppression can be reduced. Many patients take maintenance doses of cyclosporin or azathioprine, sometimes with a small supplement of corticosteroids.

There is therefore an evolution of susceptibility to infection after a transplant.

The patient will begin with an increased susceptibility to infection simply because of chronic underlying disease. Hepatic cirrhosis causes splenic dysfunction and increased risk of pneumococcal disease. Hypersplenism may cause pancytopenia. Renal failure causes a broad-spectrum susceptibility to infections.

Since the transplant itself is a surgical procedure, the patient will bear the infection risks of hospital admission, anaesthesia and surgery, and sometimes also those of intensive care.

On rare occasions, the transplanted organ contains infective agents. These may be persisting or latent viruses, such as CMV, human herpesvirus type 6 (HHV-6) or HHV8, or dormant organisms such as *Toxoplasma*. Donors from tropical countries may have migrating parasites such as *Strongyloides* in their organs. Infections originating in this way usually become evident soon after the transplant. Efforts are made to eliminate the possibility of transmissible infection in the donor, or to ensure that the recipient has antibodies to agents such as *Toxoplasma* and CMV if the donor is also positive.

The effects of strong immunosuppression are usually important for about 3 months. During this time a wide range of viral, fungal, parasitic and atypical infections may occur. Bacterial infections are also likely, and are vigorously investigated and treated. Skin and bowel suppressive treatment may be given with topical disinfectants and non-absorbable antibiotics and antifungal agents. Aciclovir prophylaxis is often given, and protects against herpes simplex infections and also against CMV disease (though the mechanism for the latter is poorly understood). Anti-CMV immunoglobulin is also used in some cases, and has an additive effect with antivirals.

Importance of cytomegalovirus in post-transplant patients

One of the most difficult infections to treat in the first 3 months after a transplant is CMV infection. Unlike AIDS

Time after transplant	Disease risk
First 2 weeks	Infectious complications of anaesthetic and surgery: *Staphylococcus aureus* or enterobacterial infection of wound, enterobacterial, staphylococcal or candidal infection of organ or deep tissues Pneumococcal or other chest infection
First 3 months	Chest infections with CMV, *Legionella*, *Aspergillus*, *Mycobacterium*, *Candida* or other fungus Skin and mucosal infections with herpes simplex, *Candida* or other yeast Infection of long-term indwelling cannula Cerebral infections with *Toxoplasma*, *Aspergillus* or *Candida*
After 3 months	Skin infection (herpes zoster) Pneumonia with *Nocardia*, *Aspergillus* or CMV Cerebral infection with *Listeria*, *Cryptococcus*, *Nocardia*, *Toxoplasma* or JC virus

CMV, cytomegalovirus.

Table 22.2 Range of infection risks following organ transplantation.

Duration of effect	Type of defect
First month	Lack of natural killer cells (and other non-CD8 cytotoxic cells)
3–4 months	Total T-cell count deficiency Decreased interleukin-2 production Decreased neutrophil chemotaxis
4–6 months	Decreased CD4 (helper/inducer) cells Decreased CD8 (cytotoxic) cells Decreased response to polysaccharide antigens and reduced immunoglobulin A responses to antigen
1 year or more	Decreased secondary antibody responses Decreased alveolar macrophage functions Reduced proliferative responses by CD4 cells

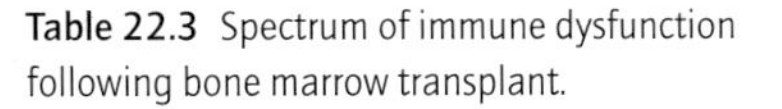

Table 22.3 Spectrum of immune dysfunction following bone marrow transplant.

patients, whose pulmonary CMV infections are rarely clinically important, more than half of transplant patients with CMV pneumonitis in the first 3 months will die. The treatment of CMV pneumonitis depends on the use of ganciclovir. This drug can cause neutropenia, thrombocytopenia and oncogenesis. If given with zidovudine, it causes bone marrow depression. Foscarnet, used for CMV retinitis in AIDS, is not recommended for other CMV infections. It is extremely toxic, causing renal impairment in half of those treated and making cyclosporin therapy difficult.

Ganciclovir treatment of cytomegalovirus pneumonitis

Ganciclovir by i.v. infusion over 1 h, 5 mg/kg 12-hourly for 14–21 days; may be continued at 5 mg/kg daily (or 6 mg/kg daily on 5 days per week), if risk of recurrence exists.

The later risks of reduced cell-mediated immunity are largely related to infections caused by reactivation of latent infections, and include herpes zoster, toxoplasmosis or progressive multifocal leucoencephalopathy (caused by JC virus). Infection with environmental agents of low pathogenicity, such as *Listeria monocytogenes* or *Nocardia asteroides*, is also a continued risk (Table 22.2).

Bone marrow transplant patients

Bone marrow transplant patients have slightly different problems because, as well as the immunosuppressive effects of their underlying disease and treatment, lymphoma, leukaemia or myeloma will cause immunosuppression before any cytotoxic therapy or transplant is undertaken. The transplant is then preceded by ablative chemotherapy and radiotherapy, causing virtually complete suppression of immune responses. The transplanted bone marrow is also often treated to suppress its cytotoxic cell population, in an effort to avoid severe graft-versus-host disease. As the transplanted bone marrow gradually becomes established, many of its functions will recover (Table 22.3).

The need for immunosuppressive maintenance therapy will depend on the source of the transplanted marrow, and its degree of tissue match with the recipient. Autologous transplants (the patient's own bone marrow, harvested during remission of malignancy) rarely need immunosuppressive support. Transplants from close relatives may require little or no antirejection treatment once they are established. Transplants from unrelated donors can require permanent immunosuppression, and may also cause graft-versus-host disease, which itself carries a risk of infection and/or rejection.

As bone marrow transplants are becoming more successful, and will be used to treat metabolic and other non-malignant disease, immunosuppressive therapy is likely to evolve, and possibly become narrower in spectrum. Expertise in long-term follow-up and the treatment of late manifestations of opportunistic infection (which may include slow infections similar to slow virus diseases) will become increasingly important.

23 Hospital Infections

Introduction, 416
Patient susceptibilities to hospital infection, 416

Infection due to intravenous cannulae, 417
Clinical features, 417
Prevention and control, 417
Treatment, 418
Diagnosis, 418

Infection associated with urinary catheters, 419

Susceptibilities of intensive therapy patients, 419
Tracheal intubation and artificial ventilation, 419
Other susceptibilities, 419

Surgery and its contribution to infection, 420
The patient, 420
Prophylaxis of surgical infections, 420
The surgical team, 421
Surgical infections in immunocompromised patients, 421

Environmental factors in hospital infection, 422
Hospital water supplies, 422
Hospital air supplies, 423
Operating theatres, 423

Hospital equipment and the spread of infection, 423

Isolation facilities in hospitals, 423
Source isolation, 423
Protective isolation, 426

Prevention of infection in laboratories, 427
Safety cabinets, 428
Containment level 4, 428

Control of infection in hospitals, 428
Control of Infection Committee, 429
Control of Infection Team, 429
Control of infection standards, 429
Control of an outbreak, 429
Hospital cleaning and disinfection, 430
Hospital waste disposal, 431

Introduction

The hospital is an ideal environment for the transmission of pathogens, because patients with similar diseases and susceptibilities are housed in an enclosed community. Patients share contact with many healthcare workers each day. In this setting, the ward, patients and workers become colonized by organisms adapted to the special environment. New susceptible individuals are frequently added to the population, and are at risk of colonization and infection. In Britain, approaching 15% of all patients admitted to hospital develop a hospital-acquired infection, with the risk increasing for a longer hospital stay.

Respiratory tract, urinary and wound infections are common in all hospital patients. In addition, immunocompromised patients readily develop infection with organisms of low virulence. Patients in the intensive therapy unit, where many antibiotics are used, can be colonized with naturally resistant organisms, which may cause pneumonia or bacteraemia.

Organisms with multiple antibiotic resistances can cause problems on general wards as well as in special units. So-called methicillin-resistant *Staphylococcus aureus* (MRSA) strains are an example. Many are resistant not only to antistaphylococcal penicillins, but also to a range of other antistaphylococcal agents. They readily colonize skin wounds, ulcers and indwelling devices such as intravenous cannulae and urinary catheters. If they cause infections, prolonged treatment may be necessary, using expensive or toxic drugs, such as teicoplanin or vancomycin.

Hospital infection not only imposes a burden of illness and prolonged admission on the patient, it also imposes the cost of investigation and treatment on the hospital, as well as preventing the use of the bed for other patients. It carries the risk of spread, particularly to other patients, and demands time-consuming and expensive control measures.

The concepts of host, organism and environment have already been discussed (see Chapter 1). The same approach can be applied to the natural history of hospital infection. In this chapter the main hospital pathogens will be described. We will also review the special susceptibilities of patient populations to organisms which often have low pathogenicity in the community, and discuss features of the hospital environment that influence the transmission of these pathogens.

Patient susceptibilities to hospital infection

Although it seems glib to say that patients are in hospital

because they are sick, the need for admission implies an alteration in host defences. This is obvious in patients immunocompromised by an illness such as leukaemia, or by treatment such as cytotoxic chemotherapy or high-dose corticosteroids. It is less obvious in fit patients admitted for routine surgery. However, the effects of anaesthesia and postoperative pain may inhibit coughing, leading to postoperative hypostatic pneumonia, or they may make micturition difficult, leading to urinary infection. The surgical wound itself presents another potential site of entry for infection. It is essential, therefore, to assess each patient carefully for such factors.

Patient predispositions to hospital infection

1 Pre-existing condition (chronic chest disease, obstructed urinary outflow or previous immunosuppression).

2 Need for invasive devices (intravenous cannulae, urinary catheters, etc.).

3 Effect of surgery (skin wound, tissue trauma, opening colonized viscus, anaesthesia, immobilization, introduction of foreign material such as joint prosthesis or arterial graft).

4 Effect of antibiotic treatment (antibiotic-associated diarrhoea, colonization by resistant organisms, predisposition to superficial fungal infections).

5 Effect of immunosuppressive treatment (corticosteroids, cancer chemotherapy or transplant immunosuppression).

6 Exposure to healthcare workers and other patients who may transmit pathogens.

7 Exposure to pathogens in the environment, especially bedding and food.

Infection due to intravenous cannulae

Intravenous devices of many kinds can be placed in the vascular system for varying periods. The time for which they can be maintained depends significantly on the likelihood of infection in each site. Peripheral venous catheters are readily colonized by organisms of the skin flora. Trivial infection, with mild inflammation, is common but more invasive disease with organisms such as *S. aureus* can cause significant morbidity and mortality (Fig. 23.1).

Many patients now have venous catheters which enter the right atrium. These may be used for intravascular monitoring, as with Swan–Ganz catheters, or to give intravenous feeding or drugs which are irritant to peripheral veins. When long-term intravenous therapy is required, long lines, such as Hickman or Portacath catheters, can be inserted via a subcutaneous track or tunnel. Infection in these devices is serious as not only must the infection and its complications be managed, but the line must also be replaced using a new tunnel. This requires a second operative procedure, and is also costly in terms of time and resources.

The common pathogens of intravenous catheters are flora from the skin of the host, particularly *S. epidermidis* and *S. aureus*. More rarely, corynebacteria may be implicated, especially the naturally multidrug-resistant species *Corynebacterium jeikeium*, which may cause line-related sepsis in leukaemic patients. *Acinetobacter* spp. can cause line infections in intensive care.

Clinical features

These are usually mild unless septicaemia supervenes. Vigilance is important in detecting this early stage, when treatment is likely to be successful, and complications few. There may be signs of inflammation at the site of the skin entry, with tenderness, cellulitis or slight purulent exudate. As in infective endocarditis, bacteraemia is usually continuous. Fever is often present but is usually mild; around 37.5–38.5°C. When line-related sepsis is likely, the patient should also be examined for signs of metastatic infection or endocarditis.

Prevention and control

The control of line-related sepsis must start with education and training in medical and nursing schools. A strict protocol of skin disinfection and sterile technique must always be used when inserting intravenous access devices. The choice of device is also important. Those with side ports are prone to colonization at this point, where no flow occurs. Giving sets may also provide a nidus for colonization. This is especially true if multiple access points are available, each with a dead space where fluids can become static. Contamination can be introduced into intravenous fluids and giving sets by repeated addition of drugs to the intravenous system. Ideally additive drugs should be incorporated into intravenous fluids in the manufacturing pharmacy, under sterile and controlled conditions.

The most important factor in preventing line-related sepsis is the regular review of inserted lines. To facilitate this the date when lines were inserted must be documented in the patient's case record. Peripheral intravenous lines should ideally be resited every 48 h. Central lines should be changed if there is evidence of infection. The life of tunnelled lines is much longer than that of

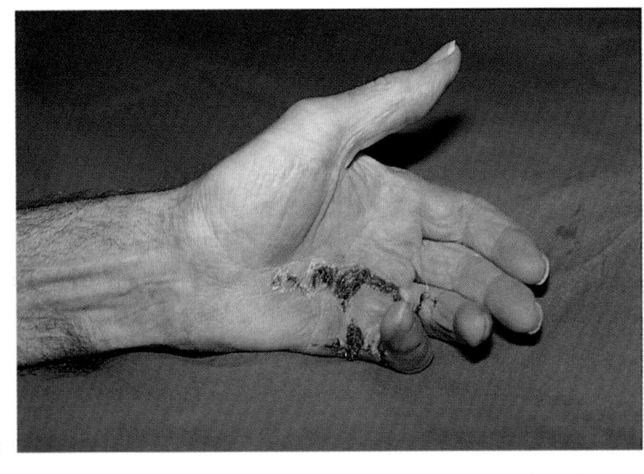
(a)

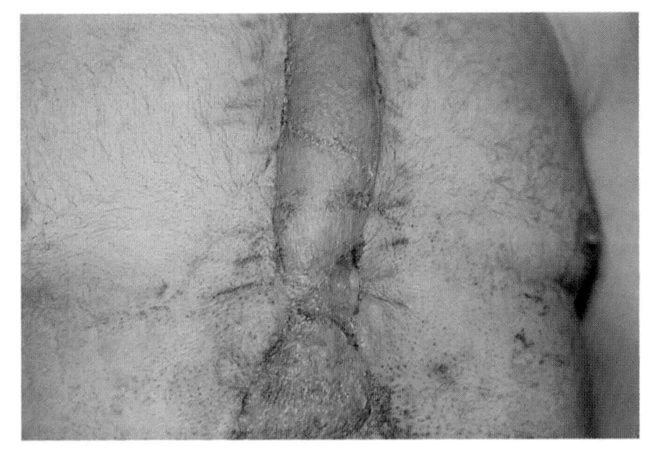
(b)

Fig. 23.1 (a) An infected minor operation wound on the hand. Methicillin-resistant *Staphylococcus aureus* (MRSA) was recovered from swabs. (b) A patient in the same ward required skin grafting after MRSA infection of a sternotomy wound.

peripheral ones, but clinicians must be aware of the risks of infection, and intervene to remove the line whenever it occurs. Special situations, such as intravenous feeding, encourage infection by providing a rich supply of nutrients within the catheter lumen. They are best managed by a specialist team which includes clinicians, pharmacists and microbiologists.

Measures to prevent line-related sepsis

1 Choice of device (excluding side ports and dead spaces).
2 Aseptic and atraumatic insertion.
3 Preparation of additive drugs and parenteral feeds in the pharmacy.
4 Maintenance of adequate hygiene and dressing of insertion site.
5 Regular review of insertion site.
6 Replacement of giving set (and cannula when indicated) at appropriate intervals.
7 Removal of cannula from inflamed site.
8 Removal or changing of cannula in a bacteraemic patient.

Treatment

When bacteraemia is associated with an intravenous access device the device should be removed. The skin insertion site and the cannula tip should both be cultured, and blood culture should be obtained, both through the infected device and via a separate, peripheral site. When the infecting organism is of low virulence such as *S. epidermidis*, these measures should be sufficient but if the infection is severe a glycopeptide antibiotic can be given for 48 h. When *S. aureus* is the infecting organism, 2 weeks' intravenous antistaphylococcal therapy is needed to minimize mortality and complications. After completion of treatment, physical review should be performed to exclude persisting focal infection, such as endocarditis or osteomyelitis. When *S. epidermidis* colonizes a 'precious' cannula, an attempt to eradicate the organisms by intracannular treatment with a glycopeptide is sometimes made. Even if successful, this should be followed by vigilant review to detect recrudescence.

Diagnosis

Line-related sepsis can be investigated by blood cultures. Ideally samples should be collected through the suspect line and also from the peripheral blood. Multiple cultures are required as the organisms are often derived from normal flora and it is only when repeated isolates of an organism with similar biochemical activity and antibiotic susceptibility profile are made that infection is confirmed. Alternatively the tip of the cannulae can be cultured and a standardized method can be used to facilitate the interpretation of the results (the Maki roll method). The main problem with this approach is that the results are semi-quantitative and the cannula must be removed to examine it. This dificulty is overcome by the introduction of bush sampling methods that allow the tip of the cannula to be sampled while it is still *in situ*. This is especially valuable for patients with a long-standing intravenous access device.

Management of line-related sepsis
1 Removal of the affected device.
2 Culture of site, cannula and blood.
3 Short course of glycopeptide antibiotic (active against staphylococci and corynebacteria).
4 Full treatment if *Staphylococcus aureus* is isolated.
5 Review of the need for insertion of new device.

Infection associated with urinary catheters

Indwelling urinary catheters provide an easy route for ascending infection of the urinary tract. After a number of days, organisms will reach the bladder, often by ascending between the catheter and the urethral wall. Permanent bladder catheterization is always associated with bacterial colonization of the urine.

Gram-negative organisms are the commonest colonizers of the catheterized bladder. *Escherichia coli, Klebsiella pneumoniae* and *Pseudomonas* spp. are often seen. *Proteus* spp. are also seen in chronically stagnant urine; its ability to metabolize urea and produce alkaline ammonia predisposes to the deposition of calcium as stones or 'sand'. In turn, these deposits can act as a reservoir of infection.

After transurethral prostatectomy, *S. aureus* or coagulase-negative staphylococci can cause urinary colonization, occasionally complicated by epididymitis.

Catheter-related urinary colonization is often asymptomatic, but there is a risk of ascending infection or bacteraemia if the catheter becomes blocked or is vigorously manipulated. Colonization is inevitable and does not require treatment unless there is evidence of infection. If fever, urinary tract symptoms or rigors occur, appropriate antibiotic therapy should be given. The catheter should usually be replaced before chemotherapy is discontinued, to ensure adequate drainage of the bladder and to remove a possible nidus of infection.

Closed drainage systems, in which the catheter is never opened directly to the environment, delay the entry and ascent of organisms, and afford a barrier to the introduction of hospital pathogens from attendants' hands. The risks can also be minimized by careful attention to aseptic technique when the catheter is inserted, and to the personal hygiene of the catheterized patient.

Avoiding urinary catheter-related sepsis
1 Sterile, atraumatic insertion.
2 Appropriate choice of catheter type and size.
3 Use of closed drainage systems.
4 Maintenance of good patient hygiene.
5 Replacement of catheter at appropriate intervals.
6 Removal of calcific deposits from the bladder (if they form).
7 Avoidance of excessive catheter manipulation or unnecessary bladder washouts.
8 Treatment of bacteriuria only when symptomatic.

Susceptibilities of intensive therapy patients

Patients in the intensive therapy unit (ITU) are susceptible to infection for several reasons. Immune responses are often diminished by the stress and metabolic effects of existing disease. Many patients in intensive care have recently undergone anaesthetic and surgical risks. Additionally, many of the barriers to infection provided by innate immunity (see Chapter 1) are breached because of the need for complex intravenous therapy, invasive monitoring, urinary catheters, artificial ventilation and extracorporeal procedures such as dialysis or haemofiltration.

Tracheal intubation and artificial ventilation

The endotracheal tube provides a means for organisms in the pharynx to bypass the mucociliary blanket defence and gain direct access to the lower respiratory tract. Where a patient has been in hospital for some time and may have received several courses of antibiotics the normal upper respiratory flora has often been replaced with Gram-negative organisms such as *Acinetobacter, Stenotrophomonas* or *Enterococcus faecium*. These organisms are naturally resistant to many first-line antibiotics and can easily invade the respiratory tract. Artificial ventilation usually imposes the need for muscular paralysis; this inhibits the normal sighing and coughing reflexes, further reducing the ability of patients to resist bacterial invasion of the lungs.

Other susceptibilities

The patient in the ITU has a multiplicity of intravenous and intra-arterial cannulae. The comments about management outlined above must be vigorously applied.

In addition to the special risks of intensive care procedures, many patients will have reduced innate and specific immunity because of organ failure, underlying malignancy, previous or pre-existing infection or chronic airways disease.

It is now known that early and adequate nutrition greatly improves the prognosis for intensive care patients. Enteral feeding often requires the insertion of percutaneous gastric cannulae, producing a further susceptible skin puncture. The feed itself must be hygienically prepared, as spoilage can introduce enteric infection. Intravenous feeding carries a high risk of cannula-related infection (see above).

Common causes of lung infections in hospital settings

1 *Streptococcus pneumoniae* (often local strains, may be penicillin-tolerant).
2 Methicillin-resistant *Staphylococcus aureus* (usually needs glycopeptide treatment).
3 *Moraxella catarrhalis* (usually produces beta-lactamase).
4 *Klebsiella pneumoniae* (always resistant to ampicillin).
5 *Escherichia coli.*
6 Enterococci (need beta-lactam/aminoglycoside, alternatively aminoglycoside–glycopeptide combination, meropenem or imipenem; a few are aminoglycoside- or glycopeptide-resistant).
7 *Pseudomonas* or related organisms (need aminoglycoside, extended-spectrum penicillin or antipseudomonal cephalosporin).
8 *Candida* (needs fluconazole as itraconazole is unpredictably distributed in critically ill patients).
9 Other yeasts (may need amphotericin treatment).

Most patients in the ITU also have an indwelling urinary catheter, and this may also act as a source of sepsis and secondary septicaemia.

The types of organisms causing local and bacteraemic infections in the ITU depend on local environmental factors and antibiotic usage. Common pathogens causing problems of treatment include *Klebsiella*, other antibiotic-resistant Gram-negative rods and enterococci. MRSA can cause intermittent outbreaks of colonization and infection.

Surgery and its contribution to infection

Modern techniques and anaesthetics make extensive and complex surgery possible, but the impact of infection greatly influences the outcome. Lister, the pioneer of antiseptic surgery, said that each operation was an experiment in bacteriology. This remains true today, though more control can be exerted over the experiment and its adverse effects. Patients must often be admitted to the ITU after a complex operation. This adds to the range of infections to which they are susceptible.

A number of factors influence the occurrence of infection in surgical patients: the patient, the operation, the antimicrobial prophylaxis, the surgical team, the hospital environment and the postoperative care.

The patient

Patients often come to surgery with pre-existing health problems. Minimizing the time between admission and the surgical procedure will limit the opportunity to acquire resistant hospital pathogens. Wherever possible, existing infection should be treated before surgery is undertaken, and preferably before hospital admission. Patients with respiratory infections should receive appropriate antibiotics and physiotherapy. Antimicrobials should not be prescribed for trivial reasons in advance of surgery, as these may allow the replacement of sensitive normal flora with multidrug-resistant hospital strains.

Prophylaxis of surgical infections

The introduction of antimicrobial prophylaxis has done much to reduce the incidence of surgical infection. There are a number of basic principles that guide its use. The agents used should be bactericidal and active against the organisms likely to be implicated in infection. To ensure that they are available at the susceptible site at the time of operation, the first dose is often given intramuscularly as part of premedication or at induction if an intravenous preparation is used. There is no evidence that additional benefit is gained by continuing prophylaxis for more than 1–3 days.

Choosing appropriate surgical prophylaxis

For this purpose, operations can be classified into three categories: clean, contaminated and infected.

Clean operations

In clean operations only the skin, or a site such as a joint which is normally bacteriologically sterile, is breached. In this case, the commonest organisms implicated in postoperative infection are staphylococci from the skin.

Postoperative infection after clean operations is almost always mild wound infection, and affects less than 2% of patients. Antimicrobial prophylaxis is not usually indicated.

The exception is when a prosthetic device, such as a vascular graft or hip prosthesis, is to be inserted. In these circumstances the consequences of infection are catastrophic and prophylaxis is indicated.

Systems designed to minimize the transmission of skin bacteria into the operation site include filtered air supplies to the operating theatre, impermeable, ventilated suits for surgeons, adequate skin preparation and sterile surgical drapes or dressings.

Attempts to eradicate those bacteria that enter the wound include the use of antibiotic-impregnated orthopaedic cement, and even antibiotic-impregnated intravascular prostheses. Nevertheless, the most common organisms infecting implanted devices remain staphylococci from the patient's own skin.

Neurosurgical operations in which the meninges are opened carry a risk of postsurgical meningitis. This is rare, but when it occurs it is often with *S. aureus* or Gram-negative organisms, including *Pseudomonas* spp. and *Acinetobacter* spp.

Contaminated operations

In these operations the surgeon opens an organ, such as the large bowel, which possesses a normal flora. Without prophylaxis the risk of infection varies between 10 and 40%. When the bowel is opened, a mixture of facultative and obligate anaerobes is released, and prophylaxis active against these organisms should include metronidazole and a broad-spectrum antibiotic such as a second-generation cephalosporin. In the upper gastrointestinal tract obligate anaerobes are uncommon and prophylaxis with a second-generation cephalosporin alone is adequate. After gastric and duodenal surgery, candidal infections occasionally occur. When obstruction is present, obligate anaerobes may accumulate, for example in the biliary tree or stomach, and the prophylaxis must be adjusted accordingly.

Instrumentation through a colonized or infected hollow organ is similar to a contaminated operation. Examples are cystocopy or ureteroscopy of the infected urinary tract and endoscopic retrograde cholangiopancreatography, when the endoscope must enter the sterile biliary tree after passing through the colonized duodenum. Prophylaxis must be given in these circumstances, as the risk of infection approaches 100%, and bacteraemia is common.

Infected operations

Infected operations are those in which infection already exists, and contamination with pathogens is inevitable. Drainage of an intraperitoneal abscess and excision of perforated bowel are good examples. Appropriate antimicrobial therapy, rather than prophylaxis, should be prescribed in this situation.

Examples of prophylactic regimens for surgical procedures

1 For upper gastrointestinal tract, endoscopic retrograde cholangiopancreatography or infected biliary tree: single dose of gentamicin *or* broad-spectrum cephalosporin 2 h before surgery.

2 For colonic and rectal surgery: single dose of gentamicin *or* cefuroxime *plus* metronidazole 2 h before surgery.

3 For hysterectomy: single dose of metronidazole rectally or i.v. 1–2 h before surgery.

4 For high amputations of the leg: benzylpenicillin i.v. or i.m. 300–600 mg 6-hourly for 5 days *or* metronidazole rectally 1 g or i.v. 500 mg 8-hourly for 5 days.

5 For joint replacement: cefuroxime or co-amoxiclav in standard parenteral doses for not more than 3 days (bone cement also contains aminoglycoside to inhibit skin-derived staphylococci).

The surgical team

The surgical team, or scrubbed team, is intimately involved with the operation. If one of the team has an infected or colonized skin or respiratory site, the pathogen involved, often a *Streptococcus* or *Staphylococcus*, may be shed into the patient's wound. The number of bacteria released into the operating theatre air depends on the number of persons in the theatre and their movement. The surgical team should therefore be as small as possible, and movements should be limited as much as possible.

The number of organisms shed can also be related to the design of surgical gowns and the type of material used. A direct relationship between this and postoperative infection is less clear. Impervious materials reduce shedding to a minimum but are very uncomfortable to wear. In the Charnley system, where the surgical team wear impervious 'space suits', these have individual air supplies and are cooled.

Surgical infections in immunocompromised patients

Modern medical practice means that there are now many patients with severe immunocompromise (see Chapter

22). These patients, however, do not form a homogeneous group. In addition to their underlying conditions which impose differing infection risks, the nature of their immunocompromise will alter the range of infections to which they are subject. Renal and liver transplant patients require less immunosuppression to prevent rejection and are therefore subject to fewer opportunist infections than cardiac or bone marrow transplant patients, in whom infection is a more important determinant of outcome. In understanding the infection risks to which patients are exposed it is convenient to discuss these under treatment considerations and immunological factors (below).

Treatment considerations

In many patients the transplantation or other therapeutic intervention imposes particular surgical risks. In cardiac transplantation the chest wall wound is a common site of infection, as is the urinary tract in renal transplant patients. For patients given radiotherapy, radiation damage to the intestinal epithelium makes bacterial escape across the gut wall more likely. After liver transplantation, postoperative infections such as cholangitis and biliary peritonitis are related to the surgery on the biliary tract.

Immune considerations

Different diseases have a differing impact on the various components of the immune system. In some, such as human immunodeficiency virus (HIV) infection, the major effect falls on T-cell function. Cyclosporin treatment also has a major immunosuppressive effect on T cells. In multiple myelomatosis the main effect is on humoral immunity. Neutrophils are principally affected in patients undergoing induction chemotherapy for leukaemia. Other conditions such as splenectomy can cause poor clearance of blood-borne parasites, and a defect in control of capsulate organisms. Immunodeficiency is seldom purely cellular or humoral: for instance, T-cell disorder also leads to humoral deficiency, because of the lack of T-helper function.

Environmental factors in hospital infection

The hospital environment is very different from the general environment, and poses particular dangers to patients ill-equipped to resist infection. Hospitals provide many opportunities for person-to-person transmission by contact. Hospital food may transmit food-borne disease if kitchen hygiene or food handling is unsatisfactory. Food is often a reservoir of local *Pseudomonas* strains. The air supply in the hospital environment is controlled and may allow for the transmission of aerosol-borne organisms such as legionellae if systems are not properly maintained. Similarly, air-borne organisms such as varicella zoster virus can spread rapidly in the hospital environment if the appropriate control measures are not taken. *Aspergillus* spores always exist in unfiltered air, and can infect neutropenic patients.

Hospital water supplies

The water supply of a hospital is complex. Unlike a domestic building or office system, there are great demands for water which must be delivered to a very large number of sites and for widely differing uses. In addition to wash-hand basins and showers, there is a need for central heating and air-conditioning. Several departments also require the delivery of steam for heating or disinfection.

In the life of a hospital, water use will evolve with changing demands, and changing use of ward and laboratory areas. This affords many opportunities for lengths of pipework to be extended, or to go out of use. Cold water may become stagnant in these areas, or hot water may lose its heat. A great danger is that *Legionella* spp. will colonize the warm water, especially when stagnant water has given up its protective chlorine. Sporadic cases or outbreaks of legionellosis may then occur. Inadequately cleaned and disinfected air-conditioning systems or cooling towers are also common sources of legionellosis (see p. 147). *Legionella pneumophila* serogroup 1 tends to cause classic legionnaires' disease, while other strains have reduced virulence, usually only causing disease in immunocompromised patients.

Any area where fast-running water may cause aerosols is a potential source of infection. Shower heads and spray-type taps may harbour legionellae in rubber washers, and disperse large numbers.

Frequent use, adequate maintenance and cleaning are all important in minimizing risk. Legionellae cannot survive at temperatures above 55°C, or replicate below 20°C. The risk of scalding by adequately hot water can be reduced by installing mixer taps. Redundant or overlong stretches of pipework, where legionellae could multiply in stagnant water, must be avoided. Pasteurizing the hot water system, by applying extra heat, may help to clear legionellae from the system. In addition it may be necessary to add additional chlorine to the cold water system. These extra measures may be necessary after pipework has been decommissioned for repair or alteration.

Control of legionellae in hospitals
1 Minimize long or redundant runs of pipework.
2 Ensure adequate chlorination of cold water.
3 Ensure adequate temperature of hot water.
4 Avoid spray taps and rubber washers.
5 Maintain taps, shower heads and cooling systems meticulously.
6 Consider pasteurization of colonized hot water systems.

Hospital air supplies

The piping and trunking of hospital air-conditioning may gradually accumulate much dust and building rubble. Sporing organisms and fungi such as *Aspergillus* spp. may thrive in these conditions. They can be discharged into the air and cause respiratory or systemic disease in susceptible individuals. Operating theatres, laboratories and individual patient isolation rooms are sites of particular risk.

Operating theatres

The construction and maintenance of operating theatres is a specialist subject. The emphasis is on easy-to-clean, impermeable surfaces, including those of movable equipment. Design is intended to minimize the need for movement of staff, and to direct the movement of patient, staff and theatre waste away from the operating or clean areas, rather than towards or through them.

There are guidelines which set out the maximum number of organisms tolerable in the air of an operating theatre. Air is supplied to the theatre through filters. Before the theatres may be used, and following repairs to the filters or other decommissioning, air quality should be tested. A special air sampler (e.g. Casella slit sampler) is used, which draws a known volume of air through a slit and deposits particles on solid bacteriological medium. Colonies of organisms can be counted after the medium is incubated, and speciation performed if indicated.

Hospital equipment and the spread of infection

Disposable hospital equipment such as syringes, needles, blood lancets, scalpels, intravenous cannulae and urinary catheters makes the introduction of infection by often-repeated invasive procedures very rare. Blood-borne infections can still be transmitted, however, by inoculation accidents with used sharp instruments. Most hospitals have strict protocols for the handling and disposal of sharps. Since the occurrence of variant Creutzfeldt–Jacob disease (CJD), and the discovery of prion protein in human tonsillar and bowel tissue, there is now a policy to use disposable surgical instruments and drapes wherever possible, and to apply strict protocols to the decontamination of all operating theatre equipment.

Some equipment is too complex and expensive for single use. Examples include endoscopes and associated biopsy equipment, complex surgical instruments, positive-pressure ventilators and high air-loss beds. These all have internal channels which can come into contact with body fluids, wound exudates or infected tissues. All have been involved in hospital outbreaks of infection. Automated cleaning systems, in which a sequence of cleaning solutions and disinfectants is pumped through the equipment for a fixed time, have reduced mishaps in decontamination after use. Such systems are expensive, and can only be used properly if the hospital possesses enough equipment to allow a satisfactory cleaning cycle before the next use.

Even such apparently simple equipment as mattresses, linen and beds can contribute to infection. The impermeable covers of mattresses can develop holes, allowing moisture and bacteria to enter. Linen which is washed at too low a temperature, or not pressed, can be contaminated by spore-bearing organisms. Low skin-pressure beds may be made of foam components, plastic beads or air-filled rubber bubbles, all of which can accumulate fluid and pathogens if not properly maintained and cleaned. All of these have caused outbreaks of skin infection, sometimes with systemic extension, often in patients with burns, skin grafts or immunosuppressive conditions.

Isolation facilities in hospitals

There are two main types of patient isolation, designed for different purposes. In source isolation the aim is to ensure that the organisms infecting or colonizing the patient are not transmitted to other patients or staff. In protective isolation the aim is to prevent organisms being transmitted to patients with special susceptibility to infection.

Source isolation

There are several types of source isolation depending on the infection that is being controlled. Until recently, isolation protocols were divided into groups depending on the routes of transmission (including blood, wound and enteric, and respiratory isolation). It is simpler, and mistakes are less likely, if a universal type of isolation is practised for all infected patients, and this is now becoming the rule in many Western hospital settings.

More stringent, high-security isolation is occasionally indicated for dangerous infections, whose treatment is difficult and which may be transmitted to medical carers. Only a few viral haemorrhagic fevers, rare bacterial pneumonias such as plague, and some fungal pneumonias fall into this category. Patients with infectious multidrug-resistant tuberculosis are also cared for in special facilities.

Blood, wound and enteric isolation

The purpose of blood, wound and enteric isolation is to prevent the transmission of organisms normally spread by contact or ingestion. Patients are nursed in a side room which contains a wash-hand basin and preferably a separate toilet facility. Nursing and medical staff remove white coats before entering the room and put on an apron or gown. The apron is discarded and hands are washed before leaving the room. Gloves are worn when handling the patient's body fluids or excreta.

Infections that can be contained by blood, wound and enteric precautions

1 Most bacteraemias.
2 Localized infections and wounds of the skin.
3 Infections with methicillin-resistant *Staphyloccus aureus*.
4 Viral hepatitides.
5 Infectious diarrhoeas.

Respiratory isolation

The precautions taken here are similar to blood, wound and enteric precautions, but the patient's room should be ventilated by a system that extracts the air to the exterior, and does not permit air flow into other ward areas. If the patient is transferred to another department of the hospital for further treatment or investigations, such as the radiology department, the patient should wear a face mask. The mask does not prevent contact with organisms suspended in droplet nuclei, but it does provide a barrier against gross contamination by fragments or large droplets of sputum which may be produced during coughing. Nurses or physiotherapists having close facial contact with a patient may wear face masks for similar reasons.

Contamination of the air outside the patient's room can be minimized by keeping the door closed. This will also ensure that the ventilation system is not overloaded by extraneous air flows.

Infections that can be contained by respiratory isolation

1 Pulmonary tuberculosis (sensitive organisms, for first 2 weeks of treatment; resistant organisms, until sputum smear is negative).
2 Varicella and herpes zoster.
3 Measles.
4 Pertussis.
5 Diphtheria.
6 Skin infections in patients with exfoliative conditions.
7 Chest infections caused by unusually resistant organisms, e.g. penicillin-resistant *Streptococcus pneumoniae*.

Control of methicillin-resistant *S. aureus*

Infection with MRSA seriously limits options for antimicrobial therapy, as methicillin resistance is usually coupled with resistance to other agents. Many studies show that the mortality rate associated with MRSA is higher, and the length of hospital stay is longer, than for patients with susceptible strains (resulting in significantly increased healthcare costs). Acquisition of MRSA carries the highest relative risk for developing a hospital infection. Staff hands are the main vehicle of transmission but some individuals ('staphylococcal dispersers') release airborne infectious particles, heavily contaminating the environment. Infected patients and, where possible, carriers should be managed in side-room accommodation with blood, wound and enteric precautions (see above). An isolation ward is often required in hospitals in which MRSA numbers are substantial. MRSA transmission is likely to occur if wards are overcrowded and antibiotic prescribing is uncontrolled. Infection control policies such as handwashing may be neglected if staffing levels are too low or many temporary staff are employed. Control activities will vary with the nature of the hospital and the degree of risk.

Hospitals with rare MRSA cases

In hospitals in which MRSA emerges rarely, efforts should be made to prevent it being established. Central to this is the universal acceptance and implementation of the infection control policies (staff must actually wash their hands between each patient contact). There is good evidence that uncontrolled antibiotic prescribing provides ideal conditions for MRSA to become established. Patients admitted from other hospitals or from overseas should be screened, and colonized or infected patients isolated.

Hospitals with endemic MRSA

When MRSA is endemic, a graded approach may be employed depending on the degree of risk to the patient.

In minimal risk areas, such as psychiatric units or long-stay care of the elderly units, isolation is not necessary.

In low-risk areas, including most medical wards, acute care of the elderly wards and non-neonatal paediatric units, colonized patients should be isolated and eradication of MRSA from the ward should be attempted.

In moderate risk areas, including general surgery, urology, dermatology, neonatal, obstetrics and gynaecology wards, screening of patients and staff should be performed if there are more than two cases, carriers should be isolated and antiseptic detergent used for patient washing.

In high-risk areas such as intensive care units, special care baby units, burns units, transplantation services, and orthopaedics, trauma and vascular surgery wards, patients should be screened on admission and on discharge; all colonized patients should be isolated, and staff should be screened if there is evidence of transmission; surgical antibiotic prophylaxis regimens may be changed to include a glycopeptide. Ward closure may be necessary to control transmission but this should only be performed in consultation with clinicians, managers and the infection control team when there is evidence of continuing transmission. Physical control measures must be supplemented with effective antimicrobial prescribing policies.

Screening procedures for MRSA

Patients may be screened for MRSA to investigate possible transmission in the hospital environment. The nose, perineum and any susceptible skin site, such as a venepuncture site or tracheostomy, as well as any infected area or burn, should be sampled. Staff should be screened if there is evidence of continuing transmission in the face of effective physical control measures. Exfoliated skin areas, due to psoriasis, dermatitis or eczema are at high risk of persistent colonization by MRSA.

There are a number of different laboratory approaches to isolation of MRSA. Selective media often use a high salt content which *S. aureus* can tolerate to reduce the growth of commensal organisms. An indicator system such as mannitol and phenol red may be used to indicate colonies for further identification. Definitive identification of MRSA is made by confirming the presence of coagulase and DNAse together with a test of methicillin resistance. Some laboratories have introduced DNA amplification techniques based on the *mecA* gene to speed up detection and identification. A simple rapid alternative is a latex agglutination technique that detects the *mecA* gene product PBP2′. It is important that MRSA are typed so that the epidemic can be monitored and new strains identified

Control of drug-resistant pulmonary tuberculosis in hospitals

Patients should be managed in single-room isolation until two or three successive sputum smears are negative for acid–alcohol bacilli. The patient's room should be at negative pressure relative to the adjacent hospital areas. To reduce the bacterial load in the room, air should be extracted to the exterior to be replaced by clean air which will dilute organisms in the room (the US Centers for Disease Control recommend six complete air changes per hour).

Susceptible visitors to the room should wear close-fitting filter masks which cover nose and mouth. Care attendants who have received bacillus Calmette–Guérin (BCG) immunization or who are tuberculin test reactors may not require routine respiratory protection. A negative-pressure room should be used for procedures such as obtaining induced sputum specimens or performing bronchial lavage, bronchoscopy or upper GI endoscopy.

Strict isolation methods have recently been applied to the control of multidrug-resistant tuberculosis, after the occurrence of outbreaks in patients and care workers (Table 23.1). Patients have been nursed in negative-pressure single rooms and this has assisted in the control of epidemics. In the UK, BCG vaccine is offered to all individuals, and this may have helped to prevent outbreaks among health workers (see Chapter 18).

High-security isolation

Highly secure isolation is provided by rooms in which the air flow is strictly controlled, and all air leaving the room is filtered to remove droplet or viral particles. Attendant staff wear protective clothing, including face protection, gloves, trousers and boots which provide a barrier to personal contamination. In exceptional cases, filtered respiratory protection is indicated, or the patient may be cared for in a filtered environment such as a bed isolator. All infectious waste from such units is decontaminated by heat or chemical methods before leaving the area.

These facilities are only available in specialist units. The only conditions which might normally require such strict isolation are viral haemorrhagic fevers transmissible from

Type of patient/contacts	Infectious	Potentially infectious*	Non-infectious
Drug-sensitive disease			
Other patients immunocompetent	Single room	Open ward	Open ward
Other patients immunocompromised	Negative-pressure room	Single room	Open ward
Drug-resistant disease			
Other patients immunocompetent	Single room	Open ward	Open ward
Other patients immunocompromised	Negative-pressure room	Single room	Open ward
Multidrug-resistant disease			
Other patients immunocompetent	Negative-pressure room	Single room	Open ward
Other patients immunocompromised	Negative-pressure room	Negative-pressure room	Single room

* Three negative consecutive smears but one or more cultures positive or cultures awaited.

Table 23.1 Minimum requirements for the isolation of patients with suspected or proven tuberculosis (Interdepartmental Working Group On Tuberculosis 1998; available from the Department of Health).

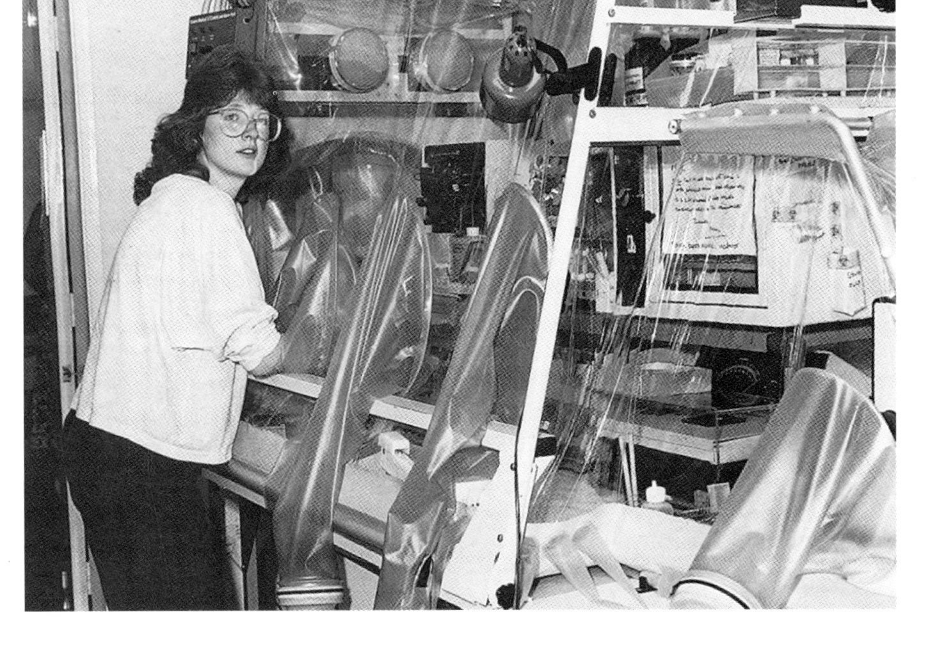

Fig. 23.2 Negative-pressure filter isolator for high-security laboratory work. This laboratory contains an incubator, an automated blood count machine, a cassette-based biochemistry analyser, a coagulometer, a blood mixture, a microscope and a centrifuge in a bench area of 2 m^2.

person to person (Crimean-Congo, Ebola, Lassa and Marburg virus infections, and some rarer, related arenavirus infections; see Chapter 24). Small-scale, negative-pressure filtered isolators are available for the laboratory handling of specimens containing dangerous pathogens (Fig. 23.2).

Protective isolation

Protective isolation is used for patients who are highly susceptible to infection (see Chapter 22). Patients with severe neutropenia are nursed in protective isolation when their neutrophil count falls below 0.5×10^9/l. Protection is not only physical, in the form of single-room isolation, and filtered air to reduce the risk of *Aspergillus* infection, but must also include arrangements to control the risk of infection from such sources as Gram-negative organisms in food or *Listeria* in soft cheeses.

Often the most difficult aspect of all these forms of isolation is ensuring that staff adhere to the effective policies. Large ward rounds, many ancillary staff, numbers of untrained voluntary workers and crowds of visitors all contribute to transmission of infection, and busy staff may often neglect the simple precautions of hand-washing when work pressures become intense.

Prevention of infection in laboratories

In European legislation, pathogens come under the heading of biological agents and are classified into categories depending on the hazard that they present to the people who work with them. Where they are handled in the workplace, an assessment of the resulting risk must be made, and protocols to minimize this are designed in accordance with the provisions of the Control of Substances Hazardous to Health (COSHH) regulations.

Definition of a biological agent
Any microorganism, cell culture or human endoparasite, including any which have been genetically modified, which may cause any infection, allergy, toxicity or otherwise create a hazard to human health.

In the UK the Advisory Committee on Dangerous Pathogens (ACDP) is a committee of the Health and Safety Executive, which advises on the design and operation of laboratories and other workplaces, depending on the type of organisms handled. Pathogens may be classified according to the recommendations of the ACDP, and other countries have similar systems. They give four hazard groups:

Category 1: An organism that is unlikely to cause disease in immune-competent humans.

Category 2: An organism that may cause human disease, and that might be a hazard to laboratory workers, but is unlikely to spread in the community. Laboratory exposure rarely produces infection and effective prophylaxis or treatment is usually available.

Category 3: An organism that may cause severe human disease and presents a serious hazard to laboratory workers. It may present a risk of spread to the community, but there is usually effective prophylaxis or effective treatment available.

Category 4: An organism that causes severe human disease and is a serious hazard to laboratory workers. It may present a high risk of spread to the community and there is usually no effective prophylaxis or treatment (Table 23.2).

Laboratory safety requires a well-designed laboratory suite. Attention must be paid to the materials employed in floors, walls and benching, and to the provision of services, including water supply and ventilation. All units where work on category 2, 3 or 4 organisms is intended must inform the Health and Safety Executive before work commences.

Containment level 3 facilities must be provided for any laboratory likely to isolate and handle organisms in hazard group 3 or which examines specimens which

Pathogen	Laboratory procedures
Category 2	
Staphylococcus aureus *Streptococcus pyogenes* *Escherichia coli* Cytomegalovirus	Open, easily cleanable bench with adequate workspace; dedicated working overalls; separate rest area; no eating, drinking, smoking, etc. in the laboratory; hand-washing facilities available in the laboratory
Category 3	
Salmonella typhi *Shigella dysenteriae* *Brucella* spp. *Mycobacterium tuberculosis* Hepatitis B virus	As above, plus separate room dedicated to this category; manipulations must be performed in class 1 safety cabinets; dedicated overalls; hand-washing facilities in room; ventilation by air extraction to exterior
Category 4	
Rabies virus Lassa virus Marburg virus Ebola virus	As above, but separate unit away from general circulation; HEPA-filtered, negative-pressure ventilation; all work in class 3 cabinets or laboratory isolator; all laboratory waste and effluent disinfected before leaving unit

HEPA, high-efficiency particulate air.

Table 23.2 Examples of the categorization of pathogens and their laboratory handling.

might contain such organisms (e.g. sputum). A laboratory suite may contain many laboratories and facilities, such as media preparation rooms, autoclave facilities, an incubator room and a cold room. Each of the individual laboratories within the clinical microbiology suite should conform to containment level 2 or containment level 3 (COSHH regulations).

A containment level 3 laboratory should be sited away from the main work of the department and access limited to authorized personnel who are trained in the use of the rooms and in the manipulation of hazard group 3 organisms. The doors should be locked when the laboratory is not in use. A continuous air flow through the laboratory must be maintained when work is in progress. A system must operate to prevent positive pressurization of the room if the extraction fans fail. Reversed air flows into the ventilation system must also be prevented. A microbiological safety cabinet of class 1 or class 2 must be available and all procedures where cultures or specimens are manipulated must be performed in this cabinet, which should be exhausted through a high-efficiency particulate air (HEPA) filter to the outside air. The laboratory should be sealable so that it can be fumigated. Ventilation, filters and cabinets must be regularly maintained and tested, to ensure continuing adequate function.

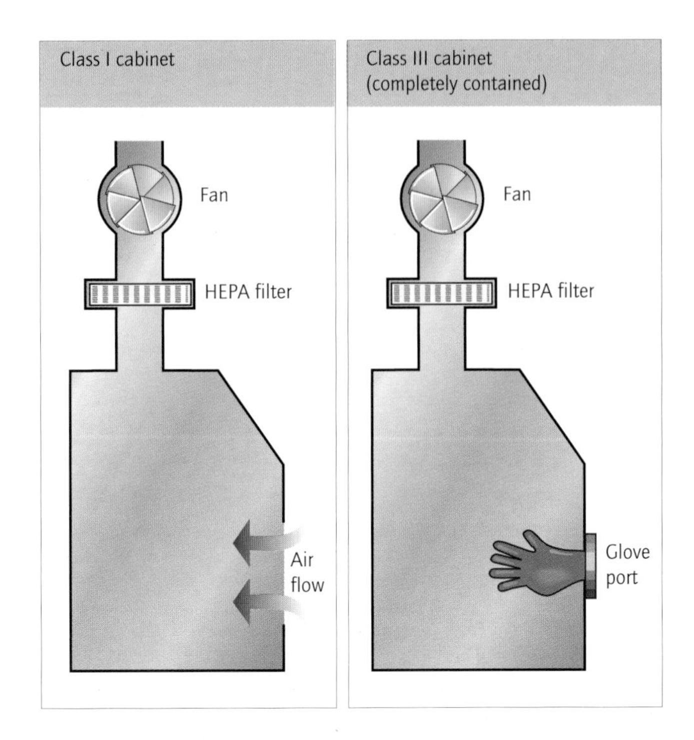

Fig. 23.3 Class I and class III exhaust-ventilated cabinets. HEPA, high-efficiency particulate air.

Safety cabinets

Class 1 cabinets

Class 1 cabinets or exhaust-protective cabinets are simple in design, with air being drawn through the face of the cabinet and out through the HEPA filter to the outside air.

Class 2 cabinets

Air is drawn into the cabinet through a HEPA filter and directed downwards on to the work surface. A portion of the filtered air is extracted. Class 2 microbiological safety cabinets provide protection to the operator and also to the work. They are therefore suitable for tissue culture, where contamination of cell lines must be minimized.

Class 3 cabinets

These are similar in clinical design to class 1 cabinets, except that they are fully enclosed and the operator works through glove ports (Fig. 23.3). Air is drawn into the cabinet and exhausted through a HEPA filter. This type of cabinet provides maximum protection to the worker from aerosol hazard. Some argue that the need to manipulate all materials and equipment with gloved hands increases the risk of accidental self-inoculation when needles and other sharp implements must be used.

Containment level 4

Containment level 4 laboratories are rare and usually sited at national reference or research laboratories. They are operated on the basis of complete security of the material used in the laboratory. Laboratory workers completely change their clothes before entering the laboratory via an air-lock while work is contained in class 3 cabinets and there is a negative pressure between the laboratory, the air-lock and the outside. Air enters through HEPA filters and is extracted through a pair of HEPA filters. A double-sided interlocked autoclave ensures that all material leaving the laboratory is rendered safe. The worker should be visible within the laboratory through glass panels and an intercom or telephone system provided with an additional competent person available to assist in emergencies. Respirators must be available for this contingency.

Control of infection in hospitals

Appropriate organization and effective management protocols are essential to the control of hospital infection.

These arrangements promote good clinical practice to minimize the occurrence of infection, promote the teaching and use of universal precautions or other infection control methods, and allow a coordinated response to outbreaks when they arise. The control of infection organization has two main strands: the Control of Infection Committee and the Control of Infection Team.

Control of Infection Committee

The Control of Infection Committee is generally a subcommittee of the senior medical committee of the hospital, and is empowered to develop and implement infection control policies and procedures. The chairperson of the committee is usually the consultant microbiologist or a consultant of equivalent expertise, such as the consultant in infectious diseases or communicable disease control. Other key staff include a hospital manager, and senior representatives from hospital services involved in infection control procedures (operating theatres, sterile supplies, nursing staff, cleaning and catering services, maintenance services and medical and surgical departments). The committee should be able to co-opt those professionals whose expertise or executive action are required.

Outbreak committee

When an outbreak occurs, the Control of Infection Committee may meet to coordinate a response, although in general a smaller group is often more practical, and can include clinicians and nursing staff from the affected area, as well as a community liaison or press officer.

Control of Infection Team

The Control of Infection Team is the core group whose role is to implement the control of infection policy, and to monitor its effectiveness. This team usually consists of the Control of Infection Officer (CIO) who is also the chairperson of the Control of Infection Committee. National authorities are increasingly insisting that the CIO has completed a diploma course in hospital infection control. In addition there are control of infection nursing staff, and these often include ward-based control of infection 'link' nurses. They work together to manage the control of infection policy on a day-to-day basis, to collect statistics on infection rates and to identify any problems. The team will monitor compliance with the policy at ward level and will implement emergency control measures when indicated. It will undertake specific surveillance of particular organisms, such as MRSA in surgical units, *Klebsiella* endemicity in ITUs and antibiotic-resistant urinary pathogens, and will screen suspected carriers. It will also provide assistance and advice to hospital staff on current policy, and on developments made necessary by changes of activity or procedures in various hospital departments.

The remit of the Control of Infection Team and Committee is necessarily wide and impacts closely on the activity of many clinical and other departments. This is because apparently trivial factors can have a profound effect on transmission of microorganisms in the hospital environment.

Control of infection standards

Examples of these have been drawn up by the Association of Medical Microbiologists, the Hospital Infection Society and the Infection Control Association in the Public Health Laboratory Service. The Department of Health has recently issued guidelines developed by a broad-based advisory group. It sets out criteria for management structure and responsibilities related to control of infection. It recommends policies and procedures for appropriate microbiological services, surveillance for the control of infection, and relevant education policies.

It suggests that an infection control structure with sufficient resources and clear lines of responsibility should be set out. This responsibility should lie directly with the senior management of the hospital. A Control of Infection Committee and effective Control of Infection Team should be present. The plans for controlling outbreaks of infection should be laid down by the Control of Infection Committee. The remit is very wide, involving all aspects of the hospital's work, including mortuary services, sterile supply services, hotel services, engineering, disposal of waste products and purchase of all equipment. None of these ideals can be serviced without an adequate microbiological laboratory. Regular surveillance for infections should be in place. Education is the main method of preventing transmission of infection and the Infection Control Team should be central to the education of medical, nursing and paramedical staff. Adequate resources and staffing should be provided by the hospital management to ensure that this important task can be carried out.

Control of an outbreak

With effective control of infection procedures in place, outbreaks ideally should be the exception. They are inevitable, however, because of the nature of the hospital environment, admitting patients from the community who may be incubating diseases such as *Salmonella* infections or chickenpox, or receiving specialist referrals from other hospitals where particular 'problem' organisms may be common. Each department must have a plan for responding to outbreaks which can be foreseen, such as *Salmonella* infections, MRSA, legionnaires' disease and multidrug-resistant Gram-negative organisms.

There are three main strands to controlling outbreaks within the hospital. Once a true outbreak has been confirmed they are: (i) to identify the source or reservoir; (ii) to halt the transmission; and (iii) to modify the host risk.

Identifying sources and reservoirs

Reservoirs of infection in hospitals are identified by screening patients within the hospital environment or before admission to hospital. Patients coming from hospitals with known epidemics, e.g. MRSA, may be screened for the presence of this organism and isolated until shown to be negative. Where the environment is a likely reservoir of infection, for example in legionnaires' disease, it may be necessary to sample the water supply at various points. Gram-negative bacteria may have their reservoir in particular hospital equipment, such as water baths, mattresses or endoscopes. When blood-borne viruses are involved, intravenous infusion, sampling or injection protocols may have been violated. Outbreaks such as MRSA infections may continue by person-to-person spread. Testing other patients and staff then becomes important in identifying the source of the problem.

Halting transmission

This is the most difficult aspect of infection control as it involves improving routine procedures such as handwashing and asepsis, isolating colonized and infected patients, and modifying the way in which patients are nursed. Cohort nursing or targeted nursing may allow individual groups of infected patients to be nursed by a team not involved with caring for other, non-infected patients. It may be necessary to stop admitting new patients until the situation is controlled.

Modifying host factors

It is more difficult to modify host risks. Some of these change naturally as patients recover from operations or can cease immunosuppressive drugs. Where epidemics of multidrug-resistant organisms are a problem, it may be possible to modify the host risk by controlling antibiotic usage. This will require liaison between the prescribing clinicians and the microbiology and infection control teams. Several studies show that strict control of antibiotic prescribing can tip the balance against MRSA and halt the spread of this organism.

Hospital cleaning and disinfection

These are two important aspects of the Control of Infection Committee's work. If the hospital is clean and its equipment and special areas adequately disinfected, staff can work safely and the risk of spreading infection is minimized.

Disinfection

Disinfection is the removal of sufficient microbial contamination from equipment to allow its safe use. This may range from the cleaning and disinfection of a vacated bed to the removal of all microbial contamination from a reusable surgical instrument.

Disinfection by cleaning

This is an extremely effective way of decontaminating floors, furniture and ordinary work surfaces. It entails the removal of dust and organic matter by wiping or washing with detergent solution, and then wiping dry with a clean cloth. Machine-washing cutlery and crockery at an adequate heat (above 80°C) is a good method of decontamination, as is washing linen at an adequate temperature (above 75°C). Steam-pressing bed sheets and towels is sufficient to remove many bacterial spores.

In some circumstances additional treatment with disinfectants (various types of antibacterial agents) is necessary to reduce bacterial contamination further. Different types of disinfectants are used for different purposes:

1 Chloros (sodium hypochlorite or bleach) is an oxidizing agent which kills most vegetative bacteria and many viruses. A 1% solution is used for cleaning surfaces. For disinfecting spillages, a 10% solution is poured over, before they are wiped up, or absorbed into granules and swept up. Chloros is toxic to humans and corrodes many metals.

2 Halogen disinfectants (chlorides and iodides) have low toxicity to human skin. They kill many bacteria, and iodine can kill spores. They are used as skin washes, scrubs and disinfectants. Chlorhexidine (containing chloride) and Betadine (containing iodine) are extensively used in intensive care units and operating theatres.

3 Alcohols act rapidly to kill vegetative bacteria, many viruses and fungi. They may be used as sprays to disinfect surfaces, such as trolley tops, and also as rubs for rapid hand disinfection. Halogen disinfectants can be dispensed as alcoholic solutions, for additional bactericidal effect. Alcohols are easily diluted to below effective concentrations by evaporation, and penetrate organic matter poorly. Their flammability makes them too dangerous to use near diathermy units and other operating theatre equipment.

4 Aldehydes are non-corrosive and kill a wide range of organisms. Glutaraldehyde (Cidex) solution is widely used in automated decontamination systems. Sealed rooms and large equipment are occasionally decontaminated by fumigation with formaldehyde dissolved in water vapour.

5 Phenol-based disinfectants such as Hycolin are non-corrosive, and highly toxic to microorganisms but are not very active against viruses. They are used to disinfect contaminated surfaces such as the floors of ambulances, bed frames and bathroom equipment. Suspensions of tarry phenolics, such as Sudol, are used for cleaning highly contaminated stone and ceramic floors, such as mortuary areas. They are highly microbicidal, but leave a residue which is hard to remove from instruments and equipment.

6 Ethylene oxide is a gas that kills bacteria, viruses, fungi and bacterial spores. It can penetrate complex instruments, and is used for cleaning laparoscopes, arthroscopes, reusable cardiac catheters and delicate surgical instruments. It is applied at low pressure, usually mixed with carbon dioxide to avoid risk of explosion, and at a temperature of 55°C. Instruments must be aired after treatment to remove the irritant gas.

Disinfectants do not work adequately in the presence of organic matter, which may degrade them. Debris can also prevent the disinfectant from reaching its target surface. All equipment should be cleaned or washed before being disinfected.

Physical methods of disinfection

Heat is the most commonly used physical disinfection method in hospitals. Superheated and pressurized steam is used in autoclaving. Porous material such as operating gowns and drapes are prepared for use by this method. Stainless steel and some plastic and rubber equipment can also be autoclaved.

Adequate sterilization is a function of temperature and time of exposure. In an autoclave, air is drawn out of the load by creating a vacuum, and is replaced by pressurized steam. This process is repeated in five to eight cycles. The temperature in the vessel is then held, usually at 134°C, for 3–20 min. An extended cycle is used for potentially contaminated material from patients with CJD.

The effectiveness of the cycle can be tested by including heat-stable bacterial spores in the load, and showing that they do not germinate after treatment. The attainment of adequate temperatures can be shown by using autoclavable tape to wrap the load; the tape shows a colour change at the correct temperature–time combination. Equipment can be prepacked in semipermeable paper or plastic wrappers, and the whole package autoclaved. The equipment is then sterile until the pack is broken.

Closed jars and bottles would rupture in the waves of heat and pressure produced by standard autoclaves. Special autoclaves with balanced pressure and heat cycles must be used for these.

Gamma radiation is widely used in industry to remove microbial contamination after manufacture from plastic, silicon and rubber equipment, usually after packaging. As the gamma rays penetrate deeply, the method can be used to sterilize fragile and complex equipment such as cardiac catheters, pacemaker controllers and complex interventional radiology equipment.

Hospital waste disposal

A hospital produces a huge amount of waste. This includes:

1 Domestic-type waste from kitchens, washrooms, dining facilities and public areas.

2 Clinical waste such as used dressings, wound drainage, used disposable equipment and even organs and limbs from the surgical department.

3 Discarded, used sharps.

Domestic waste can be removed by local authority services and disposed of in various ways without hazard to disposal workers or the public. Clinical waste must be safely packed and clearly identified. In the UK it is put into strong, yellow plastic sacks at the site where it is generated. These sacks are stored in strong bins or skips until they are removed intact for incineration locally, or by a licensed disposal firm. Used sharps are disposed of directly into strong, leak-proof bins at the site where they are generated. These bins are sealed when full, and are removed intact for storage and incineration.

Laboratory waste contains high concentrations of pathogens. All used containers, media and disposable equipment are autoclaved before they are either disposed of in the hospital waste system or recycled for laboratory use. Laboratories use the same sharps disposal system as the other hospital areas.

Case 23.1: Hernia causes a pain in the neck

History: A 69-year-old man was admitted to hospital for a routine herniorrhaphy. His general health was good, except for moderate spinal osteoarthritis. He had not smoked tobacco since giving up cigarettes 10 years earlier. The anaesthesia and operation proceeded uneventfully, but 3 days later he noticed soreness and a slight discharge of pus from the medial end of his wound. Swabs were taken, and a growth of methicillin-resistant *Staphylococcus aureus* (MRSA) was identified on culture. The wound was cleaned twice daily with chlorhexidine lotion, the discharge ceased and healing continued.

Further progress and investigation: One week later, he complained of intense aching in his lower neck, and pain in the medial side of his right upper arm. Examination showed temperature 38°C, limitation of neck rotation, and pain on neck movement. Neurological examination was normal. X-rays of the cervical spine showed only osteoarthritic changes. On the following day the patient remained febrile. He complained of weakness of the legs, and developed acute retention of urine, with bladder volume 750 ml An urgent MR scan of the cervical and thoracic spine was carried out (Fig. CS.4). The white blood cell count was $15.6 \times 10^9/l$, with 85% neutrophils.

Fig. CS.4 MR scan showing oedema and partial destruction of second cervical vertebra with loss of the T1/T2 disc space, and a pale abscess mass surrounding and compressing the spinal cord.

Questions: What is seen on the MR scan?
What bacterial aetiology must be included in the differential diagnosis?
What immediate treatment should be commenced?
What further management is indicated?

Further management and progress: The MR scan (below) showed oedema and a soft-tissue mass surrounding the second thoracic vertebra, with partial destruction of the vertebral body, and loss of the T1/T2 disc space (arrowhead). A pale mass, representing an extradural abscess, is surrounding and compressing the spinal cord (arrow). Blood cultures were taken, and antimicrobial therapy commenced with teicoplanin, cefotaxime and metronidazole. (the glycopeptide antibiotic was considered essential, because of the recent isolation of MRSA from this patient). The neurosurgical team arranged immediate surgical decompression and drainage of the abscess.

Culture of pus from the abscess cavity produced a pure, heavy growth of MRSA. The blood culture remained negative, and the patient made a complete recovery after a further 2 weeks of parenteral therapy.

Comment: Although colonization and infection of skin and small wounds is often a trivial problem, hospital-acquired resistant organisms can cause severe and life-threatening disease if they invade systemically. This patient must have acquired his vertebral abscess by blood-borne spread from a brief bacteraemia, probably with lodgement of MRSA in the mildly traumatized osteoarthritic tissues of the neck. He was fortunate that earlier investigation had revealed that he was colonized by a resistant organism, so that effective empirical treatment could be chosen.

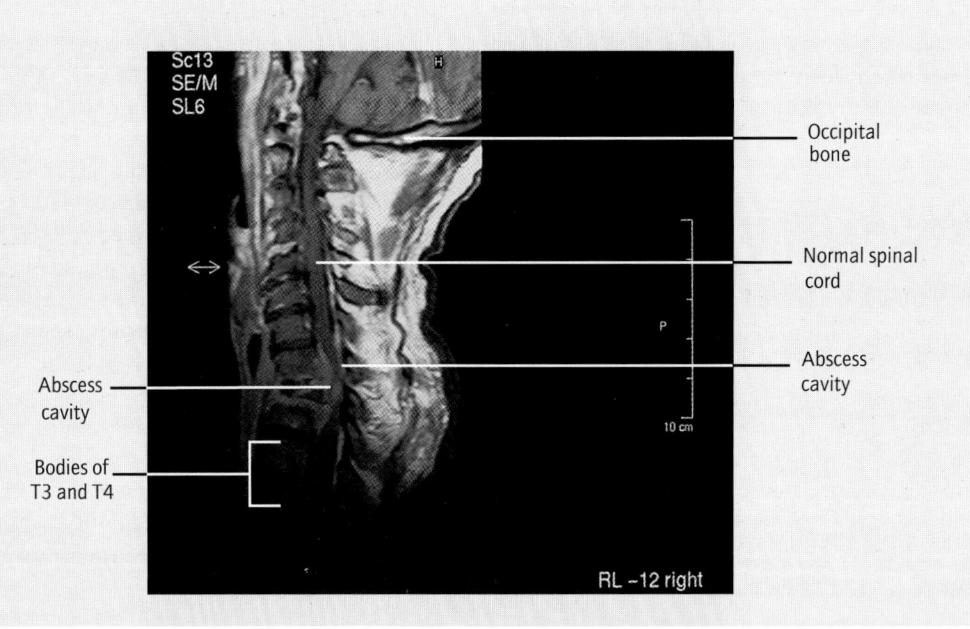

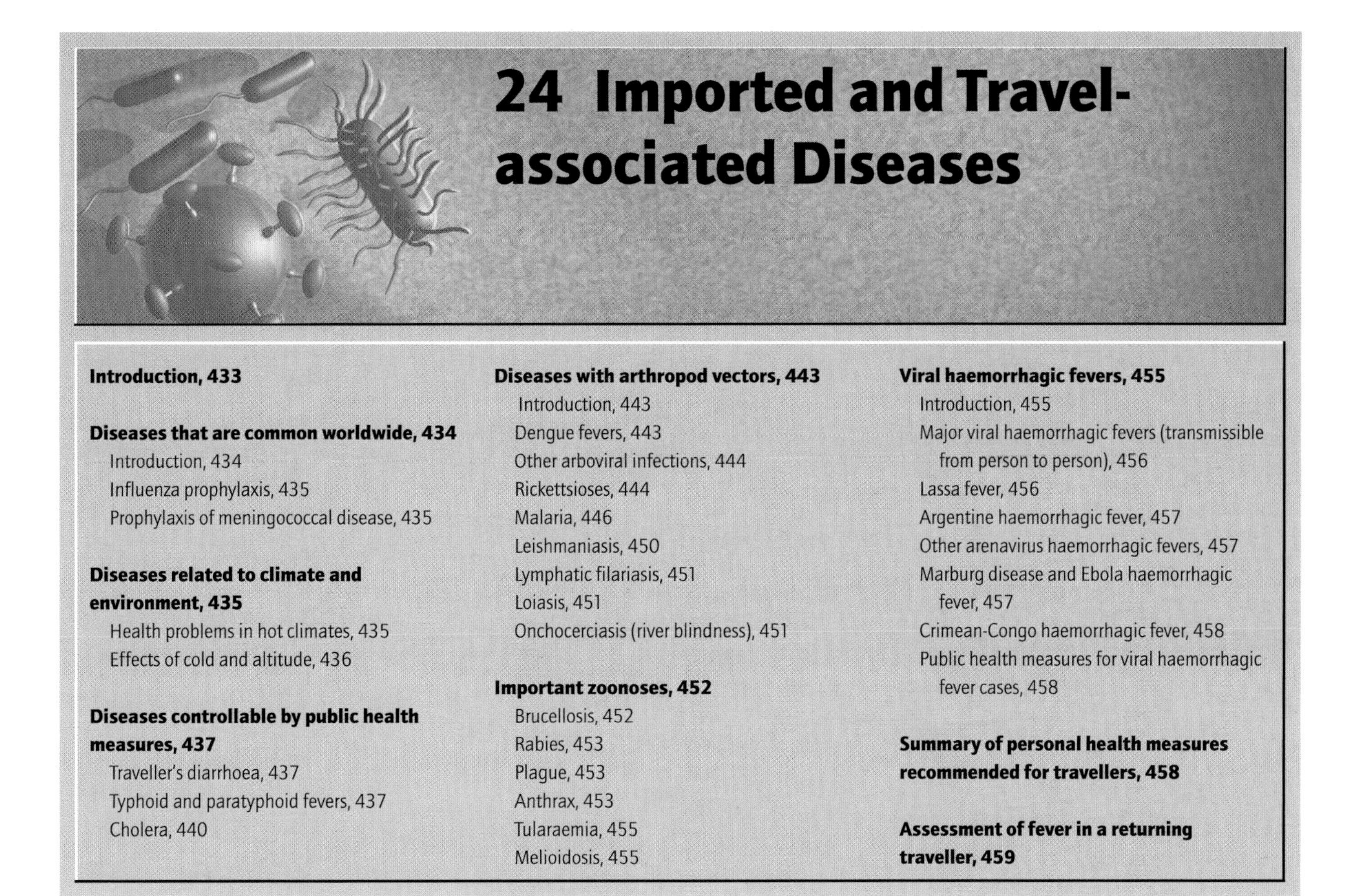

24 Imported and Travel-associated Diseases

Introduction, 433

Diseases that are common worldwide, 434
- Introduction, 434
- Influenza prophylaxis, 435
- Prophylaxis of meningococcal disease, 435

Diseases related to climate and environment, 435
- Health problems in hot climates, 435
- Effects of cold and altitude, 436

Diseases controllable by public health measures, 437
- Traveller's diarrhoea, 437
- Typhoid and paratyphoid fevers, 437
- Cholera, 440

Diseases with arthropod vectors, 443
- Introduction, 443
- Dengue fevers, 443
- Other arboviral infections, 444
- Rickettsioses, 444
- Malaria, 446
- Leishmaniasis, 450
- Lymphatic filariasis, 451
- Loiasis, 451
- Onchocerciasis (river blindness), 451

Important zoonoses, 452
- Brucellosis, 452
- Rabies, 453
- Plague, 453
- Anthrax, 453
- Tularaemia, 455
- Melioidosis, 455

Viral haemorrhagic fevers, 455
- Introduction, 455
- Major viral haemorrhagic fevers (transmissible from person to person), 456
- Lassa fever, 456
- Argentine haemorrhagic fever, 457
- Other arenavirus haemorrhagic fevers, 457
- Marburg disease and Ebola haemorrhagic fever, 457
- Crimean-Congo haemorrhagic fever, 458
- Public health measures for viral haemorrhagic fever cases, 458

Summary of personal health measures recommended for travellers, 458

Assessment of fever in a returning traveller, 459

Introduction

Worldwide travel has increased enormously since the 1950s, as it has become easier, quicker and more affordable (Fig. 24.1). Air travel is rapid, allowing movement around the world in a shorter time than the incubation period of almost any disease. Only the most hostile areas are inaccessible to tourism, but even these may be entered by geologists and other researchers. Overland travel is popular, as is the experience of sharing unfamiliar living conditions with local people. Outbreaks of respiratory or intestinal infection can occur in large hotels or cruise ships.

Travellers are vulnerable to infections unfamiliar to their home-based medical services, presenting diagnostic problems, difficulties in management and unexpected complications. Contagious diseases are also hazardous to contacts and can be serious public health problems.

Unfamiliar features of imported diseases

1 Presenting features.
2 Diagnostic methods.
3 Management requirements.
4 Unexpected complications.
5 Unexpected infectiousness.

Other factors that increase the vulnerability of travellers include:

1 The temptation to take risks with food, water, animals and sexual contacts when relaxing away from the conventions of home.
2 The different epidemiology of some diseases in different environments (e.g. heterosexual versus homosexual transmission of human immunodeficiency virus (HIV), prevalence of open pulmonary tuberculosis or existence of epidemic diseases such as diphtheria).
3 The incomplete understanding of health hazards and protective measures with which travellers often arrive at a destination.
4 The stress that accompanies long journeys across time zones which may make travellers unusually susceptible to some diseases.
5 In the case of refugees, privation, malnutrition and pre-existing disease or injury which may widen the range of infections to which they are vulnerable.

DISEASE LIST

(Rather than being listed by organism, diseases of travel can conveniently be considered in aetiological or epidemiological groups.)

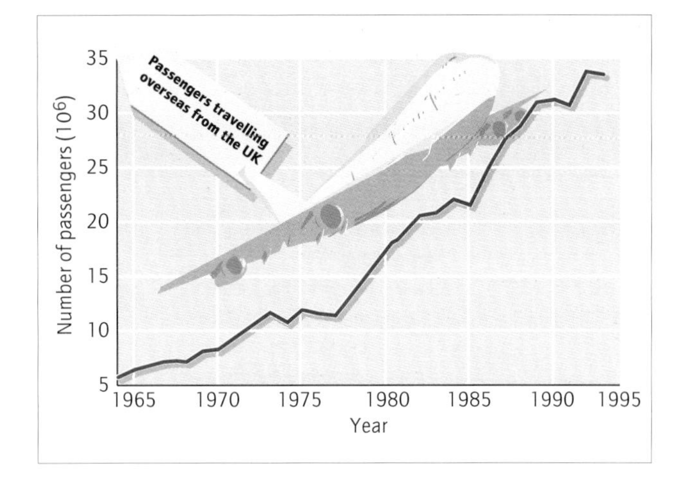

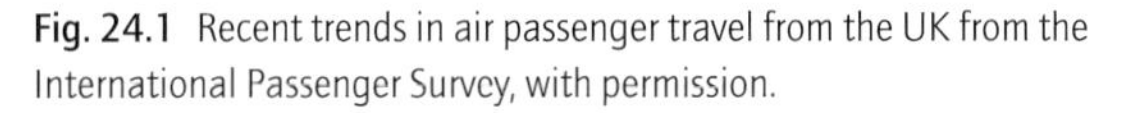
Fig. 24.1 Recent trends in air passenger travel from the UK from the International Passenger Survey, with permission.

Diseases common worldwide
- Influenza
- Community-acquired pneumonias
- Urinary tract infections
- Meningococcal disease
- Sexually transmitted diseases

Diseases related to climate and environment
- Sunburn
- Heat exhaustion and heatstroke
- Dermatophyte infections
- Folliculitis
- Cold injury
- Altitude sickness

Diseases controllable by public health measures

Sanitation, food hygiene and safe drinking water
- Hepatitis A
- Hepatitis E
- Viral gastroenteritis
- Traveller's diarrhoea
- Bacterial food poisoning
- Bacillary dysentery
- Enteric fevers
- Cholera
- Giardiasis
- Amoebiasis
- Cryptosporidiosis
- Helminth infections

Immunization
- Poliomyelitis
- Diphtheria

Education
- Sexually transmitted diseases
- HIV infection

Risks of contact with mud and water
- Leptospirosis
- Hookworms
- Strongyloidiasis
- Guinea worms (increasingly rare)
- Schistosomiasis
- Liver flukes

Diseases with arthropod vectors
- Dengue fevers
- Arboviral encephalitides
- Other arboviral infections, e.g. phlebotomus fever
- Rickettsial infections
- Plague
- Lyme disease
- Malaria
- Leishmaniasis
- Trypanosomiasis
- Filariasis
- Onchocerciasis

Some important zoonoses
- Brucellosis
- Hantavirus infections
- Rabies
- Tularaemia
- Melioidosis
- Anthrax

Viral haemorrhagic fevers
- Yellow fever
- Dengue haemorrhagic fever
- Lassa fever and other arenavirus infections
- Marburg fever
- Ebola fever
- Crimean-Congo haemorrhagic fever

Many of these diseases are discussed in some detail in other chapters. Important or common diseases not mentioned elsewhere will be described in detail in this chapter. Brief information on presentation, diagnosis and management will be given for the remainder.

Diseases that are common worldwide

Introduction

However exotic a traveller's destination, common infections still pose a risk. Indeed, crowding, stress, insect bites, altered personal hygiene facilities, fatigue and altered patterns of hydration may predispose to clinical expression of common infections.

Some epidemic diseases may be active at the destination while in abeyance at home. Influenza and meningococcal disease are good examples whose epidemic cycles wax and wane around the world.

Influenza prophylaxis

Pandemics of influenza A begin in eastern Asia and spread westwards across the world, following the winter season, first southwards to Australasia and then northwards to Europe and North America. World Health Organization laboratories maintain surveillance, to warn of the emergence of new epidemic strains. Appropriate vaccines can quickly be constructed.

Immunization is useful for travellers who will enter areas of influenza activity, and should be offered to elderly individuals, or those with special susceptibilities who would be given the vaccine routinely in the UK. It affords about 75% protection to healthy recipients. Debilitated and elderly recipients are not so well protected, but the severity of illness and the risk of death are significantly reduced.

Prophylaxis of meningococcal disease

Group B meningococci, for which there is still no effective vaccine, remain the prevalent epidemic organisms of Europe and North America, but in the Middle East, Africa and South America most epidemics are of group A. In the late 1980s a large epidemic of group A disease affected many Muslim pilgrims travelling through Mecca and the Middle East. Local outbreaks of group A and C meningococci also occur in many countries, particularly West and central Africa.

Meningococcus vaccine, containing polysaccharide antigens of group A and C, affords good protection to adults and older children for at least 3 years after a single dose. A polyvalent A/C/Y/W135 vaccine is also available in some countries. Polysaccharide vaccines are poorly immunogenic in children below the age of 18–24 months, but recently developed conjugate vaccines will avoid this problem.

Diseases related to climate and environment

Health problems in hot climates

Sunburn

Sunburn is skin damage caused by ultraviolet radiation from the sun. Most cases are superficial and heal completely by desquamation without scarring. A few are severe, and cause blistering. Prolonged exposure to sunlight causes ageing and wrinkling of the skin, while both chronic and repeated severe exposure predispose to malignant skin conditions, including melanoma.

Sunburn is prevented by avoiding exposure or covering the skin. Barriers to damaging ultraviolet B (UVB) radiation can be applied as suntan creams, whose protection number indicates the recommended duration of exposure for untanned European skin. UVA radiation is less damaging, but creams containing a suspension of titanium oxide offer protection. Tanning protects from sunburn, but must be accomplished gradually. It does not prevent the ageing or carcinogenic effects of prolonged exposure.

Heat exhaustion

Heat exhaustion is caused by excessive sodium and water loss in sweat. People unacclimatized to hot environments secrete large amounts of sweat with a high sodium content, which falls slowly during the first month of acclimatization. This adaptation process cannot be speeded by exercise or medication, so most brief holidays do not allow acclimatization.

The symptoms of heat exhaustion are malaise, nausea, headache and collapse. Replacement of sodium and water results in rapid improvement. It can be achieved with oral rehydration solutions, as for acute diarrhoea, or simply by giving plentiful dilute squash or fruit drinks to which salt has been added (about 1.5 teaspoons per pint, or 3 per litre).

Exercise in hot conditions may cause sweating of up to 4 l/h, so travellers should take plentiful fluids and add salt to their meals. Salt tablets may be of benefit if strenuous sport is played. It is sensible to rest in the midday heat, as local people usually do.

Sunstroke

Sunstroke is a condition of hyperpyrexia and shock, often induced by excessive exercise, associated with too much clothing or failure of sweating due to dehydration. It is a medical emergency which can lead to liver damage, haemolysis, renal failure and death.

The collapsed patient should be moved into the shade and excess clothing removed. The temperature should be reduced by tepid sponging or by wrapping in moist sheets. Fanning is helpful at this stage. When the core or rectal temperature is below 40°C the damage to tissue ceases, and slower cooling may be allowed to continue naturally. The patient requires frequent observation in case the temperature rises again, and haematological and biochemical assessment should be carried out without delay.

Skin infections

In hot climates the skin is constantly moist and easily becomes macerated, or traumatized by the friction of moist clothing. Hair follicles or sweat glands may be blocked by soft keratin plugs, causing hyperaemia and papular swelling, often called a sweat rash. In these circumstances dermatophyte infections easily occur. Staphylococcal folliculitis is also common. A healthy skin can be maintained by avoiding insect attack and wearing loose, light clothes, preferably made of absorbent natural fibres. Both the clothes and the skin should be regularly washed. Antiperspirants are not as effective in hot climates as in temperate ones.

Rarer skin infections such as *Acinetobacter* infections will not respond to penicillins and cephalosporins. Cuts and abrasions from coral may become infected by marine vibrios and are similarly resistant. Both will often respond to oral tetracycline or ciprofloxacin.

In areas where diphtheria exists, skin ulcers and abrasions may be colonized or infected with *Corynebacterium diphtheriae*. Infected lesions often have a greyish membrane at the base, and a slight serosanguineous discharge. In local child populations the small dose of toxin produced in the lesion often induces natural immunity. Troublesome lesions respond rapidly to treatment with penicillins or erythromycin, but it should be remembered that such lesions are infectious. Immunity declines slowly after childhood immunization, so adult travellers may be susceptible. Visitors to Western countries from overseas may never have been immunized.

Prevention of insect bites

Insect bites are common in hot climates and are easily infected by staphylococci or streptococci (such streptococcal infections make nephritis a common paediatric problem in tropical countries). Insect repellents are effective in reducing bites. Loose clothing is also a useful barrier; the arms and legs should be covered at dusk when mosquitoes are highly active. Window and door screens keep insects out of buildings; the use of 'knock-down' sprays or 'mosquito coils' kills any that have entered. In malarious areas mosquito-proof bed-nets offer important protection, which is much increased by impregnation with permethrin.

Female chigger or jigger fleas lodge in the skin of the feet and enlarge to up to 1 cm in diameter as they fill with eggs. They can be teased out with a needle. The use of footwear in sandy areas affords good protection. Tumbu flies deposit eggs on drying laundry, and emerging larvae penetrate the skin. As the larva grows and develops, a boil-like lesion results; covering the lesion with vaseline causes the maggot to emerge for air, when it can be grasped and removed. Ironing all laundry prevents this problem.

Skin infections in hot climates

1 Staphylococcal folliculitis.

2 Staphylococcal and streptococcal infections of insect bites.

3 Infections with *Acinetobacter*, pseudomonads or marine vibrios.

4 Colonization or infection with *Corynebacterium diphtheriae*.

5 Dermatophyte infections.

Effects of cold and altitude

Frostbite

Frostbite occurs when the skin is sufficiently frozen to cause tissue damage. As thawing takes place, fluid leaks from affected blood vessels and painful blisters appear. Deeper injury can cause necrotic and anaesthetic lesions, with risk of secondary infection. Temporary superficial freezing (frostnip) is completely reversible if rapidly thawed by applying a warm hand or clothing.

Altitude sickness

Altitude sickness is the result of excessive accumulation of fluid in body compartments, which occurs when rapid ascent of more than 10–15 000 feet (3000–4500 m) is made without time for physiological acclimatization. Individuals vary in their susceptibility. Mild symptoms include headache, thirst and sleeplessness. Mild diuresis and the alkalinizing effect of prophylactic acetazolamide 250 mg once or twice daily can lessen the severity of altitude sickness, but this is not a substitute for acclimatization.

Altitude-related pulmonary oedema is indicated by cough, wheeze and breathlessness. Acetazolamide and modified-release nifedipine 20 mg twice daily are often used as prophylaxis. Nifedipine 30 mg 8-hourly and, in severe cases, dexamethasone 8 mg 8-hourly can be used as emergency treatment while arranging urgent descent to below 13 000 feet (4000 m), which is the treatment of choice. An antibiotic such as co-amoxiclav or cefuroxime should also be given, as secondary bacterial chest infection is very common. Portable pressure 'cocoons' are carried on some expeditions, but are not a substitute for descent.

High-altitude cerebral oedema is a medical emergency, indicated by headache, nausea, vomiting and drowsiness.

Coma and convulsions occur in severe cases. Frusemide and dexamethasone may reduce the risk of progression, but descent is the urgent requirement.

Diseases controllable by public health measures

Traveller's diarrhoea

Introduction

This is defined as the occurrence of at least three abnormally loose stools in any day, or one or more loose stools with features such as vomiting, abdominal cramps or fever. It is common, affecting from 8 to 50% of travellers, depending on local sanitary standards. Half or more of cases are caused by enterotoxigenic *Escherichia coli* (ETEC), but other intestinal pathogens are also important (see Chapter 8). Up to half of cases are fatigued or bedbound for at least 1 day.

Clinical features

Symptoms begin most often on the third day after reaching the destination. The illness lasts an average of 4 days and few patients have more than five or six diarrhoea stools per day. An attack caused by ETEC is followed by immunity to the local ETEC strain.

Management

Most cases are brief and not significantly shortened by treatment with antibiotics.

In endemic areas 40–60% of ETEC are susceptible to agents such as amoxycillin, trimethoprim, co-trimoxazole or ciprofloxacin. Elderly and debilitated patients, who may suffer prolonged or complicated illness, may be advised to take a 2- to 3-day course of oral ciprofloxacin at the first sign of diarrhoea.

Prevention

Although several antibiotics and also bismuth preparations can reduce the incidence of diarrhoea in travellers, resistant ETEC soon emerge where antibiotics are frequently used. Chemoprophylaxis is therefore not recommended unless the necessity is exceptional. Vaccines against enterotoxins are being developed., mainly for the prevention of gastroenteritis in children, but they may become a useful safety measure for travellers. Dietary precautions can contribute to prevention, especially in the short term. Travellers who suffer from diarrhoea have more often than not taken chopped fresh fruit, sandwiches with mixed fillings, raw or lightly cooked seafood and untreated water (including ice cubes).

Water safety

It is safest to avoid untreated water, including fruits and salads washed in it and ice cubes made from it. Commercial brands of mineral water and carbonated drinks are usually safe. Tap water may be purified by boiling. Alternatively, it can be filtered in a portable filter, and boiling or purifying tablets then used to destroy viruses.

Typhoid and paratyphoid fevers

Introduction

These diseases, caused by *Salmonella typhi* and *S. paratyphi* type A or B, are uncommon but often severe infections of travellers. They occur where sanitation is poor or drinking water is insufficiently safe. Even with modern treatment, morbidity is considerable, and an increasing tendency to antibiotic resistance means that some cases are difficult to treat.

Pathology of enteric fever

For a few days after infection, salmonellae replicate in the gut, and are excreted in the faeces. This is followed by the primary bacteraemia which is usually asymptomatic. The organism then replicates in reticuloendothelial cells. A secondary bacteraemia heralds reinvasion of the gut, particularly of the Peyer's patches, and the onset of symptoms. During this period blood cultures are positive, and salmonellae can be isolated from the urine in some patients. Treatment of the infection may render the blood sterile, but bone marrow cultures may remain positive until late in the course of antimicrobial treatment.

Clinical features

Typhoid fever is the typical enteric fever. The incubation period varies from 6 days to 4 weeks, but averages 2 weeks. The insidious onset is often mistaken for 'flu'. Symptoms include fever, which increases daily, headache, abdominal discomfort, constipation and often a dry cough. The pulse rate often fails to rise with the temperature, producing the effect of a relative bradycardia (Fig. 24.2). There may be a neutrophilia at this stage. Confusion is common, varying from taciturnity or bad dreams to frank delirium or apparent psychosis. Confused patients

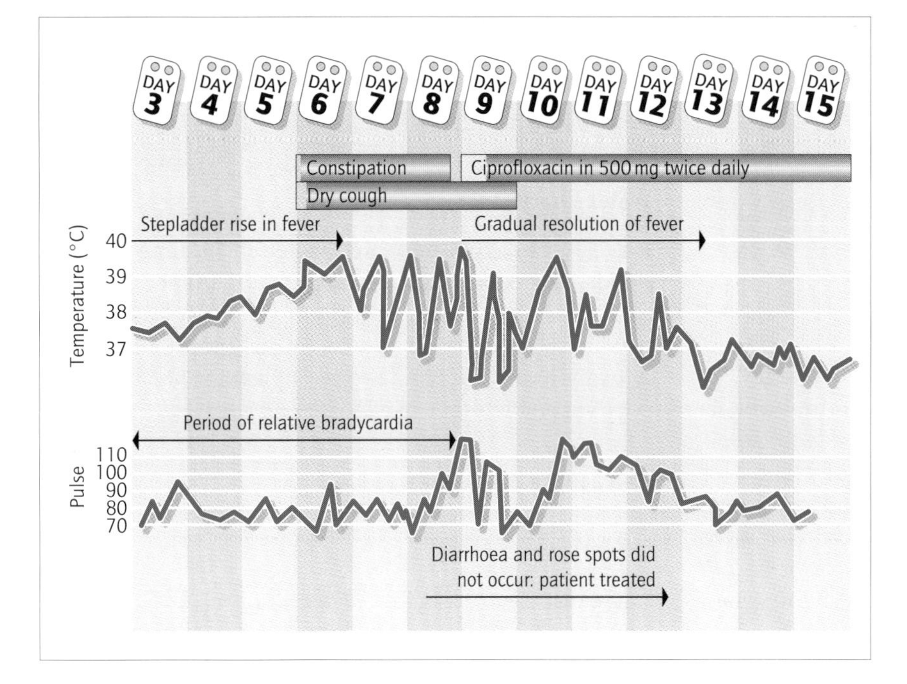

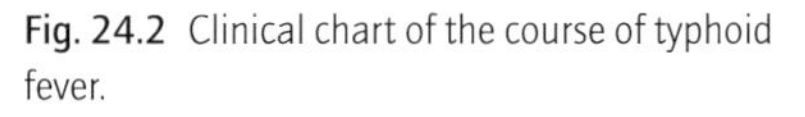
Fig. 24.2 Clinical chart of the course of typhoid fever.

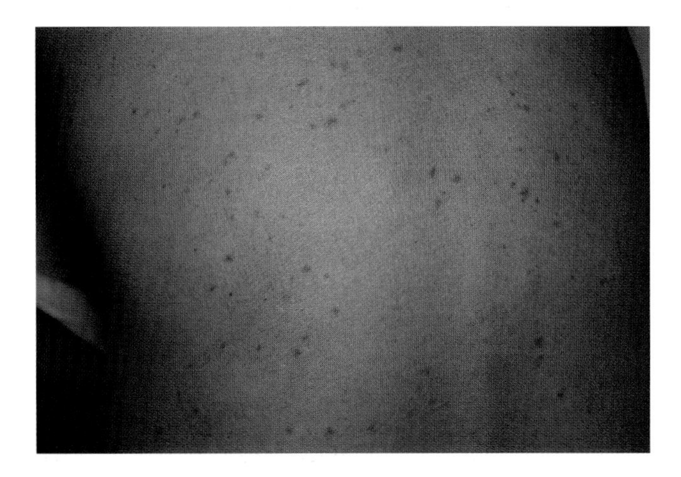
Fig. 24.3 Rose spots on the ninth day of typhoid fever.

are often very restless and may hurt themselves while attempting to escape from hospital.

After 7–10 days the fever reaches its peak; a handful of rose spots often appears on the flanks, buttocks or costal margins (Fig. 24.3), and diarrhoea begins. At this stage tachycardia develops and the white blood cell count usually shows a neutropenia.

In untreated cases complications can be expected from the second week of illness. The commonest are intestinal bleeding or perforation, usually from deeply ulcerated Peyer's patches. Bleeding may be slight, and mixed with greenish diarrhoea stools, but can be catastrophic. Similarly, small perforations can become walled off by omentum, causing temporary local signs of peritonism which resolve with continued treatment. Large or multiple perforations require emergency surgery. Bleeding and perforation are the main causes of fatalities from enteric fevers.

Late features and complications of typhoid fever
1 Bowel haemorrhage.
2 Bowel perforation.
3 Acute cholecystitis.
4 Osteomyelitis (especially spinal).
5 Other, rare metastatic infections.
6 Relapse.
7 Prolonged *Salmonella typhi* excretion.

Relapse is an important feature of typhoid, with an incidence of 10–15%. Blood cultures are again positive. It can occur after either treatment or spontaneous resolution, often being less severe than the original illness, but occasionally it is severe or fatal. It is more likely after inadequate treatment, and less likely after treatment with ciprofloxacin.

Less common presentations

Less common presentations of enteric fevers are important to recognize. Paratyphoid A closely resembles typhoid, except that rose spots are rarely seen. Paratyphoid B may have an incubation period of 4–5 days. It is usually a bacteraemic and diarrhoeal disease from its onset, the greenish watery stools becoming bloody as the feverish illness progresses. In this disease a widespread rash of rose spots often develops. Some patients have persisting fever, fatigue and malaise but remain ambulant; they are at risk of late complications.

Typhoid fever is often atypical in children. The fever does not follow the classic evolution but is often high, swinging and persistent. Bowel signs and symptoms are minimal in many cases; signs of pneumonia may predominate. Splenomegaly is common, particularly after the first week or 10 days.

Possible presentations of typhoid fever in infants and children

1 Complications of high fever.
2 Apparent acute pneumonia.
3 Persisting fever with splenomegaly.

Respiratory features are common in all age groups. Cough is a frequent symptom and the chest X-ray sometimes shows segmental or nodular pneumonitis; *S. typhi* may be isolated from sputum. Rare features of typhoid include acute cholecystitis and osteomyelitis, which particularly affects the lumbar spine.

Clinical diagnosis

The evolution of clinical features often suggests the diagnosis, which can be made with near certainty if typical rose spots appear.

Laboratory diagnosis

In common with other Enterobacteriaceae, salmonellae can be readily cultured on simple nutrient-selective media.

In cases of suspected enteric fever blood, urine and faeces should be submitted for bacteriological culture. In difficult cases, or those recently treated with antibiotics, bone marrow culture can yield positive results. Blood culture has a sensitivity of approximately 60%, and bone marrow 80%. Using conventional culture systems detection of positive isolates may be delayed, but in automated systems, positives can be detected within 24 h and subculture commenced. Isolation of *S. typhi* in faeces alone must be interpreted with caution, as it may indicate asymptomatic carriage rather than true infection.

Laboratory diagnosis of typhoid fever

1 Blood culture.
2 Bone marrow culture.
3 (Urine or faeces culture in appropriate clinical illness.)

Antigen detection

S. typhi lipopolysaccharide antigen can be detected in the serum and the urine of patients with typhoid. Techniques reported include counterimmunoelectrophoresis and enzyme-linked immunosorbent assay (ELISA). Results can be obtained more rapidly than from bacterial culture, and remain positive after chemotherapy has been initiated.

Management

Typhoid fever

The choice of antibiotic for enteric fevers lies between ciprofloxacin, high-dose amoxycillin, co-trimoxazole and chloramphenicol. Except in the case of amoxycillin, standard doses are adequate but ciprofloxacin should be continued for 7–10 days and the other drugs must be given for 2 weeks to minimize the risk of relapse. The dose of amoxycillin should be 500 mg–1.0 g 8-hourly (100 mg/kg daily in divided doses for a child). All of these agents can be given by mouth if the patient is able to take them.

Ciprofloxacin and chloramphenicol act most quickly, the fever usually falling in 2–5 days (average 3.5 days). Co-trimoxazole and amoxycillin are often recommended for typhoid in children and in natives of endemic areas. Adverse reactions to chloramphenicol are rare in the treatment of typhoid, as the duration of therapy is limited. While a small dose-related fall in the white cell count may occur, the opposite often happens as typhoid-related neutropenia resolves.

Resistance to all of these drugs is increasingly common, especially in the Indian subcontinent. Ceftriaxone in standard doses may then be effective, with improvement in fever over 2–4 days.

Severely ill patients may suffer exacerbation of fever and prostration at the start of specific treatment, with falling serum albumin and the appearance of hypotension. This will often respond to treatment with intravenous hydrocortisone 100 mg three times daily. The dose can be rapidly reduced according to the patient's response.

Antibiotic treatment of typhoid fever

1 Oral ciprofloxacin 500 mg twice daily for 10 days (child 7.5 mg/kg twice daily).

2 Intravenous ciprofloxacin may be given in a dose of 200 mg twice daily (child 5 mg/kg twice daily) until oral therapy is possible.

3 Alternatives:

(a) Co-trimoxazole orally 960–1440 mg twice daily (child 6 weeks to 5 months, 120 mg; 6 months–5 years, 240 mg; 6–12 years, 480 mg; all 12-hourly for 2 weeks).

Amoxycillin orally 500 mg–1 g 8-hourly for 2 weeks (child up to 10 years, 250 mg 8-hourly).

(b) Chloramphenicol, orally or i.v., 2–3 g daily in divided doses (child 50–100 mg/kg daily) for 2 weeks.

4 *For antibiotic-resistant infection*: ceftriaxone i.v. 1–2 g daily (can also be given i.m., if i.v. route not available).

Paratyphoid A and B are less predictable than typhoid in their response to antibiotics. Ciprofloxacin or chloramphenicol are the drugs of choice and are usually effective, though ciprofloxacin-resistant paratyphoid B has been reported.

Prevention and control

Avoiding high-risk food and drinking water can much reduce the chance of exposure to enteric fevers.

Immunization is available against typhoid fever:

1 Vi-polysaccharide vaccine (Typhim Vi or Typherix), given in a single 0.5 ml dose, provides protection equivalent to whole-cell vaccine, with fewer febrile side-effects, but sometimes irritation at the injection site. A booster is recommended after 3 years.

2 Oral Ty21a live vaccine (Vivotif), given in three doses of one capsule each on alternative days. The capsules must be kept refrigerated, and taken before food, with warm water. Booster doses are recommended at yearly intervals.

Cholera

Introduction

In spite of its rarity in developed countries, cholera is still an important infection worldwide. It may occur as part of an epidemic or arise sporadically in the developing world. It is occasionally found in travellers returning from endemic countries.

Epidemiology

Cholera is spread mainly through drinking faecally contaminated water. Food, especially shellfish, may also be a vehicle of infection. Large epidemics occur in countries lacking adequate facilities for the disposal of sewage and safe drinking water. Cholera is a pandemic infection, capable of causing epidemics that affect many countries around the globe simultaneously. Smaller epidemics and outbreaks have been reported in Mediterranean and east European countries.

During the 19th century, several cholera pandemics spread from India throughout Asia, Europe and the Americas. During the first half of the 20th century, the disease was largely confined to Asia. The seventh cholera pandemic spread from Indonesia in 1961, and has now reached all continents, including South America, where the disease has reappeared for the first time this century (Fig. 24.4). In most countries, infection is due to the E1 Tor biotype, although the classic biotype has re-emerged in Bangaladesh. More recently, the new serotype, O139, has appeared as a cause of epidemic illness in Bangladesh and India. It is now recognized as the eighth pandemic.

Cholera is rare in developed countries. There is a small focus of infection in Texas and Louisiana due to a unique strain of *Vibrio cholerae* O1. There has been no indigenous cholera infection in the UK this century. Travellers to endemic areas are only occasionally affected. Since 1981 only 45 cases of cholera have been imported into England and Wales. It has been estimated that the risk of infection in a traveller is about 1 in 500 000.

Pathogenesis

Cholera is a toxin-mediated disease caused by the O1 and O139 serotypes of *V. cholerae*. Cholera toxin is very closely related to the heat-labile toxin of *Escherichia coli* (see Chapter 8). In the future, the same vaccine may be effective against both heat-labile *E. coli* and cholera toxin.

Clinical features

The usual incubation period is 3 or 4 days. The severity of illness is extremely variable, many patients simply having a gastroenteritis-type disease. In classic cholera there is an abrupt onset of severe diarrhoea, at first watery and brown, but quickly changing to pale fluid stools containing only a little mucus and cell debris—the so-called rice-water stools. Fever is not prominent. Continuous fluid loss quickly leads to shock. The diarrhoea contains many

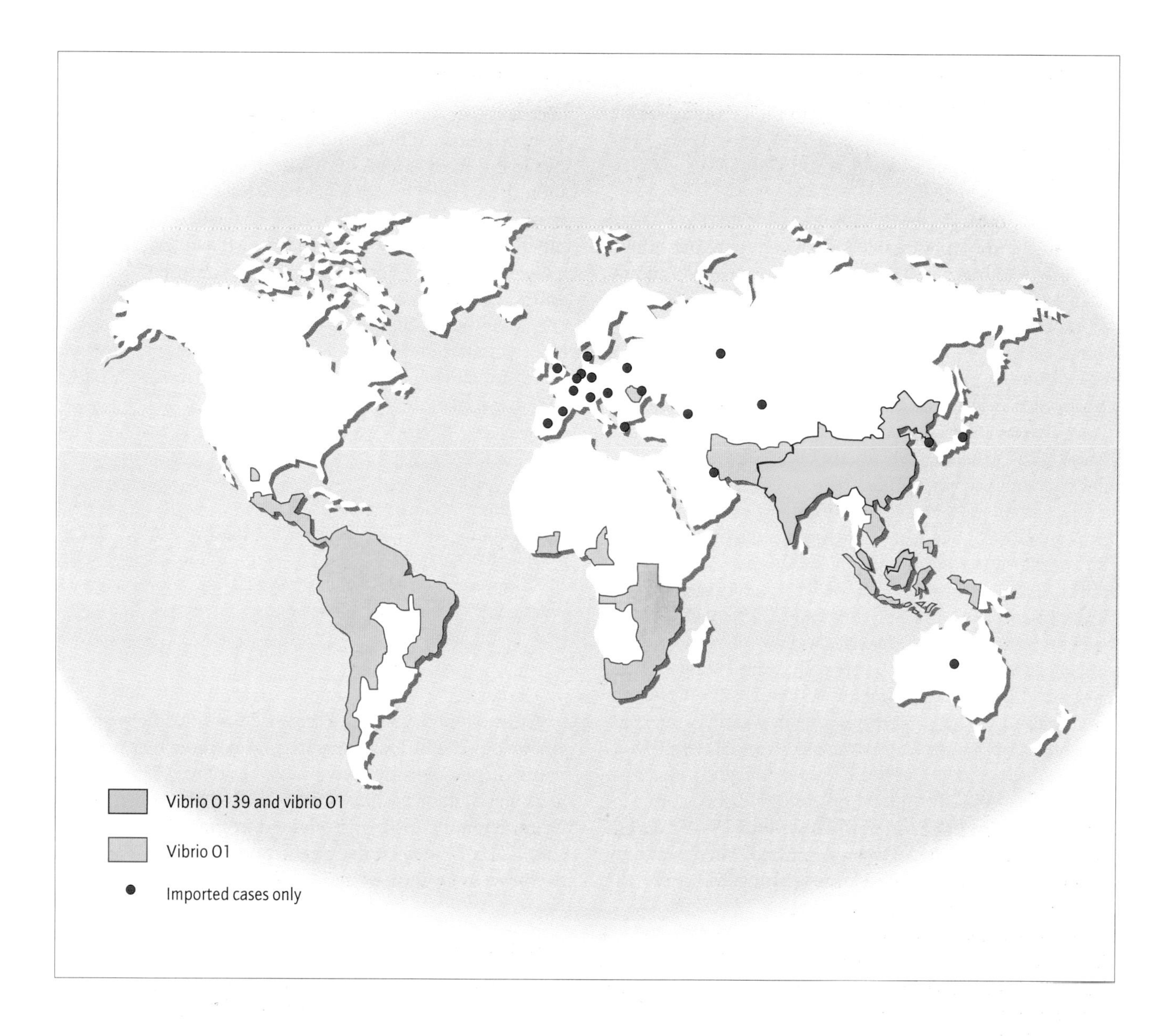

Fig. 24.4 Progress of the seventh (O1) and eighth (O139) cholera pandemics across the world: the situation in 1993.

organisms and is highly infectious. This, coupled with the difficulty of maintaining personal and domestic hygiene, contributes to the rapid spread of the disease.

Diagnosis

Clinical suspicion will be alert in an epidemic or outbreak situation. There are few diseases that cause such sudden dehydration in adults, though child cases may be hard to distinguish from severe gastroenteritis. Examination of the diarrhoea stools will give an early result but, even so, treatment should not be delayed pending diagnosis.

Laboratory diagnosis

Microscopy

V. cholerae have characteristic darting motility, and can be seen in freshly passed stool specimens from patients with acute disease. This rapid diagnosis can be confirmed by demonstrating inhibition with specific antiserum.

Isolation

Specimens should be transported to the laboratory with minimum delay but, if this cannot be avoided, specialized transport media such as that of Cary–Blair can be employed.

Media for the isolation of pathogenic vibrios have a high pH (8.6) and incorporate bile salts which together inhibit the growth of other enteric bacteria. The most commonly used medium is thiosulphate–citrate–bile–salt–sucrose (TCBS) agar. *V. cholerae* ferments sucrose after 24 h incubation producing yellow colonies due to a colour change of bromothymol blue indicator with acid production. Other vibrios, including food-poisoning species such as *V. parahaemolyticus*, grow well on this medium. Most, including *V. parahaemolyticus*, are non-sucrose fermenters whose colonies appear blue/green.

Suspect colonies are subcultured and identified using the conventional biochemical tests employed for the identification of Enterobacteriaceae (see Chapter 3). There are more than 70 serotypes of *V. cholerae* based on the lipopolysaccharide 'O' antigen. Only O1 and O139 have been associated with human disease. Confirmed colonies of *V. cholerae* are therefore serotyped, using a slide agglutination technique employing anti-O1 and anti-O139 antisera. Toxin production by the organism is then confirmed, as only toxin-producing strains cause cholera. Non-toxigenic O1 and O139 strains are not pathogenic, and require no public health action. Rapid diagnostic techniques including latex agglutination and DNA amplification techniques have been described but have only limited availability in developing countries where the disease is common.

Laboratory diagnosis of cholera

1 Comma-shaped bacteria with darting motility in the faeces.
2 Sucrose-fermenting organisms identified on thiosulphate citrate bile–salt sucrose agar.
3 *Vibrio cholerae* confirmed using biochemical tests.
4 Serotype O1 or O139 identified by slide agglutination.
5 Toxigenicity confirmed serologically.

For epidemiological purposes, the O1 and O139 strains can be typed, using agglutination, biochemical or phage reactions, as El Tor or classic types, which also exist in three biotypes (Inaba, Ogawa and Hikojima).

In areas where culture is not possible, serological surveys using techniques to detect antibodies to the O1 lipopolysaccharide antigen can be useful in epidemiological surveys.

Management

Because the mucosal cells are intact in cholera their absorptive function is undamaged. Oral rehydration is successful in over 90% of cases. Intravenous rehydration should be given to exhausted or shocked patients. Very large initial volumes of 4–6 l may be needed, followed by several litres per day while diarrhoea lasts. Spontaneous recovery is usual once hydration is controlled, but the diarrhoeal illness may last for 4–5 days. Chemotherapy may be indicated in severe or prolonged disease, or in the elderly and debilitated. Tetracycline is often effective; ciprofloxacin may also be given, and is easier to use if parenteral treatment is necessary.

Treatment of cholera

1 Oxytetracycline orally or via nasogastric tube 500 mg 6-hourly for 3–5 days.
2 Alternative: ciprofloxacin orally 500 mg 12-hourly *or* i.v. 200 mg 12-hourly, both for 3–5 days.

While asymptomatic carriage of classic cholera strains is unusual, the El Tor strains may be excreted by asymptomatic carriers and by convalescent patients. Patients and their close contacts should therefore have follow-up stool examination before being released from medical supervision. A 4- or 5-day course of chemotherapy will eradicate excretion in most cases.

Prevention and control

Cholera vaccine is of limited use. The protection afforded is less than 50% and only lasts for about 6 months. The vaccine has no effect on carriage and is therefore of no use in preventing spread. The World Health Organization has now abolished the requirements in the International Health Regulations for a certificate of vaccination against cholera. Nevertheless, evidence of vaccination in travellers from infected areas may occasionally be required. A live oral cholera vaccine is now licensed in several countries and is safer and more effective than the traditional killed vaccine.

The most useful measure in preventing the spread of cholera is provision of safe drinking water and sanitary disposal of human faeces. Food likely to be contaminated, especially fish and shellfish, should be thoroughly cooked before eating. Travel and trade restrictions between countries are not effective.

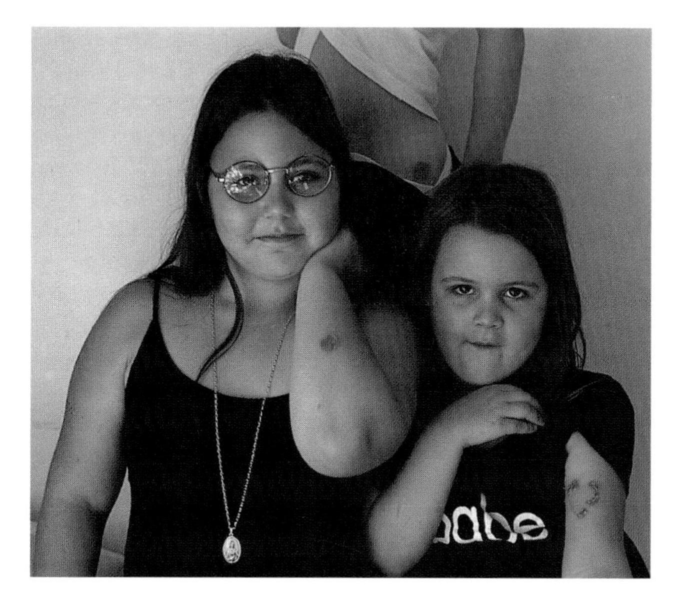

Fig. 24.5 Three sisters stayed for the summer in the Caribbean with their grandmother, who was horrified by these skin lesions. They are, however, simple secondary *Staphylococcus aureus* infection (impetigo) in insect bites.

Diseases with arthropod vectors

Introduction

Arthropod bites may transmit serious vector-borne diseases, but their most common problems are troublesome allergic reactions or secondary bacterial infections, which are extremely common in tropical areas (Fig. 24.5).

Dengue fevers

Introduction

Dengue virus is a flavivirus, whose four serotypes are widely distributed in tropical areas. Large epidemics can occur and travellers in affected areas are at considerable risk of exposure. The feverish illness can be severe and debilitating, and reinfection with a new serotype can cause enhanced disease with a substantial mortality (dengue haemorrhagic fever, DHF). DHF is an important cause of severe childhood disease in the Far East and the Caribbean. Infection is transmitted from person to person by *Aedes* sp. mosquitoes, which are common in rural and urban areas, and bite in the daytime.

Clinical features

The incubation period is 5–7 days. Illness begins abruptly with high fever, often severe arthralgia and frontal headache (this has been called breakbone fever). At this stage there is neutropenia and there may be mild hepatocellular disturbance in liver function tests. Symptoms often abate after 4–6 days but in many cases the disease is biphasic. Fever recommences, there is a generalized lymphadenopathy, and a macular rash may appear on the trunk and proximal limbs (Fig. 24.6). The platelet count often falls and the rash can then contain petechiae (Fig. 24.7), but significant bleeding does not occur and the condition is transient; fever abates in another 5 or 6 days. Convalescence is moderately rapid, taking 2 or 3 weeks.

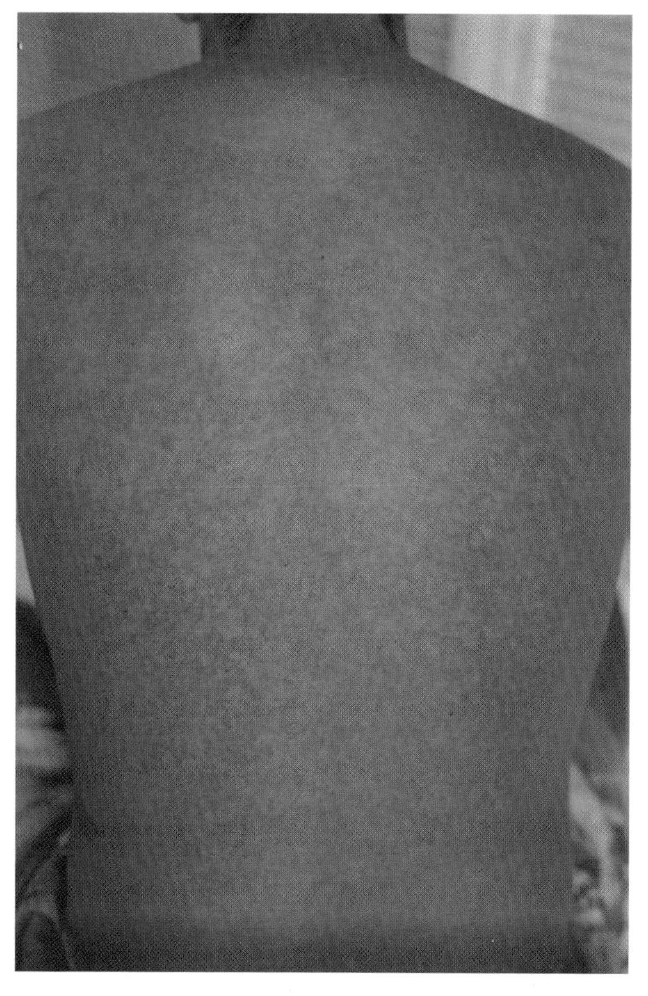

Fig. 24.6 Macular rash seen in the second feverish phase of dengue fever. Courtesy of Dr M.G. Brook.

Diagnosis and management

Typical illness is readily recognized, but many cases are mild or atypical. Virus can be demonstrated by

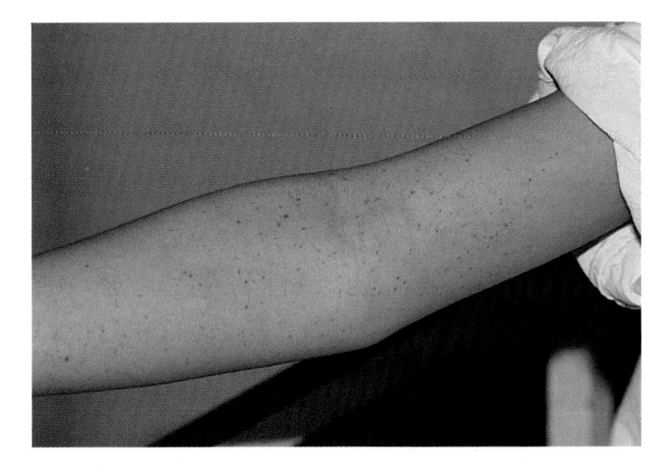

Fig. 24.7 Dengue haemorrhagic fever: positive tourniquet test. Courtesy of Dr D. Lewis.

polymerase chain reaction (PCR) techniques in blood during the acute phase. Serological testing shows high antibody titres and the presence of immunoglobulin M (IgM) antibodies. Management is symptomatic, as there is no specific treatment.

Complications

Immunity to dengue fevers is type-specific. Second attacks can occur with different serotypes, when antibody-bound virus is not neutralized, but can attach to macrophages via their Fc receptors, enhancing both viral entry into cells and cell activation. The immunopathological features are then enhanced, producing DHF. A typical onset develops into severe disease with profound thrombocytopenia. Haemorrhage, pleural effusion and associated secondary pneumonitis may occur, with up to 15% mortality. Shock, presaged by haemoconcentration, is a serious development requiring vigorous supportive treatment; the mortality of dengue shock syndrome (DSS) is near 40%. Children are most often affected, but the disease has been recorded in adults.

Other arboviral infections

Introduction

There are many arboviral infections transmitted by mosquitos, sandflies and ticks. They are all characterized by short incubation periods and intense feverish syndromes, some with lymphadenopathy or rash. Important among these are the zoonotic encephalitides, which cause severe meningoencephalitis with a high risk of long-term sequelae (see Chapter 13). The reservoirs of infection are birds, rodents or other small mammals and transmission to larger animals and humans is by mosquito bites. In the USA horses are often affected (by eastern and western equine encephalitis), or transmission may be direct from birds (as in St Louis encephalitis in Florida and central America) or rodents (as in California encephalitis). In Nepal, India and Far Eastern countries, culecine mosquitos carry Japanese B encephalitis from pigs to humans. European tick-borne encephalitis is endemic in wooded areas of central Europe and eastern Scandinavia, and an eastern strain exists in eastern Europe, Mongolia and China. West Nile fever is increasingly reported in Africa, eastern Europe and eastern United States.

Prevention and precautions

For many of these diseases the main preventive measure is control of mosquitoes in endemic areas, and avoidance of mosquito or tick bites by humans. Effective vaccines are available against tick-borne encephalitis and Japanese B encephalitis. They should be offered to travellers who will have rural exposure in endemic areas. They are not necessary unless the traveller remains close to the reservoir of infection, as the vectors have a very limited range.

Rickettsioses

Introduction

Rickettsioses are systemic diseases which are common in many countries. Rocky Mountain spotted fever (caused by *Rickettsia rickettsii*) is tick-borne, and is endemic in the Rocky Mountains and in several rural areas on the eastern seaboard of the USA. Epidemic typhus (*R. prowazekii*) affects many populations infested by lice, which are the reservoir of infection. Both of these diseases are life-threatening if untreated. Less grave but still severe illnesses are tick typhus (*R. conorii*), endemic typhus (*R. mooseri*, transmitted by fleas from mouse to humans) and scrub typhus (*R. tsutsugamushi*), transmitted by bugs. Similar but rare diseases are trench fever, caused by *Bartonella quintana* and the ehrlichioses, caused by *Ehrlichia* spp. The agents of these two diseases are members of the family Rickettsiaceae. Q fever is caused by the sheep pathogen, *Coxiella burnetii* another member of the Rickettsiaceae.

Diseases caused by Rickettsiaceae

Typhus group
Epidemic typhus (*Rickettsia prowazekii*)
Murine or endemic typhus (*R. mooseri*)

Spotted fever group
Rocky Mountain spotted fever (*R. rickettsii*)
Tick typhus (*R. conorii and R. africae*)
Scrub typhus (*R. tsutsugamushi*)
Rickettsial pox (*R. akari*)

Q fever
Coxiella burnetii

Bartonelloses
Trench fever (*Bartonella quintana*)
Cat-scratch fever (*B. henselae*)

Erlichioses
Monocytic ehrlichiosis (*Ehrlichia chaffeensis*)
Granulocytic ehrlichiosis (*Ehrlichia species*)

Clinical features

All of the rickettsial diseases have an incubation period of about 13 days. They begin abruptly with swinging fever and frontal headache. Confusion is common during the peaks of fever. The main pathology is an endovasculitis, which causes a rash and bleeding diathesis. The rash of Rocky Mountain spotted fever begins as macules on the hands and feet, then spreads over the body, becoming petechial or haemorrhagic. That of epidemic typhus begins in the axillae, and also becomes purpuric as it spreads. Rashes, often with a petechial element, may be seen in the other rickettsioses (Fig. 24.8). That of rickettsial pox is variegate, having macular, purpuric and pustular elements. It often closely resembles chickenpox.

About half of all imported rickettsial diseases in the UK are tick typhus, originating from the Mediterranean, the Arabian Gulf or Africa. The eschar of the originating tick bite may still be visible on admission as a black scab surrounded by inflammation (Fig. 24.9). The rash is generalized and the conjunctivae are suffused. The blood count shows a slight neutrophilia and transaminases are usually elevated; sometimes clinical jaundice is present. Mild abnormalities of clotting and of renal function are often detectable (a milder version of the intravascular coagulation, bleeding and systems failure of the more severe rickettsioses).

Untreated tick typhus lasts for about 3 weeks before the fever and rash resolve. Although rarely fatal, it is a severe disease followed by debility and prolonged convalescence. Endemic typhus may be more severe, causing renal failure or intravascular coagulation. The ehrlichioses cause similar disease without the rash, and with inclusion bodies in the monocytes or granulocytes.

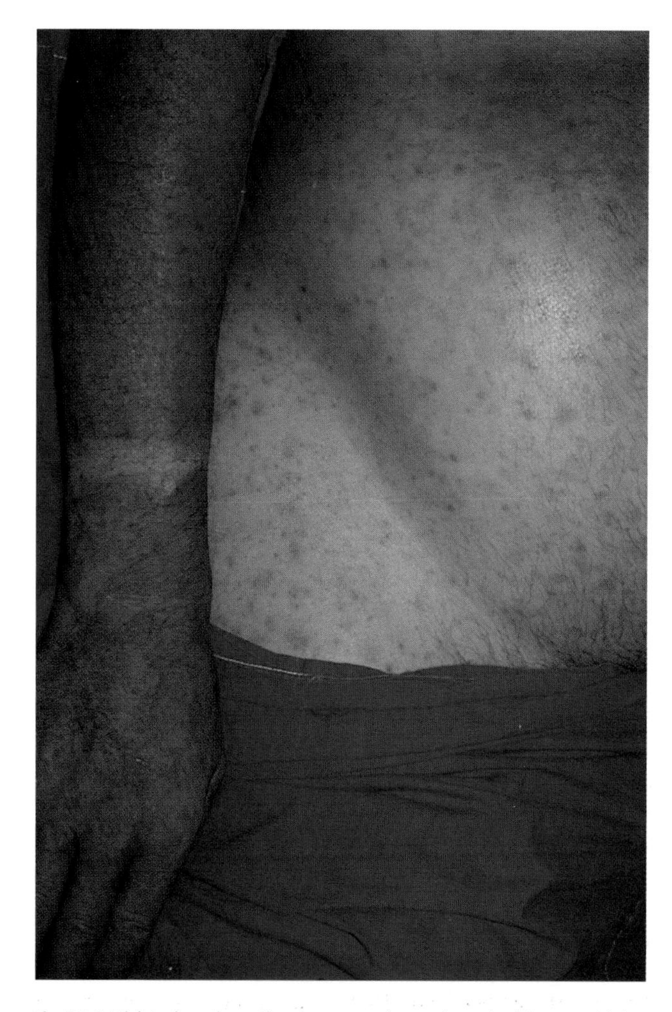

Fig. 24.8 Tick typhus: the patient has a maculopapular rash with a petechial element, conjunctival injection, severe headache and myalgia.

Diagnosis

The travel history, arthropod exposure and clinical features often suggest the diagnosis. Severe cases with purpura must be distinguished from meningococcal disease, and sometimes viral haemorrhagic fevers.

Diagnosis of rickettsial infection is usually made serologically using immunofluoresence, complement fixation or specific IgM enzyme immunoassay (EIA). PCR-based diagnosis is also available.

Q fever is usually diagnosed by a complement fixation method using acute and convalescent serum. *Coxiella bur-*

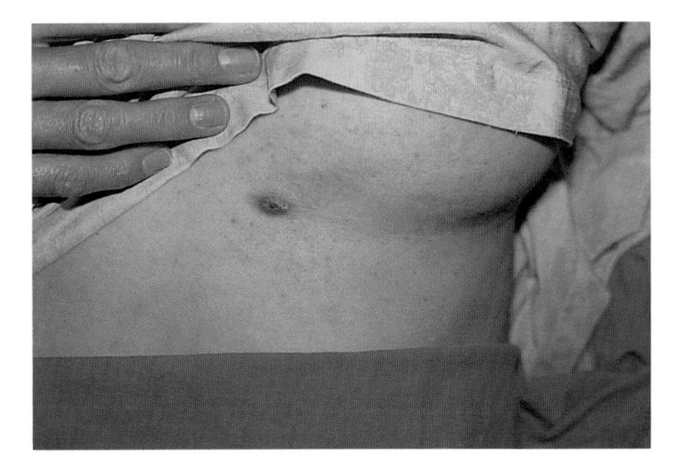

Fig. 24.9 Tick typhus: a black eschar at the site of the infecting tick bite.

netii expresses different antigens at different phases of infection. Concentrations of antibodies to antigens known as phase 2 antigens rise in acute infection but antibodies to phase 1 antigens are only elevated in chronic granulomatous Q fever or endocarditis. EIA and PCR-based methods are available in reference laboratories.

Bartonella henselae may be demonstrated by Warthin–Starry silver staining and less effectively by Gram staining. It has been isolated in culture: freshly prepared brain–heart infusion agar containing 5 or 10% rabbit or horse blood should be used and the plates should be incubated in a humid atmosphere for up to 3–4 weeks. *Bartonella* spp. grow best on solid or semisolid media and do not produce turbidity or convert enough carbon dioxide for ready detection in automated systems. Colonies on blood agar are pleomorphic. Immunofluorescence and EIA techniques have been described for the detection of IgM and IgG antibodies to *Bartonella henselae*. PCR-based techniques have proved to be the most sensitive diagnostic tests.

Treatment of rickettsial infections

1 Doxycycline 200 mg daily for 10–14 days.
2 Alternative: chloramphenicol orally or i.v. 500 mg 6-hourly for 10–14 days.

Management

The treatment of choice is doxycycline 100 mg twice daily. Chloramphenicol, orally or intravenously is an alternative. A 10–14-day course is usually required. As in typhoid fever, the temperature may not fall to normal for 3–4 days.

Malaria

Introduction and pathology

Malaria is one of the most important imported diseases. The benign malarias are debilitating diseases with a relapsing course. Falciparum (malignant) malaria may be life-threatening and should be treated as a medical emergency. The diagnosis should be actively excluded in every feverish traveller from the tropics, for two to three thousand cases are reported each year in the UK, with up to a dozen fatalities from unsuspected or late-diagnosed falciparum malaria.

Malarial sporozoites attach to and enter red blood cells, using surface antigens such as blood group determinants. There they multiply by binary fission, digesting haemoglobin to derive energy and producing a waste haem pigment, haemozoin, which is deposited in endothelium and surrounding tissues, causing an intense inflammatory response and activation of platelets and complement. The daughter parasites rupture the red blood cell to enter the plasma and continue the blood infection cycle by infecting new cells. Haemolysis is increased by an immune response to malarial antigens on the surface of infected cells.

Plasmodium falciparum parasites can enter red cells at all stages of maturity, whereas *P. vivax* infects only reticulocytes and *P. malariae* favours senescent cells. As *P. falciparum* parasites multiply to the schizont stage they cause rigidity of the infected cell, with the appearance of 'sticky' surface projections. This leads to sludging of parasitized red cells in the microcirculation, particularly of parenchymal organs. *P. falciparum* causes a more intense infection, with more vascular damage and imflammatory response than the other malaria types. Vascular shunting, tissue hypoxia, endothelial damage and disseminated intravascular coagulation occur in severe infections.

Causes of the four types of malaria

1 *Plasmodium falciparum* (malignant tertian malaria).
2 *P. vivax* (benign tertian malaria).
3 *P. ovale* (benign tertian malaria).
4 *P. malariae* (benign quartan malaria).

Clinical features

The only consistent clinical features of malaria are fever and rigors. Patients present initially with a chaotic, swingeing fever; rigors occur when the temperature rises. The fever becomes periodic when synchronous release of parasites is established after 7–14 days. Fevers occur every third day (tertian fever) in vivax and ovale malaria, or every fourth day (quartan fever) in malariae malaria. In falciparum malaria the fevers are less regular, but approximate to a tertian pattern.

Many non-specific symptoms may be present, including abdominal pain, headache, diarrhoea, dysuria and frequency, sore throat and cough. Physical examination may be normal or splenomegaly may be detectable. In chronic or relapsing malaria the spleen can be very large. Hepatomegaly and mild jaundice may also be present.

Cerebral malaria presents with encephalopathy. It is largely a disease of the non-immune and in endemic areas it mainly affects children below the age of 4, in whom it must be distinguished from childhood bacterial meningitis. Hypoglycaemia, convulsions and hypoxia readily occur and worsen the prognosis considerably. Blackwater fever results from severe intravascular haemolysis. Profound anaemia, jaundice, haemoglobinuria and acute renal failure quickly develop in untreated cases. Pulmonary oedema is common, and frequently coexists with cerebral disease. The absence of bile in the urine distinguishes haemolytic jaundice from that of viral hepatitis.

Important features of malignant malaria

1 Encephalopathy (cerebral malaria).
2 Pulmonary oedema (acute or adult respiratory distress syndrome).
3 Acute renal failure.
4 Severe intravascular haemolysis, with profound anaemia.
5 Haemoglobinuria (blackwater fever).

Diagnosis

Clinical diagnosis may be possible if periodic fever and splenomegaly exist, but the only reliable means of diagnosis is the demonstration of parasites in the red blood cells. This is best done by preparing thick and thin blood films, which are stained with Giemsa or Field's stain, respectively. Rapid Romanowsky-type stains are also suitable for thin films. Scanty parasites are easier to detect in thick films, while thin films are often additionally helpful in speciation (Fig. 24.10). Stick or card tests can be used to detect the histidine-rich surface antigen of *P. falciparum* or *Plasmodium sp.* enzymes in blood. These are very sensitive tests, but false positive results can occur in patients with rheumatoid factor. Some automated blood-counting machines can also detect malarial parasites, but parasite counting and speciation is best done by an experienced observer.

Management

Benign malarias

Benign malarias should be treated promptly with chloroquine, to which they are rarely resistant. The appropriate regimen is:

1 Chloroquine base 600 mg immediately (four tablets).
2 Next dose 300 mg 6 h later (two tablets).
3 Two more doses of 300 mg 24 and 48 h after second dose.

Equivalent doses of chloroquine preparations are chloroquine base 150 mg = chloroquine sulphate 200 mg = chloroquine phosphate 250 mg.

Chloroquine acts quickly and the fever should be abolished within 12–24 h. Nausea or intestinal irritation is rarely severe enough to prevent completion of the treatment. Pruritus particularly affects dark-skinned people and may make continued treatment intolerable, in which case quinine may be substituted.

Chloroquine exacerbates skin conditions such as psoriasis. Alternative agents such as quinine should be used in this situation.

Eradication of liver parasites

Eradication of liver parasites is necessary after treatment of acute vivax or ovale malaria (hypnozoites are not killed by chloroquine or quinine, which mainly affect developing schizonts).

Liver parasites are eradicated by a 2-week course of primaquine 15 mg daily (250 μg/kg daily for children).

Difficult cases

Chloroquine resistance has been well documented in a few cases of vivax malaria contracted in Indonesia and Papua New Guinea. Quinine is probably the alternative treatment of choice, though Malarone® (atovaquone plus proguanil) may also be used. In the same areas, primaquine tolerance has caused failure of eradication of liver parasites when standard doses are given. A further course at twice the dose should be successful.

Falciparum malaria

An increasing proportion of *P. falciparum* parasites are chloroquine resistant in all endemic areas, so that chloroquine is now unreliable therapy. Uncomplicated cases should be given quinine 600 mg 8-hourly. The dose may be reduced to 400 mg 8-hourly if nausea, tinnitus or deafness occurs. The temperature drops slowly or erratically, often

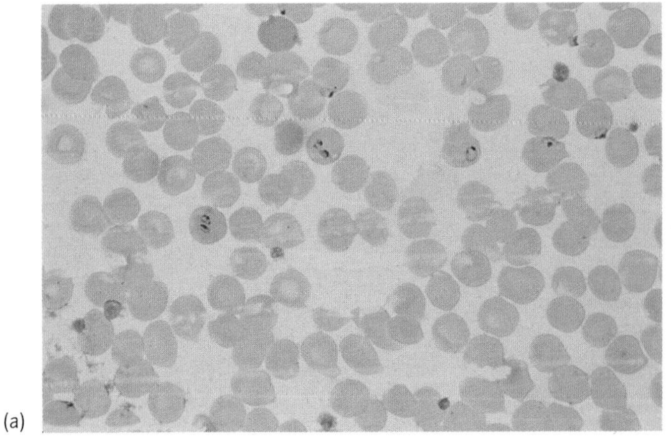
(a)

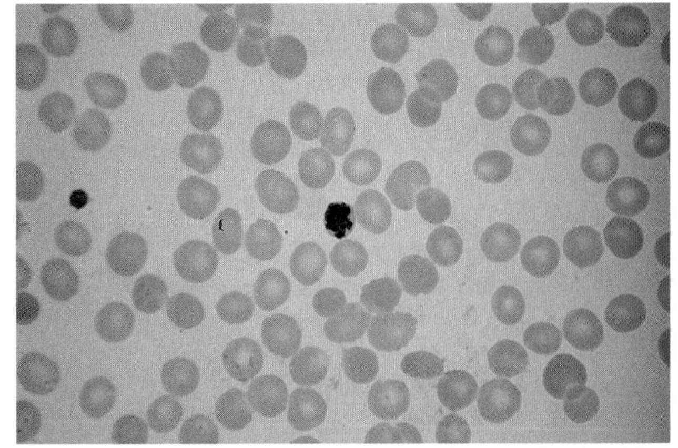
(d)

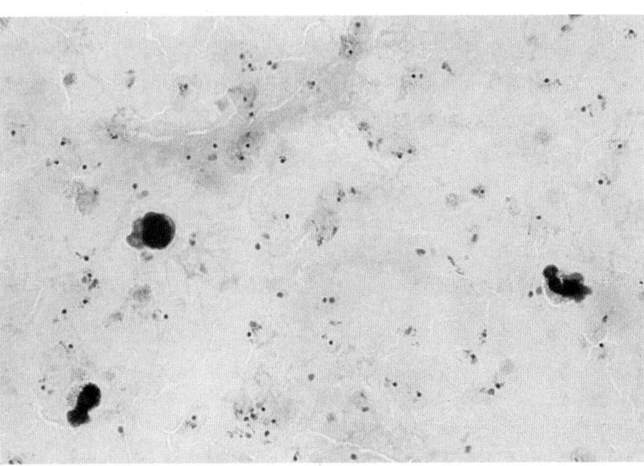
(b)

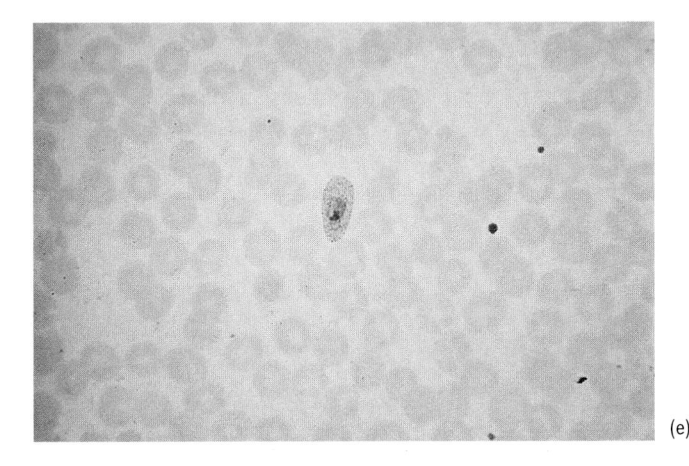
(e)

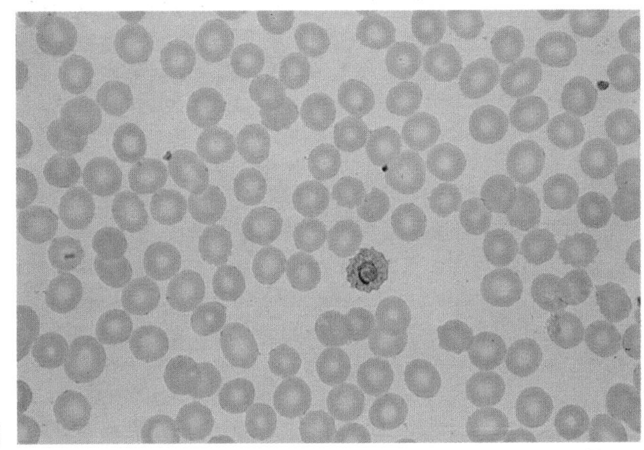
(c)

Fig. 24.10 Malaria: (a) thin blood film, stained with a Romanowsky stain, shows numerous *Plasmodium falciparum* trophozoites, some red cells with double parasitization and accroche forms (parasite applied to the rim of the red cell); (b) thick blood film, Field's stain, shows numerous *P. falciparum* trophozoites and one white blood cell; (c) thin blood film shows *P. vivax* trophozoite in a cell with typical Schuffner's dots; (d) thin blood film shows *P. vivax* schizont; (e) thin blood film shows *P. ovale* trophozoite.

remaining normal only after 2–3 days. Parasitaemia should disappear after 48 h of treatment.

With quinine monotherapy 7–10 days' treatment is needed. This can be shortened by giving Fansidar, three tablets, as a single dose on the fourth or fifth day, and discontinuing the quinine thereafter. Fansidar resistance exists in the Far East and some east African countries. If this is suspected, or if quinine–Fansidar treatment fails, quinine treatment can be completed with tetracycline 250 mg 6-hourly (or doxycycline 100 mg daily) for 2 weeks.

Severe falciparum malaria

Cerebral malaria or other severe forms of malignant malaria, including all cases with a parasitaemia above 3% of all red cells, must be treated immediately with intravenous quinine. This is not given by bolus injection because of the high risk of cardiac depression, cerebral irritation, nausea and vomiting. The safest procedure is to give an infusion of 10 mg/kg in 5% dextrose over 4 h. This dose can be repeated 12-hourly until oral therapy is possible. The first dose of quinine should be 20 mg/kg in very ill patients, and severe cases in children. Quinidine intravenously, 10–15 mg/kg 12-hourly is effective if quinine is unavailable.

Other important aspects of treatment include maintenance of adequate blood glucose levels, correction of anaemia by transfusion if necessary, and avoidance of convulsions. Many experts advocate the use of prophylactic phenytoin in patients with coma.

Pulmonary oedema may require vigorous treatment; diuretics and fluid restriction are not highly effective (and in excess could contribute to reduced cerebral blood flow). High inspired oxygen tension and intermittent or continuous positive-pressure ventilation are required in many cases.

Acute renal failure will often respond to conservative management, with a diuresis occurring in 4–7 days. A minority of patients need dialysis.

In spite of severe cerebral disturbance and difficulties with fluid handling, there is little evidence that treatment for cerebral oedema is helpful in cerebral malaria. Indeed, it has been shown that dexamethasone therapy may *prolong* coma without improving the outcome.

Other treatments for falciparum malaria

1 Malarone® (atovaquone plus proguanil) is an alternative to oral (but not intravenous) quinine. It is given in a dose of four tablets daily for 3 days (child 11–20 kg, one tablet daily; 21–30 kg, two tablets daily; 31–40 kg, three tablets daily). It is nearly as effective as quinine for *P. falciparum*.

2 Mefloquine is a substitute for oral, but not intravenous quinine. Dose: two doses of 10 mg/kg, 6 or 8 h apart; maximum dose 1500 mg. Unfortunately, psychotic side-effects are common at therapeutic doses. Mefloquine may be teratogenic in the first trimester of pregnancy. The drug is contraindicated in epileptics, in severe liver disorder and in pregnancy. Mild ataxia and nausea are common. Rare side-effects include skin rashes and cardiac conduction defects.

3 Artemesinin drugs are widely used in the Far East for severe, and multidrug-resistant malaria, with results comparable to those of quinine. They have the advantage of being available as tablets, as suppositories and as intramuscular as well as intravenous preparations. Increased cerebral irritation and insignificant prolongation of coma are problems in severe cases, but do not preclude their use.

4 Halofantrine: only for uncomplicated, chloroquine-resistant falciparum malaria. It must not be given after mefloquine prophylaxis. Give three doses of 500 mg at 6-hourly intervals. This should be repeated 1 week later. This drug is not recommended for children under 23 kg in weight. Larger children may be given 250 or 375 mg doses, depending on their weight. The adult dose may be given to children over 37 kg. Halofantrine causes prolongation of the Q-T interval. Sudden cardiac dysrhythmias have occurred in predisposed patients, in those taking other drugs which have the same effect and, rarely, in individuals with no known predisposition. It should not be given to individuals who have been taking mefloquine for this reason. If quinine treatment is unavoidable after halofantrine medication, cardiac monitoring should be considered. Halofantrine is no longer recommended for standby self-treatment of malaria.

5 Exchange transfusion will reduce high parasitaemia, but it is not known whether this improves the outcome as it cannot affect the pathological process in occluded small blood vessels where parasitized red cells are already immobilized.

Prophylaxis of malaria

The risk of malaria is greatly reduced by avoidance of mosquito bites. The antimosquito measures already described should always be used by travellers to endemic areas to enhance the effect of chemoprophylaxis.

Four main types of chemoprophylaxis are recommended by British experts:

1 *For benign malarias*: chloroquine 300 mg weekly or proguanil 100 mg daily, starting 1 week before exposure.

2 *For falciparum malaria in central, east and west Africa, and in parts of Indonesia, Malaysia and South America where resistant P. falciparum occurs*: mefloquine 250 mg (one tablet) weekly, starting 2–3 weeks before exposure. Mefloquine resistance occurs in a number of areas, but particularly in the borders of Thailand with Cambodia and Myanmar (Burma). Mefloquine in therapeutic doses can cause agitation and intrusive dreams; cases of severe depression or psychosis have been reported. At prophylactic doses severe adverse effects are uncommon (about 1 in 10 000 recipients). This drug is contraindicated in epileptics, and those with a history of psychiatric disease. There are no data on its safety in the first trimester of pregnancy.

Alternative: Malarone® (licensed in Denmark): one tablet daily starting 3 days before exposure (child 11–20 kg, one-quarter of adult dose; 21–30 kg, half adult dose; 31–40 kg, three-quarters of adult dose).

3 *For falciparum malaria in the Cambodian and Myanmar borders of Thailand*: doxycycline 100 mg daily, starting 3 days before exposure.
Alternative: Malarone® (see above).
4 *For falciparum malaria in India, Sri Lanka, most rural areas of Peru and Bolivia*: chloroquine 300 mg weekly *plus* proguanil 200 mg daily starting 1 week before exposure.

In all cases medication should be continued throughout the exposure period and for 4 weeks after leaving. There is no contraindication to malarial prophylaxis.

Malarial prophylaxis for pregnant women and for children: falciparum malaria threatens life, and in pregnancy endangers both the mother and the pregnancy itself. Neonates and infants are susceptible to severe disease. Malarial prophylaxis is much less risky than the disease itself and should never be omitted. Both chloroquine and proguanil are safe and well tolerated by infants and pregnant women. Doses for children under 12 years old are:

1 Up to 6 weeks—one-eighth of adult dose.
2 Up to 1 years—one-quarter of adult dose.
3 Up to 5 years—half of adult dose.
4 6–12 years—three-quarters of adult dose.

Even if prophylaxis has been taken continuously, malaria should be actively excluded if a traveller becomes feverish after returning home; falciparum malaria occasionally occurs up to a year later, and vivax up to 2 or 3 years.

Standby treatment for malaria

Even when prophylaxis is properly used, there is a small risk of malaria. Travellers in remote areas who cannot obtain timely investigation and treatment, may be given a supply of emergency treatment. Suitable drugs for this are quinine (600 mg 8-hourly for 5 days), *plus* Fansidar (single dose of three tablets), or Malarone® (four tablets daily for 3 days). Medical attention should be sought as soon as possible after their use, so that the blood can be checked for parasites and anaemia.

Leishmaniasis

Introduction

Leishmaniasis is disease caused by protozoa of the genus *Leishmania*. There are a number of different species, all transmitted by sandflies, causing several types of clinical disease. Cutaneous leishmaniasis is common in tropical countries, the Middle East and many Mediterranean areas. Small rodents, and sometimes dogs, are the reservoir of infection. A severe, erosive mucocutaneous disease, called espundia, occurs in tropical Latin America. Systemic leishmaniasis is less common, affecting tropical areas of Africa and India. It is a rare disease in the northern and central Mediterranean area, but more common in the eastern Mediterranean area.

Cutaneous leishmaniasis

In the old world this is caused by *L. tropica* var. *major* and var. *minor*. A typical oriental sore is caused by var. *major*, and has an incubation period of 2–6 weeks. The ragged, punched-out ulcer usually appears on the face or extremities, and is accompanied by regional lymphadenopathy. The lesion reaches up to 2.5 cm in diameter and heals slowly over 3–6 months, leaving a depressed, tissue-paper scar.

A more indolent, granulomatous type of lesion follows infection with var. *minor*, whose incubation period can be as long as a year. It appears as a purplish nodule, which gradually breaks down and slowly heals over several months. Lymphadenopathy is rare.

Diagnosis is made by demonstrating the protozoa in scrapings from the base of the lesion or in biopsy material. Specific treatment is not often required, but disfiguring facial lesions may be treated with intramuscular sodium stibogluconate for 10 days. Secondary infection may complicate the lesion and should be treated promptly to avoid further scarring.

Espundia causes destructive, scarring lesions of the oropharyngeal mucosa and surrounding structures, including the nasal septum. Secondary infection commonly exacerbates the condition. Vigorous and prolonged drug treatment is necessary, using sodium stubogluconate and/or paromomycin, and surgical aftercare may be required.

Systemic (visceral) leishmaniasis

In tropical areas the commonest cause is *L. donovani*, which is mostly transmitted from human to human. In the Mediterranean it is caused mainly by a variant, *L. infantum*, for which dogs and foxes are an important reservoir. The incubation period is variable but averages about 3 months. Fever is constant for the first few weeks, but then becomes intermittent. The typical features of hepatosplenomegaly, pancytopenia and increased skin pigmentation (which gives the disease the name 'kala-azar'—the black sickness) develop over several weeks. The erythrocyte sedimentation rate is often very high, and is related to a polyclonal elevation of immunoglobulin G (IgG).

The diagnosis may be made by demonstrating protozoa (Leishman–Donovan bodies) packed into the mononuclear phagocytes of the spleen (Fig. 24.11), liver or bone

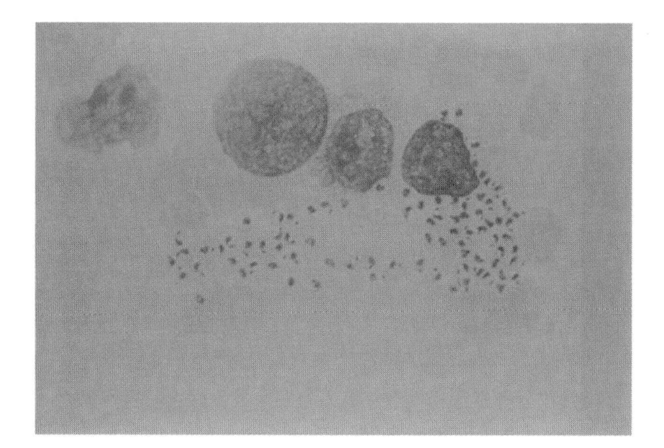

Fig. 24.11 Systemic leishmaniasis: this patient had typical hepatosplenomegaly, pancytopenia and persisting fever. Splenic aspirate revealed Leishman–Donovan bodies – macrophages packed with the protozoan parasites.

marrow. Splenic aspiration is most likely to give positive results. Serodiagnosis by ELISA test and genome detection by PCR are also possible. Culture of blood, bone marrow or splenic aspirate, or biopsy is performed in reference laboratories.

The treatment of systemic leishmaniasis is a specialist procedure. Organic pentavalent arsenic is given in the form of sodium stibogluconate 10 mg/kg daily by intravenous injection. Side-effects such as vomiting, coughing and substernal pain are common. Injections should be given intramuscularly if coughing and pain occur, but are themselves painful. The duration of treatment varies from 20 to 30 days, and depends on the type of disease and the response to treatment. Failure of response sometimes occurs. Liposomal amphotericin B 1 mg/kg (up to 3 mg/kg as initiation therapy) has shown good effect, and can be used in a 3-week course. Alternative treatments include paromomycin (an aminoglycoside), amphotericin B or mixtures of allopurinol and fluconazole.

Lymphatic filariasis

This is a disease caused by the nematode worms *Wuchereria bancrofti* and *Brugia malayi*. Adult worms live coiled together in the lymphatics of humans. Pregnant female worms release large numbers of microfilariae, which reach the peripheral blood and must then be ingested by a biting mosquito to complete their life cycle. Several species of mosquito can transmit the disease.

Clinical manifestations begin 9–12 months after infection. They are probably due to a hypersensitivity reaction to the release of larvae into the blood. There is usually a series of episodic fevers, each of which lasts a few hours. The fever is rarely high, but shivering and sweating are common. Lymphangitis may be visible if it affects lymphatics near to the skin surface; common sites are the lower leg or the thigh. Pain and redness of the skin mimic cellulitis or erysipelas. Abdominal pain or scrotal inflammation are rarer manifestations.

The filariae die within about 5 years of a single infecting episode, and no further symptoms occur. In rare cases with many reinfections, accumulating fibrosis of affected lymphatics can eventually produce lymphoedema (elephantiasis).

Diagnosis is best made serologically by an IgG antifilaria ELISA test. Microfilariae can be demonstrated in the peripheral blood of patients in endemic areas, but the sensitivity of this test is only about 70%. Midday and midnight blood films should be examined, to allow for differing periodicities of microfilaraemia.

Treatment is with diethylcarbamazine, which kills both adult worms and microfilariae. The dose is 6 mg/kg daily in divided doses. Febrile allergic reactions are common at the onset of treatment, which is commenced under close medical supervision, starting with 1 mg/kg daily and working up to a therapeutic dose. Full dosage is then continued for 21 days. Antihistamines and/or corticosteroids may be needed until febrile reactions cease.

Treatment of filariasis (excluding onchocerciasis)

1. Give diethylcarbamazine orally:
   Day 1: 1 mg/kg as a single dose.
   Day 2: 3 mg/kg as a single dose.
   Day 3: 6 mg/kg as a single dose.
   Next 20 days: 6 mg/kg daily.
2. Ivermectin, 150 mg, single dose.

Loiasis

Loiasis, caused by *Loa loa*, is often asymptomatic but may produce localized allergic skin swellings (Calabar swellings), or is visible if an adult worm migrates across the eye. If it requires treatment, diethylcarbamazine is effective.

Onchocerciasis (river blindness)

Onchocerciasis is transmitted by *Simulium* flies which deposit their eggs near fast-flowing water. Adult *Onchocerca volvulus* live in the skin, often forming macroscopic nodules which contain convoluted worms. Illness is caused by inflammatory reactions to the millions of microfilariae that invade the skin and eye. Treatment with diethylcarbamazine is dangerous because of severe inflammatory reactions with possible destructive involvement of the eye. Ivermectin 150 mg produces gradual and sustained reduction in microfilariae with little allergic reaction, and the dose can be repeated annually until infection is eradicated. Surgical excision of skin nodules has also been used as a means of reducing microfilarial numbers.

Important zoonoses

Brucellosis (see also Chapter 25)

Introduction

Human infection with *Brucella* spp. is usually acquired from the natural hosts, cattle (*B. abortus*) or goats (*B. melitensis*). Less common forms occur in sheep, pigs and, rarely, dogs. Brucellosis is an important disease worldwide, but rare in the UK because of successful eradication programmes, particularly for cattle. Imported cases arrive particularly from rural Africa, Mediterranean countries and the Middle East, where brucellosis is very common (and can also affect camels). The disease affects many body systems, causing severe morbidity and some mortality.

Clinical features

Acute brucellosis

Acute brucellosis has an incubation period of 1–3 weeks. Fever is usually an important finding. It may be continuous, swingeing or undulating (rising steadily for 7–10 days and then falling for a day or two). Weakness, shivering and sweating are prominent and most patients have arthralgia, particularly of large joints. In spite of the severity of the illness, mortality is low and untreated cases tend to recover after anything from 3 or 4 weeks to several months (average 3 or 4 months).

Physical findings are rarely impressive. They include tender splenomegaly in about half of patients, mild cervical and axillary lymphadenopathy, and tenderness of the spine.

Chronic brucellosis

Chronic brucellosis is uncommon. It may be mistaken for tuberculosis, or for chronic disease of the systems mainly involved. There may be a considerable element of hypersensitivity in cases of longer duration, but granulomatous inflammation also occurs and may cause arthralgia, uveitis, orchitis, meningoencephalitis and liver abnormalities.

Important additional features

The commonest is bone and joint disease, in which brucellae can be recovered from inflammatory effusions. Spinal brucellosis, often affecting cervical vertebrae, is seen in endemic areas and must be distinguished from tuberculosis and other chronic osteomyelitis.

Less common, but requiring urgent diagnosis and treatment, are neurobrucellosis, with fits, cranial nerve lesions or neuropsychiatric features, and *Brucella* endocarditis, which usually affects previously damaged valves.

Diagnosis

Clinical features are non-specific. There is neutrophilia in acute disease. Later there is neutropenia with a few activated mononuclear cells. Liver granulomata are often present, but are indistinguishable from those of sarcoidosis and non-caseating tuberculosis. The diagnosis must therefore be suspected from the epidemiological circumstances, and appropriate laboratory tests performed.

Diagnosis of brucellosis

1 Blood cultures (maintained for up to 6 weeks).
2 Culture of joint effusions or pus from bone.
3 Antibody detection (usually enzyme-linked immunosorbent assay).

Management

The treatment of choice is doxycycline, which should be given for at least 6 weeks. The incidence of relapse is reduced if rifampicin is added. An aminoglycoside, added for 2–3 weeks of treatment, is an alternative to rifampicin. Although the need for intramuscular injections is inconvenient, streptomycin may be the more effective drug. Other treatments, such as co-trimoxazole and ciprofloxacin, will achieve defervescence of fever and reduce local inflammation, but recurrence is common after their use. These drugs can be used for treating children.

Treatment of brucellosis

1 Doxycycline orally 200 mg 12-hourly for 6 weeks; *plus*
2 *Either* rifampicin orally 600 mg daily *or* (for first 3 weeks) gentamicin i.m. or i.v. in standard doses, with blood-level monitoring.
3 *For children under age 7*: co-trimoxazole 12-hourly; 6 weeks–5 months, 120 mg; 6 months–5 years, 240 mg; 6–12 years 480 mg; *plus* rifampicin 10 mg/kg daily.

Recurrences can occur, even after full tetracycline plus aminoglycoside treatment. They should be treated with a second course of the same therapy.

Rabies (see also Chapter 13)

Rabies is a lyssavirus infection of many warm-blooded animals including birds, squirrels, skunks, cats, horses and cattle, but is most often transmitted to humans by dogs, wolves, insectivorous bats and mongooses. The virus causes myeloencephalitis and is excreted in tears and saliva. It is inoculated by bite, scratch or mucosal contamination from animals excreting the virus. Virus enters the peripheral nerves via which it travels to the central nervous system.

The incubation period is only 1–2 weeks after bites on the head or face, and averages 60–90 days, but can be over a year for bites of the lower leg, remote from the central nervous system. Illness often begins with paraesthesiae at the inoculation site. Extreme anxiety is common at this stage. The main clinical features are then either ascending polyneuritis or rapidly developing encephalitis. The patient may be obtunded or speechless (dumb rabies), or excitable and agitated (furious rabies; Fig. 24.12) with intermittent spasms of the pharynx precipitated by stimulation of the face or mouth (hydrophobia). There is no specific treatment.

Early diagnosis depends on clinical suspicion, and exclusion of more treatable causes of myelitis or encephalitis. Rabies virus can be demonstrated by PCR or immunofluorescence, and cultured from saliva, tears, nuchal biopsies and corneal scrapings. Rising titres of antibodies can be demonstrated. Negri bodies are seen in some neural cells in both humans and animals.

Fig. 24.12 A cat with furious rabies attacks an object which has entered its cage. Courtesy of Dr D. Lewis.

Plague (see also Chapter 25)

Plague is the systemic disease caused by *Yersinia pestis*. In nature it is a disease of rodents, especially rats, transmitted by fleas, which may also attack humans. The fleas themselves are affected by the infection, for bacteria block their foregut, causing regurgitation of infected material when they attempt to bite.

> Plague is an internationally notifiable disease. Countries in which human plague cases exist are listed weekly in the World Health Organization *Weekly Epidemiological Report*.

The incubation period is 4 or 5 days, and three-quarters of all cases develop the bubonic form of the disease. After a day or two of fever the bubo appears, usually in an inguinal node, sometimes in the axilla and rarely elsewhere. The degree of swelling is variable. Many patients are bacteraemic. In those who recover untreated the bubo usually discharges offensive pus some time after the fever subsides. A minority of patients develop pneumonic plague, which rapidly evolves into severe pneumonia, often with watery, blood-stained sputum and early respiratory failure. Large numbers of bacteria are excreted and secondary cases may occur.

The diagnosis may be suspected on epidemiological and clinical findings. *Y. pestis* can be recovered from blood, pus or sputum. Work on cultures of *Y. pestis* should only be performed in laboratories with appropriate containment level 4 facilities. Tetracycline, ciprofloxacin or chloramphenicol is effective treatment. Seven to 10 days' tetracycline or ciprofloxacin prophylaxis should be offered to those in contact with pneumonic cases.

Anthrax

Introduction and epidemiology

Anthrax is a disease particularly of hoofed animals, caused by *Bacillus anthracis*, an aerobic, spore-bearing, Gram-positive rod related to *B. cereus* and *B. subtilis*, but with a wider range of antibiotic sensitivities than either. In its natural hosts it causes a fatal septicaemic disease in which the animal's blood becomes packed with bacilli. After the host's death, the bacilli form spores which remain viable in the soil for decades. Farm animals are susceptible; an outbreak in the UK in the 1980s affected pigs. Humans become infected by close contact with infected or dead animals, with bones, bonemeal, hides, hooves or meat. Badger-hair shaving brushes occasionally caused infection in the past. The usual route of infection is inoculation into the skin, less often by inhalation or ingestion of spores.

Clinical features

An oedematous skin lesion (malignant pustule) is the commonest feature. An inflamed nodule ulcerates and forms a black scab, surrounded by a halo of vesicles or pustules. A common site is the neck at the collar, the dorsum of the wrist or the arm. Local draining lymph nodes are enlarged and tender. A helpful diagnostic feature is the enormous extent of oedema that surrounds the lesion (Fig. 24.13).

Pneumonia can follow inhalation of massive spore loads, often from hides or dusty bonemeal. It is often fulminant, associated with mediastinitis and mediastinal lymphadenitis leading to death in 2 or 3 days, but is not transmitted from person to person.

Gastrointestinal anthrax is probably acquired by ingestion of large spore loads. It causes abdominal pain and severe watery diarrhoea, which contains many sporing organisms.

Septicaemia is usually fatal. It can occur with untreated skin infection, but is common with pneumonitis and gastrointestinal disease. Death is often due to pulmonary embolism or cerebral vein thrombosis, even after antibiotic treatment has reduced fever.

Diagnosis

Lesion swabs or vesicle fluid will produce colonies in blood agar, which are composed of tangled chains of

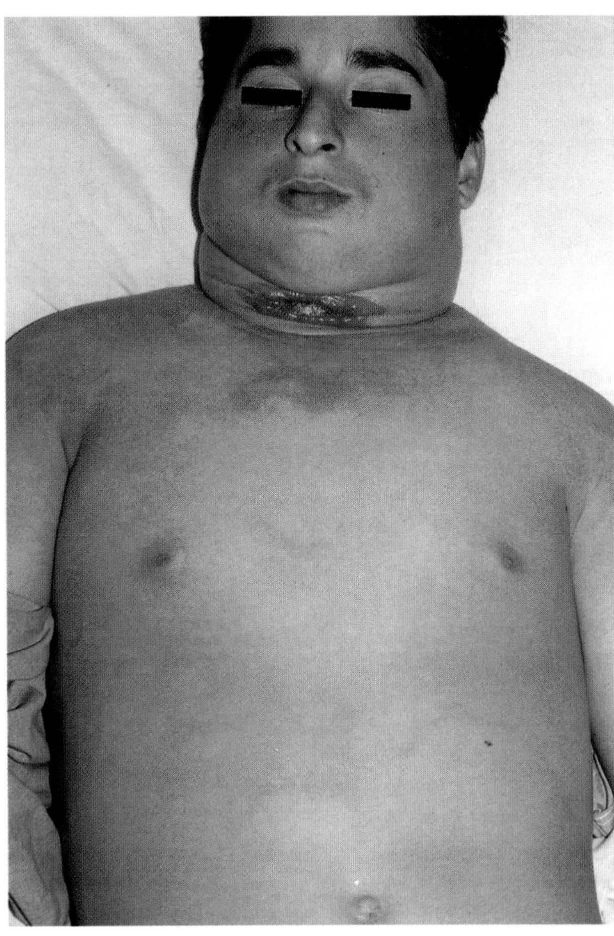

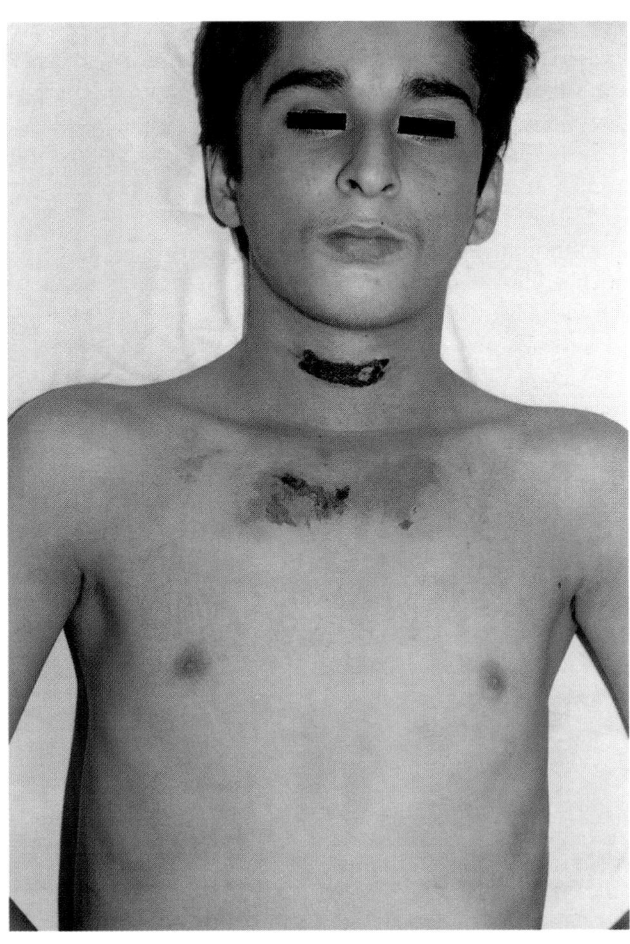

(a) (b)

Fig. 24.13 (a) Anthrax, showing the lesion with its halo of vesicles at the centre of the neck, and extensive odema, affecting the face and the whole torso. (b) After recovery, the extent of the earlier oedema is dramatically evident. (Reproduced with permission from Felek *et al.* (1999) *Journal of Infection* 1999; 38:201) square-ended Gram-positive rods. These filamentous colonies are called Medusa head colonies. Spore stains demonstrate central spores in bacteria from mature colonies. Blood cultures readily produce a heavy growth; indeed, methylene blue-stained blood smears from septicaemic animals and humans will often demonstrate bacilli (without spores, which do not form in the living host).

Treatment

This should be commenced while the diagnosis is being confirmed, to minimize the risk of bacteraemia. Intravenous benzylpencillin 2.4–3.0 g 6-hourly is the treatment of choice. Oral ampicillin 500 mg–1 g 6-hourly should be used for continuation. Ciprofloxacin intravenously 200 mg 12-hourly or orally 500 mg 12-hourly is a good alternative. Treatment should be continued until the lesion is healed, and the oedema completely resolved.

Prevention and control

Anthrax vaccines are available for animal and human use. Inactivated vaccine is offered to workers at occupational risk; few adverse effects are recorded other than local inflammation. Hides, and hoof and bone products, must be heat- or chemically treated to destroy spores before they are moved between countries or used in manufacturing processes. The tanning process renders hides and leather safe.

Tularaemia (see also Chapter 25)

This is a disease caused by *Francisella tularensis*, an organism of rodents. Hunters and woodsmen can be infected by direct contact or inoculation from live animals, skins or carcasses. The disease may be localized to the skin and local lymph nodes (ulceroglandular form) or cause systemic infection with fever, variable rashes and sometimes pneumonia (typhoidal form). Rare cases of ocular, pharyngeal or abdominal infection also occur. Patients with pneumonitis have a significant mortality.

Blood cultures are usually negative, so diagnosis rests on clinical suspicion or recovery of organisms from affected tissue sites. Serological diagnosis can be made by reference laboratories.

Aminoglycosides and tetracycline are effective in treating tularaemia.

Melioidosis

This is caused by *Burkholderia pseudomallei*, whose reservoir is probably in the water buffalo and its wetland habitat. Infection is commonest in Asian and Far Eastern farming areas. Human infection presents as pyogenic disease, such as empyema, sometimes after years of apparent latency. The organism can be recovered from pus, and is usually sensitive to third-generation cephalosporins.

Viral haemorrhagic fevers

Introduction

Viral haemorrhagic fevers (VHFs) is a general term describing several groups of severe viral infections in which haemorrhage is part of the clinical picture. These include some diseases already discussed in this and other chapters.

DISEASE LIST

Arboviruses
- Dengue haemorrhagic fever (flavivirus)
- Yellow fever (flavivirus)
- Crimean-Congo haemorrhagic fever (nairovirus)
- Others (Chikungunya, Rift Valley fever)

Hantavirus infections
- Haemorrhagic fevers with renal syndrome
- Hantavirus pulmonary syndrome

Arenaviruses
- Lassa fever
- Argentine haemorrhagic fever (Junin virus)
- Bolivian haemorrhagic fever (Machupo virus)
- Rarer pathogenic arenaviruses, occurring also in South America (Guanarito, Sabia)

Filoviruses
- Marburg disease
- Ebola virus haemorrhagic fever

> Yellow fever is an internationally notifiable disease. Countries where human cases exist are listed in the *Weekly Epidemiological Report*.

These viral infections tend to cause severe systemic diseases with fever, malaise, variable sore throat and headache, arthralgia and increasing prostration. Multiorgan damage is common, with evidence of liver dysfunction, bone marrow depression, renal impairment and evidence of widespread tissue damage (falling sodium levels, elevation of non-liver transaminases, falling blood pressure, encephalopathy and extreme lassitude). These problems may be accompanied by specific features such as rash, diarrhoea or renal failure. Haemorrhage is usually due to platelet deficiency or dysfunction, and evidence of significant disseminated intravascular coagulation is uncommon.

Haemorrhagic fevers with renal syndromes

These diseases are caused by viruses of the hantavirus family, whose reservoir is in asymptomatic mice and voles. Transmission to humans is either by inoculation or by inhalation of body fluids from host animals. Severe disease, caused by Hantaan and Seoul virus (in the Far East and Korea) and Dobrava–Belgrade virus (in eastern Europe), cause eye pain, moderate haemorrhage and severe uraemia, with a case fatality rate of up to 15%. Milder diseases of the same type occur in western Europe and forested parts of Scandinavia. These are often caused by the Puumala virus.

The disease occurs in three main phases: (i) an acute influenza-like syndrome; (ii) an intermediate stage of hypotension or shock, accompanied by haemorrhagic features and thrombocytopenia; and (iii) a late stage of oliguria and renal failure.

Diagnosis is based on demonstrating IgM antibodies, usually by ELISA. The viruses can be recovered in cell cultures of blood or serum; viral genome can be demonstrated in blood by PCR techniques.

Management is mainly supportive. Early treatment with tribavirin can abort the hypotensive and nephropathic phases of the disease.

Hantavirus pulmonary syndrome

This is a severe respiratory distress syndrome with a short incubation and fulminant course. It is caused by a variety of hantaviruses, transmitted by inhalation of dense aerosols of infected mouse excreta (typically from mouse-infested dwellings), and is endemic throughout the Americas. Many infections (caused by Sin Nombre and New York viruses) do not cause renal disease, but others (Bayou, Andes and Black Creek Canyon virus) can also cause renal failure. The case fatality rate is 40% or more, due to respiratory failure. Diagnosis is by PCR of blood, by culture of blood or respiratory secretions, or by serology in late disease. High-dose tribavirin has only a marginal benefit.

Major viral haemorrhagic fevers (transmissible from person to person)

Although all haemorrhagic fevers are serious for affected patients, most are not transmitted from person to person. However, some have caused significant nosocomial transmission and are therefore subject to special precautions when diagnosed or suspected in Western countries. These are (in approximately increasing order of infectiousness): Lassa fever, Marburg disease, Ebola virus haemorrhagic fever and Crimean-Congo haemorrhagic fever. As the maximum incubation period of viral haemorrhagic fevers is 3 weeks, they can be excluded if more than 21 days have elapsed between leaving the endemic area and the onset of fever. In the UK *The Management and Control of Viral Haemorrhagic Fevers* (HMSO 1996) sets out the recommendations of the Department of Health for obtaining advice and arranging management for suspected cases. High-security infectious diseases units (HSIDUs) have special clinical and laboratory facilities for handling patient management, and can liaise with the Central Public Health Laboratory: Virus Reference Division* to arrange diagnostic tests.

Lassa fever

Introduction

Lassa fever is an arenavirus infection whose natural reservoir is the multimammate rat *Mastomys natalensis*, which only carries Lassa virus in West Africa (other arenaviruses less pathogenic to humans exist in other parts of Africa). Transmission is by inoculation or mucosal contamination by infectious urine from asymptomatic rats. Small numbers of medical and laboratory staff have been infected when handling cases or specimens.

The disease has an insidious onset with fever, malaise, aches and pains, dry cough, sore throat, moderate gastrointestinal symptoms and increasing prostration. Many cases may be mild or self-limiting, but Lassa fever is a common cause of hospital admission in endemic areas and many patients have severe illnesses of 2–3 weeks' duration. Those who have high transaminases (more than 10 times the upper reference level) usually have high viraemias (more than 4 log tissue culture infective dose 50), and have a mortality of up to 15%.

Late and fatal cases have haematuria, blood streaking of sputum and easy bruising, due to poor platelet aggregation. Hypotension can be profound and encephalopathies may occur. Peripheral blood neutrophilia reflects extensive tissue damage. There is often extensive non-pitting oedema of the lower face and neck. Severe haemorrhage is a rare or terminal event.

Diagnosis

Diagnosis depends on clinical and epidemiological suspicion. Laboratory diagnosis is based on demonstrating

*Central Public Health Laboratory: Virus Reference Division, 61 Colindale Avenue, London NW9 5HT.

viral RNA in blood by PCR techniques, IgM or IgG antibodies by ELISA, or recovery of viruses in vero cell cultures of blood. Urine cultures become positive later than blood cultures and may remain positive for several weeks. Positive blood cultures have also been demonstrated in afebrile patients in early convalescence.

Treatment

Treatment is based on intensive support and tribavirin therapy (which ameliorates disease but does not abolish viraemia or viruria). It is important to exclude immediately life-threatening diseases such as malaria, and no patient should be referred as a case of viral haemorrhagic fever before malaria has been adequately considered and investigated.

Haemorrhagic conditions that should be considered before a diagnosis of viral haemorrhagic fever is assumed

1 Malignant malaria.
2 Meningococcal disease.
3 Severe rickettsial infections.
4 Gram-negative septicaemia with disseminated intravascular coagulation.

Supportive measures, with attention to perfusion, fluid balance and fever, are important in reducing mortality. Tribavirin for intravenous use is available in HSIDUs, and can reduce mortality in severe disease. Follow-up is necessary to confirm eventual clearance of virus from blood and urine.

Prevention and control

Prevention and control measures are concentrated on close family members and medical staff who have contact with the patient and his or her blood. Casual and social contacts are not at risk. In most cases surveillance of health and temperature during the possible incubation period is sufficient. Tribavirin has been given to high-risk contacts such as sexual partners or persons contaminated with the patient's blood, but there is no firm evidence of its effectiveness.

Argentine haemorrhagic fever

This disease is caused by Junin virus, an arenavirus whose natural hosts are harvest mice. Epidemics occur in northern Argentina, particularly during the corn harvest. The disease is like Lassa fever; petechial rashes and platelet dysfunction are common. Patients with severe disease may have encephalopathies. The mortality is 10–15%.

Convalescent plasma, containing neutralizing antibodies to Junin virus, can reduce mortality to 3%, but a late mild encephalopathy is common in treated patients. An effective vaccine is now available.

Other arenavirus haemorrhagic fevers

Machupo virus causes occasional cases and outbreaks in rural Bolivia. Guanarito virus cases outbreaks in Venezuela, and Sabia virus has been identified in Brazil, as well as in two laboratory-associated infections.

Marburg disease and Ebola haemorrhagic fever

Epidemiology

The epidemiology of these filovirus infections is not understood. The original outbreak in Marburg followed the importation of infected vervet monkeys into a scientific laboratory. Transmission occurred from monkeys and their tissues to laboratory workers, from patients to medical attendants and from one convalescent patient to his wife (the patient's semen was found to contain virus). The main route of transmission is via the blood and body fluids and the case fatality rate is about 50%. One patient in a South African outbreak had Marburg virus isolated from the anterior chamber of the eye many weeks after her illness.

Ebola infection has caused large community outbreaks with fatality rates of 75–80%. The first outbreak was complicated by extensive hospital transmission, perhaps associated with reuse of needles. Countries affected include Zaire, Sudan, Gabon and Sierra Leone (one case only). A nurse in Johannesburg was fatally infected when caring for a patient from Gabon in 1996. Most outbreaks appear to originate from human exposure to an infected monkey or chimpanzee, though there is no evidence that these are the natural reservoir of disease.

Clinical features

The clinical picture is similar in both diseases. After 3 or 4 days' insidious onset with high fever and prostration, diarrhoea occurs and a measles-like rash is common. Mild or gross haemorrhagic features may quickly develop, including bloody diarrhoea with abdominal pain. Leucocytosis occurs and transaminases rise as tissue damage proceeds. Platelet dysfunction is the main cause of haemorrhage.

Diagnosis

Diagnosis may be suspected on clinical and epidemiological grounds. Early diagnosis is by demonstration of viral RNA in blood. The viruses grow readily in cell cultures and can be demonstrated by immunofluorescence. IgM and IgG antibodies in the patient's blood can be demonstrated at appropriate stages of the disease.

No antiviral agent, including interferon, has a significant beneficial effect; vigorous supportive treatment must be given. Strict patient isolation is more urgent than with Lassa fever, as both the patient and his or her body fluids are more infectious, and there is a small possibility of droplet infection from either source.

Crimean-Congo haemorrhagic fever

Epidemiology

The natural epidemiology of this nairovirus disease depends on tick-borne transmission between warm-blooded animals. Cattle and farmed ostriches are important reservoirs in many African countries but camels and rodents may be important elsewhere. Crimean-Congo haemorrhagic fever (CCHF) is not restricted to tropical Africa, but is endemic in the Middle East and parts of Bulgaria, former Yugoslavia and the former southern USSR. It is an infectious condition. Secondary cases are relatively common, particularly among clinical care personnel.

Clinical features

The clinical picture follows an incubation of 4–10 days with a maximum of 13 days. Abrupt onset of severe viral symptoms is followed in 2 or 3 days by collapse, extensive bruising and purpura, haematemesis and melaena. A rapid rise in transaminases and profound drop in the platelet count occurs during the first week of illness. In these patients there is evidence of intravascular coagulation. Cases of intermediate severity can occur, and must be differentiated from meningococcal or rickettsial diseases or from haemorrhagic chickenpox.

Tribavirin may have some action in CCHF, but treatment is mainly supportive.

Public health measures for viral haemorrhagic fever cases

Viral haemorrhagic fevers are unfamiliar diseases in the UK, where they are extremely rare. There are several important aspects to their safe and expeditious management.

1 Ensure that common severe infections are not missed. Over 90% of viral haemorrhagic fever suspects have a final diagnosis of falciparum malaria, which can cause disseminated intravascular coagulation (see text note, p. 457) Blood films must be examined to exclude malaria before the question of viral haemorrhagic fever is seriously considered in most cases. Seriously ill patients can be commenced on empirical quinine treatment. Suspicion of and empirical treatment for meningococcal disease may also be important.
2 Obtain advice from a specialist. Regional infectious and tropical disease specialists will have up-to-date information on endemic areas, and will often be able to exclude the diagnosis of viral haemorrhagic fever on the basis of epidemiological and clinical data.

Cases where the diagnosis cannot readily be excluded fall into two main categories:

(a) Medium-risk cases, exposed to endemic areas but not to known sources of infection. If malaria studies and initial investigations do not suggest an alternative diagnosis, these patients' specimens should be handled in a specialist laboratory and the patient may require transfer to a specialist infectious diseases unit for further investigation and management.

(b) High-risk cases, when the diagnosis of viral haemorrhagic fever is known, obvious or strongly suspected, or is indicated after investigation as a medium-risk case. Such cases should be managed in an HSIDU, where special control of infection measures are employed during investigation and management. The staff at the HSIDU will accept such cases, and will liaise with ambulance services to arrange transfer.

3 Viral haemorrhagic fevers are notifiable diseases. They should be notified urgently by telephone to the appropriate Proper Officer, with follow-up in writing. The names of face-to-face healthcare contacts should be recorded for health surveillance. Contacts of high-risk cases are followed up for 3 weeks after the last contact, or until the diagnosis is disproven. Local public health specialists will advise on cleaning and disinfection of premises and equipment. Standard hospital procedures are usually adequate for this. Special disinfection or fumigation is hardly ever indicated.

Summary of personal health measures recommended for travellers

1 Vaccinations for epidemic diseases. The desirability of these will depend on the current prevalence of epidemics in the destination country. Vaccines available include:

(a) Influenza vaccine.
(b) Meningococcal vaccine.
(c) Measles mumps rubella vaccine for infants.

(Cholera vaccine is no longer required by any country.)

Other vaccines for personal protection will depend on whether a traveller will have other occupational or social exposures:

(a) Tetanus, diphtheria and polio vaccines or boosters.
(b) Hepatitis B vaccine.

2 Protection from food- and water-borne diseases is recommended for travellers to countries where efficient sanitation and safe water supplies cannot be guaranteed. The following measures may be considered:

(a) Typhoid immunization.
(b) Hepatitis A prophylaxis with inactivated vaccine.
(c) Human normal immunoglobulin can confer immediate, short-term protection from hepatitis A if urgently indicated.
(d) Chemoprophylaxis for diarrhoea (special circumstances only).
(e) Water filter/purification kits (campers and independent travellers).

3 Malaria prophylaxis is essential for everybody who travels to an endemic area. Measures to avoid insect bites should also be emphasized.

4 Yellow fever vaccination is recommended for all travellers who will enter an endemic area. Certificates of vaccination are often required for travellers passing from endemic areas to other countries. Certification is subject to International Health Regulations, and is available only from accredited centres.

5 Immunization against local infections may be advisable for travellers having rural or community exposure, especially during certain seasons. Diseases to consider include:

(a) Japanese B encephalitis (stays of more than 1 month in rural farming areas of affected countries).
(b) Tick-borne encephalitis (camping, walking or rural work in forested areas of affected countries during late spring and summer).
(c) Rabies (independent travel in remote areas of affected countries; relief work, especially if involving animal contact in those countries; caving in areas where bat rabies occurs).

Assessment of fever in a returning traveller

Different diseases may become evident at different times after a traveller returns from an overseas visit, depending on the incubation periods of the various infections (Table 24.1).

Time after return home	Disease
During first week	Viral gastroenteritis Traveller's diarrhoea Bacillary dysentery, sexually transmitted diseases, influenza, dengue (and other arboviral infections)
1–2 weeks	Malaria Hepatitis A Typhoid fever Paratyphoid fever Rickettsial infections
2–4 weeks	Typhoid fever Amoebiasis Hepatitis C Katayama fever
1–6 months	Hepatitis B HIV seroconversion illness Hepatitis E Amoebiasis Rabies Cutaneous leishmaniasis Systemic leishmaniasis
More than 6 months	Relapses of vivax or ovale malaria Reactivation of malariae malaria Strongyloidiasis (larva currens) Rabies Systemic leishmaniasis AIDS

AIDS, acquired immunodeficiency syndrome; HIV, human immunodeficiency virus.

Table 24.1 Intervals between a traveller's return home and the presentation of imported diseases.

Case 24.1: Young men with itchy feet

History: Three young men shared a 2-week holiday at a Caribbean resort, where they spent most of their days barefoot, participating in beach activities. Near the end of their holiday, they each developed localized, red, intensely itchy lesions on their feet (Fig. CS.5). These lesions persisted for 3 weeks, despite treatment with clotrimazole cream, and a referral to the travellers' clinic was made.

Questions: What is the diagnosis?
What is the pathology of the lesion?
What treatment would you recommend?

Management and progress: Based on the typical lesions, a clinical diagnosis of cutaneous larva migrans was made. This is due to skin invasion by the larvae of dog hookworms. Hookworm ova are deposited on sand or soil in the faeces of infected dogs. The ova release larvae, which undergo a soil cycle, becoming invasive, and are destined to adhere to and burrow through the skin of animals walking on the affected ground. Larvae adhering to human skin will invade locally, producing an irritating allergic reaction. They cannot, however, complete their life cycle, and die after several weeks. The larvae can be killed by a systemic course of albendazole, 400 mg twice daily for 3 days. An alternative is a paste of 10% thiabendazole (which can be made by a pharmacist), applied under occlusion to aid skin penetration. Topical corticosteroid cream may help to reduce itching during the next week or so, as larval antigens disperse.

Case 24.2: Insect bites that will not heal

History: A 26-year-old woman went for a 2-week holiday to a coastal resort in a north African country. Three weeks after returning home, she became aware that two small 'insect bites' on her right leg were enlarging and becoming nodular. During the following 2 weeks, three further lesions appeared, and the original two lesions had become ulcerated, with scabs which repeatedly separated and re-formed.

Physical examination: This showed the scabbed and nodular lesions (Fig. CS.6). There was no constitutional complaint, the temperature was normal and there was no hepatosplenomegaly or regional lymphadenopathy.

Questions:
What is the diagnosis?
How can the diagnosis be confirmed?
How should the condition be treated?

Investigation and management: A clinical diagnosis of cutaneous leishmaniasis was made, based on the slow evolution, the size (1.0–2.0 cm diameter) and the chronic ulceration and scabbing with a typical 'rolled' epithelial border. This type of lesion is probably due to *Leishmania. tropica* var. *major*. Scrapings from the granulomatous mound at the base of the lesion were examined histologically for the presence of parasites within mononuclear phagocytes. This did not prove positive. A more reliable test would be to biopsy the lesion, including part of the central area. In view of the typical clinical appearance, it was considered unjustified to produce a scar by undertaking biopsy. Serology is negative in cutaneous leishmaniasis.

There is no evidence that topical antileishmanial drugs influence the course of cutaneous leishmaniasis. Systemic therapy may be indicated for large or disfiguring lesions, but is toxic, and not recommended for uncomplicated lesions. Spontaneous healing should occur in 3–5 months,

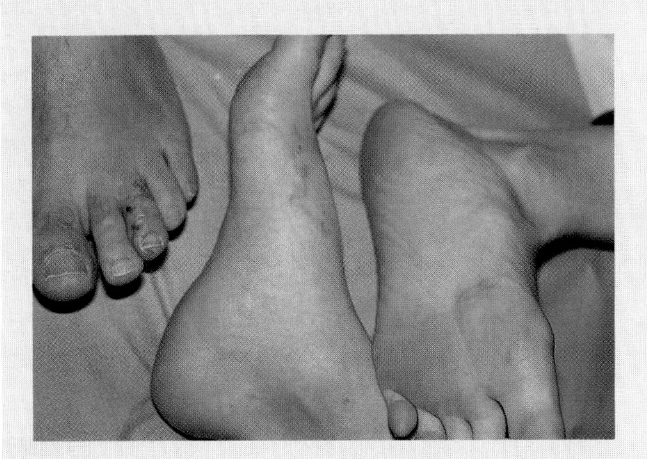

Fig. CS.5 Intensely itchy lesions in the thinner skin of the feet, one with an obvious inflammatory 'track'.

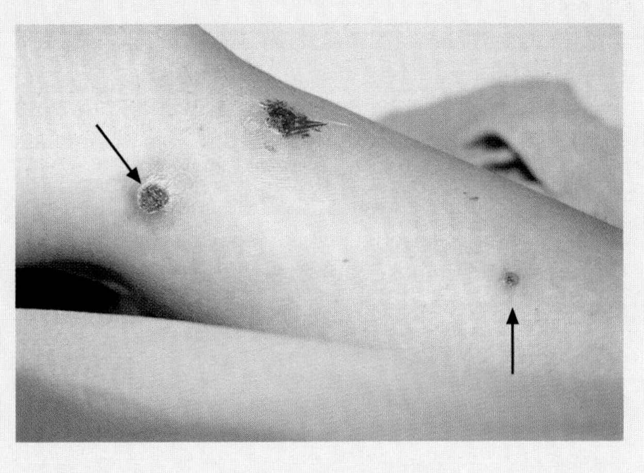

Fig. CS.6 Lesions interpreted as insect bites on the right shin and calf (one has been biopsied, and is bleeding). The large, intact lesion (arrow) shows a typical 'rolled' epithelial edge and central scab. A smaller, more recent lesion is seen on the lower calf (arrowhead)

25 Some Systemic Zoonoses

Introduction, 461

Brucellosis, 462
- Introduction and epidemiology, 462
- Clinical features, 462
- Diagnosis, 463
- Treatment, 463

Lyme disease (borreliosis), 464
- Introduction and epidemiology, 464
- Clinical features, 466
- Diagnosis, 466
- Treatment, 467
- Prevention of borreliosis, 467

Toxoplasmosis, 467
- Introduction and epidemiology, 467
- Clinical features, 468
- Diagnosis, 468
- Treatment, 469

Plague, 469
- Introduction, 469
- Clinical considerations, 470
- Diagnosis, 470
- Treatment, 470

Tularaemia, 470
- Introduction, 470
- Diagnosis, 470
- Treatment, 470

Rat bite fever (Haverhill fever), 471
- Introduction, 471
- Diagnosis, 471
- Treatment, 471

Zoonotic streptococcal infections, 471
- Introduction, 471
- *Streptococcus suis*, 471
- *Streptococcus zooepidemicus*, 471

Herpesvirus simiae infection, 471

Zoonotic paramyxoviruses, 472
- Hendra virus, 472
- Nipah virus, 472

Introduction

A zoonosis is an animal disease that can be accidentally transmitted to humans. Several such diseases, including anthrax, Q fever, ornithosis, hydatid disease, hantavirus infections and Lassa fever, have already been mentioned in other chapters of this book. A few diseases, such as *Campylobacter* infections and salmonelloses, are common to animals and humans and are relatively readily transmitted between humans as well as between animals. True zoonoses are not easily transmitted from person to person, save for cases of exceptional pulmonary disease, as in pneumonic plague.

Most zoonoses are acquired when humans intrude into the animals' environment, or handle animals or their carcasses. This occurs during farming, handling raw animal products such as hides, meat or bones, and during hunting, camping or trekking. Consumption of untreated or uncooked products such as milk, cheese or preserved meat can also be a means of acquiring a zoonosis.

Common activities which predispose to zoonoses

Consuming untreated milk, cream, yoghurt and curd cheese

1 *Salmonella*, *Escherichia coli* O157 or *Campylobacter* infections.
2 Brucellosis.
3 Q fever.
4 Haverhill fever.
5 Tick-borne encephalitis (when the animal is infected).

Hunting, trapping, skinning and butchering wild animals

1 Plague.
2 Tularaemia.
3 Rabies.

Butchering farm animals

1 Q fever.
2 Streptococcal skin infections.
3 Brucellosis.
4 *Erysipelothrix* skin infections.
5 *Streptococcus suis* systemic infections.
6 Anthrax.
7 Crimean-Congo haemorrhagic fever (from ostriches).

Eating undercooked meat
1 Toxoplasmosis.
2 Trichinellosis (from pork).
3 Pork or beef tapeworms.
4 *Salmonella, Escherichia coli* O157 or *Campylobacter* infections.

Handling dead animals, untanned hides or unpasteurized bonemeal
1 Q fever.
2 Tularaemia.
3 Plague.
4 Anthrax.

Ingestion or inoculation of animal urine
1 Leptospirosis.

Camping, hiking or forestry working in warm climates
1 Tick-borne encephalitis.
2 Borreliosis.
3 Arboviral encephalitides.
4 Hantavirus infections (haemorrhagic fever with renal syndrome (HFRS) or hantavirus pulmonary syndrome).

Brucellosis

Introduction and epidemiology

Brucella spp. are common pathogens of warm-blooded animals, particularly hoofed animals. When clinically evident in the natural host, they often cause infectious abortion, but inapparent infection and excretion of the organisms also occur. The animal's milk may be contaminated, either by excretion in the milk itself or by contamination during unhygienic milking processes.

Humans usually acquire infection by consuming unpasteurized dairy products or by contact with infected products of conception, during either delivery or abortion. The infective dose is low, and aerosol-transmitted laboratory infections can occur. Brucellosis was first described in British soldiers in Malta, who were infected by drinking raw goats' milk. Cases were subsequently recognized associated with cows' milk. Chronic brucellosis or undulant fever was described later, again after consumption of cows' milk.

Occupational exposure is important in the occurrence of brucellosis. Cowhands, veterinary practitioners and goatkeepers are at some risk. However, control programmes have made the infection extremely rare in the UK, as most herds are nowadays *Brucella*-free. Almost all cases are now imported, and are associated with consumption of untreated milk or dairy products. Brucellosis is still common in the Middle East (where the disease also affects camels) and in rural parts of Africa and Asia; a large out break occurred in goats and humans in Malta in 1995. Milk is important for its nutritious value, but is often consumed straight from the cow or goat, and in developing countries is sometimes preserved by adding the animal's urine.

Clinical features

Brucellosis has a variable clinical picture. The incubation period ranges from 5 days to several weeks, and the onset may be acute or insidious.

Acute brucellosis

Acute brucellosis has an abrupt onset, with high swingeing fever, often rigors and sometimes myalgia and arthralgia. The patient feels severely unwell, but often has few physical signs. About half have enlarged lymph nodes in the cervical chain, and about a quarter have splenomegaly. The white cell count is often, but not always, raised and the liver function tests may be slightly abnormal because of granulomatous hepatitis. Untreated acute brucellosis is rarely fatal, but causes severe morbidity for 5 or 6 weeks. Brucellae are intracellular pathogens that are able to survive inside cells of the reticuloendothelial system using a superoxide dismutase and nucleotide-like substances to inhibit the intracellular killing mechanisms of the host.

Chronic brucellosis

Chronic brucellosis is uncommon. It may follow on from an acute attack or commence insidiously. Typically, it causes malaise, depression and a fever which waxes and wanes over periods of 2 or 3 weeks (undulant fever). The white cell count is often low, because of neutropenia.

Local effects

There are many local effects of brucellosis. Almost any organ or tissue can be involved, although granulomatous inflammation of reticuloendothelial organs is the most constant feature. This causes lymphadenopathy, splenomegaly and abnormalities of the liver function tests.

Bone and joint involvement

Effusion of a large joint or chronic osteomyelitis of the

spine are the common manifestations, but any joint can be involved. Differentiation from tuberculosis is important in patients from areas where brucellosis is common, and this depends on clinical suspicion and appropriate diagnostic tests (see also Chapter 18).

Genitourinary disorders

Orchitis is relatively common. It develops after several days of fever, and may be associated with chills and increased malaise. Physical signs vary from aching pain in the testicles to acute tenderness and swelling. Renal involvement is usually evidenced by proteinuria and 'sterile' pyuria.

Neurobrucellosis

Neurobrucellosis is uncommon, but can be severe. Meningoencephalitis is the commonest feature, and can sometimes be life-threatening. Encephalopathy is a rare feature. Depression or acute psychosis can occur; only the accompanying fever, or local disease elsewhere, will suggest the diagnosis.

Endocarditis

Endocarditis is a rare manifestation of brucellosis, often causing rapid valve destruction.

Clinical features of brucellosis

1. Fever (undulant in chronic infection).
2. Pyogenic arthritis.
3. Spinal osteomyelitis.
4. Lymphadenopathy.
5. Splenomegaly (especially in acute disease).
6. Abnormal liver function.
7. Orchitis or testicular pain.
8. Endocarditis.
9. Meningoencephalitis.
10. Depression or psychosis.
11. Sterile pyuria.

Diagnosis

Diagnosis depends on epidemiological or clinical suspicion, and on appropriate laboratory tests.

Blood cultures are often positive, but *Brucella* spp. are relatively slow-growing, and cultures must be maintained for 4 weeks until they are declared negative if conventional culture methods are used. Modern automated blood culture systems allow positive detection within a few days. Culture of aspirated joint effusions, bony abscesses, or liver or bone marrow biopsies is much more sensitive than blood culture, so as many different specimens as possible should be obtained whenever the diagnosis is considered.

Serological tests are useful, and often more rapid than blood cultures. Traditional agglutination reactions (standard agglutination test, SAT) are often used but can show a strong anamnestic response after re-exposure to *Brucella* antigens, giving rise to a false-positive diagnosis. Enzyme-linked immunosorbent assay (ELISA) tests are widely performed, and can detect immunoglobulin M (IgM) antibodies in acute disease, or high levels of IgA and IgG in chronic infection.

Liver biopsy will often show multiple granulomata, especially in established disease. These are rarely caseating, unless caused by *B. suis*, and must be distinguished from the granulomata of sarcoidosis or miliary tuberculosis.

Treatment

The treatment of choice is doxycycline, which should be continued for 4–12 weeks. Relapse can occur, and is less likely if gentamicin 2–5 mg/kg daily or netilmicin 4–6 mg/kg daily (with blood-level monitoring) or rifampicin 600 mg daily is given for at least 3 weeks. Doxycycline plus aminoglycoside probably offers the best chance of relapse-free recovery, but is less convenient than doxycycline plus rifampicin because of the need for intramuscular injection and blood-level monitoring. Positive blood cultures persist after treatment in 10–15% of patients treated with older tetracyclines for 2 months combined with aminoglycoside for the first month. For endocarditis or neurobrucellosis, doxycycline plus at least one additional drug should be continued for a total of 3 months.

Co-trimoxazole produces early improvement, but is followed by relapse with positive blood cultures in 35–50% of cases. It is not recommended as monotherapy. Quinolones have not been found as effective as tetracycline–aminoglycoside. Both drugs are useful as components of multidrug therapy.

Depression or psychotic symptoms do not always respond readily to antimicrobial therapy. Psychiatric support may be required, and additional treatment with psychotropic drugs may be needed.

Treatment in pregnancy is unsatisfactory, as tetracyclines cannot be given. A compromise is to treat with co-trimoxazole plus rifampicin, and to offer retreatment after delivery if bacteraemia recurs. Children may also be treated with a combination of co-trimoxazole and rifampicin.

Treatment of brucellosis
1 Doxycycline orally 200 mg first doses, then 100–200 mg daily for 2–3 months; *or*
Demeclocycline orally 300 mg 12-hourly for 2–3 months; *plus*
Gentamicin i.m. or i.v. 2–5 mg/kg daily in three divided doses (with blood-level monitoring) for the first month; *or*
Netilmicin single dose i.m. or i.v. 4–6 mg/kg daily (with blood-level monitoring); *or*
Rifampicin orally 600 mg twice daily (or 300 mg 6-hourly) for the first month.
2 In pregnancy: co-trimoxazole orally 960–1440 mg 12-hourly for 2 weeks, then 480 mg 12-hourly for 6 more weeks; *plus*
Rifampicin orally 300–600 mg twice daily for the first 4 weeks (give first-choice treatment after delivery if indicated).
3 For a child under 12 years tetracycline is contraindicated; use co-trimoxazole orally; 6 months–5 years, 240 mg twice daily; 6–12 years, 480 mg twice daily for 8 weeks; *plus*
Rifampicin 10 mg/kg daily for the first 4 weeks.

Lyme disease (borreliosis)

Introduction and epidemiology

Lyme disease has a prolonged natural history and several stages affecting different body systems. It is caused by *Borrelia* spp., which are transmitted from animals to humans by the bite of hard ticks of the genus *Ixodes* (Fig. 25.1). The natural hosts of the *Borrelia* species are small rodents, but human disease is often acquired via deer. Dogs can also be infected. The main organism causing borreliosis in the USA is *B. burgdorferi* and in various parts of Europe is *B. afzelii* or *B. garinii*.They all cause disease with similar clinical characteristics; US strains may be more likely to produce carditis, while European strains may more often produce central nervous system manifestations.

Borreliosis is most common following occupational or leisure exposure to open forest or parkland, where ticks are plentiful. The presence of large animals such as deer seems to increase the likelihood of infection. Transmission has been recorded in the New Forest, the deer parks of London and south-east England, and in parts of Scotland. Imported cases may originate in European countries such as Germany or Austria, in Scandinavian countries and in eastern and central areas of the USA. A tick must remain attached for about 20 h to transfer an infective dose of borrelias in its saliva.

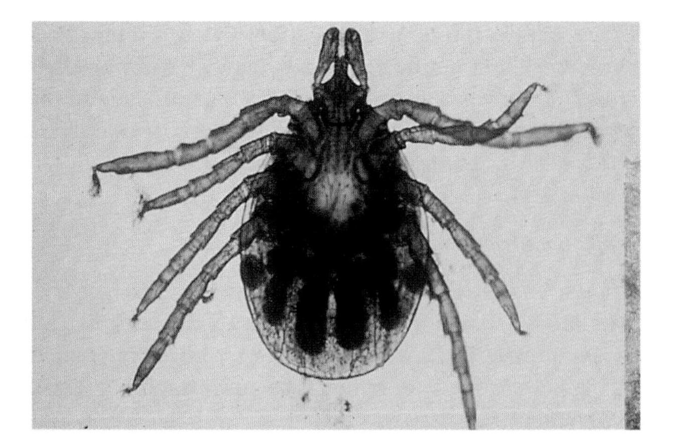

Fig. 25.1 Hard ixodid tick, the vector of borreliosis. Hard ticks are also vectors for tick-borne encephalitis and some rickettsial infections (Courtesy of the Ministry of Defence.)

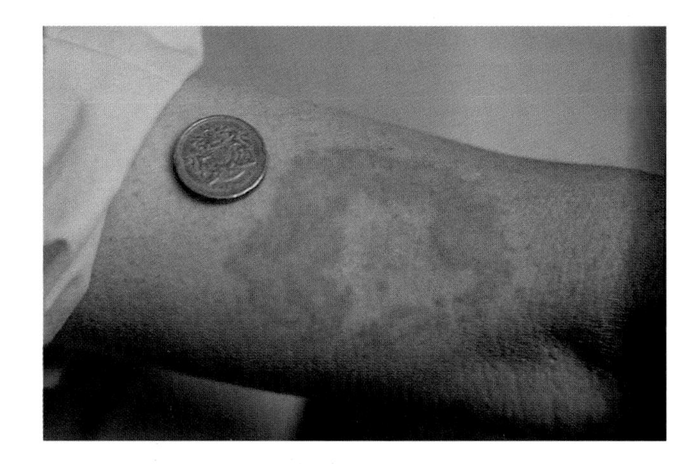

Fig. 25.2 Erythema chronicum migrans. (Courtesy of Dr M. G. Brook.)

Early Lyme disease and erythema chronicum migrans

After an incubation of 1–3 weeks, about 80% of infected patients develop a characteristic rash surrounding the tick bite. A disc of erythema expands, often clearing in the centre (Fig. 25.2). This may enlarge sufficiently to encompass a limb before gradually fading. Multiple erythema chronicum migrans (ECM) lesions are occasionally seen. Patients often have fever, aches and pains, and malaise during the eruption, which can last for 2–4 weeks. There may be a mild leucocytosis, and the erythrocyte sedimentation rate is often raised to 40–60 mm/h.

Borrelial lymphocytoma

Borrelial lymphocytoma is a rarer skin manifestation which can occur at the same time as ECM, after it, or occasionally in late disease. Although it may occur at or near the site of the original tick bite, it has a predilection for the ear (especially in children) or the breast. It is usually a dusky nodule or plaque which may last for several months if untreated, and reaches 1–5 cm in diameter. Not all patients have associated constitutional symptoms. Histology shows non-specific lymphocytic infiltration with germinal centres, and must be differentiated from other granulomatous disorders and lymphomata.

Early disseminated Lyme borreliosis

These disorders can follow the original infection after an interval varying from 2 or 3 weeks to 2 or 3 months.

Lyme arthritis

Lyme arthritis is common in the USA, affecting about half of patients whose early infection is untreated. Its onset varies from a few days to 2 years after exposure. It is usually an asymmetrical large joint arthritis, but is occasionally palindromic. There is synovitis and often moderate effusion of affected joints. The effusion has a high protein level and contains neutrophils. This, with a raised erythrocyte sedimentation rate, makes it easy to interpret the illness as seronegative rheumatoid arthritis. Untreated arthritis tends to recur progressively less frequently over 2–4 years. Few patients have permanent or erosive joint disease.

About 1 in 10 untreated patients develop intermittent arthralgias and periarticular pain without objective synovitis or effusion. In some cases these symptoms persist for 5 years or more.

Peripheral neuropathies

Peripheral neuropathies are common in many types of Lyme disease, and may accompany other manifestations. Facial paralysis is one of the most common neuropathies, but others, including unilateral phrenic nerve palsy, have been described. They tend to resolve spontaneously over a period of weeks.

Neuroborreliosis

Relapsing lymphocytic meningitis was recognized long before Lyme disease. It is now known that a significant proportion of these cases are caused by neuroborreliosis.

Polyradiculitis is a disabling and progressive feature of Lyme disease, which may be more common in European types of infection. It presents as localized pain in the affected roots, with dysfunction of the associated nerves. A typical presentation would be low back or sacral pain with a weak knee or foot drop. Paraesthesiae, loss of sensation and absent reflexes are common findings.

The radiculitis is often accompanied by meningism and lymphocytosis in the cerebrospinal fluid (CSF). Occasional plasma cells are also seen, and the CSF glucose level may be slightly lowered. This syndrome of relapsing meningoradiculitis (Bannwarth's syndrome) was described before the aetiology was understood.

Occasionally, encephalitis or encephalopathy can be a manifestation of Lyme disease.

Cardiological effects

The cardiological effects are a result of myocarditis, often with conduction defects. Complete heart block is common (Fig. 25.3). Prolonged and progressive cardiomyopathy has been described in occasional cases.

Late (chronic) Lyme disease

This occurs as a peripheral neuropathy of the glove-and-stocking type, associated with a typical violaceous inflammation of the skin called atrophic acrodermatitis. Histology of affected skin shows a lymphocytic infiltrate, often with many plasma cells. The acrodermatitis is asymmetrical, most commonly affecting a foot or heel, sometimes the elbow or hand. After months or years without treatment the lesions change from oedematous to thin and atrophic. Even though this syndrome may have existed for months or years, together with malaise and mild depression, it is amenable to treatment.

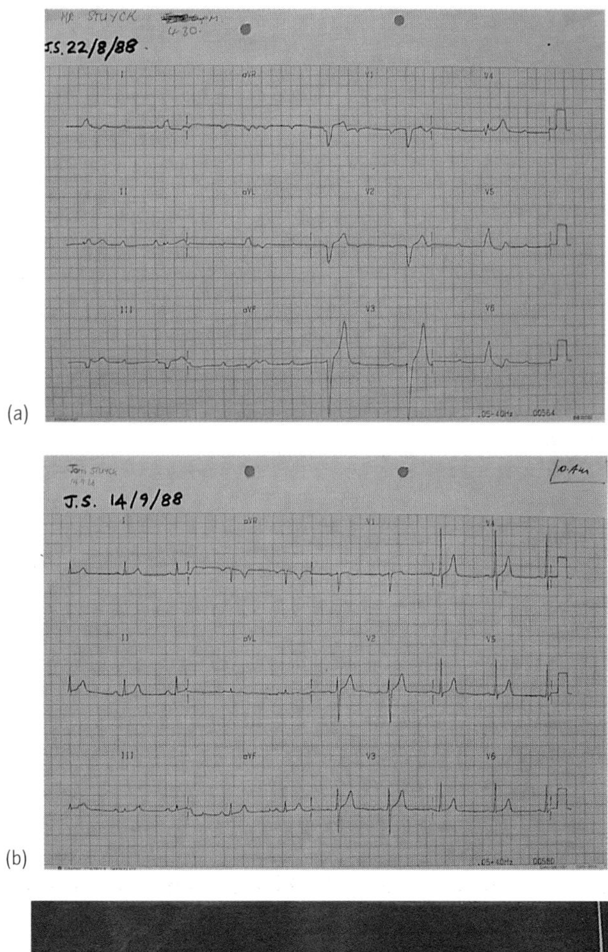

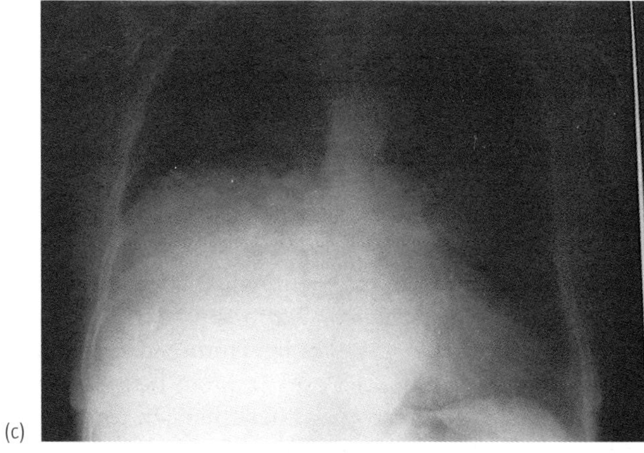

Fig. 25.3 (a) Electrocardiogram (ECG) showing a complete heart block in a young patient with secondary Lyme disease. (b) The second ECG after treatment shows that conduction is normal. (c) The same patient had a transient phrenic nerve palsy, with raised right diaphragm.

Clinical features

Clinical manifestations of borreliosis

Early localized
Flu-like illness.
Erythema chronicum migrans.
Borrelial lymphocytoma.

Early disseminated
Arthritis.
Myocarditis.
Neuropathies.
Relapsing lymphocytic meningitis.
Polyradiculitis.

Late
Peripheral neuropathy.
Atrophic acrodermatitis (acrodermatitis chronica atrophica).

Diagnosis

Clinical and epidemiological suspicion is important; unless a history of tick exposure in endemic areas is elicited, the diagnosis may be overlooked. Laboratory tests for borreliosis are still under development, as false positive and false negative tests are quite common. The diagnosis of borreliosis is therefore not straightforward, and is based on correlation of clinical and laboratory findings rather than on laboratory tests alone.

In specialist laboratories, *B. burgdorferi* can be cultured from active skin lesions. Polymerase chain reaction can demonstrate *Borrelia* spp. DNA in the synovium and CSF, and in blood in the stage of ECM, but this test is not generally available.

IgM antibodies can be demonstrated by ELISA tests using flagellar antigens, but only just over half of patients have positive tests at the earlier presentations. The antibodies persist in the blood for a variable time, but are often replaced by exclusively IgG antibodies by the time that early disseminated features appear.

Although serological tests are evolving as the infecting organism is better understood, there is currently a significant prevalence of false positive IgG antibody tests, probably due to cross-reaction with other spirochaetal antibodies. Furthermore, many patients have negative serological tests at first presentation. Confirmatory Western blot tests are therefore required when ELISA tests are positive. Isolated positive IgG tests should be interpreted with caution, unless accompanied by appropriate symptoms and physical signs. Serum antibodies may never become detectable if treatment is given early.

When there are neurological features, locally produced antibodies can be demonstrated in the CSF, and this is useful evidence of infection.

Treatment

There is still some doubt as to the best approach to treatment.

For early infections doxycycline is the most reliable antibiotic. Given in a 3-week course at the stage of ECM, this is usually curative, and will abort the late disseminated features.

For later infections more intensive and prolonged treatment is necessary, as relapse commonly follows short courses of tetracycline. Cefotaxime and ceftriaxone have both been used successfully in neuroborreliosis, but in half or more of patients symptoms persist after a 2- or 3-week course. Even longer courses are not always followed by cure, and it is worth giving oral doxycycline, or ampicillin plus probenecid to give a total of 8–12 weeks' antibiotic treatment. Longer continuation courses may have some benefit in late disease.

Penicillin is effective initially, and may be curative in early infections, but carries a risk of recrudescence if used in secondary or late disease. Erythromycin is less effective than doxycycline.

Treatment of borreliosis

1 (a) Early: doxycycline orally 200 mg, then 100 mg daily for 20 days more.
(b) Alternative, or for child: ampicillin orally 500 mg 6-hourly (child under 10 years, 250 mg 6-hourly); *or* amoxycillin orally 500 mg 8-hourly (child under 10 years, 250 mg 8-hourly) for 3 weeks.
(c) Second choice: erythromycin orally, 500 mg 6-hourly (child up to 2 years, 125 mg 6-hourly; 2–8 years, 250 mg 6-hourly) for 3 weeks.

2 Disseminated or late: cefotaxime i.v. 1–2 g 8-hourly *or* ceftriaxone i.v. 1–2 g daily, both for 2–4 weeks; *followed by* doxycycline orally 100 mg daily *or* ampicillin 250 mg 6-hourly *plus* probenecid 500 mg twice daily (child under 10 years, half adult dose) for 4–8 weeks.

Prevention of borreliosis

Avoidance of tick contact, by appropriate use of clothing and insect repellents is important. Inspection of the skin and early removal of ticks is also helpful, as 24–48 hours' attachment is necessary before an infective dose of borrelias is transferred from tick to host.

A vaccine, directed at the outer surface protein of *B. burgdorferi*, is now available (not currently licensed in Britain). It stimulates the production of antibodies which kill the tick-borne stages of the organism, so that when the tick ingests the blood of the immunized host, the borrelias in the tick's body cavity are killed before they can pass to the host.

Toxoplasmosis

Introduction and epidemiology

Toxoplasma gondii is a protozoan parasite of the family Sporozoa. It is found in the tissues of almost all warm-blooded creatures, but the only hosts for its definitive life cycle are the cat family. Cats are infected by predation on other infected creatures, or by consumption of oocysts derived from the faeces of recently infected cats.

In the cat, oocysts develop in the intestinal mucosa, and release parasites into the bloodstream and tissues. These replicate rapidly by fission, and are called tachyzoites. Tissue tachyzoites form pseudocysts: masses of organisms within expanded cells. The organisms in tissue cysts are quiescent, as they are controlled by cell-mediated immunity, and are called bradyzoites. Meanwhile some parasites re-enter the enterocytes and develop into oocysts, which are shed in the faeces. Shedding of oocysts begins about 10 days after infection and persists for 2–3 weeks.

Infection can follow ingestion of tachyzoites, bradyzoites or oocysts. Humans may be infected by contact with cats, particularly kittens, who become infected as they begin to predate. However, contact with uncooked meat is a major route of infection. Bradyzoites can be demonstrated in many types of butcher's meat, and in the types of ham which are preserved and eaten without cooking in many countries. Toxoplasmosis has occurred when a transplanted heart has contained bradyzoites. In this case, the recipient is immunosuppressed, and this may modify the presentation of the disease, as will the previous level of immunity against toxoplasmosis.

In the UK the peak age for seroconversion to toxoplasmosis is 15–35 years, and about half of all adults have evidence of past infection. Immunocompetent individuals rarely have a recognizable illness. The importance of the disease lies in its ability to cause transplacental infection (see Chapter 17) and opportunistic infection in acquired immunodeficiency syndrome (AIDS) sufferers (see Chapter 22).

Clinical features

Clinically expressed toxoplasmosis (Fig. 25.4) tends to present in one of three ways:

1 A mononucleosis syndrome is common in young adults. The features are fever, malaise and one or more enlarged lymph nodes. There is an atypical mononucleosis in the peripheral blood but the heterophil antibody test is negative. The illness is self-limiting, with a variable duration of up to many weeks.

2 A single, persistently enlarged lymph node, or occasionally a skin or soft-tissue nodule. The differential diagnosis of lymphadenitis is large, and excision biopsy is often needed to exclude tuberculosis, sarcoidosis or lymphoma.

3 Acute choroidoretinitis is a feature of late-stage toxoplasmosis. Although there may have been a feverish illness some weeks or months before the onset of eye symptoms, such a history may be remote, and is rarely elicited. Ocular features of toxoplasmosis can develop months or years after congenital infection.

Rare manifestations include myocarditis, encephalitis, encephalomyelitis and pneumonitis (all more common in the immunosuppressed).

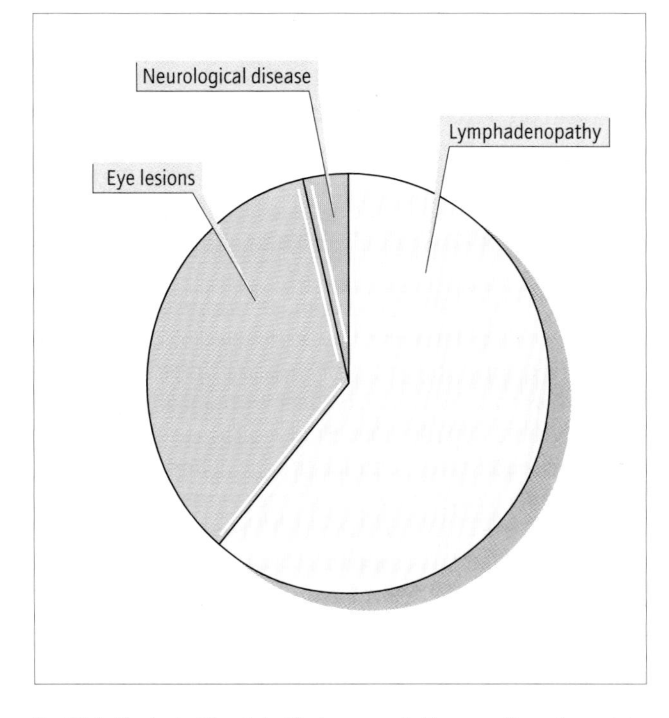

Fig. 25.4 Pie chart of the clinical features reported in cases of toxoplasmosis in England and Wales 1989–92.

Diagnosis

Several serological tests are useful in the diagnosis of toxoplasmosis. A latex agglutination test is widely used by laboratories for screening sera. Titres rise rapidly during acute disease; titres of 1:4000–1:64 000 are common. These titres fall slowly over 18–24 months to reach background seropositivity.

Confirmatory tests carried out in reference laboratories include IgM ELISA tests, the *Toxoplasma* dye test (which uses live trophozoites to measure antibody-mediated inhibition of intravital dye uptake), a complement fixation test and a haemagglutination test. The dye test is the 'gold standard' against which other tests are compared.

The *Toxoplasma* dye test rises and falls in parallel with the latex agglutination test, but is less prone to false positive results. The complement fixation test has a similar time course. Dye-test antibody titres may reach 1: 32 000–1:64 000 in acute disease.

IgM antibodies indicate recent infection. They persist for 6 months or more, suggesting that the initial infection subsides slowly, as bradyzoite pseudocysts form and are maintained inactive by immune reaction. IgM antibodies sometimes persist for over a year. This is a problem for women who wish to conceive although, if the dye test and latex agglutination titres are falling, the risks of parasitaemia and transplacental infection are almost certainly negligible. Antibody affinity studies can indicate whether the infection is at a late stage; these are reference laboratory tests.

Late toxoplasmosis: eye disease

Choroidoretinitis is a late manifestation of toxoplasmosis, and occurs when acute antibody titres have fallen considerably. Dye test and latex agglutination titres of 1: 256–1:512 (dye test) or 1:128 (latex agglutination) are not uncommon. The haemagglutination test, however, has a slower response to acute infection, and may show significantly elevated titres at the time that eye disease occurs.

Histological diagnosis

A typical histological picture in excised lymph nodes is

strongly suggestive of toxoplasmosis. Giemsa-stained tissue smears, myocardial biopsy or brain biopsy preparations can occasionally be shown to contain cysts or typical crescent-shaped trophozoites.

Culture and PCR

Culture of CSF may be helpful in acute brain infection. This is a specialist procedure, and must usually be prearranged with a reference laboratory. Positive cultures can also be obtained from infected placenta, products of conception, CSF and brain in congenital infection. PCR is now a useful adjunct to diagnosis, especially when biopsy material is available (for instance, brain biopsy material from patients with AIDS).

Diagnosis of toxoplasmosis

1 Serology: IgM ELISA, latex agglutination, complement fixation test (haemagglutination test in choroidoretinitis).
2 Histological appearance of biopsied nodes.
3 Giemsa-stained tissue specimens.
4 Culture of cerebrospinal fluid or affected tissue (rarely performed).
5 Polymerase chain reaction.

Treatment

Toxoplasmosis is self-limiting in most cases, and is often diagnosed relatively late in the acute stage. The risks of treatment must therefore be balanced against the likely benefit. However, treatment is always justified in myocardial or central nervous system disease.

The treatment of choice is a combination of sulphonamide and pyrimethamine. Sulphadimidine is the best choice of sulphonamide available in the UK. The doses should not be less than 1 g 6-hourly. Pyrimethamine is given in a dose of 50 mg daily. Both of these drugs are folate inhibitors, and in these doses can cause a significant fall in the white cell count. The blood count should be closely monitored during treatment, which must often be continued for 4 weeks or more.

Sulphonamides are contraindicated in pregnancy. The treatment of choice is then spiramycin. This is not available in the UK, and must therefore be purchased for the individual patient (see Chapter 17).

Plague

Introduction

Plague, caused by *Yersinia pestis*, is naturally a flea-borne disease of rodents, and exists in many rural and wooded areas throughout the world. When the diseased flea bites, bacteria are regurgitated through its mouthparts, inoculating a large infective dose into its host. As rats die and the fleas lose their preferred hosts, other animals and humans are increasingly bitten, and human cases of plague occur. Urban foci of transmission also exist, where feral animals, humans and rats share the environment (Fig. 25.5). Most human cases are acquired from inoculation via flea-bite transmission or by handling infected rodents. A few cases are acquired by inhalation of organisms, often from an infected human or companion animal. Plague is transmitted from rat to rat by the rat flea, whose pharynx becomes blocked by oedema and replicating bacteria.

Fig. 25.5 Plague hazard warning in an urban setting.

Clinical considerations

The incubation period is usually 2–4 days, but can be up to 12 days. The onset of disease is abrupt, with fever, prostration and rigors. Plague is a toxaemic and bacteraemic as well as a local disease. Its characteristic clinical features are enlarged, suppurating regional lymph nodes (buboes) and haemorrhagic manifestations.

Buboes

The buboes affect the lymph nodes draining the flea-bitten area, though the flea bites are rarely apparent. Inguinal nodes are more often affected than others. The mass of enlarged nodes is surrounded by boggy, often haemorrhagic oedema, and in untreated cases will often point and discharge pus after a week or two.

Skin and mucous membranes

The skin and mucous membranes are often affected by petechial and ecchymotic lesions. In bacteraemic plague, an intense haemorrhagic rash may quickly appear.

Pneumonitis

Pneumonitis is a less common feature, but is often rapidly fatal. Extensive lung involvement and respiratory failure can develop in 24–36 h. Patients with pneumonitis are highly infectious, and readily transmit pneumonic plague to family members and health workers.

Diagnosis

The diagnosis may be suspected on clinical and epidemiological grounds. Gram-stained smears of lymph-node aspirate, pus or infected sputum often show numerous small, Gram-negative rods with characteristic bipolar staining. The organism grows readily in standard cultures and blood culture media. *Y. pestis* should be handled using laboratory containment level 3 (CL 3) precautions.

Treatment

Treatment with broad-spectrum antibiotics is usually highly effective for bubonic plague, though it must be begun promptly, as the disease evolves quickly. Pneumonic plague has a high case fatality rate. Quinolones, aminoglycosides, tetracyclines and chloramphenicol are all effective. Doxycycline or ciprofloxacin is effective prophylaxis for those exposed to cases of pneumonic plague.

Tularaemia

Introduction

This is a disease of rodents and birds caused by *Francisella tularensis*. It is transmitted to humans by inoculation, either by bite or scratch, or by injuries acquired when handling or skinning carcasses. Ticks may also transmit tularemia. It is usually a sporadic disease affecting hunters, trappers and tourists to rural or forested areas.

Cutaneous–lymphatic (ulceroglandular) presentation

A cutaneous–lymphatic presentation is common. A nodular or suppurative lesion develops at the inoculation site, with extension up the lymphatic channels and marked enlargement of draining lymph nodes, which are often very tender and painful. Occasionally, the primary lesion appears as a painful conjunctival ulcer. Lymph-node pathology can exist without a detectable skin lesion.

Typhoidal presentation

A typhoidal presentation is a feature of bacteraemic disease. The patient presents with persisting high fever, but without local features. The severity of the illness varies from persisting fever to prostrating and debilitating disease. Some typhoidal cases develop a widespread pneumonitis which can lead to respiratory failure. Splenomegaly and a transient rash are sometimes seen.

Diagnosis

Diagnosis depends on suspicion, and differentiation of systemic manifestations from brucellosis and typhoid fever. Blood cultures are rarely positive, but organisms can be cultured from skin lesions and lymph-node aspirate or biopsy. Sputum may be positive in cases of pneumonitis.

Serological tests are available at reference laboratory level. False positives and cross-reactions occur, so that interpretation of results must be discriminating.

Treatment

Treatment is always warranted, as there is a mortality of 5–8% in systemic cases. Quinolones or aminoglycosides are the treatments of choice, producing a rapid response in lung and systemic disease. The skin and lymph-node lesions often heal more slowly, even when treated with antibiotics. Tetracyclines or chloramphenicol are also effective, but relapse may follow treatment with these drugs.

Rat bite fever (Haverhill fever)

Introduction

Two organisms may be transmitted by the bites of rats—*Streptobacillus moniliformis*, with an incubation period of 7–10 days, and *Spirillum minus*, with an incubation period of 1–4 weeks. *Streptobacillus moniliformis*, the cause of Haverhill fever, can also be transmitted by contaminated food or milk; milk-borne outbreaks have been described.

Both pathogens cause fever and peripheral rash. The fever tends to be relapsing when caused by *Spirillum minus* and swingeing when caused by *Streptobacillus moniliformis*. The rash may be papular or petechial and occasionally contains small pustules. Arthritis is common in *S. moniliformis* infections. Both types of infection are rare causes of pyrexia of unknown origin with a rash. They are both usually accompanied by neutrophilia. The liver function tests may be slightly abnormal, and prolonged prothrombin times can be demonstrable.

S. moniliformis is a rare cause of endocarditis.

Diagnosis

Diagnosis depends largely on cultures of pus, joint fluid and blood. *S. moniliformis* can be fastidious, but in appropriate culture media produces tangled chains of Gram-negative bacteria. *Spirillum minus* does not grow well in artificial media.

Treatment

Treatment with parenteral benzylpenicillin is effective against both organisms. A course of 1.2 g 6-hourly for 7–10 days is usually sufficient. Streptobacillary endocarditis requires 4–6 weeks' treatment.

Zoonotic streptococcal infections

Introduction

Some animals are colonized and may become infected by pyogenic streptococci which rarely affect humans. Two well-recognized examples of this are *Streptococcus suis*, of pigs, and *S. zooepidemicus*, which can affect horses.

Streptococcus suis

S. suis has an epidemiology in pigs similar to that of the meningococcus in humans. The organism is carried in the nasopharynx, particularly of piglets. When subjected to crowding, or the stress of transport, the animals may develop clinical meningitis. Humans are infected through close contact, usually with pig carcasses, and tend to develop meningitis with bacteraemia.

Gram-positive cocci may be seen in the CSF of human cases. A beta-haemolytic *Streptococcus* is demonstrated on culture, but is Lancefield group R or S rather than A (C or G), as expected. Treatment is with benzylpenicillin, with or without an aminoglycoside. A course of 10–14 days is usually required.

Streptococcus zooepidemicus

S. zooepidemicus causes bacteraemic or soft-tissue infections, usually in people who have close contact with horses. It appears in culture as a beta-haemolytic *Streptococcus* of Lancefield group C. The treatment of choice is benzylpenicillin plus an aminoglycoside.

Herpesvirus simiae infection

Herpesvirus simiae inhabits the mouth and mucocutaneous borders in monkey species, in a similar way to which herpes simplex virus affects humans. It causes cold-sore lesions in a proportion of affected animals. Monkey-handlers can become infected if the animal's saliva is inoculated via a bite or scratch.

Human infections may be local, producing persistent herpetic vesicles, but there is a severe risk of potentially fatal viral encephalitis. Treatment with aciclovir is effective in suppressing the skin lesions, but relapse often follows cessation of therapy. Affected individuals therefore often require long-term aciclovir treatment.

The nature of a herpetic lesion in a monkey-handler can be confirmed by isolation and characterization of the virus from vesicle or skin scrapings. *H. simiae* is a hazard category 4 virus for which there is no reliable treatment or prophylaxis. Diagnostic culture is therefore carried out in reference laboratories such as the Virus Reference Division of the Central Public Health Laboratory in London, or the Centers for Disease Control in Atlanta, Georgia, USA, where containment level 4 facilities exist.

Zoonotic paramyxoviruses

Hendra virus

Hendra virus caused severe systemic and respiratory disease in horses in two sites in Australia in the late 1990s. Human stable workers were infected, probably by contact with respiratory secretions of sick horses, with at least one death. The reservoir of virus was found to be in fruit bat colonies.

Nipah virus

Epidemics of encephalitic disease in western Malaysia in 1999 were at first thought to be Japanese encephalitis but, after 80% of sufferers were found negative for this virus, cultures revealed a paramyxovirus related to Hendra virus. This virus was found to be widespread in herds of pigs, on which the area's economy was based. Many patients were treated with tribavirin, though the effectiveness of this is so far unproven, and many herds of infected pigs were slaughtered, with a consequent reduction in human cases. Smaller outbreaks of Nipah virus infection have been described or suspected in Singapore and Indonesia. The reservoir of infection has not yet been identified.

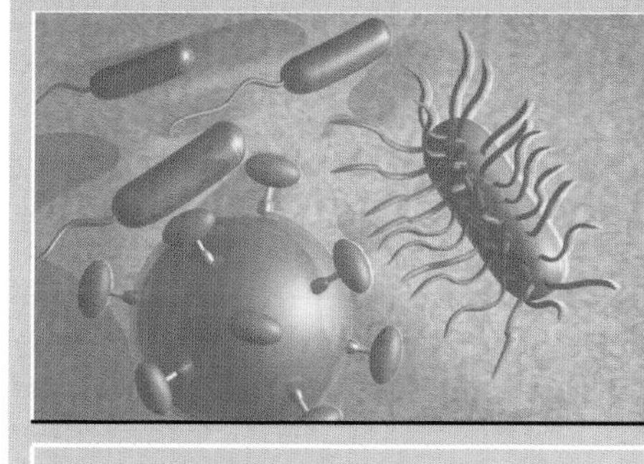

26 Control of Infection in the Community

Communicable disease and the law, 473
Role of national agencies, 473

Communicable disease surveillance, 474
Principles and practice of surveillance, 474
Sources of data, 474
Dissemination of information, 476
Surveillance in other countries, 476

Prevention and control of communicable disease, 476
Social and environmental factors, 476
Health education, 478
Food safety, 479
Vector control, 479
Immunization, 479
Contact tracing, 482
Chemoprophylaxis, 482
Screening, 482
Outbreak investigation, 483

Communicable disease and the law

The legal framework for control of infection in the community has developed over the past 150 years. In the UK, much of the old legislation was brought together under the Public Health (Control of Disease) Act (1984) and the Public Health (Infectious Diseases) Regulations (1988). These acts do not apply in Scotland, although similar legislation does exist.

Responsibility for the control of communicable disease was vested in local authorities during the 19th century, when the post of medical officer of health (MOH) was established. When the National Health Service was created, the MOH, who was employed by the local authority, retained responsibilities for control of infection and related activities such as immunization. The post of MOH was abolished in 1974; however, some residual functions including communicable disease control were retained by local authorities.

Each local authority has a designated proper officer (in Scotland, medical officer) with statutory powers for prevention and control of infection. These statutory powers relate to the notifiable diseases (see p. 475) and to certain other diseases, usually defined. The proper officer is usually (but not always) a consultant in communicable disease control (CCDC) who is a health authority employee acting on behalf of the local authority. The post of CCDC is sometimes referred to as MOEH (medical officer for environmental health). In Scotland, the equivalent post is the consultant in public health (communicable disease and environmental health).

In addition to the CCDC in the community, hospitals have a control of infection officer who is often the local microbiologist. Air and sea ports have a medical officer who liaises with the local CCDC.

Functions of the consultant in communicable disease control

1 Surveillance of communicable diseases.
2 Development of policies for control of communicable disease.
3 Investigation and control of outbreaks.
4 Coordination of immunization and acquired immunodeficiency syndrome prevention programmes.
5 Provision of advice on prevention and control.

Role of national agencies

In England and Wales, the Communicable Disease Surveillance Centre (CDSC) was established in 1979 as part of the Public Health Laboratory Service (PHLS). The laboratory network of the PHLS had existed for many years, and the government inquiries that followed two incidents of laboratory-acquired smallpox infection in the 1970s recommended the creation of an epidemiological unit within the PHLS that could coordinate the management of major incidents. The functions of the CDSC are:

1 Surveillance of communicable diseases and of immunization programmes.
2 Investigation of outbreaks.
3 Epidemiological research.
4 Training in the epidemiology and control of communicable disease.

In Scotland, the equivalent agency is the Scottish Centre for Infection and Environmental Health (SCIEH). Most countries have similar organizations, for example the Centers for Disease Control and prevention in the USA. In some countries the functions are performed by the Ministry of Health.

The Office for National Statistics (ONS) collects vital statistics (births, marriages and deaths), provides population estimates and projections and administers the notification system (see p. 475). The equivalent organization in Scotland is the General Register Office (Scotland).

The Ministry of Agriculture, Fisheries and Food (MAFF; Department of Agriculture and Fisheries in Scotland) is responsible for the control of infection in animals. Close collaboration between MAFF, CDSC/SCIEH and CCDCs is important in relation to the control of zoonotic infections. The Food Standards Agency (FSA) is responsible for overseeing the safety of food production, manufacture and handling up to the point of sale.

Communicable disease surveillance

Principles and practice of surveillance

Surveillance of disease is the continuous systematic collection and analysis of relevant morbidity and mortality data. The primary aim of surveillance is to identify trends or clusters of disease which require preventive action; thus surveillance is *information for action*. Surveillance is also used to evaluate control measures such as vaccination programmes, to plan resource allocation and to provide baseline epidemiological information for research workers.

The process of surveillance involves four basic steps:

1 Collection of data.

2 Analysis of data to provide statistics.

3 Interpretation of statistics to provide meaningful information.

4 Dissemination of narrative reports to those who need to know.

An ideal surveillance system should meet the following criteria:

1 The disease under surveillance must be of sufficient importance to justify the resources required to undertake effective monitoring.

2 The surveillance should be timely—the data should be collected and interpreted sufficiently quickly to enable effective control measures to be taken. This is particularly important for communicable diseases, where large outbreaks can occur very suddenly and cause considerable morbidity.

3 The data collected should be representative of the total population. In practice this is difficult to achieve as surveillance systems usually build on routine data sources from atypical populations such as hospital inpatients. The role of the epidemiologist is to understand and take account of the biases inherent in surveillance systems when interpreting the data.

4 It should be consistent over time and between geographical areas. If the proportion of cases detected by a surveillance system is not constant, it becomes difficult to interpret any changes in reported disease incidence. During periods of increased disease incidence the efficiency of reporting tends also to increase. A change in the criteria for reporting a disease can have a large effect on reported incidence. For example, the inclusion of patients with a CD4 count of less than 200/μl in the case definition for acquired immunodeficiency syndrome (AIDS) resulted in a 200% increase in the cumulative total of cases reported in the USA.

5 Reporting should be complete wherever possible.

6 It should be simple, in both structure and ease of operation.

7 It should be flexible, thus capable of adapting to changing information needs (for example, emergence of a new disease).

Criteria for an effective surveillance system

1 Disease under surveillance is of sufficient public health importance.

2 Timeliness.

3 Representative of the total population.

4 Consistency over time and between geographical areas.

5 Completeness of reporting.

6 Simplicity.

7 Flexibility.

A surveillance system may be active, passive or stimulated-passive. A passive system relies on routinely collected data, e.g. notifications of infectious disease, where cases are reported without any specific encouragement or inducement. When special efforts are made to improve reporting to a passive surveillance scheme, e.g. by written reminders or following up non-reporting centres, the scheme becomes stimulated-passive. An active surveillance system is where all potential reporting units are contacted at regular intervals and specifically asked to report the condition under surveillance; in addition, if no cases have been seen, a 'negative' report is requested. The British Paediatric Surveillance Unit (BPSU; see p. 476) is a good example of an active surveillance system.

In general, passive surveillance systems are used for common or less severe diseases such as food poisoning where it is not essential (or possible) to ascertain every case. Stimulated-passive or active reporting is usually only necessary for rare or serious conditions, or those where a public health programme of elimination is planned, for example poliomyelitis.

Sources of data

Reports of outbreaks and other infectious disease incidents are often provided on an *ad hoc* basis. There are however, many sources of data, routine and non-routine, that constitute the formal communicable disease surveillance network.

Routine sources of data on infectious disease in the UK

1 Statutory notifications to the consultant in communicable disease control.
2 Laboratory reports.
3 General practitioner reporting schemes.
4 Hospital inpatient and outpatient statistics.
5 Returns from sexually transmitted disease clinics.
6 Death certificates.
7 School medical officers.
8 Occupational health departments.

Routine data sources

There are several routine sources of data on infectious disease in the UK. The most important are notifications, laboratory reports, general practitioner surveillance schemes, hospital data, clinic returns and death certificates.

Notification

Notification is a statutory requirement. All doctors are required to notify any cases of specified infections seen by them to the designated proper officer for the relevant local authority. Notification is based on clinical suspicion and does not require laboratory confirmation, although if the diagnosis is altered as a result of laboratory investigations, the notification can be corrected by the notifying doctor. A small fee is payable for each notification.

Notifiable diseases in the UK (differs in Scotland):

1 Acute encephalitis.
2 Acute poliomyelitis.
3 Anthrax.
4 Cholera.
5 Diphtheria.
6 Dysentery (amoebic or bacillary).
7 Leprosy.
8 Leptospirosis.
9 Malaria.
10 Measles.
11 Meningitis.
12 Meningococcal septicaemia (without meningitis).
13 Mumps.
14 Ophthalmia neonatorum (neonatal ophthalmia).
15 Paratyphoid fever.
16 Plague.
17 Rabies.
18 Relapsing fever.
19 Rubella.
20 Scarlet fever.
21 Smallpox.
22 Tetanus.
23 Tuberculosis.
24 Typhoid fever.
25 Typhus.
26 Viral haemorrhagic fever.
27 Viral hepatitis.
28 Whooping cough.
29 Yellow fever.

Weekly summaries of notifications are forwarded from local authorities to the ONS where they are collated into weekly, quarterly and annual reports. These reports are transmitted to the national agencies for communicable disease surveillance (CDSC and SCIEH).

The notification system has developed over more than 100 years. Many of the conditions are of historical interest only, whereas others, e.g. legionellosis, are noticeable by their absence. A local authority has the discretion to add conditions to the list that occur in that authority; addition of a disease to the national list requires approval by the Secretary of State.

Laboratory reports

Laboratory reports provide reliable information on many infections. These are provided on a voluntary basis by National Health Service, public health and private laboratories to the CDSC and SCIEH. The main limitation of laboratory data is that they are biased towards patients with severe infections or where laboratory confirmation is likely to influence management. For some conditions, especially many viral infections, laboratory reports do not therefore provide a representative picture of infection in the community.

General practitioner surveillance schemes

The Royal College of General Practitioners (RCGP) operates a national network of 'spotter' practices, serving a population of approximately 750 000, who provide weekly reports of first-time consultations on a variety of infectious and non-infectious conditions. It provides a particularly sensitive index of influenza activity.

In addition to the RCGP scheme, local general practitioner surveillance networks operate in several parts of the country.

Hospital data

Hospital data provide useful information on infections that usually require admission, such as meningococcal meningitis. They tend, however, to be incomplete and out of date.

Clinic data

Clinic data are useful for certain conditions, particularly sexually transmitted diseases. Annual returns of numbers of diagnoses from sexually transmitted disease clinics are collected by CDSC.

Death certificates

Death certificates have a limited contribution to surveillance, as deaths from infection are rare. However, the ratio of deaths to cases (the case fatality ratio) can provide useful information on the changing severity of some diseases and the effectiveness of new treatment measures.

Other useful routine sources of data include the Medical Officers of Schools Association (MOSA) which reports incidents in boarding schools, NHS direct and some occupational health departments.

Special surveillance schemes

For some infections, routine data are not available and special surveillance systems have been established. Examples in the UK include the confidential reporting of AIDS and human immunodeficiency virus (HIV)-related disease to the CDSC and SCIEH, the National Congenital Rubella Registry at the Institute of Child Health, and the British Paediatric Surveillance Unit (BPSU). Under the BPSU scheme, all consultant paediatricians are sent a monthly card with a menu of about 12 reportable conditions. Any cases seen are then followed up by the lead investigator for that condition. Conditions may be added to or deleted from the menu on application to the BPSU. The BPSU scheme has been particularly useful for ascertaining cases of rare infectious disorders such as subacute sclerosing panencephalitis.

Dissemination of information

Notification data are published in the weekly *Communicable Disease Report* (CDR), which is published by the CDSC and contains both laboratory and notification data. The CDR is distributed to CCDCs, microbiologists, infectious disease physicians and others with an interest in communicable disease control. Similar bulletins are published in other countries, for example the *Morbidity and Mortality Weekly Report* (MMWR) in the USA.

Surveillance in other countries

Most countries operate a notification system for communicable diseases, although the list of reportable conditions varies widely from one country to another. In North America, the use of standardized case definitions has been widely adopted. In some countries, telephone reporting is used. In France, general practitioners report through minicomputer terminals linked via the national telephone network. Pan-European surveillance is now being developed for a number of infections including AIDS, travel-associated legionellosis, meningococcal infection and influenza.

Prevention and control of communicable disease

Prevention may be primary, secondary or tertiary. Primary prevention aims to prevent or reduce exposure to the infectious agent. This is the most effective method, but also the most difficult to achieve. The purpose of secondary prevention is to detect infection at an early stage, so that control measures to prevent further spread can be taken. Much of the day-to-day work of a CCDC is secondary prevention. In tertiary prevention, the aim is to minimize the disability arising from infection.

There is a wide range of measures that may be taken to prevent or control infection in the community and in hospitals (Table 26.1). Some of these measures may be used for more than one type of prevention. Control procedures that relate mainly to the control of infection in the community are discussed in this chapter; for the control of infection in hospitals see Chapter 23.

Social and environmental factors

Although infection is still an important clinical problem in developed societies, many of the more serious diseases that were common in the past have largely been brought under control. In contrast, infectious disease remains a major cause of morbidity and mortality in developing countries.

The main factors in reducing the burden of infectious disease in Western society were the improvements in social and environmental conditions that took place in the late 19th and early 20th century. Public health legislation forced industry and local government to spend money on sanitation and better housing. The average family size

Table 26.1 Prevention and control of infection.

Primary prevention	Secondary prevention	Tertiary prevention
Immunization (pre-exposure)	Immunization (postexposure)	Effective treatment of acute infection*
Improved housing and sanitation	Contact tracing	Management of postinfectious disorders†
Provision of safe food pasteurization of milk	Screening of food handlers, health workers, etc.	Physiotherapy, speech therapy
Vector control	Chemoprophylaxis	
Behaviour modification (sexual, hygiene, etc.)	Effective surveillance	
Isolation, barrier nursing‡	Outbreak investigation and management	
Disinfection, sterilization‡		
Laboratory safety‡		

* See Chapter 4.
† See Chapter 21.
‡See Chapter 23.

Fig. 26.1 Consumption of untreated water in a developing country: a major source of infection. Courtesy of SmithKline Beecham.

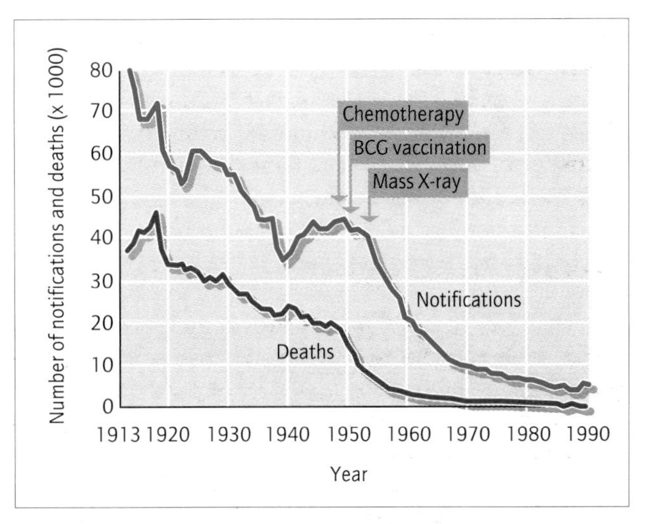

Fig. 26.2 Respiratory tuberculosis in England and Wales from 1913 to 1998. The disease started to decline long before chemotherapy and other medical interventions became available, due to improvements in living conditions.

shrank rapidly during the early part of the 20th century, resulting in less crowding. Better nutrition meant that the population was less susceptible to disease.

The most important environmental measures are provision of adequately treated drinking water and safe disposal of faeces. Water-borne infection accounts for more deaths in developing countries than any other disease (Fig. 26.1). The recent spread of cholera throughout South America illustrates the vital importance of basic sanitation. Even in the UK, water-borne outbreaks, for example of cryptosporidiosis, are relatively common, due to treatment failures and post-treatment contamination of water supplies.

Better housing conditions have made a major contribution to controlling respiratory infections such as tuberculosis that are spread by close person-to-person contact. The reduction in tuberculosis illustrates the relative importance of environmental conditions compared to more high-tech prevention methods. Cases of respiratory tuberculosis have markedly declined since reliable records began last century. The introduction of mass radiography, chemotherapy and bacillus Calmette–Guérin (BCG) vaccination in the second half of this century has made virtually no difference to the rate of decline (Fig. 26.2).

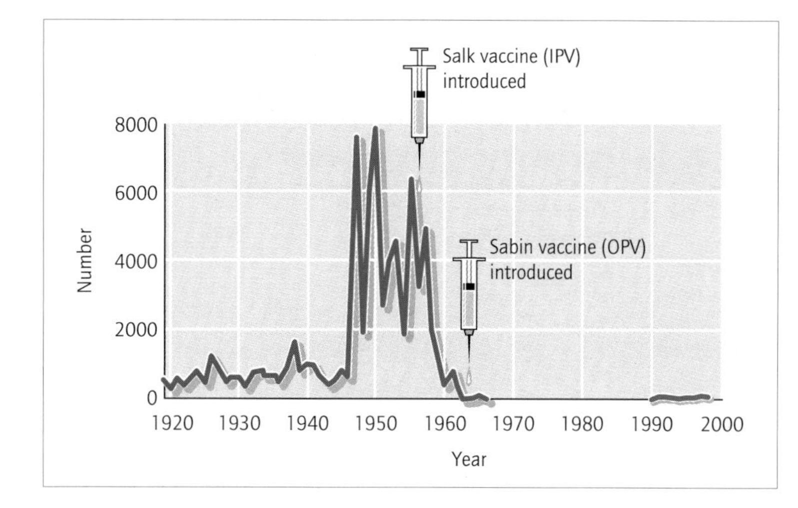

Fig. 26.3 Paralytic poliomyelitis in England and Wales from 1914 to 1998. The outbreaks in the 1940s and 1950s were associated with an increasing age at infection, with a corresponding increase in the ratio of paralytic to non-paralytic cases. IPV, inactivated poliovaccine; OPV, oral poliovaccine.

Paradoxically, the burden of some infectious diseases actually rises as living conditions improve. This applies to conditions where the rate of complications is greater in adults than in children. The epidemics of paralytic poliomyelitis that occurred in the 1940s and 1950s are attributed to improved sanitation with a consequent reduction of wild virus circulation in young children (Fig. 26.3). This led to an increase in the average age at which infection occurred. As the ratio of paralytic to non-paralytic cases rises with age, the number of paralytic cases actually increased. A similar phenomenon has recently been observed for hepatitis A in some countries, where the number of cases with jaundice is rising as the average age at infection increases.

Another infection to emerge as result of modern living is legionnaires' disease. The causal agent, *Legionella pneumophila,* thrives in microenvironments such as showerheads, cooling towers and air-conditioning units. Outbreaks of legionnaires' disease occur when bacterially contaminated aerosols are generated from these water systems, particularly when the systems are not properly maintained (Fig. 26.4).

Fig. 26.4 Inside view of a cooling tower—a potential source for *Legionella pneumophila*. This badly maintained tower was the source of a large outbreak of legionnaires' disease.

Health education

Many health education programmes have been conducted at local and national levels with the aim of reducing exposure to infectious diseases. These include safe sex campaigns, needle exchange schemes, advice to pregnant women, guidance on food hygiene and advice to travellers. These campaigns have been conducted by government-funded bodies such as the Health Education Authority, voluntary agencies and industry (especially in relation to food hygiene). At the local level many individ-

Table 26.2 Milk-borne outbreaks of salmonellosis in England, Wales and Scotland, from 1980 to 1984, showing the impact of compulsory pasteurization in Scotland in 1983.

	England and Wales		Scotland	
Time period	No. of outbreaks	No. of cases	No. of outbreaks	No. of cases
1980–1982	40	540	21	1090
1983–1984	22	518	8	46

uals and agencies may be involved, including CCDCs, general practitioners, health promotion departments, health visitors and voluntary groups.

With a few notable exceptions, health education has met with limited success in preventing exposure to communicable disease. Research has shown that campaigns often succeed in raising public awareness, but seldom result in behaviour modification. Where behaviour modification does occur, it is often short-lived. Many infectious diseases are considered by the public to be relatively unimportant. The smaller the perceived risk of infection, the less likely that an education programme will succeed. The advent of AIDS, together with the high media profile for diseases such as food poisoning, legionellosis and meningococcal meningitis, has, however, shifted the public perception of infectious disease in recent years.

In general, the education programmes most likely to succeed are those that involve the local community in their design and implementation, and are ongoing rather than one-off events.

Food safety

Salmonellosis and other bacterial causes of food poisoning have increased considerably in recent years (see Chapter 8). This has attracted considerable public attention and led to a major review of the legislation governing food safety, culminating in the Food Safety Act (1990). Food law has now been harmonized across the European Community since the introduction of the single European market.

Much of the responsibility for enforcement of food safety legislation lies with environmental health officers (EHOs) who are employed by local authorities to inspect food premises. MAFF officers carry out enforcement duties on dairy farms. This includes the extensive legislation which has recently been introduced to control *Salmonella* infection. The Food Standards Agency has been recently set up to avoid the situation where MAFF is responsible for both food production and safety.

Arthropod	Diseases
Mosquito	Malaria, dengue fever, filiariasis, yellow fever
Sandfly	Leishmaniasis, sandfly fever
Fly	Trypanosomiasis
Flea	Plague, rickettsial fevers, tungiasis
Tick	Relapsing fever
Mite	Scabies, typhus
Maggot	Myiasis
Louse	Pediculosis, relapsing fever, typhus

Table 26.3 Arthropods of medical importance.

Pasteurization of milk greatly reduces the risk of exposure to many pathogens such as *Mycobacterium bovis* and *Campylobacter* sp. In Scotland it is illegal to produce or sell unpasteurized milk, whereas in England and Wales it is not. The ban in Scotland has led to a substantial reduction in the number of milk-associated outbreaks compared to England and Wales (Table 26.2).

Vector control

This is particularly important in tropical countries where arthropods play an important role in many infections, both as vectors and as primary causal agents (Table 26.3).

Travellers to the tropics can reduce the risk of infection by taking measures to avoid insect bites using, for example, repellents such as diethyltoluamide, mosquito nets and protective clothing. Attempts to control insect populations by the use of pesticides have usually been unsuccessful due to the emergence of resistance.

Immunization

The introduction of effective immunization programmes has been one of the most significant public health achievements this century. The control of smallpox, polio and diphtheria would not have been possible without the development of immunizing agents against these diseases.

Vaccines and immunoglobulins

Immunization may be achieved passively by administration of an immunoglobulin preparation, or actively by use of a live or non-replicating vaccine.

Immunoglobulins

Immunoglobulins are prepared from plasma treated by ethanol fractionation. They provide short-term protection against certain infections and are also used in the treatment of immune disorders and to supplement antiviral therapy, especially in the immunocompromised (see Chapter 22). The most commonly used preparation is human immunoglobulin (HIG) which is prepared from pooled plasma and therefore contains antibodies to viruses that are prevalent in the general population. HIG is mainly used for postexposure prophylaxis of acute hepatitis A. It is also given to travellers to countries where hepatitis A is endemic; however, as protection is relatively short-lived and active immunization is now available, its role in pre-exposure prophylaxis is diminishing. The other main indication for HIG is for postexposure prophylaxis against measles in immunosuppressed contacts.

Other immunoglobulins are available for postexposure management of specific infections. These are prepared from hyperimmune donors. Hepatitis B immunoglobulin, in combination with active immunization, is used for postexposure prophylaxis following accidental exposure to infected blood, and for babies born to acutely or chronically infected mothers. Varicella zoster immunoglobulin is indicated for susceptible immunosuppressed and pregnant contacts of chickenpox or herpes zoster, for neonates whose mothers develop chickenpox in the period 7 days before to 7 days after delivery, and for susceptible neonates in contact with chickenpox or zoster. Tetanus immunoglobulin is used in the management of tetanus-prone wounds in patients who have not been immunized or in whom the last dose of vaccine was given more than 10 years previously. Tick-borne encephalitis immunoglobulin is available in countries where the disease is endemic, notably Austria, for prophylaxis following tick bites. Rabies immunoglobulin is indicated for prophylaxis following warm-blooded animal bites in countries where the disease is endemic in the animal population.

Immunoglobulins and their use for prophylaxis of infection

1 *Human immunoglobulin*: pre-exposure prophylaxis of hepatitis A for short-term travel abroad (2 months or less), adult, 250 mg; child under 10 years, 125 mg; longer-term travel (3–5 months) and for postexposure prophylaxis of hepatitis A 500-mg; child under 10 years, 250 mg.
Postexposure prophylaxis of measles: child under 1 year, 250 mg; 1–2 years, 500 mg; 3 years and over, 750 mg; to allow an attenuated attack, child under 1 year, 100 mg; 1 year and over, 250 mg.
2 *Hepatitis B immunoglobulin*: for postexposure prophylaxis, adult, 500 IU; child under 4 years, 200 IU; 5–9 years, 300 IU; neonate 200 IU as soon as possible after birth.
3 *Human rabies immunoglobulin*: for postexposure prophylaxis, 20 IU/kg, by infiltration around the wound and any remaining by i.m. injection.
4 *Tetanus immunoglobulin*: for postexposure prophylaxis, 250 IU, increased to 500 IU if more than 24 h have elapsed or if there is risk of heavy contamination.
5 *Varicella zoster immunoglobulin*: for postexposure prophylaxis (as soon as possible and not later than 10 days after exposure), child up to 5 years, 250 mg; 6–10 years, 500 mg; 11–14 years, 750 mg; over 15 years, 1 g; second dose required if further exposure occurs after 3 weeks.
6 *Tick-borne encephalitis immunoglobulin*: this is available for postexposure prophylaxis after a tick bite in an endemic area.

All of the above preparations should be given intramuscularly.

Vaccines

Vaccines are derived from whole viruses and bacteria, or their antigenic components. Live vaccines are prepared from attenuated strains that have minimal pathogenicity but are capable of inducing a protective immune response. They multiply in the human host and provide antigenic stimulation over a period of time. This results in durable immunity, usually after a single dose. Vaccine failures are uncommon, and are usually the result of inadequate handling or administration. The main disadvantage of live vaccines is that they occasionally cause a full-blown infection in the recipient. This is most likely to occur in the immunosuppressed patient; live vaccines are usually contraindicated for this group. Their use in pregnancy should also be avoided because of the potential risk of fetal infection.

Non-replicating vaccines contain either inactivated

whole organisms or antigenic components. Increasingly sophisticated methods are being used to produce these vaccines, such as protein–polysaccharide conjugation (*Haemophilus influenzae*) and genetic expression of protective antigens (hepatitis B). Because replication does not take place in the human host, non-replicating vaccines are safe for use in pregnancy and in the immunosuppressed. Their disadvantage is that more than one dose is usually required for protection. Local reactions are relatively common with non-replicating vaccines; this is related to the quantity of antigen they contain.

Types of vaccines, their modes of action and contraindications

Live: examples—bacillus Calmette–Guérin (BCG), measles/mumps/rubella, oral polio, yellow fever

1 Multiply inside the human host and provide continuous antigenic stimulation over a period of time.
2 Provide durable immunity, usually after a single dose.
3 Contraindicated in pregnancy and the immunosuppressed.

Non-replicating: examples—pertussis, influenza, Haemophilus influenzae type b, rabies, typhoid

1 Do not multiply inside the human host.
2 Antibody response is related to the antigen content and potency.
3 Multiple doses are often required, with subsequent booster doses.
4 No general contraindications.

Strategic aspects of immunization programmes

The aim of an immunization programme may be eradication, elimination or containment.

Eradication is total absence of the organism in humans, animals and the environment. Once a disease has been eradicated, the immunization programme can be discontinued. The only disease that has been eradicated by immunization is smallpox. Smallpox had many features that favoured eradication—an easily recognizable illness with no subclinical or latent infection, no long-term carriers, visible evidence of immunity (a characteristic scar), absence of non-human hosts, low infectivity and a long incubation period. Poliomyelitis shares many of these characteristics, and the World Health Organization now intends to eradicate polio globally during the year 2000. Elimination is where the disease has disappeared, but the organism remains in animal hosts, the environment or causing subclinical infection in humans. Unlike eradication, it is not possible to discontinue immunization. Containment is the point at which a disease, although not eliminated, is no longer considered to be a significant public health problem.

Possible aims for an immunization programme, with examples of diseases for which these have been achieved

1 Eradication: removal of the causal agent (e.g. smallpox).
2 Elimination: absence of disease, although the causal agent remains (e.g. polio).
3 Containment: reduction of disease to the point at which it is no longer a public health problem (e.g. *Haemophilus influenzae* type b).

There are two basic approaches to immunization programmes: universal or selective. Universal immunization has been adopted for most of the childhood vaccines. A selective programme aims to protect only those at risk from disease. This is less expensive than universal immunization, and tends to be used for the more costly vaccines such as hepatitis B. In practice it is often difficult to identify and immunize those who are genuinely at risk.

Immunization schedules

The ages at which vaccines are given, and the preparations used, vary considerably from one country to another. The approach in the UK has been to minimize the number of clinic visits and to secure protection as early in life as possible, without compromising efficacy. The currently recommended schedule is shown in Table 26.4. The British schedule is similar to that used in many European countries and in the USA.

Surveillance of immunization programmes

The ingredients for a successful immunization programme are a safe, effective vaccine and high coverage (uptake) in the target population. The safety and efficacy of vaccines are established in clinical trials before they are licensed. After licensing, batches of all vaccines are regularly tested for potency and toxicity at the National Institute for Biological Standards and Control (NIBSC) before release. Any severe or unusual reactions to vaccines should be reported on a yellow card to the Committee on Safety of Medicines. Additional surveillance schemes

Age	Vaccines
2 months	Diphtheria/tetanus/pertussis (DTP) *Haemophilus influenzae* type b (Hib) Meningococcus group C Oral poliovaccine (OPV)
3 months	DTP Hib Meningococcus group C OPV
4 months	DTP Hib Meningococcus group C OPV MMR
12–15 months	Measles/mumps/rubella (MMR)
4–5 years	Diphtheria/tetanus (DT) OPV
10–14 years*	Bacillus Calmette–Guérin (BCG)
15–18 years	Tetanus/low-dose diphtheria (Td) OPV

* In some parts of the country BCG is given in the neonatal period.

Table 26.4 British childhood immunization schedule (2000).

have been established to monitor the efficacy and safety of some vaccines, e.g. BCG. Annual serological surveys of age-specific antibody prevalence to measles, mumps and rubella (MMR) are undertaken by the CDSC to monitor the impact of the MMR vaccine. Coverage of vaccines is assessed from annual returns to the health departments and the COVER (cover of vaccination evaluated rapidly) scheme, which is run by the CDSC.

The target coverage for childhood vaccines is 95% at 2 years of age. Most immunizations are given by general practitioners, who are paid according to whether they achieve targets. Each health district has a designated immunization coordinator, who is usually either a community paediatrician or the CCDC, with local responsibility for management of the programme. Vaccine coverage has improved considerably in recent years although there has been a small decline in MMR coverage since the late 1990s (Fig. 26.5).

Contact tracing

The principle of contact tracing is to identify those who have been in contact with an infectious disease in order that preventive measures can be taken. These measures may include screening for evidence of infection and subsequent treatment, active or passive immunization and chemoprophylaxis (see below).

For contact tracing to be successful, speed of notification of the index case and follow-up of contacts are of the essence. This poses particular difficulties for sexually transmitted diseases, where patients may be unwilling to notify their partners. Contact tracing is most likely to be effective for infections with longer incubation periods and where adequate chemotherapy is available.

Chemoprophylaxis

This is used for control of more serious infections such as diphtheria and meningococcal disease. It is also sometimes used during influenza outbreaks among high-risk patients, such as elderly nursing-home residents. The aim of chemoprophylaxis may be either to eliminate carriage of pathogenic organisms, reducing the risk of infection in those not yet exposed, or to treat newly acquired infection in contacts who are non-immune and may be incubating the disease. The former requires prophylaxis of a large contact network, whereas the latter can be achieved by treating only those who have been in close contact with the index case. Both strategies can be difficult to implement as they require healthy individuals to take antibiotics which may produce side-effects. Single-dose regimens are the most effective. It is important to determine the antibiotic sensitivity of the strain from the index case, as this may influence the agent selected for chemoprophylaxis.

Example of a chemoprophylaxis regimen, in this case for meningococcal disease

Rifampicin 600 mg orally every 12 h for 2 days; child 10 mg/kg (3 months–1 year, 5 mg/kg) every 12 h for 2 days; contraindicated in pregnancy, liver disease and alcoholism.

Alternative: ciprofloxacin 500 mg single oral dose (adults and children over 10 years only); contraindicated in pregnancy.

Screening

A screening programme should fulfil the following criteria. First, the disease should be an important public health problem. Second, there should be a recognizable latent or early symptomatic stage. Third, the screening test should be harmless, sensitive and specific. Finally, effective treatment should be available for the condition in question, with general agreement on who should be treated.

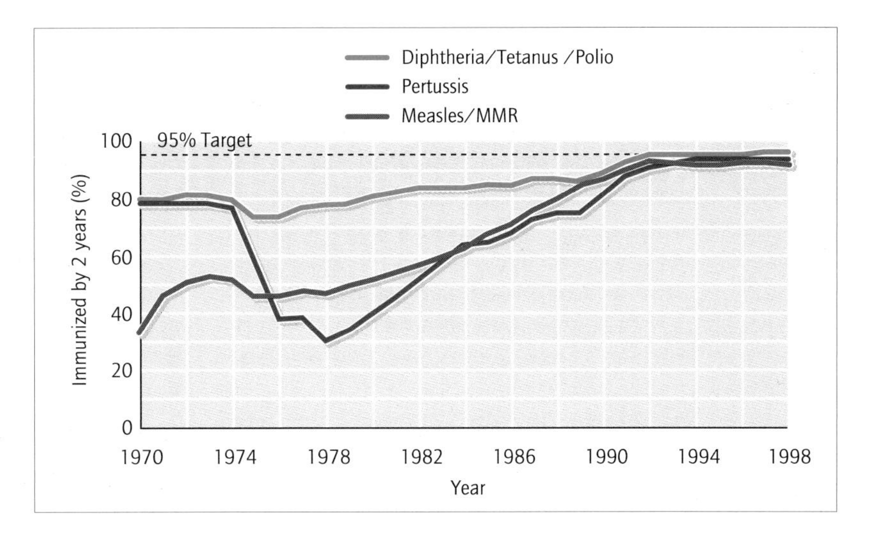

Fig. 26.5 Vaccine coverage in England and Wales from 1970 to 1998.

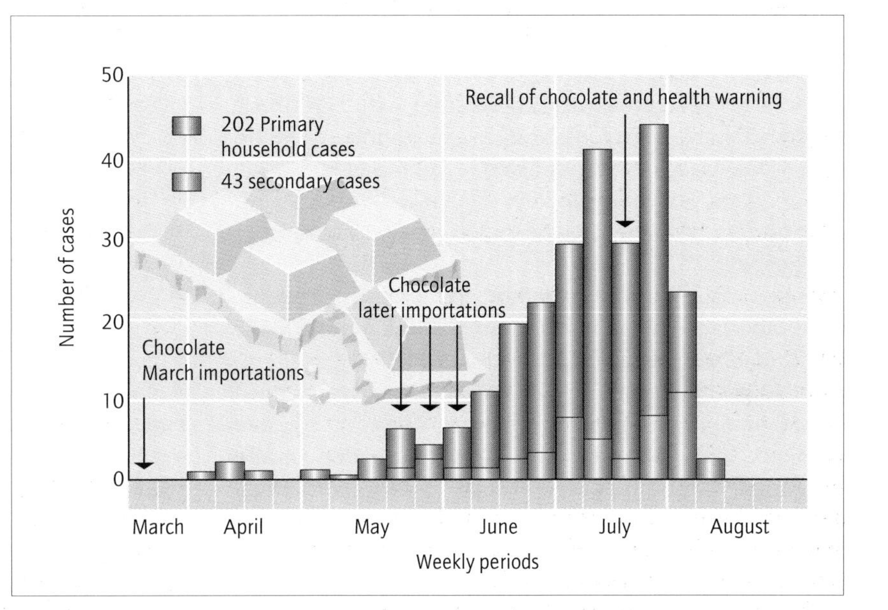

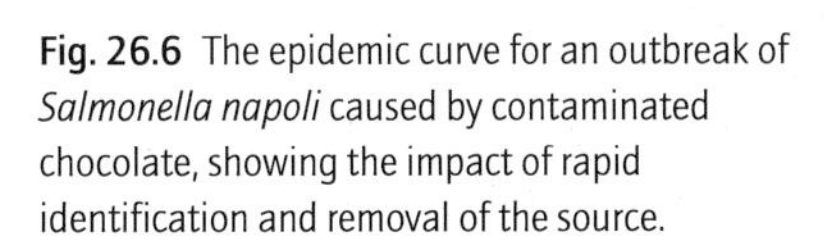

Fig. 26.6 The epidemic curve for an outbreak of *Salmonella napoli* caused by contaminated chocolate, showing the impact of rapid identification and removal of the source.

It is rare that these criteria can be met. Screening for infectious disease is often adopted in response to public and political pressure or for medicolegal reasons rather than on the basis of sound public health. For example, it is commonplace to exclude food-handlers with recent *Salmonella* infection until three consecutive negative stool samples are obtained. This results in unnecessary time off work and contributes little to the control of food poisoning. Screening for tuberculosis by mass radiography has been discontinued, although tuberculin testing is still widely used for screening of healthcare workers and teachers. The most effective screening programmes have been those aimed at preventing congenital rubella and syphilis (see Chapter 17).

Outbreak investigation

Although there is no legal requirement for CCDCs to investigate outbreaks, there is no doubt that prompt identification and control of incidents can prevent substantial morbidity. For example, it was estimated that 30 000 cases of salmonellosis were averted following the investigation of a single outbreak which was caused by contaminated imported chocolate (Fig. 26.6).

Methods

Prompt recognition of the outbreak is a prerequisite. For this, adequate surveillance must be in place. When apparent clusters are identified, it is important to distinguish between true outbreaks and those caused by reporting artefacts. A knowledge of the expected incidence for the time of year and place is necessary. Factors that may cause a pseudo-outbreak include the availability of a new or more sensitive laboratory test, the appointment of a physician with a particular disease interest and undue media involvement.

The next step is to confirm the diagnosis. This may seem obvious, but is often overlooked. In 1985, a large outbreak of legionnellosis associated with a hospital outpatient department was initially labelled as epidemic influenza until appropriate laboratory investigations were undertaken. Once the diagnosis has been confirmed, a case definition should be developed. Initially this should be non-specific so that all suspected cases can be identified and investigated; later on in the investigation a more rigorous definition, often with laboratory confirmation, will be required. This case definition should then be the basis for collecting and counting cases. Collection of cases will often involve active case finding; for example, by phoning general practitioners, laboratories and hospitals.

At this stage of the investigation, the aim is to formulate a hypothesis that explains the source of the outbreak, the mode of transmission and the duration. The basic epidemiological information to be collected on all cases must include date of onset of symptoms, age/sex and place of residence (time, person, place). This information should be tabulated and an epidemic curve drawn. Additional information required will depend on the nature of the outbreak. For example, a salmonella incident will involve obtaining food consumption histories, whereas in an outbreak of legionnaires' disease details of likely exposures should be sought, such as foreign travel. These preliminary enquiries should be carefully conducted as face-to-face interviews, ideally with cases affected early in the outbreak.

The hypothesis generated should then be tested by a formal analytical study. This may either be a case-control or cohort study. In a case-control study, exposure histories are sought from cases and healthy controls. The relative risk of exposure to the postulated source of the outbreak is then calculated for cases and controls. The controls must be drawn from the same population as the cases, so that they have had the same opportunity for exposure to the source. Case-control studies are suited to investigation of uncommon infections such as botulism. Their main disadvantage is the potential for bias arising from selection of controls.

In a cohort study, the disease outcome is compared between those exposed and not exposed to the source. Cohort studies are often used for the investigation of outbreaks with high attack rates such as food-poisoning incidents. They are more time-consuming and expensive than case-control studies, but less prone to bias.

In both case-control and cohort studies a structured questionnaire should be used. The interviews are conducted by telephone or face to face; in larger investigations a postal questionnaire may be used.

Steps in the investigation of an outbreak of infectious disease

1 Establish that an outbreak exists.
2 Confirm the diagnosis.
3 Define the population at risk.
4 Define, collect and count cases.
5 Describe the cases according to time (onset), person (age/sex) and place.
6 Formulate a hypothesis to explain the source, mode of transmission and duration of outbreak.
7 Test the hypothesis by an analytical study (case-control or cohort).
8 Plan and implement control measures.
9 Evaluate effectiveness of control measures.

A major difficulty with outbreak investigations is recall bias. The interviews may be conducted several days or even weeks after the event. Patients may not remember their exposures or bias their responses towards their perception of the source of the outbreak. For example, patients affected by food poisoning frequently ascribe their symptoms to the meal consumed immediately before the onset of illness, rather than in the period 24–48 h previously; their recall for the earlier meals may therefore be less accurate. This can be overcome by giving prompts, such as asking patients to consult their diary during the interview. It is also important to conceal the exposure variable in the questionnaire by including questions about other exposures that are not under investigation.

Organization and management

The CCDC should, at an early stage, convene an outbreak control team which should meet frequently to review progress and plan control measures. This group may include the local microbiologist, chief environmental health officer, director of public health, public health laboratory director, hospital control of infection officer, a representative of the suspected source (e.g. a water company) and, where appropriate, representatives of CDSC and MAFF. It is important that all communication with the press is through a single source; the outbreak control team should agree the contents of all press statements.

All stages of the outbreak investigation should be carefully documented. This is particularly important where legal action is likely to ensue. A preliminary report should be prepared within 48 h of the investigation, and interim reports at regular intervals thereafter. These should be approved by the outbreak control team. The final report should be a comprehensive account of the investigation and include an evaluation of the control measures that were implemented as well as recommendations for the prevention and management of future incidents.

Case study 26.1: Variant Creutzfeld–Jakob disease: an emerging infection, establishing causality

History: In 1996, the UK Creutzfeld–Jakob disease (CJD) surveillance unit in Edinburgh recognized 10 cases of a new variant of human spongiform encephalopathy disease, with a different clinical course and pathological features from classical CJD. It was thought possibly to be a human manifestation of infection with the agent of bovine spongiform encephalopathy (BSE). BSE is a transmissible spongiform encephalopathy (TSE), naturally acquired by the oral route, and transmissible to other species. Classical CJD is also transmissible from person to person; thus transmission of BSE from cattle to man seems entirely plausible.

Question: What epidemiological procedures could be used to investigate this new condition?

Epidemiological investigation: Cases were actively sought, using a case definition based on: (i) the distinct clinical course: an onset with psychiatric symptoms, followed by abnormal sensation at 2 months, ataxia at 5 months, myclonus at 8 months, akinetic mutism at 11 months and death between 12 and 24 months; and (ii) the unique pathological findings, including formation of plaques of abnormal prion protein (PrP^{Sc}) surrounded by a halo of spongiform change, distributed throughout the cerebrum and cerebellum. By the end of 1998, 35 cases had been diagnosed in the UK and one in France.

Relationships of time, place and person (personal exposure), were sought by case-note studies and interviews. A number of epidemiological criteria indicated that the association with BSE was causal:

1 All but one of the cases were found in Britain, where BSE was endemic.

2 There was a temporal relationship. Ten years elapsed between the first cases of BSE in cattle and the new disease in man. This interval is consistent with the long incubation periods observed for other human TSE's, such as kuru.

3 There is a consistency of likely exposure—although there has only been one case outside the UK, this patient, a bodybuilder, may have injected himself with a bovine equivalent of human growth hormone. The demonstration of an association with cattle in two different populations (Britain and France) lends weight to a causal association.

4 Detailed neuropathological investigation has demonstrated identical features in macaques experimentally infected with BSE.

Question: What is the next step in controlling the outbreak and confirming its cause?

Outbreak control measures: The outbreak has been controlled in cattle by: (i) banning the use of contaminated feedstuffs for all animals; (ii) controlling the preparation of carcasses, so that all possibly infected material is removed after slaughter and before entry to the human food chain; (iii) slaughter of all cattle in the age group which could be incubating the disease. It is too early to expect a fall in cases of vCJD, because the longest incubation period of human TSEs is near to 40 years. The absence of cases of vCJD in populations where BSE does not exist, however, provides supportive evidence for its aetiology.

Subject Index

Page references in *italics* refer to figures; those in **bold** refer to tables

abacavin (ABC) 75, 315
abscesses
  amoebic 186
  Bartholin's 224
  breast, during lactation 335–6
  Brodie's 288–9, 290
  cerebral and intracranial 282–5
  liver 191, 206–10
  lung 150
  peritonsillar (quinsies) 124, *125*
  retropharyngeal 130–1
  and salmonella bacteraemia 171, 173
  and sepsis 366
acetylhydrazine 61
N-acetyltransferases 68
achlorhydria 8
aciclovir 72–3
  adverse effects 72
  in pregnancy 327
acne 102
acquired immune deficiency syndrome *see* AIDS
acridine orange 37
acrodermatitis, atrophic 465
actinomycosis 102, 308, 385
acute (adult) respiratory distress syndrome (ARDS) 362, 372
acylureidopenicillins 65
adefovir 74
adenoids 111
adenovirus infection
  chest infections 140
  and conjunctivitis 112
  and sore throats 117–18
adrenoceptor agonists 371
Advisory Conmmittee on Dangerous Pathogens (ACDP) 427
aerobes, obligate 27
afimbrial adhesion (AFAI) 215
agammaglobulinaemia 412
agar 38
  Chrome 103
  crystal violet-blood 39
  deoxycholate citrate 39, 161
  iso-sensitest 58
  MacConkey's 39, *40*
  Mueller–Hinton 58
  New York City medium 39
  Sabouraud's 39
  thiosulphate citrate bile-salt sucrose (TCBS) 161, 177
agar incorporation method 58
agent factors, affecting infection 4
agglutination tests 46
agranulocytosis
  chloraphenicol-associated 69
  hepatitis-associated 201
  penicillin-associated 61
AIDS 296, 310, 313–14
  see also HIV infection/AIDS
air supplies, hospital 423
albendazole 80
algae, blue-green 182
alginate-producing bacteria 29
alkaline phosphatase 384
allopurinol, and toxic epidermal necrolysis 95
allylamines 77–8
alpha$_1$-acid glycoprotein 19
altitude sickness 436–7
alveoli 132
amantadine 72, 140
amantadine hydrochloride 140
amikacin 67, *68*, 356
aminoglycosides 52, 53, *56*, 67–8
  adverse effects 60, 68, **373**
  bacteria resistant to 62, 68
  therapeutic monitoring 60
aminopenicillins 64, 65–6
aminophylline 70
amnionitis 320
amoebae, culture 44
amoebiasis 184–6
  chronic intestinal 185–6
  diagnosis 47
amoxycillin 64
amphotericin B 62, 76–7
  adverse effects 60
ampicillin 64
  adverse effects 61, *62*, 66–7, **373**
amputation, leg, prophylactic regimens 421
amylases, blood 384
anaemia
  aplastic 61, **395**
  Coombs-positive haemolytic 61
  microangiopathic 168
anaerobes
  facultative 27
  obligate 27
  production of bacterial growth inhibitors 8
anaphylaxis, after taking antibiotics 61–2, 66
angina
  Ludwig's 130
  Vincent's 37, 131
angiomatosis, bacillary 317
angiotensin 371
animal diseases, spread to humans *see* zoonoses
anthrax 14, 453–5
  gastrointestinal 454
anti-gas gangrene serum (AGGS) 100
antibacterial agents 63–72
antibiotics
  mechanisms of resistance to 62–3
  modes of action 51, 52
  in selective media 39
  and the sites of infection 52–4
  see also antimicrobials; drugs
antibodies 365–6
  which damage tissues 21
antibody index 254
antibody-antigen complexes 21
antibody-antigen reaction, and serology 45–6
antibody-dependent cell-mediated cytotoxicity (ADCC) 11
antidiarrhoeal drugs 160
antiemetic drugs 160
antifungal therapy 76–8
antigenaemia, hepatitis E 393
antigens 8–9
  and influenza epidemics 138
  and the immune response 9–14
    antigenic drift 15
    antigenic shift 15
    antigenic variation 15
  class II human lymphocyte antigens 10
  hepatitis B surface antigen 197
  hepatitis Be antigen 197
  Lancefield group 121, 123
  shift 15
  T-cell-independent 28
  see also superantigens
antihelminthic drugs 79–80
antimalarial drugs 51, 78–9
antimicrobials 51–80
  adverse effects 60–2, **373**
  checking effectiveness of 60
  minimum bacterial concentration (MBC) 56, *57*
  minimum inhibitory concentration (MIC) 56–7, 58

antimicrobials (*Continued*)
modes of action 51–2
pharmacology 52–4
and PUO 391
resistance to *see* resistance to antimicrobials
susceptibility testing 41, 49–50, 55–63
see also antibiotics; drugs
antiprotozoal drugs 68, 78–9
antistreptolysin O titre (ASOT) 97, 387, 399
antiviral therapy 72–5
antiretroviral drugs 75–6, 315–16
resistance to 63, 76
aortitis 243
in Reiter's syndrome 402
syphilitic 251
apoptosis 10, 393
'apple-jelly' appearance 102
arbovirus encephalitis 280, 444
ARDS (acute (or adult) respiratory distress syndrome) 362, 372
Argyll Robertson pupil 305
artemesinin family 51
artemether 79
artemisinins 79
arthritis
complication of meningococcal meningitis 270
gonococcal 302, 369
Lyme 465
postinfectious 21, 173, 394, **395**, 400–1
rheumatoid 380
suppurative 293–5
tuberculous 295
arthropods, vectors **5**, 443–52, 479
artusenate 79
ascariasis 33, 187
aspartate transaminase 384
aspergillosis 155
aspirin 18, 400
atovaquone 79
attachment of organisms, to body surfaces 14
attack complex 8
attic infection 115
'atypical mononuclear cells' 119
auramine 37
autoantibodies, in connective-tissue diseases 387
autoclaving 431
auxanograms 41
auxotyping 41
azathioprine 413
azidodeoxythymidine (AZT) *see* zidovudine
azithromycin 71
azoles 77
aztreonam 65

B cells (B lymphocytes), and immunity *9*, 10–11
bacilli 26
bacillus Calmette–Guérin (BCG) vaccine 21, 344–5, 357–8, 477
Bacillus cereus food poisoning 179
bacteraemias
community acquired 364–5
defences of the blood 365–7
diagnosis 368–9
during pregnancy 320
epidemiology 364–5
follow-up of patients with 378
'hospital' **372**
and immunocompromised patients 405
management 369–73
neonatal 333–4, 365
pathogens causing 364–5
gonococci 294, 302, 369
meningococci 364–5, 366, 369
pneumococci 364, 369
salmonella 171, 313, 368
staphylococci 247, 364, 368, 369, 374–5
streptococci 124, 144, 364, 368, 369
pathological accompaniments 367–8
brain abscesses 282
suppurative arthritis 294–5
urinary tract infections 221
see also sepsis and sepsis syndromes; septicaemia
bacteria 23
atmospheric requirements 27
capsules 15, 28–9
cell walls 16, 27–8
and antimicrobials 51
classification 26–7
DNA
and quinolone antimicrobials 51, *55*, 69
transfer between bacteria 63, *64*
Gram-negative 26, 27
antimicrobial-resistant 63
Gram-negative septicaemia 376–8
and immunocompromised patients 406
toxic activity 16
Gram-positive 26, 27
bacteraemias 369
and immunocompromised patients 406
toxic activity 16, 122
protein 'capsules' 28–9
protein synthesis *56*
inhibition 52, 69, 71
spores 14, 27, *28*
structure *24*, 27–30
typing 41–4
see also antibacterial agents; bacterial infections
bacterial infections
conjunctiva/cornea 113–14
gastrointestinal tract 165–81
genital 300–7
in immunocompromised patients 405–6, 408–9
liver parenchyma 203–6
lower respiratory tract 141–53
middle ear 114–15
skin and mucosae 91–102
throat and mouth 120–31
bacteriocin typing 42
bacteriocins 8
bacteriophages, and phage typing 42
balanitis, circinate 402
Bartholin's abscess 224
bartonelloses *see* cat-scratch disease; trench fever
BCG vaccine 21, 344–5, 357–8, 477
bean haemagglutins 181
benzathine penicillin 65
benzimidazoles 79–80
benzylpenicillin 58, 61, 66, **373**
beta-lactam antibiotics 51, *54*, 63–7, 292
beta-lactamase 62, 65
testing 59
bile salts 158
bilharzia *see* schistosomes/schistosomiasis
biliary system 191
see also cholangitis
bilirubin 191
'biological agents', definition 1, 427
biopsy, and PUO 390–1
biotyping 41
birds, diseases from 5, 151
birth, infections contracted during *see* intrapartum infections
bites
animal 108–10, 366, 412
and rat bite fever 471
insect, and travellers 436, *443*
blackwater fever 447
bladder symptoms of urinary infection 216
blastomycosis 155
blood
culture 385, 410–11
defences 365–7
tests in PUO 383–4
body, human
normal flora *2*
ringworm 105
boils 93
bone
infections
osteomyelitis 287–93
suppurative arthritis 293–5
tuberculosis 353
structure and growth 286–7, *288*
bone marrow
adverse effects of antimicrobials 61
transplants 83, 405, 422
and immune dysfunction **414**, 415
borreliosis *see* Lyme disease
botulism 3, 179–80
infant 180
bovine spongiform encephalopathy (BSE) 282, 485
bowel 157
colonizing flora *2*, 158
helminth infections 186–7
immune response in 158
operations on 421
X rays 389
see also gastrointestinal tract
Bowenoid papulosis 83
brachial plexitis 120, *121*
breakpoint method 58
British Paediatric Association Surveillance Unit (BPASU) 474, 476
Brodie's abscess 288–9, 290

bronchi 132
bronchiectasis 142
bronchioles 132
bronchiolitis 135–7
bronchitis, chronic *see* chronic obstructive pulmonary disease
bronchoalveolar lavage 133, 411
brucellosis 205–6, 383, 452–3, 462–4
  and infective arthritis 295
  in pregnancy 336
  undulant fever 381
BSAC standardized disc sensitivity testing method 58
BSE *see* bovine spongiform encephalopathy
buboes, and plague 470
bullae *20*
byssinosis 380

C3 8, 365
C3a 8, 365
C3b 8, 15, 365
C5a 8
C6–C9 8
C-reactive protein (CRP) 19, 384
caeruloplasmin 19
Campylobacter infections 175–6, 397
cancrum oris 228
candidiasis 11, 103–4
  and AIDS **318**
  candidal oesophagitis 311–12
  genital 219, 308
  oral 85
  and prolonged antibiotic treatment 292
canicola fever 203
cannulae
  gastric 420
  intravenous 29, 98, 417–19
capnophiles 27
capreomycin 356
capsids, symmetry 25
capsules, bacterial 15, 28–9
carbamazepine 18
carbapenems 65
carbenicillin 65
carbol fushcin 37
carboxypenicillins 65
carbuncles 93
carcinogens, exposure to 380
cardiovascular system 243
cardiovascular system infections 243–53
carditis, of rheumatic fever 401
caruncle, urethral 220
cat bites 108–10
cat-scratch disease 101–2, 109
catheters
  intravascular 29, 406, 417
  urinary 8, 419, 420
  see also cannulae, intravenous
cats, and toxoplasmosis 467
cavernous sinus thrombosis 115
CD4 cells 11
  and HIV infection 311, 313
CD8 cells 11
CD16 cells 11
cefaclor 67
cefamandole 67
cefazolin 67
cefepime 67
cefixime 67
cefotaxime 67
ceftazidime 67
ceftriaxone 67
cefuroxime 67
cefuroxime axetil 67
cellulitis 81, 98–9
  like erysipelas 96, 98
central nervous system infections 254–85
  cerebral and intracranial abscesses 282–5
  investigations *259*, *260*
  pathogenesis 255
  poliomyelitis 262–4
  see also encephalitis; meningitis
cephalosporins 41, 53, 67–8
  adverse effects 61, 62, **373**
  structure *67*
cercariae
  liver fluke 213
  schistosome 7, 33, 210, 211
cerebral toxoplasmosis 411
cerebrospinal fluid (CSF) 254, 260
  and meningitis 256, 258, 259–60, 261–2, 265, 267, *269*
  normal composition 258
  rhinorrhoea *272*
  rising pressure 255
  staining 36
cervical intraepithelial neoplasia 297
cervicitis 300, 301, 302–3
cervix, carcinoma 297, 313
cestodes (tapeworms) 34, 284
  see also hydatid disease
chancroid 306–7
Charcot joints 305
chemoprophylaxis 482
chemotherapy, and immune deficits 405, 422
chickenpox (varicella) 21, 234–8, 408
  and HIV infection 313
  maternal 326–7
  varicella embryopathy and neonatal varicella 326
  and varicella pneumonia 140
  see also varicella zoster immunoglobulin
chigger 436
childbirth
  intrapartum infections *see* congenital and perinatal infections
  maternal infections related to 334–6
childhood infections 225
  chickenpox (varicella) 234–8
  conjunctivitis 112
  human herpesvirus type 6 240
  human herpesvirus type 7 240
  human parvovirus B19 239–40
  Kawasaki disease 240–2
  measles 225–9
  mumps 230–1
  osteomyelitis *289*, 290, 293
  rubella 232–4
  suppurative arthritis 293–4
  urinary tract 222
chlamydial infection
  and conjunctivitis 113–14, 301
  genital 300–1
  incidence 220, 296
  neonatal and infant 332–3
  and pneumonia/pneumonitis 150–2, 153, 332, 333, 412
  and urethritis 220
chloramphenicol 52, 53, 69
  adverse effects 61, 69
  bacterial resistance to 62
  for conjunctivitis 113
  for meningitis 269–70
chloroquine 78
cholangitis 54, 209–10
cholera 440–2, 477
cholesteatomas 115
cholesterol, in the liver 191
chorea, of rheumatic fever 401
chromatography, high-performance liquid (HPLC) 60
chronic obstructive pulmonary disease (COPD; chronic bronchitis) 141–2, 149
cidofovir 74, 89
ciguatera poisoning 182
cilastatin 66
cinchona bark 51, 78
ciprofloxacin 51, 69–70, 152, 291–2, 357, **373**
cirrhosis, hepatic 413
CJD *see* Creutzfeld–Jakob disease
clarithromycin 71, 356
clavulanic acid 65
clindamycin, adverse effects **373**
clinic data 476
cloxacillin 64
clubbing, of the fingers 248
Clutton's joints 330
co-trimoxazole
  adverse effects 61
  and parasitic infections 391
coagglutination 46
coagulation cascade 364
cocci 26
  anaerobic **372**
    and gas-forming infections 99
coccidia 31
coccidioidomycosis 155
cold, common 115–16
cold abscess 342
cold sores 83, *85*
coliform infections
  and cellulitis 98–9
  coliform pneumonia 156, *188*
  and osteomyelitis 287
colitis
  haemorrhagic 166, 167, 168, 190
  pseudomembranous 3, 8, 180–1, 290
  salmonella 171
colon 157
colonization resistance 7, 158
communicable disease 3
  and the law 473 4
  prevention and control 476–85

communicable disease (*Continued*)
  see also immunization
  surveillance 474–6
  see also transmission of infection
Communicable Disease Surveillance Centre (CDSC) 258, 473
complement, deficiency **405**, 412
complement systems
  alternative 8, *9*, 365, *366*, 412
  classical 8, *9*, 365, *367*, 412
computed tomographic scans, and PUO 388, *389*
condylomata acuminata 83
congenital and perinatal infections 320–1
  maternal infections related to childbirth 334–6
  transplacental, intrapartum and postnatal infections 322–31
  via intrapartum and perinatal routes 331–4
  see also pregnancy
conjugation, genetic 63
conjunctivitis 111–14
  bacterial 113
  childhood 112
  chlamydial 113–14, 301, 332, 333
  haemorrhagic 112
  herpes simplex *84*
  neonatal 332, 333
  see also keratitis; keratoconjunctivitis; ophthalmia, neonatal; trachoma
connective tissue disorders, tests for 380–1, 387–8
consultants in communicable disease control (CCDC) 473, 476, 484
contact, direct 5
contact lenses 112, 113, 114
contact tracing 482
Control of Infection Committee 429, 430
Control of Infection Officer 429
Control of Infection Team 429
Control of Substances Hazardous to Health (COSHH) 427, 428
convalescence 393–4
convulsions, febrile 17, 18
COPD *see* chronic obstructive pulmonary disease
cornea, infection *see* keratitis; keratoconjunctivitis
corns 83
corticosteroids 391, 413
corynebacterial infections 7, 417
coryza 115–16
countercurrent immunoelectrophoresis (CIE) 46
cowpox 89–90
Coxiella burnetti endocarditis 205, 249
coxsackie A viral infections 116–17, 261
coxsackie B viral infections 117, 223, 261
  and myocarditis 246
creatine kinase 384
Creutzfeld–Jakob disease 282
  variant (v-CJD) 282, 485
Crimean-Congo haemorrhagic fever (CCHF) 458
Crohn's disease 225
cross-immunity 21
croup 134–5
CRP (C-reactive protein) 19, 384
crust *20*
cryptococcosis 11, 411
cryptosporidiosis 37, 183–4, 410, 477
crystal violet 39
culture 38, 385–6
  antimicrobial susceptibility testing 41
  automation 40
  blood 385, 410–11
  limitations 40
  media 39–40
  and opportunistic infections 410–11
  of protozoa and helminths 44
  screening 40
  selective 39
  tissue culture 44
  typing 41–3
  urine 385
  viral 44–5, 386
Curtis–Fitz–Hugh syndrome 300
cutaneous reactions to antimicrobials 61–2
cyanobacterial toxins 182
cycloserine 357
cyclosporiasis 184
cyclosporin 403, 413, 422
cystic fibrosis 29, 142, 374
cysticercosis, cerebral 284
cystitis 54, 221–3
cystourethrograms 222
cytokines 9, 16
  and fever 17
cytomegalovirus infection 111, 202, 316, 397
  congenital and neonatal 324–5
  and immunocompromised patients 405, 408
    post-transplant 413–15
  retinitis **318**
  treatment 414
cytopathic effect (CPE) 44, *45*, 116
cytopenias, immunologically-mediated 61
cystoscopy, infection from 421

dairy products, unpasteurized 461, 462, 479
dalfopristin/quinupristin 52
death certificates 476
delerium 17
dengue fevers 443–4
dengue haemorrhagic fever (DHF) 443, *444*
deoxyribonuclease 93
Department of Agriculture and Fisheries 474
dermatophytes 104–6
dermis 81
  infections 81
determinants of disease 4
diagnosis, laboratory techniques 35–50
diarrhoea 158
  drug treatment 160
  and haemolytic-uraemic syndrome *169*
  infant feeding in diarrhoeal illnesses 159–60
  laboratory diagnosis 160, *161*, *164*
  traveller's 166, 168, 437
  see also stools
diarrhoetic shellfish poisoning 181
dicloxacillin 66
didanosine (DDI) 75, 315
dihydrofolate reductase 51
dihydropteroic acid 51
dinoflagellates, and shellfish poisoning 181
dip tests, urine 216, 217
diphtheria 127–30, 479–80, 482
  toxin 16, 17, 128
diseases, communicable *see* communicable disease
disinfection, in hospitals 430–1
DNA, bacterial *see* bacteria
DNA gyrase 51
dog bites 108, **109**, 412
DOPA agonists 370–1
doxycycline 68
drift, antigenic 15, 138
drugs
  and hypersensitivity with fever 380, 391
  illegal, abuse 247, 287
  see also antibiotics; antimicrobials
Duke University criteria **249**, 252
duodenal surgery 421
dysentery
  amoebic 174, 183, 185
  bacillary (shigellosis) 173–5

E-test 58
early antigen (EA) complex 118
ear(s)
  middle ear infections 114–15
  otitis externa 100
  toxic effects of aminoglycosides 60
Ebola haemorrhagic fever 457–8
ecchymoses *20*
echocardiography, and PUO 388
echovirus infections 261
ECM *see* erythema chronicum migrans
eczema
  eczema herpeticum 85
  and skin infections 82
education, health 478–9
efavirenz 76, 316
ehrlichioses 444, 445
Eimeridia 30
electrolyte requirements, children's **160**
electrophoresis
  multilocus enzyme (MLEE) 43
  polyacrylamide gel (PAGE) 43
  pulsed-field gel (PFGE) 43
elephantiasis 451
ELISAs 47–9, 330
emboli, and endocarditis 247, 250
empyema, pneumococcal *145*
encephalitis 278–82
  arbovirus 280, 444
  Californian 280, 444
  and cerebral oedema 255
  eastern equine 280, 444
  encephalitis lethargica 139
  herpes simplex 254, 279–80, 285

Japanese B 280, 444, 459
Murray Valley **280**
Nipah virus 472
postinfectious 21, **395**
post-chickenpox 237, 278
post-measles 229, 278, 394
post-rubella 234
Powassan virus **280**
St. Louis 280, 444
tick-borne **280**, 444, 459, 480
toxoplasmic **318**
Venezualan equine **280**
western equine 280, 444
West Nile fever 444
see also meningoencephalitis
encephalopathies, spongiform 281–2
endarteritis, infective 251
endemics 7
endocarditis 7
acute (aggressive) 248–9
and bacteraemia 368
chlamydial 249
Coxiella burnetti endocarditis 205, 249
culture-negative 248
infective 3, 246–51, 252–3
mycoplasmal 249
of rheumatic fever 401
staphylococcal 247, 249
endoscopic retrograde cholangiopancreatography 421
endospores *see* spores, bacterial
endothelium 243
and bacteraemia/sepsis 364, 365
endotoxins, bacterial 16, 27
inhibition 372
endotracheal tubes 8, 419
endovasculitis 243
enrichment media 39
enteral feeding 420
enteric fevers 383
see also typhoid fever
enteritis, campylobacter 397
enterocines 158
enterococcal infections **372**, 420
and cardiovascular infections 243, 247, 249, 251
and cellulitis 98–9
and gas-forming infections 99
and neutropenia 406
enterotoxins, staphylococcal 92, 178
enteroviral infections
enteroviral pharyngitis 116–17
enterovirus meningitis 261–2
see also poliomyelitis
environment, and pathogen survival 4, 14
environmental health officers (EHOs) 479
environmental pathogens 5
enzootic ovine abortion 336
enzyme immunoassay (EIA) tests 305
enzymes, lytic 121–2
eosinophilia 384
pulmonary 187
and strongyloidiasis 156, 188
epidemic methicillin-resistant *Staphylococcus aureus* (EMRSA) 40, 93
epidemics
of childhood diseases 225
cholera 440
echovirus infections 261
hepatitis A 192
influenza 138
measles 226
mumps 230
parvovirus B19 325
pneumonia 150
polio 478
respiratory syncytial virus (RSV) 135
rubella 232, 322–3
Salmonella napoli 483
worldwide *see* pandemics
epidemiology, of infections 3–7
epidermis 81
bacterial infection 81
epidermodysplasia verruciformis 83
epididymitis 300
tuberculous 352
epididyo-orchitis, acute 223–4
epiglottitis, acute 125–7
epilepsy, and fever 18
epitopes 8–9
Epstein–Barr nuclear antigen (EBNA) complex 118
Epstein–Barr viral infection 111, 118–20, 140, 394
diagnosis 386
and hepatitis 202
and HIV infection 313
ERIC-PR 43
erysipelas 81, 96–8, 376
like cellulitis 96, 98
erysipeloid 101
erythema chronicum migrans (ECM) 100–1, *464*, 465
erythema infectiosum 239
erythema marginatum *383*, 401
erythema multiforme 376, **395**, 396–7
erythema nodosum 341, 346, 376, 394, **395**
erythema nodosum leprosum (ENL) 360
erythema subitum 240
erythrasma 101
erythrocyte sedimentation rate (ESR) 20, 384, 387
erythromycin 71
adverse effects 71, **373**
erythromycin estolate, adverse effects 61
Escherichia coli infections **371**, 419
gastroenteritis 165–8
and lactose intolerance 160
neonatal bacteraemia 333
ethambutol, for tuberculosis 356
ethionamide, adverse effects 61
everninomycin 52
exotoxins 16
staphylococcal 17, 92, 95, 334, 368
streptococcal 17, 122, 368
eyes
infection 5
and toxoplasmosis 468

faeces *see* diarrhoea; stools
famciclovir 73, 87, 88
Fansidar® 79
farmer's lung 380
fasciitis, necrotizing 100, 376
fascioliasis 213, 214
feet, ringworm infection 106
feline immunodeficiency virus (FIV) infection 319
ferritin 365
fetus, infections *see* congenital and perinatal infections
fever 17–18, 368, 379
patterns *381–2*
relapsing 26
in a returning traveller **459**
treatment in neutropenic patients 407–8
see also pyrexia of unknown origin
fifth disease (slapped cheek syndrome) 24, 325
filariasis 33, 37, 451
fimbriae (pili) 14, 29
of *E. coli* 166, 215, 216
and pathogenicity 3, 166
fingers, clubbing 248
fish, food poisoning from 181, 182
flagella
bacterial 29–30
protozoan 31
flatworms *see* platyhelminthes
fleas **5**, 366
jigger 436
and plague 469
flies **5**
Simulium 451
tumbu 436
flotation methods 182
flucloxacillin 64
adverse effects 61, **373**
fluconazole 77, 407
flucytosine (5-fluorocytosine) 78
fluid requirements, children's **160**
flukes *see* trematodes
fluorescent treponemal antibody absorption (FTA-ABS) test 305, **306**, 330
5-fluorocytosine *see* flucytosine
fluoroquinolones 69–70
adverse effects 70
folate 51
folliculitis, staphylococcal 436
fomites 5, 112
food poisoning
bacterial 5, 479
Bacillus cereus 179
Clostridium perfringens 180
from vibrios 176–7
salmonella 169–73, 479, 483
staphylococcal 178–9
Yersinia 177–8
management 162–3
and non-bacterial toxins 181–2
prevention 163
shellfish 164, 165, 176–7
viral 165
see also gastrointestinal tract infections
Fort Bragg fever 148

foscarnet 74, 414
Fournier's gangrene 100
frostbite 436
fungal infections (mycoses)
  antifungal prophylaxis 407
  and immunocompromised patients 33, 408, 409
  lower respiratory tract 155
  and neutropenia 406, 407
  skin 103–8
    deep 103
    superficial 32–3, 103
  see also antifungal therapy
fungi 23
  blood-borne 361
  classification 32–3
  saprophytic 1
  systemic 33
furunculosis 91–6
  clinical features 92–3
  complications 94–6
  diagnosis 93–4
  epidemiology 91
  management 94
  microbiology 91–2
  pathogenesis 92
  pseudomonal 95
fusidic acid 71–2
  adverse effects 61, 72
Fusobacterium necrophorum infection *291*

gallbladder, mucocoele of the 242
gamma-globulin, deficiency **405**
ganciclovir 74, 414
  adverse effects 61, 74
gangrene
  diabetic, and gas-forming infections 99
  Fournier's 100
  gas gangrene 99–100
  ischaemic, and gas-forming infections 99
  synergistic 17, 122
gas-forming infections
  gas gangrene 99–100
  limited-extent 99
gastric acid 158
gastric surgery 421
gastritis, acute, and *Helicobacter pylori* 178
gastroenteritis
  E. coli 165–8
  viral 163–5
    rotavirus 163–4
gastrointestinal tract 157
  colonizing flora of the bowel 2, 158
gastrointestinal tract infections
  bacterial 165–81
  defenses against 8, 158
  management 158–63
  non-bacterial toxins and food poisoning 181–2
  parasitic 182–9
  viral 163–5
  see also food poisoning
general practitioner surveillance schemes 475
General Register Office (Scotland) 474
genes
  antibiotic-resistant 63
  pathenogenicity islands 30
  and postinfectious disorders 21
  transfer between bacteria 63, *64*
genital infections 296, 309
  bacterial 300–7
  candidiasis 219, 308
  chlamydial 300–1
  pelvic inflammatory disease (PID) 307–8
  trichomonal 36, 308–9
  tuberculosis 352–3
  viral 296–300
  see also sexually transmitted diseases
genital tract 296
genital ulcer disease 307
gentamicin 67, *68*
Gerstmann–Straussler–Scheinker syndrome 282
giardiasis 159, 183
Giemsa stains 37
gingivostomatitis, primary herpetic *84*, 85
glomerulonephritis, poststreptococcal 29, 121, 394, **395**, 398–9
  see also nephritis, postinfectious
glycolipid, bacterial capsules 29
glycopeptide antibiotics 51, 54, 70
  adverse effects 70
  resistance to 63
gonococcal bacteraemia 294
gonorrhoea 3, 302–4
  diagnosis 37, 302–3
  incidence 220, 296, *297*
Gram reaction/stain 26, 36–7
Gram-negative organisms
  cell wall *53*
  see also bacteria
Gram-positive organisms
  cell wall *53*
  see also bacteria
granulocyte counts, and PUO 383–4
granulocytopenia 61
granuloma inguinale 307
granulomata
  and fungal infections 103, 107
  and phacocyte deficiency 8
granulomatous diseases, tests for 387–8
grey baby syndrome 69, 329
griseofulvin 78
groin, ringworm 105–6
Gruinard 14
Guillain–Barré syndrome 149, 394, **395**, 397–8

HAART *see* highly active antiretroviral therapy
haemagglutinin 138
haemagglutins, bean 181
Haemagogus spp. (mosquito) 202
haemolysins 368
haemolysis 17
  postinfectious **395**
haemolytic-uraemic syndrome (HUS) 166, 167, 168, *169*, 190, 403
haemophagocytic syndrome **395**
Haemophilus influenzae infections
  granulocyte counts 383
  H. influenzae arthritis 295
  H. influenzae bacteraemic disease 365, 366
  H. influenzae meningitis 272–5
haemopoietic system, adverse effects of antimicrobials 61
haemorrhagic fevers, viral *see* viral haemorrhagic fevers
haemosporidians 30
halofantrine 79
hand, foot and mouth disease 88–9, 116–17
hands, ringworm infection 106
hantavirus pulmonary syndrome 456
haptoglobin 19
Haverhill fever 471
Haversian system 286, *287*
Heaf test 344–6, 386
health education 478–9
heart, transplants 413, 422
heat exhaustion 435
helminths 33–4
  bowel infections 186–7
  culture 44
  see also antihelminthic drugs
Henoch–Schönlein disease 268, 398, *399*
hepatitis
  congenital 320
  granulomatous 205
  hepatitis A 191–5, 384, 478, 480
  hepatitis B 7, 196–9, 214
    neonatal and perinatal 331–2
    postexposure prophylaxis 480
    treatment 197–8, **199**
  hepatitis C 200–1
    treatment 50, **199**, 200–1
  hepatitis D (delta hepatitis) 199–200
  hepatitis E 201
    antigenaemia 393
  viral
    differential diagnosis 194
    and a harmful immune response 21
hepatocellular damage, from antimicrobials 61
hepatocytes 158
'herald patch' 107
herpangina 116, *117*
herpes, genital *see* herpes simplex, genital
herpes simplex
  congenital and intrapartum 326
  encephalitis 254, 279–80, 285
  genital (genital herpes) 83, 220, 296, *297*, 298–300
  and immunocompromised patients 408
  mucocutaneous 83–6
    and AIDS **318**
  postinfectious disorders 396
  treatment 72–3, 86
  type 2 meningitis 264
  and urethritis 220
herpes zoster (zoster; shingles) 86–8, 235, 236, 238, 408, 480
  and HIV infection 86, 313
herpesvirus infection 393

herpesvirus simiae infection 471
Hib infection 144, 147, 272, 273
highly active antiretroviral therapy (HAART) 75, 315–17, 408
histoplasmosis 155, 409
HIV infection/AIDS 422
clinical features 311–14
congenital 327–8
diagnosis and staging 313, 314–15
epidemiology 7, 310, *311*
immunocompromised patients 15–16, 17, 21
in infants/children 313–14, 328
management 59, 315–17
and drug resistance 49–50, 63, 76, 316
highly active antiretroviral therapy (HAART) 75, 315–17
judging response to treatment 50
opportunistic diseases 311–12, 313, 317, **318**
Bartonella henselae infection 110
fungal 33
herpes simplex 83, 300, **318**
herpes zoster 86, 313
mycobacterial 316, 317, **318**, 339
poxviruses 89
prevention and control 317–18
surveillance schemes 476
and thrombotic thrombocytopenic purpura 403
virology 310–11
see also AIDS
HLAs *see* human leucocyte antigens
hookworms 7, 187–8, 460
treatment 79, 188
horizontal spread, of disease 4
hospital infections **372**, 405–6, 416
control 428–31
data on 476
environmental factors 422–3
and intensive therapy patients 419–20
and intravenous cannulae 29, 98, 406, 417–19
isolation facilities 423–6
patient predisposition 416–17
prevention in laboratories *426*, 427–8
spread by hospital equipment 423
and surgery 420–2
and urinary catheters 419
hosts
damage to 16
factors affecting infection 4
HTLV-associated myelopathy (HAM) 318–19
HTLV-I infection 318–19, 332
HTLV-II infection 319
human herpesvirus infections
type 6 (HHV-6) 240, 413
type 7 (HHV-7) 240
type 8 (HHV-8) 313, 413
human immunoglobulin (HIG) 480
human leucocyte antigens (HLAs) 10
B27 21, 301, 396, 402
class I 11
class II 11, 14
human normal immunoglobulin (HNIG) 195
human papillomavirus (HPV) infections 83, 296–8
human parvovirus B19 (HPV-19) infection 239–40, 384
congenital 325–6
human T-lymphotrophic virus I (HTLV-I) infection 318–19
in infancy 332
human T-lymphotrophic virus II (HTLV-II) infection 319
HUS *see* haemolytic-uraemic syndrome
Hutchinson's teeth 330
hyaluronidase 368
hydatid disease 33, 212–13
hydrocephalus, and tuberculous meningitis 352
hydrops 326
hydroxynaphthoquinones 79
hypersensitivity reactions
with fever 380, 391
type I, to antibiotics 61–2
hypersplenism 413
hypogammaglobulinaemia 411–12
hypotension, and bacteraemias/sepsis **362**, 369
hypothalamus 9, 11, 17
hypoxia, management 371
hysterectomy, prophylactic regimens 421

iceberg, epidemiological 22
ileum 157
imidazoles 77
imipenem 65, 66
immigrants, screened for tuberculosis 358
immune complex disease 394
immune response
in the bowel 158
during convalescence 394–5
harmful effects of 21
to polysaccharide capsular antigens 28
see also immunity
immunity
cell-mediated 11, *12*
depression 15
cross-immunity 21
disorders 404–5
see also immunocompromised patients
humoral 11, *13*
evasion 15
non-specific 7–8
overstimulation 15
specific 8–14
suppression 15–16
see also immune response
immunization programs 479–82, *483*
immunization/vaccination against 479–80
anthrax 455
borreliosis 467
cholera 442, 459
diphtheria 130
Haemophilus influenzae type b (Hib) 147, 274–5
hepatitis A 195
hepatitis B 198, 332
influenza 140, 435
Japanese B encephalitis 459
measles 226, 229, 314
meningococcal meningitis 271
mumps 231
overseas diseases 458–9
pertussis 155
pneumococcal pneumonia 144
polio 158, 262–3, 264, 479–80
rabies 281, 459
rotavirus infection 164
rubella 234, 323
smallpox 479–80
tick-borne encephalitis 459
tuberculosis (BCG) 21, 344–5, 357–8, 477
typhoid fever 440
varicella (chickenpox) 238
yellow fever 202, 203, 459
immunoblotting 49
immunocompromised patients 1, 21, 404
classification of infections in 404–5
complement deficiency 412
diagnosis of opportunistic infection 410–11
hypogammaglobulinaemia 411–12
neutropenia 405–8
splenectomy 412–13
surgical infections 421–2
T-cell deficiency 408–11
transplant patients 413–15
see also HIV infection/AIDS; immunosuppression
immunodeficiency *see* immunocompromised patients
immunofluorescence 38, 44, *45*
indirect 47
immunoglobulins 10, 11
administration of 480
hepatitis B (HBIG) 198, 332
human normal (HNIG) 195
human rabies (HRIG) 281
human tetanus (HTIG) 480
IgA 7, 14–15
secretory 82, 132, 158
IgG 11
IgG_2 28
IgM 11
and convalescence 394
intravenous 242, 398
tick-borne encephalitis 480
varicella zoster (VZIG) 238, 480
see also zoster immunoglobulin (ZIG)
immunological attack, microbial defence against 14–15
immunological memory 11
immunopathology 21
immunosuppression 422
immunosuppresive therapy 380
and listeriosis 277
and measles 15, 228–9
impetigo 5, 91–6, *443*
bullous *see* Lyell's syndrome
imported disease *see* travel-associated disease

India ink 37
indicator media 39–40
indinavir 75
indomethacin, and toxic epidermal necrolysis 95
infarction, myocardial 242
infection 2–3, 366
  laboratory diagnosis 35–50
  manifestations of 17–21
  see also hospital infections; pathogens
infectiousness 3
inflammation 18–20, 368
  non-specific tests for 384
influenza 5, 137–40, 482
  and croup 134
  immunization against 140, 435
ingestion, and pathogen transmission 5–6
inhalation, and pathogen transmission 5
inoculation of infection 6, 366, 423
insect bites, and travellers 436, *443*
insertion sequence typing 43
insomnia, fatal familial 282
intensive therapy unit, infections 419–20
intercellular adhesion molecules (ICAMs) 362
interferon
  harmful effects 21, 384
  interferon-alpha 17
interleukins
  IL-1 9, 11, 16, 17, 362, 372
  IL-2 11, *12*, 393–4
  IL-6 9, 372
  IL-8 362
intertrigo 103
intestines *see* bowel
intraepithelial neoplasias *see* neoplasias
intrapartum infections 320, 321, 322
intrauterine devices 308
intrinsic reproduction rate (IRR) 3
involucra 288, *291*
iron binding 365
ischaemia, myocardial 242
isolation, in hospitals 423–6
isoniazid
  adverse effects 61
  for tuberculosis 355–6
isotope scans, and PUO 388, *390*
isoxazolylpenicillins 65–6
itraconazole 77

Janeway lesions 247
Jarisch–Herxheimer reaction 251, 306
jaundice
  cholestatic 61
  and hepatitis A 193
jejunum 157
jigger fleas 436
joints
  Charcot 305
  Clutton's 330
  replacement, prophylactic regimens 421
  structure 287
  suppurative arthritis 293–4
  tuberculosis 353
  see also bone

K cells 11
kanamycin 67
Kaposi's sarcoma 313, 317
Katayama fever 211
Kawasaki disease 240–2, 384
keratitis 112
  herpes simplex 112–13
keratoconjunctivitis 112, 113
  amoebic 114
  shipyard eye 112
keratoderma blenorrhagica 402
Kernig's sign 256
ketoconazole 77
ketodeoxyoctanoic acid (KDO) 16
kidney beans, red 181
kidney infections 215
  see also urinary tract infections
killer lymphocytes 11
kinetoplasts 30
Koch, Robert 1, 38
Koplik's spots 226, 227
Kuru 282
Kyanasur forest disease **280**

laboratories
  and hospital infections *426*, 427–8
  waste 431
laboratory reports 475
laboratory techniques, infection diagnosis 35–50
lactation, and breast abscesses 335–6
lactobacilli 220, 296, 307
lactophenol blue 37
lactose intolerance, secondary acquired 160
lamivudine (3TC) 63, 75, 76, 315
Lancefield group antigens 121, 123
laparoscopy, and PUO 390
laparotomy, and PUO 390
larva currens 188
larva migrans 188, 460
lassa fever 456–7
Leeuwenhoek, Anthonie van 35
leg amputation, prophylactic regimens 421
legionellosis 148, 422
legionnaires' disease 5, 147–9, 397, 422, 478
legislation, and infection control 473, 479
Leishman–Donovan bodies 385, 451
leishmaniasis 411–12, 450–1, 460
  diagnosis 47
  and the immune system 15, 17
  visceral 385
lepromatous disease *see* leprosy
leprosy 358–60
  borderline (BB) 359
  diagnosis 37, 359
  lepromatous (LL) 15, 359
leptospirosis 5
leucocidins 368
leukaemia
  acute T-lymphoblastic 319
  chronic lymphocytic, and herpes zoster infection *87*
  immunocompromised patients 21, 405, 408, 422
levamisole 80

lice **5**
ligase chain reaction 49
lipoarabinomannan 340
lipodystrophy 75
lipopolysaccharide 16
listerial meningitis 276–7
listeriosis 282, 361
  congenital and neonatal 328–30
liver 191
  and alkaline phosphatase 384
  transplants 413, 422
liver failure
  and hepatitis A infection 194–5
  and hepatitis B infection 198
liver flukes 213–14
liver infections 191
  abscesses 191, 206
    amoebic 207–9
    cholangitis 209–10
    pyogenic 206–7
  and antimicrobials 54
    adverse effects 61
  bacterial
    miliary tuberculosis 353
    of the parenchyma 203–6
  parasites 210–14
  viral 191–203
    see also hepatitis
loiasis 451
louping ill **280**
lower respiratory tract 7, 132
  mechanical problems 132–3
lower respiratory tract infections
  acquired in hospital 420
  bacterial
    atypical pneumonias 150–3
    pertussis 153–5
    pulmonary tuberculosis 341, 349–51
    pyogenic 141–50
  fungal 155
  laboratory diagnosis 133
  non-infectious conditions mimicing 133
  viral 133–41
Ludwig's angina 130
lumbar puncture
  and cerebral abscesses 283
  and meningitis 256–8, 261, 265
lupus vulgaris 102, 339
Lyell's syndrome (scalded skin syndrome) 95, 334
Lyme disease 26, 100–1, 464–7
  diagnosis 49, 260, 466–7
  and neuroborreliosis 281
lymph nodes, aspirate 385
lymphadenitis
  and erysipelas 97
  tuberculous *344*
lymphocytes 9
  counts 384
  see also B cells; K cells; T cells
lymphocytoma, borrelial 465
lymphocytosis 384
lymphogranuloma venereum 301–2
lymphoma(s)
  and AIDS 313, 317

Burkitt's 118
lysosomes 8, 9
lysozymes, mucosal 7, 82
lytic cycle 42
lytic enzymes 121–2

M antigen 28–9
MacConkey's agar 39, *40*
maculopapular 52, 53, *56*, 71
macropapular rashes *20*
macrophages 8, 9–11
alveolar 132
role in sepsis syndromes 362
macules *20*
magnetic resonance scans, and PUO 388–9
Maki roll method 418
malaria 4, 446–50
cerebral 447, 448–9
chronic/relapsing 447
collection of parasites 35
diagnosis 36, 37, 447
falciparum 334, 446, 447–50, 458
fever 381
and the immune system 17, 412
life cycle of *Plasmodium* 31, *32*
malariae 447
management 447–9
ovale 446–7
prophylaxis 449–50, 459
vivax 446–7
see also antimalarial drugs
Malarone® 79
malignancy, and pyrexia of unknown origin 379–80
Mantoux test 345–6, 386
Marburg disease 457–8
mastoiditis 114–15
MDRTB (multidrug-resistant tuberculosis) 355
measles 5, 7, 114, 225–9
cough 134
and HIV infection 313
immunization 226, 229, 314, 408
and immunosuppression 15, 228–9
post-measles encephalitis 229, 278, 394
postexposure prophylaxis 480
mebendazole 79
medical officer for environmental health (MOEH) 473
medical officer of health (MOH) 473
medical officers 473
Medical Officers of Schools Association (MOSA) 476
mefloquine 78–9
membrane attack complex 365
meningism 255–6, *257*
mild 382
meningitis 255–60
amoebic 255
bacterial 256, 265
commonest causes *266*
cryptococcal 309, **318**
diagnosis 258–60
Haemophilus influenzae 272–5
listerial 276–7
meningococcal 4, 265–72
neonatal 166, 265
pneumococcal 275–6
Streptococcus 'milleri' 277–8
tuberculous 256, 344, 346, 351–2, 382
chronic lymphocytic 33
diagnosis 256–60
and fluid intake 255
and lumbar puncture 256–8, 261, 265
lymphocytic 281
meningism without 256
postsurgical 421
and protozoa 31
viral 256, 260–2, 263
coxsackievirus 261
diagnosis 258, *260*, 261–2
enterovirus 261
herpes simplex type 2 264
and herpes zoster infection 86, *87*
meningococcal disease 81, 364–5, 366, 412, 482
immunization against 271, 435
post-meningococcal synovitis *271*, 394
meningoencephalitis 256
congenital 320
meropenem 65, 66
metastatic infections, bacterial 368
metazoa 23
methicillin, resistance to 60, 93
methicillin-resistant *Staphylococcus aureus* (MRSA) 93, 406, 416, *418*, 420, 424–5, 430, 432
epidemic (EMRSA) 40, 93
and osteomyelitis *291*, 292
methylene blue 37
metronidazole 62, 70–1
side-effects 71, 292
microaerophilic organisms 27
microbial synergy 17
microorganisms
classification 23
see also pathogens
microscopy, for direct examination 35–8
microspora 30
milk
infections from 5–6, 176, 277
and lactose intolerance 160
minimum bacterial concentration (MBC) 56, *57*
minimum inhibitory concentration (MIC) 56, 57, 58
Ministry of Agriculture, Fisheries and Food (MAFF) 474
mites **5**
see also scabies
mitomycin 403
MLEE (multilocus enzyme electrophoresis) 43
molecular diagnostics 49–50
molecular mimicry 394
molecular susceptibility testing 59–60
molluscum contagiosum 36, 89, **318**
monkeypox 89, 91
monobactams 51, 62, 65
mononucleosis, infectious 118–20
and ampicillin 61
immune suppression 15
postinfectious disorders 395, 397
Monospot 120
mosquitoes **5**
and dengue fevers 443
and encephalitis 280, 444
and malaria 4
and yellow fever 202
mouth
colonizing flora *2*
viral infections of the throat and mouth 115–20
MRSA *see* methicillin-resistant *Staphylococcus aureus*
mucocoele, of the gallbladder 242
mucocutaneous lymph-node syndrome *see* Kawasaki disease
mucosa-associated lymphoid tissue (MALT), tumours of 178
mucosae
bacterial infections 91–102
defenses against infection 7–8, 82
gastrointestinal specimens 159
and pathogens 5, 14–15, 366
viral infections 82–91
multidrug-resistant tuberculosis (MDRTB) 355
multilocus enzyme electrophoresis (MLEE) 43
multiorgan dysfunction syndrome (MODS) 361
multiple sclerosis, and fever 18
mumps 230–1
mupirocin 94
mutations, point, detection 59
mycobacterial infections 338–9
cutaneous 102, 339
and HIV/AIDS 316, 317, **318**, 339
and immunocompromised patients 408
Mycobacterium avium-intracellulare (MAI) infection 316, **318**, 339
non-tuberculous 357
wasting effect 340
see also leprosy; tuberculosis
mycoplasmas 23
mycoses *see* fungal infections
myeloencephalitis, ascending 88
myelomatosis, multiple 422
myelopathy, HTLV-associated (HAM) 318–19
myocardial infarction, and Kawasaki disease 242
myocardial ischaemia, and Kawasaki disease 242
myocarditis 245–6
infective 243
myocardium 243

nails, fungal infection 106
candidiasis 103
nalidixic acid 51, 69
adverse effects 70
nasopharyngeal secretions, laboratory diagnosis 134

National Congenital Rubella Registry 476
National Office for Statistics (NOs) 473–4
NCCLS method 58
necrobacillosis 150
nelfinavir 75
nematodes (roundworms) 33, 34, 451
neonates, infections acquired by 320, *321*, 322
neoplasias, intraepithelial 298
  cervical 297
nephritis, postinfectious 21, 436
  poststreptococcal 394
    complications 96
  see also glomerulonephritis, poststreptococcal
netilmicin 67
neuralgia, postherpetic 88
neuritis
  ascending *see* Guillain–Barré syndrome
  postinfectious 21
neuroborreliosis 281, 465
neutral red indicator 39
neutropenia 405–8
neutrophilia 383, 387
neutrophils 8, 19, 362, *363*, 365
  and immune deficiency 405–8
  and mucosae 7
  substances released by *19*, 362
  see also granulocyte counts
nevirapine 315
niclosamide 80
Nikolsky's sign 95, 334, 397
Nipah virus 280
nocardiosis 411
nodules, of rheumatic fever 401
non-nucleoside reverse transcriptase inhibitors (NNRTIs) 76, 315
non-steroidal anti-inflammatory drugs 19
notifiable diseases 473, 475
nucleic acid typing 43
nucleoside analogues 72–4
  see also nucleoside reverse transcriptase inhibitors
nucleoside reverse transcriptase inhibitors (NRTIs) 75, 315
O-nucleotidyltransferases 68
nystatin 77

oedema
  cerebral
    and encephalitis 255
    high-altitude 436–7
  pulmonary, high-altitude 436–7
oesophagitis, candidal 311–12
ofloxacin 51, 69–70
Ogston, Alexander 35
onchocerciasis (river blindness) 451–2
operating theatres 423
ophthalmia, neonatal
  chlamydial 113, 332, 333
  gonococcal 328
opsonization 8, 365, 412
orchitis, mumps 230, 231
orf 36, 89, 90
ornithosis/psittacosis 151, 153
Osler's nodes 247
osteomyelitis *289*, 293
  acute haematogenous 286, 287–93
  childhood *289*, 290, 293
  chronic 293
  pseudomonal 287
  Salmonella 171, *290*
otitis externa 100
otitis media, acute 114–15
otosclerosis 225
outbreaks 6
  common-source *6*, *7*
  investigations 483–5
  person-to-person (propagating) *6*, *7*
  point-source 6–7
oxacillin 64
oxazolidinones 52
oxygen radicals 19

Paget's disease 225
pandemics 7, 15
  cholera 440, *441*
  influenza 138, 435
papillomavirus infections *see* human papillomavirus infections
papules *20*
para-amino benzoic acid (PABA) 51
para-amino salicylic acid (PAS) 357
paracetamol, for fever 18
parainfluenza 140
paralytic shellfish poisoning 181
paranasal sinuses, colonizing flora *2*
parapesis, tropical spastic 318–19
parasites 1–2
  and antimicrobials 391
  in the blood 384
  gastrointestinal tract 159, 182–9
  liver infections 210–14
  and overstimulation of the immune response 15, 17
  skin 108
paratyphoid fever 437, 438–9, 440
paronychia 103
parvovirus infection *see* human parvovirus B19 (HPV-19) infection
Pastia's sign 122
pathogenesis of infection 14–17
pathogenicity 3, 14
  islands 30
pathogens 1–2
  characteristics of successful 14
  dynamics of colonization and infection 21–2
  laboratory handling 427–8
  structure and classification 23–34
  see also infection
PCR *see* polymerase chain reaction
pelvic inflammatory disease (PID) 307–8
pelvis, female, tuberculosis 352–3
penems 51
penicillin-binding protein (PBP) 63
penicillin(s) 51, 53, 63–7
  adverse effects 61, 62, 66–7
  isoxazolyl 64
  penicillin G 64, 65
  penicillin V 65
  resistance to 41, 59, 62–3, 65, 93
peptidoglycan 7, 51
pericardial effusion 244, 245
pericarditis 243–5
  constrictive 243, 245
  suppurative/pyogenic 244, 245
  tuberculous 342, *343*
  viral 244, 245
pericardium 243
perihepatitis 300, 301, 303
perinatal infections *see* congenital and perinatal infections
perineum, colonizing flora *2*
periosteum 286, *290*
peritonitis, tuberculous 342–4
pertussis (whooping cough) 4, 153–5, 384
petechiae *20*
Petri dishes 38
PFGE 43
phage typing 42
phagocytes 8, *10*
  and bacteraemia 365
  and inflammation 19
phagocytosis 8, *10*, 365
  microbial defence against 15, 28
  and staphylococci 374
phagolysosomes 9, *10*, 365
phagosomes 8, 9, *10*, 365
pharyngitis, enteroviral 116–17
pharyngoconjunctival fever 117
pharynx
  pharyngeal diphtheria 128
  reservoirs of infection in 111
phenytoin, and toxic epidermal necrolysis 95
phosphodiesterase inhibitors 371
O-phosphotransferases 68
piedra 33
pigeon fancier's lung 380
pigs
  Japanese B encephalitis 444
  Nipah virus 280, 472
  retroviruses 319
  streptococcal infections 471
pili *see* fimbriae
pink eye 112
piperazine 80
pityriasis rosea 107–8, 240
pityriasis versicolor 33, 106, *107*
plague 453, 469–70
plasma, and inflammation 19–20
platelet counts 384
platyhelminthes (flatworms) 34
pleconaril 74–5, 117
pneumococcal infections 8
  neutrophil counts 383
  and pericarditis 243
  pneumococcal bacteraemia 364, 369
  pneumococcal empyema *145*
  pneumococcal meningitis 275–6
  pneumococcal pneumonia 142–4, 313
Pneumocystis carinii pneumonia (PCP) 312, 316, **318**, 409

pneumonia(s)
anaerobic 150
aspiration 149–50
atypical 150–3
chlamydial 150–2, 153, 412
coliform 156, *188*
community-acquired 142, 150
management **146**, 149
Haemophilus influenzae 144–7
Klebsiella 144
and measles 228
Mycoplasma 150, 394
pneumococcal/lobar 142–4, 313, 369
Pneumocystis carinii (PCP) 312, 316, **318**, 409
staphylococcal, and chickenpox 237
varicella 140
viral 140–1
influenza 138
see also Q fever
pneumonitis
congenital 320
lymphocytic interstitial (LIP) 313
neonatal chlamydial 332, 333
and plague 470
poliomyelitis 116, 262–4, 481
immunization against 158, 262–3, 264, 408
non-paralytic 263
paralytic 263
rise in incidence 478
polyacrylamide gel electrophoresis (PAGE) 43
polyarteritis nodosa 387
polyenes 76–7
polykaryocytes 85
polymerase chain reaction (PCR) 43, 49, 50, 59
polymicrobial infections 17
polyradiculitis 465
polysaccharide capsules 28
Pontiac fever 148
populations *see* epidemiology
porcine endogenous retrovirus (PERV) infection 319
porphyria 385
postinfectious disorders 21, 393–6, **395**
aplastic anaemia **395**
arthritis 21, 173, 394, **395**, 400–1
encephalitis 21, 229, 234, 237, 278, 394, **395**
erythema multiforme 376, **395**, 396–7
erythema nodosum 376, **395**
and fever 381, *383*
Guillain–Barré syndrome 394, **395**, 397–8
haemolysis **395**
haemophagocytic syndrome **395**
Henoch–Schönlein disease 398, *399*
postmeningococcal synovitis *271*, 394
poststreptococcal glomerulonephritis 29, 121, **395**, 398–9
Reiter's syndrome 301, 394, **395**, 402–3
Reye's syndrome **395**, 400
rheumatic fever 376, **395**, 400–2
serositis **395**
thrombocytopenia **395**
thrombotic thrombocytopenic purpura 403
Pott's disease 353
poultry
and campylobacter infection 176
and salmonella infection 169, 170
pox, rickettsial 444, 445
praziquantel 80
precipitation tests 46
pregnancy
and chickenpox 234
and genital warts 297
and herpes simplex infection 300
see also herpes simplex, congenital and intrapartum
and HIV infection 317
see also HIV infection/AIDS, congenital
and listeriosis 277
see also listeriosis, congenital and neonatal
and live vaccines 480
and malaria 450
maternal infections related to childbirth 334–6
rare zoonoses 336
and rubella 232, 234
see also rubella, congenital
urinary tract infections 215, 223
and vertical spread of disease 4
see also congenital and perinatal infections
primaquine 79
probenecid 66
procaine penicillin 65
proguanil 79
proper officers 473
properdin 8, 412
prophylaxis of infection
chemoprophylaxis 482
immunoglobulins for 480
prostaglandins
and fever 17
and inflammation 19
prostatism 222
prostatitis 300
prosthetic devices 366, 421
prosthetic valve infections 247
protease inhibitors (PIs) 75, 76, 315
proteins
bacterial *see* bacteria
iron-binding 365
plasma, and inflammation 19
typing 43
proteolytic enzymes *10*
prothionamide 356
protozoa 23
amoeboid 31
ciliate 31
classification 30–1
culture 44
encystation 31
flagellate 30
life cycle 31–2
spore-forming 30
structure 31
transmission 31, **32**
see also antiprotozoal drugs
pseudomonal osteomyelitis 287
psittacosis *see* ornithosis/psittacosis
Public Health (Control of Disease) Act (1984) 473
Public Health (Infectious Diseases) Regulations (1988) 473
puerperal fever 335, 375
pulsed-field gel electrophoresis (PFGE) 43
PUO *see* pyrexia of unknown origin
pus 35
pustules *20*
pyelonephritis 54, 216
chronic 223
pyomyositis, acute 100
pyrazinamide
adverse effects 61
for tuberculosis 356
pyrexia of unknown origin (PUO)
causes 379–80
investigations
imaging and biopsy 388–91
initial assessment 380–1
laboratory 383–5
microbiological 385–6
physical examination 381–3
preliminary results 386
serological and probe tests 386–7
tests for connective tissue and granulomatous diseases 387–8
no improvement 392
therapy, trials 391–2
pyrimethamine 51, 79
pyuria 218

Q fever 152, 153, 205, 245, 246, 444, 445–6
in pregnancy 336
quinine 78
adverse effects 61, 78
quinolones 51, *55*, 69–70
quinsies 124, *125*
quinupristin/dalfopristin 52

rabies 14, 280–1, 453, 480
radial haemolysis test, for rubella 233
radiculitis 299, 465
radioimmunoassay 47
rapid microagglutionation test (RMAT) 46
rapid plasma reagin (RPR) test 305
rashes 20–1
allergic 227
bacteraemic 368, 369
chickenpox 235, *236*
circinate balanitis 402
dengue fever 443
from antimicrobial treatment 61, *62*
gonococcal bacteraemia 302, *303*
Henoch–Schönlein disease 398
herpes zoster 86, 87
HIV seroconversion illness 311, *312*
HPV-19 239
Kawasaki disease 240, 241
measles 226–7
meningococcal meningitis 267, 268, *269*

rashes (*Continued*)
and pyrexia of unknown origin 382
rheumatic fever 401
of rickettsioses 445
rubella 227, 232
rat bite fever 471
'red man' syndrome 70
red tides 181
reflux nephropathy 223
rehydration
intravenous 159
oral 159
Reiter's syndrome 21, 301, 380, 394, **395**–6, 402–3
relapsing fever 26
renal failure 222, 223, 413
renal toxicity, of antimicrobials 60–1
renal transplants 422
renal tuberculosis 352
reservoirs, of infection 4, 14
resistance to antimicrobials 55
antibacterials 62–3, 68
antivirals 76
transmission to other organisms 63
resistance to infection 7–14
respiratory burst 15
respiratory obstruction, threatened 120
respiratory syncytial virus (RSV) infection 21, 135–7, 140
respiratory tract
lower *see* lower respiratory tract
upper *see* upper respiratory tract
restriction fragment length polymorphism (RFLP) 43, *44*
retinitis, cytomegalovirus **318**
retroviral infections
in animals 319
and antiretroviral drugs 75–6, 315–16
HTLV-I 318–19
HTLV-II 319
see also HIV infection/AIDS
reverse transcriptase (RT) 25
Reye's syndrome 18, **395**, 400
rheumatic fever 21, 29, 121, 376, **395**
rheumatic heart disease, and infective endocarditis 246
rhinorrhoea, cerebrospinal fluid *272*
ribotyping 43
rickettsial pox 236
rickettsioses 81, 444–6
rifabutin 52
rifampicin 52, *56*
adverse effects 61
resistance to 59
for tuberculosis 355
Rift Valley fever **280**
rigors, and fever 381, *383*
ringworm 105–6
ritonavir 75
Ritter's syndrome *see* scalded skin syndrome
river blindness 451–2
RMAT 46
mRNA 50
RNA polymerase 52
Rocky Mountain spotted fever 444, 445
rodents
diseases from 140, 203, 453, 456
and encephalitides 280
Romanowsky stains 37, *38*
roseola infantum 240
rotavirus gastroenteritis 163–4
Roth's spots 247
roundworms *see* nematodes
RTX haemolysins 216
rubella 232–4, 384
congenital 4, 232, 233, 320, 322–4
diagnosis 232–3, 323
and immune thrombocytopenia 21

sacroileitis 402
safety cabinets 428
salmonella infection/salmonellosis 169–73, 361, **371**, 461
and bacteraemias 171, 313, 368
blood cultures 159
and enteric fever 437
and food poisoning 169–73, 479, 483
seasonal incidence *175*
and HIV infection 313
and osteomyelitis *290*
and schistosomiasis 17
salpingitis 301, 303
salvarsan 51
sandflies **5**, 30, 280, 450
saquinavir 75
sarcoidosis 387
scabies 108
scalded skin syndrome (Lyell's syndrome; Ritter's syndrome) 95, 334
scalp, ringworm 105
scarlet fever **29**, 122–3, 376
rash 17, 21
surgical, complications 96
schistosomes/schistosomiasis (bilharzia) 1–2, 33, 210–12
immune system evasion 15
and rectal mucosae specimens 159
and *Salmonella* infection 17
and the skin 7
treatment 80, 212
and the urinary tract 223
scombrotoxin 181
Scottish Centre for Infection and Environmental Health (SCIEH) 473
screening, of patients 40
screening programmes 482–3
SDS-PAGE typing 43
secondary acquired lactose intolerance 160
secretors 21
selective media 39
selenite F 39
sensitivity, to antimicrobials 55
sepsis and sepsis syndromes 361, **362**
acute (adult) respiratory distress syndrome (ARDS) 362
defences of the blood 365–7
epidemiology 364–5
line-related 412–13, 417–19
management 369–73
pathogenesis 362–4
see also bacteraemia; septicaemia
septic shock **362**
septicaemia 7, 294
Gram-negative 376–8
staphylococcal 374–5
streptococcal 375–6
sequestra 288, 290, *291*
serology 45–6
complement fixation tests 46
ELISAs 47–9
indirect fluorescent antibody tests 47
precipitation and agglutination tests 46
for PUO 386–7
radioimmunoassay 47
Western blotting 49
serositis 380, **395**
and meningococcal meningitis 270
sexually transmitted diseases 296, *297*
and urethritis 220
see also genital infections
sheep, infections from 151, 336
shellfish, and food poisoning 164, 165, 176–7, 181
shift, antigenic 15
shigellosis 173–5
shingles *see* herpes zoster
shipyard eye 112
sialic acid 8
sickle-cell disease 412
and malaria 4
and osteomyelitis 287
and salmonella bacteraemia 171
silver sulphadiazine 94
simian immunodeficiency virus (SIV) infection 319
sinusitis, paranasal 115
skin
bacterial infections 91–102
a barrier to infection 7, 82
changes in systemic disease 81
colonizing flora 2, 82
fungal infections 103–8
infections in hot climates 436
parasites 108
penetration by pathogens 5, 82, 405
reactions to antimicrobials 61–2
and spread of disease 5
structure 81, *82*
viral infections 82–91
see also rashes
slapped cheek syndrome 24, 325
slide agglutination 46
slime, bacterial 29
smallpox 479–80, 481
soft sore (chancroid) 306–7
sources, of infection 4–5
Southern blotting 49, 59
SPEs (streptococcal pyrogenic exotoxins) 17, 122
spiral bacteria 26–7
spleen
and bacteraemia 365
and capsulate organisms 28
hypersplenism 413

and immune deficiency **405**, 412–13
rupture 120
splenectomy 412–13, 422
spongiform encephalopathies 281–2
spores, bacterial 14, 27, *28*
sporotrichosis 33, 106–7
sporozoa 30, 31, 183
sputum, and respiratory infections 133
SSPE *see* subacute sclerosing panencephalitis
stains 36–7
staphylococcal infections
acquired in hospital 417, 418, 419
MRSA *418*, 420, 424–5, 430, 432
bacteraemia/septicaemia 247, 364, 368, 374–5
food poisoning 178–9
neonatal 334
septicaemia 294
of the skin 91, 436
Staphylococcus aureus endocarditis 247
treatment **370**, 375
stavudine (D4T) 75, 315
Stevens–Johnson syndrome 61, 396–7
Stokes method 57–8
stomach 157
stools
rice-water 440
specimens 159, 160, *161*
see also diarrhoea
strawberry tongue 122, *123*
streptococcal infection **370**
poststreptococcal disorders 121, 124, 376, 394, **395**, 396, 398–9
streptococcal bacteraemia/septicaemia 124, 294, 364, 368, 375–6
streptococcal toxic shock syndrome 123
tonsillitis 28, 121–*5*
streptococcal pyrogenic exotoxins (SPEs) 17, 122
streptogramins 52
streptokinase 122
streptolysins 121–2
streptomycin 67
bacteria resistant to 62, 68
for tuberculosis 356
string test 159
strongyloidiasis 156, 159, 188–9, 410, 411
hyperinfection syndrome 410
subacute sclerosing panencephalitis (SSPE) 228, 229, 394, 476
sulbactam 65
sulphonamides 51, *52*
adverse effects 60–1
bacteria resistant to 63
and toxic epidermal necrolysis 95
sulphone therapy, adverse effects 61
sulphonylureas, and toxic epidermal necrolysis 95
'sulphur granules' 102, 385
summer flu 261
sunburn 435
sunstroke 435
superantigens 11–14, 92, 122
surgery 420–2
surveillance
disease 474–6
of immunization programs 481–2
susceptibility testing, of antimicrobials 41, 49–50, 55–63
sweat rash 436
swimmer's itch 7, 211
synergy, microbial 17
synovitis 398
postmeningococcal *271*, 394
of Reiter's syndrome 402
syphilis 243, 260, 304–6
congenital 330
diagnosis 47, 305
incidence *297*
intrapartum infections 320
meningovascular 305
syphilitic aortitis 251
systemic inflammatory response syndrome (SIRS) 361
systemic lupus erythematosus 380

T cells (T lymphocytes)
deficiency, and opportunistic infections **405**, 408–11
and immunity 9, 10, 11, 393–4
T-helper cells 10, 11
tabes 305
tamponade, pericardial 245
tanapox 89, 91
tapeworms *see* cestodes
tattoo parlours 7
tears 112
teeth
colonizing flora *2*
and congenital syphilis 330
teichoic acid 27–8
teicoplanin 51, 70, **373**
temperature, body 17
see also fever
terbinafine 77–8
tetanospasmin 3
tetanus immunoglobulin 480
tetracycline(s) 52, *56*, 68–9
adverse effects 60, 61, 68–9
bacteria resistant to 63
and parasitic infections 391
tetrahydrofolate 51
thiabendazole 80
threadworms 33, 186
three-glass test 216, *217*
throat
bacterial infections 120–31
colonizing flora *2*
viral infections of the throat and mouth 115–20
thrombocytopenia **395**
in chickenpox 237–8
immune 21, 61
thrombocytopenic purpura
congenital 320
immune 234
thrombosis, cavernous sinus 115
thrombotic thrombocytopenic purpura (TTP) 168
postinfectious 403
thymidine kinases (TKs) 72, 73
ticarcillin 65
tick typhus 444, 445, *446*
ticks **5**
and encephalitis 280, 444, 459, 480
see also Lyme disease; tick typhus
tine test 345–6
tinea infections 104–6
tissue culture 44
TNF *see* tumour necrosis factor
tobramicin 67
tongue, and scarlet fever 122, *123*
tonsillitis
follicular 122, 123
streptococcal 28, 121–*5*, 122, 123
tonsils 111
and infectious mononucleosis 119
topoisomerase 51, *55*, 69
toxic epidermal necrolysis 95
toxic shock syndrome 17, 95
and chickenpox 237, *238*
streptococcal 123
toxins (TSSTs) 14, 17, 95
toxins 16–17
bacterial, diseases caused by 178–81
haemolytic 17, 123
inactivated by the body 158
non-bacterial, and food poisoning 181–2
see also endotoxins; exotoxins
toxocariasis 189
toxoids 16
toxoplasmosis 467–9
cerebral 411
congenital 330–1
and HIV 15
in immunocompromised patients 409–10
intrapartum infections 320
treatment 51, 410, 469
trachea 132
intubation 419
tracheostomy, for diphtheria *128*, 130
trachoma 114
transaminases 384
transduction, genetic 63
transformation, genetic 63
transient cerebral ischaemic events, and fever 18
transmission of infection 4–6, 14
transplacental infections 320, *321*, 322
see also congenital and perinatal infections
transplant patients
infections 408, 413–15, **414**, 422
see also xenotransplantation
transposons 63
travel-associated diseases 387, 433–4
altitude sickness 436–7
anthrax 453–5
with arthropod vectors 443–52, 479
assessment of fever 459
brucellosis *see* brucellosis
cholera 440–2, 477
climate-related 435–7
common-worldwide 434–5

travel-associated diseases (*Continued*)
  controllable by public health measures 437–42
  Crimean-Congo haemorrhagic fever 485
  dengue fevers 443–4
  Ebola haemorrhagic fever 457–8
  frostbite 436
  hantavirus pulmonary syndrome 456
  heat exhaustion 435
  and influenza prophylaxis 435
  insect bites 436, *443*
  lassa fever 456–7
  leishmaniasis *see* leishmaniasis
  loiasis 451
  lymphatic filariasis 451
  malaria *see* malaria
  Marburg disease 457–8
  melioidosis 455
  and meningococcal disease prophylaxis 435
  onchocerciasis (river blindness) 451–2
  paratyphoid fever 437–40
  plague 453
  rabies 280–1, 453
  rickettsioses 81, 444–6
  sunburn 435
  sunstroke 435
  traveller's diarrhoea 166, 168, 437
  tularaemia 109, 366, 455
  typhoid fever 437–40
  viral haemorrhagic fevers 425–6, 455–8
  yellow fever 202–3, 455, 459
  zoonoses 452–5
    encephalitides 444
travellers, health measures for 458–9
trematodes (flukes) 34
  see also liver flukes; schistosomes/schistosomiasis
trench fever 444
Treponema pallidum haemagglutination assay (TPHA) 305, **306**
trials of therapy, for PUO 391–2
triazoles 77
tribavirin 74, 136, 140
trichinellosis 189
trichomoniasis
  incidence 220, 296
  vaginal 36, 308–9
trimethoprim 51, *52*
  bacteria resistant to 63
tripod sign 256
Trypanosomatidae 30
trypanosomiasis 411–12
  African 15, 17, 30
  cerebral 260
  South American 30
tsetse flies 30
tube agglutination 46
tuberculin tests 344–6, 358, 386, 392
tuberculoid leprosy 359
tuberculomata 284, 352
tuberculosis 337
  diagnosis 37, 344–9
  disseminated 341, 354
  drug-resistant 348–9, 354–5, 358, 425
  endobronchial 341–2
  epidemiology 337–8
  of the female pelvis 352–3
  and HIV infection 15
  and the 'iceberg of infection' 22
  and immune suppression 15
  isolation of patients with **426**
  lymph-node 342
  microbiology 338–9
  miliary 341, 353
  multidrug-resistant (MDRTB) 355
  pathogenesis 339–41
  and pericarditis 243
  pleural 342
  postprimary 340–1, 349–53
  postprimary pulmonary 349–51
  prevention and control 357–8
  primary 340, 341–4, 394
  primary pulmonary 341
  reduction in incidence 477
  renal 352
  treatment 50, 354–7
    trials 391, 392
tuberculous arthritis 295
tuberculous epididymitis 352
tuberculous lymphadenitis *344*
tuberculous meningitis 256, 344, 346, 351–2, 382
tuberculous pericarditis 342, *343*
tuberculous peritonitis 342–4
tularaemia 109, 366, 455, 470–1
tumbu flies 436
tumour necrosis factor (TNF) 10, 16, 17
  TNF-alpha 362, 372
tumours, nasopharyngeal 118
typhoid fever 437–40
typhus 444, 445
  tick 444, 445, *446*
typing, of microorganisms 41–3
Tzanck cells 84, 85

ulceroglandular tularaemia 109
ulcers
  dendritic 112–13
  genital 306–7, 317
  peptic, and *Helicobacter pylori* 178
  of skin and soft tissue
    buruli (tropical) ulcers 339
    and gas-forming infections 99
    and osteomyelitis 287, *291*
ultrasound scans, and PUO 388
umbilical infection 334
upper respiratory tract 111
  infections *see* upper respiratory tract infections
  normal flora 111
upper respiratory tract infections 111
  acquired in hospital 419
  conjunctivitis 111–14
  middle ear 114–15
  throat and mouth
    bacterial 120–31
    viral 115–20
ureteric reflux 222
ureteroscopy, infection from 421
urethra 219, 296
  caruncle 220
  colonizing flora *2*
  symptoms of urinary infection 216
urethral syndrome 220
urethritis 219–20, 300, 301, 302–3
urinary tract infections 215–24
  ascending 221–3
  in children 222
  defence mechanisms against 7
  diagnosis 216–19
  in men 222
  pathogenesis 215–16
  in pregnancy 215, 223
  prophylaxis 222
  symptoms 216
  tuberculosis 352
  and urinary catheters 8, 419
urine
  biochemical tests 384–5
  cultures 385
  non-infectious irritants in 220
  testing for urinary infection 216–19
urticaria, and larva currens 188

V-beta receptors 14
vaccines 480–1
  anthrax 455
  cholera 442
  coverage 482, *483*
  diphtheria 130, **482**
  Hib 147, 274–5, 412, **482**
  influenza 140
  Japanese encephalitis 280
  measles 229
  meningococcal 271, 435
  MMR (measles/mumps/rubella) 231, 234, 324, 408, **482**
  MMRV (measles/mumps/rubella/varicella) 238
  mumps 231
  pertussis 154–5, **482**
  pneumococcal 144
  polio 263, 264
  rabies 281, 459
  rotavirus 164
  rubella 234, 324
  tetanus 459, **482**
  tick-borne encephalitis 280, 459
  toxin 16
  typhoid fever 440
  varicella (chickenpox) 238
  yellow fever 203, 459
  see also immunization/vaccination
vaccinia 89, 90–1
vagina 296
  colonizing flora 2, 296
vaginitis, non-specfic 219
vaginosis, bacterial 307
valaciclovir 73
  for herpes simplex 86
  for herpes zoster 87, 88
valproate 18
vancomycin 51, 70
  adverse effects 70, **373**

bacterial resistance to 60, 62
therapeutic monitoring 60
variable surface glycoprotein (VSG) 15
varicella *see* chickenpox
varicella zoster immunoglobulin (VZIG) 238, 408, 480
vasoconstrictors 371
vasopressin 371
VDRL (Venereal Disease Research Laboratory) test 305, 306
vectors 5
ventilation, artificial 419
ventriculitis 278
verrucae (plantar warts) 83
vertical spread, of disease 4
vesicles *20*
Vibrionaceae 177
vibrios, 'food-poisoning' 176–7
Vincent's angina 37, 131
viraemia, during pregnancy 320
viral haemorrhagic fevers (VHFs) 425–6, 455–8
viral infections
and cell-mediated immunity 11
conjunctiva/cornea 111–13
and immune thrombocytopenia 21
in immunocompromised patients 408
intestinal tract 163–5
liver 191–203
lower respiratory tract 133–41
middle ear 114
skin and mucosal 82–91
throat and mouth 115–20
virulence 3
virus attachment proteins (VAPs) 25–6
virus neutralization 44–5
viruses
attachment 25–6
blood-borne 361
collection 35
culture 44, 386
envelopes 25, *27*
and the immune system 15
nucleic acid 24
in the pharynx 111
replication 24–5, *26*
resistance to antivirals 63
respiratory, culture 386
structure and classification 23–6
symmetry 25, *26*
vitamin A, and measles 229
vomiting, treatment 160
vomitus, tested for intesinal infections 159
von Willebrand factor 403

warts (papillomavirus infections) 83, 296–7
genital 297
waste disposal, hospital 431
water
abroad 437, 477
supplies 147, *148*, 149, 478
hospital 422–3
Waterhouse–Friderichsen syndrome 128, 270
Wegener's granulomatosis 387
Weil's disease 203
Western blotting 49
West Nile fever 444
whipworms 186
whitlows, herpetic *84*, 86
whooping cough *see* pertussis
winter vomiting disease 165
Wood's light 105

X-ray techniques, and PUO 382–3, 389–90
xenotransplantation 319

yeast infections, in the immunosuppressed 409
yellow fever 202–3, 455, 459
Yersinia infections 177–8

zalcitabine (DDC) 75, 315
zanamivir 75, 140
zidovudine (azidodeoxythymidine, AZT) 75, 76, 315
adverse effects 61, 75
Ziehl–Nielsen (ZN) stain 37
zoonoses 4–6, 461–2
anthrax 453–5
herpesvirus simiae infection 471
hydatid disease 33
melioidosis 455
paramyxoviruses 472
plague 453, 469–70
poxvirus 89–90, 91
and pregnancy 336
rabies 14, 280–1, 453
rat bite fever (Haverhill fever) 471
streptococcal infections 471
and travellers 452–5
tularaemia 109, 366, 455, 470–1
see also brucellosis; Lyme disease; toxoplasmosis
Zoonoses Order 173
zoster *see* herpes zoster
zoster immunoglobulin (ZIG) 326
zoster-associated pain (ZAP) 88
zymodeme analysis 43

Index of Organisms

Page references in *italics* refer to figures; those in **bold** refer to tables

Acanthamoeba spp. 31, 114
Acinetobacter spp. 94, 265, 364, **372**, 417, 419, 421, 436
Actinobacter actinomycetecomitans 102
Actinomyces spp. 102, 308
  A. israelii 102
adenoviruses 24, 117–18, 140, 164, 165
  type 8 112
Aedes spp. 202, 443
Aeromonas spp. 177
alphaviruses, and encephalitides **280**
Andes virus 456
arboviruses 280, 444, 455
arenaviruses 455
Ascaris spp.
  A. lumbricoides *182*, 187
  treatment of infection 79, 80, 187
Aspergillus spp. 33, 155, 407, 408, 426
  A. flavus 33, 406
  A. fumigatus 33, 406
  A. niger 33, 406
  and neutropenia 406
  and otitis externa 100
astroviruses 164

Babesia spp. 30
bacillus Calmette–Guérin (BCG) 339
Bacillus spp.
  B. anthracis 1, 28, 453
  B. cereus, food poisoning 179
  B. subtilis 179
  spores 27
Bacteroides spp. 364, **372**
  B. fragilis 39, 65, 209, 369, **372**
Balantidium coli 31
Bartonella spp.
  B. henselae 101–2, 109–10, 317, 446
  B. quintana 444
Bayou virus 456
BCG (bacillus Calmette–Guérin) 339
Black Creek Canyon virus 456
Blastocystis hominis 31
Blastomyces spp. 155
Bordetella spp.
  B. bronchoseptica 153
  B. parapertussis 153
  B. pertussis 27, 153
Borrelia spp. 26, 37, 100, 464
  B. afzelii 100, 464
  B. burgdorferi 100–1, 281, 464, 467, 469
  B. duttoni 15
  B. garinii 100, 464
  B. recurrentis 15
  B. vincenti 131
Brucella spp. 27, 206, 292
  B. abortus 452
  B. melitensis 452
Brugia malayi 451
bunyaviruses, and encephalitides **280**
Burkholderia spp. **372**
  B. cepacia 142
  B. pseudomallei 455

calciviruses 36, 164, 165
Calymmatobacterium granulomatis 307
Campylobacter spp. (campylobacters) 5, 27, *162*, 175–6, 412, 461, 479
  and body temperature 17
  C. coli 175
  C. fetus 175
  C. jejuni 175
  C. lari 175
Candida spp.
  C. albicans 33, 103, 104, 220, 406
    and drug abusers 247
    genital infections 308
  C. dublini 406
  C. glabrata 33, 406, 407
  C. krusei 33, 406, 407
  C. parasilosis 406
  in the gut 157
  and lower respiratory tract infections 150, 420
  and meningitis 265
  and neutropenia 406
  and prolonged treatment for osteomyelitis 292
Capnocytophagia canimorsus 109, 412
Chlamydia spp. 23, 151, 152
  C. pneumoniae 114, 150–2, 153
  C. psittaci 5, 150, 151, 152, 336
  C. trachomatis 151, 152
    and conjunctivitis 113–14
    and genital infections 300, 301, 302
    and gonococcal neonatal ophthalmia 328
    and pelvic inflammatory disease 307
    and urethritis 220
  defence against immunological attack 15
  and urinary infection 220
Citrobacter spp. 364
Clonorchis (opisthorchis) sinensis 213
Clostridium spp. 364, **372**
  C. botulinum 3, 179, 180
  C. difficile 3, 8, 158, 180–1
  C. perfringens 14, **372**
    bacteraemia 369
    diagnosis 160
    and food poisoning 180
    and gas gangrene 99
    and liver infections 209
    toxins 17, 99, 160, 368
  C. tetani 3, 5
  and gas gangrene 99
  spores 14, 27, *28*
Coccidiodes immitis 33, 155
coronaviruses 115, 164, *165*
Corynebacterium spp.
  C. diphtheriae 16, 40, 111, 127–30, 243, 436
    typing 41, *42*
  C. jeikeium 366, 406, 407, 417
  C. minutissimum 101
  C. ulcerans 127
Coxiella burnetii 151, 152, 205, 249, 444, 445–6
coxsackie A viruses 88, 116, 117
  type A10 116
coxsackie B viruses 117, 223
Cryptococcus neoformans 33, 46
  staining 37
Cryptosporidium spp. 162, 412
  C. parvum 30, 44, 183, *184*, 410
Cyclospora catayensis 184
cytomegalovirus (CMV) 111, 202, 386, 413

dengue virus 443
Dientamoeba fragilis 31
DNA viruses 24
Dobrava–Belgrade virus 456

Echinococcus granulosus 212
echoviruses 116, 261
Ehrlichia spp. 444
EMRSA 40
Encephalitozoon spp.
  E. cuniculi 30
  E. intestinalis 30
Entamoeba spp. 31
  E. dispar 1, 31, 185
  E. histolytica 1, 31, 70, 159, *182*, 184, 185–6, 207
    and pericarditis 243
  typing 43
enterinvasive *E. coli* (EIEC) 167
enteroaggregative *E. coli* (EAEC) 167
Enterobacter spp. 364, **371**, 406

Enterobacteriaceae *40*, 41, 43, 158, 170, 265
  and bacteraemias 364, 369, 406
  and immunocompromised patients 406
Enterobius spp.
  E. vermicularis *182*, 186
  treatment 79, 80, 186
Enterococcus spp. 406
  E. faecalis **372**
  E. faecium **372**, 419
Enterocytozoon beneusii 30
enterohaemorrhagic *E. coli* (EHEC) *see* verocytotoxic *E. coli*
enteropathogenic *E. coli* (EPEC) 167, 168
enterotoxigenic *E. coli* (ETEC) 166, 437
enteroviruses 116, 117
  cultures 386
  and meningitis 261, 262
  and myocarditis 246
  type 30 112
  type 70 261
  type 71 88, 261
Epstein–Barr virus (EPV) 111, 118–20, 140, 202
  defence against immunological attack 15
Erysipelothrix rhusiopathiae 101
Escherichia coli 160, 166, **371**
  antimicrobial-resistant 168
  and bacteraemias/septicaemia 364, 376
  and cholangitis 209
  culture *40*
  enteroaggregative (EAEC) 167
  enterohaemorrhagic (EHEC) 166–7
  enteroinvasive (EIEC) 167
  enteropathogenic (EPEC) 167, 168
  enterotoxigenic (ETEC) 166, 437
  gastroenteritis 165–8
  and hospital infections 419
  and immunocompromised patients 412
  and infective endarteritis 251
  and lower respiratory tract infections 141, 142, 156, *188*
  and osteomyelitis 293
  pathogen or colonizer 1
  pathogenesis 166–7
  uropathic 14, 29, 215, 220
  verocytotoxic (VTEC) 166–7, 168, 190
ETEC (enterotoxigenic *E. coli*) 166, 437

Fasciola hepatica 213
filoviruses 455
flaviviruses 202, **280**, 443–4
Francisella tularensis 109, 455, 470
Fusarium spp. 406
Fusiformis spp. 131
Fusobacterium spp. 26, 37, 130
  F. necrophorum *291*

Gardnerella spp. 220
  G. vaginalis 307
Giardia spp. 31, *162*, 412
  G. intestinalis 30, 70, *182*, 183
Guanarito virus 457

Haemophilus spp. 147, 364
  culture 39
  H. aegyptius 113
  H. ducreyi 306
  H. influenzae 15, 28, 273
    and acute epiglottitis 125
    antibiotic-resistant 62, 147
    and bacteraemia 365, 366
    and conjunctivitis 113
    detection 46
    immunity to 21
    immunization against 125, 127, 147, 481
    and immunocompromised patients 412
    and lower respiratory tract infections 137, 139, 141, 142, 144–7
    and meningitis 258, 272, 273, 276
    and middle-ear infections 114, 115
    and neonatal bacteraemia 333
    and osteomyelitis 290
    pathogenesis 273
    pneumonia 144–7
    and suppurative arthritis 294, 295
    type b (Hib) 144, 147, 272, 273
    typing 41
  and measles 228
Hantaan and Seoul virus 456
hantaviruses 456
Helicobacter spp. 27
  H. hominis 178
  H. pylori 157, 178
Hendra virus 472
hepatitis A virus (HAV) 116, 193
hepatitis B virus (HBV) 4, 7, 24, 196
hepatitis C virus (HCV) 200
hepatitis D virus 199
hepatitis E virus 201
hepatitis G virus 201
herpes simplex virus (HSV) 83–6, 111
  and antivirals 72–3, 279
  genital 83, *297*, 298–300
  and keratitis 113
  type 1 (HSV-1) 83, 298, 299
  type 2 (HSV-2) 83, 297, 298, 299
  and urethritis 220
herpesvirus simiae 471
herpesviruses 11, 24
  culture 386
  and immunocompromised patients 408
  negatively-stained 36
  see also herpes simplex; herpes zoster; herpesvirus simiae infection; human herpesvirus
Histoplasma capsulatum 33, 155, 409
HIV (human immunodeficiency virus)
  antiviral susceptibility 59
  attachment 25–6
  drug resistance 49–50, 63, 76, 316
  HIV-1 and HIV-2 310, 327
HPVs *see* human papillomaviruses
human herpesvirus
  type 6 (HHV-6) 111, 240
  type 7 (HHV-7) 107, 240
  type 8 (HHV-8) 313
human immunodeficiency virus *see* HIV
human papillomaviruses (HPVs) 83, 296–7
human parvovirus B19 (HPV-19) 239, 326
human T-lymphotrophic virus I (HTLV-I) 318–19, 332
human T-lymphotrophic virus II (HTLV-II) 319

influenza virus 137
  antigenic variation 15
  attachment 25
  envelope *27*
  structure *137*
  synergy with *Streptococcus pneumoniae* 17
  type A 138
    and antiviral therapy 72
  type B 138
  type C 138
Isospora belli 30, 410
Ixodes spp. 464

Junin virus 457

Klebsiella spp. 64, 364, 406
  K. oxytoca **371**
  K. pneumoniae 28, 133, 141, 150, 209, **371**
    hospital infections 419, 420
    pneumonia 144
  a urinary pathogen 221

Lactobacillus acidophilus 308
Lassa virus *45*
Legionella spp. 148, 422–3
  L. pneumophila 5, 14, 147–9, 422, 478
    detection of antibodies to 46, 149
    typing 41
Leishmania spp. 30, 450
  culture 44
  defence against immunological attack 15
  flagella 31
  L. donovani 31, 450
  L. infantum 450
  L. tropica var. *major* 450, 460
  L. tropica var. *minor* 450
  typing 43
Leptospira spp. 7, 204
  L. biflexa 203
  L. canicola 203
  L. hebdomadis 203
  L. icterohaemorrhagiae 203
  L. interrogans 203
Listeria monocytogenes 5, 276–7, 329, 330, 333, 364, 408–9
  typing 41, 42
Loa loa 451
lyssaviruses 280, 453

Machupo virus 457
Malassezia furfur 32–3
Marburg virus 457
Mastomys natalensis 456
measles virus 226, 229
  cytopathic effect 44, *45*
meningococci 8, 435
Micrococcus spp. 91
Microsporum spp. 32, 105
Mobiluncus sp. 307
Moraxella spp.

M. catarrhalis 141, 420
M. lacunata 113
mumps virus 223, 230
Mycobacterium spp. (mycobacteria) 11, 338
antimicrobial-resistant 59, 348–9
cell wall 28
classification 339
culture 347–8
defence against immunological attack 15, 340
environmental (atypical) 341, 355
and immunocompromised patients 408, 409
M. africanum 339
M. avium-intracellulare 316, **318**, 339, 342, 356, 408, 409
M. bovis 338, 339, 479
M. chelonei 102, 339, 357, 408, 409
M. cookei 339
M. fortuitum 102, 339, 357
M. genevense 339
M. haemophilum 339
M. hiberniae 339
M. kansasii 339, 408, 409
M. leprae 28, 29, 339, 340, 358–9
M. marinum 102, 339, 357
M. scrofulaceum 342
M. tuberculosis 1, 15, 338, 339, *346*, 349, 408
antimicrobial-resistant 59, 355
cutaneous infections 102, 339
and immunocompromised patients 408, 409
typing 43, 348
M. ulcerans 339, 357
M. xenopi 339
staining 28, 37
survival within macrophages 340
Mycoplasma spp.
M. hominis 307
M. pneumoniae 150, 152, 153
and harmful immune response 21, 394
and hypogammaglobulinaemia 412
and pericarditis 245
postinfectious disorders 396, 397

Naegleria spp. 114
N. fowleri 31
Neisseria spp.
culture 39
N. gonorrhoeae 3, 14, 29, 58, 303
and neonatal bacteraemia 333
and neonatal ophthalmia 113, 328
and pelvic inflammatory disease 307
penicillin-resistant 59, 62
penicillinase-producing (PPNG) 58
specimen collection 35
typing 41
and urethritis 220
N. meningitidis 15, 28, 111
and bacteraemia 364
detection 46
immunity to 21
and meningitis 258, 265, 266, 267
typing 43
New York virus 456
Nipah virus 472
Nocardia spp. 206
Norwalk virus 164
Nosema connori 30

Onchocerca volvulus 451
Orthomyxovirus 137
orthopoxviruses 89

Paracoccidoides braziliensis 33
parainfluenza viruses 134, 135
paramyxoviruses 134, 386
zoonotic 472
parapoxviruses 89, 90
parvovirus 24, 239
Pasteurella multocida 98, 109, 366
Peptococcus spp. **372**
Peptostreptococcus spp. **372**
Plasmodium spp. 30
life cycle 31, *32*
P. falciparum 4, 49, 446, 447, 449
antimalarial-resistant 59
culture 44
stained trophozoites *38*, *448*
P. malariae 446
P. ovale *448*
P. vivax 4, 446, *448*
Pleistophora spp. 30
Plesiomonas shigelloides 177
pneumococci 8, 142, 243
defence against immunological attack 15
pathogenicity 3
Pneumocystis spp., and HIV 15, 311–12, 317
Pneumovirus 136
polio/poliomyelitis viruses 21–2, 116, 262
polyomaviruses 24
porcine endogenous retroviruses (PERVs) 319
poxviruses 24, 36, 89–91
Prevotella melaninogenicus 130
Propionobacterium acnes 102
Proteus spp. 218, *219*, 364, **371**
hospital infections 419
and osteomyelitis 293
Pseudoallerchia boydii 406
Pseudomonas spp. 95, **372**, 406, 407
antimicrobial resistance 63
and cellulitis 99
hospital infections 419, 420, 421, 422
and osteomyelitis 293
and otitis externa 100
P. aeruginosa 28, 29, 64
and conjunctivitis 113
furuncle-type lesions 93
and lower respiratory tract infections 142, 150
typing 42
treatment **372**, 377
Puumala virus 456

rabies virus 281
reoviruses 25
respiratory syncytial virus (RSV) 135, 136, 140
immunopathological damage 21
retroviruses 25, 75–6
in animals 319
HTLV-I 318–19, 332
HTLV-II 319
see also HIV
rhinoviruses 115
Rickettsia spp.
R. africae 445
R. conorii 444
R. mooseri 444
R. prowazekii 444
R. rickettsii 444
R. tsutsugamushi 444
rickettsiae 23, 243, 445
RNA viruses 24–5, *26*
drug resistance 76, 316
rotaviruses 25, 36, *162*, 163, 164
RSV *see* respiratory syncytial virus
rubivirus 232

Sabia virus 457
Salmonella (salmonellae) 5, 30, 157, 158, **371**
agglutination tests 46
and bacteraemia 364, 368
culture 39, 40
drug-resistant 41, 173
and food poisoning *162*, 169–73
and osteomyelitis 287, *290*
S. arizona 170
S. cholerae-suis 170, 171
S. dublin 171
S. enteritidis 169, 170
S. napoli *483*
S. paratyphi 170, **371**, 437
S. typhi 28, 70, **371**, 437, 439
S. typhimurium 169, 170, *171*
typing 41
S. virchow 169
and schistosomiasis 17
toxicity 16
typing 41–2, 170, 172
Sarcocystis sp. 30
Sarcoptes scabiei 108
Schistosoma spp. 210–12
S. haematobium 210, 223, *385*
S. japonicum 210, 211
S. mansoni 210–11
Serratia spp. 209, 364, **371**, 406
Shigella spp. 30
agglutination tests 46
S. boydii 174
S. dysenteriae 174
S. flexneri 174
typing 41–2
S. sonneri *162*, 174
typing 42
simian immunodeficiency virus (SIV) 310, 319
Simulium spp. 451
Sin Nombre virus 456
SIV *see* simian immunodeficiency virus
small round structured viruses (SRSVs) 164, 165
Spirillum minus 471
Sporothrix schenckii 33, 107

SRSVs (small round structured viruses) 164, 165
Staphylococcus spp. (staphylococci) 91–2
after herpes zoster infection 88
coagulase-negative **370**, 375
exotoxins 17, 92, 95, 334, 368
from the skin 7, 91, 366
S. aureus 5, 17, 366, **370**
and acute pyomyositis 100
antimicrobial susceptibility 60, 62, 63, 93–4
and bacteraemias/septicemia 247, 374
and breast abscesses 335
and cardiovascular infections 247, 248
and cellulitis 98
and chickenpox 237
and conjunctivitis 113
exotoxins 17, 95
and food poisoning 178
and granulocyte counts 383
and hospital infections 417, 418, 419, 420, 424–5
and impetigo/furunculosis 91–2, 93, *443*
and lower respiratory tract infections 137, 139, 142
and Ludwig's angina 130
and measles 228
methicillin-resistant (MRSA) 40, 93, *291*, 292, 416, *418*, 420, 424, 425, 430, 432
microbiological diagnosis 93
and middle-ear infections 114, 115
neonatal infections 334
and neutropenia 406
and osteomyelitis 290, 293
typing 41, 42, 43, 93
S. epidermidis 1, 29, 248, 366, 405, 406, 407
and hospital infections 417, 418
typing 41, *42*
S. haemolyticus 70
S. saprophyticus 220, 221
and superantigens 14
toxins 17, 178
typing 42
Stenotrophomonas spp. 419
Streptobacillus moniliformis 471
Streptococcus spp. (streptococci)
and bacteraemia/septicaemia 364, 368, 375, 376
culture 39
exotoxins 17, 122, 368
group A **370**
group B **370**
group C and G **370**
S. agalactiae 28, **124**
S. anginosis 277–8
S. bovis **124**
S. constellatus 277–8
S. dysgalactiae **124**
S. equi **124**
S. equisimilis **124**
S. intermedius 277–8
S. 'milleri' 115, **124**, 249, 277–8, 368, **370**
S. mitis 406
S. oralis 406
S. pneumoniae 3, 14–15, 16, 27–8, 28, 142–4, 364, **370**
antimicrobial-resistant 41, 59, 62–3
and conjunctivitis 113
detection 46
and immunocompromised patients 412
and influenza virus 17
and lower respiratory tract infections 137, 139, 141, 142–4, 420
and measles 228
and meningitis 258
and middle-ear infections 114, 115
and neonatal bacteraemia 333
pathogenesis 142–3
pneumococcal/lobar pneumonia 142–4
type b 46, 333
typing 43
S. pyogenes 17, 40, **124**
and acute epiglottitis 125
capsular antigens (M-types) 28–9, 121
and cellulitis 98
and chickenpox 237
detection 46, 123–4
and endocarditis 248
and erysipelas 96
and glomerular nephritis 29, 121
and granulocyte counts 383
and impetigo 93, 94
and Ludwig's angina 130
and middle-ear infections 114
and necrotizing fasciitis 100, 376
nephritogenic 394
pathogenicity factors 121
and puerperal infections **29**
and rheumatic fever 21, 29, 121, 400
and scarlet fever **29**, 122–3, 376
and septicaemia 375
and tonsillitis 121–4
S. sanguis 3
S. suis 471
S. zooepidemicus **124**, 471
toxins 16, 17, 121–2
Strongyloides spp.
and coliform pneumonia 156, *188*
and organ transplants 413
S. stercoralis *36*, 44, 183, 188, 410

Taenia spp.
T. saginata, treatment of infection 80
T. solium 284
treatment of infection 80
Toxocara spp.
T. canis 189
T. cati 189
Toxoplasma spp. 31, 413
T. gondii 15, 30, 409–10, 467
transfusion-transmitted virus (TTV) 24, 201
Treponema pallidum 26, 304, 305
Trichinella spiralis 189
Trichomonas spp. 30
T. vaginalis 70, 220, 308–9
Trichophyton spp. 32, 105
Trichuris spp.
T. trichiura 186
treatment 79
Trypanosoma brucei var. *rhodesiense*, evading the immune response 15
TTV (transfusion-transmitted virus) 24, 201

Ureaplasma urealyticum 301

varicella zoster virus (VZV) 86, 235
and antivirals 72
and immunocompromised patients 408
verocytotoxic *E. coli* (VTEC) 166–7, 168, 190
Vibrio spp.
and cellulitis 98
culture 39
and food poisoning 176–7
skin lesions 94
V. alginolyticus 177
V. cholerae 30, 177, 440, 441, 442
V. fluvialis 177
V. mimicus 177
V. parahaemolyticus 176–7, 442
Vittaforma corneum 30

Wuchereria bancrofti 451

Yersinia spp.
Y. enterocolitica 177
Y. pestis 453, 469
Y. pseudotuberculosis 177–8